Bimonoidal Categories, E_n-Monoidal Categories, and Algebraic K-Theory

Volume I: Symmetric Bimonoidal Categories and Monoidal Bicategories

Mathematical
Surveys
and
Monographs

Volume 283

Bimonoidal Categories, E_n-Monoidal Categories, and Algebraic K-Theory

Volume I: Symmetric Bimonoidal Categories and Monoidal Bicategories

Donald Yau

AMERICAN
MATHEMATICAL
SOCIETY
Providence, Rhode Island

2020 *Mathematics Subject Classification.* Primary 18D20, 18F25, 18M05, 18M15, 18M50, 18M60, 18M65, 19D23, 55P43, 55P48.

For additional information and updates on this book, visit
www.ams.org/bookpages/surv-283

Library of Congress Cataloging-in-Publication Data

Names: Yau, Donald Y. (Donald Ying), 1977– author.
Title: Bimonoidal categories, E_n-monoidal categories, and algebraic K-theory : volume I : symmetric bimonoidal categories and monoidal bicategories / Donald Yau.
Description: Providence : American Mathematical Society, 2024. | Series: Mathematical surveys and monographs, 0076-5376 ; 283 | Includes bibliographical references and index.
Identifiers: LCCN 2024025153 | ISBN 9781470478094 (paperback) | ISBN 9781470478469 (ebook)
Subjects: LCSH: Categories (Mathematics) | K-theory. | AMS: Category theory; homological algebra – Categories with structure – Enriched categories (over closed or monoidal categories). | Category theory; homological algebra – Categories and geometry – Algebraic K-theory and L-theory. | K-theory – Higher algebraic K-theory – Symmetric monoidal categories. | Algebraic topology – Homotopy theory – Spectra with additional structure ($E\infty$, $A\infty$, ring spectra, etc.). | Algebraic topology – Homotopy theory – Loop space machines, operads.
Classification: LCC QA169 .Y37 2024 | DDC 512/.66–dc23/eng20240823
LC record available at https://lccn.loc.gov/2024025153
Mathematical Surveys and Monographs ISSN: 0076-5376 (print); 2331-7159 (online)
DOI: https://doi.org/10.1090/surv/283

∞ The paper used in this book is acid-free and falls within the guidelines established to ensure permanence and durability.
Visit the AMS home page at `https://www.ams.org/`

10 9 8 7 6 5 4 3 2 1 29 28 27 26 25 24

The author dedicates this book to Jacqueline.

Contents

This is a three-volume work; each volume consists of two parts. The preface and backmatter are identical across all three volumes. The table of contents for each volume is followed by a copy of the tables of contents for the other two volumes.

Reference Convention. In each of Volumes II and III, the numbering of parts, chapters, and pages restarts at 1. Page numbers are prefixed with a volume number so that, for example, page n of Volume II is printed II.n. References (chapters, definitions, theorems, equations, etc.) to a different volume are preceded by I, II, or III according to the volume. For example, $a.b.c$ in Volume I is denoted by I.$a.b.c$ in Volumes II and III, and Chapter N in Volume III is denoted by Chapter III.N in Volumes I and II.

Main Contents of Volume II

Main Contents of Volume III

Preface

Bimonoidal categories are categorical analogues of rings without additive inverses. They have been actively studied in category theory, homotopy theory, and algebraic K-theory since around 1970. There is an abundance of new applications and questions of bimonoidal categories in mathematics and other sciences. This work provides the first unified treatment of bimonoidal and higher ring-like categories, their connection with algebraic K-theory and homotopy theory, and applications to quantum groups and topological quantum computation. With ample background material, extensive coverage, detailed presentation of both well-known and new theorems, and a list of open questions, this work is a user-friendly resource for beginners and experts alike.

Bimonoidal and E_n-Monoidal Categories

A *bimonoidal category* C is a categorical analogue of a rig, which is a ring without additive inverses. In this categorification, the addition, multiplication, 0, and 1 of a rig are replaced by functors and objects in a bimonoidal category. Rig axioms are replaced by natural structure morphisms, along with suitable coherence axioms of their own.

More specifically, in place of the rig addition and multiplication, C has two monoidal structures

$$\big(\mathsf{C}, \oplus, \mathbb{0}, \alpha^{\oplus}, \lambda^{\oplus}, \rho^{\oplus}, \xi^{\oplus}\big) \quad \text{and} \quad \big(\mathsf{C}, \otimes, \mathbb{1}, \alpha^{\otimes}, \lambda^{\otimes}, \rho^{\otimes}\big).$$

The first is symmetric monoidal, and called the *additive structure*. The second is plain monoidal, and called the *multiplicative structure*. As with plain monoidal categories, there are variants with braided or symmetric multiplicative structure, and a variety of intermediate multiplicative structures parametrized by E_n-operads.

In place of distributivity relations in a rig, a bimonoidal category has natural *distributivity monomorphisms* for objects A, B, and C:

$$A \otimes (B \oplus C) \xrightarrow{\delta^l_{A,B,C}} (A \otimes B) \oplus (A \otimes C)$$

$$(A \oplus B) \otimes C \xrightarrow{\delta^r_{A,B,C}} (A \otimes C) \oplus (B \otimes C).$$

These data are required to satisfy a finite list of axioms that (i) are checkable in practice and (ii) ensure that (symmetric/braided) bimonoidal categories have good coherence and other categorical properties. An important special case is a *tight* bimonoidal category, in which the distributivity monomorphisms δ^l and δ^r are *isomorphisms*.

A number of examples, arising in both algebraic and homotopical contexts, are discussed throughout the text. Here, we summarize three important ones. More examples are discussed in the next section about quantum science.

(1) The category of finite dimensional complex vector spaces, $\mathsf{Vect}^{\mathbb{C}}$, is a tight symmetric bimonoidal category with its additive and multiplicative structures given by the usual direct sum and tensor product of vector spaces. More generally, each distributive symmetric monoidal category is a tight symmetric bimonoidal category.
(2) The nonnegative integers and permutations form the objects and the morphisms of a tight symmetric bimonoidal category Σ, called the finite ordinal category.
(3) May's bipermutative categories, with the additional axiom $\xi^{\otimes}_{-,\mathbb{0}} = \mathrm{Id}$, are tight symmetric bimonoidal categories.

The definition and coherence theorems for symmetric bimonoidal categories are due to Laplaza [**Lap72a, Lap72b**]. These theorems and their plain bimonoidal analogues are discussed in detail in Part 1. In addition to providing completely detailed proofs, we also correct some subtle and nontrivial inaccuracies in the original statements and proofs. See Sections 3.11 and 4.7 for related discussion. Just as applications of monoidal categories heavily depend on Mac Lane's coherence theorem, Laplaza's two coherence theorems for symmetric bimonoidal categories, as well as their plain and braided analogues, are crucial to their applications.

Part 2 applies Laplaza's coherence theorems to prove a number of theorems about bimonoidal categories in the context of 2-dimensional categories. These include existence of a bi-initial object, confirming a conjecture of Baez [**Bae18**], and a symmetric monoidal bicategory of matrices, $\mathsf{Mat}^{\mathsf{C}}$, constructed from a tight symmetric bimonoidal category C. (Note the unfortunately subtle confluence of terminology that "symmetric monoidal *bicategory*" and "symmetric *bimonoidal* category" refer to wildly distinct algebraic structures.) In the case $\mathsf{C} = \mathsf{Vect}^{\mathbb{C}}$, $\mathsf{Mat}^{\mathsf{C}}$ is the symmetric monoidal bicategory of *coordinatized 2-vector spaces*, one version of the 2-vector spaces introduced by Kapranov and Voevodsky in [**KV94**].

Braided bimonoidal categories, along with their corresponding coherence and strictification theorems, are discussed in Part II.1. These structures are of interest for applications in quantum science, discussed below. The relevant coherence results are new, confirming a conjecture of Blass and Gurevich [**BG20a**].

Part II.2 introduces a similar but distinct categorification of rigs, called E_n-*monoidal categories*. These have *factorization morphisms* in place of distributivity monomorphisms, and are significant for the K-theoretic applications in Part III.2. The E_n-monoidal structure is a generalization of n-fold monoidal structure due to Balteanu, Fiedorowicz, Schwänzel, and Vogt [**BFSV03**]. Special cases of E_n-monoidal categories include, or are closely related to, the *bipermutative categories* of May [**May77**], the *ring categories* of Elmendorf and Mandell [**EM06**], and the *braided bimonoidal categories* of Richter [**Ric10**].

Applications in Quantum Science

Due to the ubiquity of ring-like structures and categories, bimonoidal categories are increasingly applied in a variety of disciplines in mathematics and other formal sciences related to quantum algebra. To support readers with a variety of interests, this work includes summary and introduction to several such applications.

Quantum Groups. The first part of Chapter II.3 extends a well-known fact in quantum group theory. We observe that the category of left modules over a braided bialgebra, which is also known as a quasitriangular bialgebra in the literature, is a tight braided bimonoidal category.

Topological Quantum Computation. The second part of Chapter II.3 discusses applications of braided bimonoidal categories to topological quantum computation (TQC). We prove that the Fibonacci anyons and the Ising anyons, which are two of the most important models in TQC, are both tight braided bimonoidal categories.

Centers. Monoidal, braided monoidal, and symmetric monoidal categories are connected by the Drinfeld center and the symmetric center. Kassel [**Kas95**] and Majid [**Maj91**] explain the relationship between the Drinfeld center construction, due to Drinfeld in unpublished work, and modules over the Drinfeld double of a a finite dimensional Hopf algebra. (See Note II.1.7.2 for further explanation and context.) Bimonoidal and ring categorical analogues of these center constructions are discussed in Chapters II.4 and II.9.

Reversible Programming. Section 2.6 is a brief illustration that symmetric bimonoidal categories naturally arise in reversible programming of finite types. We observe that there is a symmetric bimonoidal groupoid whose objects are syntax of finite types. Note 2.7.5 directs the reader to further applications in the sheet diagrams of [**CDH∞**] and the work of [**Hin13**] on quantum circuits.

Applications in Algebraic K-Theory

Uses of additive and multiplicative categorical structure in homotopy theory are among the earliest and most well-developed applications of the material from Volumes I and II of this work. Volume III focuses on those applications, beginning with the work of Segal [**Seg74**] that assembles structured ring spectra from permutative categories. Under the Segal K-theory functor K^{Se}, the symmetric monoidal structure of a permutative category C results in the additive structure in the cohomology theory represented by the spectrum $\mathsf{K}^{\mathsf{Se}}(\mathsf{C})$.

The computational importance of multiplicative structure in cohomology motivates significant interest in multiplicative structure for the representing spectra, leading to the highly-structured spectra that are presaged in work of Adams [**Ada95**] and realized in the S-modules of Elmendorf-Kriz-Mandell-May [**EKMM97**], the symmetric spectra of Hovey-Shipley-Smith [**HSS00**], and the orthogonal spectra of Mandell-May-Schwede-Shipley [**MMSS01**], among other equivalent models. Thus, there is corresponding interest in bimonoidal structures for the input categories.

One difficulty, however, is that Segal K-theory does *not* preserve the multiplicative aspect of such structures. As a resolution, the work of Elmendorf and Mandell [**EM06, EM09**] introduces an alternative construction that (a) is suitably equivalent to Segal's construction and (b) preserves multiplicative structure. This is known as Elmendorf-Mandell K-theory, K^{EM}.

The formalism in which these statements can be made precise is that of multicategories, where multilinear functors between permutative categories encode the relevant multiplicative structure. In these terms, the essential difference between the Segal and Elmendorf-Mandell constructions is that the latter is *multifunctorial*, while the former is merely functorial.

Part III.1 develops the necessary supporting theory of enriched monoidal categories and multicategories, but is also of independent interest. Part III.2 contains the applications to algebraic K-theory, including a review of the relevant background in homotopy theory and detailed treatments of both the Segal and Elmendorf-Mandell constructions.

The second half of Part III.2 applies the Elmendorf-Mandell K-theory multifunctor K^{EM} to the E_n-monoidal categories in Part II.2 to produce structured ring spectra. We prove in detail that K^{EM} sends

- small ring categories to strict ring symmetric spectra,
- small bipermutative categories to E_∞-symmetric spectra,
- small braided ring categories to E_2-symmetric spectra, and
- small E_n-monoidal categories to E_n-symmetric spectra for $2 \leq n < \infty$.

The strict ring and E_∞ cases are from [**EM06, EM09**]. The $1 < n < \infty$ cases are new results.

Audience and Features

This work is aimed at graduate students and advanced researchers with an interest in category theory, homotopy theory, algebraic K-theory, and their applications. Below are some features that make this work a unique and user-friendly resource.

Unified Presentation: The literature on bimonoidal categories, higher ring-like categories, enriched monoidal categories, multicategories, and their connection with algebraic K-theory, homotopy theory, and the sciences is scattered across many journal articles over several decades, with varying definitions, notations, and terminology. This work presents these topics in a unified manner, with both well-known and new theorems.

Background Material: To make this work self-contained and to bring the reader quickly up to speed, there is extensive background material on
- basic category theory (Chapter 1),
- 2-dimensional categories (Chapter 6),
- braided structures (Chapter II.1),
- abelian categories (Section II.2.3),
- braided, also known as quasitriangular, bialgebras (Section II.3.1),
- enriched monoidal categories (Chapters III.1, III.2, and III.3),
- pointed objects and pointed diagram categories (Chapter III.4),
- enriched multicategories (Chapters III.5 and III.6), and
- homotopy theory (Chapter III.7).

These chapters and sections form a substantial portion of this work.

Open Questions: Appendix A discusses open questions related to the topics of this work. The reader is encouraged to take advantage of these open questions and use them as a springboard to read the main text.

Detailed Discussion: This work contains many highly detailed and carefully structured proofs for both known and new theorems. For each major result, our discussion has much more detail than one would normally find in the literature. Our detailed discussion has several purposes.

Exercises with Solutions. Our detailed presentation makes the material accessible to a diverse audience, including those who are new to bimonoidal and higher ring-like categories and algebraic K-theory. Students are encouraged to regard the numerous detailed proofs as exercises with full solutions. Each result whose proof has many different parts has been carefully structured to make it easy for the reader to jump forward and backward.

Axioms. Symmetric bimonoidal categories are defined by 24 axioms, and the list of axioms for (braided) bimonoidal categories is similarly substantial. Our detailed discussion helps the reader see exactly where these axioms are used and why they are needed.

Laplaza's Theorems. The Coherence Theorems 3.9.1 and 4.4.3 for symmetric bimonoidal categories are central results in this subject that have been cited and used numerous times in the literature. Their original proofs given by Laplaza in [**Lap72a, Lap72b**] were written in outline form, with much detail and some cases in the proofs completely omitted. Moreover, Laplaza's original proofs and statements of these theorems have several subtle and nontrivial inaccuracies that have never been made explicit before and are not easy to spot. For both archival and educational purposes, we present fully detailed proofs of these theorems and correct the inaccuracies. Sections 3.11 and 4.7 have more related discussion.

K-Theory Multifunctors. The K-theory multifunctors in Chapters III.9 and III.10, due to Elmendorf-Mandell [**EM06, EM09**], are fundamental constructions for multiplicative structure of algebraic K-theory spectra. They are essential for our development of E_n-monoidal symmetric spectra from corresponding structure on small permutative categories. We use the theory of enriched monoidal categories and enriched multicategories from Part III.1 to give complete explanations of the constructions and their properties. This treatment corrects an inaccuracy in the statement of [**EM09**, Theorem 1.3] and some other statements about expanding the domain of the K-theory multifunctor. The basic issue has to do with monoidal units and, to the authors' knowledge, has not been previously explained. See Note III.10.8.2 for further discussion.

Reading Guides: In addition to a detailed introduction, almost every chapter has a brief *Reading Guide* that provides an alternative to reading that chapter linearly. Our presentation in the main text follows a straightly logical order and has a lot of detail. By following the reading guide, it is possible to first obtain a bird's-eye view of that chapter before digesting all the detail. The end of this Preface also includes several thematic reading guides for salient topics that span multiple chapters.

Motivation and Explanation: Main definitions and results are often preceded by discussion that motivates the upcoming definitions and proofs. Whenever useful, definitions and results are followed by a detailed explanation that interprets and unpacks the various components. In the text, these are clearly marked as *Motivation* and *Explanation*, respectively. Examples include Motivation 2.1.1, Explanation 2.4.7, and Section 4.1.

Organization: There are extensive cross-references throughout the text. In addition to a detailed index, there are lists of main facts and notations, each organized by chapters. While the text follows a strictly logical order, it is not necessary to read the chapters in a linear order. The reader can jump straight to a section and use the extensive cross-references to fill in the necessary definitions and facts.

Part and Chapter Summaries

Part 1: Symmetric Bimonoidal Categories

This part studies symmetric bimonoidal categories and bimonoidal categories (Chapter 2). It presents highly detailed proofs of Laplaza's Coherence Theorems for symmetric bimonoidal categories (Chapters 3 and 4), May's Strictification Theorem for tight symmetric bimonoidal categories (Chapter 5), and their nonsymmetric analogues for bimonoidal categories. The only prerequisite for this part is some basic knowledge of category theory, which is summarized in Chapter 1.

Part 2: Bicategorical Aspects of Symmetric Bimonoidal Categories

Applying Laplaza's Coherence Theorems, this part proves several new theorems on the connection between symmetric bimonoidal categories and bicategories. All the necessary definitions of 2-dimensional category theory are summarized in Chapter 6. The first main result is a confirmation of Baez's Conjecture (Chapter 7) that proves the existence of a bi-initial object in a 2-category of symmetric bimonoidal categories. Chapter 8 proves that a matrix construction $\mathsf{Mat}^{\mathsf{C}}$ sends each tight symmetric bimonoidal category to a symmetric monoidal bicategory.

Part II.1: Braided Bimonoidal Categories

Starting with a preliminary chapter on the braid groups and braided monoidal categories, this part is a detailed study of braided bimonoidal categories (Chapter II.2), which are strictly more general than Richter's [**Ric10**] and the BD categories of Blass-Gurevich [**BG20a**]. This part discusses applications to quantum groups and topological quantum computation (Chapter II.3), bimonoidal centers (Chapter II.4), coherence and strictification of braided bimonoidal categories (Chapters II.5 and II.6), and the braided versions of Baez's Conjecture and the matrix construction (Chapters II.7 and II.8). Our coherence and strictification theorems confirm the Blass-Gurevich Conjecture. The main theorems in Parts 1 and 2 are used in this part.

Part II.2: E_n-Monoidal Categories

This part studies a closely related variant of bimonoidal categories, called ring categories, and their bipermutative, braided, and higher analogues, called E_n-monoidal categories. Ring and bipermutative categories are due to Elmendorf-Mandell [**EM06, EM09**]. An E_n-monoidal category combines n ring categories with a common additive structure and an n-fold monoidal category as in [**BFSV03**]. The categories in this part are applied in Part III.2 to obtain E_n-symmetric spectra via algebraic K-theory. This part is independent of the earlier parts, except for some definitions and statements of theorems.

Part III.1: Enriched Monoidal Categories and Multicategories

To prepare for Part III.2, this part lays the groundwork on enriched monoidal categories (Chapters III.1, III.2, and III.3), smash products (Chapter III.4), and multicategories (Chapters III.5 and III.6). In addition to their roles in the Segal K-theory functor and the Elmendorf-Mandell K-theory multifunctor, the detailed discussion of enriched monoidal categories—including change of enrichment, coherence, self-enrichment, and the Enriched Yoneda Lemma—and multicategories is also of independent interest. These chapters assume only a basic knowledge of monoidal categories, as summarized in Section III.1.1.

Part III.2: Algebraic K-Theory

This part studies the interconnection between E_n-monoidal categories (Part II.2), homotopy theory (Chapter III.7), and algebraic K-theory. The first half discusses in detail the Segal K-theory functor (Chapter III.8) and the Elmendorf-Mandell K-theory multifunctor (Chapters III.9 and III.10) from small permutative categories to symmetric spectra. The second half (Chapters III.11, III.12, and III.13) applies the K-theory multifunctor to small ring, bipermutative, braided ring, and E_n-monoidal categories to obtain, respectively, strict ring, E_∞-, E_2-, and E_n-symmetric spectra. These structured ring spectra are fundamental objects in homotopy theory. Our discussion shows how they arise from E_n-monoidal categories via algebraic K-theory.

In the main text, each chapter starts with a detailed introduction. A summary of each chapter follows.

Part 1: Symmetric Bimonoidal Categories

Chapter 1: Basic Category Theory

To make this book self-contained, this chapter reviews the basics of category theory, starting from the definitions of categories, functors, and natural transformations. Then it discusses adjunctions, equivalences of categories, (co)limits, (co)ends, and Kan extensions. The remaining sections review (symmetric) monoidal categories, (symmetric) monoidal functors, monoidal natural transformations, and their coherence theorems.

Chapter 2: Symmetric Bimonoidal Categories

This chapter introduces symmetric bimonoidal categories and bimonoidal categories. Then we prove Laplaza's Theorem 2.2.13 that says that half of the 24 symmetric bimonoidal category axioms are formal consequences of the other 12 axioms. The weaker bimonoidal analogue is Proposition 2.2.14. The remaining sections discuss examples of symmetric bimonoidal categories, including distributive symmetric monoidal categories, the finite ordinal category Σ, a variant Σ', and left and right bipermutative categories. The finite ordinal category Σ is an important part of (i) the distortion category $\mathcal{D}$ (Chapter 4) used in Laplaza's Second Coherence Theorem 4.4.3, (ii) Baez's Conjecture (Chapter 7), and (iii) the braided version of Baez's Conjecture (Chapter II.7). Section 2.6 contains an application of symmetric bimonoidal categories to reversible programming of finite types.

Chapter 3: Coherence of Symmetric Bimonoidal Categories

This chapter proves Laplaza's First Coherence Theorem 3.9.1 for symmetric bimonoidal categories that satisfy a monomorphism assumption. This assumption is automatically satisfied if tightness—that is, the invertibility of the distributivity morphisms δ^l and δ^r—is assumed, but the general form of this theorem only requires that the distributivity morphisms be natural monomorphisms. The analogue of this coherence theorem for bimonoidal categories is Theorem 3.10.7. Section 3.11 discusses the main differences between this chapter and Laplaza's original work in [**Lap72a**].

Chapter 4: Coherence of Symmetric Bimonoidal Categories II

This chapter proves Laplaza's Second Coherence Theorem 4.4.3 for symmetric bimonoidal categories that satisfy the same monomorphism assumption as in Theorem 3.9.1. The analogue of this coherence theorem for bimonoidal categories is Theorem 4.5.8. Section 4.7 discusses the main differences between this chapter and Laplaza's original work in [**Lap72b**]. Both Coherence Theorems 3.9.1 and 4.4.3 say that some formal diagrams in certain symmetric bimonoidal categories commute. The first theorem has an assumption called *regularity* on the common domain of the two paths involved, which is analogous to Mac Lane's Coherence Theorem 1.3.3 for monoidal categories. The second theorem has an assumption about the two paths themselves, which is reminiscent of the Joyal-Street Coherence Theorem II.1.6.3 for braided monoidal categories. In Chapter II.5, we observe that the second, but not the first, theorem has a braided analogue.

Chapter 5: Strictification of Tight Symmetric Bimonoidal Categories

This chapter proves May's Strictification Theorem 5.4.6 of *tight* symmetric bimonoidal categories to right bipermutative categories. The latter are tight symmetric bimonoidal categories whose additive structures and multiplicative structures are both permutative categories, and whose structure morphisms $\lambda^{\bullet}$, $\rho^{\bullet}$, δ^r, and $\xi^{\otimes}_{-,\mathbb{0}}$ are identities. Unlike the Coherence Theorems 3.9.1 and 4.4.3, the strictification theorem requires the tightness assumption. Our detailed proofs show exactly where the invertibility of δ^l and δ^r is used. Theorem 5.4.7 is another version of the strictification theorem involving *left* bipermutative categories, in which δ^l, instead of δ^r, is the identity. Theorems 5.5.11 and 5.5.12 are the corresponding strictification results for tight bimonoidal categories. Section 5.6 briefly discusses the history of related strictification theorems and claims. The proofs in this chapter are repurposed in Chapter II.6 to prove the strictification form of the Blass-Gurevich conjecture for braided bimonoidal categories.

Part 2: Bicategorical Aspects of Symmetric Bimonoidal Categories

Chapter 6: Definitions from Bicategory Theory

Without assuming any knowledge of 2-dimensional categories, in this chapter we review the basics of 2-/bicategories, pasting diagrams, lax functors, lax transformations, modifications, and adjunctions in bicategories. Then it reviews multiplicative structures, including monoidal bicategories, their braided, sylleptic, and symmetric analogues, the Gray tensor product for 2-categories, (permutative) Gray monoids, and permutative 2-categories. Most of these topics are discussed in detail in the book [**JY21**].

Chapter 7: Baez's Conjecture

This chapter proves Baez's Conjecture (Theorems 7.8.1 and 7.8.3). Section 7.1 defines a 2-category $\mathsf{Bi}_{\mathsf{r}}^{\mathsf{fsy}}$ with *flat* small symmetric bimonoidal categories as objects. Flatness (Definition 3.9.9) is much weaker than tightness, and it guarantees that the Coherence Theorems 3.9.1 and 4.4.3 are applicable. The first version of Baez's Conjecture (Theorem 7.8.1) says that the finite ordinal category Σ is a lax bicolimit of the 2-functor $\varnothing \longrightarrow \mathsf{Bi}_{\mathsf{r}}^{\mathsf{fsy}}$. Another version is Theorem 7.8.3, which says that the variant Σ' of Σ is also such a lax bicolimit. We emphasize that our proof of Baez's Conjecture does *not* use the Strictification Theorems 5.4.6 and 5.4.7. This allows us to use flat small symmetric bimonoidal categories in the 2-category $\mathsf{Bi}_{\mathsf{r}}^{\mathsf{fsy}}$, instead of the smaller class of tight ones. Section 7.9 discusses the relationship between our version of Baez's Conjecture and the more restricted version in [**CDH∞, Elg21**] for rig categories, which are multiplicatively nonsymmetric and tight.

Chapter 8: Symmetric Monoidal Bicategorification

This chapter proves Theorem 8.15.4. It says that, for each tight symmetric bimonoidal category C, a matrix construction $\mathsf{Mat}^{\mathsf{C}}$ is a symmetric monoidal bicategory, with no strict structure morphisms in general. Therefore, the construction $\mathsf{Mat}^{\mathsf{C}}$ is a direct connection between tight symmetric bimonoidal categories and symmetric monoidal bicategories. The objects in $\mathsf{Mat}^{\mathsf{C}}$ are nonnegative integers. Its 1-/2-cells are matrices whose entries are objects/morphisms in C. The horizontal composition in the bicategory $\mathsf{Mat}^{\mathsf{C}}$ uses the usual matrix product. The monoidal composition in its monoidal bicategory structure uses the matrix tensor product, which is also known as the Kronecker product. The category of coordinatized 2-vector spaces, which is $\mathsf{Mat}^{\mathsf{C}}$ with $\mathsf{C} = \mathsf{Vect}^{\mathbb{C}}$, is such a symmetric monoidal bicategory. This chapter uses the Coherence Theorems 3.9.1 and 3.10.7 and the graph theoretic machinery in Chapter 3, but neither the Coherence Theorem 4.4.3 nor the Strictification Theorems 5.4.6 and 5.4.7.

Part II.1: Braided Bimonoidal Categories

Chapter II.1: Preliminaries on Braided Structures

To prepare for the rest of Part II.1, this chapter discusses the braid groups and braided monoidal categories. First it defines the braid groups and proves some useful properties for sum braids and block braids. Then it reviews braided monoidal categories and proves some basic properties, including two manifestations of the third Reidemeister move. Next it proves in detail that the Drinfeld center of a monoidal category is a braided monoidal category and that the symmetric center of a braided monoidal category is a symmetric monoidal category. Then it recalls the Joyal-Street Coherence Theorem II.1.6.3 for braided monoidal categories.

Chapter II.2: Braided Bimonoidal Categories

This chapter defines braided bimonoidal categories. They are defined using 12 of the 24 Laplaza axioms of a symmetric bimonoidal category, together with two additional axioms that are variants of the only two Laplaza axioms involving the braiding $\xi^{\otimes}$. In a symmetric bimonoidal category, each of these two variant axioms is equivalent to the original Laplaza axiom. This is reminiscent of the fact that a braided monoidal category has two hexagon axioms, which are equivalent to each other in a symmetric monoidal category. A *tight* braided bimonoidal category—that is, one with invertible distributivity morphisms δ^l and δ^r—is equivalent to a BD category in the sense of Blass and Gurevich [**BG20a**]. The first main observation in this chapter is Theorem II.2.2.1, which says that each braided bimonoidal category satisfies all 24 Laplaza axioms. Therefore, a symmetric bimonoidal category is precisely a braided bimonoidal category whose braiding satisfies the symmetry axiom. The second main result in this chapter says that an abelian category with a compatible (symmetric/braided) monoidal structure is a tight (symmetric/braided) bimonoidal category. The additive structure comes from the abelian structure, and the multiplicative structure comes from the monoidal structure. The braided case of this result is due to Blass and Gurevich [**BG20a**].

Chapter II.3: Applications to Quantum Groups and Topological Quantum Computation

This chapter shows that braided bimonoidal categories arise naturally in quantum groups and topological quantum computation (TQC). The first main observation is Theorem II.3.2.19. It says that for a (symmetric/braided) bialgebra A, the category $\mathsf{Mod}(A)$ of left A-modules, equipped with the usual direct sum and tensor product, is a tight (symmetric/braided) bimonoidal category. This is an extension of the important fact in quantum group theory that, for a braided bialgebra A, $\mathsf{Mod}(A)$ is a braided monoidal category. Next we prove in detail that Fibonacci anyons and Ising anyons, which are two central models in TQC, are both tight braided bimonoidal categories. In each case, the additive structure comes from an abelian category structure, and the multiplicative structure comes from the fusion rules of anyons.

Chapter II.4: Bimonoidal Centers

This chapter generalizes the Drinfeld center of a monoidal category and the symmetric center of a braided monoidal category (Sections II.1.4 and II.1.5) to the bimonoidal setting. Generalizing the Drinfeld center, Theorem II.4.4.3 says that, for each tight bimonoidal category C, the bimonoidal Drinfeld center $\overline{\mathsf{C}}^{\mathsf{bi}}$ is a tight braided bimonoidal category. Tightness is required for this theorem because the invertibility of δ^l and δ^r is used in the construction of $\overline{\mathsf{C}}^{\mathsf{bi}}$. The proof of this theorem is another good illustration of the axioms of a braided bimonoidal category, since we will use all 24 Laplaza axioms and the two variant axioms in the braided case. Generalizing the symmetric center, Theorem II.4.5.3 says that, for each braided bimonoidal category C, the bimonoidal symmetric center $\mathsf{C}^{\mathsf{sym}}$ is a symmetric bimonoidal category.

Chapter II.5: Coherence of Braided Bimonoidal Categories

This chapter proves the Coherence Theorem II.5.4.4 for braided bimonoidal categories that satisfy a monomorphism assumption. As in the symmetric case (Theorems 3.9.1 and 4.4.3), the monomorphism assumption in Theorem II.5.4.4 is automatically satisfied if tightness is assumed. This theorem is the braided analogue of Laplaza's Second Coherence Theorem 4.4.3. It uses a braided version $\mathcal{D}^{\mathsf{br}}$ of the distortion category that involves the symmetric groups and the braid groups to keep track of, respectively, the additive symmetry $\xi^{\oplus}$ and the braiding $\xi^{\otimes}$. Reminiscent of the Joyal-Street Coherence Theorem II.1.6.3 for braided monoidal categories, Theorem II.5.4.4 says that, if two paths have the same image in the braided distortion category $\mathcal{D}^{\mathsf{br}}$, then they have the same value in the braided bimonoidal category in question. This condition of having the same image in $\mathcal{D}^{\mathsf{br}}$ is very much checkable in practice. In fact, the proofs of the main results in Chapters II.6, II.7, and II.8 all use Theorem II.5.4.4 and involve checking this condition many times. In [**BG20a**], Blass and Gurevich conjectured the existence of a coherence theorem for their BD categories, which are equivalent to our tight braided bimonoidal categories. Theorem II.5.4.4 confirms the Blass-Gurevich Conjecture in the form of commutative formal diagrams.

Chapter II.6: Strictification of Tight Braided Bimonoidal Categories

This chapter proves two Strictification Theorems II.6.3.6 and II.6.3.7 for tight braided bimonoidal categories. As in the symmetric case (Theorems 5.4.6 and 5.4.7), strictification requires the tightness assumption because the construction of the strictification uses the invertibility of the distributivity morphisms δ^l and δ^r. A *right permbraided category* is a tight braided bimonoidal category with both the additive and the multiplicative structures strict monoidal, and with identities for the structure morphisms $\lambda^\bullet$, $\rho^\bullet$, δ^r, $\xi^\otimes_{-,\mathbb{0}}$, and $\xi^\otimes_{\mathbb{0},-}$. Theorem II.6.3.6 says that each tight braided bimonoidal category is adjoint equivalent to a right permbraided category via strong braided bimonoidal functors. Theorem II.6.3.7 is the analogue that strictifies each tight braided bimonoidal category to a *left* permbraided category, in which δ^l, instead of δ^r, is the identity. Theorems II.6.3.6 and II.6.3.7 are two further positive answers to the Blass-Gurevich Conjecture [**BG20a**] in the form of strictification.

Chapter II.7: The Braided Baez Conjecture

This chapter proves the braided version of Baez's Conjecture. Section II.7.1 defines the 2-category $\mathsf{Bi}_\mathsf{r}^{\mathsf{fbr}}$ with *flat* small braided bimonoidal categories as objects. As in the symmetric case, flatness (Definition II.5.4.5) is much weaker than tightness, and it guarantees that the Braided Bimonoidal Coherence Theorem II.5.4.4 is applicable. The first version of the Braided Baez Conjecture (Theorem II.7.3.4) says that the finite ordinal category Σ is a lax bicolimit of the 2-functor $\varnothing \longrightarrow \mathsf{Bi}_\mathsf{r}^{\mathsf{fbr}}$. Another version is Theorem II.7.3.6, which says that the variant Σ' of Σ is also such a lax bicolimit. Also like the symmetric case, the proofs of the Braided Baez Conjecture do *not* use the Strictification Theorems II.6.3.6 and II.6.3.7. This allows us to use flat small braided bimonoidal categories in the 2-category $\mathsf{Bi}_\mathsf{r}^{\mathsf{fbr}}$, instead of the smaller class of tight ones. The reader may wonder why the finite ordinal category Σ and its variant Σ' are bi-initial in both the symmetric case (Theorems 7.8.1 and 7.8.3) and the braided case. This is analogous to the fact that the ring of integers is initial in both the category of rings and the category of commutative rings.

Chapter II.8: Monoidal Bicategorification

The main Theorem II.8.4.7 in this chapter says that, for each tight braided bimonoidal category C, the matrix construction Mat^C is a monoidal bicategory. While most of the definitions for Mat^C are the same as in the symmetric case in Chapter 8, there are two subtleties. First, in the current braided case, the lax functoriality constraint $\boxtimes^2$ of the monoidal composition $\boxtimes$ in Mat^C has two additional conditions about the braided distortions of the two paths involved; see (II.8.2.15) and (II.8.2.16). These conditions about the braided distortions are necessary because a braid is not determined by its underlying permutation, and the braided distortion category $\mathcal{D}^{\mathsf{br}}$ involves the braid groups. The second subtle point is that, even if C is a tight braided bimonoidal category, the monoidal bicategory Mat^C does *not* seem to have any reasonable braided monoidal bicategory structure in general. We will explain this point in more detail near the end of Section II.8.4. The difficulty once again comes from the fact that the braided distortion category $\mathcal{D}^{\mathsf{br}}$ involves the braid groups, and a braid with an identity underlying permutation is usually not the identity braid.

Part II.2: E_n-Monoidal Categories

Chapter II.9: Ring, Bipermutative, and Braided Ring Categories

This chapter discusses ring and bipermutative categories in the sense of Elmendorf-Mandell and the braided version. The main difference between these categorical notions and their bimonoidal counterparts in Parts 1 and II.1 is that ring categories have generally non-invertible *factorization morphisms*

$$(A \otimes C) \oplus (B \otimes C) \xrightarrow{\partial^l_{A,B,C}} (A \oplus B) \otimes C$$

$$(A \otimes B) \oplus (A \otimes C) \xrightarrow{\partial^r_{A,B,C}} A \otimes (B \oplus C)$$

that go in the opposite direction as the distributivity morphisms δ^r and δ^l. Ring categories with invertible factorization morphisms are special cases of tight bimonoidal categories, so the latter's strictification theorems in Chapter 5 also apply to such ring categories. The bipermutative and braided analogues are also true. Similar to the endomorphism rig of a commutative monoid, each small permutative category C yields an endomorphism ring category $\mathsf{Perm}^{\mathsf{su}}(\mathsf{C};\mathsf{C})$. Similar to the reduction of Laplaza's axioms in symmetric bimonoidal categories in Section 2.2 and the braided version in Theorem II.2.2.1, about half of the ring category axioms are redundant in a bipermutative category and a braided ring category. This is an extension of an observation in [**EM06**, Fig. 1]. Moreover, the Drinfeld center and the symmetric center have natural analogues for these ring-like categories. As we will discuss in Chapters III.11 and III.12, the Elmendorf-Mandell K-theory multifunctor sends small ring, braided ring, and bipermutative categories to, respectively, strict ring, E_2-, and E_∞-symmetric spectra. The strict ring and E_∞ cases are due to Elmendorf-Mandell [**EM06, EM09**], and the E_2 case is new.

Chapter II.10: Iterated and E_n-Monoidal Categories

Keeping in mind that the ring-like categories in Chapter II.9 correspond to E_n-symmetric spectra for $n \in \{1, 2, \infty\}$ via algebriac K-theory, this chapter discusses the categorical structure for the general E_n cases. An n-fold monoidal category in the sense of [**BFSV03**] has n monoidal structures $\otimes_1, \ldots, \otimes_n$ that are strictly associative and unital and interact via the exchange natural transformations

$$(A \otimes_j B) \otimes_i (C \otimes_j D) \xrightarrow{\eta^{i,j}_{A,B,C,D}} (A \otimes_i C) \otimes_j (B \otimes_i D)$$

for $1 \leq i < j \leq n$. Monoids in the monoidal category of small n-fold monoidal categories are precisely small $(n+1)$-fold monoidal categories. We introduce the notion of an *E_n-monoidal category* as a permutative category $(\mathsf{C}, \oplus)$ equipped with an n-fold monoidal structure $\{\otimes_i, \eta^{i,j}\}$ and factorization morphisms $\{\partial^{l,i}, \partial^{r,i}\}$ for each monoidal structure $\otimes_i$, such that (i) each $(\oplus, \otimes_i, \partial^{l,i}, \partial^{r,i})$ is a ring category and (ii) several axioms relating $\eta^{i,j}$, $\partial^{l,i}$, and $\partial^{r,i}$ hold. Ring categories are E_1-monoidal categories. Braided ring categories and bipermutative categories are special cases of, respectively, E_2- and E_n-monoidal categories for $n \geq 2$. Moreover, each small category generates a free E_n-monoidal category. In Chapter III.13, we will show that the Elmendorf-Mandell K-theory of a small E_n-monoidal category is an E_n-symmetric spectrum for $n \geq 1$.

Part III.1: Enriched Monoidal Categories and Multicategories

Chapter III.1: Enriched Monoidal Categories

This chapter gives the basic definitions and properties for enriched monoidal categories, including plain, braided, and symmetric variants. Definition III.1.4.25 describes 2-categories of each, with 1- and 2-cells given by appropriately monoidal enriched functors and natural transformations, respectively. For our applications to K-theory in Part III.2, the enriching category V is symmetric monoidal closed.

However, our treatment in this chapter addresses the more general case that V is merely monoidal, with additional assumptions about braided or symmetric monoidal structure stated as necessary.

Section III.1.5 discusses the important special case V = Cat, the category of small categories with its Cartesian product. Explanation III.1.5.3 describes how the monoidal V-categories in this case are strict versions of monoidal bicategories. The braided and symmetric cases are similarly compared.

Chapter III.2: Change of Enrichment

This chapter describes change of enriching category induced by a symmetric monoidal functor, showing that monoidal structures are preserved. Sections III.2.1 through III.2.4 give a thorough treatment of 2-functoriality results. As an application, Corollary III.2.4.17 shows that taking underlying categories gives a 2-functor from small monoidal V-categories, V-functors, and V-natural transformations to ordinary monoidal categories, functors, and natural transformations. Similar statements hold for the braided and symmetric cases.

A partial reverse of Corollary III.2.4.17 is given in Theorem III.2.5.1. The theorem shows that, for given V-enriched data, various enriched monoidal axioms are satisfied if and only if the corresponding monoidal axioms for the underlying data are satisfied. This provides a mechanism to lift ordinary monoidal structures to enriched monoidal structures.

Sections III.2.5 and III.2.6 apply Theorem III.2.5.1 to lift coherence and strictification results for ordinary monoidal, braided, and symmetric monoidal categories to their enriched counterparts. The Enriched Monoidal Coherence Theorem III.2.5.6 and Enriched Epstein's Coherence Theorem III.2.5.8 play a significant role in subsequent chapters.

Chapter III.3: Self-Enrichment and Enriched Yoneda

This chapter restricts to the case that V is a symmetric monoidal closed category. Theorem III.3.3.2 shows, via Theorem III.2.5.1, that the canonical enrichment of V over itself is symmetric monoidal as a V-category. The next several sections develop the theory of V-enriched co/ends followed by the V-Yoneda Lemma (Theorem III.3.6.9) and an equivalent form called the V-Yoneda Density Theorem III.3.7.8. These are applied to develop the Day convolution and internal hom for enriched diagram categories (Theorem III.3.7.22). The remainder of the chapter discusses additional theory of enriched diagram categories and tensor/cotensor structures that will be important for the development of enriched K-theory functors in Part III.2.

Chapter III.4: Pointed Objects, Smash Products, and Pointed Homs

This chapter gives the definitions and properties of smash products and pointed homs. These will be used throughout Part III.2, and the smash product of pointed multicategories, developed in Chapter III.5, will be particularly important.

Section III.4.3 uses the Day convolution and internal hom to develop symmetric monoidal closed structure for pointed diagram categories. The results are summarized in Theorem III.4.3.37. Applications of this material appear in Chapters III.8, III.9, and III.10, where the Segal and Elmendorf-Mandell K-theory constructions are given via certain pointed diagram categories.

Chapter III.5: Multicategories

This chapter gives relevant background on multicategories, multifunctors, and multinatural transformations. Theorem III.5.5.14 shows that the category of small multicategories is complete and cocomplete. The Boardman-Vogt tensor product of multicategories, and the associated smash product for pointed multicategories, are developed in Section III.5.6. The corresponding internal hom and its pointed variant are developed in Section III.5.7.

Chapter III.6: Enriched Multicategories

This chapter develops basic definitions and properties for enriched multicategories. One of our important applications, developed in Section III.6.3, is the enriched multicategory associated to an enriched symmetric monoidal category. Our first use of this is in Section III.6.4 where we describe the Cat-enriched multicategory structure on $\mathsf{Multicat}$, the category of small multicategories. It is induced by showing that the tensor product makes $\mathsf{Multicat}$ symmetric monoidal as a Cat-enriched category (Theorem III.6.4.3). The pointed variant, with the smash product of small pointed multicategories, is given in Theorem III.6.4.4 and will be essential for Part III.2.

Sections III.6.5 and III.6.6 cover our second important application of enriched multicategories. The category $\mathsf{PermCat}^{\mathsf{su}}$, consisting of small permutative categories and strictly unital symmetric monoidal functors, has a Cat-enriched multicategory structure given by multilinear functors and multilinear transformations (Definitions III.6.5.4 and III.6.5.11). Propositions III.6.5.10 and III.6.5.13 show that this Cat-enriched multicategory structure is induced from that of small pointed multicategories and their smash product. Section III.6.6 gives a second, direct proof of the Cat-enriched multicategory axioms.

Part III.2: Algebraic K-Theory

Chapter III.7: Homotopy Theory Background

This chapter gives relevant background from homotopy theory. Sections III.7.1 and III.7.2 introduce simplicial sets and simplicial homotopy, along with the nerve and geometric realization functors. The category of symmetric spectra, with its symmetric monoidal closed structure, is presented in Sections III.7.3 through III.7.6. Then, Sections III.7.7 and III.7.8 give a short review of Quillen model categories and a number of key examples.

Chapter III.8: Segal K-Theory of Permutative Categories

This chapter presents the K-theory functor K^{Se} due to Segal [**Seg74**]. Its inputs are small permutative categories and its outputs are symmetric spectra. Section III.8.3 describes the key construction as given by Segal. Sections III.8.4 and III.8.5 describe an equivalent construction that compares more readily with the K-theory multifunctor of Elmendorf-Mandell, K^{EM}.

Chapter III.9: Categories of $\mathcal{G}_*$-Objects

This chapter is the first of two that replace the Segal K-theory functor with a simplicially-enriched multifunctor due to Elmendorf-Mandell [**EM06, EM09**]. This chapter focuses on the replacement of Γ-categories and Γ-simplicial sets with pointed diagrams out of a larger indexing category $\mathcal{G}$. The construction of symmetric spectra from such diagram categories is given in Section III.9.3 and is denoted $\mathsf{K}^{\mathcal{G}}$. Sections III.9.2 and III.9.4 use the material from Part III.1 to explain that the new diagram categories and the new functor K^g are symmetric monoidal, in the enriched sense of Chapter III.1, over the category of pointed simplicial sets.

Chapter III.10: Elmendorf-Mandell K-Theory of Permutative Categories

This chapter is the second of two that replace the Segal K-theory functor with a simplicially enriched multifunctor due to Elmendorf-Mandell [**EM06, EM09**]. This chapter focuses on the construction of $\mathcal{G}_*$-categories from small permutative categories, replacing Segal's construction of Γ-categories from the same. Additional material from Part III.1 is used throughout the chapter to explain that the multi/categories and multi/functors are enriched either in the symmetric monoidal sense of Chapter III.1 or in the multicategorical and multifunctorial sense of Chapter III.6. Section III.10.6 contains the proof that the Segal and Elmendorf-Mandell K-theory symmetric spectra associated to a small permutative category C are level equivalent (Theorem III.10.6.10). Because K^{EM} is an enriched multifunctor, it preserves operad actions. We state this result as Theorem III.10.3.33 and apply it in Chapters III.11, III.12, and III.13.

Chapter III.11: K-Theory of Ring and Bipermutative Categories

This is the first of three chapters that contain algebraic K-theory applications of the ring-like categories in Part II.2. The main K-theory results in this chapter, Corollaries III.11.3.16 and III.11.6.12, are from [**EM06, EM09**], and they are the E_1 and the E_∞ cases. These results state that the Elmendorf-Mandell K-theory multifunctor K^{EM} sends (i) small ring categories to strict ring symmetric spectra and (ii) small bipermutative categories to E_∞-symmetric spectra. They are obtained by combining the multifunctor K^{EM} and the fact that the associative operad and the Barratt-Eccles operad parametrize, respectively, ring and bipermutative category structures on small permutative categories. Since the associative operad has monoids as algebras and the Barratt-Eccles operad is an E_∞-operad, the K-theory results follow.

Chapter III.12: K-Theory of Braided Ring Categories

This chapter contains the E_2 analogues of the results in Chapter III.11. The first part of this chapter discusses the braid operad Br, which generalizes the Barratt-Eccles operad. This is a categorical E_2-operad (Theorem III.12.2.4) whose algebras in Cat are small braided strict monoidal categories (Proposition III.12.3.22). The main categorical input is Theorem III.12.4.5, which says that Br parametrizes braided ring category structures, as in Chapter II.9, on small permutative categories. Applying the K-theory multifunctor K^{EM}, it follows that K^{EM} sends small braided ring categories to E_2-symmetric spectra (Corollary III.12.5.3). The K-theory result, Corollary III.12.5.3, and the main categorical input, Theorem III.12.4.5, are new results.

Chapter III.13: K-Theory of E_n-Monoidal Categories

This chapter contains the general E_n analogues for $n \geq 1$ of the categorical and K-theory results in Chapters III.11 and III.12. The first part of this chapter discusses the n-fold monoidal category operad Mon^n. This is a categorical E_n-operad (Theorem III.13.2.1) whose algebras in Cat are small n-fold monoidal categories (Proposition III.13.3.18) as in Chapter II.10. The main categorical input is Theorem III.13.4.12, which says that Mon^n parametrizes E_n-monoidal category structures on small permutative categories. Applying the K-theory multifunctor

K^{EM}, it follows that K^{EM} sends small E_n-monoidal categories to E_n-symmetric spectra for $n \geq 1$ (Corollary III.13.5.2). As in Chapter III.12, the K-theory result, Corollary III.13.5.2, and the main categorical input, Theorem III.13.4.12, are new results.

Appendix A: Open Questions

This chapter discusses open questions related to the topics of this work. We encourage the reader to read these open questions at any time and use them as additional motivation for the main text.

Reading Guides

Supplementing the chapter introductions and individual reading guides therein, the following guides describe themes that span multiple chapters. For especially broad topics, we include a selection of general references for background or further reading. The Notes section at the end of each chapter provides additional references relevant to the content of that chapter.

Category Theory. For a refresher of basic category theory, including braided and symmetric monoidal categories, read Chapters 1 and II.1. For bicategories and 2-categories, read Chapter 6. For abelian categories, read Section II.2.3. For enriched category theory, read Section III.1.2 and Chapter III.3. Bimonoidal categories are built upon monoidal categories. Thus, a thorough understanding of the definitions and coherence of monoidal categories is necessary to understand bimonoidal categories and their coherence.

References for basic category theory include [**Awo10, BK00, Gra18, Lei14, Rie16, Rom17, Sim11**]. References for more advanced category theory include [**Bor94a, Bor94b, Bor94c, ML98, Mit65, Sch72**]. References for Abelian categories include [**EGNO15, Fre03, Mit65**]. References for enriched categories include [**Bor94b, Cru09, For04, Kel05**]. References for ends and coends include [**Day70, DK69, Lor21**]. For further reference on 2-dimensional categories, we highly recommend [**JY21**].

Symmetric Bimonoidal Categories. To review the axioms of a symmetric bimonoidal category, read Sections 2.1 and 2.2. For Laplaza's Coherence Theorems for symmetric bimonoidal categories, read Theorems 3.9.1 and 4.4.3. For strictification theorems, read Theorems 5.4.6, 5.4.7, 5.5.11, and 5.5.12. For their bimonoidal analogues, read Theorems 3.10.7 and 4.5.8. For Baez's Conjecture, read Definition 7.1.8 and Theorem 7.8.1. The introductions of Chapters 3, 4, 5, and 7 have more detailed description and reading suggestions. This material on (symmetric) bimonoidal categories is used extensively in Volume II in the discussion of braided bimonoidal categories.

Braided Bimonoidal Categories. To review the axioms of a braided bimonoidal category, read Definition II.2.1.29 and Theorem II.2.2.1. For their coherence and strictification (i.e., the Blass-Gurevich Conjecture), read Theorems II.5.4.4, II.6.3.6, and II.6.3.7. For the braided version of Baez's Conjecture, read Definition II.7.1.5 and Theorem II.7.3.4. The introductions of these chapters have more detailed discussion and reading guides. This material on braided bimonoidal categories is not heavily used in Volume III in our discussion of algebraic K-theory.

and

- an assignment on morphisms

$$\begin{aligned} \mathsf{C}(X,Y) &\longrightarrow \mathsf{D}(FX,FY) \\ f &\longmapsto Ff. \end{aligned}$$

These data are required to satisfy the following two conditions.

Composition: $F(gf) = (Fg)(Ff)$, provided that the composite gf is defined.
Identities: $F1_X = 1_{FX}$ for each object $X \in \mathsf{C}$.

Moreover, we define the following.

- Functors are composed by composing the assignments on objects and on morphisms.
- The *identity functor* of a category C is determined by the identity assignments on objects and morphisms, and is written as either Id_{C} or 1_{C}. ◇

Example 1.1.5. We let $\mathbb{1}$ denote the terminal category; it has a unique object $*$ and unique morphism 1_*. ◇

Example 1.1.6. We write Cat for the category with small categories as objects, functors as morphisms, identity functors as identity morphisms, and composition of functors as composition. ◇

Definition 1.1.7. Suppose $F, G: \mathsf{C} \longrightarrow \mathsf{D}$ are functors. A *natural transformation* $\theta: F \longrightarrow G$ consists of a morphism $\theta_X: FX \longrightarrow GX$ in D for each object $X \in \mathsf{C}$ such that the diagram

$$\begin{array}{ccc} FX & \xrightarrow{\theta_X} & GX \\ {\scriptstyle Ff}\downarrow & & \downarrow{\scriptstyle Gf} \\ FY & \xrightarrow{\theta_Y} & GY \end{array}$$

in D is commutative for each morphism $f: X \longrightarrow Y$ in C.

Moreover, we define the following.

- Each morphism θ_X is called a *component* of θ.
- The *identity natural transformation* $1_F: F \longrightarrow F$ of a functor F has each component an identity morphism.
- A *natural isomorphism* is a natural transformation in which every component is an isomorphism.
- A *natural monomorphism* is a natural transformation in which every component is a monomorphism. ◇

Definition 1.1.8. Suppose $\theta: F \longrightarrow G$ is a natural transformation for functors $F, G: \mathsf{C} \longrightarrow \mathsf{D}$.

- Suppose $\phi: G \longrightarrow H$ is a natural transformation for another functor $H: \mathsf{C} \longrightarrow \mathsf{D}$. The *vertical composition*

$$\phi\theta: F \longrightarrow H$$

is the natural transformation with components

$$(\phi\theta)_X = \phi_X \circ \theta_X: FX \longrightarrow HX \quad \text{for} \quad X \in \mathsf{C}.$$

- Suppose $\theta': F' \longrightarrow G'$ is a natural transformation for functors $F', G': \mathsf{D} \longrightarrow \mathsf{E}$. The *horizontal composition*

$$\theta' * \theta: F'F \longrightarrow G'G$$

is the natural transformation whose component $(\theta' * \theta)_X$ for an object $X \in \mathsf{C}$ is defined as either composite in the commutative diagram

$$\begin{array}{ccc} F'FX & \xrightarrow{\theta'_{FX}} & G'FX \\ {\scriptstyle F'\theta_X}\downarrow & & \downarrow{\scriptstyle G'\theta_X} \\ F'GX & \xrightarrow{\theta'_{GX}} & G'GX \end{array}$$

in D. ◇

Adjunctions and Equivalences.

Definition 1.1.9. For categories C and D, an *adjunction* from C to D is a triple (L, R, ϕ) consisting of

- a pair of functors

$$\mathsf{C} \underset{R}{\overset{L}{\rightleftarrows}} \mathsf{D}$$

  and
- a bijection

$$\mathsf{D}(LX, Y) \xrightarrow[\cong]{\phi_{X,Y}} \mathsf{C}(X, RY)$$

  that is natural in the objects $X \in \mathsf{C}$ and $Y \in \mathsf{D}$.

Such an adjunction is also called an *adjoint pair*, with L the *left adjoint* and R the *right adjoint*. We also denote such an adjunction by $L \dashv R$. ◇

Unless otherwise specified, we always display the left adjoint on top, pointing to the right. In an adjunction $L \dashv R$ as above, setting $Y = LX$ or $X = RY$, the natural bijection ϕ yields natural transformations

$$\begin{aligned} 1_{\mathsf{C}} &\xrightarrow{\eta} RL \\ LR &\xrightarrow{\varepsilon} 1_{\mathsf{D}}, \end{aligned} \tag{1.1.10}$$

which are called, respectively, the *unit* and the *counit*. The vertically composed natural transformations

$$\begin{aligned} R &\xrightarrow{\eta R} RLR \xrightarrow{R\varepsilon} R \\ L &\xrightarrow{L\eta} LRL \xrightarrow{\varepsilon L} L \end{aligned} \tag{1.1.11}$$

are equal to 1_R and 1_L, respectively. Here

$$\eta R = \eta * 1_R$$
$$R\varepsilon = 1_R * \varepsilon$$

and similarly for $L\eta$ and εL. The identities in (1.1.11) are called the *triangle identities*. Conversely, an adjunction is completely determined by functors (L, R) and natural transformations (η, ε) that satisfy the triangle identities.

Definition 1.1.12. A functor $F: \mathsf{C} \longrightarrow \mathsf{D}$ is called an *equivalence* if there exist

- a functor $G: \mathsf{D} \longrightarrow \mathsf{C}$ and
- natural isomorphisms

$$1_{\mathsf{C}} \xrightarrow[\cong]{\eta} GF \quad \text{and} \quad FG \xrightarrow[\cong]{\varepsilon} 1_{\mathsf{D}}.$$

If, in addition, F is left adjoint to G with unit η and counit ε, then $(F, G, \eta, \varepsilon)$ is called an *adjoint equivalence*. ◇

A functor F is an equivalence if and only if it is both

- *fully faithful*, which means that each function $\mathsf{C}(X,Y) \longrightarrow \mathsf{D}(FX,FY)$ on morphism sets is a bijection, and
- *essentially surjective*, which means that for each object $Y \in \mathsf{D}$, there exists an isomorphism $FX \xrightarrow{\cong} Y$ for some object $X \in \mathsf{C}$.

Colimits and Limits.

Definition 1.1.13. Suppose $F : \mathsf{D} \longrightarrow \mathsf{C}$ is a functor. A *colimit of* F, if it exists, is a pair $(\mathsf{colim}\, F, \delta)$ consisting of

- an object $\mathsf{colim}\, F \in \mathsf{C}$ and
- a morphism $\delta_d : Fd \longrightarrow \mathsf{colim}\, F$ in C for each object $d \in \mathsf{D}$

such that the following two conditions are satisfied.

Naturality: For each morphism $f : d \longrightarrow d'$ in D, the diagram

$$\begin{array}{ccc} Fd & \xrightarrow{\delta_d} & \mathsf{colim}\, F \\ {\scriptstyle Ff}\downarrow & & \| \\ Fd' & \xrightarrow{\delta_{d'}} & \mathsf{colim}\, F \end{array}$$

in C is commutative. A pair $(\mathsf{colim}\, F, \delta)$ with this property is called a *cocone of* F.

Universality: The pair $(\mathsf{colim}\, F, \delta)$ is *universal* among cocones of F. This means that if (X, δ') is another cocone of F, then there exists a unique morphism $h : \mathsf{colim}\, F \longrightarrow X$ in C such that the diagram

$$\begin{array}{ccc} Fd & \xrightarrow{\delta_d} & \mathsf{colim}\, F \\ \| & & \downarrow{\scriptstyle h} \\ Fd & \xrightarrow{\delta'_d} & X \end{array}$$

is commutative for each object $d \in \mathsf{D}$. ◇

A *limit of* F, which is denoted by $(\lim F, \delta)$ if it exists, is defined dually by turning the morphisms δ_d for $d \in \mathsf{D}$ and h backward.

- For a functor $F : \mathsf{D} \longrightarrow \mathsf{C}$, its colimit, if it exists, is also denoted by $\mathsf{colim}_{x \in \mathsf{D}}\, Fx$ and $\mathsf{colim}_{\mathsf{D}}\, F$, and similarly for limits.
- A *small (co)limit* is a (co)limit of a functor whose domain category is a small category.
- A *finite (co)limit* is a (co)limit of a functor whose domain category is a finite category.
- A category C is *(co)complete* if it has all small (co)limits.

A left adjoint $F : \mathsf{C} \longrightarrow \mathsf{D}$ preserves all the colimits that exist in C. In other words, if $H : \mathsf{E} \longrightarrow \mathsf{C}$ has a colimit, then $FH : \mathsf{E} \longrightarrow \mathsf{D}$ also has a colimit, and the natural morphism

$$\underset{e \in E}{\mathsf{colim}}\, FHe \longrightarrow F\Big(\underset{e \in E}{\mathsf{colim}}\, He\Big) \tag{1.1.14}$$

is an isomorphism. Similarly, a right adjoint $G : \mathsf{D} \longrightarrow \mathsf{C}$ preserves all the limits that exist in D.

Example 1.1.15. Here are some special types of colimits in a category C.

(1) An *initial object* $\varnothing^{\mathsf{C}}$ in C is a colimit of the functor $\varnothing \longrightarrow \mathsf{C}$, where $\varnothing$ is the empty category with no objects and no morphisms. It is characterized by the universal property that for each object X in C, there is a unique morphism $\varnothing^{\mathsf{C}} \longrightarrow X$ in C.

(2) A *coproduct* is a colimit of a functor whose domain category is a discrete category, that is, a category with only identity morphisms. We use the symbols $\coprod$ and $\sqcup$ to denote coproducts.

(3) A *pushout* is a colimit of a functor whose domain category has the form

$$\bullet \longleftarrow \bullet \longrightarrow \bullet$$

with three objects and two nonidentity morphisms.

(4) A *coequalizer* is a colimit of a functor whose domain category has the form

$$\bullet \rightrightarrows \bullet$$

with two objects and two nonidentity morphisms.

Terminal objects, products, pullbacks, and equalizers are the corresponding limit concepts. A *zero object* is an object that is both initial and terminal. $\diamond$

Coends and Ends. For a category C, its *opposite category* C^{op} is the category with the same objects as C and with morphism sets

$$\mathsf{C}^{\mathsf{op}}(A, B) = \mathsf{C}(B, A).$$

The identity morphisms and composites in C^{op} are defined by those in C.

Definition 1.1.16. Suppose $F : \mathsf{C}^{\mathsf{op}} \times \mathsf{C} \longrightarrow \mathsf{D}$ is a functor.

(1) A *cowedge* of F is a pair (X, ζ) consisting of
- an object $X \in \mathsf{D}$ and
- morphisms

$$\zeta_c : F(c,c) \longrightarrow X \quad \text{for} \quad c \in \mathsf{C}$$

such that the diagram

$$\begin{array}{ccc} F(d,c) & \xrightarrow{F(1_d, g)} & F(d,d) \\ {\scriptstyle F(g,1_c)}\downarrow & & \downarrow{\scriptstyle \zeta_d} \\ F(c,c) & \xrightarrow{\zeta_c} & X \end{array}$$

is commutative for each morphism $g : c \longrightarrow d \in \mathsf{C}$.

(2) A *coend* of F is an initial cowedge $\left(\int^{c \in \mathsf{C}} F(c,c), \omega\right)$. $\diamond$

More explicitly, a coend of F is a cowedge $\left(\int^{c \in \mathsf{C}} F(c,c), \omega\right)$ such that, given any cowedge (X, ζ) of F, there exists a unique morphism

$$\int^{c \in \mathsf{C}} F(c,c) \xrightarrow{h} X \in \mathsf{D}$$

that renders the diagram

$$\begin{array}{ccc} F(c,c) & \xrightarrow{\omega_c} & \int^{c\in\mathsf{C}} F(c,c) \\ & {\scriptstyle \zeta_c}\searrow \quad \swarrow{\scriptstyle h} & \\ & X & \end{array}$$

commutative for each object $c \in \mathsf{C}$. Dual to a coend is the notion of an *end*, which is originally due to Yoneda [**Yon60**]. The following observation follows from the definitions of a coend and a coequalizer.

Proposition 1.1.17. *Suppose $F: \mathsf{C}^{\mathsf{op}} \times \mathsf{C} \longrightarrow \mathsf{D}$ is a functor with C small and D cocomplete. Then a coend of F exists and is given by a coequalizer*

$$\int^{c\in\mathsf{C}} F(c,c) = \mathsf{coeq}\left(\coprod_{g\in\mathsf{Mor}(\mathsf{C})} F(d,c) \underset{i_c\circ F(g,1_c)}{\overset{i_d\circ F(1_d,g)}{\rightrightarrows}} \coprod_{c\in\mathsf{C}} F(c,c) \right)$$

in which $g: c \longrightarrow d$ runs through all the morphisms in C. The natural morphism ω_c is the composite

$$\begin{array}{ccc} F(c,c) & \xrightarrow{\omega_c} & \int^{c\in\mathsf{C}} F(c,c) \\ & {\scriptstyle i_c}\searrow \quad \nearrow{\scriptstyle \text{natural}} & \\ & \coprod_{c\in\mathsf{C}} F(c,c) & \end{array}$$

for each object $c \in \mathsf{C}$, with i_c the natural inclusion.

In the above setting, a coend is a colimit.

Kan Extensions.

Definition 1.1.18. Suppose

$$F: \mathsf{C} \longrightarrow \mathsf{D} \quad \text{and} \quad G: \mathsf{C} \longrightarrow \mathsf{E}$$

are functors. A *left Kan extension of F along G* is a pair

$$(\mathsf{Lan}_G F, \theta)$$

consisting of

- a functor

$$\mathsf{Lan}_G F: \mathsf{E} \longrightarrow \mathsf{D}$$

  and
- a natural transformation

$$\theta: F \longrightarrow (\mathsf{Lan}_G F)G$$

that is initial among such pairs. ◇

Explanation 1.1.19. The universal property that defines a left Kan extension in Definition 1.1.18 is that for each

- functor $H: \mathsf{E} \longrightarrow \mathsf{D}$ and
- natural transformation $\varepsilon: F \longrightarrow HG$,

there exists a unique natural transformation

$$\phi: \mathsf{Lan}_G F \longrightarrow H$$

such that ε factors as

$$\varepsilon = (\phi * 1_G)\theta.$$

This equality means that ε is equal to the following pasting of natural transformations, where $L = \mathsf{Lan}_G F$:

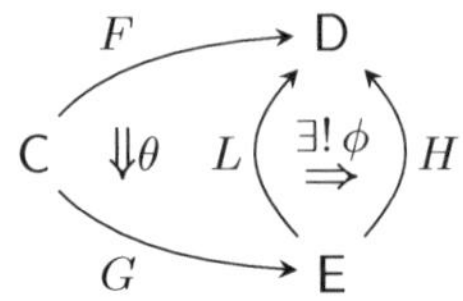

A left Kan extension is given by the coend

$$\mathsf{Lan}_G F = \int^{c \in \mathsf{C}} \mathsf{E}(Gc, -) \cdot Fc,$$

assuming that this coend and the coproduct

$$\mathsf{E}(Gc, -) \cdot Fc = \coprod_{\mathsf{E}(Gc,-)} Fc$$

exist in D. This is [**ML98**, X.4 Theorem 1]. ◇

The dual concept of a *right Kan extension of F along G* is a pair $(\mathsf{Ran}_G F, \theta)$ consisting of

- a functor

$$\mathsf{Ran}_G F: \mathsf{E} \longrightarrow \mathsf{D}$$

  and
- a natural transformation

$$\theta: (\mathsf{Ran}_G F)G \longrightarrow F$$

that is terminal among such pairs.

1.2. Monoidal Categories

In this section, we recall the definitions of a (symmetric) monoidal category, a (symmetric) monoidal functor, and a monoidal natural transformation.

Monoidal Categories and Functors.

Definition 1.2.1. A *monoidal category* is a tuple

$$(\mathsf{C}, \otimes, \mathbb{1}, \alpha, \lambda, \rho)$$

consisting of

- a category C;
- a functor $\otimes: \mathsf{C} \times \mathsf{C} \longrightarrow \mathsf{C}$, which is called the *monoidal product*;
- an object $\mathbb{1} \in \mathsf{C}$, which is called the *monoidal unit*;
- a natural isomorphism

$$(X \otimes Y) \otimes Z \xrightarrow[\cong]{\alpha_{X,Y,Z}} X \otimes (Y \otimes Z)$$

  for all objects $X, Y, Z \in \mathsf{C}$, which is called the *associativity isomorphism*;

and

- natural isomorphisms

$$\mathbb{1} \otimes X \xrightarrow[\cong]{\lambda_X} X \quad \text{and} \quad X \otimes \mathbb{1} \xrightarrow[\cong]{\rho_X} X$$

  for all objects $X \in \mathsf{C}$, which are called the *left unit isomorphism* and the *right unit isomorphism*, respectively.

These data are required to satisfy the following two axioms.

The Unity Axiom: The *middle unity diagram*

$$\begin{array}{ccc} (X \otimes \mathbb{1}) \otimes Y & \xrightarrow{\alpha_{X,\mathbb{1},Y}} & X \otimes (\mathbb{1} \otimes Y) \\ {\scriptstyle \rho_X \otimes 1_Y}\downarrow & & \downarrow{\scriptstyle 1_X \otimes \lambda_Y} \\ X \otimes Y & = = = = & X \otimes Y \end{array} \tag{1.2.2}$$

is commutative for all objects $X, Y \in \mathsf{C}$.

The Pentagon Axiom: The pentagon

$$\begin{array}{ccccc} & & (W \otimes X) \otimes (Y \otimes Z) & & \\ & {\scriptstyle \alpha_{W \otimes X,Y,Z}}\nearrow & & \searrow{\scriptstyle \alpha_{W,X,Y \otimes Z}} & \\ ((W \otimes X) \otimes Y) \otimes Z & & & & W \otimes (X \otimes (Y \otimes Z)) \\ {\scriptstyle \alpha_{W,X,Y} \otimes 1_Z}\downarrow & & & & \uparrow{\scriptstyle 1_W \otimes \alpha_{X,Y,Z}} \\ (W \otimes (X \otimes Y)) \otimes Z & & \xrightarrow{\alpha_{W,X \otimes Y,Z}} & & W \otimes ((X \otimes Y) \otimes Z) \end{array} \tag{1.2.3}$$

is commutative for all objects $W, X, Y, Z \in \mathsf{C}$.

A *strict monoidal category* is a monoidal category in which the components of α, λ, and ρ are all identity morphisms. ◇

Convention 1.2.4. In a monoidal category, we sometimes use concatenation as an abbreviation for the monoidal product. For example,

$$XY = X \otimes Y, \qquad (XY)Z = (X \otimes Y) \otimes Z,$$

and similarly for morphisms. We sometimes denote a monoidal category by $(\mathsf{C}, \otimes, \mathbb{1})$ or C. On the other hand, to emphasize the monoidal category C, we decorate the monoidal structure accordingly as $\otimes^{\mathsf{C}}$, $\mathbb{1}^{\mathsf{C}}$, α^{C}, λ^{C}, and ρ^{C}. ◇

Remark 1.2.5. By the results in [**Kel64**], in a monoidal category the equality

$$\lambda_{\mathbb{1}} = \rho_{\mathbb{1}} : \mathbb{1} \otimes \mathbb{1} \longrightarrow \mathbb{1} \tag{1.2.6}$$

holds, and the diagrams

(1.2.7)

$$\begin{array}{ccc} (\mathbb{1} \otimes X) \otimes Y & \xrightarrow{\alpha_{\mathbb{1},X,Y}} & \mathbb{1} \otimes (X \otimes Y) \\ {\scriptstyle \lambda_X \otimes 1_Y}\downarrow & & \downarrow{\scriptstyle \lambda_{X \otimes Y}} \\ X \otimes Y & = = = = & X \otimes Y \end{array} \qquad \begin{array}{ccc} (X \otimes Y) \otimes \mathbb{1} & \xrightarrow{\alpha_{X,Y,\mathbb{1}}} & X \otimes (Y \otimes \mathbb{1}) \\ {\scriptstyle \rho_{X \otimes Y}}\downarrow & & \downarrow{\scriptstyle 1_X \otimes \rho_Y} \\ X \otimes Y & = = = = & X \otimes Y \end{array}$$

are commutative. They are called the *left unity diagram* and the *right unity diagram*, respectively. Proofs of these unity properties in the more general setting of bicategories can be found in [**JY21**, Section 2.2]. ◇

Definition 1.2.8. A *monoid* in a monoidal category C is a triple (X, μ, η) with

- X an object in C;
- $\mu : X \otimes X \longrightarrow X$ a morphism, which is called the *multiplication*; and

- $\eta : \mathbb{1} \longrightarrow X$ a morphism, which is called the *unit*.

These data are required to make the following associativity and unity diagrams commutative.

$$\begin{array}{ccc} (X \otimes X) \otimes X & \xrightarrow{\alpha} & X \otimes (X \otimes X) \\ & & \big\downarrow{\scriptstyle 1_X \otimes \mu} \\ {\scriptstyle \mu \otimes 1_X}\big\downarrow & & X \otimes X \\ & & \big\downarrow{\scriptstyle \mu} \\ X \otimes X & \xrightarrow{\mu} & X \end{array} \qquad \begin{array}{ccc} \mathbb{1} \otimes X & \xrightarrow{\lambda_X} & X \\ {\scriptstyle \eta \otimes 1_X}\big\downarrow & & \big\| \\ X \otimes X & \xrightarrow{\mu} & X \\ {\scriptstyle 1_X \otimes \eta}\big\uparrow & & \big\| \\ X \otimes \mathbb{1} & \xrightarrow{\rho_X} & X \end{array}$$

A *morphism* of monoids

$$f : (X, \mu^X, \eta^X) \longrightarrow (Y, \mu^Y, \eta^Y)$$

is a morphism $f : X \longrightarrow Y$ in C that preserves the multiplications and the units in the sense that the diagrams

$$\begin{array}{ccc} X \otimes X & \xrightarrow{f \otimes f} & Y \otimes Y \\ {\scriptstyle \mu^X}\big\downarrow & & \big\downarrow{\scriptstyle \mu^Y} \\ X & \xrightarrow{f} & Y \end{array} \qquad \begin{array}{ccc} \mathbb{1} & \xrightarrow{\eta^X} & X \\ \big\| & & \big\downarrow{\scriptstyle f} \\ \mathbb{1} & \xrightarrow{\eta^Y} & Y \end{array}$$

are commutative. ◇

Definition 1.2.9. A *comonoid* in a monoidal category C is a triple (Y, Δ, ε) with

- Y an object in C;
- $\Delta : Y \longrightarrow Y \otimes Y$ a morphism, which is called the *comultiplication*; and
- $\varepsilon : Y \longrightarrow \mathbb{1}$ a morphism, which is called the *counit*.

These data are required to make the following coassociativity and counity diagrams commutative.

$$\begin{array}{ccc} Y & \xrightarrow{\Delta} & Y \otimes Y \\ & & \big\downarrow{\scriptstyle \Delta \otimes 1_Y} \\ {\scriptstyle \Delta}\big\downarrow & & (Y \otimes Y) \otimes Y \\ & & {\scriptstyle \alpha}\big\downarrow \\ Y \otimes Y & \xrightarrow{1_Y \otimes \Delta} & Y \otimes (Y \otimes Y) \end{array} \qquad \begin{array}{ccc} \mathbb{1} \otimes Y & \xrightarrow{\lambda_Y} & Y \\ {\scriptstyle \varepsilon \otimes 1_Y}\big\uparrow & & \big\| \\ Y \otimes Y & \xleftarrow{\Delta} & Y \\ {\scriptstyle 1_Y \otimes \varepsilon}\big\downarrow & & \big\| \\ Y \otimes \mathbb{1} & \xrightarrow{\rho} & Y \end{array} \tag{1.2.10}$$

A *morphism* of comonoids is a morphism of the underlying objects that preserves the comultiplications and the counits. ◇

Definition 1.2.11. For monoidal categories C and D, a *monoidal functor*

$$(F, F^2, F^0) : \mathsf{C} \longrightarrow \mathsf{D}$$

consists of

- a functor $F : \mathsf{C} \longrightarrow \mathsf{D}$;
- a natural transformation

$$FX \otimes FY \xrightarrow{F^2_{X,Y}} F(X \otimes Y) \in \mathsf{D} \tag{1.2.12}$$

for objects X and Y in C, which is called the *monoidal constraint*; and

- a morphism

(1.2.13) $$\mathbb{1}^{\mathsf{D}} \xrightarrow{F^0} F\mathbb{1}^{\mathsf{C}} \in \mathsf{D},$$

 which is called the *unit constraint.*

These data are required to satisfy the following associativity and unity axioms.

Associativity: The diagram

(1.2.14) $$\begin{array}{ccc} (FX \otimes FY) \otimes FZ & \xrightarrow{\alpha^{\mathsf{D}}} & FX \otimes (FY \otimes FZ) \\ \downarrow {\scriptstyle F^2 \otimes 1_{FZ}} & & \downarrow {\scriptstyle 1_{FX} \otimes F^2} \\ F(X \otimes Y) \otimes FZ & & FX \otimes F(Y \otimes Z) \\ \downarrow {\scriptstyle F^2} & & \downarrow {\scriptstyle F^2} \\ F((X \otimes Y) \otimes Z) & \xrightarrow{F\alpha^{\mathsf{C}}} & F(X \otimes (Y \otimes Z)) \end{array}$$

is commutative for all objects $X, Y, Z \in \mathsf{C}$.

Unity: The diagrams

(1.2.15) $$\begin{array}{ccc} \mathbb{1}^{\mathsf{D}} \otimes FX & \xrightarrow{\lambda^{\mathsf{D}}_{FX}} & FX \\ \downarrow {\scriptstyle F^0 \otimes 1_{FX}} & & \uparrow {\scriptstyle F\lambda^{\mathsf{C}}_X} \\ F\mathbb{1}^{\mathsf{C}} \otimes FX & \xrightarrow{F^2} & F(\mathbb{1}^{\mathsf{C}} \otimes X) \end{array} \qquad \begin{array}{ccc} FX \otimes \mathbb{1}^{\mathsf{D}} & \xrightarrow{\rho^{\mathsf{D}}_{FX}} & FX \\ \downarrow {\scriptstyle 1_{FX} \otimes F^0} & & \uparrow {\scriptstyle F\rho^{\mathsf{C}}_X} \\ FX \otimes F\mathbb{1}^{\mathsf{C}} & \xrightarrow{F^2} & F(X \otimes \mathbb{1}^{\mathsf{C}}) \end{array}$$

are commutative for all objects $X \in \mathsf{C}$. They are called the *left unity diagram* and the *right unity diagram*, respectively.

A monoidal functor (F, F^2, F^0) is sometimes abbreviated to F.

Moreover, a monoidal functor (F, F^2, F^0) is said to be

- *unital* if F^0 is an isomorphism;
- *strictly unital* if F^0 is the identity morphism;
- *strong* if F^0 and the components of F^2 are isomorphisms; and
- *strict* if F^0 and the components of F^2 are identity morphisms. ◇

Definition 1.2.16. For monoidal functors $F, G : \mathsf{C} \longrightarrow \mathsf{D}$, a *monoidal natural transformation* $\theta : F \longrightarrow G$ is a natural transformation between the underlying functors such that the diagrams

(1.2.17) $$\begin{array}{ccc} FX \otimes FY & \xrightarrow{\theta_X \otimes \theta_Y} & GX \otimes GY \\ \downarrow {\scriptstyle F^2} & & \downarrow {\scriptstyle G^2} \\ F(X \otimes Y) & \xrightarrow{\theta_{X \otimes Y}} & G(X \otimes Y) \end{array} \qquad \begin{array}{ccc} \mathbb{1}^{\mathsf{D}} & \xrightarrow{F^0} & F\mathbb{1}^{\mathsf{C}} \\ \| & & \downarrow {\scriptstyle \theta_{\mathbb{1}^{\mathsf{C}}}} \\ \mathbb{1}^{\mathsf{D}} & \xrightarrow{G^0} & G\mathbb{1}^{\mathsf{C}} \end{array}$$

are commutative for all objects $X, Y \in \mathsf{C}$. ◇

Symmetric Monoidal Categories and Functors. Next we consider symmetric analogues of the above concepts.

Definition 1.2.18. A *symmetric monoidal category* is a pair (C, ξ) consisting of the following data.

- $\mathsf{C} = (\mathsf{C}, \otimes, \mathbb{1}, \alpha, \lambda, \rho)$ is a monoidal category as in Definition 1.2.1.
- ξ is a natural isomorphism

$$X \otimes Y \xrightarrow[\cong]{\xi_{X,Y}} Y \otimes X \tag{1.2.19}$$

  for objects $X, Y \in \mathsf{C}$, which is called the *symmetry isomorphism* or the *braiding*.

These data are required to satisfy the following three axioms.

The Symmetry Axiom: The diagram

$$\begin{array}{ccc} X \otimes Y & \xrightarrow{\xi_{X,Y}} & Y \otimes X \\ & \diagdown\!\!\diagdown & \downarrow{\scriptstyle \xi_{Y,X}} \\ & & X \otimes Y \end{array} \tag{1.2.20}$$

is commutative for all objects $X, Y \in \mathsf{C}$.

The Unit Axiom: The diagram

$$\begin{array}{ccc} X \otimes \mathbb{1} & \xrightarrow{\xi_{X,\mathbb{1}}} & \mathbb{1} \otimes X \\ {\scriptstyle \rho_X}\downarrow & & \downarrow{\scriptstyle \lambda_X} \\ X & = & X \end{array} \tag{1.2.21}$$

is commutative for all objects $X \in \mathsf{C}$.

The Hexagon Axiom: The diagram

$$\begin{array}{ccccc} & X \otimes (Z \otimes Y) & \xrightarrow{1_X \otimes \xi_{Z,Y}} & X \otimes (Y \otimes Z) & \\ {\scriptstyle \alpha}\nearrow & & & & \searrow{\scriptstyle \alpha^{-1}} \\ (X \otimes Z) \otimes Y & & & & (X \otimes Y) \otimes Z \\ {\scriptstyle \xi_{X \otimes Z, Y}}\searrow & & & & \nearrow{\scriptstyle \xi_{Y,X} \otimes 1_Z} \\ & Y \otimes (X \otimes Z) & \xrightarrow{\alpha^{-1}} & (Y \otimes X) \otimes Z & \end{array} \tag{1.2.22}$$

is commutative for all objects $X, Y, Z \in \mathsf{C}$.

A symmetric monoidal category is said to be *strict* if the underlying monoidal category is strict. A symmetric strict monoidal category is also called a *permutative category*. ◇

Definition 1.2.23. Suppose (C, ξ) is a symmetric monoidal category.

- A *commutative monoid* is a monoid (X, μ, η) such that the diagram

$$\begin{array}{ccc} X \otimes X & \xrightarrow{\xi_{X,X}} & X \otimes X \\ {\scriptstyle \mu}\downarrow & & \downarrow{\scriptstyle \mu} \\ X & = & X \end{array}$$

  is commutative.

- A *cocommutative comonoid* is a comonoid (Y, Δ, ε) such that the diagram

$$\begin{array}{ccc} Y & = & Y \\ {\scriptstyle\Delta}\downarrow & & \downarrow{\scriptstyle\Delta} \\ Y \otimes Y & \xrightarrow{\xi_{Y,Y}} & Y \otimes Y \end{array} \tag{1.2.24}$$

is commutative.

A morphism of (co)commutative (co)monoids is a morphism of the underlying (co)monoids. ◇

Definition 1.2.25. For symmetric monoidal categories C and D, a *symmetric monoidal functor* $(F, F^2, F^0): \mathsf{C} \longrightarrow \mathsf{D}$ is a monoidal functor between the underlying monoidal categories such that the diagram

$$\begin{array}{ccc} FX \otimes FY & \xrightarrow[\cong]{\xi_{FX,FY}} & FY \otimes FX \\ {\scriptstyle F^2}\downarrow & & \downarrow{\scriptstyle F^2} \\ F(X \otimes Y) & \xrightarrow[\cong]{F\xi_{X,Y}} & F(Y \otimes X) \end{array} \tag{1.2.26}$$

is commutative for all objects $X, Y \in \mathsf{C}$. A symmetric monoidal functor is said to be *strong* (respectively, *strict*, *unital*, or *strictly unital*) if the underlying monoidal functor is so. ◇

Example 1.2.27. Here are some examples of symmetric monoidal categories.

- $(\mathsf{Set}, \times, *)$ is the category of sets and functions. A monoid in Set is a monoid in the usual sense.
- $(\mathsf{Cat}, \times, \mathbb{1})$ is the category of small categories and functors.
- $(\mathsf{Vect}^{\Bbbk}, \otimes, \Bbbk)$ is the category of $\Bbbk$-vector spaces and $\Bbbk$-linear maps over a field $\Bbbk$, with $\otimes$ the tensor product of $\Bbbk$-vector spaces. ◇

Definition 1.2.28. A symmetric monoidal category C is *closed* if, for each object X, the functor

$$- \otimes X : \mathsf{C} \longrightarrow \mathsf{C}$$

admits a right adjoint, which is denoted by $[X, -]$ and is called the *internal hom*. ◇

1.3. Coherence

In this section, we recall several coherence results for monoidal categories, symmetric monoidal categories, and symmetric monoidal functors. Each of them states either that some formal diagrams commute, or that a monoidal structure can be strictified. See the references in Notes 1.4.4 and 1.4.5 for detailed discussion of these topics.

Monoidal Categories. To state the first type of coherence result, we use the next two definitions.

Definition 1.3.1. A *word* of length $n \geq 0$ is defined inductively as follows.

- The only word of length 0 is the symbol e.
- The only word of length 1 is the symbol $-$.
- If u and v are words of lengths m and n, respectively, then $u \,\square\, v$ is a word of length $m + n$.

Moreover, we define the following.

- A *left normalized word* is the word e, $-$, $u \mathbin{\square} e$, or $u \mathbin{\square} -$, with u a left normalized word.
- A *right normalized word* is the word e, $-$, $e \mathbin{\square} u$, or $- \mathbin{\square} u$, with u a right normalized word.
- For a monoidal category C, each word w of length n determines a functor $w\colon \mathsf{C}^n \longrightarrow \mathsf{C}$ by interpreting
  - the length 0 word e as the constant functor at $\mathbb{1}$;
  - the length 1 word $-$ as the identity functor 1_{C}; and
  - $\square$ as the monoidal product in C.

  We also call this functor a *word*. ◇

Definition 1.3.2. For a monoidal category $(\mathsf{C}, \otimes, \mathbb{1}, \alpha, \lambda, \rho)$, *canonical maps* are natural isomorphisms between words of the same length, defined inductively by the following four conditions.

- The identity morphism of $\mathbb{1}$ is a canonical map.
- The identity natural transformation of 1_{C} is a canonical map.
- α, λ, ρ, and their inverses are canonical maps.
- Canonical maps are closed under $\otimes$ and vertical composites. ◇

Theorem 1.3.3 (Mac Lane's Coherence). *Suppose u and v are words $\mathsf{C}^n \longrightarrow \mathsf{C}$ of the same length in a monoidal category C. Then there exists a unique canonical map $u \longrightarrow v$.*

Example 1.3.4. The word

$$u = \big((- \mathbin{\square} -) \mathbin{\square} -\big) \mathbin{\square} -$$

is left normalized, and the word

$$v = - \mathbin{\square} \big(- \mathbin{\square} (- \mathbin{\square} -)\big)$$

is right normalized, both of length 4. Regarding the words u and v as functors $\mathsf{C}^4 \longrightarrow \mathsf{C}$, they are given by

$$\begin{aligned} u(x_1, x_2, x_3, x_4) &= \big((x_1 \otimes x_2) \otimes x_3\big) \otimes x_4 \quad \text{and} \\ v(x_1, x_2, x_3, x_4) &= x_1 \otimes \big(x_2 \otimes (x_3 \otimes x_4)\big). \end{aligned}$$

Each of the two paths in the pentagon axiom (1.2.3) is a canonical map from u to v. Mac Lane's Coherence Theorem 1.3.3 extends the unity axiom (1.2.2) and the pentagon axiom (1.2.3). ◇

Theorem 1.3.5 (Mac Lane's Strictification). *For each monoidal category C, there exist a strict monoidal category C_{st} and an adjoint equivalence*

$$\mathsf{C}_{\mathsf{st}} \underset{R}{\overset{L}{\rightleftarrows}} \mathsf{C}$$

with (i) both L and R strong monoidal functors, and (ii) $LR = 1_{\mathsf{C}}$.

Symmetric Monoidal Categories. The symmetric versions of the above coherence results involve the following concepts.

Definition 1.3.6. The symmetric group on n letters is denoted by Σ_n. Suppose C is a monoidal category.

- For a word w of length n and a permutation $\sigma \in \Sigma_n$, the *permuted word* $w\sigma\colon \mathsf{C}^n \longrightarrow \mathsf{C}$ is the composite functor $w \circ \sigma$, where $\sigma\colon \mathsf{C}^n \longrightarrow \mathsf{C}^n$ is given by
$$\sigma(x_1, \ldots, x_n) = \big(x_{\sigma^{-1}(1)}, \ldots, x_{\sigma^{-1}(n)}\big)$$
with the x_i's all objects, or all morphisms, in C.
- For a symmetric monoidal category (C, ξ), a *permuted canonical map* is a natural isomorphism between permuted words of the same length, defined as in Definition 1.3.2 by also allowing the symmetry isomorphism ξ. ◇

Example 1.3.7. For the left normalized word
$$u = \big((- \,\square\, -) \,\square\, -\big) \,\square\, -$$
and the cycle
$$\sigma = (1, 4, 2, 3) \in \Sigma_4,$$
the permuted word $u\sigma\colon \mathsf{C}^4 \longrightarrow \mathsf{C}$ is the functor given by
$$(u\sigma)(x_1, x_2, x_3, x_4) = \big((x_3 \otimes x_4) \otimes x_2\big) \otimes x_1. \qquad \diamond$$

Theorem 1.3.8 (Symmetric Coherence). *Suppose $u\sigma$ and $v\tau$ are two permuted words of the same length in a symmetric monoidal category C. Then there exists a unique permuted canonical map $u\sigma \longrightarrow v\tau$.*

Example 1.3.9. The Symmetric Coherence Theorem 1.3.8 extends the symmetry axiom (1.2.20), the unit axiom (1.2.21), and the hexagon axiom (1.2.22). ◇

Theorem 1.3.10 (Symmetric Strictification). *For each symmetric monoidal category C, there exist a permutative category C_{st} and an adjoint equivalence*
$$\mathsf{C}_{\mathsf{st}} \underset{R}{\overset{L}{\rightleftarrows}} \mathsf{C}$$
with (i) both L and R strong symmetric monoidal functors, and (ii) $LR = 1_{\mathsf{C}}$.

Monoidal Functors. Next we recall a coherence theorem for (symmetric) monoidal functors. The following definition and theorem have a plain monoidal version and a symmetric monoidal version. We will state them together using Definitions 1.3.1, 1.3.2, and 1.3.6.

Definition 1.3.11. Suppose $(F, F^2, F^0)\colon \mathsf{C} \longrightarrow \mathsf{D}$ is a (symmetric) monoidal functor between (symmetric) monoidal categories.

(1) The set of *F-iterates* is the set of functors $\mathsf{C}^n \longrightarrow \mathsf{D}$ for $n \geq 1$ defined inductively by the following two conditions.
  - $Fw\colon \mathsf{C}^n \longrightarrow \mathsf{D}$ is an F-iterate for each (permuted) word $w\colon \mathsf{C}^n \longrightarrow \mathsf{C}$ of length n.
  - If $G\colon \mathsf{C}^m \longrightarrow \mathsf{D}$ and $H\colon \mathsf{C}^n \longrightarrow \mathsf{D}$ are F-iterates, then so is the composite
$$\mathsf{C}^{m+n} = \mathsf{C}^m \times \mathsf{C}^n \xrightarrow{G \times H} \mathsf{D} \times \mathsf{D} \xrightarrow{\otimes} \mathsf{D}.$$

(2) The set of F-*coherent maps* is the set of natural transformations between F-iterates defined inductively as follows.
- Suppose θ is a (permuted) canonical map in C that does not involve $\mathbb{1}^{\mathsf{C}}$, λ^{C}, or ρ^{C}. Then $1_F * \theta$ is an F-coherent map.
- The identity natural transformation, α^{D}, their inverses, and ξ^{D} in the symmetric case, applied to F-iterates are F-coherent maps.
- F^2 is an F-coherent map.
- F-coherent maps are closed under vertical composites and $\otimes^{\mathsf{D}}$. ◇

Theorem 1.3.12 (Epstein's Coherence). *Suppose $F\colon \mathsf{C} \longrightarrow \mathsf{D}$ is a (symmetric) monoidal functor between (symmetric) monoidal categories, and $G, H\colon \mathsf{C}^n \longrightarrow \mathsf{D}$ are F-iterates. Then there exists at most one F-coherent map from G to H.*

Example 1.3.13. Consider the associativity axiom (1.2.14) for a monoidal functor $F\colon \mathsf{C} \longrightarrow \mathsf{D}$. All six functors $\mathsf{C}^3 \longrightarrow \mathsf{D}$ defined by the vertices in (1.2.14) are F-iterates. Each path along the boundary of that diagram is a component of an F-coherent map. Epstein's Coherence Theorem 1.3.12 extends the associativity axiom (1.2.14) and the compatibility axiom (1.2.26) in the symmetric case. ◇

1.4. Notes

1.4.1 (Categories). For more detailed discussion of basic category theory, we refer the reader to the books [**Awo10, BK00, Gra18, Lei14, Rie16, Rom17, Sim11**]. For higher level discussion of categories, the reader may consult the books [**Bor94a, Bor94b, Bor94c, Ehr65, ML98, Mit65, Sch72**]. The founding paper [**EML45**] of category theory is very readable and contains applications to topology. ◇

1.4.2 (Ends and Coends). More discussion about (co)ends and enriched diagram categories is in [**Day70, DK69**] and Chapter III.3. A systematic discussion about (co)ends is in [**Lor21**] and [**ML98**, IX.5–8]. ◇

1.4.3 (Terminology). What we call a (symmetric) monoidal category is what Joyal and Street [**JS93**] called a *(symmetric) tensor category*. A monoidal functor is sometimes called a *lax monoidal functor* in the literature to emphasize the fact that the structure morphisms F^2 and F^0 are not necessarily invertible. A strong monoidal functor is also known as a *tensor functor*. ◇

1.4.4 (Coherence). Mac Lane's Coherence Theorem 1.3.3 is [**ML63**, Th. 5.2]; see also [**ML98**, VII.2 Cor.]. Mac Lane's Strictification Theorem 1.3.5 is [**ML98**, XI.3 Th. 1], which is a consequence of the coherence theorem. The Symmetric Coherence Theorem 1.3.8 is [**ML98**, XI.1 Th. 1]. The Symmetric Strictification Theorem 1.3.10 is proved by adapting the proof of Theorem 1.3.5 in the nonsymmetric case, by incorporating the symmetry and using Theorem 1.3.8. ◇

1.4.5 (Coherence of Monoidal Functors). Theorem 1.3.12 is [**Eps66**, Th. 1.6]. The proof of the plain monoidal version of Theorem 1.3.12 is obtained from Epstein's proof by systematically removing all instances of the symmetry isomorphism. Another coherence result for monoidal functors is [**JS93**, Th. 1.7]. ◇

CHAPTER 2

Symmetric Bimonoidal Categories

In this chapter, we introduce symmetric bimonoidal categories in the sense of Laplaza and discuss some examples and one application to computer science. The definition is given in Section 2.1. A symmetric bimonoidal category is a categorical analogue of a commutative rig, in which one can add, multiply, and distribute products over sums. A symmetric bimonoidal category is a category equipped with

- two symmetric monoidal structures, one additive and one multiplicative,
- two multiplicative zero natural isomorphisms, and
- two distributivity natural monomorphisms,

along with a list of carefully chosen axioms. A symmetric bimonoidal category is called *tight* if the distributivity morphisms are natural isomorphisms. We also define bimonoidal categories, which do not have a multiplicative symmetry.

In Section 2.2, we prove Laplaza's theorem, with one improvement as discussed in Note 2.7.1, that half of the 24 axioms of a symmetric bimonoidal category are formal consequences of the others. Therefore, in practice, to check that one has a symmetric bimonoidal category, one has to check 12 axioms instead of 24. This reduction Theorem 2.2.13 does not apply to bimonoidal categories because its proof heavily uses the multiplicative symmetry. The only reduction of the list of axioms in the nonsymmetric case is in Proposition 2.2.14.

The next few sections contain examples of symmetric bimonoidal categories. In Section 2.3, we observe that distributive symmetric monoidal categories are symmetric bimonoidal categories whose sums are coproducts. In particular, distributive categories and the category of modules over a commutative ring are symmetric bimonoidal categories.

Section 2.4 contains an important example of a symmetric bimonoidal category Σ with finite ordinals as objects and their bijections as morphisms.

- In Section 4.2, we will construct a symmetric bimonoidal category, called the distortion category, that extends Σ. The distortion category is an essential ingredient in the formulation of the Coherence Theorem 4.4.3.
- In Chapter 7, we will show that Σ is bicategorically an initial object in a 2-category of small symmetric bimonoidal categories.
- In Chapter II.7, we will show that Σ is bicategorically an initial object in a 2-category of small braided bimonoidal categories.

This section also contains a variant of Σ denoted by Σ'.

In Section 2.5, we observe that right bipermutative categories, with Σ' being one example, are tight symmetric bimonoidal categories. A right bipermutative category has two permutative structures, one additive $\oplus$ and one multiplicative $\otimes$, with identities for both multiplicative zeros, the right distributivity, and $\xi^{\otimes}_{-,\mathbb{0}}$, along with several carefully chosen axioms. In Section 5.4, we will show that *tight* symmetric bimonoidal categories can be strictified to equivalent right bipermutative categories. In the second half of this section, we discuss left bipermutative categories, in which the left distributivity is the identity. The symmetric bimonoidal category Σ is an example of a left bipermutative category.

In Section 2.6, we discuss an application of symmetric bimonoidal categories to computer science from [**CS16**]. The main observation is that there is a symmetric bimonoidal groupoid Π with syntax of finite types as objects. The symmetric bimonoidal groupoid Π provides a variant of the Curry-Howard-Lambek correspondence for reversible programming of finite types with sums and products. Applications of *braided* bimonoidal categories to quantum group theory and topological quantum computation are discussed in Chapter II.3.

The following table summarizes basic properties of (symmetric) bimonoidal categories, along with their coherence theorems, strictification theorems, and bicategorical properties.

	bimonoidal categories	symmetric bimonoidal categories
axioms	22	24
reduced axioms	21 (2.2.14)	12 (2.2.13)
coherence	3.10.7, 4.5.8	3.9.1, 4.4.3
strictification	5.5.11, 5.5.12	5.4.6, 5.4.7
Baez's Conjecture		7.8.1, 7.8.3
bicategorification	8.4.12	8.15.4

Braided bimonoidal categories will be discussed in Part II.1. For open questions related to bimonoidal categories, see Appendix A.1.

2.1. Definitions

In this section, we define symmetric bimonoidal categories and bimonoidal categories, and provide some basic examples. Many more examples will be given in the coming sections.

Motivation 2.1.1. To motivate the definition of a symmetric bimonoidal category, consider a commutative rig $(R, +, 0, \times, 1)$, which is a commutative ring without additive inverses, such as the ring $\mathbb{Z}$ of integers. Both the addition $+$ and the multiplication $\times$ are associative and commutative. They are 2-sided unital with respect to 0 and 1, respectively. For elements $x, y, z \in R$, the following multiplicative zero and distributivity properties hold:

$$0x = 0 = x0$$
$$x(y+z) = xy + xz$$
$$(x+y)z = xz + yz$$

A symmetric bimonoidal category is a categorified version of a commutative rig, with the commutative monoids $(R, +, 0)$ and $(R, \times, 1)$ replaced by two symmetric monoidal structures. The above four rig axioms are replaced by natural transformations that satisfy a list of coherence axioms. In the next several chapters, we will see that these coherence axioms yield good coherence properties of symmetric bimonoidal categories. ◇

Definition 1.2.18 of a symmetric monoidal category is used in the next definition. In Theorem 2.2.13, we show that in each of the following twelve groups of axioms, only the first axiom is needed. However, for later reference and usage, it is more convenient to have all of them in one place.

Definition 2.1.2. A *symmetric bimonoidal category* is a tuple

$$\big(\mathsf{C}, (\oplus, \mathbb{0}, \alpha^{\oplus}, \lambda^{\oplus}, \rho^{\oplus}, \xi^{\oplus}), (\otimes, \mathbb{1}, \alpha^{\otimes}, \lambda^{\otimes}, \rho^{\otimes}, \xi^{\otimes}), (\lambda^{\bullet}, \rho^{\bullet}), (\delta^{l}, \delta^{r})\big)$$

consisting of the following data.

- $(\mathsf{C}, \oplus, \mathbb{0}, \alpha^{\oplus}, \lambda^{\oplus}, \rho^{\oplus}, \xi^{\oplus})$ is a symmetric monoidal category, which is called the *additive structure*.
- $(\mathsf{C}, \otimes, \mathbb{1}, \alpha^{\otimes}, \lambda^{\otimes}, \rho^{\otimes}, \xi^{\otimes})$ is a symmetric monoidal category, which is called the *multiplicative structure*.
- $\lambda^{\bullet}$ and $\rho^{\bullet}$ are natural isomorphisms

$$\mathbb{0} \otimes A \xrightarrow[\cong]{\lambda^{\bullet}_A} \mathbb{0} \xleftarrow[\cong]{\rho^{\bullet}_A} A \otimes \mathbb{0} \quad \text{for} \quad A \in \mathsf{C}, \tag{2.1.3}$$

  which are called the *left multiplicative zero* and the *right multiplicative zero*, respectively.
- δ^{l} and δ^{r} are natural monomorphisms

$$\begin{aligned} A \otimes (B \oplus C) &\xrightarrow{\delta^{l}_{A,B,C}} (A \otimes B) \oplus (A \otimes C) \\ (A \oplus B) \otimes C &\xrightarrow{\delta^{r}_{A,B,C}} (A \otimes C) \oplus (B \otimes C) \end{aligned} \tag{2.1.4}$$

  for objects $A, B, C \in \mathsf{C}$, which are called the *left distributivity morphism* and the *right distributivity morphism*, respectively.

To simplify the presentation, we often abbreviate $\otimes$ using concatenation. In the absence of parentheses, $\otimes$ always takes precedence over $\oplus$. For example, the left distributivity morphism is abbreviated to $A(B \oplus C) \longrightarrow AB \oplus AC$.

The above data are required to make the following 24 diagrams in C commutative for all objects A, B, C, and D in C. They are collectively known as *Laplaza's Axioms*.

Distributivity and Multiplicative Symmetry:

$$\begin{array}{ccc} (A \oplus B)C & \xrightarrow{\delta^{r}_{A,B,C}} & AC \oplus BC \\ {\scriptstyle \xi^{\otimes}_{A\oplus B,C}} \downarrow & & \downarrow {\scriptstyle \xi^{\otimes}_{A,C} \oplus \xi^{\otimes}_{B,C}} \\ C(A \oplus B) & \xrightarrow{\delta^{l}_{C,A,B}} & CA \oplus CB \end{array} \tag{2.1.5}$$

Distributivity and Additive Symmetry:

$$\begin{array}{ccc} A(B \oplus C) & \xrightarrow{\delta^{l}_{A,B,C}} & AB \oplus AC \\ {\scriptstyle 1_A \xi^{\oplus}_{B,C}} \downarrow & & \downarrow {\scriptstyle \xi^{\oplus}_{AB,AC}} \\ A(C \oplus B) & \xrightarrow{\delta^{l}_{A,C,B}} & AC \oplus AB \end{array} \tag{2.1.6}$$

$$
\begin{array}{ccc}
(A \oplus B)C & \xrightarrow{\delta^r_{A,B,C}} & AC \oplus BC \\
{\scriptstyle \xi^{\oplus}_{A,B} 1_C} \big\downarrow & & \big\downarrow {\scriptstyle \xi^{\oplus}_{AC,BC}} \\
(B \oplus A)C & \xrightarrow{\delta^r_{B,A,C}} & BC \oplus AC
\end{array}
\tag{2.1.7}
$$

Distributivity and Additive Associativity:

$$
\begin{array}{ccccc}
[(A \oplus B) \oplus C]D & \xrightarrow{\delta^r_{A\oplus B,C,D}} & (A \oplus B)D \oplus CD & \xrightarrow{\delta^r_{A,B,D} \oplus 1_{CD}} & (AD \oplus BD) \oplus CD \\
{\scriptstyle \alpha^{\oplus}_{A,B,C} 1_D} \big\downarrow & & & & \big\downarrow {\scriptstyle \alpha^{\oplus}_{AD,BD,CD}} \\
[A \oplus (B \oplus C)]D & \xrightarrow{\delta^r_{A,B\oplus C,D}} & AD \oplus (B \oplus C)D & \xrightarrow{1_{AD} \oplus \delta^r_{B,C,D}} & AD \oplus (BD \oplus CD)
\end{array}
\tag{2.1.8}
$$

$$
\begin{array}{ccccc}
A[(B \oplus C) \oplus D] & \xrightarrow{\delta^l_{A,B\oplus C,D}} & A(B \oplus C) \oplus AD & \xrightarrow{\delta^l_{A,B,C} \oplus 1_{AD}} & (AB \oplus AC) \oplus AD \\
{\scriptstyle 1_A \alpha^{\oplus}_{B,C,D}} \big\downarrow & & & & \big\downarrow {\scriptstyle \alpha^{\oplus}_{AB,AC,AD}} \\
A[B \oplus (C \oplus D)] & \xrightarrow{\delta^l_{A,B,C\oplus D}} & AB \oplus A(C \oplus D) & \xrightarrow{1_{AB} \oplus \delta^l_{A,C,D}} & AB \oplus (AC \oplus AD)
\end{array}
\tag{2.1.9}
$$

Distributivity and Multiplicative Associativity:

$$
\begin{array}{ccccc}
(AB)(C \oplus D) & & \xrightarrow{\qquad\quad \delta^l_{AB,C,D} \qquad\quad} & & (AB)C \oplus (AB)D \\
{\scriptstyle \alpha^{\otimes}_{A,B,C\oplus D}} \big\downarrow & & & & \big\downarrow {\scriptstyle \alpha^{\otimes}_{A,B,C} \oplus \alpha^{\otimes}_{A,B,D}} \\
A[B(C \oplus D)] & \xrightarrow{1_A \delta^l_{B,C,D}} & A(BC \oplus BD) & \xrightarrow{\delta^l_{A,BC,BD}} & A(BC) \oplus A(BD)
\end{array}
\tag{2.1.10}
$$

$$
\begin{array}{ccccc}
[(A \oplus B)C]D & \xrightarrow{\delta^r_{A,B,C} 1_D} & (AC \oplus BC)D & \xrightarrow{\delta^r_{AC,BC,D}} & (AC)D \oplus (BC)D \\
{\scriptstyle \alpha^{\otimes}_{A\oplus B,C,D}} \big\downarrow & & & & \big\downarrow {\scriptstyle \alpha^{\otimes}_{A,C,D} \oplus \alpha^{\otimes}_{B,C,D}} \\
(A \oplus B)(CD) & & \xrightarrow{\qquad\quad \delta^r_{A,B,CD} \qquad\quad} & & A(CD) \oplus B(CD)
\end{array}
\tag{2.1.11}
$$

$$
\begin{array}{ccccc}
[A(B \oplus C)]D & \xrightarrow{\delta^l_{A,B,C} 1_D} & (AB \oplus AC)D & \xrightarrow{\delta^r_{AB,AC,D}} & (AB)D \oplus (AC)D \\
{\scriptstyle \alpha^{\otimes}_{A,B\oplus C,D}} \big\downarrow & & & & \big\downarrow {\scriptstyle \alpha^{\otimes}_{A,B,D} \oplus \alpha^{\otimes}_{A,C,D}} \\
A[(B \oplus C)D] & \xrightarrow{1_A \delta^r_{B,C,D}} & A(BD \oplus CD) & \xrightarrow{\delta^l_{A,BD,CD}} & A(BD) \oplus A(CD)
\end{array}
\tag{2.1.12}
$$

2-By-2 Distributivity:

$$
\begin{array}{ccc}
(A \oplus B)(C \oplus D) & \xrightarrow{\delta^r_{A,B,C\oplus D}} & A(C \oplus D) \oplus B(C \oplus D) \\
{\scriptstyle \delta^l_{A\oplus B,C,D}} \big\downarrow & & \big\downarrow {\scriptstyle \delta^l_{A,C,D} \oplus \delta^l_{B,C,D}} \\
(A \oplus B)C \oplus (A \oplus B)D & & (AC \oplus AD) \oplus (BC \oplus BD) \\
{\scriptstyle \delta^r_{A,B,C} \oplus \delta^r_{A,B,D}} \big\downarrow & & \big\downarrow {\scriptstyle \alpha^{\oplus}_{AC,AD,BC\oplus BD}} \\
(AC \oplus BC) \oplus (AD \oplus BD) & & AC \oplus [AD \oplus (BC \oplus BD)] \\
{\scriptstyle \alpha^{\oplus}_{AC,BC,AD\oplus BD}} \big\downarrow & & \big\downarrow {\scriptstyle 1_{AC} \oplus (\alpha^{\oplus})^{-1}} \\
AC \oplus [BC \oplus (AD \oplus BD)] & & AC \oplus [(AD \oplus BC) \oplus BD] \\
{\scriptstyle 1_{AC} \oplus (\alpha^{\oplus})^{-1}} \big\downarrow & & \big\downarrow {\scriptstyle 1_{AC} \oplus (\xi^{\oplus}_{AD,BC} \oplus 1_{BD})} \\
AC \oplus [(BC \oplus AD) \oplus BD] & = & AC \oplus [(BC \oplus AD) \oplus BD]
\end{array}
\tag{2.1.13}
$$

Multiplicative Zero of 0:

$$\mathbb{0}\otimes\mathbb{0} \underset{\rho^{\bullet}_{\mathbb{0}}}{\overset{\lambda^{\bullet}_{\mathbb{0}}}{\rightrightarrows}} \mathbb{0} \tag{2.1.14}$$

Multiplicative Zero of a Sum:

$$\begin{array}{ccc} \mathbb{0}(A\oplus B) & \xrightarrow{\delta^{l}_{\mathbb{0},A,B}} & \mathbb{0}A\oplus\mathbb{0}B \\ {\scriptstyle\lambda^{\bullet}_{A\oplus B}}\big\downarrow & & \big\downarrow{\scriptstyle\lambda^{\bullet}_{A}\oplus\lambda^{\bullet}_{B}} \\ \mathbb{0} & \xleftarrow{\lambda^{\oplus}_{\mathbb{0}}} & \mathbb{0}\oplus\mathbb{0} \end{array} \tag{2.1.15}$$

$$\begin{array}{ccc} (A\oplus B)\mathbb{0} & \xrightarrow{\delta^{r}_{A,B,\mathbb{0}}} & A\mathbb{0}\oplus B\mathbb{0} \\ {\scriptstyle\rho^{\bullet}_{A\oplus B}}\big\downarrow & & \big\downarrow{\scriptstyle\rho^{\bullet}_{A}\oplus\rho^{\bullet}_{B}} \\ \mathbb{0} & \xleftarrow{\lambda^{\oplus}_{\mathbb{0}}} & \mathbb{0}\oplus\mathbb{0} \end{array} \tag{2.1.16}$$

Multiplicative Zero and Multiplicative Unit:

$$\mathbb{0}\otimes\mathbb{1} \underset{\rho^{\otimes}_{\mathbb{0}}}{\overset{\lambda^{\bullet}_{\mathbb{1}}}{\rightrightarrows}} \mathbb{0} \tag{2.1.17}$$

$$\mathbb{1}\otimes\mathbb{0} \underset{\lambda^{\otimes}_{\mathbb{0}}}{\overset{\rho^{\bullet}_{\mathbb{1}}}{\rightrightarrows}} \mathbb{0} \tag{2.1.18}$$

Symmetry of Multiplicative Zero:

$$\begin{array}{ccc} A\otimes\mathbb{0} & \xrightarrow{\xi^{\otimes}_{A,\mathbb{0}}} & \mathbb{0}\otimes A \\ & {\scriptstyle\rho^{\bullet}_{A}}\searrow \quad \swarrow{\scriptstyle\lambda^{\bullet}_{A}} & \\ & \mathbb{0} & \end{array} \tag{2.1.19}$$

Multiplicative Zero and Multiplicative Associativity:

$$\begin{array}{ccc} (AB)\mathbb{0} & \xrightarrow{\alpha^{\otimes}_{A,B,\mathbb{0}}} & A(B\mathbb{0}) \\ {\scriptstyle\rho^{\bullet}_{AB}}\big\downarrow & & \big\downarrow{\scriptstyle 1_A\rho^{\bullet}_{B}} \\ \mathbb{0} & \xleftarrow{\rho^{\bullet}_{A}} & A\mathbb{0} \end{array} \tag{2.1.20}$$

$$\begin{array}{ccc} (A\mathbb{0})B & \xrightarrow{\alpha^{\otimes}_{A,\mathbb{0},B}} & A(\mathbb{0}B) \\ {\scriptstyle\rho^{\bullet}_{A}1_B}\big\downarrow & & \big\downarrow{\scriptstyle 1_A\lambda^{\bullet}_{B}} \\ \mathbb{0}B & & A\mathbb{0} \\ & {\scriptstyle\lambda^{\bullet}_{B}}\searrow \quad \swarrow{\scriptstyle\rho^{\bullet}_{A}} & \\ & \mathbb{0} & \end{array} \tag{2.1.21}$$

$$\begin{array}{ccc} (\mathbb{0}A)B & \xrightarrow{\lambda^{\bullet}_{A}1_B} & \mathbb{0}B \\ {\scriptstyle\alpha^{\otimes}_{\mathbb{0},A,B}}\big\downarrow & & \big\downarrow{\scriptstyle\lambda^{\bullet}_{B}} \\ \mathbb{0}(AB) & \xrightarrow{\lambda^{\bullet}_{AB}} & \mathbb{0} \end{array} \tag{2.1.22}$$

Additive and Multiplicative Zero:

(2.1.23)
$$\begin{CD} A(\mathbb{0} \oplus B) @>{\delta^l_{A,\mathbb{0},B}}>> A\mathbb{0} \oplus AB \\ @V{1_A \lambda^{\oplus}_B}VV @VV{\rho^{\bullet}_A \oplus 1_{AB}}V \\ AB @<{\lambda^{\oplus}_{AB}}<< \mathbb{0} \oplus AB \end{CD}$$

(2.1.24)
$$\begin{CD} (\mathbb{0} \oplus B)A @>{\delta^r_{\mathbb{0},B,A}}>> \mathbb{0}A \oplus BA \\ @V{\lambda^{\oplus}_B 1_A}VV @VV{\lambda^{\bullet}_A \oplus 1_{BA}}V \\ BA @<{\lambda^{\oplus}_{BA}}<< \mathbb{0} \oplus BA \end{CD}$$

(2.1.25)
$$\begin{CD} A(B \oplus \mathbb{0}) @>{\delta^l_{A,B,\mathbb{0}}}>> AB \oplus A\mathbb{0} \\ @V{1_A \rho^{\oplus}_B}VV @VV{1_{AB} \oplus \rho^{\bullet}_A}V \\ AB @<{\rho^{\oplus}_{AB}}<< AB \oplus \mathbb{0} \end{CD}$$

(2.1.26)
$$\begin{CD} (B \oplus \mathbb{0})A @>{\delta^r_{B,\mathbb{0},A}}>> BA \oplus \mathbb{0}A \\ @V{\rho^{\oplus}_B 1_A}VV @VV{1_{BA} \oplus \lambda^{\bullet}_A}V \\ BA @<{\rho^{\oplus}_{BA}}<< BA \oplus \mathbb{0} \end{CD}$$

Distributivity and Multiplicative Unit:

(2.1.27)
$$\begin{array}{ccc} \mathbb{1}(A \oplus B) & \xrightarrow{\delta^l_{\mathbb{1},A,B}} & \mathbb{1}A \oplus \mathbb{1}B \\ & {\scriptstyle \lambda^{\otimes}_{A \oplus B}} \searrow \quad \swarrow {\scriptstyle \lambda^{\otimes}_A \oplus \lambda^{\otimes}_B} & \\ & A \oplus B & \end{array}$$

(2.1.28)
$$\begin{array}{ccc} (A \oplus B)\mathbb{1} & \xrightarrow{\delta^r_{A,B,\mathbb{1}}} & A\mathbb{1} \oplus B\mathbb{1} \\ & {\scriptstyle \rho^{\otimes}_{A \oplus B}} \searrow \quad \swarrow {\scriptstyle \rho^{\otimes}_A \oplus \rho^{\otimes}_B} & \\ & A \oplus B & \end{array}$$

This finishes the definition of a symmetric bimonoidal category.

A *bimonoidal category* has the same definition as a symmetric bimonoidal category except for the following two conditions.

- The multiplicative symmetry $\xi^{\otimes}$ is omitted, and $(\mathsf{C}, \otimes, \mathbb{1}, \alpha^{\otimes}, \lambda^{\otimes}, \rho^{\otimes})$ is a monoidal category.
- The axioms (2.1.5) and (2.1.19) are omitted.

Moreover, we define the following.

- A (symmetric) bimonoidal category is *small* if its class of objects is a set.
- A (symmetric) bimonoidal category is *tight* if both δ^l and δ^r are natural isomorphisms.
- A *(symmetric) bimonoidal groupoid* is a (symmetric) bimonoidal category in which each morphism is an isomorphism.
- The objects $\mathbb{0}$ and $\mathbb{1}$ are called the *additive zero* and the *multiplicative unit*, respectively.

- $\oplus$ and $\otimes$ are called the *sum* and the *product*, respectively.
- $\alpha^{\oplus}$, $\lambda^{\oplus}$, $\rho^{\oplus}$, and $\xi^{\oplus}$ are called the *additive associativity isomorphism*, the *left additive zero*, the *right additive zero*, and the *additive symmetry isomorphism*, respectively.
- $\alpha^{\otimes}$, $\lambda^{\otimes}$, $\rho^{\otimes}$, and $\xi^{\otimes}$ are called the *multiplicative associativity isomorphism*, the *left multiplicative unit*, the *right multiplicative unit*, and the *multiplicative symmetry isomorphism*, respectively. ◇

Explanation 2.1.29. Consider the definition of a symmetric bimonoidal category.

(1) The 24 axioms are purposely divided into 12 groups. We will see in Section 2.2 that in each group, only the first axiom is actually needed. Therefore, a symmetric bimonoidal category is defined by 12 axioms. See Theorem 2.2.13.
(2) The left distributivity and the right distributivity (2.1.4) are only required to be natural monomorphisms. Laplaza's coherence results in Chapters 3 and 4 do not need the invertibility of the distributivity morphisms.
(3) The 2-by-2 distributivity axiom (2.1.13) is the only axiom that involves the inverse of a structure morphism, namely $(\alpha^{\oplus})^{-1}$. ◇

Explanation 2.1.30. Consider the definition of a bimonoidal category.

(1) The additive structure $(\mathsf{C}, \oplus, \mathbb{0}, \alpha^{\oplus}, \lambda^{\oplus}, \rho^{\oplus}, \xi^{\oplus})$ is still a symmetric monoidal category.
(2) The axioms
    - (2.1.5) relating δ^l and δ^r, and
    - (2.1.19) relating $\lambda^{\bullet}$ and $\rho^{\bullet}$

    are omitted because they are the only ones among Laplaza's 24 axioms that involve the multiplicative symmetry $\xi^{\otimes}$.

Due to the absence of $\xi^{\otimes}$, the reduction of Laplaza's axioms in Theorem 2.2.13 from 24 to 12 does *not* apply to bimonoidal categories. The only exception is in Proposition 2.2.14. Therefore, a bimonoidal category is defined by 21 Laplaza axioms. ◇

Example 2.1.31 (Commutative Rigs)**.** Suppose

$$(R, +, \times, 0_R, 1_R)$$

is a *commutative rig*, that is, a commutative ring without additive inverses. Then R becomes a small and tight symmetric bimonoidal category with only identity morphisms and

$$(+, \times, 0_R, 1_R) = (\oplus, \otimes, \mathbb{0}, \mathbb{1}).$$

In particular, the structure morphisms $\alpha^{\oplus}$, $\lambda^{\oplus}$, $\rho^{\oplus}$, $\xi^{\oplus}$, $\alpha^{\otimes}$, $\lambda^{\otimes}$, $\rho^{\otimes}$, $\xi^{\otimes}$, $\lambda^{\bullet}$, $\rho^{\bullet}$, δ^l, and δ^r are all identities.

Similarly, if R is a *rig*, that is, a ring without additive inverses, then R becomes a small and tight bimonoidal category as above with only identity morphisms. Therefore, a (symmetric) bimonoidal category is a categorification of a (commutative) rig. ◇

Example 2.1.32 (Vector Spaces)**.** The category $\mathsf{Vect}^{\mathbb{C}}$ of finite dimensional complex vector spaces is a tight symmetric bimonoidal category with

- the usual direct sum $\oplus$ and tensor product $\otimes$ of finite dimensional complex vector spaces;

- $\mathbb{0}$ the 0 vector space; and
- $\mathbb{1} = \mathbb{C}$ as a 1-dimensional complex vector space.

The structure morphisms $\alpha^{\oplus}$, $\lambda^{\oplus}$, $\rho^{\oplus}$, $\xi^{\oplus}$, $\alpha^{\otimes}$, $\lambda^{\otimes}$, $\rho^{\otimes}$, $\xi^{\otimes}$, $\lambda^{\bullet}$, $\rho^{\bullet}$, δ^l, and δ^r are the canonical isomorphisms. The axioms in Definition 2.1.2 are satisfied because, in each case, the two composites are both given by the canonical isomorphism. ◇

2.2. Reduction of Axioms

In this section, we observe that in a symmetric bimonoidal category, only 12 of the 24 axioms are necessary; the other 12 axioms are formal consequences. We emphasize that this reduction of the list of axioms applies only to symmetric bimonoidal categories and *not* bimonoidal categories. The reason is that the proofs in this section use the multiplicative symmetry $\xi^{\otimes}$, with only one exception as in Proposition 2.2.14. The braided analogue of the reduction Theorem 2.2.13 is Theorem II.2.2.1.

Motivation 2.2.1. To motivate these reductions, recall that there are 12 groups of axioms in Definition 2.1.2, four of which have only one axiom each. In the other eight groups of axioms, we will show that only the first axiom is necessary. The axiom (2.1.5) describes how the left distributivity δ^l and the right distributivity δ^r determine each other. Once this axiom is assumed, axioms involving δ^l have analogues for δ^r. Moreover, the axiom (2.1.19) states that the left multiplicative zero $\lambda^{\bullet}$ and the right multiplicative zero $\rho^{\bullet}$ determine each other. Once this axiom is assumed, axioms involving $\lambda^{\bullet}$ have analogues for $\rho^{\bullet}$. ◇

Convention 2.2.2. The following list of conventions is used throughout the rest of this book.

(1) The subscripts of a natural transformation are omitted if there is no danger of confusion. For example, we often write $\delta^l_{A,B,C}$ as δ^l.
(2) The inverse of $\alpha^{\oplus}$ is abbreviated to $\alpha^{-\oplus}$, and similarly for other invertible structure morphisms.
(3) In each diagram, if an isomorphism, such as $\xi^{\otimes}$ or $\alpha^{\otimes}$, is replaced by its inverse, then we denote the resulting equivalent version by the same reference without further comment.
(4) If a diagram is commutative because of the naturality of some structure, such as δ^l, $\xi^{\oplus}$, $\xi^{\otimes}$, or $\oplus$, then it is denoted by nat.
(5) If a diagram is commutative because of naturality and some other property P, then only P is displayed. ◇

To simplify the presentation, we adopt the following in most of this section.

Convention 2.2.3. Suppose

$$\big(\mathsf{C}, (\oplus, \mathbb{0}, \alpha^{\oplus}, \lambda^{\oplus}, \rho^{\oplus}, \xi^{\oplus}), (\otimes, \mathbb{1}, \alpha^{\otimes}, \lambda^{\otimes}, \rho^{\otimes}, \xi^{\otimes}), (\lambda^{\bullet}, \rho^{\bullet}), (\delta^l, \delta^r)\big)$$

consists of the same data as in Definition 2.1.2, in which Laplaza's axioms are *not* assumed, with A, B, C, and D objects in C. ◇

Lemma 2.2.4. *Under Convention 2.2.3, if* (2.1.5) *is satisfied, then* (2.1.6) *is equivalent to* (2.1.7).

Proof. The outer diagram below is (2.1.6), and the inner rectangle is (2.1.7).

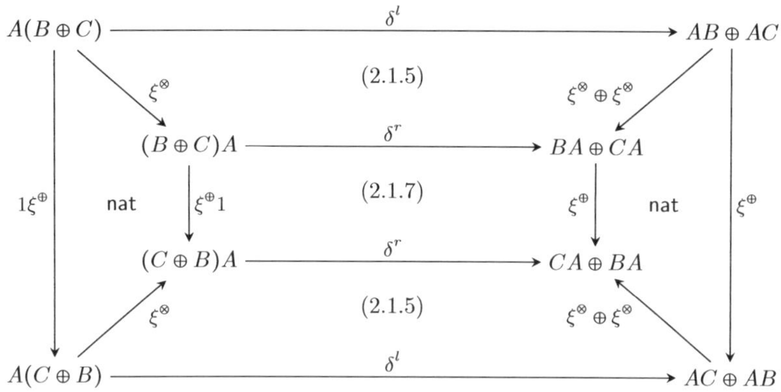

Since $\xi^{\otimes}$ is a natural isomorphism, the outer diagram is commutative if and only if the inner rectangle is commutative. □

Lemma 2.2.5. *Under Convention 2.2.3, if* (2.1.5) *is satisfied, then* (2.1.8) *is equivalent to* (2.1.9).

Proof. The outer diagram below is (2.1.9), and the middle rectangle is (2.1.8).

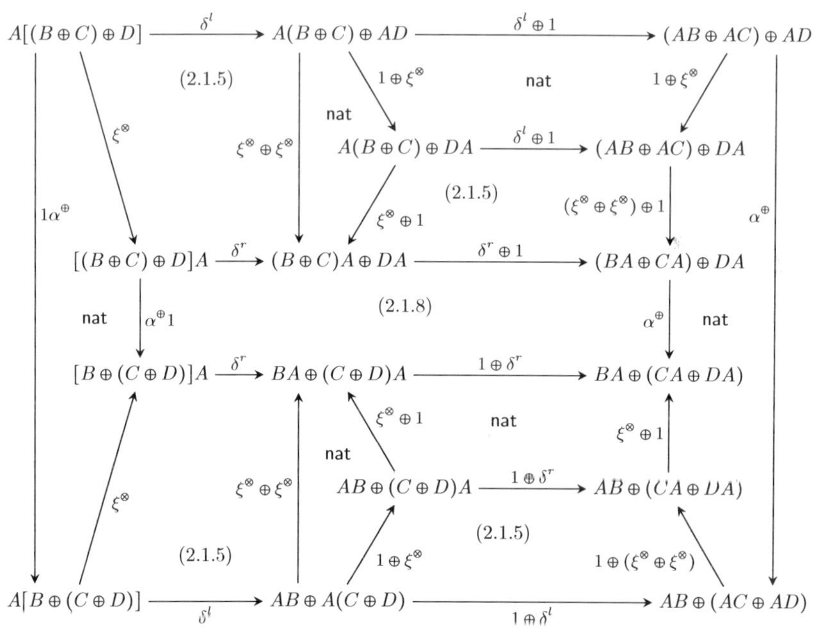

Since $\xi^{\otimes}$ is a natural isomorphism, the outer diagram is commutative if and only if the middle rectangle is commutative. □

Lemma 2.2.6. *Under Convention 2.2.3, if* (2.1.5) *is satisfied, then* (2.1.10) *is equivalent to* (2.1.11).

Proof. The outer diagram below is (2.1.10), and the middle rectangle is (2.1.11).

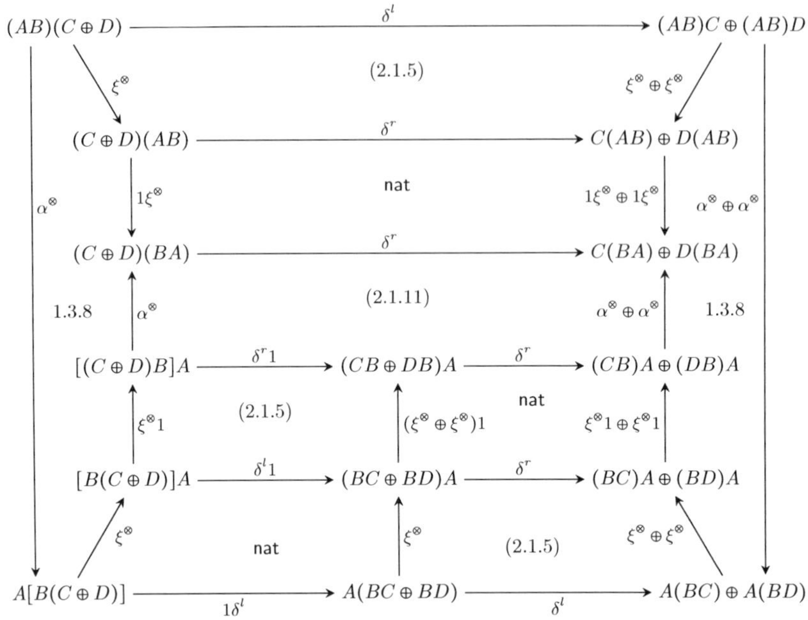

The left and the right trapezoids are commutative by Theorem 1.3.8. Since $\alpha^{\otimes}$ and $\xi^{\otimes}$ are natural isomorphisms, the outer diagram (2.1.10) is commutative if and only if the middle rectangle (2.1.11) is commutative. $\square$

Lemma 2.2.7. *Under Convention 2.2.3, if* (2.1.5) *and* (2.1.10) *are satisfied, then so is* (2.1.12).

Proof. The outer diagram below is (2.1.12).

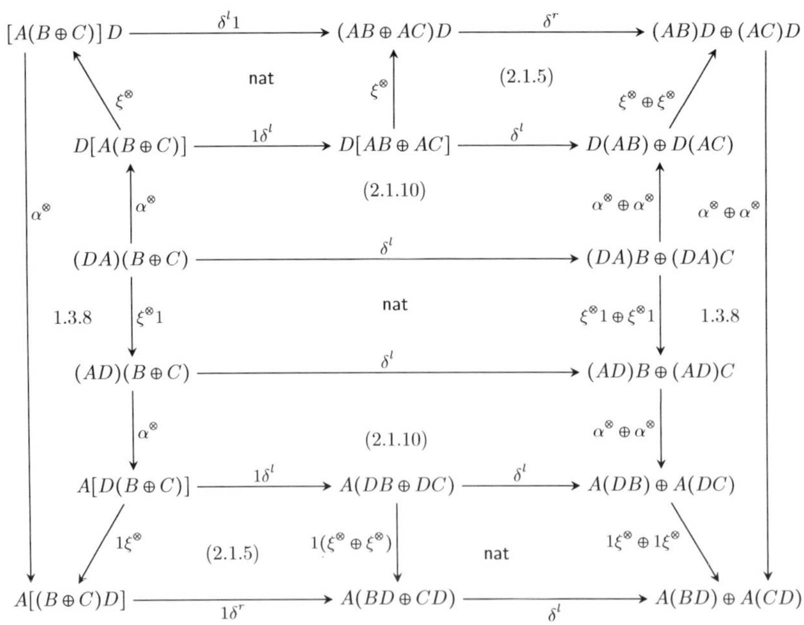

The left and the right trapezoids are commutative by Theorem 1.3.8. Since $\alpha^{\otimes}$ and $\xi^{\otimes}$ are invertible, if (2.1.10) is commutative, then so is the outer diagram (2.1.12). □

Lemma 2.2.8. *Under Convention 2.2.3, if* (2.1.5) *and* (2.1.19) *are satisfied, then* (2.1.15) *is equivalent to* (2.1.16).

Proof. Consider the following diagram.

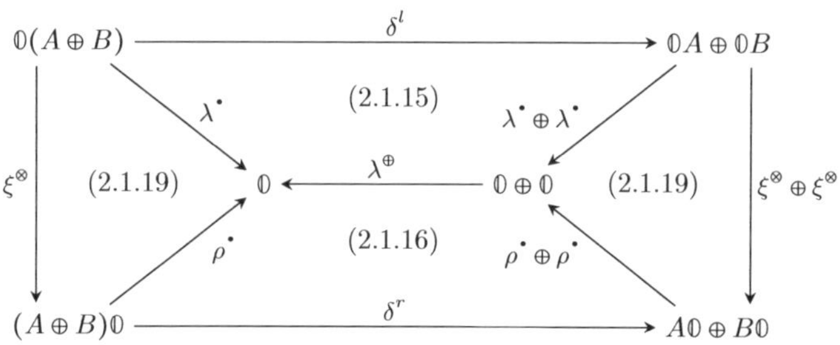

The outer diagram is commutative by (2.1.5). Since $\xi^{\otimes}$, $\lambda^{\oplus}$, and $\rho^{\bullet}$ are natural isomorphisms, the top trapezoid is commutative if and only if the bottom trapezoid is commutative. □

Lemma 2.2.9. *Under Convention 2.2.3, if* (2.1.19) *is satisfied, then* (2.1.17) *is equivalent to* (2.1.18).

Proof. Consider the following diagram.

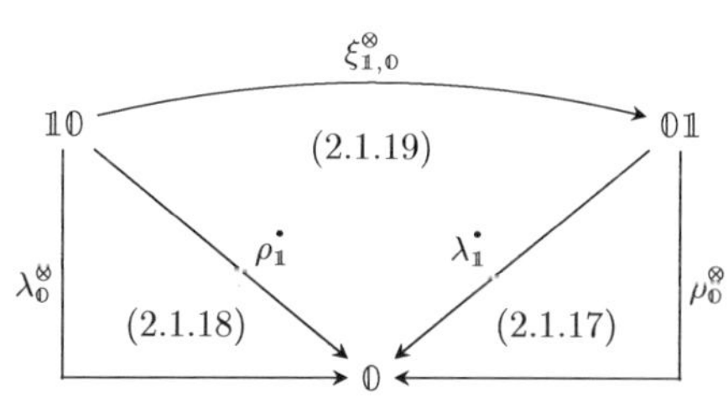

The outer diagram is commutative by the axioms (1.2.20) and (1.2.21) in the symmetric monoidal category $(\mathsf{C}, \xi^{\otimes})$. Since $\xi^{\otimes}$ is invertible, the left triangle is commutative if and only if the right triangle is commutative. □

Lemma 2.2.10. *Under Convention 2.2.3, if* (2.1.19) *is satisfied, then* (2.1.20) *implies both* (2.1.21) *and* (2.1.22).

Proof. The outer diagram below is (2.1.21).

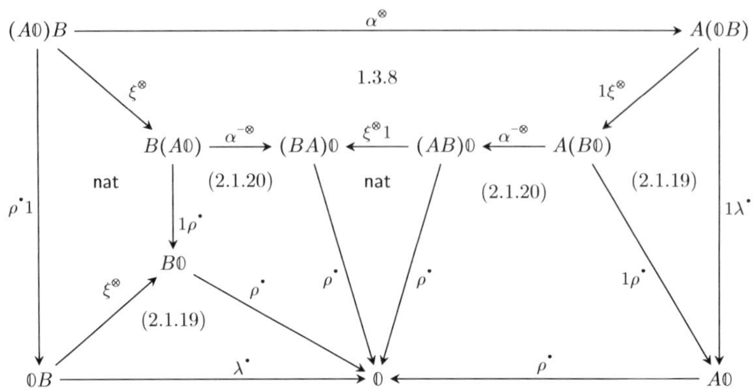

The top trapezoid is commutative by Theorem 1.3.8. Therefore, assuming (2.1.19), (2.1.20) implies (2.1.21).

For the other assertion, consider the following diagram.

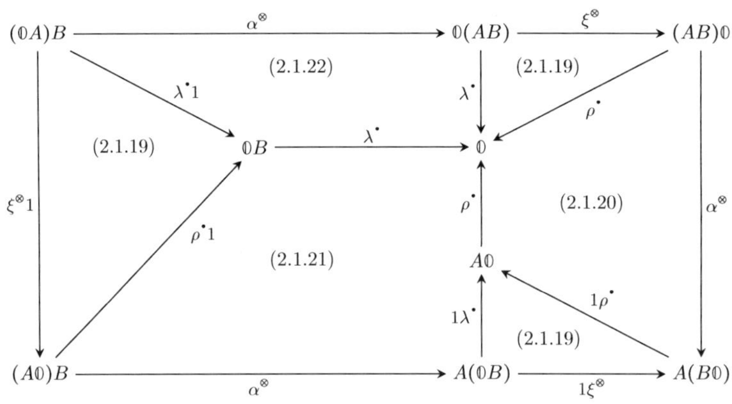

The outer diagram is commutative by Theorem 1.3.8. In the previous paragraph, we showed that assuming (2.1.19), (2.1.20) implies (2.1.21). Since $\lambda^{\bullet}$, $\rho^{\bullet}$, and $\xi^{\otimes}$ are natural isomorphisms, we infer that (2.1.20) also implies (2.1.22), which is the upper left trapezoid. □

Lemma 2.2.11. *Under Convention 2.2.3, if* (2.1.5), (2.1.6), *and* (2.1.19) *are satisfied, then the four axioms* (2.1.23), (2.1.24), (2.1.25), *and* (2.1.26) *are equivalent to each other.*

Proof. The outer diagram below is commutative by (2.1.5).

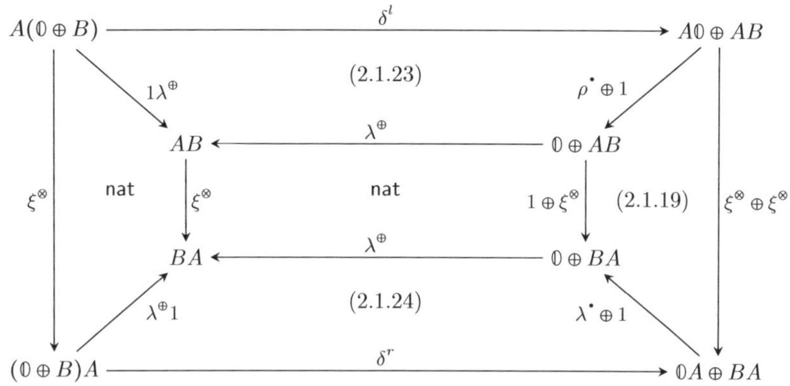

Since $\lambda^{\oplus}$, $\lambda^{\bullet}$, and $\xi^{\otimes}$ are natural isomorphisms, the above diagram shows the equivalence between (2.1.23) and (2.1.24).

The outer diagram below is commutative by (2.1.6).

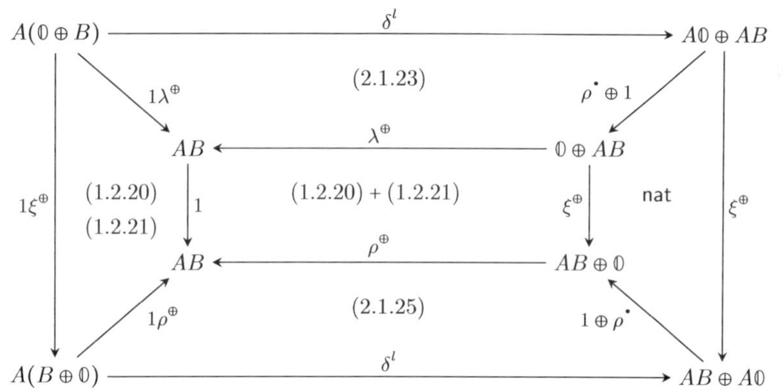

Since $\lambda^{\oplus}$, $\rho^{\oplus}$, $\xi^{\oplus}$, and $\rho^{\bullet}$ are natural isomorphisms, the above diagram shows the equivalence between (2.1.23) and (2.1.25). Note that this diagram is valid even without assuming (2.1.5), (2.1.19), and the existence of $\xi^{\otimes}$.

The outer diagram below is commutative by (2.1.5).

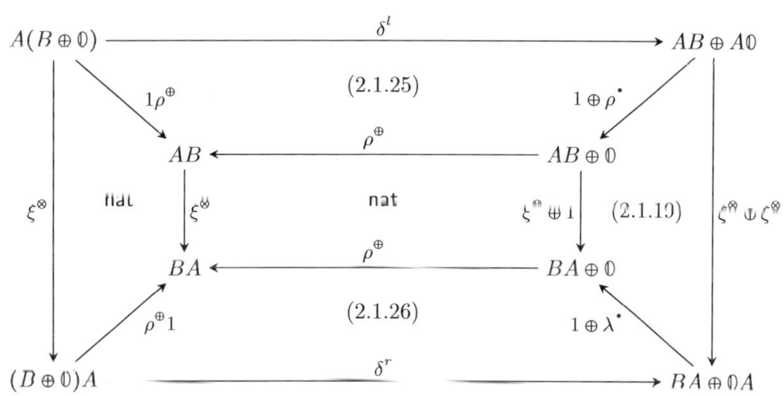

Since $\rho^{\oplus}$, $\lambda^{\bullet}$, and $\xi^{\otimes}$ are natural isomorphisms, the above diagram shows the equivalence between (2.1.25) and (2.1.26). □

Lemma 2.2.12. *Under Convention 2.2.3, if* (2.1.5) *is satisfied, then* (2.1.27) *is equivalent to* (2.1.28).

Proof. The outer diagram below is commutative by (2.1.5).

$$\begin{array}{ccccc}
\mathbb{1}(A \oplus B) & & \xrightarrow{\quad \delta^l \quad} & & \mathbb{1}A \oplus \mathbb{1}B \\
 & \searrow^{\lambda^{\otimes}} & (2.1.27) & \swarrow_{\lambda^{\otimes} \oplus \lambda^{\otimes}} & \\
\Big\downarrow \xi^{\otimes} & & A \oplus B & & \Big\downarrow \xi^{\otimes} \oplus \xi^{\otimes} \\
 & \nearrow^{\rho^{\otimes}} & (2.1.28) & \nwarrow_{\rho^{\otimes} \oplus \rho^{\otimes}} & \\
(A \oplus B)\mathbb{1} & & \xrightarrow{\quad \delta^r \quad} & & A\mathbb{1} \oplus B\mathbb{1}
\end{array}$$

The left and right triangles are commutative by the axioms (1.2.20) and (1.2.21) in the symmetric monoidal category $(\mathsf{C}, \xi^{\otimes})$. Since $\lambda^{\otimes}$ and $\rho^{\otimes}$ are natural isomorphisms, the above diagram shows the equivalence between (2.1.27) and (2.1.28). □

Combining Lemmas 2.2.4 through 2.2.12, we obtain the following main observation of this section.

Theorem 2.2.13. *In Definition 2.1.2 of a symmetric bimonoidal category, it is sufficient to assume the first axiom in each of the twelve groups of axioms, namely,* (2.1.5), (2.1.6), (2.1.8), (2.1.10), (2.1.13), (2.1.14), (2.1.15), (2.1.17), (2.1.19), (2.1.20), (2.1.23), *and* (2.1.27).

As noted at the beginning of this section, Theorem 2.2.13 does not apply to bimonoidal categories because its proof uses the multiplicative symmetry $\xi^{\otimes}$. There is only one exception as follows.

Proposition 2.2.14. *In a bimonoidal category, the two axioms* (2.1.23) *and* (2.1.25) *are equivalent, so only one of them is necessary.*

Proof. The second diagram in the proof of Lemma 2.2.11, which establishes the equivalence between (2.1.23) and (2.1.25), is valid in a bimonoidal category because it does not involve the multiplicative symmetry. □

2.3. Distributive Symmetric Monoidal Categories

In this section, we discuss a class of examples of symmetric bimonoidal categories whose sums are coproducts.

Definition 2.3.1. A symmetric monoidal category C is said to be *distributive* if the following two conditions hold.

(1) C has finite coproducts. Fix an initial object $\varnothing \in \mathsf{C}$.
(2) The natural morphisms

$$\textstyle\coprod_{i=1}^{n}(A \otimes B_i) \xrightarrow{\quad d_l \quad} A \otimes \big(\coprod_{i=1}^{n} B_i\big)$$

$$\textstyle\coprod_{i=1}^{n}(B_i \otimes A) \xrightarrow{\quad d_r \quad} \big(\coprod_{i=1}^{n} B_i\big) \otimes A$$

are isomorphisms for $n = 0, 2$ and for objects $A, B_i \in \mathsf{C}$, with $\coprod$ denoting coproduct, and with an empty coproduct denoting $\varnothing$. ◇

Recall that a symmetric bimonoidal category is *tight* if both distributivity δ^l and δ^r are natural isomorphisms.

Proposition 2.3.2. *Suppose C is a distributive symmetric monoidal category. Then C yields a tight symmetric bimonoidal category with the following data.*

The Multiplicative Structure: *It is the given symmetric monoidal structure on C.*

The Additive Structure: *It is $(\mathsf{C}, \sqcup, \varnothing)$, with $\alpha^{\oplus}$, $\lambda^{\oplus}$, $\rho^{\oplus}$, and $\xi^{\oplus}$ given by the universal properties of coproducts and $\varnothing$.*

The Multiplicative Zeros: $\lambda^{\bullet} = d_r^{-1}$ *and* $\rho^{\bullet} = d_l^{-1}$ *in the case* $n = 0$.

The Distributivity Morphisms: $\delta^l = d_l^{-1}$ *and* $\delta^r = d_r^{-1}$ *in the case* $n = 2$.

Proof. The universal properties of coproducts and $\varnothing$ imply that $(\mathsf{C}, \sqcup, \varnothing)$ is a symmetric monoidal category. By definition, δ^l and δ^r are natural isomorphisms. The twelve axioms in Theorem 2.2.13 all follow from the university properties of coproducts and $\varnothing$, and the naturality of the multiplicative structure. □

Below are some examples of distributive symmetric monoidal categories, to which Proposition 2.3.2 applies to yield tight symmetric bimonoidal categories.

Example 2.3.3 (Symmetric Monoidal Closed Categories). If C is a symmetric monoidal closed category as in Definition 1.2.28 with finite coproducts, then the natural morphisms d_l and d_r are isomorphisms. Therefore, C is a distributive symmetric monoidal category. ◇

Example 2.3.4 (Modules). For each commutative ring R, the category $\mathsf{Mod}(R)$ of R-modules is symmetric monoidal with

- the usual tensor product of R-modules;
- coproducts given by direct sums $\oplus$ of R-modules; and
- internal hom given by R-linear maps.

As a special case of Example 2.3.3, $\mathsf{Mod}(R)$ is a distributive symmetric monoidal category, hence also a symmetric bimonoidal category with $(\sqcup, \varnothing) = (\oplus, 0)$. ◇

Example 2.3.5 (Distributive Categories). A *distributive category* is a category with finite products $\times$ and finite coproducts such that the natural morphism d_l is an isomorphism for $n = 0, 2$ and $\otimes = \times$. The symmetry isomorphism of $\times$ implies that d_r is also an isomorphism for $n = 0, 2$. Therefore, a distributive category is a distributive symmetric monoidal category, hence also a symmetric bimonoidal category with $\otimes = \times$, $\mathbb{1} = *$ (a chosen terminal object), $\oplus = \sqcup$, and $\mathbb{0} = \varnothing$ (a chosen initial object). ◇

The symmetric bimonoidal categories in Proposition 2.3.2 and Examples 2.3.3 through 2.3.5 all have $(\oplus, \mathbb{0}) = (\sqcup, \varnothing)$. In other words, the additive structure is given by coproducts. For the symmetric bimonoidal categories Σ and Σ' in Section 2.4, their sums are *not* given by coproducts.

2.4. Finite Ordinal Category

In this section, we discuss a symmetric bimonoidal category Σ that will play an important role in Chapters 4, 7, and II.7.

(1) In Chapter 4, Σ is extended to the distortion category $\mathcal{D}$ that is a crucial ingredient in formulating and proving the Coherence Theorem 4.4.3.
(2) In Chapter 7, we show that Σ is bicategorically an initial object in a suitable 2-category of small symmetric bimonoidal categories.
(3) In Chapter II.7, we show that Σ is bicategorically an initial object in a suitable 2-category of small braided bimonoidal categories.

A variation of Σ is discussed in the second half of this section. Further explanation of the formulas in the next definition is given in Explanation 2.4.7.

Definition 2.4.1. Define Σ by the following data.

Category: As a category, Σ is defined as follows.

Objects: They are nonnegative integers $n \geq 0$.

Morphisms: For objects m and n, the morphism set is

$$\Sigma(m,n) = \begin{cases} \Sigma_n & \text{if } m = n, \text{ and} \\ \varnothing & \text{if } m \neq n. \end{cases}$$

Here Σ_n is the symmetric group on n letters.

Composition is that of bijections, and $\mathrm{id}_n \in \Sigma_n$ is the identity morphism on n.

The Additive Structure: The functor

$$\Sigma \times \Sigma \xrightarrow{\oplus} \Sigma$$

is defined on objects by

$$m \oplus n = m + n,$$

and on morphisms

$$\Sigma_m \times \Sigma_n \xrightarrow{\oplus} \Sigma_{m+n}$$

by the *block sum*

(2.4.2) $$(\sigma \oplus \tau)(i) = \begin{cases} \sigma(i) & \text{if } 1 \leq i \leq m, \text{ and} \\ \tau(i-m) + m & \text{if } m+1 \leq i \leq m+n \end{cases}$$

for $(\sigma, \tau) \in \Sigma_m \times \Sigma_n$.

The additive zero $\mathbb{0}$ is the object 0, which is a strict two-sided unit for $\oplus$. The additive structure isomorphisms $\alpha^{\oplus}$, $\lambda^{\oplus}$, and $\rho^{\oplus}$ are identities. The additive symmetry isomorphism

$$m \oplus n = m + n \xrightarrow{\xi^{\oplus}_{m,n}} n + m = n \oplus m$$

is the bijection defined by

(2.4.3) $$\xi^{\oplus}_{m,n}(j) = \begin{cases} j+n & \text{if } 1 \leq j \leq m, \text{ and} \\ j-m & \text{if } m+1 \leq j \leq m+n. \end{cases}$$

The Multiplicative Structure: The functor

$$\Sigma \times \Sigma \xrightarrow{\otimes} \Sigma$$

is defined on objects by

$$m \otimes n = mn,$$

and on morphisms

$$\Sigma_m \times \Sigma_n \xrightarrow{\ \otimes\ } \Sigma_{mn}$$

by

$$(\sigma \otimes \tau)\big(i + (j-1)m\big) = \sigma(i) + \big(\tau(j) - 1\big)m \tag{2.4.4}$$

for $(\sigma,\tau) \in \Sigma_m \times \Sigma_n$, $1 \le i \le m$, and $1 \le j \le n$.

The multiplicative unit $\mathbb{1}$ is the object 1, which is a strict two-sided unit for $\otimes$. The multiplicative structure isomorphisms $\alpha^{\otimes}$, $\lambda^{\otimes}$, and $\rho^{\otimes}$ are identities.

The multiplicative symmetry isomorphism

$$m \otimes n = mn \xrightarrow{\ \xi^{\otimes}_{m,n}\ } nm = n \otimes m$$

is the bijection defined by

$$\xi^{\otimes}_{m,n}\big(i + (j-1)m\big) = j + (i-1)n \tag{2.4.5}$$

for $1 \le i \le m$ and $1 \le j \le n$.

The Multiplicative Zeros: The left multiplicative zero $\lambda^{\bullet}$ and the right multiplicative zero $\rho^{\bullet}$ are both identity natural transformations. This is well defined because

$$m \otimes 0 = 0 = 0 \otimes m.$$

Distributivity: The left distributivity morphism

$$m \otimes \big(n \oplus p\big) \xrightarrow{\ \delta^{l}_{m,n,p}\ } \big(m \otimes n\big) \oplus \big(m \otimes p\big)$$

is the identity permutation in $\Sigma_{m(n+p)}$. The right distributivity morphism

$$\big(m \oplus n\big) \otimes p \xrightarrow{\ \delta^{r}_{m,n,p}\ } \big(m \otimes p\big) \oplus \big(n \otimes p\big)$$

is the permutation in $\Sigma_{(m+n)p}$ defined by

$$\begin{aligned} \delta^r\big(i + (k-1)(m+n)\big) &= i + (k-1)m \\ \delta^r\big(j + m + (k-1)(m+n)\big) &= j + (k-1)n + pm \end{aligned} \tag{2.4.6}$$

for $1 \le i \le m$, $1 \le j \le n$, and $1 \le k \le p$.

This finishes the definition of Σ. ◇

Explanation 2.4.7. Here we provide geometric intuition of the constructions in Definition 2.4.1.

Objects and morphisms: The object $m \in \Sigma$ is regarded as an interval with m objects, arranged from left to right. A morphism in Σ permutes the objects within such an interval.

Sums: The object $m \oplus n = m + n$ is the combined interval with m on the left and n on the right.

Block sums: In the block sum $\sigma \oplus \tau$ in (2.4.2), σ acts on the interval on the left, and τ acts on the interval on the right.

Additive symmetry: $\xi^{\oplus}_{m,n}$ in (2.4.3) interchanges the two intervals and leaves the order within each interval unchanged.

Products: The object $m \otimes n = mn$ consists of n intervals, each with m objects. The elements in n are now used as indices for these n intervals. Alternatively, it is an $n \times m$ matrix with each row a copy of m and with the rows indexed by n.

Product of morphisms: In the bijection $\sigma \otimes \tau \in \Sigma_{mn}$ in (2.4.4), $\tau \in \Sigma_n$ permutes the n intervals, and $\sigma \in \Sigma_m$ permutes within each interval. In the matrix description, τ permutes the n rows, and σ permutes the m columns.

Multiplicative symmetry: $\xi^{\otimes}_{m,n}$ in (2.4.5) redistributes n intervals of m objects each to m intervals of n objects each. Specifically, it sends the ith object in the jth interval in the domain to the jth object in the ith interval in the codomain. In the matrix description, this corresponds to taking the transpose of an $n \times m$ matrix.

Left distributivity: $\delta^l_{m,n,p}$ is the identity because $(n+p)$ intervals of m objects each is already equal to n intervals of m objects each followed by p intervals of m objects each.

Right distributivity: $\delta^r_{m,n,p}$ in (2.4.6) redistributes p intervals of $(m+n)$ objects each to p intervals of m objects each followed by p intervals of n objects each. Alternatively, in the matrix description, we first arrange $(m+n)p$ objects into a $p \times (m+n)$ matrix of the form $[A|B]$ with
- A the $p \times m$ matrix consisting of the first m columns and
- B the $p \times n$ matrix consisting of the last n columns.

Then this component of δ^r rearranges the matrix $[A|B]$ to the array $\left[\frac{A}{B}\right]$, leaving the order of the entries in each of A and B unchanged. The two rows in (2.4.6) correspond to the action of δ^r on, respectively, the (k,i)-entry in A and the (k,j)-entry in B. ◇

Recall from Definitions 1.2.1 and 1.2.18 that a *permutative category* is a symmetric strict monoidal category.

Proposition 2.4.8. *Σ in Definition 2.4.1 is a small and tight symmetric bimonoidal category whose additive structure and multiplicative structure are both permutative categories.*

Proof. That the additive structure is a strict monoidal category follows from the fact that $\alpha^{\oplus}$, $\lambda^{\oplus}$, and $\rho^{\oplus}$ are identities. The unit axiom (1.2.21) holds because $\xi^{\oplus}_{m,0}$ is the identity map. The symmetry axiom (1.2.20) and the hexagon axiom (1.2.22) both follow from the description in Explanation 2.4.7 of $\xi^{\oplus}_{m,n}$ as interchanging two consecutive intervals. So the additive structure is a symmetric strict monoidal category.

To see that the multiplicative associativity can be defined as the identity, suppose $\sigma \in \Sigma_m$, $\tau \in \Sigma_n$, and $\pi \in \Sigma_p$. Then both $(\sigma \otimes \tau) \otimes \pi$ and $\sigma \otimes (\tau \otimes \pi)$ in Σ_{mnp} are given by the bijection

$$i + (j-1)m + (k-1)(mn) \longmapsto \sigma(i) + \big(\tau(j) - 1\big)m + \big(\pi(k) - 1\big)(mn) \tag{2.4.9}$$

for $1 \le i \le m$, $1 \le j \le n$, and $1 \le k \le p$. See Explanation 2.4.14. Therefore, the identity natural transformation $\alpha^{\otimes}$ is well defined. Both identity natural transformations $\lambda^{\otimes}$ and $\rho^{\otimes}$ are well defined by (2.4.4) and that $\mathbb{1} = 1$ is a strict two-sided unit for $\otimes$. That the multiplicative structure is a strict monoidal category follows from the fact that $\alpha^{\otimes}$, $\lambda^{\otimes}$, and $\rho^{\otimes}$ are identities.

Next we check the symmetric monoidal category axioms.

- The unit axiom (1.2.21) holds because $\xi^{\otimes}_{m,1}$ is the identity map.
- The symmetry axiom (1.2.20) holds by the description of $\xi^{\otimes}_{m,n}$ as redistributing n intervals of m objects each to m intervals of n objects each. Alternatively, regarding $\xi^{\otimes}_{m,n}$ as taking the transpose of an $n \times m$ matrix, its inverse is the transpose given by $\xi^{\otimes}_{n,m}$.
- The hexagon axiom (1.2.22) is equivalent to the commutativity of the following diagram for $m, n, p \geq 0$.

$$\begin{array}{ccc} m \otimes n \otimes p & \xrightarrow{1_m \otimes \xi^{\otimes}_{n,p}} & m \otimes p \otimes n \\ & {\scriptstyle \xi^{\otimes}_{mn,p}} \searrow \quad \swarrow {\scriptstyle \xi^{\otimes}_{m,p} \otimes 1_n} & \\ & p \otimes m \otimes n & \end{array}$$

This diagram is commutative because both composites are given by the bijection

$$i + (j-1)m + (k-1)(mn) \longmapsto k + [i - 1 + (j-1)m]p \tag{2.4.10}$$

for $1 \leq i \leq m$, $1 \leq j \leq n$, and $1 \leq k \leq p$. See Explanation 2.4.14.

Therefore, the multiplicative structure is a symmetric strict monoidal category.

For each of the twelve axioms listed in Theorem 2.2.13 for Σ, we check that the two relevant permutations are equal. First, for the axiom (2.1.6) with $A = m$, $B = n$, and $C = p$, each of the two composites is the permutation in $\Sigma_{m(n+p)}$ given by

$$\begin{cases} i + (j-1)m \longmapsto i + (p+j-1)m \\ i + (n+k-1)m \longmapsto i + (k-1)m \end{cases} \tag{2.4.11}$$

for $1 \leq i \leq m$, $1 \leq j \leq n$, and $1 \leq k \leq p$. See Explanation 2.4.15.

For the axiom (2.1.5), with the same notation as above, each of the two composites is the permutation in $\Sigma_{(m+n)p}$ given by

$$\begin{cases} i + (k-1)(m+n) \longmapsto k + (i-1)p \\ j + m + (k-1)(m+n) \longmapsto k + (m+j-1)p \end{cases} \tag{2.4.12}$$

for $1 \leq i \leq m$, $1 \leq j \leq n$, and $1 \leq k \leq p$. See Explanation 2.4.16.

For the 2-by-2 distributivity axiom (2.1.13) with $D = q$, each of the two composites is the permutation in $\Sigma_{(m+n)(p+q)}$ given by

$$\begin{cases} i + (k-1)(m+n) \longmapsto i + (k-1)m \\ j + m + (k-1)(m+n) \longmapsto j + (k-1)n + mp \\ i + (m+n)(p+l-1) \longmapsto i + (l-1)m + (m+n)p \\ j + m + (m+n)(p+l-1) \longmapsto j + (l-1)n + (m+n)p + mq \end{cases} \tag{2.4.13}$$

for $1 \leq i \leq m$, $1 \leq j \leq n$, $1 \leq k \leq p$, and $1 \leq l \leq q$. See Explanation 2.4.17.

The other nine axioms—(2.1.9), which is equivalent to (2.1.8) by Lemma 2.2.5, (2.1.10), (2.1.14), (2.1.15), (2.1.17), (2.1.19), (2.1.20), (2.1.23), and (2.1.27)—hold because each permutation involved is the identity. □

We now provide geometric description of the bijections that appeared in the proof of Proposition 2.4.8.

Explanation 2.4.14. For the bijection in (2.4.10), suppose

$$A = [\bullet, \ldots, \bullet]$$

is a $1 \times m$ matrix with m objects, and M is the $p \times mn$ matrix

$$M = \begin{bmatrix} A_{11} & \cdots & A_{1n} \\ \vdots & \ddots & \vdots \\ A_{p1} & \cdots & A_{pn} \end{bmatrix}$$

with each A_{kj} a copy of A for $1 \leq j \leq n$ and $1 \leq k \leq p$. Then the bijection in (2.4.10), namely $\xi^{\otimes}_{mn,p}$, sends M to its transpose

$$M^T = \begin{bmatrix} A^T_{11} & \cdots & A^T_{p1} \\ \vdots & \ddots & \vdots \\ A^T_{1n} & \cdots & A^T_{pn} \end{bmatrix}$$

with the superscript T denoting transpose. The formula (2.4.10) is the action of this bijection on the ith object in A_{kj}.

For the bijection in (2.4.9), namely

$$(\sigma \otimes \tau) \otimes \pi = \sigma \otimes (\tau \otimes \pi) \in \Sigma_{mnp},$$

regard M as a $p \times n$ matrix with (k, j)-entry A_{kj}. Then each of these two bijections permutes

- the p rows of M via $\pi \in \Sigma_p$;
- the n columns of M via $\tau \in \Sigma_n$; and
- the objects in each $1 \times m$ matrix $A_{kj} = A$ via $\sigma \in \Sigma_m$. ◇

Explanation 2.4.15. For the bijection in (2.4.11), first arrange $m(n+p)$ objects into an $(n+p) \times m$ matrix of the form $M = \left[\frac{A}{B}\right]$ with

- A the $n \times m$ matrix consisting of the top n rows and
- B the $p \times m$ matrix consisting of the bottom p rows.

Then the bijection in (2.4.11) rearranges the matrix $\left[\frac{A}{B}\right]$ to $\left[\frac{B}{A}\right]$ by swapping the blocks A and B. The two rows in (2.4.11) correspond to the action of this bijection on, respectively,

- the (j, i)-entry in A in M and
- the (k, i)-entry in B in M. ◇

Explanation 2.4.16. For the bijection in (2.4.12), first arrange $(m+n)p$ objects into a $p \times (m+n)$ matrix of the form $M = [A|B]$ with

- A the $p \times m$ matrix consisting of the first m columns and
- B the $p \times n$ matrix consisting of the last n columns.

Then the bijection in (2.4.12) sends the matrix $[A|B]$ to its transpose $\left[\frac{A^T}{B^T}\right]$, with A^T and B^T denoting the transposes of, respectively, A and B. The two rows in (2.4.12) correspond to the action of this bijection on, respectively,

- the (k, i)-entry in A in M and
- the (k, j)-entry in B in M. ◇

Explanation 2.4.17. For the bijection in (2.4.13), first arrange $(m+n)(p+q)$ objects into a $(p+q)\times(m+n)$ matrix M of the form

$$M = \left[\begin{array}{c|c} A & B \\ \hline C & D \end{array}\right]$$

with

- A a $p \times m$ matrix,
- B a $p \times n$ matrix,
- C a $q \times m$ matrix, and
- D a $q \times n$ matrix.

Then the bijection in (2.4.13) rearranges M to the array

$$\begin{bmatrix} A \\ B \\ C \\ D \end{bmatrix}.$$

The four rows in (2.4.13) correspond to the action of this bijection on, respectively,

- the (k,i)-entry in A in M,
- the (k,j)-entry in B in M,
- the (l,i)-entry in C in M, and
- the (l,j)-entry in D in M. ◇

Next we describe a variant of Σ in which the right distributivity is an identity.

Definition 2.4.18. Define Σ' by the following data.

The Additive Structure: The additive structure

$$(\Sigma', \oplus, \mathbb{0}, \alpha^{\oplus}, \lambda^{\oplus}, \rho^{\oplus}, \xi^{\oplus})$$

is the same as that of Σ in Definition 2.4.1.

The Multiplicative Structure: The functor

$$\Sigma' \times \Sigma' \xrightarrow{\ \otimes\ } \Sigma'$$

is again defined by $m \otimes n = mn$ on objects, and by

$$(\sigma \otimes \tau)\big(j + (i-1)n\big) = \tau(j) + \big(\sigma(i) - 1\big)n \tag{2.4.19}$$

on morphisms for $(\sigma, \tau) \in \Sigma_m \times \Sigma_n$, $1 \le i \le m$, and $1 \le j \le n$.

The multiplicative unit $\mathbb{1}$ is the object 1, which is a strict two-sided unit for $\otimes$. The multiplicative structure isomorphisms $\alpha^{\otimes}$, $\lambda^{\otimes}$, and $\rho^{\otimes}$ are identities.

The multiplicative symmetry isomorphism

$$mn \xrightarrow{\ \xi'^{\otimes}_{m,n}\ } mn$$

is defined by

$$\xi'^{\otimes}_{m,n}\big(j + (i-1)n\big) = i + (j-1)m \tag{2.4.20}$$

for $1 \le i \le m$ and $1 \le j \le n$.

The Multiplicative Zeros: Both $\lambda^{\bullet}$ and $\rho^{\bullet}$ are identity natural transformations.

Distributivity: The right distributivity morphism δ^r is the identity natural transformation. The left distributivity morphism

$$m \otimes (n \oplus p) \xrightarrow{\delta^l_{m,n,p}} (m \otimes n) \oplus (m \otimes p)$$

is the permutation in $\Sigma_{m(n+p)}$ defined by

$$\begin{aligned} \delta^l\big(j + (i-1)(n+p)\big) &= j + (i-1)n \\ \delta^l\big(k + n + (i-1)(n+p)\big) &= k + (i-1)p + mn \end{aligned} \tag{2.4.21}$$

for $1 \le i \le m$, $1 \le j \le n$, and $1 \le k \le p$.

This finishes the definition of Σ'. ◇

Explanation 2.4.22. The additive structure of Σ' has the same geometric interpretation as that of Σ in Explanation 2.4.7. Consider the rest of Σ' in Definition 2.4.18.

Products: The object $m \otimes n = mn$ is now regarded as consisting of m intervals, each with n objects. Alternatively, it is an $m \times n$ matrix with each row a copy of n and with the rows indexed by m.

Product of morphisms: In the bijection $\sigma \otimes \tau \in \Sigma_{mn}$ in (2.4.19), $\sigma \in \Sigma_m$ permutes the m intervals, and $\tau \in \Sigma_n$ permutes within each interval. In the matrix description, σ permutes the m rows, and τ permutes the n columns.

Multiplicative symmetry: $\xi'^{\otimes}_{m,n}$ in (2.4.20) redistributes m intervals of n objects each to n intervals of m objects each by sending the jth object in the ith interval in the domain to the ith object in the jth interval in the codomain. In the matrix description, this corresponds to taking the transpose of an $m \times n$ matrix.

Right distributivity: $\delta^r_{m,n,p}$ is the identity because $(m+n)$ intervals of p objects each is already equal to m intervals of p objects each followed by n intervals of p objects each.

Left distributivity: $\delta^l_{m,n,p}$ in (2.4.21) redistributes m intervals of $(n+p)$ objects each to m intervals of n objects each followed by m intervals of p objects each. Alternatively, in the matrix description, we first arrange $m(n+p)$ objects into an $m \times (n+p)$ matrix of the form $[A|B]$ with
- A the $m \times n$ matrix consisting of the first n columns and
- B the $m \times p$ matrix consisting of the last p columns.

Then this component of δ^l rearranges the matrix $[A|B]$ to the array $\left[\frac{A}{B}\right]$, leaving the order of the entries in each of A and B unchanged. The two rows in (2.4.21) correspond to the action of δ^l on, respectively, the (i,j)-entry in A and the (i,k)-entry in B. ◇

A minor modification of the proof of Proposition 2.4.8 yields the following.

Proposition 2.4.23. *Σ' in Definition 2.4.18 is a small and tight symmetric bimonoidal category whose additive structure and multiplicative structure are both permutative categories.*

2.5. Bipermutative Categories

In this section, we introduce strict analogues of a symmetric bimonoidal category, which are called a right, respectively left, bipermutative category. We observe that right and left bipermutative categories are tight symmetric bimonoidal categories. In Chapter 5, we will show that tight symmetric bimonoidal categories can be strictified to right and left bipermutative categories. First we define right bipermutative categories.

Right Bipermutative Categories.

Motivation 2.5.1. To motivate the definition of a right bipermutative category, recall from Section 2.4 the two symmetric bimonoidal categories, Σ and Σ', whose additive and multiplicative structures are both permutative categories, that is, symmetric strict monoidal categories. Their left and right multiplicative zeros, $\lambda^\bullet$ and $\rho^\bullet$, are both identities. However, only one of the two distributivity morphisms is an identity: δ^l in Σ and δ^r in Σ'. These two examples suggest that, in defining a strict analogue of a symmetric bimonoidal category, we should demand that the multiplicative zeros and only one of the two distributivity morphisms be the identities. In a right bipermutative category, the right distributivity morphism δ^r is the identity. ◇

Definition 2.5.2. A *right bipermutative category* is a tuple

$$\big(\mathsf{C}, (\oplus, \mathbb{0}, \xi^\oplus), (\otimes, \mathbb{1}, \xi^\otimes), (\lambda^\bullet, \rho^\bullet), (\delta^l, \delta^r)\big)$$

consisting of the following data.

- $(\mathsf{C}, \oplus, \mathbb{0}, \xi^\oplus)$ and $(\mathsf{C}, \otimes, \mathbb{1}, \xi^\otimes)$ are permutative categories.
- $\lambda^\bullet$, $\rho^\bullet$, δ^l, and δ^r are natural transformations as in (2.1.3) and (2.1.4).

The above data are required to satisfy the following four conditions.

(1) $\lambda^\bullet$ and $\rho^\bullet$ are both equal to the identity natural transformation of the constant functor $\mathsf{C} \longrightarrow \mathsf{C}$ at $\mathbb{0}$.
(2) δ^r is an identity natural transformation.
(3) $\xi^\otimes_{A,\mathbb{0}} : A \otimes \mathbb{0} \longrightarrow \mathbb{0} \otimes A$ is the identity morphism of $\mathbb{0}$ for each object A.
(4) The axioms (2.1.5), (2.1.7), and (2.1.13) are satisfied.

This finishes the definition of a right bipermutative category. ◇

Explanation 2.5.3. Consider Definition 2.5.2.

- The axiom (2.1.5) is equivalent to the following commutative diagram.

(2.5.4)
$$\begin{array}{ccc}
A(B\oplus C) & \xrightarrow{\delta^l} & AB\oplus AC \\
{\scriptstyle \xi^\otimes}\big\downarrow & & \big\uparrow{\scriptstyle \xi^\otimes \oplus \xi^\otimes} \\
(B\oplus C)A & \xrightarrow[=]{\delta^r} & BA\oplus CA
\end{array}$$

In particular, the left distributivity morphism δ^l is a natural isomorphism.

- The axiom (2.1.7) means the commutativity of the following diagram.

(2.5.5)
$$\begin{array}{ccc}
(A\oplus B)C & \xrightarrow[=]{\delta^r} & AC\oplus BC \\
{\scriptstyle \xi^\oplus 1}\big\downarrow & & \big\downarrow{\scriptstyle \xi^\oplus} \\
(B\oplus A)C & \xrightarrow[=]{\delta^r} & BC\oplus AC
\end{array}$$

- The axiom (2.1.13) means the commutativity of the following diagram.

$$
\begin{array}{ccccc}
(A \oplus B)(C \oplus D) & & \xrightarrow[=]{\delta^r} & & A(C \oplus D) \oplus B(C \oplus D) \\
\delta^l \downarrow & & & & \downarrow \delta^l \oplus \delta^l \\
(A \oplus B)C \oplus (A \oplus B)D & & & & AC \oplus AD \oplus BC \oplus BD \\
& \underset{\delta^r \oplus \delta^r}{\overset{=}{\searrow}} & & \swarrow_{1 \oplus \xi^{\oplus} \oplus 1} & \\
& & AC \oplus BC \oplus AD \oplus BD & &
\end{array}
\tag{2.5.6}
$$

◇

Recall from Definition 2.1.2 that a symmetric bimonoidal category is *tight* if both δ^l and δ^r are natural isomorphisms.

Proposition 2.5.7. *Each right bipermutative category is a tight symmetric bimonoidal category.*

Proof. The right distributivity morphism δ^r is the identity by definition. The left distributivity morphism δ^l is a natural isomorphism by (2.5.4). Consider the 24 axioms in Definition 2.1.2 for a right bipermutative category.

- The axioms (2.1.5), (2.1.7), and (2.1.13) hold by assumption.
- The axiom (2.1.6) holds by Lemma 2.2.4.
- The axiom (2.1.19) holds by the identity assumptions on $\lambda^{\bullet}$, $\rho^{\bullet}$, and $\xi^{\otimes}_{-,\mathbb{0}}$.
- The axioms (2.1.8), (2.1.11), (2.1.14), (2.1.16), (2.1.17), (2.1.20), (2.1.24), and (2.1.28) hold by the assumption that $\lambda^{\bullet}$, $\rho^{\bullet}$, and δ^r are identities.

We finish the proof by applying Lemmas 2.2.5 through 2.2.12 to obtain all 24 axioms of a symmetric bimonoidal category. □

Example 2.5.8. The symmetric bimonoidal category Σ' in Proposition 2.4.23 is a right bipermutative category. ◇

Example 2.5.9 (Coordinatized Vector Spaces)**.** A variation of the tight symmetric bimonoidal category $\mathsf{Vect}^{\mathbb{C}}$ in Example 2.1.32 is the right bipermutative category $\mathsf{Vect}^{\mathbb{C}}_{\mathsf{c}}$ of *coordinatized* finite dimensional complex vector spaces defined as follows.

- Its objects are nonnegative integers.
- A morphism $m \longrightarrow n$ is an $n \times m$ complex matrix.
- Categorical composition is given by matrix multiplication.
- The identity morphism $1_n : n \longrightarrow n$ is the $n \times n$ identity matrix.

It becomes a right bipermutative category with the following data.

- $m \oplus n = m + n$ on objects, with 0 as the additive zero.
- For complex matrices A and B, $A \oplus B$ is the matrix direct sum

  $$A \oplus B = \left[\begin{array}{c|c} A & 0 \\ \hline 0 & B \end{array}\right]$$

  with each 0 representing a 0 matrix of appropriate size.
- $\xi^{\oplus}_{m,n} : m + n \longrightarrow n + m$ is the permutation matrix

  $$\xi^{\oplus}_{m,n} = \left[\begin{array}{c|c} 0 & 1_n \\ \hline 1_m & 0 \end{array}\right]$$

This permutation matrix is obtained from the identity matrix 1_{m+n} by permuting its $m+n$ rows using the block permutation in (2.4.3).

  - For a matrix C with $n+m$ columns, $C\xi^{\oplus}_{m,n}$ is obtained from C by swapping the first n columns with the last m columns.
  - For a matrix D with $m+n$ rows, $\xi^{\oplus}_{m,n}D$ is obtained from D by swapping the first m rows with the last n rows.

- $m \otimes n = mn$ on objects, with 1 as the multiplicative unit.
- For complex matrices
  - $A = (A_{ji}): m \longrightarrow n$ and
  - $B = (B_{lk}): p \longrightarrow q$,

  $A \otimes B: mp \longrightarrow nq$ is the following matrix tensor product.

$$A \otimes B = \begin{bmatrix} A_{11}B & \cdots & A_{1m}B \\ \vdots & \ddots & \vdots \\ A_{n1}B & \cdots & A_{nm}B \end{bmatrix}$$

  Each $A_{ji}B = (A_{ji}B_{lk})_{l,k}$ is the scalar product.
- $\xi^{\otimes}_{m,n}: mn \longrightarrow nm$ is the permutation matrix obtained from the identity matrix 1_{mn} by permuting its mn rows using the permutation in (2.4.20).
  - For a matrix C with mn columns, $C\xi^{\otimes}_{m,n}$ is obtained from C by permuting its columns using the permutation in (2.4.5).
  - For a matrix D with mn rows, $\xi^{\otimes}_{m,n}D$ is obtained from D by permuting its rows using the permutation in (2.4.20).
- The structure morphisms $\lambda^{\bullet}$, $\rho^{\bullet}$, and δ^r are identities.
- δ^l is defined as the composite in (2.5.4).

This finishes the right bipermutative categorical data of $\mathsf{Vect}^{\mathbb{C}}_{\mathsf{c}}$.

The naturality of $\xi^{\oplus}$ follows from the matrix equalities

$$\begin{bmatrix} B & 0 \\ 0 & A \end{bmatrix}\begin{bmatrix} 0 & 1_n \\ 1_m & 0 \end{bmatrix} = \begin{bmatrix} 0 & B \\ A & 0 \end{bmatrix} = \begin{bmatrix} 0 & 1_{n'} \\ 1_{m'} & 0 \end{bmatrix}\begin{bmatrix} A & 0 \\ 0 & B \end{bmatrix}$$

for $A: m \longrightarrow m'$ and $B: n \longrightarrow n'$. The naturality of $\xi^{\otimes}$ means the equality

$$\xi^{\otimes}_{m',n'}(A \otimes B) = (B \otimes A)\xi^{\otimes}_{m,n}$$

that expresses $B \otimes A$ as a row and column permutation of $A \otimes B$. The naturality of δ^r follows from the following matrix equalities.

$$(A \oplus B) \otimes C = \left[\begin{array}{c|c} A \otimes C & 0 \\ \hline 0 & B \otimes C \end{array}\right] = (A \otimes C) \oplus (B \otimes C)$$

On the other hand, δ^l is not the identity because

$$C \otimes (A \oplus B) \neq (C \otimes A) \oplus (C \otimes B)$$

for general matrices A, B, and C. For example, if

$$C = \begin{bmatrix} 0 & 1 \\ 1 & 0 \end{bmatrix} \quad \text{and} \quad A = B = (1), \tag{2.5.10}$$

then

$$C \otimes \big((1) \oplus (1)\big) = C \otimes 1_2 = \begin{bmatrix} 0 & 1_2 \\ 1_2 & 0 \end{bmatrix},$$

while

$$\big(C \otimes (1)\big) \oplus \big(C \otimes (1)\big) = C \oplus C = \begin{bmatrix} C & 0 \\ 0 & C \end{bmatrix}.$$

Next we check the right bipermutative category axioms.

- The diagram (2.5.4) is commutative by the definition of δ^l in terms of δ^r and $\xi^{\otimes}$.
- The diagram (2.5.5) is commutative because
$$\xi^{\oplus}_{m,n} \otimes 1_p = \begin{bmatrix} 0 & 1_{np} \\ 1_{mp} & 0 \end{bmatrix} = \xi^{\oplus}_{mp,np}$$
for $m,n,p \geq 0$.
- The diagram (2.5.6) is commutative for the following reasons:
  - The axiom (2.1.13) holds in Σ', which is equivalent to (2.5.6).
  - The two composites in (2.5.6) for $\mathsf{Vect}^{\mathbb{C}}_{\mathsf{c}}$ are the two permutation matrices obtained from $1_{(m+n)(p+q)}$ by permuting its rows using the two corresponding permutations in (2.5.6) for Σ'.

A more conceptual way to think about $\mathsf{Vect}^{\mathbb{C}}_{\mathsf{c}}$ is to regard each object m as the complex vector space $\mathbb{C}^m$ equipped with the standard Kronecker basis. A morphism $m \longrightarrow n$ is then a $\mathbb{C}$-linear map $\mathbb{C}^m \longrightarrow \mathbb{C}^n$, regarded as a complex $n \times m$ matrix via the standard bases. Composition of $\mathbb{C}$-linear maps

$$\mathbb{C}^m \longrightarrow \mathbb{C}^n \longrightarrow \mathbb{C}^p$$

corresponds to matrix multiplication. The identity matrix 1_n corresponds to the identity map on $\mathbb{C}^n$. The rest of the structures in $\mathsf{Vect}^{\mathbb{C}}_{\mathsf{c}}$ are similarly interpreted as in $\mathsf{Vect}^{\mathbb{C}}$ for the vector spaces $\mathbb{C}^n$ with the standard bases.

We remark that there is an incorrect claim in [**KV94**, Ex. 5.6] that says that in $\mathsf{Vect}^{\mathbb{C}}_{\mathsf{c}}$ both distributivity morphisms are the identities. As we illustrated with a simple example (2.5.10) above, δ^l in $\mathsf{Vect}^{\mathbb{C}}_{\mathsf{c}}$ is not the identity. ◇

Left Bipermutative Categories. Next we define left bipermutative categories, in which the left distributivity morphism δ^l is the identity.

Definition 2.5.11. A *left bipermutative category* is a tuple

$$\big(\mathsf{C}, (\oplus, \mathbb{0}, \xi^{\oplus}), (\otimes, \mathbb{1}, \xi^{\otimes}), (\lambda^{\bullet}, \rho^{\bullet}), (\delta^l, \delta^r)\big)$$

consisting of the following data.

- $(\mathsf{C}, \oplus, \mathbb{0}, \xi^{\oplus})$ and $(\mathsf{C}, \otimes, \mathbb{1}, \xi^{\otimes})$ are permutative categories.
- $\lambda^{\bullet}$, $\rho^{\bullet}$, δ^l, and δ^r are natural transformations as in (2.1.3) and (2.1.4).

The above data are required to satisfy the following four conditions.

(1) $\lambda^{\bullet}$ and $\rho^{\bullet}$ are both equal to the identity natural transformation of the constant functor $\mathsf{C} \longrightarrow \mathsf{C}$ at $\mathbb{0}$.
(2) δ^l is an identity natural transformation.
(3) $\xi^{\otimes}_{A,\mathbb{0}} : A \otimes \mathbb{0} \longrightarrow \mathbb{0} \otimes A$ is the identity morphism of $\mathbb{0}$ for each object A.
(4) The axioms (2.1.5), (2.1.6), and (2.1.13) are satisfied.

This finishes the definition of a left bipermutative category. ◇

Explanation 2.5.12. Consider Definition 2.5.11.

- The axiom (2.1.5) is equivalent to the following commutative diagram.

(2.5.13)
$$\begin{array}{ccc} (A \oplus B)C & \xrightarrow{\delta^r} & AC \oplus BC \\ {\scriptstyle \xi^{\otimes}} \downarrow & & \uparrow {\scriptstyle \xi^{\otimes} \oplus \xi^{\otimes}} \\ C(A \oplus B) & \xrightarrow[=]{\delta^l} & CA \oplus CB \end{array}$$

In particular, the right distributivity morphism δ^r is a natural isomorphism.

- The axiom (2.1.6) means the commutativity of the following diagram.

$$\begin{array}{ccc} A(B\oplus C) & \xrightarrow[=]{\delta^l} & AB\oplus AC \\ {\scriptstyle 1\xi^{\oplus}}\big\downarrow & & \big\downarrow{\scriptstyle \xi^{\oplus}} \\ A(C\oplus B) & \xrightarrow[=]{\delta^l} & AC\oplus AB \end{array} \tag{2.5.14}$$

- The axiom (2.1.13) means the commutativity of the following diagram.

$$\begin{array}{ccccc} (A\oplus B)(C\oplus D) & & \xrightarrow{\quad\delta^r\quad} & & A(C\oplus D)\oplus B(C\oplus D) \\ {\scriptstyle \delta^l}\big\downarrow{\scriptstyle =} & & & & {\scriptstyle =}\big\downarrow{\scriptstyle \delta^l\oplus\delta^l} \\ (A\oplus B)C\oplus(A\oplus B)D & & & & AC\oplus AD\oplus BC\oplus BD \\ & {\scriptstyle \delta^r\oplus\delta^r}\searrow & & \swarrow{\scriptstyle 1\oplus\xi^{\oplus}\oplus 1} & \\ & & AC\oplus BC\oplus AD\oplus BD & & \end{array} \tag{2.5.15}$$

◇

Proposition 2.5.16. *Each left bipermutative category is a tight symmetric bimonoidal category.*

Proof. The left distributivity morphism δ^l is the identity by definition. The right distributivity morphism δ^r is a natural isomorphism by (2.5.13). Consider the 24 axioms in Definition 2.1.2 for a left bipermutative category.

- The axioms (2.1.5), (2.1.6), and (2.1.13) hold by assumption.
- The axiom (2.1.19) holds by the identity assumptions on $\lambda^{\bullet}$, $\rho^{\bullet}$, and $\xi^{\otimes}_{-,0}$.
- The axioms (2.1.9), (2.1.10), (2.1.14), (2.1.15), (2.1.17), (2.1.20), (2.1.23), and (2.1.27) hold by the assumption that $\lambda^{\bullet}$, $\rho^{\bullet}$, and δ^r are identities.

We finish the proof by applying Lemmas 2.2.4 through 2.2.12 to obtain all 24 axioms of a symmetric bimonoidal category. □

Example 2.5.17. The symmetric bimonoidal category Σ in Proposition 2.4.8 is a left bipermutative category. ◇

2.6. Application: Reversible Programming of Finite Types

In this section, we discuss an application of symmetric bimonoidal categories to reversible programming of finite types with sums and products. This section is based on the paper [**CS16**]. The main observation [**CS16**, Th. 3] is Theorem 2.6.2, which asserts the existence of a symmetric bimonoidal groupoid Π with syntax of finite types as objects. This is actually true by the construction of Π. Theorem 2.6.2 is a bimonoidal manifestation of the theme of the paper [**BS11**], in which the close connection between category theory, physics, and computation is described. Here the categorical concept is a symmetric bimonoidal groupoid. Sums and products of objects in such a category model syntax of finite types. Near the end of this section, we point out a slight improvement of Π. More discussion about [**CS16**] and Π is in Note 2.7.3 and Examples 3.9.10 and 4.4.5. In particular, Laplaza's Coherence Theorems 3.9.1 and 4.4.3 apply to Π.

Motivation 2.6.1. The Curry-Howard-Lambek correspondence relates the following concepts:

Type theory: types, programs (or type equivalences), and program transformations (or equivalences of type equivalences);

Propositional logic: propositions, proofs, and transformations of proofs; and

Cartesian closed categories: objects, morphisms, and commutative diagrams.

Extending the work of [**BJS11, JS12**], in [**CS16**] a variant of the Curry-Howard-Lambek correspondence is proposed for finite type reversible programming. In categorical language, *finite type* means that there are functors representing sums and products, along with appropriate distributivity and units. Therefore, the concept of finite type is naturally associated with bimonoidal categories.

Reversibility can be interpreted in several ways.

Category theory: In a category, reversibility means that each morphism is invertible, so we are dealing with groupoids.

Physics: The conservation of information in physics is another motivation for reversibility. The No-Hiding Theorem [**BP07, KGB⁺19, SPK11**] states essentially that no physical processes can destroy quantum information. If computation is regarded as a physical process, then it makes sense that this process can be reversed.

Homotopy type theory: A *type isomorphism* is a fundamental concept in homotopy type theory [**Pro13**], where a proof of an equality between two terms is thought of as a path between two points in a topological space. Since each path has an inverse, so should a proof.

In summary, bimonoidal groupoids form the appropriate framework for finite type reversible programming. ◇

Recall from Definition 2.1.2 that a *symmetric bimonoidal groupoid* is a symmetric bimonoidal category in which each morphism is invertible.

Theorem 2.6.2. *There is a symmetric bimonoidal groupoid* Π *with*

- *syntax of finite types as objects and*
- Π*-terms and* Π*-combinators as morphisms.*

In the rest of this section, we explain the category Π as defined in [**CS16**] and point out one improvement.

Syntax of Finite Types as Objects. The objects in the groupoid Π are defined inductively as follows.

- Each Agda type is an object in Π. Agda is the dependently typed functional programming language available at `https://wiki.portal.chalmers.se/agda/pmwiki.php`. It is an extension of Martin-Löf type theory.
- Π is equipped with two distinguished objects $\mathbb{0}$ and $\mathbb{1}$.
- Inductively, if A and B are objects in Π, then so are their *sum* $A \oplus B$ and *product* $A \otimes B$.

In the language of type theory and propositional logic:

- $\mathbb{0}$ is the empty type $\bot$, which corresponds to inconsistency.
- $\mathbb{1}$ is the unit type $\top$, which corresponds to the trivially true proposition.
- $\oplus$ is the sum type $\uplus$, which corresponds to the disjunction of propositions.
- $\otimes$ is the product type $\times$, which corresponds to the conjunction of propositions.

In [**CS16**], the objects $A \oplus B$ and $A \otimes B$ are denoted by, respectively, $A + B$ and $A * B$. Instead of starting with Agda types, we can, in fact, start with any class of objects in the above definition, and the construction below is still valid.

Π-Terms. There are two kinds of morphisms in Π. Morphisms of the first kind are called Π-*terms* in [**CS16**, Fig. 1 and Section 6.1]. Using the notation in Definition 2.1.2, Π-terms include the following isomorphisms for objects A, B, and C in Π.

Identity Morphisms:

$$A \xrightarrow{1_A} A$$

The Additive Structure:

$$(A \oplus B) \oplus C \xrightleftarrows[\alpha^{-\oplus}_{A,B,C}]{\alpha^{\oplus}_{A,B,C}} A \oplus (B \oplus C) \qquad\qquad A \oplus B \xrightleftarrows[\xi^{-\oplus}_{A,B}]{\xi^{\oplus}_{A,B}} B \oplus A$$

$$\mathbb{0} \oplus A \xrightleftarrows[\lambda^{-\oplus}_{A}]{\lambda^{\oplus}_{A}} A \xleftrightarrows[\rho^{-\oplus}_{A}]{\rho^{\oplus}_{A}} A \oplus \mathbb{0}$$

The Multiplicative Structure:

$$(A \otimes B) \otimes C \xrightleftarrows[\alpha^{-\otimes}_{A,B,C}]{\alpha^{\otimes}_{A,B,C}} A \otimes (B \otimes C) \qquad\qquad A \otimes B \xrightleftarrows[\xi^{-\otimes}_{A,B}]{\xi^{\otimes}_{A,B}} B \otimes A$$

$$\mathbb{1} \otimes A \xrightleftarrows[\lambda^{-\otimes}_{A}]{\lambda^{\otimes}_{A}} A \xleftrightarrows[\rho^{-\otimes}_{A}]{\rho^{\otimes}_{A}} A \otimes \mathbb{1}$$

The Multiplicative Zeros:

$$\mathbb{0} \otimes A \xrightleftarrows[\lambda^{-\bullet}_{A}]{\lambda^{\bullet}_{A}} \mathbb{0} \xleftrightarrows[\rho^{-\bullet}_{A}]{\rho^{\bullet}_{A}} A \otimes \mathbb{0}$$

Distributivity:

$$A \otimes (B \oplus C) \xrightleftarrows[\delta^{-l}_{A,B,C}]{\delta^{l}_{A,B,C}} (A \otimes B) \oplus (A \otimes C)$$

$$(A \oplus B) \otimes C \xrightleftarrows[\delta^{-r}_{A,B,C}]{\delta^{r}_{A,B,C}} (A \otimes C) \oplus (B \otimes C)$$

The distributivity morphisms δ^l and δ^r are also equipped with inverses because Π is a groupoid. These isomorphisms are denoted by different symbols in [**CS16**]. For example, $\lambda^{\oplus}$, $\alpha^{\otimes}$, $\lambda^{\bullet}$, and δ^r are denoted by, respectively, $unite_{+}l$, $assocr_{*}$, $absorbr$, and $dist$ in [**CS16**, Fig. 1].

Π-Combinators. Suppose $f: A \longrightarrow B$, $g: B \longrightarrow C$, and $h: A' \longrightarrow B'$ are morphisms in Π. Then, inductively, so are the following.

Composition: $gf: A \longrightarrow C$, whose inverse is $f^{-1}g^{-1}$.
Sum: $f \oplus h: A \oplus A' \longrightarrow B \oplus B'$, whose inverse is $f^{-1} \oplus h^{-1}$.
Product: $f \otimes h: A \otimes A' \longrightarrow B \otimes B'$, whose inverse is $f^{-1} \otimes h^{-1}$.

In [**CS16**, Fig. 2], composition, sum, and product are denoted by, respectively, $\odot$, $\oplus$, and $\otimes$, and are called Π-*combinators*. The above data are subject to the relations that Π is a symmetric bimonoidal groupoid as in Definition 2.1.2. To be more explicit, the relations ensure the following.

- Π with its identity morphisms, composition, and inverses is a groupoid.
- The additive structure $(\Pi, \oplus, \mathbb{0}, \alpha^\oplus, \lambda^\oplus, \rho^\oplus, \xi^\oplus)$ is a symmetric monoidal category.
- The multiplicative structure $(\Pi, \otimes, \mathbb{1}, \alpha^\otimes, \lambda^\otimes, \rho^\otimes, \xi^\otimes)$ is a symmetric monoidal category.
- $\lambda^\bullet$, $\rho^\bullet$, δ^l, and δ^r are natural isomorphisms.
- The Laplaza axioms (2.1.5)–(2.1.28) are satisfied.

This finishes the construction of Π.

An Improvement of Π. The relations above are stated in [**CS16**, Fig. 3–12] with the following 13 Laplaza axioms, where Fig. N refers to that figure in [**CS16**].

- (2.1.5) and (2.1.6) in Fig. 12;
- (2.1.8), (2.1.10), and (2.1.13) in Fig. 11;
- (2.1.14), (2.1.15), (2.1.17), (2.1.19), (2.1.21), (2.1.22), and (2.1.23) in Fig. 10; and
- (2.1.27) in Fig. 6.

It is possible to shorten the above list as follows. By Lemma 2.2.10, the axiom (2.1.20) implies both (2.1.21) and (2.1.22), as long as (2.1.19) is assumed. Therefore, in [**CS16**, Fig. 10], the two relations (2.1.21) and (2.1.22) may be replaced by the single relation (2.1.20). Using the notation in [**CS16**], the relation (2.1.20) is stated as follows.

$$absorbl \quad \Leftrightarrow \quad (assocr_*) \odot (id{\leftrightarrow} \otimes absorbl) \odot absorbl$$

By Theorem 2.2.13, this shortened list of relations defines a symmetric bimonoidal groupoid.

2.7. Notes

2.7.1 (Symmetric Bimonoidal Categories)**.** The 24 axioms in Definition 2.1.2 of a symmetric bimonoidal category are due to Laplaza [**Lap72a**]. The only differences between our definition and Laplaza's is that his associativity isomorphisms $\alpha^\oplus$ and $\alpha^\otimes$, which he denoted by α' and α, move parentheses from right to left, instead of left to right. Since these are natural isomorphisms, these differences are only cosmetic. All the lemmas in Section 2.2 are also due to Laplaza, with one exception. In Lemma 2.2.10, the assertion that (2.1.20) implies (2.1.21) is not included in [**Lap72a**]. ◇

2.7.2 (Terminology)**.** The name *symmetric bimonoidal category* goes back to at least [**May77**], but the literature contains several different names for this concept.

- In [**Lap72a, Lap72b**] Laplaza called such a category *coherent*.
- They are also called *symmetric ring categories* or *symmetric rig categories* elsewhere, including [**CS16, Elg21**].
- In other places, including [**BG20a, Hin13**], the word *distributive* is used instead of bimonoidal, ring, or rig.

One main reason we use the word *bimonoidal* is that it aligns better with bipermutative categories, including those in Section 2.5 and those of Elmendorf-Mandell in Chapters II.9 and III.11.

The word *bimonoidal* in categorical probability theory [**FP18**] means something different. In that paper, a *bimonoidal monad* is a monad on a monoidal category equipped with both a monoidal functor structure and an opmonoidal functor structure that are compatible in an appropriate sense. $\diamond$

2.7.3 (Symmetric Rig Categories)**.** There are two remarks concerning some definitions in [**CS16**], on which Section 2.6 is based.

- [**CS16**, Def. 8] of a braided monoidal category includes only the left hexagon axiom and omits the right hexagon axiom in (II.1.3.17). In a symmetric monoidal category, these two hexagon axioms are equivalent to each other by the symmetry axiom (1.2.20), so only one of them is needed. However, in a general braided monoidal category, the two hexagon axioms are not equivalent to each other, so both of them are needed.
- In [**CS16**, Def. 10], a symmetric rig (= tight bimonoidal) category is defined as a rig (= tight bimonoidal) category in which the multiplicative structure is symmetric. This is not correct because a bimonoidal category is not equipped with a multiplicative symmetry $\xi^{\otimes}$, so the Laplaza axioms (2.1.5) and (2.1.19) are not included. A symmetric bimonoidal category is a bimonoidal category in which
  - the multiplicative structure is symmetric, *and*
  - the axioms (2.1.5) and (2.1.19) are satisfied. $\diamond$

2.7.4 (Bipermutative Categories)**.** Our right bipermutative category is almost the same as May's bipermutative category [**May77**], except that May did not include the axiom that $\xi^{\otimes}_{-,\mathbb{0}}$ be the identity. The material in Section 2.5 on right bipermutative categories is due to May [**May77**, Section 6.3]. The symmetric bimonoidal category Σ' in Proposition 2.4.23 appeared in [**May77**, Example 6.5.1]. In Chapter 5 we will show that tight symmetric bimonoidal categories can be strictified to right bipermutative categories. In the literature, the name *bipermutative category* refers to either May's version or a different version due to Elmendorf-Mandell [**EM06**], which we will discuss in Chapters II.9 and III.11. See also Note 5.6.3. $\diamond$

2.7.5 (Applications to Diagrammatic Calculus)**.** In addition to Section 2.6 on reversible programming of finite types, bimonoidal categories have other applications in computer science. Sheet diagrams for tight bimonoidal categories that generalize string diagrams for monoidal categories [**JS91a, Sel11**] are discussed in [**CDH**∞]; see also Example 3.10.9 and Note 7.9.2. Sheet diagrams are three-dimensional, two for the additive structure $\oplus$ and an additional one for the multiplicative structure $\otimes$.

- Precursors of sheet diagrams appeared in [**Sta15**] to diagrammatically separate the quantum parts of quantum circuits from the classical wires.
- Sheet diagrams in [**Del20**] provide a diagrammatic calculus for faceted dataflow in OpenRefine (`https://openrefine.org`), which is a popular data wrangling software.

For open questions related to sheet diagrams, see Questions A.1.6 and A.2.8. Moreover, in [**Hin13**], tight symmetric bimonoidal categories are used to provide a categorical proof of the operational equivalence of two quantum circuits. Chapter II.3 discusses applications of braided bimonoidal categories to quantum groups and topological quantum computation. ◇

CHAPTER 3

Coherence of Symmetric Bimonoidal Categories

The purpose of this chapter is to prove the Coherence Theorem 3.9.1 for symmetric bimonoidal categories, which is due to Laplaza [**Lap72a**]. This coherence result says that certain formal diagrams in symmetric bimonoidal categories commute, provided that a condition about monomorphisms is satisfied. We emphasize that this coherence theorem, like Definition 2.1.2, only requires the distributivity morphisms δ^l and δ^r in (2.1.4) to be natural monomorphisms, instead of natural isomorphisms. The end of Section 3.9 contains examples where Theorem 3.9.1 is applicable. Theorem 3.10.7 is the multiplicatively nonsymmetric analogue of Theorem 3.9.1 for bimonoidal categories. For open questions related to coherence of symmetric bimonoidal categories, see Question A.1.6.

Outline of the Coherence Theorem. The proof of Theorem 3.9.1 involves a series of reductions until the Symmetric Coherence Theorem 1.3.8, for both the additive structure $\oplus$ and the multiplicative structure $\otimes$, can be applied to finish the proof. The reduction steps for Theorem 3.9.1 are roughly analogous to the proof of Mac Lane's Coherence Theorem 1.3.3 as given in [**ML63**, Theorem 5.2] and [**ML98**, VII.2 Corollary]. For coherence of monoidal categories, one first reduces to the situation where the monoidal unit, the left unit isomorphism, and the right unit isomorphism are not involved. One is then left with the associativity isomorphisms. A further argument reduces the proof to applying the monoidal category axioms and the naturality of some structure morphisms.

Since a symmetric bimonoidal category has more structure and many more axioms than a monoidal category, the proof of Theorem 3.9.1 has many more steps and cases than that of Theorem 1.3.3. Here is a brief outline of the reduction steps for the Coherence Theorem 3.9.1 for symmetric bimonoidal categories.

(1) Reduce away the additive zeros $\lambda^{\oplus}$ and $\rho^{\oplus}$, and the multiplicative zeros $\lambda^{\bullet}$ and $\rho^{\bullet}$.
   - In Section 3.3, we first prove the existence and the uniqueness of such reductions for objects.
   - In Sections 3.4 and 3.5, we show the existence of such reductions for paths.

   Section 3.5 ends with a special case of the Coherence Theorem 3.9.1; see Proposition 3.5.33.
(2) Reduce away the distributivity morphisms δ^l and δ^r.
   - The existence of such reductions for objects is proved in Section 3.6.
   - The existence of such reductions for paths is proved in Lemma 3.6.12 and Section 3.7.
(3) Reduce away the multiplicative units $\lambda^{\otimes}$ and $\rho^{\otimes}$ in Section 3.8.

(4) In Section 3.9, we assemble the results in the previous sections to reduce the proof of Theorem 3.9.1 to the situation where only the structure morphisms $\alpha^{\oplus}$, $\xi^{\oplus}$, $\alpha^{\otimes}$, $\xi^{\otimes}$, and their inverses (that is, the additive and the multiplicative associativity and symmetry) are left. Then we apply the Symmetric Coherence Theorem 1.3.8 both additively and multiplicatively to finish the proof of Theorem 3.9.1.

The proof of the Coherence Theorem 3.10.7 for bimonoidal categories is obtained from that of Theorem 3.9.1 by removing $\xi^{\pm\otimes}$, and by using Theorem 1.3.3 instead of Theorem 1.3.8 for the multiplicative structure.

Mac Lane's Coherence Theorems 1.3.3 and 1.3.8 are stated in terms of words, canonical maps, and their symmetric variants. Section 3.1 contains the symmetric bimonoidal analogues of these concepts. Many proofs in this chapter involve inductions on formal words. Section 3.2 contains several concepts on formal words that will be used in later sections to perform inductions.

As we mentioned above, the Coherence Theorem 3.9.1 is due to Laplaza, and we generally follow the broad outline of his original proof. However, there are some nontrivial differences between this chapter and Laplaza's original proof, which will be discussed in Section 3.11. The relation to a 2-monad approach is discussed in Note 3.11.7.

Reading Guide. The proof of Theorem 3.9.1 is presented straightly linearly in this chapter. Since this proof has many steps and cases, as a possible alternative to reading this chapter linearly, we suggest the following.

- First read Theorem 3.9.1, whose proof is only about three pages long by that point, without worrying about the terms and results from earlier sections. The point is to first obtain a bird's-eye view of the structure of the proof.
- Then read the earlier sections, but skip all the proofs, or at least the longer ones.
- With all the necessary concepts and statements of preliminary results in mind, read Theorem 3.9.1 again. The outlined proof above should now make sense.
- After enough rest and some mental preparation, read the proofs in the earlier sections.

We deliberately divided the proof of Theorem 3.9.1 into many lemmas and cases to clarify the overall structure of the proof, and to make jumping forward and backward easier. Students are encouraged to regard the lemmas, the cases within each lemma, and their detailed proofs as exercises with full solutions.

Detail. In addition to proving Theorem 3.9.1, the many detailed proofs in this chapter have several additional purposes.

Axioms: Some of the proofs in this chapter are where the axioms of a symmetric bimonoidal category in Definition 2.1.2 are used. For example, Lemmas 3.4.12, 3.5.9, and 3.6.12 use, respectively, 2, 3, and 11 symmetric bimonoidal category axioms. Seeing these axioms arising naturally in these proofs helps demystify them.

Corrections: Laplaza's original proof contains some inaccuracies that have never been made explicit before. For both educational and archival purposes, it is important to rectify these issues. However, without going into detail, it would be difficult to (i) pinpoint these subtle issues, and (ii) discuss the corrections and where they fit in the big picture. Section 3.11 contains more detailed discussion of the differences between this chapter and Laplaza's original proof, and the necessary corrections for the latter.

Other Theorems: Several other main theorems in this book use the preliminary results in this chapter and Theorem 3.9.1.

- The proof of the Coherence Theorem 4.4.3 uses the same reduction steps as in the proof of Theorem 3.9.1. So it uses essentially all the preliminary results in this chapter before Section 3.9.
- The proof of the Strictification Theorem 5.4.6 of tight symmetric bimonoidal categories to equivalent right bipermutative categories involves Theorem 3.9.1 many times. In fact, both the definition of the associated right bipermutative category and the proofs of its properties use coherence. See Explanation 5.2.31, Lemmas 5.3.1, 5.3.4, 5.3.8, and 5.4.4, and Note 5.6.2.
- The proofs of Baez's Conjecture (Theorem 7.8.1) and the Bicategorification Theorem 8.15.4 use Theorem 3.9.1.
- The proof of the Coherence Theorem II.5.4.4 of braided bimonoidal categories uses many of the proofs in this chapter. A detailed treatment here will allow us to be both precise and concise at the same time in the braided case.

Throughout this chapter, as in Definition 2.1.2, we often abbreviate $\otimes$ using concatenation, with $\otimes$ taking precedence over $\oplus$ in the absence of parentheses. For example,

$$(ab)c \oplus a'(b'c') = \big((a \otimes b) \otimes c\big) \oplus \big(a' \otimes (b' \otimes c')\big).$$

3.1. Regularity

The main objective of this section is to introduce the language and notation that are needed to state and prove the first coherence theorem for symmetric bimonoidal categories.

Motivation 3.1.1. In the Coherence Theorems 1.3.3 and 1.3.8 for (symmetric) monoidal categories, the assertions are stated in terms of

- formal variables in the forms of (permuted) words in Definitions 1.3.1 and 1.3.6; and
- natural isomorphisms in the forms of (permuted) canonical maps in Definitions 1.3.2 and 1.3.6.

To state and prove the first Coherence Theorem 3.9.1 for symmetric bimonoidal categories, we first need to develop symmetric bimonoidal analogues of words and canonical maps.

The formal alphabets are taken from a set X with two distinguished elements 0^X and 1^X. The analogues of words are elements in the free $\{\oplus, \otimes\}$-algebra X^{fr} in Definition 3.1.2. The analogues of morphisms between words are paths consisting of prime edges in Definition 3.1.8. These two concepts together constitute the graph of X, which is denoted by $\mathsf{Gr}(X)$, in Definition 3.1.9. The first half of this section contains all the definitions necessary to define the graph of X.

To apply these concepts to symmetric bimonoidal categories C, we use a certain graph morphism $\varphi\colon \mathsf{Gr}(X) \longrightarrow \mathsf{C}$ in Definition 3.1.14. This is analogous to interpreting a word as a functor in Definition 1.3.1. With this graph morphism, we can define the *value* of a path in $\mathsf{Gr}(X)$ as the corresponding composite morphism in C. The first coherence theorem of symmetric bimonoidal categories then takes on a familiar form: two paths from a to b in $\mathsf{Gr}(X)$ have the same value in C, provided a satisfies a regularity condition.

The regularity condition ensures that we do *not* assert generally false statements such as this: $1_{x\oplus x}$ and $\xi^{\oplus}_{x,x}\colon x \oplus x \longrightarrow x \oplus x$ are equal for all objects x. The second half of this section is devoted to the concept of regularity. ◇

Elementary Graph. To define the graph of X in Definition 3.1.9, we first define its vertices in the next definition.

Definition 3.1.2. Suppose S is a set. The *free* $\{\oplus,\otimes\}$*-algebra* of S is the set S^{fr} defined inductively by the following two conditions.

- $S \subset S^{\mathsf{fr}}$.
- If $a, b \in S^{\mathsf{fr}}$, then the symbols

$$a \oplus b \quad \text{and} \quad a \otimes b$$

  also belong to S^{fr}. They are called, respectively, the *sum* and the *product* of a and b.

To simplify the presentation, we sometimes abbreviate $a \otimes b$ to ab. In the absence of clarifying parentheses, $\otimes$ takes precedence over $\oplus$. ◇

Example 3.1.3. For $a, \ldots, f \in S$,

$$(ab \oplus cd)e \oplus f = \big[\big((a \otimes b) \oplus (c \otimes d)\big) \otimes e\big] \oplus f$$

in S^{fr}. ◇

We also need the following concept of a graph.

Definition 3.1.4. A *graph* $G = (V, E)$ is a pair consisting of the following data.

- V is a class. An element in V is called a *vertex* in G.
- E assigns to each ordered pair (u, v), with $u, v \in V$, a set $E(u, v)$, an element of which is called an *edge* with *domain* u and *codomain* v. We also denote such an edge by

$$u \longrightarrow v, \qquad e\colon u \longrightarrow v, \qquad \text{or} \qquad u \xrightarrow{\;e\;} v$$

  if e is the name of the edge.

A *path* in such a graph is a nonempty finite sequence of edges $(e_n, \ldots, e_1)$ as in

$$v_0 \xrightarrow{\;e_1\;} v_1 \xrightarrow{\;e_2\;} \cdots \xrightarrow{\;e_n\;} v_n.$$

Such a path is denoted by $v_0 \longrightarrow\!\!\!\!\rightarrow v_n$ and is said to have *length* n, domain v_0, and codomain v_n. ◇

Example 3.1.5. Each category C has an associated graph (V, E), with V the class of objects in C, and $E(u, v) = \mathsf{C}(u, v)$ for objects $u, v \in \mathsf{C}$. A nonempty finite sequence of composable morphisms in C yields a path in the associated graph. ◇

The edges in the graph of X in Definition 3.1.9 are built from the elementary edges in the next definition.

Definition 3.1.6. Suppose X is a set with two distinguished elements 0^X and 1^X, called the *additive zero* and the *multiplicative unit*, respectively. The *elementary graph of* X, denoted by $\mathsf{Gr}^{\mathsf{el}}(X)$, is the graph defined as follows.

Vertices: The set of vertices in $\mathsf{Gr}^{\mathsf{el}}(X)$ is the free $\{\oplus,\otimes\}$-algebra X^{fr} of X.
Edges: Edges in $\mathsf{Gr}^{\mathsf{el}}(X)$ are of the following types for all $x,y,z \in X^{\mathsf{fr}}$.

The Additive Structure:

$$(x\oplus y)\oplus z \underset{\alpha^{-\oplus}_{x,y,z}}{\overset{\alpha^{\oplus}_{x,y,z}}{\rightleftarrows}} x\oplus(y\oplus z) \qquad\qquad x\oplus y \underset{\xi^{-\oplus}_{x,y}}{\overset{\xi^{\oplus}_{x,y}}{\rightleftarrows}} y\oplus x$$

$$0^X\oplus x \underset{\lambda^{-\oplus}_{x}}{\overset{\lambda^{\oplus}_{x}}{\rightleftarrows}} x \underset{\rho^{-\oplus}_{x}}{\overset{\rho^{\oplus}_{x}}{\leftrightarrows}} x\oplus 0^X$$

The Multiplicative Structure:

$$(x\otimes y)\otimes z \underset{\alpha^{-\otimes}_{x,y,z}}{\overset{\alpha^{\otimes}_{x,y,z}}{\rightleftarrows}} x\otimes(y\otimes z) \qquad\qquad x\otimes y \underset{\xi^{-\otimes}_{x,y}}{\overset{\xi^{\otimes}_{x,y}}{\rightleftarrows}} y\otimes x$$

$$1^X\otimes x \underset{\lambda^{-\otimes}_{x}}{\overset{\lambda^{\otimes}_{x}}{\rightleftarrows}} x \underset{\rho^{-\otimes}_{x}}{\overset{\rho^{\otimes}_{x}}{\leftrightarrows}} x\otimes 1^X$$

The Multiplicative Zeros:

$$0^X\otimes x \underset{\lambda^{-\bullet}_{x}}{\overset{\lambda^{\bullet}_{x}}{\rightleftarrows}} 0^X \underset{\rho^{-\bullet}_{x}}{\overset{\rho^{\bullet}_{x}}{\leftrightarrows}} x\otimes 0^X$$

Distributivity:

$$x\otimes(y\oplus z) \xrightarrow{\delta^{l}_{x,y,z}} (x\otimes y)\oplus(x\otimes z)$$

$$(x\oplus y)\otimes z \xrightarrow{\delta^{r}_{x,y,z}} (x\otimes z)\oplus(y\otimes z)$$

Identity:

$$x \xrightarrow{1_x} x$$

This finishes the definition of $\mathsf{Gr}^{\mathsf{el}}(X)$.

Moreover, we define the following.

- The set of edges in $\mathsf{Gr}^{\mathsf{el}}(X)$ is denoted by $\mathsf{E}_{\mathsf{el}}(X)$, the elements of which are called *elementary edges*.
- $\alpha^{\oplus}$ and $\alpha^{-\oplus}$ are *formal inverses* of each other, and similarly for the other 9 pairs of elementary edges in the first three groups above.
- 1_x is called the *identity* of x.
- The names in Definition 2.1.2 are reused for elementary edges. For example, $\lambda^{\bullet}$ is called the left multiplicative zero, and δ^r is called the right distributivity. ◇

Graph. The next two definitions will be combined to yield the graph of X in Definition 3.1.9.

Definition 3.1.7. With $(X, 0^X, 1^X)$ as in Definition 3.1.6, consider the free $\{\oplus,\otimes\}$-algebra $\mathsf{E}^{\mathsf{fr}}_{\mathsf{el}}(X)$ of the set $\mathsf{E}_{\mathsf{el}}(X)$ of elementary edges. The *domain* and *codomain* of an element $f \in \mathsf{E}^{\mathsf{fr}}_{\mathsf{el}}(X)$ are elements in X^{fr} defined inductively as follows.

- For an elementary edge $f \in \mathsf{E}_{\mathsf{el}}(X)$, its (co)domain are those of f in the elementary graph $\mathsf{Gr}^{\mathsf{el}}(X)$.
- Suppose $f_1, f_2 \in \mathsf{E}^{\mathsf{fr}}_{\mathsf{el}}(X)$ with
  - $u_i \in X^{\mathsf{fr}}$ the domain of f_i and
  - $v_i \in X^{\mathsf{fr}}$ the codomain of f_i

  already defined for $i = 1, 2$. Then:
  - $f_1 \oplus f_2$ has domain $u_1 \oplus u_2$ and codomain $v_1 \oplus v_2$.
  - $f_1 \otimes f_2$ has domain $u_1 \otimes u_2$ and codomain $v_1 \otimes v_2$. ◇

Definition 3.1.8. Continuing Definition 3.1.7, *identity prime edges* and *nonidentity prime edges* are elements in $\mathsf{E}^{\mathsf{fr}}_{\mathsf{el}}(X)$ defined inductively by the following four conditions.

- Elementary edges of the type 1_x for $x \in X^{\mathsf{fr}}$ are identity prime edges.
- Elementary edges not of the type 1_x for $x \in X^{\mathsf{fr}}$ are nonidentity prime edges.
- If $e_1, e_2 \in \mathsf{E}^{\mathsf{fr}}_{\mathsf{el}}(X)$ are identity prime edges, then so are $e_1 \oplus e_2$ and $e_1 \otimes e_2$.
- If f is a nonidentity prime edge, and if e is an identity prime edge, then

$$f \oplus e, \quad e \oplus f, \quad f \otimes e, \quad \text{and} \quad e \otimes f$$

  are nonidentity prime edges.

A *prime edge* means either an identity prime edge or a nonidentity prime edge. The set of prime edges is denoted by $\mathsf{E}^{\mathsf{pr}}(X)$. An identity prime edge is also called an *identity*. ◇

Definition 3.1.9. With $(X, 0^X, 1^X)$ as in Definition 3.1.6, the *graph of* X, which is denoted by $\mathsf{Gr}(X)$, is the graph defined as follows.

Vertices: The set of vertices in $\mathsf{Gr}(X)$ is the free $\{\oplus, \otimes\}$-algebra X^{fr} of X.
Edges: The set of edges in $\mathsf{Gr}(X)$ is the set $\mathsf{E}^{\mathsf{pr}}(X)$ of prime edges as in Definition 3.1.8, with (co)domain as in Definition 3.1.7. ◇

Definition 3.1.10. Consider $\mathsf{Gr}(X)$.

- Suppose $f: a \longrightarrow b$ is a prime edge that does not involve δ^l and δ^r. Its *formal inverse* $g: b \longrightarrow a$ is the prime edge obtained from f as follows.
  - Each identity 1_x for $x \in X^{\mathsf{fr}}$ in f is replaced by 1_x in the opposite direction.
  - If f involves an elementary edge ϵ that is not an identity, then replace ϵ by its formal inverse.
- Suppose $P: a \twoheadrightarrow b$ is a path in which each prime edge does not involve δ^l and δ^r. Its *formal inverse* $Q: b \twoheadrightarrow a$ is the path obtained from P by replacing each of its prime edges by its formal inverse. ◇

Example 3.1.11. Suppose $x, y, z \in X^{\mathsf{fr}}$. With $\otimes$ abbreviated to concatenation, below are two paths in $\mathsf{Gr}(X)$.

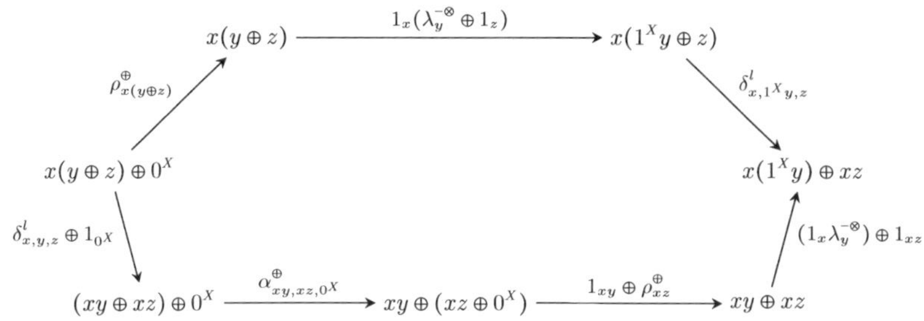

The top path has length 3, with the outer edges elementary edges, and the middle one a nonidentity prime edge. The bottom path has length 4, with $\alpha^{\oplus}$ an elementary edge, and the other three edges nonidentity prime edges. ◇

Example 3.1.12. For elements $w, x, y, z \in X^{\mathsf{fr}}$,

$$w \oplus x(y \oplus z) \xrightarrow{1_w \oplus 1_x 1_{y \oplus z}} w \oplus x(y \oplus z)$$

is an identity. The formal inverse of the prime edge

$$w \oplus x(y \oplus z) \xrightarrow{1_w \oplus \lambda_x^{-\otimes} 1_{y \oplus z}} w \oplus (1^X x)(y \oplus z)$$

is the prime edge

$$w \oplus (1^X x)(y \oplus z) \xrightarrow{1_w \oplus \lambda_x^{\otimes} 1_{y \oplus z}} w \oplus x(y \oplus z),$$

and vice versa. ◇

The Value of a Path. The graph of X is interpreted in a symmetric bimonoidal category via the following concept.

Definition 3.1.13. Suppose $G_1 = (V_1, E_1)$ and $G_2 = (V_2, E_2)$ are graphs. A *graph morphism* $f: G_1 \longrightarrow G_2$ consists of functions

- $f_V: V_1 \longrightarrow V_2$ and
- $f_E: E_1(u, v) \longrightarrow E_2\big(f_V u, f_V v\big)$ for $u, v \in V_1$.

To simplify the notation, both f_V and f_E are denoted by f below. ◇

Definition 3.1.14. Suppose given the data (X, C, φ) as follows.

- X is a set with two distinguished elements 0^X and 1^X as in Definition 3.1.6.
- C is an arbitrary symmetric bimonoidal category as in Definition 2.1.2, equipped with the graph structure in Example 3.1.5.
- $\varphi: X \longrightarrow \mathsf{Ob}(\mathsf{C})$ is any function such that

$$\varphi(0^X) = \mathbb{0} \quad \text{and} \quad \varphi(1^X) = \mathbb{1}. \tag{3.1.15}$$

Extend φ to a graph morphism

$$\mathsf{Gr}(X) \xrightarrow{\varphi} \mathsf{C}$$

as follows.

Vertices: For $x, y \in X^{\mathsf{fr}}$ such that $\varphi x, \varphi y \in \mathsf{Ob}(\mathsf{C})$ are already defined, we define

$$\begin{aligned} \varphi(x \oplus y) &= \varphi x \oplus \varphi y \quad \text{and} \\ \varphi(x \otimes y) &= \varphi x \otimes \varphi y. \end{aligned} \tag{3.1.16}$$

Elementary Edges: φ sends each elementary edge to the structure morphism in C with the same name, and with the subscripts replaced by their images under φ.

Prime Edges: If $f_1, f_2 \in \mathsf{E}^{\mathsf{pr}}(X)$ are prime edges with at most one of them non-identity, and with $\varphi(f_1)$ and $\varphi(f_2)$ already defined, then we define

$$\begin{aligned}\varphi(f_1 \oplus f_2) &= \varphi(f_1) \oplus \varphi(f_2) \quad \text{and}\\ \varphi(f_1 \otimes f_2) &= \varphi(f_1) \otimes \varphi(f_2).\end{aligned} \tag{3.1.17}$$

This finishes the definition of the graph morphism φ.

Moreover, we define the following.

- For a path $P = (f_n, \ldots, f_1)$ in $\mathsf{Gr}(X)$ with domain u and codomain v, its *value* in C is the composite

$$\varphi P = \varphi(f_n) \circ \cdots \circ \varphi(f_1) \in \mathsf{C}(\varphi u; \varphi v). \tag{3.1.18}$$

- A diagram with vertices and edges in $\mathsf{Gr}(X)$ is *commutative in* C if its image under the graph morphism φ is a commutative diagram in C.
- A diagram with vertices and edges in $\mathsf{Gr}(X)$ is *commutative* if it is commutative in each symmetric bimonoidal category C and for each function $\varphi\colon X \longrightarrow \mathsf{Ob}(\mathsf{C})$ satisfying (3.1.15). ◇

Remark 3.1.19. In Definition 3.1.14, the 24 symmetric bimonoidal category axioms are not needed. Therefore, the definition still makes sense if C only has the data portion of a symmetric bimonoidal category. ◇

Example 3.1.20. The graph morphism φ sends elementary edges to structure morphisms in C. Below are two examples.

$$\begin{aligned}\varphi(\alpha^{\oplus}_{x,y,z}) &= \alpha^{\oplus}_{\varphi x, \varphi y, \varphi z}\colon (\varphi x \oplus \varphi y) \oplus (\varphi z) \longrightarrow \varphi x \oplus (\varphi y \oplus \varphi z)\\ \varphi(\lambda^{\bullet}_{x}) &= \lambda^{\bullet}_{\varphi x}\colon \mathbb{0} \otimes \varphi x \longrightarrow \mathbb{0}.\end{aligned}$$

◇

Example 3.1.21. The image under φ of an identity prime edge is an identity morphism by the functoriality of $\oplus$ and $\otimes$ in C. If f is a prime edge with formal inverse g, then

$$\varphi(f)^{-1} = \varphi(g),$$

and similarly for a path with a formal inverse. ◇

Example 3.1.22. For elements $x, y, z \in X^{\mathsf{fr}}$, the diagram

$$\begin{array}{ccccc} (x \oplus y)z & \xrightarrow{\delta^r_{x,y,z}} & xz \oplus yz & \xrightarrow{1_{xz} \oplus \xi^{\otimes}_{y,z}} & xz \oplus zy\\ {\scriptstyle \xi^{\otimes}_{x\oplus y,z}}\big\downarrow & & & & \big\downarrow{\scriptstyle \xi^{\otimes}_{x,z} \oplus 1_{zy}}\\ z(x \oplus y) & & \xrightarrow{\qquad\delta^l_{z,x,y}\qquad} & & zx \oplus zy \end{array}$$

in $\mathsf{Gr}(X)$ is commutative. This is true because its image under the graph morphism $\varphi\colon \mathsf{Gr}(X) \longrightarrow \mathsf{C}$ is commutative by

- the axiom (2.1.5) in C applied to the objects φx, φy, and φz; and
- the functoriality of $\oplus$ in C, which implies

$$\big(\xi^{\otimes}_{x,z} \oplus 1_{zy}\big)\big(1_{xz} \oplus \xi^{\otimes}_{y,z}\big) = \xi^{\otimes}_{x,z} \oplus \xi^{\otimes}_{y,z}.$$

Similarly, each of the other 23 symmetric bimonoidal category axioms in C gives a commutative diagram in $\mathsf{Gr}(X)$. ◇

Regularity. Next we introduce a restriction, called regularity, on the elements in X^{fr} for the first coherence theorem of symmetric bimonoidal categories. As discussed in Motivation 3.1.1, this restriction is needed to avoid incoherent situations. Regularity is defined in terms of the following concept.

Definition 3.1.23. Suppose $(X, 0^X, 1^X)$ is as in Definition 3.1.6. Define its *strict* $\{\oplus, \otimes\}$*-algebra* X^{st} as the quotient set of X^{fr} in Definition 3.1.2 by the smallest relation that contains the following identifications for all elements $x, y, z \in X^{\mathsf{fr}}$.

The Additive Structure:

$$\begin{aligned}(x \oplus y) \oplus z &= x \oplus (y \oplus z)\\ 0^X \oplus x = x &= x \oplus 0^X\\ x \oplus y &= y \oplus x\end{aligned}$$

The Multiplicative Structure:

$$\begin{aligned}(x \otimes y) \otimes z &= x \otimes (y \otimes z)\\ 1^X \otimes x = x &= x \otimes 1^X\\ x \otimes y &= y \otimes x\end{aligned}$$

The Multiplicative Zeros:

$$0^X \otimes x = 0^X = x \otimes 0^X$$

Distributivity:

$$\begin{aligned}x \otimes (y \oplus z) &= (x \otimes y) \oplus (x \otimes z)\\ (x \oplus y) \otimes z &= (x \otimes z) \oplus (y \otimes z)\end{aligned}$$

This finishes the definition of X^{st}.

Moreover, denote by

$$X^{\mathsf{fr}} \xrightarrow{\ \mathsf{supp}\ } X^{\mathsf{st}} \tag{3.1.24}$$

the quotient map, called the *support*. ◇

Definition 3.1.25. An element $x \in X^{\mathsf{fr}}$ is *regular* if there exist elements $x^i_j \in X$ for $1 \le i \le m$ and $1 \le j \le k_i$ for each i with $m, k_1, \ldots, k_m > 0$, such that the following three conditions hold.

(i) The equality

$$\mathsf{supp}(x) = \mathsf{supp}\left(\bigoplus_{i=1}^{m} \left(x^i_1 \otimes \cdots \otimes x^i_{k_i}\right)\right) \tag{3.1.26}$$

holds in X^{st}. The iterated sum $\bigoplus_{i=1}^m$ and each of the m iterated products $x^i_1 \otimes \cdots \otimes x^i_{k_i}$ have some bracketings. By the definition of X^{st}, different bracketings yield the same support.

(ii) For each $1 \le i \le m$, the elements $x^i_1, \ldots, x^i_{k_i} \in X$ are distinct.

(iii) The m elements

$$\mathsf{supp}\left(x^i_1 \otimes \cdots \otimes x^i_{k_i}\right) \in X^{\mathsf{st}}$$

for $1 \le i \le m$ are distinct. ◇

The rest of this section contains basic observations and examples about support and regular elements.

Example 3.1.27. For elements $a, \ldots, f \in X$,

$$\begin{aligned}\mathsf{supp}\big((ab \oplus cd)e \oplus f\big) &= \mathsf{supp}\big(\, [(ab)e \oplus (cd)e] \oplus f\big)\\ &= \mathsf{supp}\big(a(be) \oplus [c(de) \oplus f]\,\big).\end{aligned}$$

In general, because of the relations that define X^{st}, when computing the support of an element in X^{fr}, we may

- distributive $\otimes$ over $\oplus$ as much as possible and
- ignore the additive bracketings of the summands and the multiplicative bracketings within each summand. ◇

Example 3.1.28. Elements in X are regular. If $x_1, \ldots, x_m$ are distinct elements in X, then the elements

$$x_1 \oplus \cdots \oplus x_m \quad \text{and} \quad x_1 \otimes \cdots \otimes x_m \in X^{\mathsf{fr}}$$

with any bracketings are regular. On the other hand, $x_1 \oplus x_1$ and $x_1 \otimes x_1$ are not regular. ◇

Lemma 3.1.29. *The following statements hold.*

(1) *If two elements in X^{fr} have the same support, then one of them is regular if and only if the other one is regular.*
(2) *If $x \twoheadrightarrow y$ is a path in $\mathsf{Gr}(X)$ as in Definition 3.1.9, then*

$$\mathsf{supp}(x) = \mathsf{supp}(y).$$

Proof. The first assertion holds because regularity depends only on the support of an element in X^{fr}.

If $x \longrightarrow y$ is an elementary edge as in Definition 3.1.6, then x and y are connected by one of the relations that define X^{st} as a quotient set of X^{fr}. So they have the same support. This implies that if $x \longrightarrow y$ is a prime edge as in Definition 3.1.8, then x and y have the same support. The second assertion now follows from an induction on the length of a path in $\mathsf{Gr}(X)$, with the initial case being a single prime edge. □

For the next observation, consider the 24 axioms (2.1.5)–(2.1.28) of a symmetric bimonoidal category with

- A, B, C, and D taken to be distinct elements in X;
- $\mathbb{0}$ and $\mathbb{1}$ interpreted as 0^X and 1^X, respectively; and
- each vertex in each diagram regarded as an element in X^{fr}.

The next observation provides many examples of regular elements.

Proposition 3.1.30. *In each of the 24 axioms in the setting of the previous paragraph:*

(1) *All the vertices have the same support.*
(2) *All the vertices are regular.*

Proof. All the vertices in each axiom have the same support because they are all connected by the relations that define X^{st} as a quotient set of X^{fr}.

For the second assertion, it suffices to show that in each axiom, one vertex is regular. Together with the first assertion and Lemma 3.1.29(1), one vertex being regular would imply that all the vertices are regular.

For the first axiom (2.1.5) consider the vertex $AC \oplus BC$ and the following four cases.

(1) If $A, B, C \in X \smallsetminus \{0^X\}$ then

$$\mathsf{supp}(AC) \neq \mathsf{supp}(BC).$$

(2) If $A = 0^X$, then $B, C \neq 0^X$, and

$$\mathsf{supp}(AC \oplus BC) = \mathsf{supp}(BC).$$

(3) If $B = 0^X$, then $A, C \neq 0^X$, and

$$\mathsf{supp}(AC \oplus BC) = \mathsf{supp}(AC).$$

(4) If $C = 0^X$, then

$$\mathsf{supp}(AC \oplus BC) = \mathsf{supp}(0^X \oplus 0^X) = \mathsf{supp}(0^X).$$

In each of the four cases, we conclude that $AC \oplus BC$ is regular because A, B, and C are distinct in X. The same kind of case-by-case analysis applies to the next 8 axioms (2.1.6)–(2.1.13).

Each of the axioms (2.1.14)–(2.1.22) has a vertex equal to 0^X, which is regular. Each of the axioms (2.1.23)–(2.1.26) has a vertex equal to AB or BA, which are both regular. Each of the axioms (2.1.27) and (2.1.28) has a vertex equal to $A \oplus B$, which is regular. □

Example 3.1.31. In Example 3.1.11, the seven vertices in X^{fr} have the same support, which is that of $xy \oplus xz$. If x, y, and z are distinct elements in X, then $xy \oplus xz$ is regular by the same kind of case-by-case proof in Proposition 3.1.30. ◇

Example 3.1.32. The three elements in X^{fr} appearing in Example 3.1.27 have the same support. If $a, \ldots, f \in X$ are distinct, then these three elements in X^{fr} are all regular by the same kind of case-by-case proof in Proposition 3.1.30. ◇

3.2. Induction Devices

The purpose of this section is to introduce some tools for performing induction arguments in the graph of X as in Definition 3.1.9. These concepts will be used in subsequent sections to prove the Coherence Theorem 3.9.1 for symmetric bimonoidal categories. We will need the following three concepts for induction arguments.

Definition 3.2.1. Denote by $\mathbb{Z}_+$ the set of positive integers. Suppose $(X, 0^X, 1^X)$ is a set with two distinguished elements as in Definition 3.1.6, and with X^{fr} as in Definition 3.1.2.

(1) The *norm*

(3.2.2)
$$X^{\mathsf{fr}} \xrightarrow{\mathsf{norm}} \mathbb{Z}_+$$

is the function defined inductively by the following two conditions.

- $\mathsf{norm}(x) = 1$ for $x \in X$.
- For $a, b \in X^{\mathsf{fr}}$ with $\mathsf{norm}(a)$ and $\mathsf{norm}(b)$ already defined, define

$$\mathsf{norm}(a \oplus b) = \mathsf{norm}(a \otimes b) = \mathsf{norm}(a) + \mathsf{norm}(b).$$

(2) The *rank*

$$X^{\mathsf{fr}} \xrightarrow{\mathsf{rank}} \mathbb{Z}_+ \tag{3.2.3}$$

is the function defined by

$$\mathsf{rank}(a) = 2 \cdot \mathsf{norm}(a) \quad \text{for} \quad a \in X^{\mathsf{fr}}.$$

(3) The *size*

$$X^{\mathsf{fr}} \xrightarrow{\mathsf{size}} \mathbb{Z}_+ \tag{3.2.4}$$

is the function defined inductively by the following two conditions:

- $\mathsf{size}(x) = 2$ for $x \in X$.
- For $a, b \in X^{\mathsf{fr}}$ with $\mathsf{size}(a)$ and $\mathsf{size}(b)$ already defined, define

$$\begin{aligned} \mathsf{size}(a \oplus b) &= \mathsf{size}(a) + \mathsf{size}(b) \quad \text{and} \\ \mathsf{size}(a \otimes b) &= \mathsf{size}(a) \cdot \mathsf{size}(b). \end{aligned}$$

We call $\mathsf{norm}(a)$, $\mathsf{rank}(a)$, and $\mathsf{size}(a)$, respectively, the norm of a, the rank of a, and the size of a. ◇

Explanation 3.2.5. The norm of $a \in X^{\mathsf{fr}}$ is computed by counting the number of elements in X appearing in the expression of a, counting multiplicity. So if an element $x \in X$ appears n times in a, then it is counted as n in the norm of a.

The size of a is computed by replacing

- each element of X appearing in a by the number 2 and
- $\{\oplus, \otimes\}$ in X^{fr} by $\{+, \times\}$ of integers. ◇

Example 3.2.6. Suppose $x_j^i \in X$ for $1 \leq i \leq m$ and $1 \leq j \leq k_i$ for each i with $m, k_1, \ldots, k_m > 0$. Consider the element

$$a = \bigoplus_{i=1}^{m} \left(x_1^i \cdots x_{k_i}^i \right) \in X^{\mathsf{fr}} \tag{3.2.7}$$

appearing in (3.1.26), in which the iterated sum $\bigoplus_{i=1}^m$ and each of the m iterated products $x_1^i \cdots x_{k_i}^i$ have some bracketings. Then its norm, rank, and size are as follows.

$$\begin{aligned} \mathsf{norm}(a) &= k_1 + \cdots + k_m \\ \mathsf{rank}(a) &= 2(k_1 + \cdots + k_m) \\ \mathsf{size}(a) &= 2^{k_1} + \cdots + 2^{k_m} \end{aligned}$$

Furthermore, the equality

$$\mathsf{size}(a) = \mathsf{rank}(a)$$

holds if and only if $k_i \leq 2$ for each $1 \leq i \leq m$. ◇

Example 3.2.8. For the element

$$b = (x_1 \oplus x_2)(x_3 \oplus x_4) \in X^{\mathsf{fr}}$$

with $x_1, x_2, x_3, x_4 \in X$, we have

$$\mathsf{norm}(b) = 4, \quad \mathsf{rank}(b) = 8, \quad \text{and} \quad \mathsf{size}(b) = 16.$$ ◇

Below are some basic observations that we will later use for induction arguments.

Lemma 3.2.9. *The following statements hold for elements $a \in X^{\mathsf{fr}}$.*

(1) $\mathsf{norm}(a) = 1$ *if and only if* $a \in X$.
(2) $\mathsf{size}(a) - \mathsf{rank}(a) \geq 0$.

Proof. For assertion (1), if $a \in X$, then $\mathsf{norm}(a) = 1$ by definition. If a does not lie in X, then a has the form $a_1 \oplus a_2$ or $a_1 \otimes a_2$ for some $a_1, a_2 \in X^{\mathsf{fr}}$. So $\mathsf{norm}(a) \geq 2$.

For assertion (2), first observe that the rank can be equivalently defined as follows.

- $\mathsf{rank}(x) = 2$ for $x \in X$.
- For $a, b \in X^{\mathsf{fr}}$ with $\mathsf{rank}(a)$ and $\mathsf{rank}(b)$ already defined, define

$$\mathsf{rank}(a \oplus b) = \mathsf{rank}(a \otimes b) = \mathsf{rank}(a) + \mathsf{rank}(b).$$

Now we prove the second assertion by induction on the norm. For norm 1, that is, for an element $x \in X$, there are equalities

$$\mathsf{size}(x) = \mathsf{rank}(x) = 2.$$

The induction step follows from the inequalities for $a, b \in X^{\mathsf{fr}}$:

$$\begin{aligned}
\mathsf{rank}(a \oplus b) &= \mathsf{rank}(a) + \mathsf{rank}(b) \\
&\leq \mathsf{size}(a) + \mathsf{size}(b) && (\text{since } \mathsf{norm}(a), \mathsf{norm}(b) < \mathsf{norm}(a \oplus b)) \\
&= \mathsf{size}(a \oplus b); \\
\mathsf{rank}(a \otimes b) &= \mathsf{rank}(a) + \mathsf{rank}(b) \\
&\leq \mathsf{rank}(a) \cdot \mathsf{rank}(b) && (\text{since } \mathsf{rank}(a), \mathsf{rank}(b) \geq 2) \\
&\leq \mathsf{size}(a) \cdot \mathsf{size}(b) && (\text{since } \mathsf{norm}(a), \mathsf{norm}(b) < \mathsf{norm}(a \oplus b)) \\
&= \mathsf{size}(a \otimes b).
\end{aligned}$$

This proves assertion (2). □

Lemma 3.2.10. *Suppose $f : a \longrightarrow b$ is a prime edge that is either an identity or involves a single instance of $\alpha^{\pm\oplus}$, $\xi^{\pm\oplus}$, $\alpha^{\pm\otimes}$, or $\xi^{\pm\otimes}$. Then the following statements hold.*

(1) $\mathsf{norm}(a) = \mathsf{norm}(b)$.
(2) $\mathsf{rank}(a) = \mathsf{rank}(b)$.
(3) $\mathsf{size}(a) = \mathsf{size}(b)$.

Proof. If f is an identity, then $a = b$, which implies the desired equalities.

For $x, y, z \in X^{\mathsf{fr}}$ and each $\odot \in \{\oplus, \otimes\}$, there are equalities as follows.

$$\begin{aligned}
\mathsf{norm}((x \odot y) \odot z) &= \mathsf{norm}(x) + \mathsf{norm}(y) + \mathsf{norm}(z) \\
&= \mathsf{norm}(x \odot (y \odot z)) \\
\mathsf{norm}(x \odot y) &= \mathsf{norm}(x) + \mathsf{norm}(y) \\
&= \mathsf{norm}(y \odot x) \\
\mathsf{size}((x \oplus y) \oplus z) &= \mathsf{size}(x) + \mathsf{size}(y) + \mathsf{size}(z) \\
&= \mathsf{size}(x \oplus (y \oplus z)) \\
\mathsf{size}((x \otimes y) \otimes z) &= \mathsf{size}(x) \cdot \mathsf{size}(y) \cdot \mathsf{size}(z) \\
&= \mathsf{size}(x \otimes (y \otimes z)) \\
\mathsf{size}(x \oplus y) &= \mathsf{size}(x) + \mathsf{size}(y) \\
&= \mathsf{size}(y \oplus x) \\
\mathsf{size}(x \otimes y) &= \mathsf{size}(x) \cdot \mathsf{size}(y) \\
&= \mathsf{size}(y \otimes x)
\end{aligned}$$

They imply the assertions for norm and size. The assertion for rank follows from the definition $\mathsf{rank} = 2 \cdot \mathsf{norm}$. □

The next few assertions are about the quantity size − rank.

Lemma 3.2.11. *For elements $a, b \in X^{\mathsf{fr}}$ and each $\odot \in \{\oplus, \otimes\}$, the inequality*

$$\mathsf{size}(a \odot b) - \mathsf{rank}(a \odot b) \geq \mathsf{size}(c) - \mathsf{rank}(c)$$

holds for each $c \in \{a, b\}$.

Proof. The inequality for $\oplus$ follows from Lemma 3.2.9 (2) and the following computation.

$$\begin{aligned}
&\mathsf{size}(a \oplus b) - \mathsf{rank}(a \oplus b) \\
&= \mathsf{size}(a) + \mathsf{size}(b) - \mathsf{rank}(a) - \mathsf{rank}(b) \\
&= \big(\mathsf{size}(a) - \mathsf{rank}(a)\big) + \big(\mathsf{size}(b) - \mathsf{rank}(b)\big)
\end{aligned}$$

Similarly, the inequality for $\otimes$ follows from the last equality above and the following computation.

$$\begin{aligned}
&\mathsf{size}(a \otimes b) - \mathsf{rank}(a \otimes b) \\
&= \mathsf{size}(a) \cdot \mathsf{size}(b) - \mathsf{rank}(a) - \mathsf{rank}(b) \\
&\geq \mathsf{size}(a) + \mathsf{size}(b) - \mathsf{rank}(a) - \mathsf{rank}(b)
\end{aligned}$$

The inequality uses the fact that $\mathsf{size}(a), \mathsf{size}(b) \geq 2$. □

Next we want to describe elements in X^{fr} that satisfy $\mathsf{size}(a) = \mathsf{rank}(a)$. The following are two preliminary observations in this direction, and use the concept of a prime edge in Definition 3.1.8.

Lemma 3.2.12. *Suppose $a \longrightarrow b$ is a prime edge that involves either δ^l or δ^r. Then the inequality*

$$\mathsf{size}(a) - \mathsf{rank}(a) > \mathsf{size}(b) - \mathsf{rank}(b)$$

holds.

Proof. For $x, y, z \in X^{\mathsf{fr}}$, there are (in)equalities as follows.

$$\begin{aligned}\mathsf{size}(x(y \oplus z)) &= \mathsf{size}(x) \cdot \big(\mathsf{size}(y) + \mathsf{size}(z)\big) \\ &= \mathsf{size}(xy \oplus xz) \\ \mathsf{rank}(x(y \oplus z)) &= \mathsf{rank}(x) + \mathsf{rank}(y) + \mathsf{rank}(z) \\ &< \mathsf{rank}(x) + \mathsf{rank}(y) + \mathsf{rank}(x) + \mathsf{rank}(z) \\ &= \mathsf{rank}(xy \oplus xz)\end{aligned}$$

These (in)equalities imply that in going from the domain to the codomain of the elementary edge

$$x(y \oplus z) \xrightarrow{\delta^l_{x,y,z}} xy \oplus xz,$$

and similarly any prime edge involving δ^l, the quantity $(\mathsf{size} - \mathsf{rank})$ strictly decreases. An almost identical computation proves the case for δ^r. □

Lemma 3.2.13. *If an element $a \in X^{\mathsf{fr}}$ satisfies*

$$\mathsf{size}(a) = \mathsf{rank}(a),$$

then a is not the domain of any prime edge that involves either δ^l or δ^r.

Proof. Suppose to the contrary that there exists a prime edge $a \longrightarrow b$ involving δ^l or δ^r. Then the hypothesis on a and Lemma 3.2.12 imply that

$$0 = \mathsf{size}(a) - \mathsf{rank}(a) > \mathsf{size}(b) - \mathsf{rank}(b).$$

This cannot happen by Lemma 3.2.9(2). □

Example 3.2.14. The converse of Lemma 3.2.13 is *not* true. For example, for elements $x, y, z \in X$, the product $(x \otimes y) \otimes z$ is not the domain of any prime edge that involves either δ^l or δ^r. However, it has rank 6 and size 8. In other words, in order for an element $a \in X^{\mathsf{fr}}$ to not be the domain of any prime edge involving either δ^l or δ^r, the property

$$\mathsf{size}(a) = \mathsf{rank}(a)$$

is sufficient but *not* necessary. ◇

The following observation characterizes elements in X^{fr} with size equal to rank.

Proposition 3.2.15. *For elements $a \in X^{\mathsf{fr}}$, the equality*

$$\mathsf{size}(a) = \mathsf{rank}(a)$$

holds if and only if a has the form

$$a = a_1 \oplus \cdots \oplus a_m$$

for some additive bracketing and some $m \geq 1$ such that for each $1 \leq i \leq m$, either

- $a_i \in X$, *or*
- $a_i = a_i^1 \otimes a_i^2$ *for some elements* $a_i^1, a_i^2 \in X$.

Proof. First suppose $a = a_1 \oplus \cdots \oplus a_m$ with some additive bracketing and with each a_i either in X or the product of two elements in X. Then $\mathsf{size}(a) = \mathsf{rank}(a)$ by Example 3.2.6.

Conversely, suppose $\mathsf{size}(a) = \mathsf{rank}(a)$. To show that a has the desired form, consider the three possible cases.

(i) If $a \in X$ then we are done.

(ii) Next suppose

$$a = a_1 \otimes \cdots \otimes a_n$$

for some $n \geq 2$ and some multiplicative bracketing, and with each $a_i \in X^{\mathsf{fr}}$ not of the form $a_i^1 \otimes a_i^2$ for any $a_i^1, a_i^2 \in X^{\mathsf{fr}}$. So each a_i is either in X or has the form $a_i^1 \oplus a_i^2$. Lemma 3.2.13 implies that each $a_i \in X$. Since

$$\mathsf{rank}(a) = 2n \leq 2^n = \mathsf{size}(a),$$

the assumption $\mathsf{size}(a) = \mathsf{rank}(a)$ implies $n = 2$. So a is a product of two elements in X, which is of the desired form.

(iii) For the remaining case, we may assume that a has the form

$$a = a_1 \oplus \cdots \oplus a_m$$

for some $m \geq 2$ and some additive bracketing, and with each $a_i \in X^{\mathsf{fr}}$ not of the form $a_i^1 \oplus a_i^2$ for any $a_i^1, a_i^2 \in X^{\mathsf{fr}}$. For each $1 \leq i \leq m$, either

- $a_i \in X$, or
- by Lemma 3.2.13 a_i is a finite product of elements in X.

Therefore, a has the form (3.2.7), which is a finite sum of finite products of elements in X. By Example 3.2.6, there are equalities

$$\mathsf{rank}(a) = 2(k_1 + \cdots + k_m) \quad \text{and}$$
$$\mathsf{size}(a) = 2^{k_1} + \cdots + 2^{k_m}.$$

The assumption $\mathsf{size}(a) = \mathsf{rank}(a)$ implies $k_i \leq 2$ for $1 \leq i \leq m$. So a is a finite sum with each summand either in X, or is a product of two elements in X.

This finishes the proof of the other direction. □

Lemma 3.2.16. *Suppose $f : a \longrightarrow b$ is a prime edge that involves either $\lambda^{\otimes}$ or $\rho^{\otimes}$.*

(1) *The following inequality holds.*

$$\mathsf{size}(a) - \mathsf{rank}(a) \geq \mathsf{size}(b) - \mathsf{rank}(b). \tag{3.2.17}$$

(2) (3.2.17) *is an equality if and only if f is a sum of identities and an elementary edge $\lambda_x^{\otimes}$ or $\rho_x^{\otimes}$ with $x \in X$.*

Proof. Suppose f involves $\lambda^{\otimes}$; the proof for $\rho^{\otimes}$ is almost identical.

For assertion (1), first consider the following (in)equalities for $c \in X^{\mathsf{fr}}$.

$$\begin{aligned}
&\mathsf{size}(1^X \otimes c) - \mathsf{rank}(1^X \otimes c)\\
&= \mathsf{size}(1^X) \cdot \mathsf{size}(c) - \mathsf{rank}(1^X) - \mathsf{rank}(c)\\
&= \big(\mathsf{size}(c) - \mathsf{rank}(c)\big) + \big(\mathsf{size}(c) - 2\big)\\
&\geq \mathsf{size}(c) - \mathsf{rank}(c)
\end{aligned} \tag{3.2.18}$$

This is an equality if and only if $\mathsf{size}(c) = 2$, that is, if $c \in X$. Since the prime edge f must involve one elementary edge of the form $\lambda_c^{\otimes}$ for some $c \in X^{\mathsf{fr}}$ and identities, a computation similar to (3.2.18) proves the inequality (3.2.17).

For assertion (2), (3.2.18) with $c \in X$ proves the "if" direction. Conversely, if (3.2.17) is an equality, then the elementary edge in f has the form $\lambda_x^{\otimes}$ for some $x \in X$ by (3.2.18). Moreover, this $\lambda_x^{\otimes}$ is a summand of f, with all other summands identities. This is true because, by a computation similar to (3.2.18), for a product of $1^X \otimes x$ with some other elements in X^{fr}, the quantity $\mathsf{size} - \mathsf{rank}$ strictly decreases when 1^X is removed. □

3.3. Reduction of Additive and Multiplicative Zeros

As the first reduction step for the Coherence Theorem 3.9.1, in this section we show that each element in the free $\{\oplus, \otimes\}$-algebra X^{fr} has a unique 0^X-reduction, in which all instances of 0^X that can be added or multiplied have been eliminated. Although a path from an element to its 0^X-reduction is not unique, we observe that its value in C is unique.

Convention 3.3.1. For the rest of this chapter, the following conventions are in effect.

(1) $(X, 0^X, 1^X)$ is as in Definition 3.1.6, so X is a set with two distinguished elements 0^X and 1^X. The free $\{\oplus, \otimes\}$-algebra X^{fr} of X is as in Definition 3.1.2.
(2) C is an arbitrary symmetric bimonoidal category as in Definition 2.1.2.
(3) $\varphi\colon X \longrightarrow \mathsf{Ob}(\mathsf{C})$ is an arbitrary function such that

$$\varphi(0^X) = \mathbb{0} \quad \text{and} \quad \varphi(1^X) = \mathbb{1},$$

and

$$\mathsf{Gr}(X) \xrightarrow{\quad\varphi\quad} \mathsf{C}$$

is the associated graph morphism in Definition 3.1.14. The *value* in C of a path (3.1.18) and the *commutativity* of a diagram in $\mathsf{Gr}(X)$ are defined in that definition. ◇

Definition 3.3.2. A 0^X*-prime edge* is a nonidentity prime edge in the sense of Definition 3.1.8 that involves $\lambda^{\oplus}$, $\rho^{\oplus}$, $\lambda^{\bullet}$, or $\rho^{\bullet}$. ◇

Example 3.3.3. In Example 3.1.11, there are two 0^X-prime edges, namely, $\rho^{\oplus}_{x(y\oplus z)}$ and $1_{xy} \oplus \rho^{\oplus}_{xz}$. ◇

Definition 3.3.4. Suppose $a \in X^{\mathsf{fr}}$.

(1) The element a is 0^X*-reduced* if either
- $a = 0^X$, or
- the expression of a in terms of elements of X does not involve 0^X.

(2) A 0^X*-reduction* of a is a path $P\colon a \twoheadrightarrow b$ in $\mathsf{Gr}(X)$ such that the following two statements hold.
- b is 0^X-reduced.
- Each edge in P is either an identity or a 0^X-prime edge. ◇

The following is basically the definition, but it is useful to state it precisely for later usage.

Lemma 3.3.5. *The following statements hold.*

(1) *Each element in X is 0^X-reduced.*
(2) *An element $a \in X^{\mathsf{fr}}$ is 0^X-reduced if and only if a is not the domain of any 0^X-prime edges.*

Proof. An element in X is either 0^X or in $X \setminus \{0^X\}$. In either case, it is 0^X-reduced.

For the second assertion, the domain of each 0^X-prime edge contains 0^X and at least one other element in X^{fr}, which may also be 0^X. Such an element is not 0^X-reduced.

Conversely, if $a \in X^{\mathsf{fr}}$ is not 0^X-reduced, then, by assertion (1), its expression in terms of elements of X involves at least two elements, and 0^X is among them. So a is the domain of some 0^X-prime edge. □

Next we show the existence of 0^X-reductions.

Lemma 3.3.6. *Each element in X^{fr} has a 0^X-reduction.*

Proof. Suppose $a \in X^{\mathsf{fr}}$. The proof is by induction on the number n of instances of 0^X in the expression of a in terms of elements of X. If $n = 0$, then a is 0^X-reduced, and the identity 1_a is a 0^X-reduction of a.

Inductively, suppose $n > 0$. If $a = 0^X$, then it is 0^X-reduced, and 1_a is again a 0^X-reduction.

If $a \neq 0^X$, then a is not 0^X-reduced. By Lemma 3.3.5 (2), a is the domain of some 0^X-prime edge $f\colon a \longrightarrow a'$. The number of 0^X in the expression of a' in terms of elements of X is $n-1$. By the induction hypothesis, a' has a 0^X-reduction $P'\colon a' \twoheadrightarrow b$. The path

$$a \xrightarrow{\ f\ } a' \xrightarrow{\ P'\ } \!\!\!\!\twoheadrightarrow b$$

in $\mathsf{Gr}(X)$ is a 0^X-reduction of a. □

Lemma 3.3.7. *Suppose $P\colon a \twoheadrightarrow b$ is a 0^X-reduction of an element $a \in X^{\mathsf{fr}}$. Then*

$$\mathsf{supp}(a) = \mathsf{supp}(0^X) \quad \textit{if and only if} \quad b = 0^X.$$

Proof. By Lemma 3.1.29 (2), there is an equality

$$\mathsf{supp}(a) = \mathsf{supp}(b).$$

So if $b = 0^X$, then

$$\mathsf{supp}(a) = \mathsf{supp}(b) = \mathsf{supp}(0^X). \tag{3.3.8}$$

Conversely, since b is 0^X-reduced, it is either 0^X or contains no 0^X. Therefore, if (3.3.8) holds, then $b = 0^X$, since otherwise its support cannot be the support of 0^X. □

Motivation 3.3.9. Uniqueness of 0^X-reductions requires more care than existence. For example, the diagram

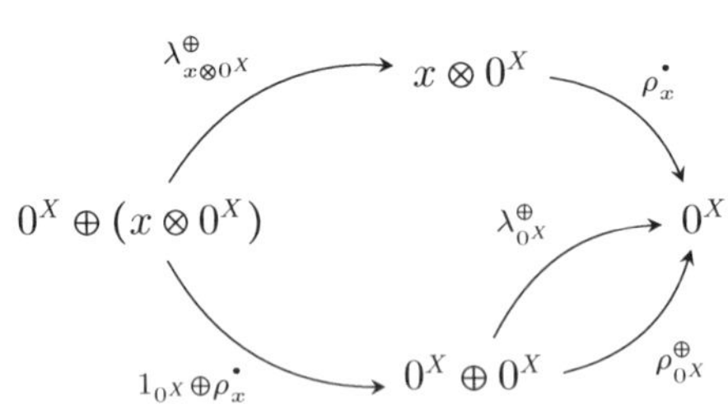

contains three different 0^X-reductions of $0^X \oplus (x \otimes 0^X) \in X^{\mathsf{fr}}$. Therefore, uniqueness should *not* refer to the paths themselves, but rather (i) the codomain and (ii) the values of the paths in the symmetric bimonoidal category C. Uniqueness of 0^X-reductions in these two senses are proved in the next two results. ◇

Notation 3.3.10. For a path $P: a \twoheadrightarrow b$ in $\mathsf{Gr}(X)$ and an element $c \in X^{\mathsf{fr}}$, the path

$$a \oplus c \xrightarrow{P \oplus 1_c} b \oplus c$$

in $\mathsf{Gr}(X)$ is obtained from P by replacing each of its edges h by $h \oplus 1_c$. The paths

- $1_c \oplus P: c \oplus a \twoheadrightarrow c \oplus b$,
- $P \otimes 1_c: a \otimes c \twoheadrightarrow b \otimes c$, and
- $1_c \otimes P: c \otimes a \twoheadrightarrow c \otimes b$

are defined analogously by replacing each edge h in P by $1_c \oplus h$, $h \otimes 1_c$, and $1_c \otimes h$, respectively. ◇

Lemma 3.3.11. *For any two 0^X-reductions*

$$\begin{array}{ccc} & \overset{P_1}{\nearrow\!\!\!\twoheadrightarrow} & b_1 \\ a & & \\ & \underset{P_2}{\searrow\!\!\!\twoheadrightarrow} & b_2 \end{array}$$

of an element $a \in X^{\mathsf{fr}}$, the equality $b_1 = b_2$ holds.

Proof. The proof is by induction on the norm of a. If $\mathsf{norm}(a) = 1$, then $a \in X$. By Lemma 3.3.5, the only possible 0^X-reductions of a are paths consisting of identity prime edges. They all have codomains equal to a.

For the induction step, suppose $\mathsf{norm}(a) > 1$. So a has the form $a_1 \oplus a_2$ or $a_1 \otimes a_2$ for some $a_1, a_2 \in X^{\mathsf{fr}}$ with strictly lower norms than a.

If a contains no 0^X in its expression, then the only possible 0^X-reductions of a are paths consisting of identity prime edges. They all have codomains equal to a.

We now assume that a contains at least one 0^X in its expression in terms of elements of X. Adding or removing an identity in a path do not change the (co)domain. Therefore, to show that $b_1 = b_2$, we may assume that for $i = 1, 2$, P_i has the form

$$a \xrightarrow{f_i} c_i \xrightarrow{P_i'} b_i$$

with

- the first edge f_i a 0^X-prime edge and
- $P_i': c_i \twoheadrightarrow b_i$ a 0^X-reduction of c_i, whose norm is strictly less than that of a.

The induction hypothesis applied to c_i implies that the codomains of any two 0^X-reductions of c_i are equal. To show that $b_1 = b_2$, we consider the two possible cases.

First, if $f_1 = f_2$, then $c_1 = c_2$. This implies that $b_1 = b_2$.

If $f_1 \neq f_2$, then up to renaming them, they fall into one of the following cases.

(1) f_1 and f_2 act on different summands of a, as in the following diagram.

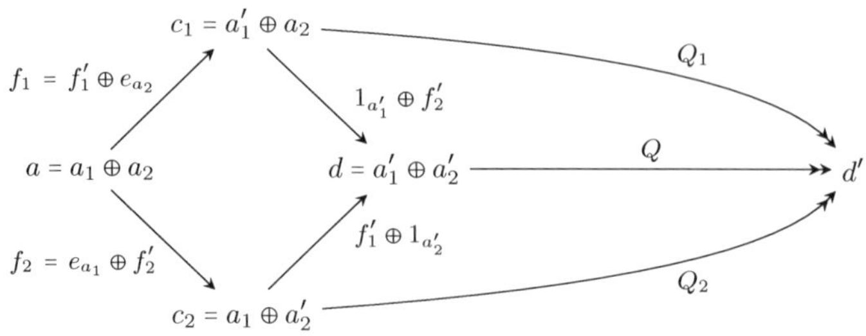

Here:

- $e_{a_i}\colon a_i \longrightarrow a_i$ is an identity prime edge for $i = 1, 2$.
- $f_1 = f'_1 \oplus e_{a_2}$ for some 0^X-prime edge $f'_1\colon a_1 \longrightarrow a'_1$.
- $f_2 = e_{a_1} \oplus f'_2$ for some 0^X-prime edge $f'_2\colon a_2 \longrightarrow a'_2$.
- By Lemma 3.3.6, $d = a'_1 \oplus a'_2$ has a 0^X-reduction $Q\colon d \twoheadrightarrow d'$.

Define the paths

$$Q_1 = \big(Q, 1_{a'_1} \oplus f'_2\big)\colon c_1 \twoheadrightarrow d'$$
$$Q_2 = \big(Q, f'_1 \oplus 1_{a'_2}\big)\colon c_2 \twoheadrightarrow d'$$

in $\mathsf{Gr}(X)$ as displayed above. Since these are 0^X-reductions of c_1 and c_2, respectively, the induction hypothesis implies that $b_1 = d' = b_2$.

(2) Case (1) has an analogue involving $\otimes$ instead of $\oplus$ in the diamond in the previous diagram. The proof for $b_1 = b_2$ in this case is the same as above after replacing $\oplus$ with $\otimes$.

(3) f_1 and f_2 act on the same summand of a, as in the following diagram.

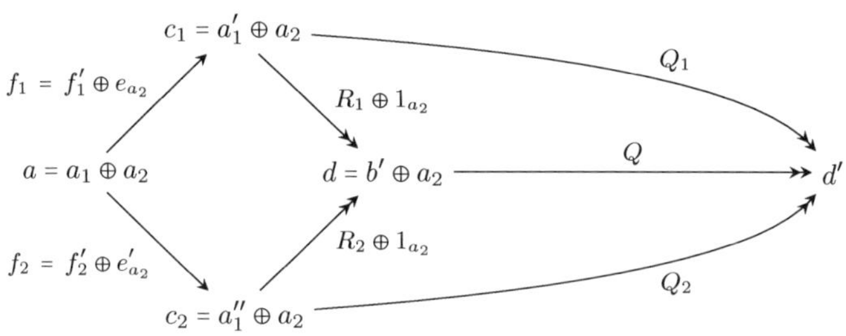

Here:

- f_1 is as in case (1).
- $e'_{a_2}\colon a_2 \longrightarrow a_2$ is an identity prime edge.
- $f_2 = f'_2 \oplus e'_{a_2}$ for some 0^X-prime edge $f'_2\colon a_1 \longrightarrow a''_1$.
- Since

$$\mathsf{norm}(a_1) < \mathsf{norm}(a),$$

the induction hypothesis applied to a_1, and Lemma 3.3.6 applied to a'_1 and a''_1, together imply that there exist 0^X-reductions

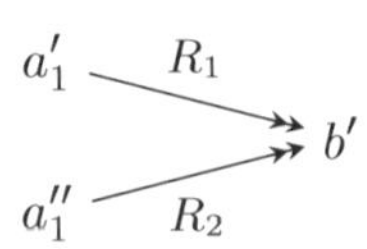

with a common codomain b'.

- By Lemma 3.3.6, $d = b' \oplus a_2$ has a 0^X-reduction $Q: d \twoheadrightarrow d'$.
- Each path $R_i \oplus 1_{a_2}: c_i \twoheadrightarrow d$ is defined in Notation 3.3.10.

Define the paths

$$Q_1 = \big(Q, R_1 \oplus 1_{a_2}\big): c_1 \twoheadrightarrow d'$$
$$Q_2 = \big(Q, R_2 \oplus 1_{a_2}\big): c_2 \twoheadrightarrow d'$$

in $\mathsf{Gr}(X)$ as displayed above. Since these are 0^X-reductions of c_1 and c_2, respectively, the induction hypothesis implies that $b_1 = d' = b_2$.

(4) There are three variants of case (3) obtained by the following modifications in the diamond in the previous diagram.
- Permute the two summands in each vertex and each edge or path.
- Replace $\oplus$ with $\otimes$.
- Permute the two summands in each vertex and each edge or path, and replace $\oplus$ with $\otimes$.

In each of these three cases, after the modifications, the proof is the same as in case (3).

(5) f_1 acts on a summand, and f_2 acts on all of a, as in the following diagram.

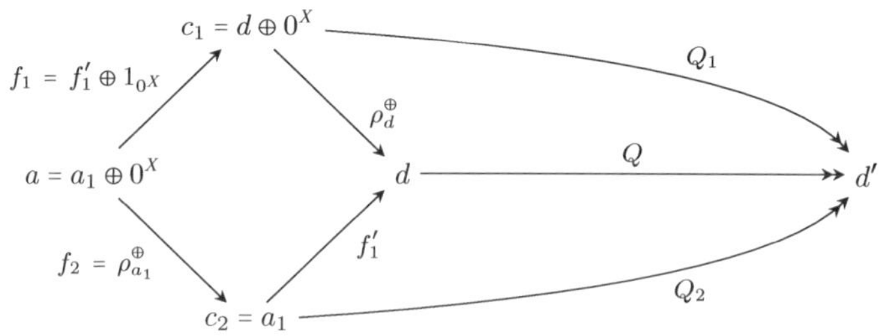

Here:
- $f_1 = f_1' \oplus 1_{0^X}$ for some 0^X-prime edge $f_1': a_1 \longrightarrow d$.
- By Lemma 3.3.6, there exists a 0^X-reduction $Q: d \twoheadrightarrow d'$.

Define the paths

$$Q_1 = \big(Q, \rho_d^{\oplus}\big): c_1 \twoheadrightarrow d'$$
$$Q_2 = \big(Q, f_1'\big): c_2 \twoheadrightarrow d'$$

in $\mathsf{Gr}(X)$ as displayed above. Since these are 0^X-reductions of c_1 and c_2, respectively, the induction hypothesis implies that $b_1 = d' = b_2$.

(6) Case (5) has a variant with $a = 0^X \oplus a_1$. The proof for this case is obtained from that of case (5) by replacing $(\rho^{\oplus}, f_1' \oplus 1_{0^X})$ with $(\lambda^{\oplus}, 1_{0^X} \oplus f_1')$.

(7) f_1 acts on a factor, and f_2 acts on all of a, as in the following diagram.

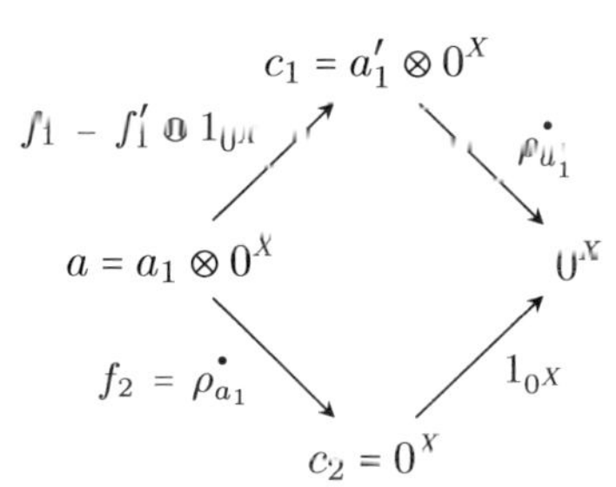

Since $\rho^{\bullet}_{a_1'}$ and 1_{0^X} are 0^X-reductions of c_1 and c_2, respectively, the induction hypothesis implies that $b_1 = 0^X = b_2$.

(8) Case (7) has a variant with $a = 0^X \otimes a_1$. The proof for this case is obtained from that of case (7) by replacing $(\rho^{\bullet}, f_1' \otimes 1_{0^X})$ with $(\lambda^{\bullet}, 1_{0^X} \otimes f_1')$.

(9) Both f_1 and f_2 act on all of a, as in the following diagram.

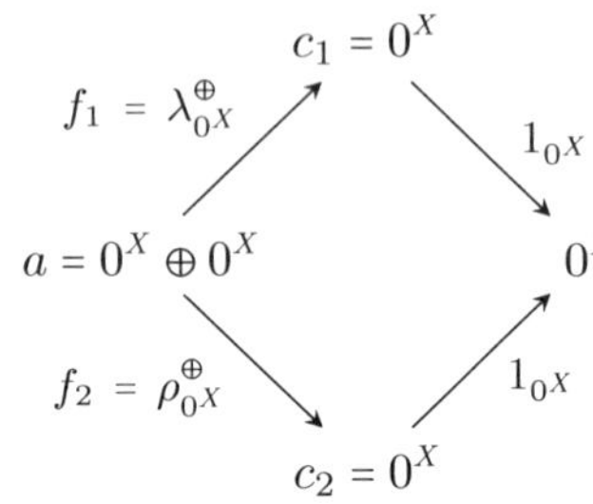

Since 1_{0^X} is a 0^X-reduction of each of c_1 and c_2, the induction hypothesis implies that $b_1 = 0^X = b_2$.

(10) Case (9) has a variant with $(\oplus, \lambda^{\oplus}_{0^X}, \rho^{\oplus}_{0^X})$ replaced by $(\otimes, \lambda^{\bullet}_{0^X}, \rho^{\bullet}_{0^X})$. Again the induction hypothesis implies that $b_1 = 0^X = b_2$.

This finishes the proof of the induction step. □

Lemma 3.3.12. *Any two 0^X-reductions*

$$a \underset{P_2}{\overset{P_1}{\rightrightarrows}} b$$

of an element $a \in X^{\mathsf{fr}}$ have the same value in C.

Proof. We use Lemma 3.3.11 and its inductive proof. If either $\mathsf{norm}(a) = 1$, or inductively $\mathsf{norm}(a) > 1$ with a containing no 0^X, then a 0^X-reduction of a consists of identities. So its value in C must be the identity morphism.

Following the proof of Lemma 3.3.11, we may assume that each P_i takes the form

$$a \xrightarrow{f_i} c_i \overset{P_i'}{\twoheadrightarrow} b$$

with

- f_i a 0^X-prime edge and
- $P_i' \colon c_i \twoheadrightarrow b$ a 0^X-reduction of c_i.

The induction hypothesis applied to c_i says that every 0^X-reduction of c_i has the same value in C as P_i'.

If $f_1 = f_2$, then $c_1 = c_2$. The induction hypothesis implies that P_1' and P_2', and therefore also P_1 and P_2, have the same value in C.

If $f_1 \neq f_2$, then we follow the proof of Lemma 3.3.11 and reuse the diagrams there with $d' = b$.

- In cases (1) and (2), the diamond is commutative by, respectively, the functoriality of $\oplus$ and $\otimes$ in C.
- In cases (3) and (4), the diamond is commutative by functoriality and the induction hypothesis applied to a_1.
- In cases (5)–(8), the diamond is commutative by, respectively, the naturality of $\rho^{\oplus}$, $\lambda^{\oplus}$, $\rho^{\bullet}$, and $\lambda^{\bullet}$.

- In case (9), the diamond is commutative by the unity property (1.2.6), that is, $\lambda^{\oplus}_{\mathbb{0}} = \rho^{\oplus}_{\mathbb{0}}$ in the symmetric monoidal category $(\mathsf{C}, \oplus, \mathbb{0})$.
- In case (10), the diamond is commutative by the axiom (2.1.14), that is, $\lambda^{\bullet}_{\mathbb{0}} = \rho^{\bullet}_{\mathbb{0}}$ in C.

Now we observe that P_1 and P_2 have the same value in C in each case. In each of cases (1)–(6), the other two subdiagrams are commutative by the definitions of the paths Q_1 and Q_2. The induction hypothesis applied to each c_i implies that P'_i and Q_i have the same value in C. The commutativity of the diagram in C then implies that P_1 and P_2 have the same value in C.

In each of cases (7)–(10), the diagram is the diamond with $b = 0^X$.

- In case (7), the induction hypothesis implies that the values of P'_1 and P'_2 in C are, respectively, $\rho^{\bullet}_{ga'_1}$ and $1_{\mathbb{0}}$. The commutative diamond in C implies that P_1 and P_2 have the same value in C.
- The proof for case (8) is obtained from that of case (7) by replacing $\rho^{\bullet}_{ga'_1}$ with $\lambda^{\bullet}_{ga'_1}$.
- In cases (9) and (10), the value in C of any 0^X-reduction of $c_1 = c_2 = 0^X$ is $1_{\mathbb{0}}$. Therefore, P_1 and P_2 have values, respectively,
  - $\lambda^{\oplus}_{\mathbb{0}}$ and $\rho^{\oplus}_{\mathbb{0}}$ in case (9) and
  - $\lambda^{\bullet}_{\mathbb{0}}$ and $\rho^{\bullet}_{\mathbb{0}}$ in case (10).

  They are equal by, respectively, (1.2.6) and (2.1.14), as explained above.

This finishes the induction step. □

Example 3.3.13. In Motivation 3.3.9, the three 0^X-reductions of $0^X \oplus (x \otimes 0^X) \in X^{\mathsf{fr}}$ have the same value in C for the following reasons.

- The upper left sub-diagram is commutative by the naturality of $\lambda^{\oplus}$ in C.
- The lower right sub-diagram is commutative by axiom (1.2.6) in the symmetric monoidal category $(\mathsf{C}, \oplus, \mathbb{0})$, which says that $\lambda^{\oplus}_{\mathbb{0}} = \rho^{\oplus}_{\mathbb{0}}$. ◇

3.4. Zero Reduction of Paths

A reduction step in the proof of the Coherence Theorem 3.9.1 involves eliminating additive zeros, multiplicative zeros, and their inverses in paths. In this section, we first give a precise definition of such a reduction; see Definition 3.4.5. In such a reduction, the (co)domain is a 0^X reduction of the (co)domain of the original path. Then we prove several basic properties about them. The proof of the existence of such a reduction is given in Section 3.5. Recall elementary edges, prime edges, and 0^X-prime edges in Definitions 3.1.6, 3.1.8, and 3.3.2. Recall that Convention 3.3.1 is in effect.

Definition 3.4.1. Consider the graph $\mathsf{Gr}(X)$ of X in Definition 3.1.9.

- An *inverse 0^X-prime edge* is a nonidentity prime edge that involves $\lambda^{-\oplus}$, $\rho^{-\oplus}$, $\lambda^{-\bullet}$, or $\rho^{-\bullet}$.
- A *0^X-free path* is a path that does not contain any 0^X-prime edges and inverse 0^X-prime edges.
- A *0^X-free edge* is a 0^X-free path of length 1. ◇

Explanation 3.4.2. In a 0^X-free path, every prime edge is either an identity or involves a single instance of $\alpha^{\pm\oplus}$, $\xi^{\pm\oplus}$, $\alpha^{\pm\otimes}$, $\xi^{\pm\otimes}$, $\lambda^{\pm\otimes}$, $\rho^{\pm\otimes}$, δ^l, or δ^r. ◇

Example 3.4.3. For elements $a, b \in X^{\mathsf{fr}}$, the prime edges

$$a \xrightarrow{\lambda_a^{-\oplus}} 0^X \oplus a \xrightarrow{\rho_b^{-\bullet} \oplus 1_a} (b \otimes 0^X) \oplus a$$

are inverse 0^X-prime edges. ◇

Example 3.4.4. Suppose $P: a \twoheadrightarrow b$ is a 0^X-free path. Then the paths

- $P \oplus 1_c: a \oplus c \twoheadrightarrow b \oplus c$,
- $1_c \oplus P: c \oplus a \twoheadrightarrow c \oplus b$,
- $P \otimes 1_c: a \otimes c \twoheadrightarrow b \otimes c$, and
- $1_c \otimes P: c \otimes a \twoheadrightarrow c \otimes b$

in Notation 3.3.10 are 0^X-free paths for each element $c \in X^{\mathsf{fr}}$. ◇

The next definition uses the concepts of commutativity near the end of Definition 3.1.14, and 0^X-reductions of an element in X^{fr} in Definition 3.3.4.

Definition 3.4.5. Suppose $P: a \twoheadrightarrow b$ is a path in $\mathsf{Gr}(X)$. A *0^X-reduction* of P is a 0^X-free path $R: a' \twoheadrightarrow b'$ such that for any 0^X-reductions

- $Q_a: a \twoheadrightarrow a'$ of a and
- $Q_b: b \twoheadrightarrow b'$ of b,

the diagram

$$\begin{array}{ccc} a & \xrightarrow{P} & b \\ {\scriptstyle Q_a}\downarrow & & \downarrow{\scriptstyle Q_b} \\ a' & \xrightarrow{R} & b' \end{array} \tag{3.4.6}$$

is commutative in the sense of Definition 3.1.14. ◇

The goal of this section and Section 3.5 is to show that each path in $\mathsf{Gr}(X)$ has a 0^X-reduction; see Proposition 3.5.32.

Explanation 3.4.7. Consider Definition 3.4.5.

(1) The domain a' of R must be the codomain of a 0^X-reduction of a, and the codomain b' of R must be the codomain of a 0^X-reduction of b. The elements a' and b' are well defined, that is, uniquely determined by a and b, by Lemmas 3.3.6 and 3.3.11. In particular, R is a 0^X-free path whose domain and codomain are both 0^X-reduced in the sense of Definition 3.3.4.

(2) For the commutativity of the diagram (3.4.6), the equality

$$(R, Q_a) = (Q_b, P)$$

of paths in $\mathsf{Gr}(X)$ is sufficient but *not* necessary. The definition only requires that the image of the diagram under the graph morphism $\varphi: \mathsf{Gr}(X) \longrightarrow \mathsf{C}$ in Definition 3.1.14 be a commutative diagram in C for arbitrary C and φ.

The following observation says that for a 0^X-free path to be a 0^X-reduction of a given path, it suffices to check the commutativity of the diagram (3.4.6) for one pair of 0^X-reductions (Q_a, Q_b). ◇

Lemma 3.4.8. *Suppose the following statements hold:*

- $P: a \twoheadrightarrow b$ *is a path in* $\mathsf{Gr}(X)$.
- $Q_a: a \twoheadrightarrow a'$ *is a* 0^X*-reduction of* a.
- $Q_b: b \twoheadrightarrow b'$ *is a* 0^X*-reduction of* b.
- $R: a' \twoheadrightarrow b'$ *is a* 0^X*-free path.*

- *The diagram*

$$\begin{array}{ccc} a & \xrightarrow{\;\;P\;\;} & b \\ {\scriptstyle Q_a}\downarrow & & \downarrow{\scriptstyle Q_b} \\ a' & \xrightarrow{\;\;R\;\;} & b' \end{array}$$

is commutative.

Then R is a 0^X-reduction of P.

Proof. By Lemmas 3.3.11 and 3.3.12, for each element in X^{fr}, the codomain and the value in C of a 0^X-reduction are unique. Therefore, in the given diagram, if Q_a and Q_b are replaced by any other 0^X-reductions of a and b, respectively, then the resulting diagram is still commutative. □

Lemma 3.4.8 will be used below to check that a given path has a 0^X-reduction. Recall from Definition 3.3.4 the concept of an element in X^{fr} being 0^X-reduced.

Lemma 3.4.9. *Suppose $P: a \twoheadrightarrow b$ is a 0^X-free path.*

(1) *If a contains no 0^X, then every vertex in P contains no 0^X and is 0^X-reduced.*
(2) *If $a = 0^X$, then the following statements hold.*
  (i) *Each edge in P is either an identity or a prime edge involving a single instance of $\alpha^{\pm\otimes}$, $\xi^{\pm\otimes}$, $\lambda^{\pm\otimes}$, or $\rho^{\pm\otimes}$.*
  (ii) *Each vertex in P is a finite product with precisely one 0^X and all other factors 1^X.*

Proof. If the expression of a in terms of elements of X contains no 0^X, then the same is true for all other vertices in P because edges in P, as described in Explanation 3.4.2, do not introduce any 0^X from the domain to the codomain.

For assertion (2), consider the subset $S \subset X^{\mathsf{fr}}$ in which each element is a finite product with precisely one 0^X and all other factors 1^X. Then $a = 0^X$ is in S as the special case with no factors of 1^X. Consider an edge

$$c \xrightarrow{\;\;f\;\;} b$$

in $\mathsf{Gr}(X)$ with

- $c \in S$ and
- f a prime edge that is also a 0^X-free path.

See Explanation 3.4.2 for the possibilities of f. We consider all possible cases below.

- If f is an identity, then $b = c \in S$.
- Suppose f is not an identity, and $c = 0^X$. Then f must be either

$$0^X \xrightarrow{\;\lambda^{-\otimes}_{0^X}\;} 1^X \otimes 0^X \quad \text{or} \quad 0^X \xrightarrow{\;\rho^{-\otimes}_{0^X}\;} 0^X \otimes 1^X.$$

  Moreover, its codomain is also in S.
- Suppose f is not an identity, and $c \neq 0^X$. Since $c \in S$, the prime edge f can only involve a single instance of $\alpha^{\pm\otimes}$ (if $\mathsf{norm}(c) \geq 3$), $\xi^{\pm\otimes}$, $\lambda^{\pm\otimes}$, or $\rho^{\pm\otimes}$. Moreover, its codomain is also in S.

Therefore, an induction on the length of P proves assertion (2), with the previous paragraph proving both the initial case and the first edge of the induction step. □

Example 3.4.10. In the context of Lemma 3.4.9 (2), for a 0^X-free path $P: 0^X \twoheadrightarrow b$, it does *not* follow that the vertices in P other than the domain are 0^X-reduced. For example, in the 0^X-free path

$$\begin{array}{ccccc} 0^X & & & & (1^X \otimes 1^X) \otimes (0^X \otimes 1^X) \\ \downarrow{\scriptstyle \lambda^{-\otimes}_{0^X}} & & & & \uparrow{\scriptstyle \lambda^{-\otimes}_{1^X} \otimes 1_{0^X \otimes 1^X}} \\ 1^X \otimes 0^X & \xrightarrow{\rho^{-\otimes}_{1^X \otimes 0^X}} & (1^X \otimes 0^X) \otimes 1^X & \xrightarrow{\alpha^{\otimes}_{1^X, 0^X, 1^X}} & 1^X \otimes (0^X \otimes 1^X) \end{array}$$

all the vertices except 0^X are *not* 0^X-reduced. ◇

Recall from (3.1.18) the value in C of a path in $\mathsf{Gr}(X)$. The next two observations are about the values in C of paths with 0^X as the domain. They provide further information in the setting of Lemma 3.4.9 (2).

Lemma 3.4.11. *Suppose $P: 0^X \twoheadrightarrow 0^X$ is a 0^X-free path. Then the value of P in C is $1_{\mathbb{0}}$.*

Proof. This follows from Lemma 3.4.9 (2) and the uniqueness part of Theorem 1.3.8 for the symmetric monoidal category $(\mathsf{C}, \otimes, \mathbb{1}, \alpha^{\otimes}, \lambda^{\otimes}, \rho^{\otimes}, \xi^{\otimes})$. □

The next observation will be needed in the proof of Lemma 3.4.14.

Lemma 3.4.12. *Suppose*

- *$P: 0^X \twoheadrightarrow b$ is a 0^X-free path, and*
- *$Q_b: b \twoheadrightarrow 0^X$ is a 0^X-reduction of b.*

Then the value in C of the path

$$0^X \xrightarrow{P} b \xrightarrow{Q_b} 0^X \tag{3.4.13}$$

is the identity morphism $1_{\mathbb{0}}$.

Proof. By Lemma 3.4.9 (2), b is a finite product with precisely one 0^X and all other factors 1^X. An induction on $\mathsf{norm}(b)$, starting with $\mathsf{norm}(b) = 1$ and $b = 0^X$, implies that there is a 0^X-reduction

$$b \xrightarrow{Q_b} 0^X$$

of b such that the following statements hold.

- $Q_b = 1_{0^X}$ if $b = 0^X$.
- If $b \neq 0^X$, then Q_b consists of prime edges, each involving $\lambda^{\bullet}$ or $\rho^{\bullet}$ with subscript a finite product of copies of 1^X.

One example of Q_b is the path below.

$$\begin{array}{ccc} b = 1^X \otimes \big(0^X \otimes (1^X \otimes 1^X)\big) & \xrightarrow{1_{1^X} \otimes \lambda^{\bullet}_{1^X \otimes 1^X}} & 1^X \otimes 0^X \\ & & \downarrow{\scriptstyle \rho^{\bullet}_{1^X}} \\ & & 0^X \end{array}$$

In general, the value in C of each prime edge in Q_b is one of the following:

- $1_{\mathbb{0}}$ if $Q_b = 1_{0^X}$;
- a canonical map as in Definition 1.3.2 involving a single instance of $\lambda^{\bullet}_{\mathbb{1}} = \rho^{\otimes}_{\mathbb{0}}$ (2.1.17) and identities, if f has a copy of $\lambda^{\bullet}_{1^X}$;
- a canonical map involving a single instance of $\rho^{\bullet}_{\mathbb{1}} = \lambda^{\otimes}_{\mathbb{0}}$ (2.1.18) and identities, if f has a copy of $\rho^{\bullet}_{1^X}$; or

- a canonical map involving identities, $\lambda^{\otimes}$, and $\rho^{\otimes}$, if f has a copy of $\lambda^{\bullet}$ or $\rho^{\bullet}$ whose subscript is a product of at least two copies of 1^X.

For an example of the last case, consider the prime edge

$$1^X \otimes \big(0^X \otimes (1^X \otimes 1^X)\big) \xrightarrow{1_{1^X}\otimes\lambda^{\bullet}_{1^X\otimes 1^X}} 1^X \otimes 0^X$$

with an instance of $\lambda^{\bullet}_{1^X\otimes 1^X}$. Its value in C factors as the composite below by the naturality of $\lambda^{\bullet}$ in C.

$$\begin{array}{ccc}
\mathbb{1}\otimes\big(\mathbb{0}\otimes(\mathbb{1}\otimes\mathbb{1})\big) & \xrightarrow{1_{\mathbb{1}}\otimes\lambda^{\bullet}_{\mathbb{1}\otimes\mathbb{1}}} & \mathbb{1}\otimes\mathbb{0} \\
{\scriptstyle 1_{\mathbb{1}}\otimes(1_{\mathbb{0}}\otimes\lambda^{\otimes}_{\mathbb{1}})} \searrow & & \nearrow {\scriptstyle 1_{\mathbb{1}}\otimes\lambda^{\bullet}_{\mathbb{1}} = 1_{\mathbb{1}}\otimes\rho^{\otimes}_{\mathbb{0}}} \\
& \mathbb{1}\otimes(\mathbb{0}\otimes\mathbb{1}) &
\end{array}$$

The axiom (2.1.17) in C, which says that $\lambda^{\bullet}_{\mathbb{1}} = \rho^{\otimes}_{\mathbb{0}}$, implies that the above composite is a canonical map involving identities, $\lambda^{\otimes}$, and $\rho^{\otimes}$.

For a general finite product I of at least two copies of $\mathbb{1}$, there is a finite composite $I \longrightarrow \mathbb{1}$ consisting of identities, $\lambda^{\otimes}$, and $\rho^{\otimes}$. In this case:

- The left slanted morphism in the previous diagram is a canonical map involving identities, $\lambda^{\otimes}$, and $\rho^{\otimes}$.
- The right slanted morphism uses either
  - $\lambda^{\bullet}_{\mathbb{1}} = \rho^{\otimes}_{\mathbb{0}}$ as above, if the prime edge involves $\lambda^{\bullet}$; or
  - the axiom (2.1.18), which says that $\rho^{\bullet}_{\mathbb{1}} = \lambda^{\otimes}_{\mathbb{0}}$, if the prime edge involves $\rho^{\bullet}$.

Therefore, in general the value of Q_b in C is the $\mathbb{0}$-component of a canonical map involving identities, $\lambda^{\otimes}$, and $\rho^{\otimes}$.

By Lemma 3.4.9 (2), every edge in P is either an identity or a prime edge involving a single instance of $\alpha^{\pm\otimes}$, $\xi^{\pm\otimes}$, $\lambda^{\pm\otimes}$, or $\rho^{\pm\otimes}$. Therefore, the value of the path (3.4.13) in C is the $\mathbb{0}$-component of a permuted canonical map $- \longrightarrow -$. It is the identity morphism $1_{\mathbb{0}}$ by the uniqueness part of the Symmetric Coherence Theorem 1.3.8 for the symmetric monoidal category $(\mathsf{C}, \otimes, \mathbb{1}, \alpha^{\otimes}, \lambda^{\otimes}, \rho^{\otimes}, \xi^{\otimes})$.

We have proved the assertion for one choice of a 0^X-reduction of b. Therefore, by Lemma 3.3.12, the assertion remains true for any other 0^X-reduction of b. □

The next observation will be used in the proofs of several lemmas in Section 3.5.

Lemma 3.4.14. *Suppose*

- $P\colon a \longrightarrow\!\!\!\!\!\rightarrow b$ *is a* 0^X*-free path with* a *and* b *both* 0^X*-reduced, and*
- $c \in X^{\mathrm{fr}}$ *is* 0^X*-reduced.*

Then the paths

$$P \oplus 1_c, \quad 1_c \oplus P, \quad P \otimes 1_c, \quad \textit{and} \quad 1_c \otimes P$$

in Notation 3.3.10 have 0^X*-reductions.*

Proof. Each of a and c has two possibilities. It is either 0^X, or its expression in terms of elements of X contains no 0^X. We consider the possible cases separately.

First, if a contains no 0^X, then b also contains no 0^X by Lemma 3.4.9 (1). There are now two sub-cases depending on the form of c.

(1) If c also contains no 0^X, then $a \oplus c$ and $b \oplus c$ contain no 0^X, and are 0^X-reduced. Since $P \oplus 1_c$ is a 0^X-free path, it is a 0^X-reduction of itself. Essentially the same argument works for the other three types of paths.

(2) If $c = 0^X$, then the diagram

$$\begin{array}{ccc} a \oplus 0^X & \overset{P\oplus 1_{0^X}}{\twoheadrightarrow} & b \oplus 0^X \\ {\scriptstyle \rho_a^{\oplus}}\big\downarrow & & \big\downarrow{\scriptstyle \rho_b^{\oplus}} \\ a & \overset{P}{\twoheadrightarrow} & b \end{array}$$

is commutative by the naturality of $\rho^{\oplus}$ in C. The left and the right vertical edges are 0^X-reductions of $a \oplus 0^X$ and $b \oplus 0^X$, respectively. Since P is a 0^X-free path, it is a 0^X-reduction of $P \oplus 1_c$.

Similarly, for the paths $1_c \oplus P$, $P \otimes 1_c$, and $1_c \otimes P$, the diagrams below are commutative by the naturality of, respectively, $\lambda^{\oplus}$, $\rho^{\bullet}$, and $\lambda^{\bullet}$ in C.

$$\begin{array}{ccc} 0^X \oplus a & \overset{1_{0^X}\oplus P}{\twoheadrightarrow} & 0^X \oplus b \\ {\scriptstyle \lambda_a^{\oplus}}\big\downarrow & & \big\downarrow{\scriptstyle \lambda_b^{\oplus}} \\ a & \overset{P}{\twoheadrightarrow} & b \end{array} \qquad \begin{array}{ccc} a \otimes 0^X & \overset{P\otimes 1_{0^X}}{\twoheadrightarrow} & b \otimes 0^X \\ {\scriptstyle \rho_a^{\bullet}}\big\downarrow & & \big\downarrow{\scriptstyle \rho_b^{\bullet}} \\ 0^X & \overset{1_{0^X}}{\twoheadrightarrow} & 0^X \end{array} \qquad \begin{array}{ccc} 0^X \otimes a & \overset{1_{0^X}\otimes P}{\twoheadrightarrow} & 0^X \otimes b \\ {\scriptstyle \lambda_a^{\bullet}}\big\downarrow & & \big\downarrow{\scriptstyle \lambda_b^{\bullet}} \\ 0^X & \overset{1_{0^X}}{\twoheadrightarrow} & 0^X \end{array}$$

They show that P, 1_{0^X}, and 1_{0^X}, respectively, are the corresponding 0^X-reductions.

Next if $a = 0^X$, then b is a finite product with precisely one 0^X and all other factors 1^X by Lemma 3.4.9 (2). We consider the following diagram with $Q_b \colon b \twoheadrightarrow 0^X$ any 0^X-reduction of b.

$$\begin{array}{ccc} 0^X \oplus c & \overset{P\oplus 1_c}{\twoheadrightarrow} & b \oplus c \\ {\scriptstyle 1_{0^X}\oplus 1_c}\big\downarrow & & \big\downarrow{\scriptstyle Q_b\oplus 1_c} \\ 0^X \oplus c & \overset{1_{0^X}\oplus 1_c}{\twoheadrightarrow} & 0^X \oplus c \\ {\scriptstyle \lambda_c^{\oplus}}\big\downarrow & & \big\downarrow{\scriptstyle \lambda_c^{\oplus}} \\ c & \xrightarrow{1_c} & c \end{array}$$

The bottom rectangle is commutative by the functoriality of $\oplus$ in C. The top rectangle is commutative by the functoriality of $\oplus$ in C and Lemma 3.4.12. Since c is 0^X-reduced, the left and the right vertical paths are 0^X-reductions of $0^X \oplus c$ and $b \oplus c$, respectively. Therefore, 1_c is a 0^X-reduction of $P \oplus 1_c$.

For $1_c \oplus P$, we modify the previous diagram by (i) switching the two summands in each vertex, each edge, and each path in the top rectangle, and (ii) replacing $\lambda_c^{\oplus}$ with $\rho_c^{\oplus}$.

For $P \otimes 1_c$, we modify the previous diagram by replacing (i) $\oplus$ with $\otimes$, and (ii) the bottom rectangle with the rectangle below.

$$\begin{array}{ccc} 0^X \otimes c & \overset{1_{0^X}\otimes 1_c}{\twoheadrightarrow} & 0^X \otimes c \\ {\scriptstyle \lambda_c^{\bullet}}\big\downarrow & & \big\downarrow{\scriptstyle \lambda_c^{\bullet}} \\ 0^X & \xrightarrow{1_{0^X}} & 0^X \end{array}$$

This is commutative by the functoriality of $\otimes$ in C. In this case, 1_{0^X} is a 0^X-reduction of $P \otimes 1_c$.

For $1_c \otimes P$, we modify the proof for $P \otimes 1_c$ by (i) switching the two factors in each vertex, each edge, and each path in the top rectangle, and (ii) replacing $\lambda_c^{\bullet}$ with $\rho_c^{\bullet}$. $\square$

3.5. Existence of Zero Reduction of Paths

In this section, we give a detailed proof that every path in $\mathsf{Gr}(X)$ as in Definition 3.1.9 has a 0^X-reduction in the sense of Definition 3.4.5; see Proposition 3.5.32. A path in $\mathsf{Gr}(X)$ consists of prime edges as in Definition 3.1.8, and prime edges are built from elementary edges as in Definition 3.1.6. The proof of Proposition 3.5.32 is broken down into a series of lemmas covering the various types of prime edges. This section ends with a preliminary version of the Coherence Theorem 3.9.1 for paths whose domains have the same support as 0^X; see Proposition 3.5.33. Recall that Convention 3.3.1 is in effect.

Lemma 3.5.1. *Suppose $f: a \longrightarrow b$ is a prime edge of one of the following types:*

- *an identity,*
- *a 0^X-prime edge as in Definition 3.3.2, or*
- *an inverse 0^X-prime edge as in Definition 3.4.1.*

Then f has a 0^X-reduction.

Proof. Recall from Lemma 3.3.6 that each element in X^{fr} has a 0^X-reduction.

- If f is an identity, then its value in C is an identity morphism. If $Q_a: a \twoheadrightarrow a'$ is a 0^X-reduction of a, then the diagram

$$\begin{array}{ccc} a & \xrightarrow{\;f\;} & a \\ {\scriptstyle Q_a}\big\downarrow & & \big\downarrow{\scriptstyle Q_a} \\ a' & \xrightarrow{\;1_{a'}\;}\!\!\!\!\rightarrow & a' \end{array}$$

  is commutative. So the identity $1_{a'}$ is a 0^X-reduction of f.
- If f is a 0^X-prime edge, and if $Q_b : b \twoheadrightarrow b'$ is a 0^X-reduction of b, then $(Q_b, f): a \twoheadrightarrow b'$ is a 0^X-reduction of a. Since the diagram

$$\begin{array}{ccc} a & \xrightarrow{\;f\;} & b \\ {\scriptstyle (Q_b,f)}\big\downarrow & & \big\downarrow{\scriptstyle Q_b} \\ b' & \xrightarrow{\;1_{b'}\;}\!\!\!\!\rightarrow & b' \end{array}$$

  is commutative, the identity $1_{b'}$ is a 0^X-reduction of f.
- If f is an inverse 0^X-prime edge, denote by $f^{-1}: b \longrightarrow a$ the 0^X-prime edge obtained from f by replacing its unique copy of $\lambda^{-\oplus}$, $\rho^{-\oplus}$, $\lambda^{-\bullet}$, or $\rho^{-\bullet}$ with its formal inverse. If $Q_a: a \twoheadrightarrow a'$ is a 0^X-reduction of a, then $(Q_a, f^{-1}): b \twoheadrightarrow a'$ is a 0^X-reduction of b. Since the diagram

$$\begin{array}{ccc} a & \xrightarrow{\;f\;} & b \\ {\scriptstyle Q_a}\big\downarrow & & \big\downarrow{\scriptstyle (Q_a,f^{-1})} \\ a' & \xrightarrow{\;1_{a'}\;}\!\!\!\!\rightarrow & a' \end{array}$$

  is commutative, the identity $1_{a'}$ is a 0^X-reduction of f.

Therefore, by Lemma 3.4.8, f has a 0^X-reduction. □

Lemma 3.5.2. *Suppose $f: a \longrightarrow b$ is a prime edge whose domain is 0^X-reduced in the sense of Definition 3.3.4. Then f has a 0^X-reduction.*

Proof. If f satisfies the hypothesis of Lemma 3.5.1, then we are done. Otherwise, f is a 0^X-free path. There are two cases.

First suppose that the expression of a in terms of elements of X contains no 0^X. By Lemma 3.4.9(1), the codomain b is also 0^X-reduced. The identities $1_a: a \longrightarrow a$ and $1_b: b \longrightarrow b$ are 0^X-reductions of a and b, respectively, and the diagram

$$\begin{array}{ccc} a & \xrightarrow{\ f\ } & b \\ {\scriptstyle 1_a}\downarrow & & \downarrow{\scriptstyle 1_b} \\ a & \xrightarrow{\ f\ } & b \end{array}$$

is commutative. Therefore, f is a 0^X-free path, and is a 0^X-reduction of f.

Next suppose $a = 0^X$. This implies that 1_a is a 0^X-reduction of a, and that f is either

$$a = 0^X \xrightarrow{\lambda^{-\otimes}_{0^X}} 1^X \otimes 0^X = b \quad \text{or} \quad a = 0^X \xrightarrow{\rho^{-\otimes}_{0^X}} 0^X \otimes 1^X = b.$$

In these two cases, a 0^X-reduction of b is $\rho^{\bullet}_{1^X}$ on the left and $\lambda^{\bullet}_{1^X}$ on the right. The diagrams

$$\begin{array}{ccc} 0^X & \xrightarrow{\lambda^{-\otimes}_{0^X}} & 1^X \otimes 0^X \\ {\scriptstyle 1_{0^X}}\downarrow & & \downarrow{\scriptstyle \rho^{\bullet}_{1^X}} \\ 0^X & \xrightarrow{\ 1_{0^X}\ } & 0^X \end{array} \quad \text{and} \quad \begin{array}{ccc} 0^X & \xrightarrow{\rho^{-\otimes}_{0^X}} & 0^X \otimes 1^X \\ {\scriptstyle 1_{0^X}}\downarrow & & \downarrow{\scriptstyle \lambda^{\bullet}_{1^X}} \\ 0^X & \xrightarrow{\ 1_{0^X}\ } & 0^X \end{array}$$

are commutative by the axioms (2.1.18) and (2.1.17), which state that $\rho^{\bullet}_{\mathbb{1}} = \lambda^{\otimes}_{\mathbb{0}}$ and $\lambda^{\bullet}_{\mathbb{1}} = \rho^{\otimes}_{\mathbb{0}}$, respectively. Therefore, 1_{0^X} is a 0^X-reduction of f. □

Lemma 3.5.3. *Suppose $f: a \longrightarrow b$ is a prime edge such that*

- *a is not 0^X-reduced, and*
- *f involves an instance of $\alpha^{\pm\oplus}$.*

Then f has a 0^X-reduction.

Proof. By assumption, $\mathsf{norm}(a) \geq 3$, and a contains at least one 0^X in its expression in terms of elements of X. First assume that f involves an instance of $\alpha^{\oplus}$. We proceed by induction on $\mathsf{norm}(a)$.

If $\mathsf{norm}(a) = 3$, then

$$a = (x \oplus y) \oplus z \xrightarrow{f = \alpha^{\oplus}_{x,y,z}} x \oplus (y \oplus z) = b \tag{3.5.4}$$

with $x, y, z \in X$, at least one of which is 0^X. Depending on whether each of x, y, and z is 0^X, there are seven cases. Consider, for example, the case $x = z = 0^X \neq y$, as in the following diagram.

$$\begin{array}{ccc} a = (0^X \oplus y) \oplus 0^X & \xrightarrow{\alpha^{\oplus}_{0^X,y,0^X}} & 0^X \oplus (y \oplus 0^X) = b \\ {\scriptstyle \lambda^{\oplus}_y \oplus 1_{0^X}}\downarrow & & \downarrow{\scriptstyle \lambda^{\oplus}_{y \oplus 0^X}} \\ y \oplus 0^X & \xrightarrow{\ 1_{y \oplus 0^X}\ } & y \oplus 0^X \\ {\scriptstyle \rho^{\oplus}_y}\downarrow & & \downarrow{\scriptstyle \rho^{\oplus}_y} \\ y & \xrightarrow{\ 1_y\ } & y \end{array} \tag{3.5.5}$$

- The top rectangle is commutative by the left unity property (1.2.7) in the monoidal category $(\mathsf{C}, \oplus, 0^X, \alpha^{\oplus}, \lambda^{\oplus}, \rho^{\oplus})$.
- The bottom rectangle is commutative by definition.
- The left and the right vertical paths are, respectively, 0^X-reductions of a and b.
- 1_y is a 0^X-free path with $y \in X$ 0^X-reduced by Lemma 3.3.5 (1).

Therefore, 1_y is a 0^X-reduction of f in this case. The other six cases are proved in the same way, using either the unity axiom (1.2.2) or the unity properties (1.2.7) in the monoidal category $(\mathsf{C}, \oplus)$. In each case, an identity is a 0^X-reduction of f.

For the induction step, suppose $\mathsf{norm}(a) > 3$. Either

$$a = a_1 \oplus a_2 \quad \text{or} \quad a = a_1 \otimes a_2$$

for some $a_1, a_2 \in X^{\mathsf{fr}}$. We consider these two cases separately below.

First suppose a is a sum $a = a_1 \oplus a_2$, and $f = f_1 \oplus e$ splits as a sum for some

- prime edge $f_1 : a_1 \longrightarrow b_1$ involving an instance of $\alpha^{\oplus}$ and
- identity $e : a_2 \longrightarrow a_2$.

There are two sub-cases depending on a_1.

- If a_1 is 0^X-reduced, then by Lemma 3.5.2, f_1 has a 0^X-reduction R_1.
- If a_1 is not 0^X-reduced, then, since $\mathsf{norm}(a_1) < \mathsf{norm}(a)$, the induction hypothesis implies that f_1 has a 0^X-reduction R_1.

In either case, there is a diagram

$$\begin{array}{ccc} a_1 & \xrightarrow{\;f_1\;} & b_1 \\ {\scriptstyle Q_{a_1}}\big\downarrow & & \big\downarrow{\scriptstyle Q_{b_1}} \\ a_1' & \xrightarrow{\;R_1\;} & b_1' \end{array} \tag{3.5.6}$$

that is commutative, with Q_{a_1} and Q_{b_1} any 0^X-reductions of a_1 and b_1, respectively, and with R_1 a 0^X-reduction of f_1, hence in particular a 0^X-free path.

Consider the following diagram.

$$\begin{array}{ccc} a = a_1 \oplus a_2 & \xrightarrow{\;f = f_1 \oplus e\;} & b_1 \oplus a_2 = b \\ {\scriptstyle Q_{a_1} \oplus 1_{a_2}}\big\downarrow & & \big\downarrow{\scriptstyle Q_{b_1} \oplus 1_{a_2}} \\ a_1' \oplus a_2 & \xrightarrow{\;R_1 \oplus 1_{a_2}\;} & b_1' \oplus a_2 \\ {\scriptstyle 1_{a_1'} \oplus Q_{a_2}}\big\downarrow & & \big\downarrow{\scriptstyle 1_{b_1'} \oplus Q_{a_2}} \\ a_1' \oplus a_2' & \xrightarrow{\;R_1 \oplus 1_{a_2'}\;} & b_1' \oplus a_2' \\ {\scriptstyle Q_1}\big\downarrow & & \big\downarrow{\scriptstyle Q_2} \\ a' & \xrightarrow{\;R\;} & b' \end{array} \tag{3.5.7}$$

- $Q_{a_2} : a_2 \longrightarrow a_2'$ is any 0^X-reduction of a_2.
- The top two rectangles are commutative by the functoriality of $\oplus$ in C.
- $Q_1 : a_1' \oplus a_2' \longrightarrow a'$ is any 0^X-reduction of $a_1' \oplus a_2'$.
- $Q_2 : b_1' \oplus a_2' \longrightarrow b'$ is any 0^X-reduction of $b_1' \oplus a_2'$.

- R is a 0^X-reduction of $R_1 \oplus 1_{a_2'}$, which exists by Lemma 3.4.14. So R is a 0^X-free path, and the bottom rectangle is commutative.
- The left and the right vertical paths are 0^X-reductions of a and b, respectively.

Therefore, by Lemma 3.4.8, R is a 0^X-reduction of f.

If f has the form $e \oplus f_1$, then we slightly modify the argument in the previous two paragraphs by switching the two summands in the top two rectangles in (3.5.7).

If f does not split as a sum, then $f = \alpha^{\oplus}_{x,y,z}$ as in (3.5.4) for some $x, y, z \in X^{\mathsf{fr}}$. For each $w \in \{x, y, z\}$, suppose $Q_w\colon w \twoheadrightarrow w'$ is a 0^X-reduction of w. The diagram

$$
\begin{array}{ccc}
a = (x \oplus y) \oplus z & \xrightarrow{f = \alpha^{\oplus}_{x,y,z}} & x \oplus (y \oplus z) = b \\
{\scriptstyle (Q_x \oplus 1_y) \oplus 1_z} \big\downarrow & & \big\downarrow {\scriptstyle Q_x \oplus (1_y \oplus 1_z)} \\
(x' \oplus y) \oplus z & \xrightarrow{\alpha^{\oplus}_{x',y,z}} & x' \oplus (y \oplus z) \\
{\scriptstyle (1_{x'} \oplus Q_y) \oplus 1_z} \big\downarrow & & \big\downarrow {\scriptstyle 1_{x'} \oplus (Q_y \oplus 1_z)} \\
(x' \oplus y') \oplus z & \xrightarrow{\alpha^{\oplus}_{x',y',z}} & x' \oplus (y' \oplus z) \\
{\scriptstyle (1_{x'} \oplus 1_{y'}) \oplus Q_z} \big\downarrow & & \big\downarrow {\scriptstyle 1_{x'} \oplus (1_{y'} \oplus Q_z)} \\
a' = (x' \oplus y') \oplus z' & \xrightarrow{\alpha^{\oplus}_{x',y',z'}} & x' \oplus (y' \oplus z') = b'
\end{array}
\tag{3.5.8}
$$

is commutative by the naturality of $\alpha^{\oplus}$ in C. Each w' is 0^X-reduced, so it is either 0^X or contains no 0^X.

- If x', y', and z' all contain no 0^X, then a' and b' also contain no 0^X, and are 0^X-reduced. The left and the right vertical paths in (3.5.8) are 0^X-reductions of a and b, respectively. Therefore, $\alpha^{\oplus}_{x',y',z'}$ is a 0^X-reduction of f in this case.
- Otherwise, at least one of x', y', and z' is 0^X. We reuse the argument in (3.5.5) applied to $\alpha^{\oplus}_{x',y',z'}$, and append the resulting diagram to the bottom of (3.5.8), and similarly for the other six cases. For each $w \in \{x, y, z\}$, the condition "$w \neq 0^X$" is replaced by "w' contains no 0^X." In each case, the combined diagram is commutative, and shows that an identity is a 0^X-reduction of f.

So far we have proved the induction step when a is a sum.

Next suppose $a = a_1 \otimes a_2$. Then f, being a prime edge involving $\alpha^{\oplus}$, must split as a product, say $f_1 \otimes e$ with

- $f_1\colon a_1 \longrightarrow b_1$ a prime edge involving $\alpha^{\oplus}$ and
- $e\colon a_2 \longrightarrow a_2$ an identity.

The diagram (3.5.6) still applies. We reuse the argument in (3.5.7) by replacing $\oplus$ with $\otimes$ to conclude that f has a 0^X-reduction. If f has the form $e \otimes f_1$, then we modify the argument further by switching the two factors in the top two rectangles. This finishes the induction and the case with f involving an instance of $\alpha^{\oplus}$.

If f involves an instance of $\alpha^{-\oplus}$, then we reuse the above argument by replacing $\alpha^{\oplus}$ with $\alpha^{-\oplus}$. $\square$

Lemma 3.5.9. *Suppose $f\colon a \longrightarrow b$ is a prime edge such that*

- *a is not 0^X-reduced, and*
- *f involves an instance of $\alpha^{\pm\otimes}$.*

Then f has a 0^X-reduction.

Proof. First suppose that f involves an instance of $\alpha^{\otimes}$. We proceed by induction on $\mathsf{norm}(a) \geq 3$ as in the proof of Lemma 3.5.3, most of which is reused here by replacing $\oplus$ with $\otimes$.

In the initial case of the induction with $\mathsf{norm}(a) = 3$ and $f = \alpha^{\otimes}_{x,y,z}$, the diagram (3.5.5) becomes the following diagram.

$$\begin{array}{ccc}
a = (0^X \otimes y) \otimes 0^X & \xrightarrow{\;f = \alpha^{\otimes}_{0^X,y,0^X}\;} & 0^X \otimes (y \otimes 0^X) = b \\
{\scriptstyle \lambda^{\bullet}_y \otimes 1_{0^X}} \big\downarrow & & \\
0^X \otimes 0^X & & \big\downarrow {\scriptstyle \lambda^{\bullet}_{y \otimes 0^X}} \\
{\scriptstyle \lambda^{\bullet}_{0^X}} \big\downarrow & & \\
0^X & \xrightarrow{\quad 1_{0^X} \quad} & 0^X
\end{array} \tag{3.5.10}$$

This diagram is commutative by the axiom (2.1.22), so 1_{0^X} is a 0^X-reduction of f. The other six cases follow similarly by the axioms (2.1.20)–(2.1.22).

For the induction step, we first consider the case with $a = a_1 \otimes a_2$ and $f = f_1 \otimes e$ for some

- prime edge $f_1\colon a_1 \longrightarrow b_1$ involving an instance of $\alpha^{\otimes}$ and
- identity $e\colon a_2 \longrightarrow a_2$.

We reuse the argument in the two paragraphs containing (3.5.6) and (3.5.7) by changing $\oplus$ to $\otimes$. The case with $f = e \otimes f_1$ is similar, with the two factors switched.

If f does not split as a product, then $f = \alpha^{\otimes}_{x,y,z}$ for some $x, y, z \in X^{\mathsf{fr}}$. We reuse the paragraph containing (3.5.8) by changing $\oplus$ to $\otimes$, (3.5.5) to (3.5.10), and similarly for the other six cases. So far we have proved the induction step when a is a product.

Next suppose $a = a_1 \oplus a_2$. Then f, being a prime edge involving $\alpha^{\otimes}$, must split as a sum, say $f_1 \oplus e$ with f_1 and e as above. We reuse the argument in (3.5.7), and conclude that f has a 0^X-reduction. If f has the form $e \oplus f_1$, then we modify the argument by switching the two summands in the top two rectangles. This finishes the induction and the case with f involving an instance of $\alpha^{\otimes}$.

If f involves an instance of $\alpha^{-\otimes}$, then we reuse the above argument by replacing $\alpha^{\otimes}$ with $\alpha^{-\otimes}$. □

Lemma 3.5.11. *Suppose $f\colon a \longrightarrow b$ is a prime edge such that*

- *a is not 0^X-reduced, and*
- *f involves an instance of $\xi^{+\oplus}$.*

Then f has a 0^X-reduction.

Proof. By assumption, $\mathsf{norm}(a) \geq 2$, and a contains at least one 0^X in its expression in terms of elements of X. First assume that f involves an instance of $\xi^{\oplus}$. We proceed by induction on $\mathsf{norm}(a)$.

If $\mathsf{norm}(a) = 2$, then

$$a = x \oplus y \xrightarrow{\;f = \xi^{\oplus}_{x,y}\;} y \oplus x = b$$

with $x, y \in X$, at least one of which is 0^X. Depending on whether each of x and y is 0^X, there are three cases corresponding to the following three diagrams.

$$\begin{array}{ccc} 0^X \oplus y & \xrightarrow{\xi^{\oplus}_{0^X,y}} & y \oplus 0^X \\ {\scriptstyle \lambda^{\oplus}_y}\downarrow & & \downarrow{\scriptstyle \rho^{\oplus}_y} \\ y & \xrightarrow{1_y} & y \end{array} \qquad \begin{array}{ccc} x \oplus 0^X & \xrightarrow{\xi^{\oplus}_{x,0^X}} & 0^X \oplus x \\ {\scriptstyle \rho^{\oplus}_x}\downarrow & & \downarrow{\scriptstyle \lambda^{\oplus}_x} \\ x & \xrightarrow{1_x} & x \end{array} \qquad \begin{array}{ccc} 0^X \oplus 0^X & \xrightarrow{\xi^{\oplus}_{0^X,0^X}} & 0^X \oplus 0^X \\ {\scriptstyle \rho^{\oplus}_{0^X}}\downarrow & & \downarrow{\scriptstyle \lambda^{\oplus}_{0^X}} \\ 0^X & \xrightarrow{1_{0^X}} & 0^X \end{array} \tag{3.5.12}$$

Each of these diagrams is commutative by the symmetry axiom (1.2.20) and the unit axiom (1.2.21) in the symmetric monoidal category $(\mathsf{C}, \oplus, \mathbb{0}, \alpha^{\oplus}, \lambda^{\oplus}, \rho^{\oplus}, \xi^{\oplus})$. In each case, an identity is a 0^X-reduction of f.

For the induction step, suppose $\mathsf{norm}(a) > 2$. Either

$$a = a_1 \oplus a_2 \quad \text{or} \quad a = a_1 \otimes a_2$$

for some $a_1, a_2 \in X^{\mathsf{fr}}$. We consider these two cases separately below.

First suppose a is a sum $a = a_1 \oplus a_2$, and $f = f_1 \oplus e$ splits as a sum for some

- prime edge $f_1 \colon a_1 \longrightarrow b_1$ involving an instance of $\xi^{\oplus}$ and
- identity $e \colon a_2 \longrightarrow a_2$.

We obtain a 0^X-reduction of f by reusing the two paragraphs containing (3.5.6) and (3.5.7). If f has the form $e \oplus f_1$, then we slightly modify the argument by switching the two summands in the top two rectangles in (3.5.7).

If f does not split as a sum, then $f = \xi^{\oplus}_{a_1,a_2}$. For each $w \in \{a_1, a_2\}$, suppose $Q_w \colon w \twoheadrightarrow w'$ is a 0^X-reduction of w. The diagram

$$\begin{array}{ccc} a = a_1 \oplus a_2 & \xrightarrow{f = \xi^{\oplus}_{a_1,a_2}} & a_2 \oplus a_1 = b \\ {\scriptstyle Q_{a_1} \oplus 1_{a_2}}\downarrow & & \downarrow{\scriptstyle 1_{a_2} \oplus Q_{a_1}} \\ a_1' \oplus a_2 & \xrightarrow{\xi^{\oplus}_{a_1',a_2}} & a_2 \oplus a_1' \\ {\scriptstyle 1_{a_1'} \oplus Q_{a_2}}\downarrow & & \downarrow{\scriptstyle Q_{a_2} \oplus 1_{a_1'}} \\ a' = a_1' \oplus a_2' & \xrightarrow{\xi^{\oplus}_{a_1',a_2'}} & a_2' \oplus a_1' = b' \end{array} \tag{3.5.13}$$

is commutative by the naturality of $\xi^{\oplus}$ in C. Each w' is 0^X-reduced, so it is either 0^X or contains no 0^X.

- If a_1' and a_2' contain no 0^X, then a' and b' also contain no 0^X, and are 0^X-reduced. The left and the right vertical paths are 0^X-reductions of a and b, respectively. Therefore, $\xi^{\oplus}_{a_1',a_2'}$ is a 0^X-reduction of f in this case.
- Otherwise, at least one of a_1' and a_2' is 0^X. We reuse the argument in (3.5.12) applied to $\xi^{\oplus}_{a_1',a_2'}$, and append the resulting diagram to the bottom of (3.5.13). In each case, the combined diagram is commutative, and shows that an identity is a 0^X-reduction of f.

So far we have proved the induction step when a is a sum.

Next suppose $a = a_1 \otimes a_2$. Then f, being a prime edge involving $\xi^{\oplus}$, must split as a product, say $f_1 \otimes e$ with

- $f_1 \colon a_1 \longrightarrow b_1$ a prime edge involving $\xi^{\oplus}$ and
- $e \colon a_2 \longrightarrow a_2$ an identity.

We reuse (3.5.6) and the argument in (3.5.7) by replacing $\oplus$ with $\otimes$ to conclude that f has a 0^X-reduction. If f has the form $e \otimes f_1$, then we modify the argument further by switching the two factors in the top two rectangles. This finishes the induction and the case with f involving an instance of $\xi^{\oplus}$.

If f involves an instance of $\xi^{-\oplus}$, then we reuse the above argument by replacing $\xi^{\oplus}$ with $\xi^{-\oplus}$. □

Lemma 3.5.14. *Suppose $f\colon a \longrightarrow b$ is a prime edge such that*

- *a is not 0^X-reduced, and*
- *f involves an instance of $\xi^{\pm\otimes}$.*

Then f has a 0^X-reduction.

Proof. First suppose that f involves an instance of $\xi^{\otimes}$. We proceed by induction on $\mathsf{norm}(a) \geq 2$ as in the proof of Lemma 3.5.11, most of which is reused here by replacing $\oplus$ with $\otimes$.

In the initial case of the induction with $\mathsf{norm}(a) = 2$ and $f = \xi^{\otimes}_{x,y}$, the diagrams in (3.5.12) become the following diagrams.

$$(3.5.15)\qquad \begin{array}{ccc} 0^X \otimes y & \xrightarrow{\xi^{\otimes}_{0^X,y}} & y \otimes 0^X \\ {\scriptstyle \lambda^{\bullet}_y}\downarrow & & \downarrow{\scriptstyle \rho^{\bullet}_y} \\ 0^X & \xrightarrow{1_{0^X}} & 0^X \end{array} \qquad \begin{array}{ccc} x \otimes 0^X & \xrightarrow{\xi^{\otimes}_{x,0^X}} & 0^X \otimes x \\ {\scriptstyle \rho^{\bullet}_x}\downarrow & & \downarrow{\scriptstyle \lambda^{\bullet}_x} \\ 0^X & \xrightarrow{1_{0^X}} & 0^X \end{array} \qquad \begin{array}{ccc} 0^X \otimes 0^X & \xrightarrow{\xi^{\otimes}_{0^X,0^X}} & 0^X \otimes 0^X \\ {\scriptstyle \rho^{\bullet}_{0^X}}\downarrow & & \downarrow{\scriptstyle \lambda^{\bullet}_{0^X}} \\ 0^X & \xrightarrow{1_{0^X}} & 0^X \end{array}$$

Each of these diagrams is commutative by the axiom (2.1.19), and also the symmetry axiom (1.2.20) in $(\mathsf{C}, \otimes)$ for the leftmost diagram. In each case, 1_{0^X} is a 0^X-reduction of f.

For the induction step, we first consider the case with $a = a_1 \otimes a_2$, and $f = f_1 \otimes e$ for some

- prime edge $f_1\colon a_1 \longrightarrow b_1$ involving an instance of $\xi^{\otimes}$ and
- identity $e\colon a_2 \longrightarrow a_2$.

We reuse the argument in the two paragraphs containing (3.5.6) and (3.5.7) by changing $\oplus$ to $\otimes$. The case with $f = e \otimes f_1$ is similar, with the two factors switched.

If f does not split as a product, then $f = \xi^{\otimes}_{a_1,a_2}$ for some $a_1, a_2 \in X^{\mathsf{fr}}$. We reuse the paragraph containing (3.5.13) by changing $\oplus$ to $\otimes$ and (3.5.12) to (3.5.15). So far we have proved the induction step when a is a product.

Next suppose $a = a_1 \oplus a_2$. Then f, being a prime edge involving $\xi^{\otimes}$, must split as a sum, say $f_1 \oplus e$ with f_1 and e as above. We reuse the argument in (3.5.7), and conclude that f has a 0^X-reduction. If f has the form $e \oplus f_1$, then we modify the argument by switching the two summands in the top two rectangles. This finishes the induction and the case with f involving an instance of $\xi^{\otimes}$.

If f involves an instance of $\xi^{-\otimes}$, then we reuse the above argument by replacing $\xi^{\otimes}$ with $\xi^{-\otimes}$. □

Lemma 3.5.16. *Suppose $f\colon a \longrightarrow b$ is a prime edge such that*

- *a is not 0^X-reduced, and*
- *f involves an instance of $\lambda^{\otimes}$.*

Then f has a 0^X-reduction.

Proof. By assumption, $\mathsf{norm}(a) \geq 2$, and a contains at least one of each of 0^X and 1^X in its expression in terms of elements of X. We proceed by induction on $\mathsf{norm}(a)$.

If $\mathsf{norm}(a) = 2$, then

$$a = 1^X \otimes 0^X \xrightarrow{f = \lambda^{\otimes}_{0^X}} 0^X = b.$$

The diagram

$$\begin{array}{ccc} 1^X \otimes 0^X & \xrightarrow{\lambda^{\otimes}_{0^X}} & 0^X \\ {\scriptstyle \rho^{\bullet}_{1^X}} \downarrow & & \downarrow {\scriptstyle 1_{0^X}} \\ 0^X & \xrightarrow{1_{0^X}} & 0^X \end{array} \tag{3.5.17}$$

is commutative by the axiom (2.1.18), which says that $\lambda^{\otimes}_{\mathbb{0}} = \rho^{\bullet}_{\mathbb{1}}$. Therefore, 1_{0^X} is a 0^X-reduction of f.

For the induction step, suppose $\mathsf{norm}(a) > 2$. Either

$$a = a_1 \oplus a_2 \quad \text{or} \quad a = a_1 \otimes a_2$$

for some $a_1, a_2 \in X^{\mathsf{fr}}$. We consider these two cases separately below.

First suppose $a = a_1 \oplus a_2$. Then f, being a prime edge with an instance of $\lambda^{\otimes}$, must split as a sum, say $f = f_1 \oplus e$ for some

- prime edge $f_1 : a_1 \longrightarrow b_1$ involving an instance of $\lambda^{\otimes}$ and
- identity $e : a_2 \longrightarrow a_2$.

We reuse the argument in the two paragraphs containing (3.5.6) and (3.5.7) to conclude that f has a 0^X-reduction. If f has the form $e \oplus f_1$, then we slightly modify the argument by switching the two summands in the top two rectangles in (3.5.7).

Next suppose $a = a_1 \otimes a_2$. If f splits as a product $f_1 \otimes e$ with f_1 and e as above, then we reuse the argument in the two paragraphs containing (3.5.6) and (3.5.7) by changing $\oplus$ to $\otimes$. If $f = e \otimes f_1$, then we modify the argument further by switching the two factors in the top two rectangles in (3.5.7).

If f does not split as a product, then

$$a = 1^X \otimes x \xrightarrow{f = \lambda^{\otimes}_x} x = b$$

with $x = a_2 \in X^{\mathsf{fr}}$ having at least one 0^X. Suppose $Q_x : x \twoheadrightarrow x'$ is any 0^X-reduction of x. The diagram

$$\begin{array}{ccc} a = 1^X \otimes x & \xrightarrow{f = \lambda^{\otimes}_x} & x = b \\ {\scriptstyle 1_{1^X} \otimes Q_x} \downarrow & & \downarrow {\scriptstyle Q_x} \\ 1^X \otimes x' & \xrightarrow{\lambda^{\otimes}_{x'}} & x' \end{array}$$

is commutative by the naturality of $\lambda^{\otimes}$ in C. The element x' is 0^X-reduced, so it is either 0^X or contains no 0^X.

- If x' contains no 0^X, then $1^X \otimes x'$ contains no 0^X, and is 0^X-reduced. Since the left vertical path is a 0^X-reduction of a, the 0^X-free path $\lambda^{\otimes}_{x'}$ is a 0^X-reduction of f in this case.
- If $x' = 0^X$, then we append the diagram (3.5.17) to the bottom of the previous diagram to conclude that 1_{0^X} is a 0^X-reduction of f.

This finishes the induction. □

Lemma 3.5.18. *Suppose $f\colon a \longrightarrow b$ is a prime edge such that*

- *a is not 0^X-reduced, and*
- *f involves an instance of $\lambda^{-\otimes}$.*

Then f has a 0^X-reduction.

Proof. By assumption, $\mathsf{norm}(a) \geq 2$, and a contains at least one 0^X in its expression in terms of elements of X. We proceed by induction on $\mathsf{norm}(a)$.

If $\mathsf{norm}(a) = 2$, then $a = a_1 \oplus a_2$ or $a_1 \otimes a_2$ for some $a_1, a_2 \in X$, at least one of which is 0^X. The prime edge f has one of the following six forms.

$$(3.5.19)\qquad \begin{array}{lll} a_1 \oplus a_2 \xrightarrow{\lambda^{-\otimes}_{a_1} \oplus 1_{a_2}} (1^X \otimes a_1) \oplus a_2 & & a_1 \otimes a_2 \xrightarrow{\lambda^{-\otimes}_{a_1} \otimes 1_{a_2}} (1^X \otimes a_1) \otimes a_2 \\ a_1 \oplus a_2 \xrightarrow{1_{a_1} \oplus \lambda^{-\otimes}_{a_2}} a_1 \oplus (1^X \otimes a_2) & & a_1 \otimes a_2 \xrightarrow{1_{a_1} \otimes \lambda^{-\otimes}_{a_2}} a_1 \otimes (1^X \otimes a_2) \\ a_1 \oplus a_2 \xrightarrow{\lambda^{-\otimes}_{a_1 \oplus a_2}} 1^X \otimes (a_1 \oplus a_2) & & a_1 \otimes a_2 \xrightarrow{\lambda^{-\otimes}_{a_1 \otimes a_2}} 1^X \otimes (a_1 \otimes a_2) \end{array}$$

First consider the upper left case in (3.5.19). There are two subcases as follows.

- Suppose $a_1 = 0^X$. The diagram

$$(3.5.20)\qquad \begin{array}{ccc} a = 0^X \oplus a_2 & \xrightarrow{\lambda^{-\otimes}_{0^X} \oplus 1_{a_2}} & (1^X \otimes 0^X) \oplus a_2 = b \\ {\scriptstyle 1_{0^X \oplus a_2}} \downarrow & & \downarrow {\scriptstyle \rho^{\bullet}_{1^X} \oplus 1_{a_2}} \\ 0^X \oplus a_2 & \xrightarrow{1_{0^X \oplus a_2}} & 0^X \oplus a_2 \\ {\scriptstyle \lambda^{\oplus}_{a_2}} \downarrow & & \downarrow {\scriptstyle \lambda^{\oplus}_{a_2}} \\ a_2 & \xrightarrow{1_{a_2}} & a_2 \end{array}$$

  is commutative by the axiom (2.1.18), which says that $\lambda^{\otimes}_{\mathbb{0}} = \rho^{\bullet}_{\mathbb{1}}$. The left and the right vertical paths are 0^X-reductions of a and b, respectively. Therefore, 1_{a_2} is a 0^X-reduction of f.
- Suppose $a_2 = 0^X \neq a_1$. The diagram

$$\begin{array}{ccc} a = a_1 \oplus 0^X & \xrightarrow{\lambda^{-\otimes}_{a_1} \oplus 1_{0^X}} & (1^X \otimes a_1) \oplus 0^X = b \\ {\scriptstyle \rho^{\oplus}_{a_1}} \downarrow & & \downarrow {\scriptstyle \rho^{\oplus}_{1^X \otimes a_1}} \\ a_1 & \xrightarrow{\lambda^{-\otimes}_{a_1}} & 1^X \otimes a_1 \end{array}$$

  is commutative by the naturality of $\rho^{\oplus}$ in C. The left and the right vertical edges are 0^X-reductions of a and b, respectively. Therefore, $\lambda^{-\otimes}_{a_1}$ is a 0^X-reduction of f.

The left middle case in (3.5.19) is proved by almost the same argument.

Consider the left bottom case in (3.5.19). There are three sub-cases as follows.

- If $a_1 = 0^X \neq a_2$, then the diagram

$$
\begin{array}{ccc}
0^X \oplus a_2 & \xrightarrow{\lambda^{-\otimes}_{0^X \oplus a_2}} & 1^X \otimes (0^X \oplus a_2) \\
{\scriptstyle \lambda^{\oplus}_{a_2}} \downarrow & & \downarrow {\scriptstyle 1_{1^X} \otimes \lambda^{\oplus}_{a_2}} \\
a_2 & \xrightarrow{\lambda^{-\otimes}_{a_2}} & 1^X \otimes a_2
\end{array}
\tag{3.5.21}
$$

  is commutative by the naturality of $\lambda^{\otimes}$ in C. Therefore, $\lambda^{-\otimes}_{a_2}$ is a 0^X-reduction of f.
- Similarly, if $a_1 \neq 0^X = a_2$, then the diagram

$$
\begin{array}{ccc}
a_1 \oplus 0^X & \xrightarrow{\lambda^{-\otimes}_{a_1 \oplus 0^X}} & 1^X \otimes (a_1 \oplus 0^X) \\
{\scriptstyle \rho^{\oplus}_{a_1}} \downarrow & & \downarrow {\scriptstyle 1_{1^X} \otimes \rho^{\oplus}_{a_1}} \\
a_1 & \xrightarrow{\lambda^{-\otimes}_{a_1}} & 1^X \otimes a_1
\end{array}
\tag{3.5.22}
$$

  is commutative by the naturality of $\lambda^{\otimes}$ in C. It shows that $\lambda^{-\otimes}_{a_1}$ is a 0^X-reduction of f.
- If $a_1 = a_2 = 0^X$, then the diagram

$$
\begin{array}{ccc}
0^X \oplus 0^X & \xrightarrow{\lambda^{-\otimes}_{0^X \oplus 0^X}} & 1^X \otimes (0^X \oplus 0^X) \\
{\scriptstyle \lambda^{\oplus}_{0^X}} \downarrow & & \downarrow {\scriptstyle 1_{1^X} \otimes \lambda^{\oplus}_{0^X}} \\
0^X & \xrightarrow{\lambda^{-\otimes}_{0^X}} & 1^X \otimes 0^X \\
{\scriptstyle 1_{0^X}} \downarrow & & \downarrow {\scriptstyle \rho^{\bullet}_{1^X}} \\
0^X & \xrightarrow{1_{0^X}} & 0^X
\end{array}
\tag{3.5.23}
$$

  is commutative by the naturality of $\lambda^{\otimes}$ for the top rectangle, and the axiom (2.1.18) in C for the bottom rectangle. Therefore, 1_{0^X} is a 0^X-reduction of f.

The three right cases in (3.5.19) are proved by almost the same argument. This finishes the proof for the initial case.

For the induction step, suppose $\mathsf{norm}(a) > 2$. Then $a = a_1 \oplus a_2$ or $a_1 \otimes a_2$ for some $a_1, a_2 \in X^{\mathsf{fr}}$, at least one of which contains 0^X. The prime edge f has one of the six forms in (3.5.19).

First consider the upper left case in (3.5.19). There are two subcases as follows.

- Suppose a_1 contains 0^X. If, furthermore, a_1 is 0^X-reduced, then $a_1 = 0^X$. The diagram

$$\begin{CD}
a = 0^X \oplus a_2 @>{\lambda^{-\otimes}_{0^X} \oplus 1_{a_2}}>> (1^X \otimes 0^X) \oplus a_2 = b\\
@V{1_{0^X \oplus a_2}}VV @VV{\rho^{\bullet}_{1^X} \oplus 1_{a_2}}V\\
0^X \oplus a_2 @>{1_{0^X \oplus a_2}}>> 0^X \oplus a_2\\
@V{Q}VV @VV{Q}V\\
a' @>{1_{a'}}>> a'
\end{CD}$$

  is commutative by the axiom (2.1.18), where Q is any 0^X-reduction of $0^X \oplus a_2$. Since the left and the right vertical paths are 0^X-reductions of a and b, respectively, $1_{a'}$ is a 0^X-reduction of f.

  Next suppose a_1 is not 0^X-reduced. Since $\mathsf{norm}(a_1) < \mathsf{norm}(a)$, the induction hypothesis implies that the prime edge $\lambda^{-\otimes}_{a_1}$ has a 0^X-reduction $R_1 \colon a_1' \twoheadrightarrow b_1'$. The diagram

(3.5.24)
$$\begin{CD}
a_1 @>{\lambda^{-\otimes}_{a_1}}>> 1^X \otimes a_1\\
@V{Q_{a_1}}VV @VV{Q_{b_1}}V\\
a_1' @>{R_1}>> b_1'
\end{CD}$$

  is commutative, with Q_{a_1} and Q_{b_1} any 0^X-reductions of, respectively, a_1 and $b_1 = 1^X \otimes a_1$. With $f = \lambda^{-\otimes}_{a_1} \oplus 1_{a_2}$, we now reuse the argument in the paragraph containing (3.5.7) to conclude that f has a 0^X-reduction.
- Suppose a_1 does not contain 0^X. Since a_1 is 0^X-reduced, the prime edge $\lambda^{-\otimes}_{a_1}$ has a 0^X-reduction $R_1 \colon a_1' \twoheadrightarrow b_1'$ by Lemma 3.5.2 (whose proof shows that $R_1 = \lambda^{-\otimes}_{a_1}$). The diagram (3.5.24) exists in this case, so again the argument in the paragraph containing (3.5.7) shows that f has a 0^X-reduction.

The left middle case in (3.5.19) is proved by almost the same argument.

For the left bottom case in (3.5.19) consider the following diagram, where $Q_i \colon a_i \twoheadrightarrow a_i'$ is a 0^X-reduction of a_i for $i = 1, 2$.

(3.5.25)
$$\begin{CD}
a = a_1 \oplus a_2 @>{\lambda^{-\otimes}_{a_1 \oplus a_2}}>> 1^X \otimes (a_1 \oplus a_2) = b\\
@V{Q_1 \oplus 1_{a_2}}VV @VV{1_{1^X} \otimes (Q_1 \oplus 1_{a_2})}V\\
a_1' \oplus a_2 @>{\lambda^{-\otimes}_{a_1' \oplus a_2}}>> 1^X \otimes (a_1' \oplus a_2)\\
@V{1_{a_1'} \oplus Q_2}VV @VV{1_{1^X} \otimes (1_{a_1'} \oplus Q_2)}V\\
a_1' \oplus a_2' @>{\lambda^{-\otimes}_{a_1' \oplus a_2'}}>> 1^X \otimes (a_1' \oplus a_2')
\end{CD}$$

This diagram is commutative by the naturality of $\lambda^{\otimes}$ in C. Each a_i' is 0^X-reduced, so it is either 0^X or contains no 0^X. There are four subcases as follows.

- If both a_1' and a_2' contain no 0^X, then $a_1' \oplus a_2'$ and $1^X \otimes (a_1' \oplus a_2')$ contain no 0^X, and are 0^X-reduced. Since the left and the right vertical paths are 0^X-reductions of a and b, respectively, the 0^X-free path $\lambda^{-\otimes}_{a_1' \oplus a_2'}$ is a 0^X-reduction of f.
- If $a_1' = 0^X$, and if a_2' contains no 0^X, then we append the diagram (3.5.21) with a_2' in place of a_2 to the bottom of the diagram (3.5.25). The combined diagram shows that $\lambda^{-\otimes}_{a_2'}$ is a 0^X-reduction of f.
- If a_1' contains no 0^X, and if $a_2' = 0^X$, then we append the diagram (3.5.22) with a_1' in place of a_1 to the bottom of the diagram (3.5.25). The combined diagram shows that $\lambda^{-\otimes}_{a_1'}$ is a 0^X-reduction of f.
- If $a_1' = a_2' = 0^X$, then we append the diagram (3.5.23) to the bottom of the diagram (3.5.25). The combined diagram shows that 1_{0^X} is a 0^X-reduction of f.

The three right cases in (3.5.19) are proved by almost the same argument.

This finishes the induction. □

Lemma 3.5.26. *Suppose $f: a \longrightarrow b$ is a prime edge such that*

- *a is not 0^X-reduced, and*
- *f involves an instance of $\rho^{\pm\otimes}$.*

Then f has a 0^X-reduction.

Proof. The proofs are obtained from those of Lemmas 3.5.16 and 3.5.18 by replacing $(\lambda^{\otimes}, \lambda^{-\otimes})$ with $(\rho^{\otimes}, \rho^{-\otimes})$, and the axiom (2.1.18) by the axiom (2.1.17), which states that $\rho^{\otimes}_{\mathbb{0}} = \lambda^{\bullet}_{\mathbb{1}}$. □

Lemma 3.5.27. *Suppose $f: a \longrightarrow b$ is a prime edge such that*

- *a is not 0^X-reduced, and*
- *f involves an instance of δ^l or δ^r.*

Then f has a 0^X-reduction.

Proof. By assumption, $\mathsf{norm}(a) \geq 3$, and a contains at least one 0^X in its expression in terms of elements of X. First assume that f involves an instance of δ^l. We proceed by induction on $\mathsf{norm}(a)$. For the rest of this proof, we abbreviate $\otimes$ to concatenation.

If $\mathsf{norm}(a) = 3$, then

$$a = x(y \oplus z) \xrightarrow{f = \delta^l_{x,y,z}} xy \oplus xz = b \tag{3.5.28}$$

for some $x, y, z \in X$, at least one of which is 0^X. Depending on whether each of x, y, and z is 0^X, there are seven cases as follows.

(1) If $y = 0^X$, then the diagram

(3.5.29)
$$\begin{array}{ccc}
a = x(0^X \oplus z) & \xrightarrow{f = \delta^l_{x,0^X,z}} & x0^X \oplus xz = b \\
& & \Big\downarrow \rho^\bullet_x \oplus 1_{xz} \\
\Big\downarrow 1_x \lambda^\oplus_z & & 0^X \oplus xz \\
& & \Big\downarrow \lambda^\oplus_{xz} \\
xz & \xrightarrow{1_{xz}} & xz
\end{array}$$

is commutative by the axiom (2.1.23).

(i) If $x, z \neq 0^X$, then xz is 0^X-reduced. Therefore, 1_{xz} is a 0^X-reduction of f in this case.

(ii) If either $x = 0^X$ or $z = 0^X$, then we append the diagram

$$\begin{array}{ccc}
0^X z & \xrightarrow{1_{0^X z}} & 0^X z \\
\lambda^\bullet_z \Big\downarrow & & \Big\downarrow \lambda^\bullet_z \\
0^X & \xrightarrow{1_{0^X}} & 0^X
\end{array}
\quad \text{or} \quad
\begin{array}{ccc}
x0^X & \xrightarrow{1_{x0^X}} & x0^X \\
\rho^\bullet_x \Big\downarrow & & \Big\downarrow \rho^\bullet_x \\
0^X & \xrightarrow{1_{0^X}} & 0^X
\end{array}$$

to the bottom of the diagram (3.5.29). In each case, the combined diagram is commutative, and shows that 1_{0^X} is a 0^X-reduction of f. If $x = z = 0^X$, then we can use either one of these two diagrams.

(2) If $y \neq 0^X$, then at least one of x and z is 0^X.

(i) If $z = 0^X$, then the diagram

(3.5.30)
$$\begin{array}{ccc}
a = x(y \oplus 0^X) & \xrightarrow{f = \delta^l_{x,y,0^X}} & xy \oplus x0^X = b \\
& & \Big\downarrow 1_{xy} \oplus \rho^\bullet_x \\
\Big\downarrow 1_x \rho^\oplus_y & & xy \oplus 0^X \\
& & \Big\downarrow \rho^\oplus_{xy} \\
xy & \xrightarrow{1_{xy}} & xy
\end{array}$$

is commutative by the axiom (2.1.25).

- If $x \neq 0^X$, then xy is 0^X-reduced. The diagram (3.5.30) shows that 1_{xy} is a 0^X-reduction of f.
- If $x = 0^X$, then we append the diagram

$$\begin{array}{ccc}
0^X y & \xrightarrow{1_{0^X y}} & 0^X y \\
\lambda^\bullet_y \Big\downarrow & & \Big\downarrow \lambda^\bullet_y \\
0^X & \xrightarrow{1_{0^X}} & 0^X
\end{array}$$

to the bottom of the diagram (3.5.30). The combined diagram is commutative, and shows that 1_{0^X} is a 0^X-reduction of f.

(ii) If $z \neq 0^X$, then $x = 0^X$. The diagram

$$\begin{array}{ccc}
0^X(y \oplus z) & \xrightarrow{\;f = \delta^l_{0^X,y,z}\;} & 0^X y \oplus 0^X z \\
 & & \Big\downarrow {\scriptstyle \lambda^{\bullet}_y \oplus 1_{0^X z}} \\
 & & 0^X \oplus 0^X z \\
{\scriptstyle \lambda^{\bullet}_{y\oplus z}} \Big\downarrow & & \Big\downarrow {\scriptstyle 1_{0^X} \oplus \lambda^{\bullet}_z} \\
 & & 0^X \oplus 0^X \\
 & & \Big\downarrow {\scriptstyle \lambda^{\oplus}_{0^X}} \\
0^X & \xrightarrow{\;1_{0^X}\;} & 0^X
\end{array}$$

is commutative by the axiom (2.1.15) and the functoriality of $\oplus$ in C. Therefore, 1_{0^X} is a 0^X-reduction of f.

This finishes the proof of the initial case.

For the induction step, suppose $\mathsf{norm}(a) > 3$. Either $a = a_1 \oplus a_2$ or $a = a_1 \otimes a_2$ for some $a_1, a_2 \in X^{\mathsf{fr}}$, at least one of which contains 0^X. We consider these cases separately below.

If $a = a_1 \oplus a_2$, then f, being a prime edge involving an instance of δ^l, must split as a sum, say $f = f_1 \oplus e$ for some

- prime edge $f_1 : a_1 \longrightarrow b_1$ involving an instance of δ^l and
- identity $e : a_2 \longrightarrow a_2$.

We reuse the argument in the two paragraphs containing (3.5.6) and (3.5.7) to conclude that f has a 0^X-reduction. If f has the form $e \oplus f_1$, then we slightly modify the argument by switching the two summands in the top two rectangles in (3.5.7).

Next suppose $a = a_1 \otimes a_2$. If $f = f_1 \otimes e$ with f_1 and e as above, then we reuse the argument in the two paragraphs containing (3.5.6) and (3.5.7) by changing $\oplus$ to $\otimes$. If $f = e \otimes f_1$, then we modify the argument further by switching the two factors in the top two rectangles in (3.5.7).

If f does not split as a product, then $f = \delta^l_{x,y,z}$ as in (3.5.28) for some $x, y, z \in X^{\mathsf{fr}}$, at least one of which contains 0^X. For each $w \in \{x, y, z\}$, suppose $Q_w : w \twoheadrightarrow w'$ is a 0^X-reduction of w. Consider the following diagram.

$$\begin{array}{ccc}
a = x(y \oplus z) & \xrightarrow{\;f = \delta^l_{x,y,z}\;} & xy \oplus xz = b \\
 & & \Big\downarrow {\scriptstyle Q_x 1_y \oplus 1_{xz}} \\
{\scriptstyle Q_x 1_{y\oplus z}} \Big\downarrow & & x'y \oplus xz \\
 & & \Big\downarrow {\scriptstyle 1_{x'y} \oplus Q_x 1_z} \\
x'(y \oplus z) & \xrightarrow{\;\delta^l_{x',y,z}\;} & x'y \oplus x'z \\
{\scriptstyle 1'_x(Q_y \oplus 1_z)} \Big\downarrow & & \Big\downarrow {\scriptstyle 1_{x'} Q_y \oplus 1_{x'z}} \\
x'(y' \oplus z) & \xrightarrow{\;\delta^l_{x',y',z}\;} & x'y' \oplus x'z \\
{\scriptstyle 1_{x'}(1_{y'} \oplus Q_z)} \Big\downarrow & & \Big\downarrow {\scriptstyle 1_{x'y'} \oplus 1_{x'} Q_z} \\
x'(y' \oplus z') & \xrightarrow{\;\delta^l_{x',y',z'}\;} & x'y' \oplus x'z'
\end{array} \tag{3.5.31}$$

This diagram is commutative by the naturality of δ^l in C. Each w' is 0^X-reduced, so it is either 0^X or contains no 0^X.

- If x', y', and z' all contain no 0^X, then $x'(y' \oplus z')$ and $x'y' \oplus x'z'$ also contain no 0^X, and are 0^X-reduced. Since the left and the right vertical paths in (3.5.31) are 0^X-reductions of a and b, respectively, the 0^X-free path $\delta^l_{x',y',z'}$ is a 0^X-reduction of f in this case.
- Otherwise, at least one of x', y', and z' is 0^X. In each of the seven sub-cases, we (i) reuse the argument for the initial case above applied to $\delta^l_{x',y',z'}$, and (ii) append the resulting diagram to the bottom of (3.5.31). For each $w \in \{x, y, z\}$, the condition "$w \neq 0^X$" is replaced by "w' contains no 0^X." In each case, we conclude that an identity is a 0^X-reduction of f.

This finishes the induction and the case with f containing an instance of δ^l.

The proof for the case with f containing an instance of δ^r is obtained from the above proof by replacing

- $\delta^l_{x,y,z}$ with $\delta^r_{y,z,x}\colon (y \oplus z)x \longrightarrow yx \oplus zx$ and
- the axioms (2.1.23), (2.1.25), and (2.1.15) in C with, respectively, the axioms (2.1.24), (2.1.26), and (2.1.16).

This finishes the proof. □

Now we combine the preliminary lemmas above to prove the following main result of this section.

Proposition 3.5.32. *Each path in* $\mathsf{Gr}(X)$ *has a* 0^X*-reduction.*

Proof. Suppose $P = (f_n, \ldots, f_1)$ is a path in $\mathsf{Gr}(X)$ for some $n \geq 1$ as displayed below.

$$a_0 \xrightarrow{\;f_1\;} a_1 \xrightarrow{\;f_2\;} \cdots \xrightarrow{\;f_n\;} a_n$$

- For each $0 \leq j \leq n$, suppose $Q_j\colon a_j \twoheadrightarrow a'_j$ is any 0^X-reduction of a_j, which exists by Lemma 3.3.6. Moreover, a'_j is uniquely determined by a_j by Lemma 3.3.11.
- For each $1 \leq i \leq n$, the prime edge $f_i\colon a_{i-1} \longrightarrow a_i$ has a 0^X-reduction $R_i\colon a'_{i-1} \twoheadrightarrow a'_i$ by one of Lemmas 3.5.1 through 3.5.3, 3.5.9, 3.5.11, 3.5.14, 3.5.16, 3.5.18, 3.5.26, and 3.5.27.

Each square in the diagram

$$\begin{array}{ccccccc} a_0 & \xrightarrow{\;f_1\;} & a_1 & \xrightarrow{\;f_2\;} & \cdots & \xrightarrow{\;f_n\;} & a_n \\ {\scriptstyle Q_0}\downarrow & & \downarrow{\scriptstyle Q_1} & & & & \downarrow{\scriptstyle Q_n} \\ a'_0 & \xrightarrow{\;R_1\;} & a'_1 & \xrightarrow{\;R_2\;} & \cdots & \xrightarrow{\;R_n\;} & a'_n \end{array}$$

is commutative by Definition 3.4.5. Therefore, the 0^X-free path

$$a'_0 \xrightarrow{\;(R_n,\ldots,R_1)\;} a'_n$$

is a 0^X-reduction of P by Lemma 3.4.8. □

Recall from (3.1.24) the support of an element in X^{fr}. Next is a preliminary form of the Coherence Theorem 3.9.1.

Proposition 3.5.33. *Suppose*

$$a \underset{P_2}{\overset{P_1}{\rightrightarrows}} b$$

are two paths in $\mathsf{Gr}(X)$ *such that*

$$\mathsf{supp}(a) = \mathsf{supp}(0^X).$$

Then the values of P_1 *and* P_2 *in* C *are equal.*

Proof. By Lemma 3.1.29 (2),

$$\mathsf{supp}(a) = \mathsf{supp}(b) = \mathsf{supp}(0^X).$$

By Lemmas 3.3.6 and 3.3.7 and Proposition 3.5.32, there exist 0^X-reductions

- $Q_a: a \twoheadrightarrow 0^X$ of a,
- $Q_b: b \twoheadrightarrow 0^X$ of b, and
- $R_i: 0^X \twoheadrightarrow 0^X$ of P_i for each $i = 1, 2$.

For each $i = 1, 2$, the diagram

$$\begin{array}{ccc} a & \overset{P_i}{\twoheadrightarrow} & b \\ {\scriptstyle Q_a}\downarrow & & \downarrow{\scriptstyle Q_b} \\ 0^X & \underset{R_i}{\twoheadrightarrow} & 0^X \end{array}$$

is commutative by Definition 3.4.5. By Lemma 3.4.11, the value of R_i in C is the identity morphism $1_{\mathbb{0}}$. Therefore, the value of P_i in C is the value of Q_a in C followed by the inverse of the value of Q_b in C. □

3.6. Reduction of Distributivity

The main observation in this and the next sections is Proposition 3.7.19. It says that each path whose vertices do not involve 0^X can be replaced by a 0^X-free path that does not involve the distributivity edges δ^l and δ^r. This will constitute another reduction step in the Coherence Theorem 3.9.1, where we eliminate δ^l and δ^r. In this section, we first define the preliminary concept of a δ-reduction of an element in X^{fr}, and show its existence in Lemma 3.6.9. The main technical result of this section is Lemma 3.6.12, which says that, under some conditions, a 0^X-free edge may be replaced by a 0^X-free path that contains no δ^l and δ^r. This observation is used repeatedly in the proof of Proposition 3.7.19.

Some of the definitions and results in this section are distributivity analogues of those in Sections 3.3 through 3.5. Recall that Convention 3.3.1 is in effect.

Motivation 3.6.1. The next definition contains the distributivity analogues of 0^X-prime edges, 0^X-reduced, and 0^X-reductions in Definitions 3.3.2 and 3.3.4. We observed in Lemma 3.3.5 that an element in X^{fr} is 0^X-reduced if and only if it is not the domain of any 0^X-prime edges. The distributivity analogue of a 0^X-reduced element in X^{fr}, which is called δ-reduced, is modeled after this lemma. In Lemma 3.6.5, we will provide a more intrinsic characterization of δ-reduced elements. ◇

Definition 3.6.2. Consider the graph $\mathsf{Gr}(X)$ of X in Definition 3.1.9.

- A *δ-prime edge* is a prime edge that involves either δ^l or δ^r.
- An element $a \in X^{\mathsf{fr}}$ is *δ-reduced* if it is not the domain of any δ-prime edges.
- A *δ-reduction* of $a \in X^{\mathsf{fr}}$ is a path $P: a \twoheadrightarrow b$ such that the following two statements hold.
  - b is δ-reduced.
  - Each edge in P is either an identity or a δ-prime edge. ◇

To characterize δ-reduced elements, we use the following concepts.

Definition 3.6.3. Suppose $a \in X^{\mathsf{fr}}$.

- The element a is a *monomial* if either
  - $a \in X$, or
  - $a = a_1 \otimes \cdots \otimes a_k$ for some multiplicative bracketing and some elements $a_j \in X$ for $1 \leq j \leq k$ with $k \geq 2$. Each a_j is called a *factor* of the monomial a.
- The element a is a *polynomial* if either
  - a is a monomial, or
  - $a = a^1 \oplus \cdots \oplus a^m$ for some additive bracketing and some monomials $a^i \in X^{\mathsf{fr}}$ for $1 \leq i \leq m$ with $m \geq 2$. Each a^i is called a monomial in the polynomial a. ◇

Explanation 3.6.4. A polynomial is a nonempty sum in which each summand is either an element in X or a product of several elements in X. ◇

Lemma 3.6.5. *For an element $a \in X^{\mathsf{fr}}$, the following statements are equivalent.*

(1) *a is a polynomial*
(2) *$1_a: a \longrightarrow a$ is a δ-reduction of a.*
(3) *a is δ-reduced.*

Proof. First observe that the domain of a δ-prime edge must involve an expression of the form

- $x \otimes (y \oplus z)$ for δ^l or
- $(y \oplus z) \otimes x$ for δ^r

for some $x, y, z \in X^{\mathsf{fr}}$. A polynomial does not contain these expressions, so it is δ-reduced. Therefore, if a is a polynomial, then the identity 1_a is a δ-reduction of a, proving (1) $\Rightarrow$ (2).

If 1_a is a δ-reduction of a, then its codomain a is δ-reduced by definition, proving (2) $\Rightarrow$ (3).

For (3) $\Rightarrow$ (1), suppose a is δ-reduced. A general element $a \in X^{\mathsf{fr}}$ can always be expressed in the form

$$a = a^1 \oplus \cdots \oplus a^m$$

for some additive bracketing and $m \geq 1$ such that each a^i is not a sum. In other words, each a^i is either in X or has the form $a_1^i \otimes a_2^i$. Since a is δ-reduced, expressions of the forms $x \otimes (y \oplus z)$ and $(y \oplus z) \otimes x$ do not appear in a. So each a^i is a monomial, and a is a polynomial. □

Example 3.6.6. For elements $a_1, \ldots, a_n \in X$ for $n \geq 1$, the elements

$$a_1 \oplus \cdots \oplus a_n \quad \text{and} \quad a_1 \otimes \cdots \otimes a_n \in X^{\mathsf{fr}}$$

with any additive, respectively multiplicative, bracketing are δ-reduced. ◇

Example 3.6.7. In Definition 2.1.2 of a symmetric bimonoidal category, consider the 24 axioms, and interpret A, B, C, and D as elements in X, $\mathbb{0}$ as 0^X, and $\mathbb{1}$ as 1^X.

- In each of (2.1.5)–(2.1.13), (2.1.15)–(2.1.16), and (2.1.23)–(2.1.28), the common domain is *not* δ-reduced, and the common codomain is δ-reduced.
- In each of (2.1.14) and (2.1.17)–(2.1.22), each vertex is δ-reduced. ◇

Example 3.6.8. Consider elements $a, b, c, d \in X$.

- $\delta^l_{a,b,c}\colon a(b \oplus c) \longrightarrow ab \oplus ac$ is a δ-reduction of $a(b \oplus c)$.
- $\delta^r_{a,b,c}\colon (a \oplus b)c \longrightarrow ac \oplus bc$ is a δ-reduction of $(a \oplus b)c$.
- The following two paths are two δ-reductions of $(a \oplus b)(c \oplus d)$.

$$\begin{array}{ccc}
 & (a\oplus b)(c\oplus d) & \\
\swarrow {\scriptstyle \delta^l_{a\oplus b,c,d}} & & {\scriptstyle \delta^r_{a,b,c\oplus d}} \searrow \\
(a\oplus b)c \oplus (a\oplus b)d & & a(c\oplus d)\oplus b(c\oplus d) \\
\downarrow {\scriptstyle \delta^r_{a,b,c}\oplus 1_{(a\oplus b)d}} & & \downarrow {\scriptstyle \delta^l_{a,c,d}\oplus 1_{b(c\oplus d)}} \\
(ac\oplus bc)\oplus(a\oplus b)d & & (ac\oplus ad)\oplus b(c\oplus d) \\
\downarrow {\scriptstyle 1_{ac\oplus bc}\oplus \delta^r_{a,b,d}} & & \downarrow {\scriptstyle 1_{ac\oplus ad}\oplus \delta^l_{b,c,d}} \\
(ac\oplus bc)\oplus(ad\oplus bd) & & (ac\oplus ad)\oplus(bc\oplus bd)
\end{array}$$

This example shows that the codomain of a δ-reduction of a given element in X^{fr} is *not* unique in general. This is different from the situation for 0^X-reductions, where the codomain is uniquely determined, as shown in Lemma 3.3.11. ◇

In Lemma 3.3.6, we observed that each element in X^{fr} has a 0^X-reduction. Next is the distributivity analogue of that assertion. Recall rank and size in Definition 3.2.1.

Lemma 3.6.9. *Each element in X^{fr} has a δ-reduction.*

Proof. We proceed by induction on $\mathsf{size}(a) - \mathsf{rank}(a)$ for elements $a \in X^{\mathsf{fr}}$. If

$$\mathsf{size}(a) - \mathsf{rank}(a) = 0,$$

then by Proposition 3.2.15, a is a polynomial in the sense of Definition 3.6.3, in which each monomial is either an element in X or a product of two elements in X. Therefore, by Lemma 3.6.5, the identity 1_a is a δ-reduction of a.

For the induction step, suppose $\mathsf{size}(a) - \mathsf{rank}(a) > 0$. If a is δ-reduced, then by Lemma 3.6.5 again, 1_a is a δ-reduction of a.

If a is not δ-reduced, then there exists a δ-prime edge $f\colon a \longrightarrow b$ with domain a. The induction hypothesis applies to b by Lemma 3.2.12, so there exists a δ-reduction $P\colon b \longrightarrow\!\!\!\!\rightarrow b'$ of b. The combined path

$$a \xrightarrow{\ f\ } b \xrightarrow{\ P\ }\!\!\!\!\rightarrow b'$$

is a δ-reduction of a. This finishes the induction. □

The next definition contains the distributivity analogue of a 0^X-free path in Definition 3.4.1.

Definition 3.6.10. Consider $\mathsf{Gr}(X)$.

- A *δ-free path* is a path that does not contain any δ-prime edges.
- A *δ-free edge* is a δ-free path of length 1.
- A *$(0^X,\delta)$-free path* is a path that is both a 0^X-free path and a δ-free path.
- A *$(0^X,\delta)$-free edge* is a $(0^X,\delta)$-free path of length 1. $\diamond$

Explanation 3.6.11. In a δ-free path, every prime edge is either an identity or involves a single instance of $\alpha^{\pm\oplus}$, $\xi^{\pm\oplus}$, $\lambda^{\pm\oplus}$, $\rho^{\pm\oplus}$, $\alpha^{\pm\otimes}$, $\xi^{\pm\otimes}$, $\lambda^{\pm\otimes}$, $\rho^{\pm\otimes}$, $\lambda^{\pm\bullet}$, or $\rho^{\pm\bullet}$.

In a $(0^X,\delta)$-free path, every prime edge is either an identity or involves a single instance of $\alpha^{\pm\oplus}$, $\xi^{\pm\oplus}$, $\alpha^{\pm\otimes}$, $\xi^{\pm\otimes}$, $\lambda^{\pm\otimes}$, or $\rho^{\pm\otimes}$. $\diamond$

The goal for the rest of this section is to prove the following preliminary version of Proposition 3.7.19. Recall from Definition 3.4.1 that a 0^X-free edge is a prime edge that does not involve $\lambda^{\oplus}$, $\rho^{\oplus}$, $\lambda^{\bullet}$, $\rho^{\bullet}$, and their formal inverses.

Lemma 3.6.12. *In $\mathsf{Gr}(X)$, suppose*

- *$f\colon a \longrightarrow b$ is a 0^X-free edge, and*
- *$g\colon a \longrightarrow c$ is a δ-prime edge.*

Then there exists a diagram

(3.6.13)

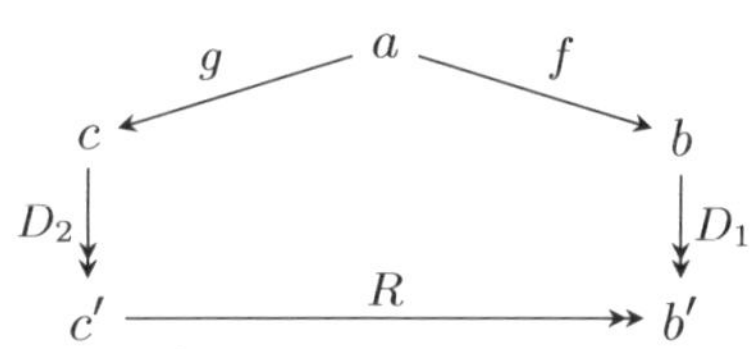

such that the following statements hold.

1. *The above diagram is commutative.*
2. *Each path D_i is either a single identity or consists of δ-prime edges.*
3. *R is a $(0^X,\delta)$-free path. Moreover, R involves only instances of $\lambda^{\pm\otimes}$ or $\rho^{\pm\otimes}$ if and only if f involves an elementary edge of the same type.*

Proof. Recall the possibilities of a 0^X-free edge in Explanation 3.4.2.

- If f is an identity, then $D_1 = g$, and each of D_2 and R is a single identity.
- If $f = g$, then each of D_1, D_2, and R is a single identity.

In the rest of this proof, we may assume that f is not an identity, and that $f \neq g$.

The proof proceeds by induction on $\mathsf{norm}(a) \geq 3$. If $\mathsf{norm}(a) = 3$, then g is either

$$a = x(y \oplus z) \xrightarrow{\delta^l_{x,y,z}} xy \oplus xz = c \quad \text{or} \quad a = (y \oplus z)x \xrightarrow{\delta^r_{y,z,x}} yx \oplus zx = c$$

for some $x,y,z \in X$. First suppose $g = \delta^l_{x,y,z}$. Depending on what f is, there are the following subcases. In what follows, *functoriality* means the functoriality of $\oplus$, $\otimes$, or both in C.

(1) If $f = \lambda^{\otimes}_{y \oplus z}$ with $x = 1^X$, then D_1, D_2, and R in the diagram (3.6.13) are defined as follows.

(3.6.14)

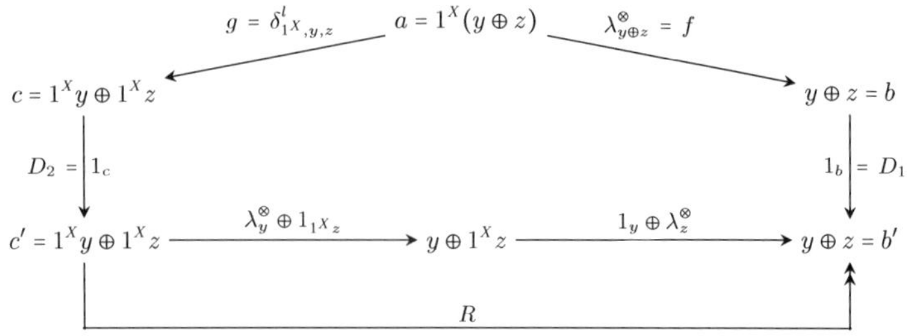

This diagram is commutative by the axiom (2.1.27) and functoriality.

(2) If f involves $\lambda^{-\otimes}$, then there are five subcases depending on whether $\lambda^{-\otimes}$ is applied to a, x, y, z, or $y \oplus z$.

(i) If $f = \lambda^{-\otimes}_a$, then D_1, D_2, and R in (3.6.13) are defined as follows.

(3.6.15)

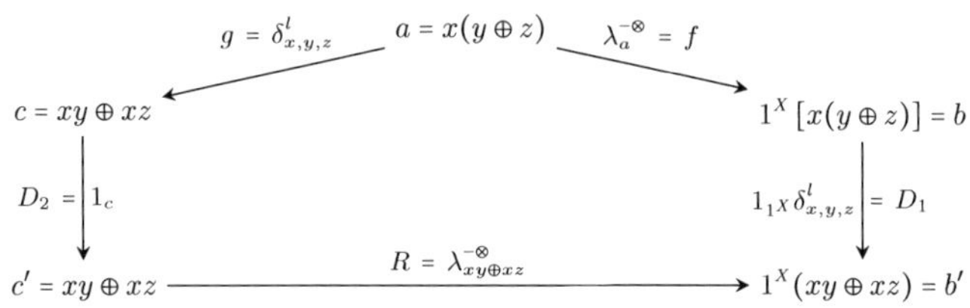

This diagram is commutative by the naturality of $\lambda^{\otimes}$ in C.

(ii) If f involves $\lambda^{-\otimes}_x$, then we consider the following diagram.

(3.6.16)

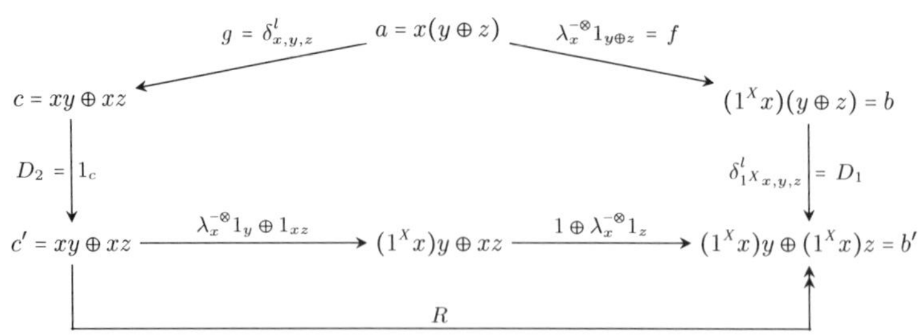

In the lower right horizontal edge, 1 means $1_{(1^X x)y}$. This diagram is commutative by the naturality of δ^l and functoriality.

(iii) If f involves $\lambda_y^{-\otimes}$ or $\lambda_z^{-\otimes}$, then there are two diagrams similar to the one in case (ii) that are commutative for the same reasons.

(iv) If f involves $\lambda_{y\oplus z}^{-\otimes}$, then we consider the following diagram.

(3.6.17)

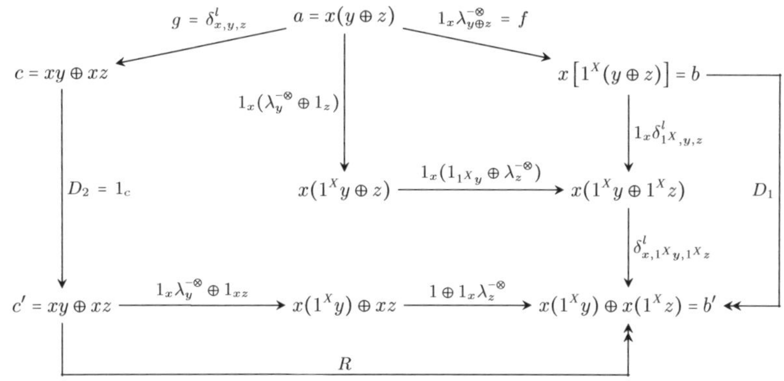

In the lower right horizontal edge, 1 means $1_{x(1^X y)}$. The upper right trapezoid is commutative by the axiom (2.1.27) and functoriality. The left subdiagram is commutative by the naturality of δ^l and functoriality.

(3) If f involves $\rho^{-\otimes}$, then as in case (2), there are five further subcases depending on whether $\rho^{-\otimes}$ is applied to a, x, y, z, or $y \oplus z$. The proofs for these five subcases are obtained from those in case (2) by
- replacing $\lambda^{-\otimes}$ with $\rho^{-\otimes}$ throughout and
- replacing $\delta^l_{1^X,y,z}$ and (2.1.27) with, respectively, $\delta^r_{y,z,1^X}$ and (2.1.28) in the upper right trapezoid in the diagram (3.6.17).

(4) If $f = \xi^{\otimes}_{x,y\oplus z}$, then we consider the following diagram.

(3.6.18)

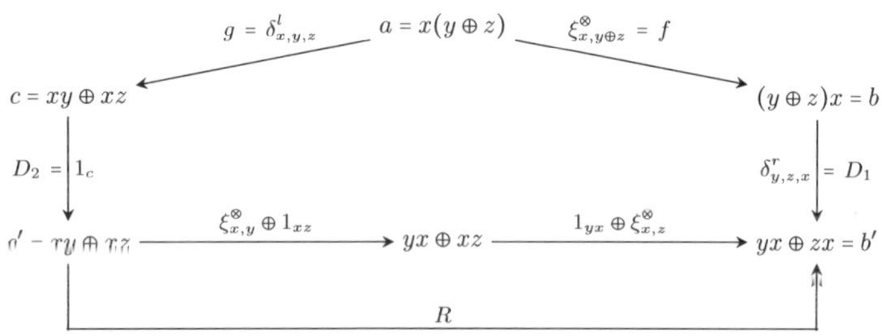

This diagram is commutative by functoriality, the axiom (2.1.5), and the symmetry axiom (1.2.20) in $(\mathsf{C}, \otimes)$.

(5) If $f = \xi^{-\otimes}_{y\oplus z,x}$, then we reuse the diagram in case (4) by replacing each $\xi^{\otimes}_{x,?}$ with $\xi^{-\otimes}_{?,x}$. The resulting diagram is commutative by functoriality and the axiom (2.1.5).

(6) If $f = 1_x\xi^{\oplus}_{y,z}$, then we consider the following diagram.

(3.6.19)

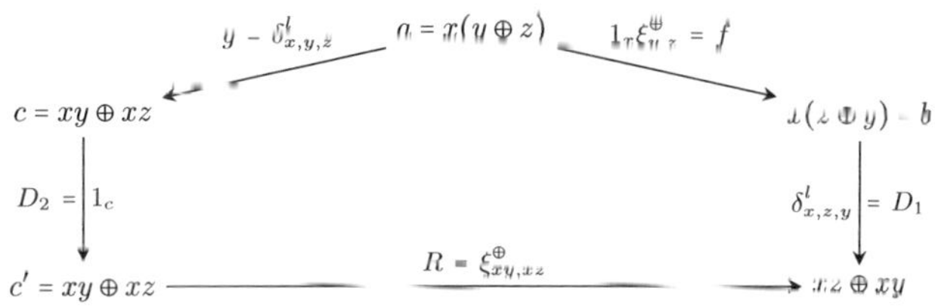

This diagram is commutative by the axiom (2.1.6).

(7) If $f = 1_x \xi^{-\oplus}_{z,y}$, then we reuse the diagram in case (6) by replacing each $\xi^{\oplus}_{*,?}$ with $\xi^{-\oplus}_{?,*}$. The resulting diagram is commutative by the axiom (2.1.6).

This finishes the initial case when $g = \delta^l_{x,y,z}$.

The proof for the case $g = \delta^r_{y,z,x}$ is obtained by slightly modifying the diagrams in cases (1)–(7) above, and using the symmetric bimonoidal category axioms

- (2.1.5), (2.1.27), (2.1.28), and
- (2.1.7) instead of (2.1.6) in the analogues of cases (6) and (7) if $f = \xi^{\oplus}_{y,z} 1_x$ or $\xi^{-\oplus}_{z,y} 1_x$.

This finishes the initial case of the induction.

For the induction step, suppose $\mathsf{norm}(a) > 3$. Either

$$a = a_1 \oplus a_2 \quad \text{or} \quad a = a_1 \otimes a_2$$

for some $a_1, a_2 \in X^{\mathsf{fr}}$ with $\mathsf{norm}(a_i) < \mathsf{norm}(a)$ for $i = 1, 2$. There are cases corresponding to the following situations.

(1) f and g act on the same summand or factor of a. To say that f acts on a summand, say a_1 of $a = a_1 \oplus a_2$, means that $f = f_1 \oplus e_2$ for some prime edge $f_1 : a_1 \longrightarrow b_1$ and some identity $e_2 : a_2 \longrightarrow a_2$. The meaning of f acting on a factor of a is similar with $\otimes$ instead of $\oplus$.

(2) f and g act on different summands or factors of a.

(3) f acts on all of a, and g acts on a summand or factor of a. To say that f acts on all of a means that f does not split as a sum or a product.

(4) f acts on a summand or factor of a, and g acts on all of a.

(5) f and g both act on all of a.

For the rest of this proof $\odot$ denotes either $\oplus$ or $\otimes$, but not both in the same diagram. We automatically adjust the symbols to avoid too many subscripts and superscripts.

(1) f and g act on the same $\odot$-factor of a, as in the following diagram.

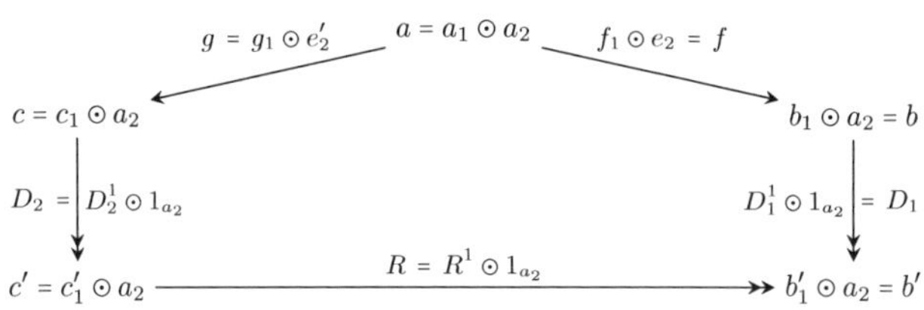

- $e_2, e_2' : a_2 \longrightarrow a_2$ are identities.
- $f_1 : a_1 \longrightarrow b_1$ is a 0^X-free edge.
- $g_1 : a_1 \longrightarrow c_1$ is a δ-prime edge.
- The paths

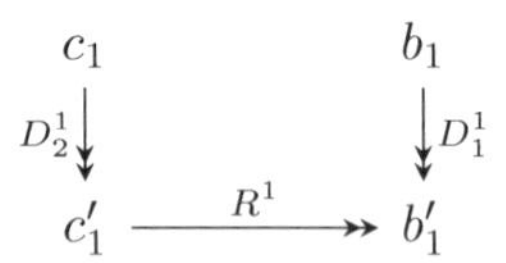

as in (3.6.13) are obtained from the induction hypothesis applied to a_1.
- The above diagram is commutative by functoriality.

If f and g both act on a_2 instead of a_1, then there is a similar diagram obtained from the induction hypothesis applied to a_2 and functoriality.

(2) f and g act on different $\odot$-factors of a.

(i) If f is a δ-prime edge, then we consider the following diagram.

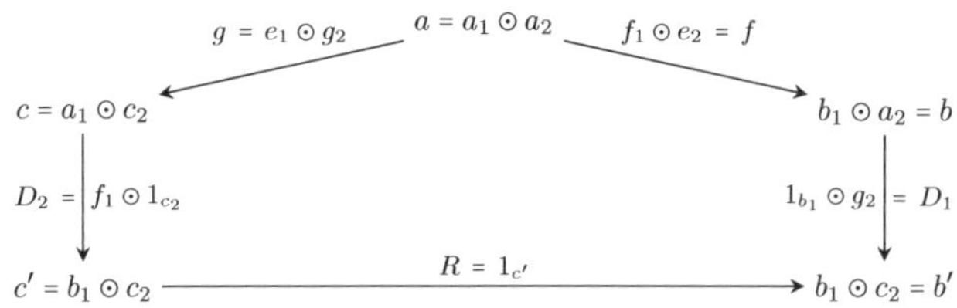

- $e_1 \colon a_1 \longrightarrow a_1$ and $e_2 \colon a_2 \longrightarrow a_2$ are identities.
- $f_1 \colon a_1 \longrightarrow b_1$ and $g_2 \colon a_2 \longrightarrow c_2$ are δ-prime edges.

This diagram is commutative by functoriality.

(ii) If f is not a δ-prime edge, then we consider the following diagram.

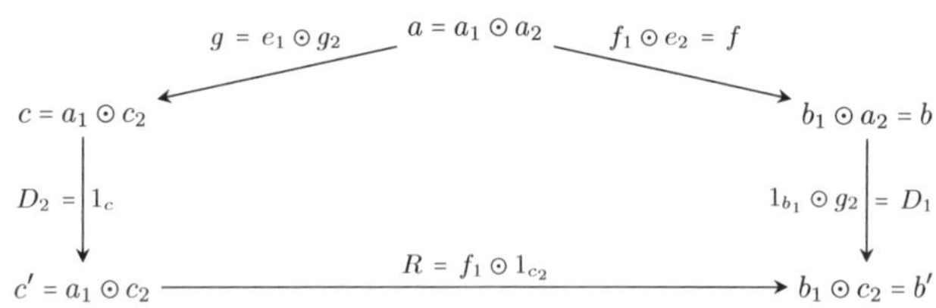

- e_1, e_2, and g_2 are as in case (i).
- $f_1 \colon a_1 \longrightarrow b_1$ is a $(0^X, \delta)$-free edge.

This diagram is commutative by functoriality.

(iii) If f and g act on a_2 and a_1, respectively, then there are two similar diagrams obtained from those in cases (i) and (ii) by switching the two $\odot$-factors.

(3) f acts on all of a, and g acts on a $\odot$-factor of a.

(i) If $f = \xi^{\odot}_{a_1,a_2}$, then we consider the following diagram.

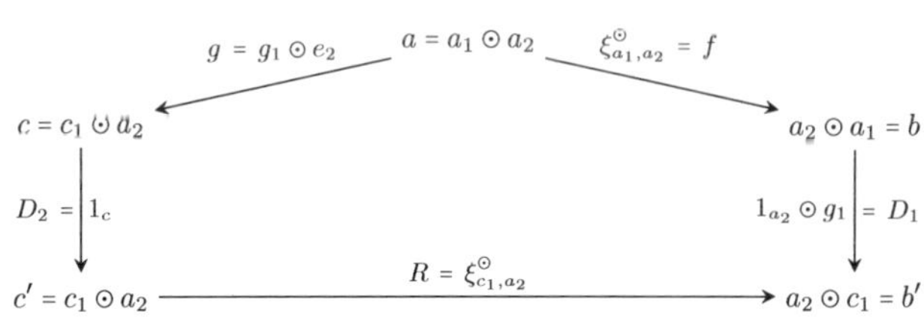

- $e_2 \colon a_2 \longrightarrow a_2$ is an identity.
- $g_1 \colon a_1 \longrightarrow c_1$ is a δ-prime edge.

This diagram is commutative by the naturality of $\xi^{\odot}$ in C. There is a similar diagram if g acts on the $\odot$-factor a_2 instead of a_1.

(ii) If $f = \xi^{-\odot}_{a_2,a_1}$, then we reuse the two diagrams in case (i) by replacing $\xi^{\odot}_{?,a_2}$ with $\xi^{-\odot}_{a_2,?}$. These diagrams are commutative by the naturality of $\xi^{\odot}$ in C.

(iii) If $f = \alpha^{\odot}_{a_1,a_2,a_3}$, and if g acts on one of the three $\odot$-factors of a, then we consider the following diagram.

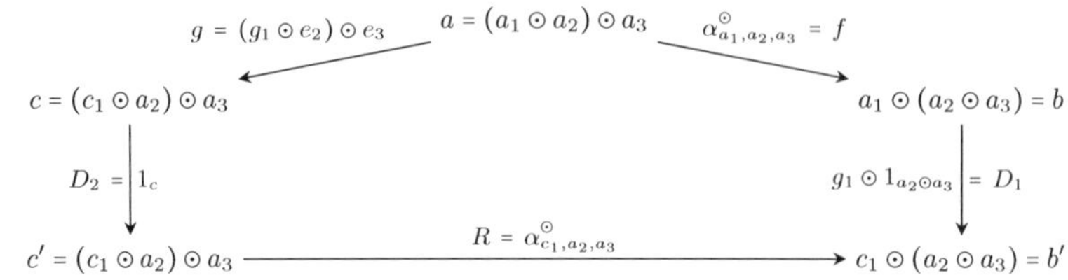

- $e_2 \colon a_2 \longrightarrow a_2$ and $e_3 \colon a_3 \longrightarrow a_3$ are identities.
- $g_1 \colon a_1 \longrightarrow c_1$ is a δ-prime edge.

This diagram is commutative by the naturality of $\alpha^{\odot}$ in C. There are two similar diagrams if g acts on a_2 or a_3 instead of a_1.

(iv) If $f = \alpha^{-\odot}_{a_1,a_2,a_3}$, and if g acts on one of the three $\odot$-factors of a, then we reuse the diagrams in case (iii) by replacing $\alpha^{\odot}$ with $\alpha^{-\odot}$. These diagrams are commutative by the naturality of $\alpha^{\odot}$ in C.

(v) If $f = \alpha^{\otimes}_{a_1,a_2,a_3}$, and if g involves δ^l on the first two $\otimes$-factors of a with $a_2 = x \oplus y$, then we consider the following diagram.

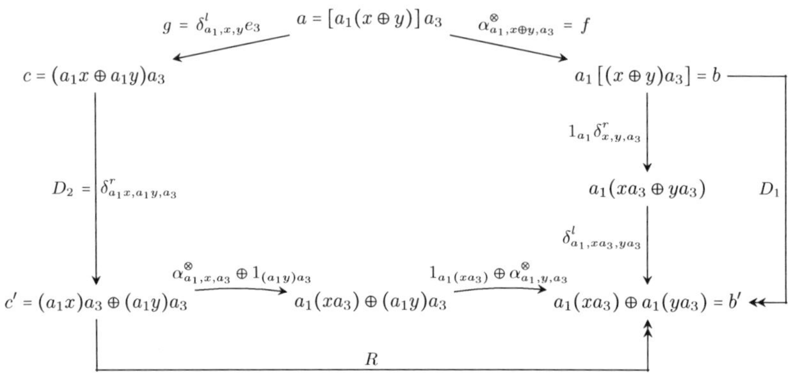

Here $e_3 \colon a_3 \longrightarrow a_3$ is an identity. This diagram is commutative by the axiom (2.1.12).

(vi) If $f = \alpha^{\otimes}_{a_1,a_2,a_3}$, and if g involves δ^r on the first two $\otimes$-factors of a with $a_1 = x \oplus y$, then there is a diagram similar to the one in case (v). This diagram is commutative by the axiom (2.1.11).

(vii) If $f = \alpha^{-\otimes}_{a_1,a_2,a_3}$, and if g involves δ^l on the last two $\otimes$-factors of a with $a_3 = y \oplus z$, then we consider the following diagram.

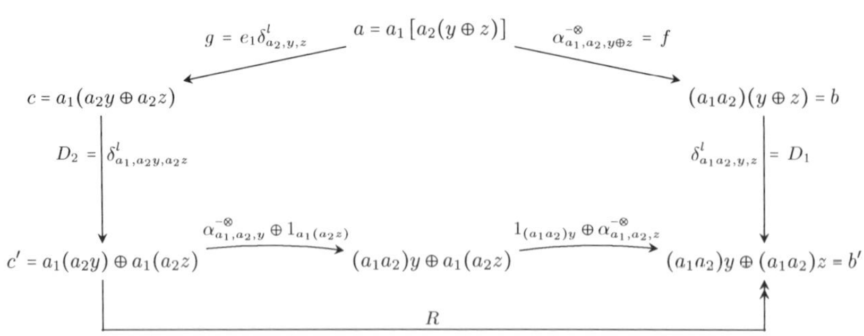

Here $e_1: a_1 \longrightarrow a_1$ is an identity. This diagram is commutative by the axiom (2.1.10).

(viii) If $f = \alpha^{-\otimes}_{a_1,a_2,a_3}$, and if g involves δ^r on the last two $\otimes$-factors of a with $a_2 = x \oplus y$, then there is a diagram similar to the one in case (vii). This diagram is commutative by the axiom (2.1.12).

(ix) If $f = \lambda^{\otimes}_{a_2}: 1^X \otimes a_2 \longrightarrow a_2$, then we consider the following diagram.

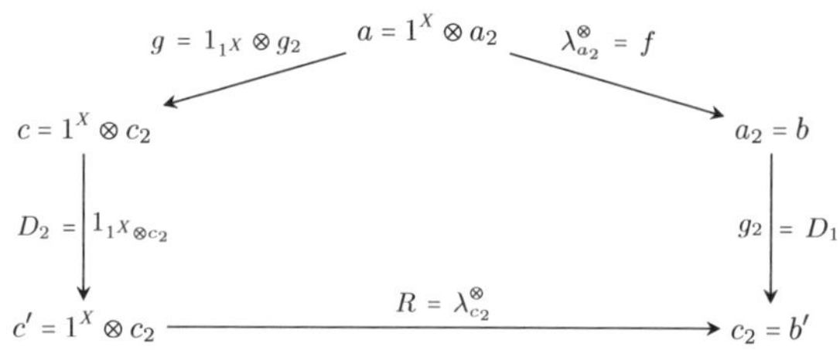

Here $g_2: a_2 \longrightarrow c_2$ is a δ-prime edge. This diagram is commutative by the naturality of $\lambda^{\otimes}$ in C.

(x) If $f = \lambda^{-\otimes}_a$, and if g acts on the $\odot$-factor a_1, then we consider the following diagram.

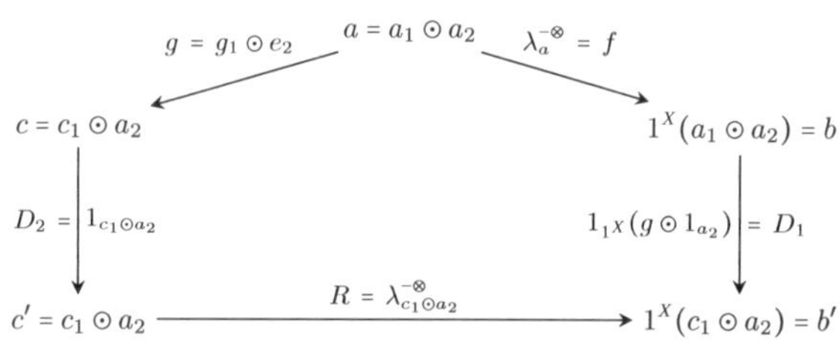

- $e_2: a_2 \longrightarrow a_2$ is an identity.
- $g_1: a_1 \longrightarrow c_1$ is a δ-prime edge.

This diagram is commutative by the naturality of $\lambda^{\otimes}$ in C.
If g acts on the $\odot$-factor a_2, then we slightly modify the above diagram by switching the two $\odot$-factors of a.

(xi) If

$$a_1 \otimes 1^X \xrightarrow{f = \rho^{\otimes}_{a_1}} a_1 \quad \text{or} \quad a \xrightarrow{f = \rho^{-\otimes}_a} a \otimes 1^X,$$

then we slightly modify the diagrams in cases (ix) and (x). The resulting diagrams are commutative by the naturality of $\rho^{\otimes}$ in C.

(xii) If $f = \delta^l_{a_1,x,y}$ with $a_2 = x \oplus y$, and if g acts on a_1, then we consider the following diagram.

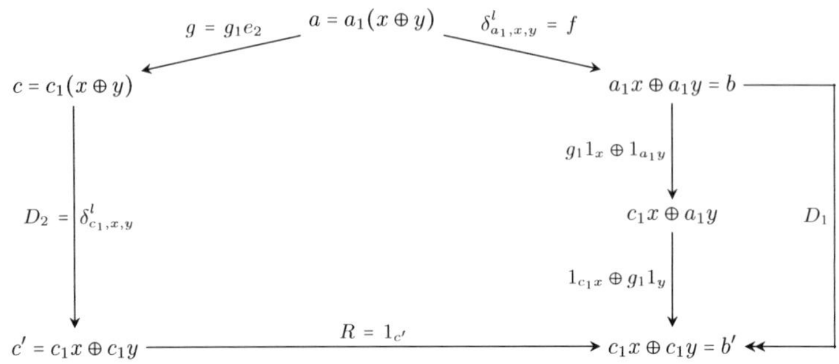

- $e_2 \colon a_2 \longrightarrow a_2$ is an identity.
- $g_1 \colon a_1 \longrightarrow c_1$ is a δ-prime edge.

This diagram is commutative by functoriality and the naturality of δ^l in C.

If g acts on x or y, then there are two diagrams similar to the one above that are commutative for the same reasons.

(xiii) If

$$(w \oplus x)a_2 \xrightarrow{f = \delta^r_{w,x,a_2}} wa_2 \oplus xa_2$$

with $a_1 = w \oplus x$, and if g acts on one of w, x, and a_2, then we slightly modify the diagrams in case (xii). The resulting diagrams are commutative by functoriality and the naturality of δ^r in C.

(4) f acts on a $\odot$-factor of a, and g acts on all of a.

First suppose $g = \delta^l_{a_1,y,z}$ with $a_2 = y \oplus z$. Then f may act on a_1, y, z, or a_2, leading to the following cases.

(i) If f acts on a_1, and if f is not a δ-prime edge, then we consider the following diagram.

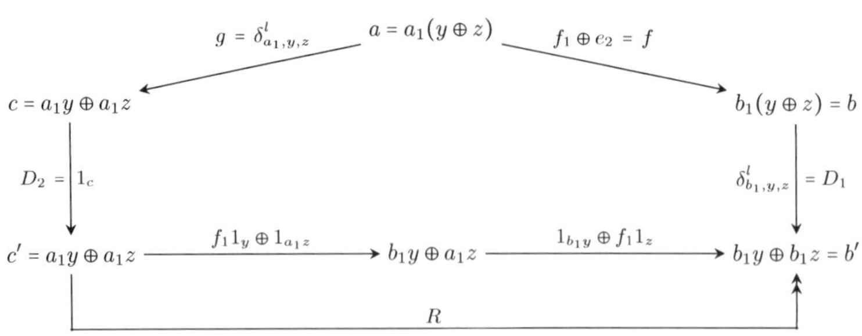

- $e_2{:}\, a_2 \longrightarrow a_2$ is an identity.
- $f_1{:}\, a_1 \longrightarrow b_1$ is a $(0^X, \delta)$-free edge.

This diagram is commutative by functoriality and the naturality of δ^l in C.

If f acts on a_1, and if f is a δ-prime edge, then we modify the previous diagram by defining

- $c' = b'$,
- $D_2{:}\, c \twoheadrightarrow c' = b'$ as the bottom composite in the previous diagram, and
- $R = 1_{c'}$.

The resulting diagram is commutative for the same reasons.

(ii) If f acts on y or z, then there are diagrams similar to those in case (i) that are commutative for the same reasons.

(iii) If

$$a_1(y \oplus z) \xrightarrow{f = e_1 \xi^{\oplus}_{y,z}} a_1(z \oplus y)$$

for some identity $e_1{:}\, a_1 \longrightarrow a_1$, then we reuse the argument in the paragraph containing (3.6.19) and the axiom (2.1.6).

Similarly, if $f = e_1 \xi^{-\oplus}_{z,y}$, then we reuse the argument in case (7) in the initial case.

(iv) If

- $a = w[(x \oplus y) \oplus z]$ with $a_1 = w$ and $a_2 = (x \oplus y) \oplus z$, and
- $f = e\alpha^{\oplus}_{x,y,z}$ for some identity $e{:}\, w \longrightarrow w$,

then we consider the following diagram.

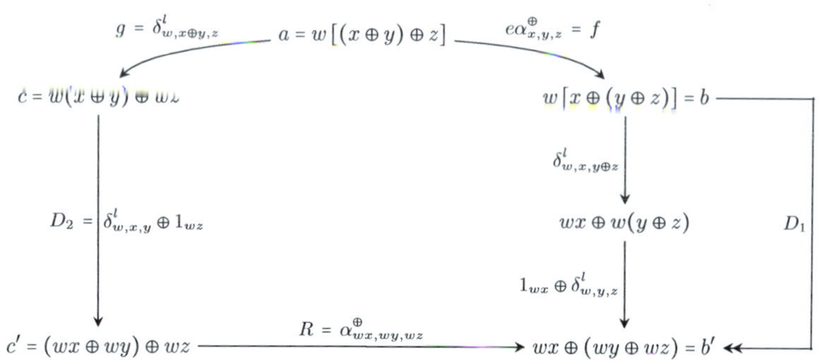

This diagram is commutative by the axiom (2.1.9).

(v) If

- $a = w[x \oplus (y \oplus z)]$ with $a_1 = w$ and $a_2 = x \oplus (y \oplus z)$, and
- $f = e\alpha^{-\oplus}_{x,y,z}$ for some identity $e{:}\, w \longrightarrow w$,

then we consider the following diagram.

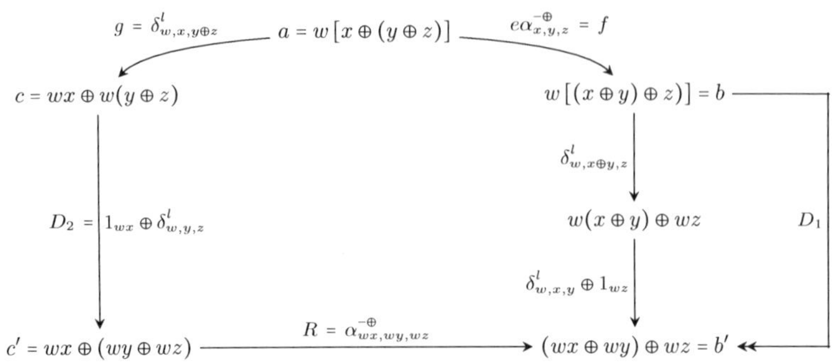

This diagram is commutative by the axiom (2.1.9).

Cases (i)–(v) account for all the cases when $g = \delta^l$.

If $a_1 = x \oplus y$, and if

$$(x \oplus y)a_2 \xrightarrow{g = \delta^r_{x,y,a_2}} xa_2 \oplus ya_2,$$

then f may act on a_2, x, y, or a_1, leading to analogues of cases (i)–(v) above. We slightly modify the diagrams above using

- functoriality and the naturality of δ^r in the analogues of cases (i) and (ii) when f acts on one of a_2, x, and y;
- (2.1.7) instead of (2.1.6) in the analogue of case (iii) when $f = \xi^{\oplus}_{x,y}e_2$ or $\xi^{-\oplus}_{y,x}e_2$ for some identity $e_2 \colon a_2 \longrightarrow a_2$; and
- (2.1.8) instead of (2.1.9) in the analogues of cases (iv) and (v) when $f = \alpha^{\oplus}_{w,x,y}e$ or $f = \alpha^{-\oplus}_{w,x,y}e$ for some identity $e \colon z \longrightarrow z$.

(5) f and g both act on all of a.

First suppose $g = \delta^l$. Depending on what f is, there are the following cases.

(i) If $f = \alpha^{\otimes}$, then we consider the following diagram.

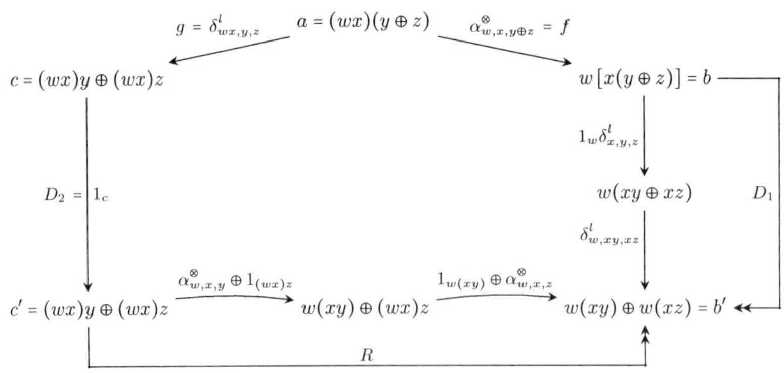

This diagram is commutative by functoriality and (2.1.10).

(ii) If $f = \xi^{\otimes}_{x,y\oplus z}$ and $g = \delta^l_{x,y,z}$, then we reuse the diagram (3.6.18).

(iii) If $f = \xi^{-\otimes}_{y\oplus z,x}$ and $g = \delta^l_{x,y,z}$, then we reuse case (5) in the initial case.

(iv) If $f = \lambda^{\otimes}_{y\oplus z}$ and $g = \delta^l_{1^X,y,z}$, then we reuse the diagram (3.6.14).

(v) If $f = \lambda^{-\otimes}_a$, then we reuse the diagram (3.6.15).

(vi) If $f = {\rho_a}^{\otimes}$, then we reuse the diagram (3.6.15) by replacing $\lambda^{-\otimes}$ with $\rho^{-\otimes}$, and using the naturality of $\rho^{\otimes}$ in C.

(vii) If $f = \delta^r$, then we consider the following diagram, where some subscripts are omitted to save space.

(3.6.20)

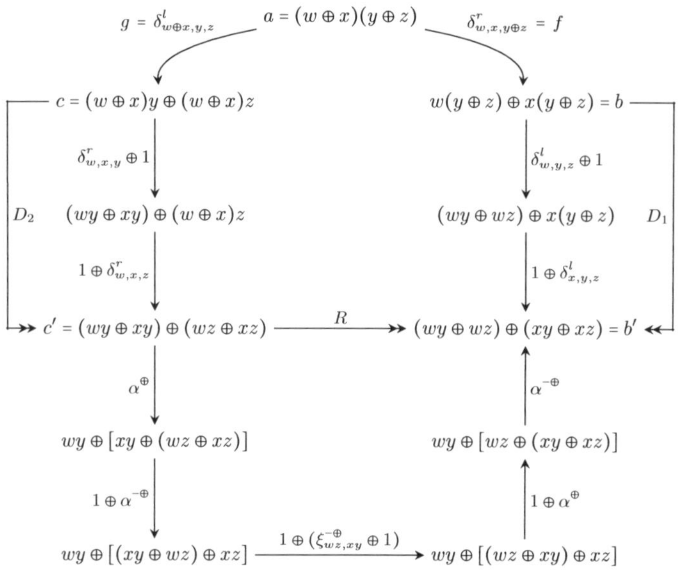

The bottom half of this diagram is the definition of the $(0^X, \delta)$-free path R, which consists of five prime edges. This diagram is commutative by the axiom (2.1.13).

Cases (i)–(vii) above account for all the cases when $g = \delta^l$.

Similarly, if $g = \delta^r$, then there are the following cases.

(i) If $f = \alpha^{-\otimes}$, then there is a diagram similar to the one in case (i) above, which is commutative by functoriality and (2.1.11).

(ii) If $f = \xi^{\otimes}$, then we use functoriality and (2.1.5).

(iii) If $f = \xi^{-\otimes}$, then we use functoriality, (2.1.5), and the symmetry axiom (1.2.20) in $(\mathsf{C}, \otimes)$.

(iv) If $f = \rho^{\otimes}$, then we use functoriality and (2.1.28).

(v) If $f = \lambda^{-\otimes}$, then we use the δ^r analogue of the diagram (3.6.15) and the naturality of $\lambda^{\otimes}$ in C.

(vi) If $f = \rho^{-\otimes}$, then we use the $(\delta^r, \rho^{-\otimes})$ analogue of the diagram (3.6.15) and the naturality of $\rho^{\otimes}$ in C.

(vii) If $f = \delta^l$, then we reuse the diagram in case (vii) above by
- switching (f, D_1) with (g, D_2) and
- reversing R and every edge in it and replacing $(\alpha^{\oplus}, \alpha^{-\oplus}, \xi^{-\oplus})$ with $(\alpha^{-\oplus}, \alpha^{\oplus}, \xi^{\oplus})$.

This finishes case (5) and the induction. The proof of Lemma 3.6.12 is now complete. □

3.7. Zero and Delta Reduction of Paths

Recall from Definition 3.6.2 that a δ-reduction of an element $a \in X^{\mathsf{fr}}$ is a path with domain a and codomain a δ-reduced element in which each edge is either an identity or a δ-prime edge. The next definition is the path analogue of a δ-reduction. In Proposition 3.5.32, we showed that each path in $\mathsf{Gr}(X)$ has a 0^X reduction. The goal of this section is to extend that observation by also eliminating the distributivity morphisms; see Proposition 3.7.19. This will constitute another reduction step in the proof of the Coherence Theorem 3.9.1. Recall that Convention 3.3.1 is in effect.

Definition 3.7.1. Suppose

- $P: a \twoheadrightarrow b$ is a path in $\mathsf{Gr}(X)$ whose vertices do not contain 0^X, and
- for each $x \in \{a, b\}$, $D_x: x \twoheadrightarrow x'$ is a δ-reduction of x.

A *$(0^X, \delta)$-reduction* of (P, D_a, D_b) is a $(0^X, \delta)$-free path $R: a' \twoheadrightarrow b'$ such that the diagram

$$(3.7.2) \qquad \begin{array}{ccc} a & \xrightarrow{P} & b \\ {\scriptstyle D_a}\downarrow & & \downarrow{\scriptstyle D_b} \\ a' & \xrightarrow{R} & b' \end{array}$$

is commutative in the sense of Definition 3.1.14. We also call R a $(0^X, \delta)$-reduction of P, suppressing D_a and D_b from the notation. ◇

The main result of this section, Proposition 3.7.19, asserts the existence of a $(0^X, \delta)$-reduction.

Explanation 3.7.3. Consider Definition 3.7.1.

(1) Since the vertices in P contain no 0^X, P is a 0^X-free path as in Definition 3.4.1. Its edges do not contain $\lambda^{\pm\oplus}$, $\rho^{\pm\oplus}$, $\lambda^{\pm\bullet}$, and $\rho^{\pm\bullet}$. In other words, as described in Explanation 3.4.2, each prime edge in P is either an identity or contains a single instance of $\alpha^{\pm\oplus}$, $\xi^{\pm\oplus}$, $\alpha^{\pm\otimes}$, $\xi^{\pm\otimes}$, $\lambda^{\pm\otimes}$, $\rho^{\pm\otimes}$, δ^l, or δ^r.
(2) The (co)domain of R is a δ-reduction of the (co)domain of the original path P. Each element in X^{fr} has a δ-reduction by Lemma 3.6.9. However, neither its codomain nor its value in C is unique, as we explained in Example 3.6.8.
(3) For the commutativity of the diagram (3.7.2), the equality
$$(R, D_a) = (D_b, P)$$
of paths in $\mathsf{Gr}(X)$ is sufficient but *not* necessary. The definition only requires that the image of the diagram under the graph morphism $\varphi: \mathsf{Gr}(X) \longrightarrow \mathsf{C}$ in Definition 3.1.14 be a commutative diagram in C for arbitrary C and φ. ◇

Example 3.7.4. Suppose P is the path

$$(w \oplus x)(y \oplus z) \xrightarrow{P = \delta^r_{w,x,y\oplus z}} w(y \oplus z) \oplus x(y \oplus z)$$

for some elements $w, x, y, z \in X \smallsetminus \{0^X\}$. Reusing the diagram (3.6.20):

- $D_a = (D_2, g): a \twoheadrightarrow c'$ is a δ-reduction of a.
- $D_b = D_1: b \twoheadrightarrow b'$ is a δ-reduction of b.
- $R: c' \twoheadrightarrow b'$ is a $(0^X, \delta)$-reduction of (P, D_a, D_b).

The commutativity of the diagram (3.7.2) follows from the axiom (2.1.13). ◇

Lemma 3.7.5. *Suppose $P: a \twoheadrightarrow b$ is a $(0^X, \delta)$-free path with a and b both δ-reduced. Then P is a $(0^X, \delta)$-reduction of P for any choices of δ-reductions of a and b.*

Proof. Since a and b are not the domains of any δ-prime edges, any δ-reduction $D_x: x \twoheadrightarrow x'$ of $x \in \{a, b\}$, which exists by Lemma 3.6.9, is a sequence of identities with $x' = x$. Therefore, the diagram (3.7.2) is commutative if $R = P$. □

The proof of the existence of a $(0^X, \delta)$-reduction is split into several cases, the first of which uses the following two facts about symmetric bimonoidal categories.

Lemma 3.7.6. *For objects $Y_1, \ldots, Y_n \in \mathsf{C}$ with $n \geq 2$, the diagram*

$$\begin{array}{ccc} & \overset{\lambda^{-\otimes}_Y}{\nearrow} & \mathbb{1}(Y_1 \oplus \cdots \oplus Y_n) = W \\ Y = Y_1 \oplus \cdots \oplus Y_n & & \Big\downarrow D \\ & \underset{L}{\searrow} & \mathbb{1}Y_1 \oplus \cdots \oplus \mathbb{1}Y_n = Z \end{array} \tag{3.7.7}$$

defined as follows is commutative:

- *Y has some additive bracketing, and $W = \mathbb{1}Y$.*
- *Z is obtained from Y by replacing each summand Y_k with $\mathbb{1}Y_k$.*
- *$L = \lambda^{-\otimes}_{Y_1} \oplus \cdots \oplus \lambda^{-\otimes}_{Y_n}$ with the same additive bracketing as Y.*
- *D is a composite of morphisms, each being a sum of*
  - *identity morphisms, for which there are zero of them for the first morphism in D; and*
  - *one $\delta^l_{\mathbb{1},Y',Y''}$ with each of Y' and Y'' a sum of consecutive Y_k's with bracketings inherited from Y.*

Proof. The proof is by induction on n. The $n = 2$ case of (3.7.7) is commutative by the axiom (2.1.27).

For the induction step, suppose $n > 2$, and $Y = Y^1 \oplus Y^2$ with

- $Y^1 = Y_1 \oplus \cdots \oplus Y_k$ and
- $Y^2 = Y_{k+1} \oplus \cdots \oplus Y_n$

with bracketings inherited from Y and some $1 \leq k \leq n-1$. Moreover, $L = L^1 \oplus L^2$ with

- $L^1 = \lambda^{-\otimes}_{Y_1} \oplus \cdots \oplus \lambda^{-\otimes}_{Y_k}$ and
- $L^2 = \lambda^{-\otimes}_{Y_{k+1}} \oplus \cdots \oplus \lambda^{-\otimes}_{Y_n}$

with bracketings inherited from Y.

The diagram (3.7.7) factors as follows.

$$\begin{array}{ccccc} & \overset{\lambda^{-\otimes}_Y}{\nearrow} & \mathbb{1}(Y^1 \oplus Y^2) = W & & \\ & & \Big\downarrow \delta^l_{\mathbb{1},Y^1,Y^2} & & \\ Y = Y^1 \oplus Y^2 & \xrightarrow{\lambda^{-\otimes}_{Y^1} \oplus \lambda^{-\otimes}_{Y^2}} & \mathbb{1}Y^1 \oplus \mathbb{1}Y^2 & & \Big\downarrow D \\ & & \Big\downarrow D^1 \oplus D^2 & & \\ & \underset{L^1 \oplus L^2}{\searrow} & \left(\mathbb{1}Y_1 \oplus \cdots \oplus \mathbb{1}Y_k\right) \oplus \left(\mathbb{1}Y_{k+1} \oplus \cdots \oplus \mathbb{1}Y_n\right) = Z & & \end{array}$$

The top half is commutative by (2.1.27). Consider the bottom half.

- If $k = 1$, then $Y^1 = Y_1$, and $D^1 = 1_{\mathbb{1}Y_1}$. The induction hypothesis is applied to Y^2.

- If $k = n-1$, then $Y^2 = Y_n$, and $D^2 = 1_{\mathbb{1}Y_n}$. The induction hypothesis is applied to Y^1.
- If $1 < k < n-1$, then the induction hypothesis is applied to Y^1 and Y^2.

In each case, the bottom half of the previous diagram is commutative. $\square$

Next is the $\rho^{\otimes}$ analogue of Lemma 3.7.6.

Lemma 3.7.8. *For objects $Y_1, \ldots, Y_n \in \mathsf{C}$ with $n \geq 2$, the diagram*

$$\begin{array}{ccc} & \overset{\rho_Y^{-\otimes}}{\nearrow} & (Y_1 \oplus \cdots \oplus Y_n)\mathbb{1} = W \\ Y = Y_1 \oplus \cdots \oplus Y_n & & \Big\downarrow D \\ & \underset{L}{\searrow} & Y_1\mathbb{1} \oplus \cdots \oplus Y_n\mathbb{1} = Z \end{array}$$

defined as follows is commutative:

- *Y has some additive bracketing, and $W = Y\mathbb{1}$.*
- *Z is obtained from Y by replacing each summand Y_k with $Y_k\mathbb{1}$.*
- *$L = \rho_{Y_1}^{-\otimes} \oplus \cdots \oplus \rho_{Y_n}^{-\otimes}$ with the same additive bracketing as Y.*
- *D is a composite of morphisms, each being a sum of*
  - *identity morphisms, for which there are zero of them for the first morphism in D; and*
  - *one $\delta^r_{Y',Y'',\mathbb{1}}$ with each of Y' and Y'' a sum of consecutive Y_k's with bracketings inherited from Y.*

Proof. This is a slight modification of the proof of Lemma 3.7.6, using the axiom (2.1.28) instead of (2.1.27). $\square$

Next is the first preliminary case of Proposition 3.7.19.

Lemma 3.7.9. *Suppose $f: a \longrightarrow b$ is a prime edge such that*

- *a and b do not contain 0^X, and*
- *a is δ-reduced.*

Then f has a $(0^X, \delta)$-reduction for any choices of δ-reductions of a and b.

Proof. For each $x \in \{a, b\}$, suppose $D_x: x \twoheadrightarrow x'$ is a δ-reduction of x, which exists by Lemma 3.6.9. First we make a few observations.

- By Lemma 3.6.5, a being δ-reduced means that it is a polynomial in the sense of Definition 3.6.3.
- Since a, being δ-reduced, is not the domain of any δ-prime edge, the δ-reduction D_a is a sequence of identities of a, and $a' = a$.
- The absence of 0^X in a and b, and that a is δ-reduced, imply that f is a $(0^X, \delta)$-free edge as in Definition 3.6.10.

The possibilities of the $(0^X, \delta)$-free edge f are listed in Explanation 3.6.11.

First suppose f is either an identity or involves an instance of

- $\alpha^{\pm\oplus}$, $\xi^{\pm\oplus}$, $\alpha^{\pm\otimes}$, $\xi^{\pm\otimes}$, $\lambda^{\otimes}$, $\rho^{\otimes}$, or
- $\lambda^{-\otimes}$ or $\rho^{-\otimes}$ applied to either a monomial in a or some $\otimes$-factor of a monomial in a.

Then b is also a polynomial, which means that it is δ-reduced. The δ-reduction D_b is a sequence of identities of b, and $b' = b$. Therefore, $R = f$ is a $(0^X, \delta)$-reduction of f.

If f is not as in the previous paragraph, then f involves $\lambda^{-\otimes}$ or $\rho^{-\otimes}$ applied to a sum of at least two monomials in a. Consider the $\lambda^{-\otimes}$ case first. In other words:

- $a \in X^{\mathsf{fr}}$ has the form

$$a = \underbrace{a^1 \oplus \cdots \oplus a^{i-1}}_{\text{empty if } i=1} \oplus A \oplus \underbrace{a^{j+1} \oplus \cdots \oplus a^m}_{\text{empty if } j=m}$$

  for some additive bracketing and some

$$A = a^i \oplus \cdots \oplus a^j \in X^{\mathsf{fr}}$$

  with some additive bracketing, $1 \le i < j \le m$ with $m \ge 2$, and each a^k a monomial as in Definition 3.6.3.
- $b \in X^{\mathsf{fr}}$ has the form

$$b = a^1 \oplus \cdots \oplus a^{i-1} \oplus 1^X A \oplus a^{j+1} \oplus \cdots \oplus a^m$$

  with the same additive bracketing as a.
- The prime edge f involves the elementary edge $\lambda_A^{-\otimes}: A \longrightarrow 1^X A$.

Note that b is not δ-reduced because δ^l can be applied to $1^X A$. Since $D_b: b \twoheadrightarrow b'$ is a δ-reduction of b, the codomain b' has the form

$$b' = a^1 \oplus \cdots \oplus a^{i-1} \oplus A' \oplus a^{j+1} \oplus \cdots \oplus a^m$$

with the same additive bracketing as a, where

$$A' = 1^X a^i \oplus \cdots \oplus 1^X a^j$$

with the same additive bracketing as A. Moreover, each prime edge in D_b is either an identity or involves δ^l.

The desired path $R: a = a' \twoheadrightarrow b'$ is defined using the following objects and edges.

- For $0 \le l \le j-i+1$, define $A_l \in X^{\mathsf{fr}}$ as the element obtained from A by replacing each of the l leftmost summands $a^?$ by $1^X a^?$. Here are some examples.

$$\begin{aligned} A_0 &= a^i \oplus \cdots \oplus a^j = A \\ A_1 &= 1^X a^i \oplus a^{i+1} \oplus \cdots \oplus a^j \\ A_{j-i} &= 1^X a^i \oplus \cdots \oplus 1^X a^{j-1} \oplus a^j \\ A_{j-i+1} &= 1^X a^i \oplus \cdots \oplus 1^X a^j = A' \end{aligned}$$

- For $0 \le l \le j-i+1$, define $a_l \in X^{\mathsf{fr}}$ as the element obtained from a by replacing A with A_l. Here are some examples.

$$\begin{aligned} a_0 &= a^1 \oplus \cdots \oplus a^{i-1} \oplus A \oplus a^{j+1} \oplus \cdots \oplus a^m = a \\ a_1 &= a^1 \oplus \cdots \oplus a^{i-1} \oplus A_1 \oplus a^{j+1} \oplus \cdots \oplus a^m \\ a_{j-i+1} &= a^1 \oplus \cdots \oplus a^{i-1} \oplus A' \oplus a^{j+1} \oplus \cdots \oplus a^m = b' \end{aligned}$$

- For $1 \le l \le j-i+1$, define $r_l: a_{l-1} \longrightarrow a_l$ as the prime edge involving the elementary edge

$$a^{i+l-1} \xrightarrow{\lambda^{-\otimes}_{a^{i+l-1}}} 1^X a^{i+l-1}$$

and the identity elementary edge of every other summand in a_{l-1}. For example, the prime edge

- $r_1: a = a_0 \longrightarrow a_1$ involves $\lambda^{-\otimes}_{a^i}: a^i \longrightarrow 1^X a^i$, and
- $r_{j-i+1}: a_{j-i} \longrightarrow a_{j-i+1} = b'$ involves $\lambda^{-\otimes}_{a^j}: a^j \longrightarrow 1^X a^j$.

Define the $(0^X, \delta)$-free path

$$R = \big(r_{j-i+1}, \ldots, r_1\big)$$

in $\mathsf{Gr}(X)$, that is,

$$a = a_0 \xrightarrow{r_1} a_1 \xrightarrow{r_2} \cdots \xrightarrow{r_{j-i+1}} a_{j-i+1} = b'.$$

We aim to show that R is a δ-reduction of f.

To show that the diagram (3.7.2) is commutative, it suffices to concentrate on the summands A in a, $1^X A$ in b, and A' in b', since f, D_b, and R in other summands are identities. The relevant diagram is commutative by Lemma 3.7.6 applied as follows:

- $Y_l = \varphi a^{i+l-1}$ for $1 \le l \le j - i + 1 = n$, with φ the graph morphism in Definition 3.1.14. For example, $Y_1 = \varphi a^i$, and $Y_n = \varphi a^j$.
- $Y = Y_1 \oplus \cdots \oplus Y_n$ has the same additive bracketing as A, and $Y = \varphi A$.
- $W = \mathbb{1}Y = \varphi(1^X A)$.
- $Z = \mathbb{1}Y_1 \oplus \cdots \oplus \mathbb{1}Y_n = \varphi A'$.
- $\lambda^{-\otimes}_Y: Y \longrightarrow \mathbb{1}Y$ is the nonidentity summand of φf.
- $L = \lambda^{-\otimes}_{Y_1} \oplus \cdots \oplus \lambda^{-\otimes}_{Y_n}$ is the nonidentity summand of the composite of φR.
- D is the nonidentity summand of the composite of φD_b.

This shows that R is a δ-reduction of f when f involves $\lambda^{-\otimes}$.

If f involves $\rho^{-\otimes}$, then we slightly modify the argument above by replacing

- $\big(\lambda^{-\otimes}, \delta^l, 1^X A, 1^X a^k\big)$ with $\big(\rho^{-\otimes}, \delta^r, A1^X, a^k 1^X\big)$ and
- Lemma 3.7.6 with Lemma 3.7.8.

All the possible cases of f, as in Explanation 3.6.11, have been considered. □

Next is the second preliminary case of Proposition 3.7.19.

Lemma 3.7.10. *Under the hypotheses of Definition 3.7.1, suppose that each edge in the path $P: a \twoheadrightarrow b$ is of one of the following types:*

- *an identity or*
- *a prime edge involving $\alpha^{\pm\oplus}$, $\xi^{\pm\oplus}$, $\alpha^{\pm\otimes}$, $\xi^{\pm\otimes}$, δ^l, or δ^r.*

Then P has a $(0^X, \delta)$-reduction.

Proof. The proof is by induction on $\mathsf{size}(a) - \mathsf{rank}(a)$, which is always nonnegative by Lemma 3.2.9 (2). First suppose

$$\mathsf{size}(a) - \mathsf{rank}(a) = 0.$$

By Lemmas 3.2.9, 3.2.10, and 3.2.12, P does not contain δ^l and δ^r, and each vertex in P satisfies $\mathsf{size} = \mathsf{rank}$. By Proposition 3.2.15, each vertex in P is a polynomial in the sense of Definition 3.6.3, which is δ-reduced by Lemma 3.6.5. Since P is a $(0^X, \delta)$-free path with a and b both δ-reduced, P is a $(0^X, \delta)$-reduction of P by Lemma 3.7.5.

For the induction step, suppose

$$\mathsf{size}(a) - \mathsf{rank}(a) > 0$$

and that the path P consists of the prime edges $(f_n, \ldots, f_1)$ as in

$$a = a_0 \xrightarrow{\;f_1\;} a_1 \xrightarrow{\;f_2\;} \cdots \xrightarrow{\;f_n\;} a_n = b \tag{3.7.11}$$

for some $n \geq 1$. The δ-reductions $D_x\colon x \twoheadrightarrow x'$ of $x \in \{a, b\}$ are already chosen in Definition 3.7.1. If $n > 1$, then for each $1 \leq i \leq n-1$, choose a δ-reduction $D_i\colon a_i \twoheadrightarrow a_i'$ of a_i, which exists by Lemma 3.6.9. If for each $1 \leq i \leq n$, the prime edge f_i has a $(0^X, \delta)$-reduction $R_i\colon a_{i-1}' \twoheadrightarrow a_i'$ with respect to these δ-reductions, then the concatenated path

$$R = (R_n, \ldots, R_1)$$

is a $(0^X, \delta)$-reduction of P, as in the following diagram.

$$\begin{array}{ccccccc}
a = a_0 & \xrightarrow{\;f_1\;} & a_1 & \xrightarrow{\;f_2\;} & \cdots & \xrightarrow{\;f_n\;} & a_n = b \\
{\scriptstyle D_a}\big\downarrow & & \big\downarrow{\scriptstyle D_1} & & & & \big\downarrow{\scriptstyle D_b} \\
a' = a_0' & \overset{R_1}{\twoheadrightarrow} & a_1' & \overset{R_2}{\twoheadrightarrow} & \cdots & \overset{R_n}{\twoheadrightarrow} & a_n' = b'
\end{array}$$

Therefore, to show that P has a $(0^X, \delta)$-reduction, it suffices to show that each prime edge $f\colon a \longrightarrow b$ of the form stated in the lemma, with a and b containing no 0^X, has a $(0^X, \delta)$-reduction.

Moreover, if a is δ-reduced, then f has a $(0^X, \delta)$-reduction by Lemma 3.7.9. Therefore, we may assume that a is not δ-reduced, which by Lemma 3.6.5 means that a is not a polynomial. The rest of the induction step is divided into two cases depending on f.

First suppose that the prime edge $f\colon a \longrightarrow b$ does not involve δ^l and δ^r. In other words, either

- f is an identity, or
- f involves $\alpha^{\pm\oplus}$, $\xi^{\pm\oplus}$, $\alpha^{\pm\otimes}$, or $\xi^{\pm\otimes}$.

Lemma 3.2.10 implies that there is an equality

$$\mathsf{size}(a) - \mathsf{rank}(a) = \mathsf{size}(b) - \mathsf{rank}(b). \tag{3.7.12}$$

Moreover, that a is not a polynomial and the current hypothesis on f imply that b is not a polynomial, so b is also not δ-reduced by Lemma 3.6.5. Since a and b are not δ-reduced, each δ-reduction $D_x\colon x \twoheadrightarrow x'$ of $x \in \{a, b\}$ has at least one δ-prime edge. We are concerned about constructing a diagram (3.7.2) that is commutative, and identity prime edges are sent by the graph morphism $\varphi\colon \mathsf{Gr}(X) \longrightarrow \mathsf{C}$ in Definition 3.1.14 to identity morphisms. Therefore, we may assume that the first edge in each δ-reduction D_x is a δ-prime edge. In the diagram (3.7.13) below, these δ-prime edges are denoted by $a \longrightarrow a_1$ and $b \longrightarrow b_1$.

Using Lemma 3.6.12 and the induction hypothesis (IH), a $(0^X, \delta)$-reduction $R\colon a' \twoheadrightarrow b'$ of f is obtained in the following diagram, which we will explain in detail.

(3.7.13)

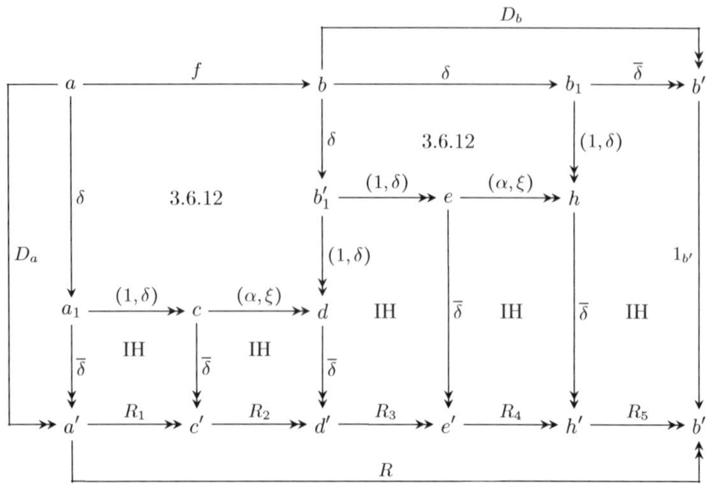

The following statements are about the diagram (3.7.13):

- Each subdiagram labeled by 3.6.12 is obtained by applying that lemma.
- Each subdiagram labeled by IH is obtained by applying the induction hypothesis.
- Each subdiagram is commutative.
- Each edge labeled by δ is a δ-prime edge.
- Each path labeled by:
  - $\overline{\delta}$ is a δ-reduction.
  - $(1, \delta)$ is either a single identity or consists of δ-prime edges.
  - (α, ξ) consists of identities and prime edges involving $\alpha^{\pm\oplus}$, $\xi^{\pm\oplus}$, $\alpha^{\pm\otimes}$, or $\xi^{\pm\otimes}$.
- The δ-reductions D_a and D_b are as displayed. The respective first edges $a \longrightarrow a_1$ and $b \longrightarrow b_1$ are both δ-prime edges by the last sentence of the previous paragraph.
- Each R_i is a $(0^X, \delta)$-reduction of the horizontal path above it. Their concatenation R is a $(0^X, \delta)$-reduction of f.

With more detail, we begin with the prime edge f, and the δ-reductions D_a and D_b, and perform the following steps.

(1) Using the current hypothesis on f and that $a \longrightarrow a_1$ is a δ-prime edge, we first apply Lemma 3.6.12 to obtain the upper left square in (3.7.13). The resulting path $b \longrightarrow\!\!\!\!\rightarrow d$ has at least one δ-prime edge $b \longrightarrow b'_1$ by the following (in)equalities.

$$
\begin{aligned}
&\mathsf{size}(d) - \mathsf{rank}(d) \\
&= \mathsf{size}(c) - \mathsf{rank}(c) && \text{(by Lemma 3.2.10)} \\
&\leq \mathsf{size}(a_1) - \mathsf{rank}(a_1) && \text{(by Lemma 3.2.12)} \\
&< \mathsf{size}(a) - \mathsf{rank}(a) && \text{(by Lemma 3.2.12)} \\
&= \mathsf{size}(b) - \mathsf{rank}(b) && \text{(by (3.7.12))}
\end{aligned}
$$

Therefore, $b \longrightarrow\!\!\!\!\rightarrow d$ cannot be a single identity, and must contain a δ-prime edge $b \longrightarrow b'_1$.

(2) Next we apply Lemma 3.6.12 to the δ-prime edges $b \longrightarrow b_1$ and $b \longrightarrow b_1'$ to obtain the top middle rectangle in (3.7.13). Since a and b contain no 0^X, all other vertices involved so far—namely, a_1, a', b_1, b', b_1', c, d, e, and h—also contain no 0^X.

(3) For each $y \in \{c, d, e, h\}$, choose a δ-reduction $y \twoheadrightarrow y'$, which exists by Lemma 3.6.9. The concatenated paths

$$b_1' \twoheadrightarrow d \twoheadrightarrow d'$$

$$b_1 \twoheadrightarrow h \twoheadrightarrow h'$$

are δ-reductions of b_1' and b_1, respectively. Moreover, since b' is δ-reduced, the identity $1_{b'}$ is a δ-reduction.

(4) We observed in step (1) that a_1 and c both have $\mathsf{size} - \mathsf{rank}$ less than that of a. Furthermore, by Lemma 3.2.12 and (3.7.12), b_1, b_1', and e also have $\mathsf{size} - \mathsf{rank}$ less than that of a. The induction hypothesis is applied five times as indicated by IH in (3.7.13) to obtain $(0^X, \delta)$-reductions
- $R_1 \colon a' \twoheadrightarrow c'$ of $a_1 \twoheadrightarrow c$;
- $R_2 \colon c' \twoheadrightarrow d'$ of $c \twoheadrightarrow d$;
- $R_3 \colon d' \twoheadrightarrow e'$ of $b_1' \twoheadrightarrow e$;
- $R_4 \colon e' \twoheadrightarrow h'$ of $e \twoheadrightarrow h$; and
- $R_5 \colon h' \twoheadrightarrow b'$ of $b_1 \twoheadrightarrow b'$.

The concatenated path

$$R = (R_5, R_4, R_3, R_2, R_1) \colon a' \twoheadrightarrow b'$$

is $(0^X, \delta)$-free, and the diagram (3.7.2) with $P = f$ is commutative by (3.7.13). Therefore, R is a $(0^X, \delta)$-reduction of f.

Next we consider the case when the prime edge f involves a single instance of δ^l or δ^r. Using Lemma 3.6.12 and the induction hypothesis (IH), a $(0^X, \delta)$-reduction $R \colon a' \twoheadrightarrow b'$ of f is obtained in the following diagram.

(3.7.14)

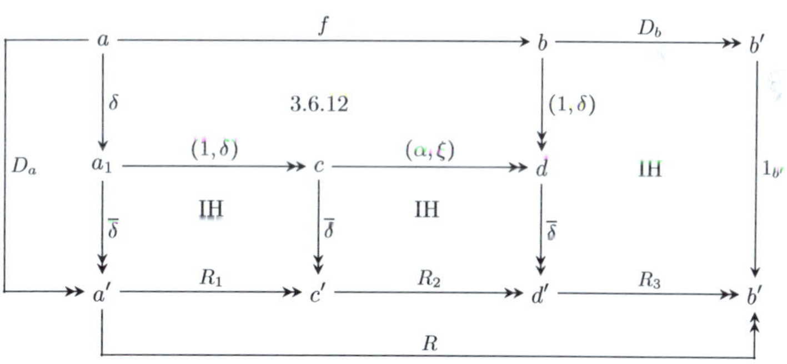

The steps for obtaining this diagram are essentially those for (3.7.13). In more detail:

(1) Using the fact that f and $a \longrightarrow a_1$ are δ-prime edges, we apply Lemma 3.6.12 to obtain the upper left rectangle in (3.7.14).

(2) For each $y \in \{c, d\}$, choose a δ-reduction $y \twoheadrightarrow y'$ using Lemma 3.6.9. The path

$$b \twoheadrightarrow d \twoheadrightarrow d'$$

is a δ-reduction of b, and $1_{b'}$ is a δ-reduction of b', since b' is δ reduced.

(3) By Lemma 3.2.12, the elements a_1, b, and c all have $\mathsf{size} - \mathsf{rank}$ less than that of a. The induction hypothesis is applied three times as indicated by IH. Each R_i is a $(0^X, \delta)$-reduction of the horizontal path above it.

The bottom concatenated path

$$R = (R_3, R_2, R_1)\colon a' \longrightarrow\!\!\!\!\!\rightarrow b'$$

is a $(0^X, \delta)$-reduction of f. This finishes the induction. □

Next is the preliminary case of Proposition 3.7.19 involving $\lambda^{\otimes}$ and $\rho^{\otimes}$.

Lemma 3.7.15. *Under the hypotheses of Definition 3.7.1, suppose that each prime edge in* $P\colon a \longrightarrow\!\!\!\!\!\rightarrow b$ *involves a single instance of* $\lambda^{\otimes}$ *or* $\rho^{\otimes}$. *Then* P *has a* $(0^X, \delta)$*-reduction.*

Proof. The proof is by induction on $\mathsf{size}(a) - \mathsf{rank}(a)$, which is always nonnegative by Lemma 3.2.9 (2). First suppose

$$\mathsf{size}(a) - \mathsf{rank}(a) = 0.$$

By Lemma 3.2.16 (1), each vertex in P satisfies $\mathsf{size} = \mathsf{rank}$. By Proposition 3.2.15, each vertex in P is a polynomial in the sense of Definition 3.6.3, which is δ-reduced by Lemma 3.6.5. Since P is a $(0^X, \delta)$-free path with a and b both δ-reduced, P is a $(0^X, \delta)$-reduction of P by Lemma 3.7.5.

For the induction step, suppose

$$\mathsf{size}(a) - \mathsf{rank}(a) > 0.$$

Reusing the reasoning in the second and the third paragraphs in the proof of Lemma 3.7.10, it suffices to show that each prime edge $f\colon a \longrightarrow b$ involving $\lambda^{\otimes}$ or $\rho^{\otimes}$, with a and b containing no 0^X and with a not δ-reduced, has a $(0^X, \delta)$-reduction. Using Lemmas 3.6.12 and 3.7.10 and the induction hypothesis (IH), we modify the diagram (3.7.14) as follows to obtain a $(0^X, \delta)$-reduction of f.

(3.7.16)

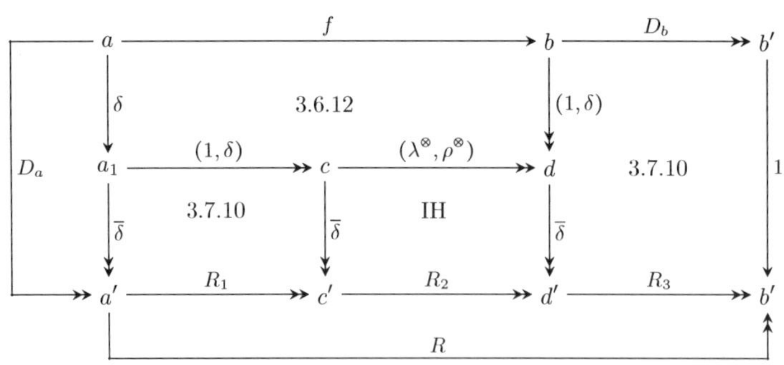

The steps for obtaining this diagram are as follows.

(1) Since a is not δ-reduced, the δ-reduction D_a has at least one δ-prime edge. We may assume that it is the first edge in D_a, as explained in the paragraph containing (3.7.12). This δ-prime edge is $a \longrightarrow a_1$ in (3.7.16).

(2) Starting with the prime edge f, which involves either $\lambda^{\otimes}$ or $\rho^{\otimes}$, and the δ-prime edge $a \longrightarrow a_1$, apply Lemma 3.6.12 to obtain the upper left rectangle in (3.7.16). By that lemma, the path $c \longrightarrow\!\!\!\!\!\rightarrow d$ consists of only prime edges involving either $\lambda^{\otimes}$ or $\rho^{\otimes}$, according to f.

(3) For each $y \in \{c, d\}$, choose a δ-reduction $y \twoheadrightarrow y'$ using Lemma 3.6.9. The path

$$b \twoheadrightarrow d \twoheadrightarrow d'$$

is a δ-reduction of b, and $1_{b'}$ is a δ-reduction of b', since b' is δ-reduced.

(4) Since the paths

$$\begin{array}{ccc} a_1 & \twoheadrightarrow & c \\ b & \xrightarrow{D_b} & b' \end{array}$$

consist of identities and δ-prime edges, we can apply Lemma 3.7.10 to each of them to obtain the $(0^X, \delta)$-reductions R_1 and R_3, respectively.

(5) By Lemma 3.2.12, c has $\mathsf{size} - \mathsf{rank}$ less than that of a. The induction hypothesis applies to the path $c \twoheadrightarrow d$ to yield the $(0^X, \delta)$-reduction R_2.

The bottom concatenated path

$$R = (R_3, R_2, R_1) \colon a' \twoheadrightarrow b'$$

is a $(0^X, \delta)$-reduction of f. This finishes the induction. □

Remark 3.7.17. The reason that Lemmas 3.7.10 and 3.7.15 are not proved by one overall induction is that in the diagram (3.7.16), b may have the same $\mathsf{size} - \mathsf{rank}$ as a by Lemma 3.2.16. If the two inductions are combined into one, then we would have to separately consider the case

$$\mathsf{size}(a) - \mathsf{rank}(a) = \mathsf{size}(b) - \mathsf{rank}(b)$$

with f as in Lemma 3.2.16 (2). The argument involved would be no easier than the proof of Lemma 3.7.15. ◇

Next is the preliminary case of Proposition 3.7.19 involving $\lambda^{-\otimes}$ and $\rho^{-\otimes}$.

Lemma 3.7.18. *Under the hypotheses of Definition 3.7.1, suppose that each prime edge in* $P \colon a \twoheadrightarrow b$ *involves a single instance of* $\lambda^{-\otimes}$ *or* $\rho^{-\otimes}$. *Then* P *has a* $(0^X, \delta)$-*reduction.*

Proof. Suppose $Q \colon b \twoheadrightarrow a$ is the formal inverse of P as in Definition 3.1.10. Each prime edge in Q involves either $\lambda^{\otimes}$ or $\rho^{\otimes}$. By Lemma 3.7.15, Q has a $(0^X, \delta)$-reduction $S \colon b' \twoheadrightarrow a'$, making the diagram

$$\begin{array}{ccc} a & \overset{Q}{\twoheadleftarrow} & b \\ {\scriptstyle D_a}\downarrow & & \downarrow{\scriptstyle D_b} \\ a' & \overset{S}{\twoheadleftarrow} & b' \end{array}$$

commutative. The formal inverse $R \colon a' \twoheadrightarrow b'$ of S is a $(0^X, \delta)$-reduction of P because

$$\varphi(P) = \varphi(Q)^{-1} \quad \text{and} \quad \varphi(R) = \varphi(S)^{-1}$$

for the graph morphism $\varphi \colon \mathsf{Gr}(X) \longrightarrow \mathsf{C}$ in Definition 3.1.14. □

Next is the main result of this section.

Proposition 3.7.19. *Under the hypotheses of Definition 3.7.1, the path* $P \colon a \twoheadrightarrow b$ *has a* $(0^X, \delta)$-*reduction.*

Proof. By the reasoning in the paragraph containing (3.7.11), it suffices to show that each prime edge in P has a $(0^X, \delta)$-reduction. By assumption, each prime edge $f{:}\, c \longrightarrow d$ in P has the property that c and d contain no 0^X. The possibilities of these 0^X-free edges are listed in Explanation 3.4.2. By Lemmas 3.7.10, 3.7.15, and 3.7.18, each prime edge in P has a $(0^X, \delta)$-reduction. □

3.8. Reduction of Multiplicative Units

In Proposition 3.7.19, we showed that each path whose vertices contain no 0^X has a $(0^X, \delta)$-reduction. In this section, we go one step further and eliminate the multiplicative units; see Proposition 3.8.14. In the first half of this section, we define a 1^X-reduction of an element in X^{fr}, and prove its existence and uniqueness in a suitable sense. In the second half of this section, we define a 1^X-reduction of a path and establish its existence under suitable conditions. Recall that Convention 3.3.1 is in effect.

Next is the multiplicative unit analogue of a 0^X-reduction and a δ-reduction in Definitions 3.3.4 and 3.6.2.

Definition 3.8.1. Consider the graph $\mathsf{Gr}(X)$ of X in Definition 3.1.9.

- A 1^X*-prime edge* is a prime edge that involves either $\lambda^{\otimes}$ or $\rho^{\otimes}$.
- An element $a \in X^{\mathsf{fr}}$ is 1^X*-reduced* if it is not the domain of any 1^X-prime edge.
- A 1^X*-reduction* of $a \in X^{\mathsf{fr}}$ is a path $P{:}\, a \longrightarrow\!\!\!\!\!\!\rightarrow b$ such that the following two statements hold.
  - b is 1^X-reduced.
  - Each edge in P is either an identity or a 1^X-prime edge. ◇

Example 3.8.2. Here are some examples of 1^X-prime edges.

- In Example 3.1.12, $1_w \oplus \lambda^{\otimes}_x 1_{y \oplus z}$ is a 1^X-prime edge.
- In Lemma 3.6.12, if f is a 1^X-prime edge, then every edge in the path R is a 1^X-prime edge.
- In the diagram (3.7.16), every edge in the path $c \longrightarrow\!\!\!\!\!\!\rightarrow d$ is a 1^X-prime edge. ◇

Example 3.8.3. The domain of a 1^X-prime edge must contain at least one instance of 1^X. Therefore, each element in X^{fr} that contains no 1^X is 1^X-reduced. However, the converse is not true. For example, if $x, y \in X$ with $y \neq 1^X$, then the elements

$$1^X \oplus x \quad \text{and} \quad y \otimes (x \oplus 1^X)$$

contain 1^X and are 1^X-reduced. ◇

Recall the concept of a polynomial in Definition 3.6.3. The following result provides a characterization of a δ-reduced element in X^{fr} that is also 1^X-reduced.

Lemma 3.8.4. *Suppose $a \in X^{\mathsf{fr}}$ is δ-reduced. Then a is 1^X-reduced if and only if it is a polynomial with each monomial either equal to 1^X or containing no instances of 1^X.*

Proof. By Lemma 3.6.5, being δ-reduced is equivalent to being a polynomial. If each monomial in a is either 1^X or contains no 1^X, then a is 1^X-reduced by the following two facts:

- The domain of $\lambda^{\otimes}$ contains an expression of the form $1^X \otimes ?$.
- The domain of $\rho^{\otimes}$ contains an expression of the form $? \otimes 1^X$.

Conversely, suppose it is not the case that each monomial in a is either 1^X or contains no 1^X. Then a contains a monomial with at least two elements in X that contains 1^X. Therefore, a is the domain of some 1^X-prime edge. □

Next we establish the existence of 1^X-reductions.

Lemma 3.8.5. *Each element in X^{fr} has a 1^X-reduction.*

Proof. We proceed by induction on $\mathsf{norm}(a)$ for elements $a \in X^{\mathsf{fr}}$, where norm is as in Definition 3.2.1. If $\mathsf{norm}(a) = 1$, then $a \in X$. So a is 1^X-reduced, and the identity 1_a is a 1^X-reduction of a.

For the induction step, suppose $\mathsf{norm}(a) > 0$. If a is 1^X-reduced, then the identity 1_a is a 1^X-reduction of a. If a is not 1^X-reduced, then there exists a 1^X-prime edge $f: a \longrightarrow b$ with domain a. Since an instance of 1^X is removed when going from a to b, the inequality

$$\mathsf{norm}(a) > \mathsf{norm}(b)$$

holds. So the induction hypothesis applies to b to yield a 1^X-reduction $P: b \twoheadrightarrow b'$ of b. The combined path

$$a \xrightarrow{f} b \xrightarrow{P} b'$$

is a 1^X-reduction of a. This finishes the induction. □

Next is the 1^X-reduction analogue of Lemma 3.3.11. It says that for a δ-reduced element in X^{fr}, the codomain of a 1^X-reduction is uniquely determined.

Lemma 3.8.6. *For any two 1^X-reductions*

$$b_1 \overset{P_1}{\twoheadleftarrow} a \overset{P_2}{\twoheadrightarrow} b_2$$

of a δ-reduced element $a \in X^{\mathsf{fr}}$, the equality $b_1 = b_2$ holds.

Proof. The element a, being δ-reduced, is a polynomial by Lemma 3.6.5. Each 1^X-prime edge in each P_i eliminates one copy of 1^X in a monomial that contains at least one other element in X. So both b_1 and b_2 are still polynomials, hence δ-reduced, and 1^X-reduced. By Lemma 3.8.4, the following statements hold.

- If a monomial in a is either a single copy of 1^X or contains no 1^X, then it is still in b_i.
- If a monomial in a contains at least two copies of 1^X and no other elements of X, then it becomes 1^X in b_i.
- If a monomial in a contains at least one copy of 1^X and at least one element in $X \setminus \{1^X\}$, then the corresponding monomial in b_i has all the copies of 1^X removed.

Therefore, $b_1 = b_2$. □

Next is the 1^X-reduction analogue of Lemma 3.3.12. Recall the graph morphism $\varphi: \mathsf{Gr}(X) \longrightarrow \mathsf{C}$ in Definition 3.1.14, with respect to which values in C are defined.

Lemma 3.8.7. *Any two 1^X-reductions*

$$a \overset{P_1}{\underset{P_2}{\rightrightarrows}} b$$

of a δ-reduced element $a \in X^{\mathrm{fr}}$ have the same value in C.

Proof. The element a, being δ-reduced, is a polynomial by Lemma 3.6.5. Suppose the polynomial a has the form

$$a = a^1 \oplus \cdots \oplus a^m$$

for some $m \geq 1$, with each a^j a monomial, and for some additive bracketing. The polynomial b has the form

$$b = b^1 \oplus \cdots \oplus b^m$$

with the same additive bracketing as a and with each monomial b^j obtained from a^j as described in the proof of Lemma 3.8.6. In particular, each monomial b^j is 1^X or contains no 1^X.

By the functoriality of $\oplus$ in C, for each $i = 1, 2$, the 1^X-reduction P_i has value in C a sum

$$\varphi P_i = P_i^1 \oplus \cdots \oplus P_i^m$$

with the same additive bracketing as a and with

$$\varphi a^j \xrightarrow{\;P_i^j\;} \varphi b^j$$

for $1 \leq j \leq m$. To show that P_1 and P_2 have the same values in C, it suffices to show the equality

$$P_1^j = P_2^j \colon \varphi a^j \longrightarrow \varphi b^j.$$

By the proof of Lemma 3.8.6, P_i^j is a composite of morphisms, each being a $\otimes$ of identity morphisms and a component of either $\lambda^{\otimes}$ or $\rho^{\otimes}$. Mac Lane's Coherence Theorem 1.3.3 in the monoidal category $(\mathsf{C}, \otimes, \mathbb{1}, \alpha^{\otimes}, \lambda^{\otimes}, \rho^{\otimes})$ implies that $P_1^j = P_2^j$ for each $1 \leq j \leq n$. □

We now define the 1^X analogue of a $(0^X, \delta)$-free path in Definition 3.6.10.

Definition 3.8.8. A 1^X-*free path* is a $(0^X, \delta)$-free path that contains no $\lambda^{\pm\otimes}$ and $\rho^{\pm\otimes}$. ◇

Explanation 3.8.9. In a 1^X-free path, every prime edge is either an identity or involves a single instance of $\alpha^{\pm\oplus}$, $\xi^{\pm\oplus}$, $\alpha^{\pm\otimes}$, or $\xi^{\pm\otimes}$. ◇

Motivation 3.8.10. In the next definition, the $(0^X, \delta)$-free path P should be thought of as a $(0^X, \delta)$-reduction of some other path as in Definition 3.7.1. The aim is to further eliminate multiplicative units, thereby replacing P by a 1^X-free path. ◇

Definition 3.8.11. Suppose

- $P \colon a \twoheadrightarrow b$ is a $(0^X, \delta)$-free path, and
- for each $x \in \{a, b\}$, $L_x \colon x \twoheadrightarrow x'$ is a 1^X-reduction of x.

A 1^X-*reduction* of (P, L_a, L_b) is a 1^X-free path $R: a' \twoheadrightarrow b'$ such that the diagram

$$\begin{array}{ccc} a & \overset{P}{\twoheadrightarrow} & b \\ {\scriptstyle L_a}\downarrow & & \downarrow{\scriptstyle L_b} \\ a' & \underset{R}{\twoheadrightarrow} & b' \end{array} \tag{3.8.12}$$

is commutative in the sense of Definition 3.1.14. We also call R a 1^X-reduction of P, suppressing L_a and L_b from the notation. ◇

The main result of this section, Proposition 3.8.14, asserts the existence of a 1^X-reduction when a and b are δ-reduced. First we need the following preliminary observation.

Lemma 3.8.13. *Suppose:*

- *$f: a \longrightarrow b$ is a $(0^X, \delta)$-free edge whose domain a is δ-reduced.*
- *For each $x \in \{a, b\}$, $L_x: x \twoheadrightarrow x'$ is a 1^X-reduction of x.*
- *If f contains $\lambda^{-\otimes}$ or $\rho^{-\otimes}$, then its subscript is a monomial or a factor of a monomial in a.*

Then a 1^X-reduction of f exists.

Proof. By Lemma 3.6.5, a is a polynomial as in Definition 3.6.3. The possibilities of the $(0^X, \delta)$-free edge f in Explanation 3.6.11 and the last assumption on f imply that b is also a polynomial and δ-reduced. By the proof of Lemma 3.8.6, the following statements hold:

- There is a canonical bijection between the set of monomials in a and the set of monomials in a'. Each monomial in a' is either equal to one in a, or is obtained from one in a by removing copies of 1^X. If the monomial started with some element in $X \setminus \{1^X\}$, then all the copies of 1^X are removed. If the monomial only has copies of 1^X, then all but one copies are removed.
- The additive bracketing of a' is equal to that of a.
- The analogous statements also hold for b and b'.

We now consider the possible cases of f.

(1) Suppose f is an identity or involves an instance of $\alpha^{\pm\oplus}$ or $\xi^{\pm\oplus}$. By the remarks above, the monomials in a and b are the same, and the monomials in a' and b' are the same. There is a prime edge $f': a' \longrightarrow b'$ of the same type as f whose subscripts are the corresponding monomials in a'. The path $R = f'$ is a 1^X-reduction of f because the diagram (3.8.12) is commutative by
   - the naturality of $\alpha^{\oplus}$ or $\xi^{\oplus}$ and
   - Lemma 3.8.7 applied to each monomial in a.

(2) Suppose f involves an instance of $\alpha^{\pm\otimes}$. Then a and b differ only by the multiplicative bracketing within a single monomial, say m_a in a and m_b in b. Denote by $m_{a'}$ and $m_{b'}$ the corresponding monomials in a' and b', respectively. Then a' and b' differ by at most the multiplicative bracketings in $m_{a'}$ and $m_{b'}$. Define $R: a' \longrightarrow b'$ as consisting of a single prime edge that is either an identity or contains an instance of $\alpha^{\pm\otimes}$ that moves the multiplicative brackets from $m_{a'}$ to $m_{b'}$ in precisely the same way as f with copies of 1^X removed, as described earlier in this proof.

The diagram (3.8.12) is commutative by

- the functoriality of $\oplus$ and
- the uniqueness in Mac Lane's Coherence Theorem 1.3.3 in $(\mathsf{C}, \otimes)$ applied to each monomial in a.

(3) Suppose f involves an instance of $\xi^{\pm\otimes}$. Then a and b differ only by the multiplicative bracketing, and the order of the factors within a single monomial. We slightly modify the argument in the previous case, using instead the uniqueness in the Symmetric Coherence Theorem 1.3.8 in the symmetric monoidal category $(\mathsf{C}, \otimes)$. The path $R: a' \longrightarrow b'$ consists of a single prime edge that is either an identity or contains an instance of $\xi^{\pm\otimes}$.

(4) If f involves an instance of $\lambda^{\otimes}$ or $\rho^{\otimes}$, then the concatenated path

$$a \xrightarrow{\quad f \quad} b \xrightarrow{\quad L_b \quad}\!\!\!\!\twoheadrightarrow b'$$

is a 1^X-reduction of a. There is an equality $a' = b'$, and the paths L_a and (L_b, f) have the same values in C by Lemmas 3.8.6 and 3.8.7. Therefore, with $R = 1_{a'}$ the diagram (3.8.12) is commutative.

(5) Suppose f involves an instance of $\lambda^{-\otimes}$ or $\rho^{-\otimes}$ with subscript a monomial or a factor of a monomial in a. Then a and b only differ by an extra copy of 1^X in a monomial in b. This implies that $a' = b'$. With $R = 1_{a'}$ the diagram (3.8.12) is commutative by

- the functoriality of $\oplus$ and
- the uniqueness in Mac Lane's Coherence Theorem 1.3.3 in $(\mathsf{C}, \otimes)$ applied to each monomial in a.

The above cases account for all the possibilities of f. □

The following main result of this section is a 1^X-reduction analogue of Propositions 3.5.32 and 3.7.19, which assert the existence of a 0^X-reduction and a $(0^X, \delta)$-reduction, respectively, under suitable conditions. Recall from Lemma 3.6.5 that an element in X^{fr} is δ-reduced if and only if it is a polynomial.

Proposition 3.8.14. *Suppose*

- $P: a \twoheadrightarrow b$ *is a* $(0^X, \delta)$*-free path with* a *and* b *both* δ*-reduced, and*
- *for each* $x \in \{a, b\}$, $L_x: x \twoheadrightarrow x'$ *is a* 1^X*-reduction of* x.

Then a 1^X*-reduction of* P *exists.*

Proof. Since P contains no δ-prime edges, and since a and b are δ-reduced,

- the naturality of $\alpha^{\otimes}$, $\lambda^{\otimes}$, $\rho^{\otimes}$, and $\xi^{\otimes}$, and
- the Symmetric Coherence Theorem 1.3.8 for $(\mathsf{C}, \otimes, \mathbb{1}, \alpha^{\otimes}, \lambda^{\otimes}, \rho^{\otimes}, \xi^{\otimes})$

imply the following statement:

> P has the same value in C as a $(0^X, \delta)$-free path $Q: a \twoheadrightarrow b$ that does *not* contain $\lambda^{-\otimes}$ and $\rho^{-\otimes}$ with subscript a sum.

For example, for elements $x, y \in X^{\mathsf{fr}}$, the two paths below have the same value in C by naturality and coherence as stated above.

$$\begin{array}{ccc}
x \oplus y & \xrightarrow{\xi^{\oplus}_{x,y}} & y \oplus x \\
{\scriptstyle \lambda^{-\otimes}_{x\oplus y}}\downarrow & & \uparrow{\scriptstyle \lambda^{\otimes}_{y\oplus x}} \\
1^X(x \oplus y) & & 1^X(y \oplus x) \\
{\scriptstyle 1_{1^X}\xi^{\oplus}_{x,y}}\downarrow & & \uparrow{\scriptstyle \lambda^{\otimes}_{1^X} 1_{y\oplus x}} \\
1^X(y \oplus x) & & (1^X 1^X)(y \oplus x) \\
{\scriptstyle \rho^{-\otimes}_{1^X(y\oplus x)}}\downarrow & & \uparrow{\scriptstyle \alpha^{-\otimes}_{1^X,1^X,y\oplus x}} \\
[1^X(y \oplus x)]\,1^X & \xrightarrow{\xi^{\otimes}_{1^X(y\oplus x),1^X}} & 1^X\,[1^X(y \oplus x)]
\end{array}$$

Therefore, we may assume that P does not contain $\lambda^{-\otimes}$ and $\rho^{-\otimes}$ with subscript a sum.

Suppose the path P consists of the prime edges $(f_n, \ldots, f_1)$ as in

$$a = a_0 \xrightarrow{f_1} a_1 \xrightarrow{f_2} \cdots \xrightarrow{f_n} a_n = b$$

for some $n \geq 1$. Observe the following:

- Since a is a polynomial, the reduction in the previous paragraph implies that each vertex in P is a polynomial, that is, δ-reduced.
- If $n > 1$, then for each $1 \leq i \leq n-1$, choose a 1^X-reduction $L_i\colon a_i \twoheadrightarrow a_i'$ of a_i, which exists by Lemma 3.8.5. Then for each $1 \leq i \leq n$, the prime edge f_i has a 1^X-reduction $R_i\colon a_{i-1}' \twoheadrightarrow a_i'$ by the reduction in the previous paragraph and Lemma 3.8.13.

The concatenated path

$$R = (R_n, \ldots, R_1)$$

is a 1^X-reduction of P because the diagram

$$\begin{array}{ccccccc}
a = a_0 & \xrightarrow{f_1} & a_1 & \xrightarrow{f_2} & \cdots & \xrightarrow{f_n} & a_n = b \\
{\scriptstyle L_a}\downarrow & & \downarrow{\scriptstyle L_1} & & & & \downarrow{\scriptstyle L_b} \\
a' = a_0' & \overset{R_1}{\twoheadrightarrow} & a_1' & \overset{R_2}{\twoheadrightarrow} & \cdots & \overset{R_n}{\twoheadrightarrow} & a_n' = b'
\end{array}$$

is commutative in the sense of Definition 3.1.14. □

3.9. The First Coherence Theorem

In this section, we use the results in the previous sections to prove the following coherence theorem for symmetric bimonoidal categories. Some examples of symmetric bimonoidal categories where this theorem can be applied are given after the proof. See Section 3.11 for further discussion about this theorem. The statement of the following theorem uses Definitions 2.1.2, 3.1.9, 3.1.14, 3.1.25, and 3.6.2 and Convention 3.3.1.

Theorem 3.9.1 (Laplaza's First Coherence). *Suppose* C *is a symmetric bimonoidal category in which the value of each* δ*-prime edge is a monomorphism. If*

$$a \underset{P_2}{\overset{P_1}{\rightrightarrows}} b$$

are two paths in $\mathsf{Gr}(X)$ *with* $a \in X^{\mathsf{fr}}$ *regular, then the values of* P_1 *and* P_2 *in* C *are equal.*

Proof. This proof involves a series of reductions. Since a is regular, the existence of a path from a to b implies that b is also regular by Lemma 3.1.29.

Reduction step 1

Choose 0^X-reductions $Q_x : x \twoheadrightarrow x'$ of $x \in \{a, b\}$, which exist by Lemma 3.3.6. The codomain x' and the value of the path Q_x in C are both uniquely determined by x by Lemmas 3.3.11 and 3.3.12. Since each $x \in \{a, b\}$ is regular, by Lemma 3.1.29 each x' is regular. By Proposition 3.5.32, each path P_i for $i = 1, 2$ has a 0^X-reduction R_i such that the diagram

$$\begin{array}{ccc} a & \xrightarrow{P_i} & b \\ {\scriptstyle Q_a}\big\downarrow & & \big\downarrow{\scriptstyle Q_b} \\ a' & \xrightarrow{R_i} & b' \end{array}$$

is commutative. Since $\lambda^{\oplus}$, $\rho^{\oplus}$, $\lambda^{\bullet}$, and $\rho^{\bullet}$ are natural isomorphisms in C, the values of Q_a and Q_b in C are isomorphisms. Therefore, to show that P_1 and P_2 have the same value in C, it suffices to show that R_1 and R_2 have the same value in C. In other words:

> *We may assume that each* $P_i : a \twoheadrightarrow b$ *is a* 0^X*-free path with* a *and* b *both* 0^X*-reduced and regular.*

Reduction step 2

If

$$\mathsf{supp}(a) = \mathsf{supp}(0^X),$$

then Proposition 3.5.33 implies that P_1 and P_2 have the same value in C. If

$$\mathsf{supp}(a) \neq \mathsf{supp}(0^X),$$

then a being 0^X-reduced as in Definition 3.3.4 implies that a contains no 0^X. That b is 0^X-reduced and Lemma 3.1.29(2) imply that b also contains no 0^X. In other words:

> *We may assume that each* $P_i : a \twoheadrightarrow b$ *is a* 0^X*-free path with* a *and* b *containing no* 0^X *and regular.*

Reduction step 3

Next suppose $D_b : b \twoheadrightarrow b'$ is a δ-reduction of b, which exists by Lemma 3.6.9. By Lemma 3.1.29, b' is regular. Observe the following:

- In each category, monomorphisms are closed under composition.
- Each edge in the δ-reduction D_b is either an identity or a δ-prime edge.
- The value of each δ-prime edge is a monomorphism in C by assumption.

These remarks imply that the value of D_b in C is a monomorphism. Therefore, it suffices to show that the two concatenated paths

$$a \underset{P_2}{\overset{P_1}{\rightrightarrows}} b \xrightarrow{D_b} b' \tag{3.9.2}$$

have the same value in C. In other words:

> *We may assume that each $P_i\colon a \longrightarrow b$ is a 0^X-free path with a and b containing no 0^X and regular, and with b δ-reduced.*

In particular, the identity 1_b is a δ-reduction of b.

Reduction step 4

Suppose $D_a\colon a \longrightarrow a'$ is a δ-reduction of a, which exists by Lemma 3.6.9. By Lemma 3.1.29, a' is regular. By Proposition 3.7.19, each P_i has a $(0^X, \delta)$-reduction R_i such that the diagram

$$\begin{array}{ccc} a & \xrightarrow{P_i} & b \\ {\scriptstyle D_a}\downarrow & & \downarrow{\scriptstyle 1_b} \\ a' & \xrightarrow{R_i} & b \end{array}$$

is commutative in the sense of Definition 3.1.14. To show that P_1 and P_2 have the same value in C, it suffices to show that R_1 and R_2 have the same value in C. In other words:

> *We may assume that each $P_i\colon a \longrightarrow b$ is a $(0^X, \delta)$-free path with a and b containing no 0^X, δ-reduced, and regular.*

Reduction step 5

Choose 1^X-reductions $L_x\colon x \longrightarrow x'$ of $x \in \{a, b\}$, which exist by Lemma 3.8.5. By Lemma 3.1.29, each x' is regular. Since each $x \in \{a, b\}$ is δ-reduced, by Lemmas 3.8.6 and 3.8.7 the codomain x' and the value of L_x in C are uniquely determined by x. By Proposition 3.8.14, each P_i has a 1^X-reduction R_i such that the diagram

$$\begin{array}{ccc} a & \xrightarrow{P_i} & b \\ {\scriptstyle L_a}\downarrow & & \downarrow{\scriptstyle L_b} \\ a' & \xrightarrow{R_i} & b' \end{array}$$

is commutative in the sense of Definition 3.1.14. Since $\lambda^{\otimes}$ and $\rho^{\otimes}$ are natural isomorphisms in C, the values of L_a and L_b in C are isomorphisms. So it suffices to show that R_1 and R_2 have the same value in C. In other words:

> *We may assume that each $P_i\colon a \longrightarrow b$ is a 1^X-free path with a and b containing no 0^X, δ-reduced, 1^X-reduced, and regular.*

Let us explain this reduction step more explicitly.

(1) By Definition 3.1.25 and Lemmas 3.6.5 and 3.8.4, the conditions on $x \in \{a, b\}$—namely, containing no 0^X, δ-reduced, 1^X-reduced, and regular—mean that x is a polynomial as in Definition 3.6.3 such that the following statements hold.

(i) Each monomial in x either
- is equal to 1^X, or
- contains no 0^X and 1^X.

(ii) Different monomials in x have different supports as in (3.1.24).

(iii) Within each monomial in x, the factors are all distinct elements in X.

For example, for distinct elements $p, q, r, s \in X \smallsetminus \{0^X, 1^X\}$, the polynomial

$$\big[1^X \oplus (pq)r\big] \oplus p(qs)$$

satisfies all three conditions stated above.

(2) As stated in Explanation 3.8.9, that each P_i is a 1^X-free path means that each of its prime edges is either an identity or involves an instance of $\alpha^{\pm\oplus}$, $\xi^{\pm\oplus}$, $\alpha^{\pm\otimes}$, or $\xi^{\pm\otimes}$. It follows that each vertex in each P_i is a polynomial that satisfies (i)–(iii) in (1) above.

When applied to a polynomial, prime edges involving:
- $\alpha^{\pm\oplus}$ move additive brackets, but do not change the monomials.
- $\xi^{\pm\oplus}$ permute monomials, but do not change the monomials.
- $\alpha^{\pm\otimes}$ move multiplicative brackets within a monomial, but do not change the additive bracketing and the order of the monomials.
- $\xi^{\pm\otimes}$ permute factors within a monomial, but do not change the additive bracketing and the order of the monomials.

These remarks and the naturality of $\alpha^{\oplus}$ and $\xi^{\oplus}$ in C imply that for each $i = 1, 2$, there is a diagram in $\mathsf{Gr}(X)$

(3.9.3)

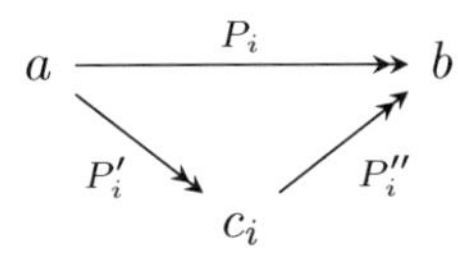

that is commutative in the sense of Definition 3.1.14 such that the following two statements hold.
- $P_i' \colon a \longrightarrow\!\!\!\!\!\rightarrow c_i$ consists of identities and prime edges involving $\alpha^{\pm\oplus}$ or $\xi^{\pm\oplus}$.
- $P_i'' \colon c_i \longrightarrow\!\!\!\!\!\rightarrow b$ consists of identities and prime edges involving $\alpha^{\pm\otimes}$ or $\xi^{\pm\otimes}$.

Moreover, there is an equality

$$c_1 = c_2$$

in X^{fr} because of the following facts.
- P_1' and P_2' only move additive brackets and permute monomials.
- P_1'' and P_2'' only move multiplicative brackets and permute factors within each monomial.
- Different monomials in a have different supports by condition (1)(ii).
- If there are paths $p \longrightarrow\!\!\!\!\!\rightarrow r$ and $q \longrightarrow\!\!\!\!\!\rightarrow r$ for elements $p, q, r \in X^{\mathsf{fr}}$, then

  $$\mathsf{supp}(p) = \mathsf{supp}(r) = \mathsf{supp}(q)$$

  by Lemma 3.1.29.

Finally, with c denoting $c_1 = c_2$, the following two statements hold.
- The Symmetric Coherence Theorem 1.3.8 for the additive structure

$$(\mathsf{C}, \oplus, \mathbb{0}, \alpha^{\oplus}, \lambda^{\oplus}, \rho^{\oplus}, \xi^{\oplus})$$

and condition (1)(ii) imply that the paths P_1' and $P_2': a \twoheadrightarrow c$ have the same value in C.

- The Symmetric Coherence Theorem 1.3.8 for the multiplicative structure

$$(\mathsf{C}, \otimes, \mathbb{1}, \alpha^{\otimes}, \lambda^{\otimes}, \rho^{\otimes}, \xi^{\otimes})$$

and condition (1)(iii), when applied to each monomial in c, imply that the paths P_1'' and $P_2'': c \twoheadrightarrow b$ have the same value in C.

Therefore, $P_1 = (P_1'', P_1')$ and $P_2 = (P_2'', P_2')$ have the same value in C. □

Explanation 3.9.4. In Theorem 3.9.1, there are two assumptions:

(i) The value of each δ-prime edge is a monomorphism in C.
(ii) $a \in X^{\mathsf{fr}}$ is regular.

These assumptions are only used in the following ways.

(1) Assumption (i) is only used in the reduction step (3.9.2) to make sure that the δ-reduction $D_b: b \twoheadrightarrow b'$, which consists of identities and δ-prime edges, has value in C a monomorphism. By Definition 3.6.2, a δ-prime edge is an iterated $\oplus$ and $\otimes$ of identities and one elementary edge δ^l or δ^r as in Definition 3.1.6. Its value in C has the same description, with identity morphisms instead of identities. However, even though $\delta^l_{x,y,z}$ and $\delta^r_{x,y,z}$ are monomorphisms in C, without further assumptions on C, it does *not* automatically follow that morphisms such as

$$\delta^l_{x,y,z} \oplus 1_w, \quad \delta^l_{x,y,z} \otimes 1_w, \quad \text{and} \quad (\delta^l_{x,y,z} \oplus 1_w) \otimes 1_v \tag{3.9.5}$$

are monomorphisms. Therefore, assumption (i) is imposed to ensure that the value of the δ-reduction D_b is a monomorphism in C.

(2) Even without the regularity assumption (ii), the diagrams (3.9.3) still exist, and each vertex in each path P_i still satisfies (1)(i) in the proof of Theorem 3.9.1. The regularity assumption is used to obtain conditions (1)(ii) and (1)(iii) in that proof. As explained there, these conditions then imply:
   - $c_1 = c_2$.
   - The paths P_1' and P_2' have the same value in C.
   - The paths P_1'' and P_2'' have the same value in C. ◇

Example 3.9.6 (Tight Symmetric Bimonoidal Categories)**.** If C is a tight symmetric bimonoidal category, then Theorem 3.9.1 applies to C for each choice of a graph morphism $\varphi: \mathsf{Gr}(X) \longrightarrow \mathsf{C}$ as in Definition 3.1.14. Recall that *tight* means that the distributivity morphisms δ^l and δ^r are natural isomorphisms, not just natural monomorphisms. Therefore, the value of each δ-prime edge is an isomorphism, hence a monomorphism, in C. In particular, Theorem 3.9.1 applies to the tight symmetric bimonoidal categories Σ and Σ' in Propositions 2.4.8 and 2.4.23. ◇

Example 3.9.7 (Distributive Symmetric Monoidal Categories)**.** These are also tight symmetric bimonoidal categories by Proposition 2.3.2. As explained in Example 3.9.6, Theorem 3.9.1 applies to each distributive symmetric monoidal category. In particular, it applies to

- symmetric monoidal closed categories with finite coproducts by Example 2.3.3;
- the category of modules over a commutative ring by Example 2.3.4; and
- distributive categories by Example 2.3.5. ◇

Example 3.9.8 (Bipermutative Categories). Right and left bipermutative categories in Definitions 2.5.2 and 2.5.11 are tight symmetric bimonoidal categories by Propositions 2.5.7 and 2.5.16, respectively. By Example 3.9.6, Theorem 3.9.1 applies to each right, respectively left, bipermutative category. ◇

Definition 3.9.9. A (symmetric) bimonoidal category C is called *flat* if each iterated sum and product of a component of δ^l or δ^r with a finite number of identity morphisms is a monomorphism. ◇

Example 3.9.10. Tight (symmetric) bimonoidal categories—that is, those with δ^l and δ^r natural isomorphisms—are flat. In particular, all the symmetric bimonoidal categories in Sections 2.3 through 2.6 are flat. Suppose C is a flat symmetric bimonoidal category. Then the morphisms in (3.9.5) and the values of δ-prime edges are monomorphisms. Therefore, Theorem 3.9.1 applies to C. ◇

Example 3.9.11. Suppose $x, y, z \in X$ are distinct elements, and that the value of each δ-prime edge is a monomorphism in C. Then the two paths in Example 3.1.11 have the same value in C by Theorem 3.9.1 because the domain $x(y \oplus z) \oplus 0^X$ is regular, as we mentioned in Example 3.1.31. ◇

Example 3.9.12. The proof of Theorem 3.9.1 will be reused in Section 4.4 to prove the second coherence theorem for symmetric bimonoidal categories. ◇

Example 3.9.13. Theorem 3.9.1 will be applied many times in Chapter 5 in the construction of a right bipermutative category that is equivalent in a suitable sense to a given tight symmetric bimonoidal category. See Explanation 5.2.31, Example 5.2.32, and Lemmas 5.3.4, 5.3.7, 5.3.8, and 5.4.4. ◇

3.10. Coherence of Bimonoidal Categories

In this section, we discuss the multiplicatively nonsymmetric analogue of the Coherence Theorem 3.9.1 that applies to bimonoidal categories instead of symmetric bimonoidal categories. It is also the bimonoidal analogue of Mac Lane's Coherence Theorem 1.3.3. As in the symmetric case, the Bimonoidal Coherence Theorem 3.10.7 does *not* require the invertibility of the distributivity morphisms δ^l and δ^r. Instead, it assumes a much weaker monomorphism condition.

The statement and the proof of Theorem 3.10.7 are obtained from those of Theorem 3.9.1 by

- omitting the multiplicative symmetry $\xi^{\otimes}$ and its inverse and
- using Mac Lane's Coherence Theorem 1.3.3 instead of Theorem 1.3.8 for the multiplicative structure $(\mathsf{C}, \otimes)$.

Below we will state the key definitions in the nonsymmetric context.

Convention 3.10.1. Throughout this section, suppose the triple (X, C, φ) consists of the following data.

- X is a set of formal variables with two distinguished elements $\{0^X, 1^X\}$ as in Definition 3.1.6.
- C is a bimonoidal category as in Definition 2.1.2, equipped with the graph structure in Example 3.1.5. In particular, its multiplicative structure

$$(\mathsf{C}, \otimes, \mathbb{1}, \alpha^{\otimes}, \lambda^{\otimes}, \rho^{\otimes})$$

  is a monoidal category, and C satisfies the 22 Laplaza axioms excluding (2.1.5) and (2.1.19).
- $\varphi\colon X \longrightarrow \mathsf{Ob}(\mathsf{C})$ is a function that satisfies $\varphi(0^X) = \mathbb{0}$ and $\varphi(1^X) = \mathbb{1}$. ◇

Definition 3.10.2. Under Convention 3.10.1, define the following.

- In Definition 3.1.6, an elementary edge in $\mathsf{Gr}^{\mathsf{el}}(X)$ is *nonsymmetric* if it is not of the form $\xi^{\otimes}_{x,y}$ and $\xi^{-\otimes}_{x,y}$ for any $x, y \in X^{\mathsf{fr}}$.
- In Definition 3.1.8, a prime edge is *nonsymmetric* if it only involves nonsymmetric elementary edges.
- The *nonsymmetric graph of* X, which is denoted by $\mathsf{Gr}^{\mathsf{ns}}(X)$, is obtained from Definition 3.1.9 by restricting to nonsymmetric prime edges. An edge or a path in $\mathsf{Gr}^{\mathsf{ns}}(X)$ is said to be *nonsymmetric*.
- The notions of
  - the graph morphism $\varphi\colon \mathsf{Gr}^{\mathsf{ns}}(X) \longrightarrow \mathsf{C}$,
  - the *value* in C of a path in $\mathsf{Gr}^{\mathsf{ns}}(X)$, and
  - *commutativity* in C

  are obtained from those in Definition 3.1.14 by restricting to $\mathsf{Gr}^{\mathsf{ns}}(X)$.
- The *nonsymmetric strict* $\{\oplus, \otimes\}$*-algebra* X^{ns} is obtained from X^{st} in Definition 3.1.23 by omitting the relation

$$x \otimes y = y \otimes x \quad \text{for} \quad x, y \in X^{\mathsf{fr}}.$$

- The quotient map

$$X^{\mathsf{fr}} \xrightarrow{\ \mathsf{nsupp}\ } X^{\mathsf{ns}}$$

  is called the *nonsymmetric support.*
- An element $x \in X^{\mathsf{fr}}$ is *nonsymmetric regular* if it satisfies the conditions (i) and (iii) in Definition 3.1.25, but with
  - the nonsymmetric support instead of the support and
  - X^{ns} instead of X^{st}. ◇

Lemma 3.10.3. *The following statements hold.*

(1) *If two elements in* X^{fr} *have the same nonsymmetric support, then one of them is nonsymmetric regular if and only if the other one is nonsymmetric regular.*

(2) *If* $x \twoheadrightarrow y$ *is a nonsymmetric path in* $\mathsf{Gr}^{\mathsf{ns}}(X)$, *then*

$$\mathsf{nsupp}(x) = \mathsf{nsupp}(y).$$

Proof. Reuse the proof for the symmetric case in Lemma 3.1.29 by replacing support, regularity, paths, and prime edges by their nonsymmetric counterparts in Definition 3.10.2. □

Example 3.10.4.

- A path in $\mathsf{Gr}(X)$ that is
  - a 0^X-reduction as in Definition 3.3.4,
  - a δ-reduction as in Definition 3.6.2, or
  - a 1^X-reduction as in Definition 3.8.1,

  is nonsymmetric. In other words, it belongs to $\mathsf{Gr}^{\mathsf{ns}}(X)$.
- Each path in Examples 3.1.11, 3.1.12, and 3.6.7 and Motivation 3.3.9 is nonsymmetric.
- Each path in (3.6.14)–(3.6.17), (3.6.19), and (3.6.20) is nonsymmetric.

On the other hand, each of the two paths in Example 3.1.22 is *not* nonsymmetric because it involves $\xi^{\otimes}$. ◇

Example 3.10.5. Regularity implies nonsymmetric regularity. Indeed, the support $\mathsf{supp}\colon X^{\mathsf{fr}} \longrightarrow X^{\mathsf{st}}$ in (3.1.24) factors into two projection maps

$$X^{\mathsf{fr}} \xrightarrow{\mathsf{nsupp}} X^{\mathsf{ns}} \longrightarrow X^{\mathsf{st}}.$$

Therefore, if an element $w \in X^{\mathsf{fr}}$ is regular as in Definition 3.1.25, then it is also nonsymmetric regular as in Definition 3.10.2. The next example demonstrates the difference between regularity and its nonsymmetric analogue. ◇

Example 3.10.6. Nonsymmetric regularity does *not* imply regularity. For example, suppose $x, y \in X \smallsetminus \{0^X, 1^X\}$ are distinct elements, and

$$z = (x \otimes y) \oplus (y \otimes x) \in X^{\mathsf{fr}}.$$

Then the equality

$$\mathsf{supp}(x \otimes y) = \mathsf{supp}(y \otimes x) \in X^{\mathsf{st}}$$

implies that z is *not* regular as in Definition 3.1.25. However, since

$$\mathsf{nsupp}(x \otimes y) \neq \mathsf{nsupp}(y \otimes x) \in X^{\mathsf{ns}},$$

$z \in X^{\mathsf{fr}}$ is nonsymmetric regular in the sense of Definition 3.10.2. ◇

For the statement of the following coherence theorem for bimonoidal categories, recall from Definition 3.6.2 that a *δ-prime edge* is a prime edge in $\mathsf{Gr}(X)$ that involves either δ^l or δ^r. Since a prime edge has at most one nonidentity elementary edge, a δ-prime edge is nonsymmetric in the sense of Definition 3.10.2.

Theorem 3.10.7 (Bimonoidal Coherence). *Under Convention 3.10.1, suppose that the value of each δ-prime edge is a monomorphism. If*

$$a \underset{P_2}{\overset{P_1}{\rightrightarrows}} b$$

are two paths in $\mathsf{Gr}^{\mathsf{ns}}(X)$ with $a \in X^{\mathsf{fr}}$ nonsymmetric regular, then the values of P_1 and P_2 in C are equal.

Proof. The proof of the current nonsymmetric case is obtained from the proofs of Theorem 3.9.1 and the preliminary results in the earlier sections of this chapter by the following procedure:

- We systematically remove all $\xi^{\pm\otimes}$ in Section 3.1–Section 3.9.
- We use the notions in Definition 3.10.2 in place of their counterparts in Section 3.1.
- We use Mac Lane's Coherence Theorem 1.3.3 instead of the Symmetric Coherence Theorem 1.3.8 for the monoidal category $(\mathsf{C}, \otimes)$ in
  - Lemmas 3.4.11 and 3.4.12 and Proposition 3.8.14 and
  - the second bullet item in the last paragraph in the proof of Theorem 3.9.1.

In particular, in the diagram (3.9.3), each path $P_i''\colon c \twoheadrightarrow b$ consists of identities and $\alpha^{\pm\otimes}$, so it only moves multiplicative brackets within each monomial in c. This is the reason why condition (ii) in Definition 3.1.25 is not included in the definition of nonsymmetric regularity. □

Example 3.10.8. Recall from Definition 3.9.9 that a bimonoidal category C is *flat* if each iterated sum and product of a component of δ^l or δ^r with a finite number of identity morphisms is a monomorphism. Tight bimonoidal categories—that is, those with δ^l and δ^r natural isomorphisms—are flat. If C is flat, then the value of each δ-prime edge is a monomorphism, so Theorem 3.10.7 applies to C. ◇

Example 3.10.9 (Sheet Diagrams). We mentioned in Note 2.7.5 that sheet diagrams for tight bimonoidal categories are discussed in [**CDH∞**]. Just as string diagrams for monoidal categories [**JS91a**] use Mac Lane's Coherence Theorem 1.3.3, sheet diagrams in [**CDH∞**] use the Bimonoidal Coherence Theorem 3.10.7. In particular, regularity in [**CDH∞**, Def. 13] and the coherence theorem [**CDH∞**, Th. 4] may be replaced by, respectively, nonsymmetric regularity in Definition 3.10.2 and Theorem 3.10.7. See Note 7.9.2 for the use of strictification in [**CDH∞**]. ◇

3.11. Notes

As we mentioned in the introduction of this chapter, the Coherence Theorem 3.9.1 for symmetric bimonoidal categories is due to Laplaza [**Lap72a**, Proposition 10], and the proof presented in this chapter follows the general outline of Laplaza's original proof. Here we point out some of the main differences between Laplaza's proof and this chapter, and explain how some inaccuracies in Laplaza's proof are fixed. In Note 3.11.7, we briefly discuss a possible 2-monad approach.

3.11.1 (Terminology and Notation). The table below shows the correspondence of terminology and notation between this chapter and [**Lap72a**].

Definition	Concept	Laplaza [**Lap72a**]
3.1.6	0^X, 1^X, X^{fr}, $\mathsf{Gr}^{\mathsf{el}}(X)$	n, u, $\underline{A}$, $\underline{G}$
3.1.7	$\mathsf{E}^{\mathsf{fr}}_{\mathsf{el}}(X)$	$\underline{H}$
3.1.9	$\mathsf{Gr}(X)$	$\underline{T}$
3.1.23	X^{st}	$\underline{A}^*$
3.1.6	elementary edge $\longrightarrow$	$\searrow$
3.1.9	path $\twoheadrightarrow$ in $\mathsf{Gr}(X)$	$\multimap$
3.3.4	0^X-reduction	reduction
3.6.2	δ-reduction	rappel
3.8.1	1^X-reduction	normalization

Moreover, the concept of a graph in Definition 3.1.4 is not explicitly defined in [**Lap72a**]. ◇

3.11.2 (Level of Detail). Laplaza's original proof is given partly in outline form, and some of the cases and proofs are omitted entirely. In this chapter, we provided all the detail and explained all the cases in full. For example:

- Section 3.5 and the main result Proposition 3.5.32, with about 15 pages total, correspond to [**Lap72a**, Prop. 5], which is condensed to about 2 pages there. Some cases, such as the one in Lemma 3.5.11, are not explicitly mentioned there.
- Lemma 3.6.12 corresponds to [**Lap72a**, Prop. 6], where the initial case of the induction with $\mathsf{norm}(a) = 3$ and some cases in the induction step are not explicitly mentioned.
- The proofs of the assertions in Section 3.8 are omitted in [**Lap72a**]. ◇

In this chapter, we corrected a few inaccurate statements in [**Lap72a**]. These are discussed in Notes 3.11.3 through 3.11.6 below.

3.11.3 (Zero-Reduction of Paths). In the proof of [**Lap72a**, Prop. 5, page 50, last paragraph], it is stated that, in our terminology,

> if $P\colon a \twoheadrightarrow b$ is a 0^X-free path with $a \in X^{\mathsf{fr}}$ 0^X-reduced, then each vertex in P is 0^X-reduced.

This statement is only partially correct because it does not account for the case $a = 0^X$, which is 0^X-reduced. For example, the elementary edge

$$0^X \xrightarrow{\lambda^{-\otimes}_{0^X}} 1^X \otimes 0^X$$

is a 0^X-free path, since it does not involve $\lambda^{\pm\oplus}$, $\rho^{\pm\oplus}$, $\lambda^{\pm\bullet}$, and $\rho^{\pm\bullet}$. However, the codomain $1^X \otimes 0^X$ is not 0^X-reduced. Example 3.4.10 has another example of a 0^X-free path whose vertices other than the domain are not 0^X-reduced.

These nontrivial cases involving 0^X-free paths with domain 0^X must be taken into account in the proofs of Propositions 3.5.32 and 3.5.33. An explicit characterization of 0^X-free paths with domain 0^X is given in Lemma 3.4.9(2). Subsequently, Lemmas 3.4.11 and 3.4.12 deal with the relevant cases, which are used in the proofs of Proposition 3.5.33 and Lemma 3.4.14, respectively. ◇

3.11.4 (Size and Rank). Just before [**Lap72a**, Prop. 6], it is stated that, in our terminology, for an element $a \in X^{\mathsf{fr}}$ the following are equivalent.

(i) a is a polynomial.
(ii) The identity 1_a is a δ-reduction of a.
(iii) $\mathsf{rank}(a) = \mathsf{size}(a)$.

This is only partially correct. While (i) and (ii) are equivalent by Lemma 3.6.5, (iii) is strictly stronger than (i) and (ii). Indeed, by Proposition 3.2.15, if $\mathsf{rank}(a) = \mathsf{size}(a)$, then a is a special kind of polynomial in which each monomial is either an element in X or a product of two elements in X. On the other hand, for elements $x, y, z \in X$, the monomial

$$a = (xy)z \in X^{\mathsf{fr}}$$

has $\mathsf{rank}(a) = 6$ and $\mathsf{size}(a) = 8$. In other words, (iii) implies (i), but (i) does not imply (iii). ◇

3.11.5 (Zero and Delta Reduction of Paths). The proof of [**Lap72a**, Prop. 7], which corresponds to our Proposition 3.7.19, requires the following corrections.

(1) The incorrect identification of (ii) and (iii) discussed in Note 3.11.4 implies that the proof of [**Lap72a**, Prop. 7] should have contained the case in Lemma 3.7.9, which in turn uses Lemmas 3.7.6 and 3.7.8. As we explained with the monomial $a = (xy)z$, even if $\mathsf{size}(a) > \mathsf{rank}(a)$, the identity 1_a may still be a δ-reduction of a. In this case, there are no δ-prime edges with domain a. The induction diagram in [**Lap72a**, page 58] relies on the existence of such a δ-prime edge with domain a, which is denoted by $a \rightharpoonup a_1$ there.
(2) In the proof of [**Lap72a**, Prop. 7], Laplaza first reduced to the case where the path P is a single prime edge, and then considered three separate cases, each proved by an induction. However, in the diagrams in [**Lap72a**, pages 58-59], the induction steps apply the induction hypothesis for a general path. The correct proof uses Lemmas 3.7.10, 3.7.15, and 3.7.18, each considering a path with some restrictions on its prime edges. The induction diagrams are (3.7.13), (3.7.14), and (3.7.16). ◇

3.11.6 (Monomorphisms in Theorem 3.9.1)**.** The statement of [**Lap72a**, Prop. 10], which corresponds to Theorem 3.9.1, does not include the hypothesis that the value of each δ-prime edge is a monomorphism in C. However, this condition is actually used implicitly in the first sentence of the second paragraph in the proof of [**Lap72a**, Prop. 10]. In the proof of Theorem 3.9.1, this hypothesis is used in the reduction step in (3.9.2). This point was elaborated in Explanation 3.9.4(1). ◇

3.11.7 (2-Monads)**.** For symmetric bimonoidal categories with δ^l and δ^r natural isomorphisms (not just monomorphisms), that is, *tight* symmetric bimonoidal categories, we mentioned in Example 3.9.6 that Theorem 3.9.1 applies. This special case of Theorem 3.9.1 is also claimed to follow from the 2-monad coherence result in [**Kel74**, pages 371–373]. This 2-monad approach has the advantage of being more conceptual in the sense that it is part of a more general 2-monad coherence result. On the other hand, there are additional issues to take into account.

First, this 2-monad approach relies on a substantial body of 2-category theory, including [**Kel74**], a number of papers in its references, and the references in those papers. Some basic 2-dimensional category theory is discussed at an elementary level in the book [**JY21**]. Coherence theory of 2-monad algebras goes beyond that book, and a systematic and highly detailed treatment comparable to that book does not exist.

Furthermore, to show that the results in [**Kel74**] are applicable to tight symmetric bimonoidal categories is itself a coherence theorem that requires a careful proof. Among other things, one has to (i) construct the relevant 2-monad and (ii) show that its pseudo algebras are precisely tight symmetric bimonoidal categories. Detailed proofs of these statements would be long and involved. An example of such a detailed proof is in [**JY21**, Ch. 9], where the authors constructed a 2-monad whose pseudo algebras are precisely cloven fibrations.

There is also a set-theoretic issue. In Theorem 3.9.1, there is no restriction on the size of the underlying category C, as long as it is a category. On the other hand, the 2-monad approach would involve the category Cat of all small categories, thereby restricting C to a small category. One may deal with this issue by assuming Grothendieck's Axiom of Universes, as in [**JY21**, Section 1.1]. Nevertheless, this is an extra set-theoretic axiom to deal with a nontrivial issue. ◇

CHAPTER 4

Coherence of Symmetric Bimonoidal Categories II

This chapter contains the second coherence theorem for symmetric bimonoidal categories, which is also due to Laplaza [**Lap72b**]. As in Theorem 3.9.1, the main Coherence Theorem 4.4.3 in this chapter states that some formal diagrams commute in symmetric bimonoidal categories that satisfy a monomorphism condition. In Theorem 3.9.1, the regularity condition is imposed on the common domain of the two paths. In Theorem 4.4.3, the regularity hypothesis is replaced by an assumption about the paths themselves. Roughly speaking, this new hypothesis says that the two paths permute the formal variables involved in the same way. Theorem 4.5.8 is the multiplicatively nonsymmetric analogue of Theorem 4.4.3 for bimonoidal categories. Moreover, Theorem II.5.4.4 is a braided version of this coherence result. For open questions related to this chapter, see Question A.5.6.

Organization. In Section 4.1, we explain the intuition and motivate the constructions and proofs in later sections.

The new hypothesis that replaces regularity involves a certain sequence of permutations, called the *distortion* of a path. This notion is defined in two stages. In Section 4.2, we define the distortion category $\mathcal{D}$ whose morphisms encode the sequences of permutations. In Theorem 4.2.29, we carefully prove that the distortion category is a left bipermutative category in the sense of Definition 2.5.11. By Proposition 2.5.16, the distortion category is a tight symmetric bimonoidal category. The distortion category extends the left bipermutative category Σ in Definition 2.4.1. Parts of the proof of Proposition 2.4.8 are reused in the proof of Theorem 4.2.29.

In Section 4.3, we associate to each path in $\mathsf{Gr}(X)$ in Definition 3.1.9 a morphism in the distortion category. By definition this morphism is the distortion of the path; see Definition 4.3.1. The distortion is the value of a path in the sense of (3.1.18), applied to the distortion category. The rest of this section contains observations and examples that illustrate the concept of distortion.

In Section 4.4, we prove the Coherence Theorem 4.4.3. This proof reuses much of the proof of Theorem 3.9.1, in particular its five reduction steps without the regularity condition. In the proof of Theorem 3.9.1, the last part after the five reduction steps was dealt with using the regularity assumption. In the proof of Theorem 4.4.3, the last part after the reduction steps is dealt with using the distortion assumption on the paths, see Lemma 4.4.1. This section ends with examples where Theorem 4.4.3 may be applied.

In Section 4.5, we discuss the multiplicatively nonsymmetric version of Theorem 4.4.3. We first define the multiplicatively nonsymmetric analogue of the distortion category, which is called the additive distortion category and is denoted by $\mathcal{D}^{\mathsf{ad}}$. Using $\mathcal{D}^{\mathsf{ad}}$, we define the additive distortion of a nonsymmetric path in Definition 3.10.2. The main Coherence Theorem 4.5.8 of this section states that, if two parallel nonsymmetric paths have the same additive distortion, then they have the same value in the bimonoidal category under consideration. Its proof is obtained from that of Theorem 4.4.3 by removing $\xi^{\pm\otimes}$ and using Theorem 1.3.3 instead of Theorem 1.3.8 for the multiplicative structure.

Section 4.6 provides a conceptual description of the distortion category and the additive distortion category as Grothendieck constructions over the finite ordinal category Σ. This observation is a repackaging of the definitions of $\mathcal{D}$ and $\mathcal{D}^{\text{ad}}$ that provides a better perspective about the relationship between $\mathcal{D}$, $\mathcal{D}^{\text{ad}}$, and Σ. The braided version for the braided distortion category is Proposition II.5.5.3.

Although we partly follow the outline of Laplaza's original proof in Theorem 4.4.3, there are some significant differences between this chapter and Laplaza's original proof. These issues are discussed in Section 4.7. In particular, we correct some inaccuracies in Laplaza's original proof; see Notes 4.7.3 and 4.7.4.

Reading Guide. As a possible alternative to reading this chapter linearly, we offer the following suggestion.

- First read Section 4.1, which motivates the distortion category, the distortion of a path, and their roles in the proof of Theorem 4.4.3.
- In Section 4.2, read Definitions 4.2.1, 4.2.7, 4.2.14, and 4.2.23, the explanation after them, and the statement of Theorem 4.2.29. Together they define the distortion category $\mathcal{D}$. Save the proofs for a second reading.
- In Section 4.3, first read Definition 4.3.1, which defines the distortion of a path. Then read Lemmas 4.3.5 and 4.3.6, which describe the images of polynomials in the distortion category, and Examples 4.3.7 and 4.3.8. Save the rest of this section for a second reading.
- In Section 4.4, first read Theorem 4.4.3, whose proof is only about half a page long. Then read Lemma 4.4.1, which is the key step where distortion replaces regularity.
- With a good grasp of the structure of the proof of Theorem 4.4.3, go back and read the proofs in Section 4.2, which mainly consist of computation involving permutations, and the second half of Section 4.3 starting with Convention 4.3.9.

As in Chapter 3, we divided the proof of Theorem 4.4.3 into a number of lemmas to clarify the overall structure of the proof and to make jumping forward and backward easier. Students are encouraged to regard the detailed proofs in Section 4.2 as exercises with full solutions.

Detail. In addition to proving Theorem 4.4.3, the detailed proofs in this chapter have the following additional purposes.

- As in Chapter 3, Laplaza's original proof of Theorem 4.4.3 contains some inaccuracies that have never been made explicit before. To pinpoint these subtle issues and discuss the corrections, it is necessary to dive into the detail. Section 4.7 contains more detailed discussion of the differences between this chapter and Laplaza's original proof and necessary corrections for the latter.
- The proof of Baez's Conjecture (Theorem 7.8.1) uses Theorem 4.4.3 a number of times.

- The proof of the Coherence Theorem II.5.4.4 of braided bimonoidal categories uses modified versions of the proofs in this chapter. A detailed treatment here will allow us to be both precise and concise at the same time in the braided case.

Concepts in Chapter 3 are also used in this chapter. Before Section 4.5, Convention 3.3.1 is in effect. In particular, C is a symmetric bimonoidal category, and X is a set with two distinguished elements 0^X and 1^X, with $\varphi\colon \mathsf{Gr}(X) \longrightarrow \mathsf{C}$ the graph morphism in Definition 3.1.14.

4.1. Motivation

In this section, we motivate

(i) the distortion category to be defined in Section 4.2,
(ii) the distortion of a path to be defined in Section 4.3, and
(iii) the proof of Theorem 4.4.3 in Section 4.4.

The distortion category is a crucial component in the formulation and proof of Theorem 4.4.3, the main coherence result of this chapter.

The Coherence Theorem 3.9.1 says that two parallel paths in $\mathsf{Gr}(X)$ whose domain is regular have the same value in the symmetric bimonoidal category C, in which the values of δ-prime edges are assumed to be monomorphisms. The assumption is imposed on the domain and not on the two paths. The regularity assumption excludes domains such as $x \oplus x$ and $x \otimes x$, which are not regular, for any $x \in X$. However, there should be some coherence properties even for parallel paths whose domains are not regular. For example, the composite

$$x \otimes x \xrightarrow{\xi^{\otimes}_{x,x}} x \otimes x \xrightarrow{\xi^{\otimes}_{x,x}} x \otimes x$$

is equal to $1_{x\otimes x}$. We seek an alternative coherence result that imposes a suitable condition on the paths and not on the domain.

The Coherence Theorem 4.4.3 is a variant of Theorem 3.9.1 that replaces the regularity assumption on the domain with an assumption about the two paths in question. As we pointed out in Explanation 3.9.4 (2), even without the regularity assumption, the diagrams (3.9.3) still exist along with condition (1)(i) in that proof for each vertex in each path P_i. The regularity assumption is used to obtain conditions (1)(ii) and (1)(iii) there. These conditions are used to conclude that the component paths P_1' and P_2' have the same value in C, and similarly for P_1'' and P_2''.

An alternative to regularity should be a concept such that the following statements hold.

(1) The reduction steps, without the regularity condition, leading up to the diagrams (3.9.3) are still valid.
(2) From the diagram (3.9.3), the new assumption on the paths P_1 and P_2 implies that they have the same value in C.

In the diagram (3.9.3), for either path $P_1 = (P_1'', P_1')$ or $P_2 = (P_2'', P_2')$, the following statements hold.

- The elements a, b, and c_i are polynomials as in Definition 3.6.3, with each monomial either equal to 1^X or containing no 0^X and 1^X.
- The first path P_i' only moves additive brackets and permutes the set of monomials.

- The second path P_i'' only moves multiplicative brackets and permutes the factors within each monomial.

Assuming step (1) above is possible, to achieve step (2), we would want the paths P_1 and P_2 to

(i) permute the set of monomials in the same way and
(ii) permute the factors in each monomial in the same way.

Suppose

$$a = a^1 \oplus \cdots \oplus a^m \in X^{\mathsf{fr}}$$

is a polynomial with some additive bracketing and with $m \geq 1$ monomials, and each monomial

$$a^j = a_1^j \otimes \cdots \otimes a_{r_j}^j \in X^{\mathsf{fr}}$$

for $1 \leq j \leq m$ has some multiplicative bracketing and $r_j \geq 1$ factors in X.

- To record the number of monomials in a and the number of factors in each monomial, we use the finite sequence of nonnegative integers
$$\underline{r} = (r_1, \ldots, r_m).$$
- To achieve (i), we need to associate to each path P_i a permutation
$$\sigma_i \in \Sigma_m \quad \text{such that} \quad \sigma_1 = \sigma_2.$$
Here Σ_m denotes the symmetric group on m letters as in Definition 2.4.1. The permutation σ_i permutes the m monomials $a^1, \ldots, a^m$ in a, ignoring its additive bracketing.
- To achieve (ii), we need to associate to each path P_i a permutation
$$\tau_i^j \in \Sigma_{r_j} \quad \text{such that} \quad \tau_1^j = \tau_2^j$$
for each $1 \leq j \leq m$. The permutation τ_i^j permutes the r_j factors in the monomial a^j, ignoring its multiplicative bracketing.

Therefore:

- To each object $a \in X^{\mathsf{fr}}$, we want to associate a finite sequence $\underline{r}$ of integers that, if a is a polynomial, records the number of monomials and the number of factors in each monomial in a.
- To each path $P: a \longrightarrow\!\!\!\!\rightarrow b$ in $\mathsf{Gr}(X)$, we want to associate a finite sequence of permutations
$$\big(\sigma; \tau^1, \ldots, \tau^m\big)$$
that, if a and b are polynomials with the same number of monomials, records
  - how P permutes the monomials in a to those in b via σ and
  - how P permutes the factors in each monomial in a to those in b via the τ^j.

This discussion suggests a category with

- finite sequences of nonnegative integers $\underline{r} = (r_1, \ldots, r_m)$ as objects and
- finite sequences of permutations $(\sigma; \tau^1, \ldots, \tau^m)$ as morphisms.

Furthermore, this should be a symmetric bimonoidal category whose sum $\oplus$ and product $\otimes$ correspond to those in X^{fr}. This is the distortion category $\mathcal{D}$ to be defined in Section 4.2 below. In Section 2.4, we already constructed a symmetric bimonoidal category Σ that corresponds to m and $\sigma \in \Sigma_m$ above. By Definition 2.4.1 and Proposition 2.4.8, Σ is a left bipermutative category as in Definition 2.5.11.

In particular, its structure morphisms $\alpha^{\oplus}$, $\lambda^{\oplus}$, $\rho^{\oplus}$, $\alpha^{\otimes}$, $\lambda^{\otimes}$, $\rho^{\otimes}$, $\xi^{\otimes}_{-,\mathbb{0}}$, $\lambda^{\bullet}$, $\rho^{\bullet}$, and δ^l are identities. We will show that the distortion category is a left bipermutative category that extends Σ in Theorem 4.2.29.

The Coherence Theorem 4.4.3 states that if the paths P_1 and P_2 have the same value in the distortion category, then they have the same value in the symmetric bimonoidal category C, in which the value of each δ-prime edge is a monomorphism. The proof is a precise version of the discussion above involving steps (1) and (2). Step (1) is in the proof of Theorem 4.4.3, and step (2) is in Lemma 4.4.1.

4.2. The Distortion Category

In this section, we construct the distortion category as motivated in Section 4.1, and observe that it is a left bipermutative category as in Definition 2.5.11. In Section 4.3, we will use the distortion category to define the distortion of a path in $\mathsf{Gr}(X)$ in Definition 3.1.9. The notion of the distortion of a path is part of the statement of the Coherence Theorem 4.4.3. Here is an outline of this section.

- The underlying category of the distortion category is defined in Definition 4.2.1 and verified in Lemma 4.2.5.
- The additive structure of the distortion category is defined in Definition 4.2.7 and verified in Lemma 4.2.12.
- The multiplicative structure of the distortion category is defined in Definition 4.2.14 and verified in Lemma 4.2.19.
- The rest of the left bipermutative category structure $(\lambda^{\bullet}, \rho^{\bullet}, \delta^l, \delta^r)$ for the distortion category is given in Definition 4.2.23 and shown to be well defined in Lemma 4.2.25.
- Theorem 4.2.29 proves that the distortion category is a left bipermutative category.

The Underlying Category. First we define the underlying category of the distortion category. Recall that Σ_m denotes the symmetric group on m letters. Its identity permutation is denoted by id_m.

Definition 4.2.1. Define the *distortion category* $\mathcal{D}$ as follows.

Objects: An object in $\mathcal{D}$ is a finite sequence

$$\underline{r} = (r_1, \ldots, r_m)$$

with $m \geq 0$ and with each r_j for $1 \leq j \leq m$ a nonnegative integer. We call m the *length* of $\underline{r}$, which is denoted by $|\underline{r}|$. The unique sequence with length 0 is denoted by $\varnothing$.

Morphisms: Suppose $\underline{s} = (s_1, \ldots, s_n)$ is an object in $\mathcal{D}$. With $\underline{r}$ as above, the morphism set $\mathcal{D}(\underline{r}; \underline{s})$ is defined as follows.

- If $m \neq n$, then $\mathcal{D}(\underline{r}; \underline{s})$ is empty.
- If $m = n$, then $\mathcal{D}(\underline{r}; \underline{s})$ is the set of finite sequences of permutations

$$\underline{\sigma} = (\sigma; \sigma_1, \ldots, \sigma_m) \in \Sigma_m \times \Sigma_{r_1} \times \cdots \times \Sigma_{r_m}$$

such that

$$\sigma\underline{r} = \big(r_{\sigma^{-1}(1)}, \ldots, r_{\sigma^{-1}(m)}\big) = \underline{s}. \tag{4.2.2}$$

The last equality means that

$$r_{\sigma^{-1}(j)} = s_j \quad \text{for} \quad 1 \leq j \leq m.$$

Identities: The identity morphism of an object $\underline{r}$ as above is the sequence

$$1_{\underline{r}} = (\mathrm{id}_m; \mathrm{id}_{r_1}, \ldots, \mathrm{id}_{r_m}) \tag{4.2.3}$$

of identity permutations, with $1_{\varnothing} = (\mathrm{id}_0;)$.

Composition: Suppose given morphisms

$$\underline{r} \xrightarrow{\ \underline{\sigma}\ } \underline{s} \xrightarrow{\ \underline{\tau}\ } \underline{t}$$

in $\mathcal{D}$ with $\underline{\sigma}$ as above and with

$$\underline{\tau} = (\tau; \tau_1, \ldots, \tau_m) \in \Sigma_m \times \Sigma_{r_{\sigma^{-1}(1)}} \times \cdots \times \Sigma_{r_{\sigma^{-1}(m)}}.$$

Their composite

$$\left(\underline{r} \xrightarrow{\ \underline{\tau\sigma}\ } \underline{t} \right) \in \mathcal{D}(\underline{r}; \underline{t})$$

is defined as

$$\underline{\tau\sigma} = \big(\tau\sigma; \tau_{\sigma(1)}\sigma_1, \ldots, \tau_{\sigma(m)}\sigma_m\big) \in \Sigma_m \times \Sigma_{r_1} \times \cdots \times \Sigma_{r_m}. \tag{4.2.4}$$

This finishes the definition of the distortion category. ◇

Lemma 4.2.5. *The distortion category $\mathcal{D}$ in Definition 4.2.1 is a groupoid.*

Proof. The unity axiom of a category in Definition 1.1.1 holds because $1_{\underline{r}}$ is a sequence of identity permutations. For the associativity axiom, suppose $\underline{\sigma}$ and $\underline{\tau}$ are as in Definition 4.2.1, and $\underline{\pi} \in \mathcal{D}(\underline{t}; \underline{u})$ is given by

$$\underline{\pi} = (\pi; \pi_1, \ldots, \pi_m) \in \Sigma_m \times \Sigma_{t_1} \times \cdots \times \Sigma_{t_m}$$

with each $t_j = r_{(\tau\sigma)^{-1}(j)}$. Then

$$\begin{aligned}
\underline{\pi}(\underline{\tau\sigma}) &= \big(\pi\tau\sigma; \pi_{(\tau\sigma)(1)}\tau_{\sigma(1)}\sigma_1, \ldots, \pi_{(\tau\sigma)(m)}\tau_{\sigma(m)}\sigma_m\big)\\
&= \big(\pi\tau\sigma; \pi_{\tau(\sigma(1))}\tau_{\sigma(1)}\sigma_1, \ldots, \pi_{\tau(\sigma(m))}\tau_{\sigma(m)}\sigma_m\big)\\
&= (\underline{\pi\tau})\underline{\sigma}.
\end{aligned}$$

Therefore, $\mathcal{D}$ is a category.

To see that $\mathcal{D}$ is a groupoid, suppose

$$\underline{\sigma} = (\sigma; \sigma_1, \ldots, \sigma_m) \in \mathcal{D}(\underline{r}; \underline{s})$$

is a morphism. Then the morphism

$$\underline{\sigma}^{-1} = \left(\sigma^{-1}; (\sigma_{\sigma^{-1}(1)})^{-1}, \ldots, (\sigma_{\sigma^{-1}(m)})^{-1}\right) \in \mathcal{D}(\underline{s}; \underline{r})$$

is the inverse of $\underline{\sigma}$. □

Explanation 4.2.6. Consider the distortion category $\mathcal{D}$.

(1) An object $\underline{r} = (r_1, \ldots, r_m)$ may be thought of as consisting of m consecutive intervals, with the jth interval having r_j objects.

(2) A morphism $\underline{\sigma} = (\sigma; \sigma_1, \ldots, \sigma_m) \in \mathcal{D}(\underline{r}; \underline{s})$ consists of
- a permutation $\sigma \in \Sigma_m$ that permutes the m intervals and leaves the order within each interval unchanged; and
- for each $1 \le j \le m$, a permutation $\sigma_j \in \Sigma_{r_j}$ that permutes the r_j objects in the jth interval in $\underline{r}$.

We think of $\underline{s}$ as the result of first applying the permutation σ_j to the jth interval in $\underline{r}$ for each $1 \le j \le m$, followed by the permutation σ that permutes the m resulting intervals.

The identity morphisms and composition are then what one would expect using the above geometric interpretation of objects and morphisms. ◇

The Additive Structure. Next we define the additive structure in the distortion category. We continue to use the notations in Definition 4.2.1. In the next two definitions, we reuse some of the structure in the left bipermutative category Σ in Definition 2.4.1 and Explanation 2.4.7.

Definition 4.2.7. Define the additive structure

$$\left(\oplus, \mathbb{0}, \alpha^{\oplus}, \lambda^{\oplus}, \rho^{\oplus}, \xi^{\oplus}\right)$$

in the distortion category $\mathcal{D}$ as follows.

The Sum: The functor

$$\mathcal{D} \times \mathcal{D} \xrightarrow{\ \oplus\ } \mathcal{D}$$

is defined as follows.

Objects: For objects $\underline{r} = (r_1, \ldots, r_m)$ and $\underline{r}' = (r'_1, \ldots, r'_k)$ in $\mathcal{D}$, define the object

$$\underline{r} \oplus \underline{r}' = \left(r_1, \ldots, r_m, r'_1, \ldots, r'_k\right) \tag{4.2.8}$$

with length $|\underline{r}| + |\underline{r}'|$.

Morphisms: For morphisms $\underline{\sigma} = (\sigma; \sigma_1, \ldots, \sigma_m) \in \mathcal{D}(\underline{r}; \underline{s})$ and

$$\underline{\sigma}' = (\sigma'; \sigma'_1, \ldots, \sigma'_k) \in \Sigma_k \times \Sigma_{r'_1} \times \cdots \times \Sigma_{r'_k}$$

in $\mathcal{D}(\underline{r}'; \underline{s}')$ with $|\underline{r}'| = |\underline{s}'| = k$, define the morphism

$$\underline{\sigma} \oplus \underline{\sigma}' = \left(\sigma \oplus \sigma'; \sigma_1, \ldots, \sigma_m, \sigma'_1, \ldots, \sigma'_k\right) \in \mathcal{D}\left(\underline{r} \oplus \underline{r}'; \underline{s} \oplus \underline{s}'\right). \tag{4.2.9}$$

Here $\sigma \oplus \sigma' \in \Sigma_{m+k}$ is their block sum in (2.4.2).

The Additive Zero: The object $\mathbb{0}$ is defined as the empty sequence $\varnothing \in \mathcal{D}$.

Associativity and Unity: The natural transformations $\alpha^{\oplus}$, $\lambda^{\oplus}$, and $\rho^{\oplus}$, with components, respectively,

$$(\underline{r} \oplus \underline{r}') \oplus \underline{r}'' \xrightarrow{\alpha^{\oplus}_{\underline{r}, \underline{r}', \underline{r}''}} \underline{r} \oplus (\underline{r}' \oplus \underline{r}'')$$

$$\varnothing \oplus \underline{r} \xrightarrow{\lambda^{\oplus}_{\underline{r}}} \underline{r}$$

$$\underline{r} \oplus \varnothing \xrightarrow{\rho^{\oplus}_{\underline{r}}} \underline{r}$$

are defined as the identities.

The Additive Symmetry: The natural transformation $\xi^{\oplus}$ has components

$$\underline{r} \oplus \underline{r}' \xrightarrow{\xi^{\oplus}_{\underline{r}, \underline{r}'}} \underline{r}' \oplus \underline{r}$$

defined as

$$\xi^{\oplus}_{\underline{r}, \underline{r}'} = \left(\xi^{\oplus}_{m,k}; \mathrm{id}_{r_1}, \ldots, \mathrm{id}_{r_m}, \mathrm{id}_{r'_1}, \ldots, \mathrm{id}_{r'_k}\right) \tag{4.2.10}$$

with $\xi^{\oplus}_{m,k} \in \Sigma_{m+k}$ the block permutation defined in (2.4.3).

This finishes the definition of $\left(\oplus, \mathbb{0}, \alpha^{\oplus}, \lambda^{\oplus}, \rho^{\oplus}, \xi^{\oplus}\right)$ in $\mathcal{D}$. ◇

Explanation 4.2.11. We continue Explanation 4.2.6.

(1) The object $\underline{r} \oplus \underline{r}'$ in (4.2.8) is the concatenation of the m intervals in $\underline{r}$ with the k intervals in $\underline{r}'$.

(2) Consider the morphism $\underline{\sigma} \oplus \underline{\sigma}'$ in (4.2.9).
- The block sum $\sigma \oplus \sigma'$ permutes
  - the m intervals in $\underline{r}$ via σ and
  - the k intervals in $\underline{r}'$ via σ'.
- The permutations σ_j for $1 \leq j \leq m$ and σ'_l for $1 \leq l \leq k$ permute, respectively,
  - the r_j objects in the jth interval in $\underline{r}$ and
  - the r'_l objects in the lth interval in $\underline{r}'$.

Starting with $\underline{r} \oplus \underline{r}'$, applying all the permutations σ_j and σ'_l, followed by the block sum $\sigma \oplus \sigma'$, the result is $\underline{s} \oplus \underline{s}'$.

(3) Regarding each of $\underline{r}$ and $\underline{r}'$ as a single block in $\underline{r} \oplus \underline{r}'$, the additive symmetry $\xi^{\oplus}_{\underline{r},\underline{r}'}$ in (4.2.10) swaps these two blocks and leaves the order in each block unchanged. ◇

Recall from Definitions 1.2.1 and 1.2.18 that a permutative category means a symmetric strict monoidal category.

Lemma 4.2.12. *With the additive structure in Definition 4.2.7, $\mathcal{D}$ is a permutative category.*

Proof. We already observed in Lemma 4.2.5 that $\mathcal{D}$ is a groupoid. To show that $\oplus$ is a functor, first observe that the sum in (4.2.9) of two identity morphisms as in (4.2.3) is another identity morphism because the block sum preserves identity morphisms.

The sum in (4.2.9) preserves composition as in (4.2.4) because for morphisms

$$\begin{aligned} \underline{\tau} &= (\tau; \tau_1, \ldots, \tau_m) \in \mathcal{D}(\underline{s}; \underline{t}) \quad \text{and} \\ \underline{\tau}' &= (\tau'; \tau'_1, \ldots, \tau'_k) \in \mathcal{D}(\underline{s}'; \underline{t}'), \end{aligned}$$

there are equalities in $\mathcal{D}(\underline{r} \oplus \underline{r}'; \underline{t} \oplus \underline{t}')$ as follows.

$$\begin{aligned} &(\underline{\tau} \oplus \underline{\tau}')(\underline{\sigma} \oplus \underline{\sigma}') \\ &= \big(\tau \oplus \tau'; \tau_1, \ldots, \tau_m, \tau'_1, \ldots, \tau'_k\big)\big(\sigma \oplus \sigma'; \sigma_1, \ldots, \sigma_m, \sigma'_1, \ldots, \sigma'_k\big) \\ &= \big((\tau \oplus \tau')(\sigma \oplus \sigma'); \tau_{\sigma(1)}\sigma_1, \ldots, \tau_{\sigma(m)}\sigma_m, \tau'_{\sigma'(1)}\sigma'_1, \ldots, \tau'_{\sigma'(k)}\sigma'_k\big) \\ &= \big((\tau\sigma) \oplus (\tau'\sigma'); \tau_{\sigma(1)}\sigma_1, \ldots, \tau_{\sigma(m)}\sigma_m, \tau'_{\sigma'(1)}\sigma'_1, \ldots, \tau'_{\sigma'(k)}\sigma'_k\big) \\ &= \big(\tau\sigma; \tau_{\sigma(1)}\sigma_1, \ldots, \tau_{\sigma(m)}\sigma_m\big) \oplus \big(\tau'\sigma'; \tau'_{\sigma'(1)}\sigma'_1, \ldots, \tau'_{\sigma'(k)}\sigma'_k\big) \\ &= (\underline{\tau}\underline{\sigma}) \oplus (\underline{\tau}'\underline{\sigma}') \end{aligned}$$

Therefore, $\oplus \colon \mathcal{D} \times \mathcal{D} \longrightarrow \mathcal{D}$ is a functor.

The identity natural transformations $\alpha^{\oplus}$, $\lambda^{\oplus}$, and $\rho^{\oplus}$ are well defined, with the last two by the definitions of the empty sequence $\varnothing$ and its identity morphism $1_{\varnothing} = (\mathrm{id}_0;)$. The unity axiom (1.2.2) and the pentagon axiom (1.2.3) are satisfied because each morphism involved is an identity. Therefore, $\mathcal{D}$ is a strict monoidal category.

The naturality of $\xi^{\oplus}$ in (4.2.10) means the commutativity of the following diagram for morphisms $\sigma \in \mathcal{D}(\underline{r};\underline{s})$ and $\sigma' \in \mathcal{D}(\underline{r}';\underline{s}')$.

$$\begin{array}{ccc} \underline{r} \oplus \underline{r}' & \xrightarrow{\xi^{\oplus}_{\underline{r},\underline{r}'}} & \underline{r}' \oplus \underline{r} \\ {\scriptstyle \underline{\sigma}\oplus\underline{\sigma}'}\downarrow & & \downarrow{\scriptstyle \underline{\sigma}'\oplus\underline{\sigma}} \\ \underline{s} \oplus \underline{s}' & \xrightarrow{\xi^{\oplus}_{\underline{s},\underline{s}'}} & \underline{s}' \oplus \underline{s} \end{array}$$

- The respective first component of the two composites, namely
$$\xi^{\oplus}_{m,k}(\sigma \oplus \sigma') \quad \text{and} \quad (\sigma' \oplus \sigma)\xi^{\oplus}_{m,k} \in \Sigma_{m+k},$$
are both given by the permutation below for $1 \le j \le m$ and $1 \le l \le k$.
$$\begin{cases} j \longmapsto k + \sigma(j) \\ m + l \longmapsto \sigma'(l) \end{cases}$$
- In each of the two composites, the remaining components are given by the permutations
$$\big(\sigma_1, \ldots, \sigma_m, \sigma'_1, \ldots, \sigma'_k\big).$$

Therefore, $\xi^{\oplus}$ is a natural transformation.

Next we check the symmetric monoidal category axioms for $\mathcal{D}$.

- The symmetry axiom (1.2.20) holds because the block permutation $\xi^{\oplus}_{m,k} \in \Sigma_{m+k}$ has inverse $\xi^{\oplus}_{k,m}$.
- The unit axiom (1.2.21) holds because $\xi^{\oplus}_{\underline{r},\varnothing}$ is the identity morphism $1_{\underline{r}}$.
- The hexagon axiom (1.2.22) is satisfied because, with $\alpha^{\oplus}$ being the identity, block permutations (2.4.3) satisfy the same hexagon axiom, that is,

$$\xi^{\oplus}_{m+k,p} = \big(\xi^{\oplus}_{m,p} \oplus \mathrm{id}_k\big)\big(\mathrm{id}_m \oplus \xi^{\oplus}_{k,p}\big) \in \Sigma_{m+k+p}. \tag{4.2.13}$$

Therefore, $\mathcal{D}$ is a permutative category with the additive structure. □

The Multiplicative Structure. Next we define the multiplicative structure in the distortion category. We continue to reuse the structure in the left bipermutative category Σ in Definition 2.4.1 and Explanation 2.4.7.

Definition 4.2.14. Define the multiplicative structure
$$\big(\otimes, \mathbb{1}, \alpha^{\otimes}, \lambda^{\otimes}, \rho^{\otimes}, \xi^{\otimes}\big)$$
in the distortion category $\mathcal{D}$ as follows.

The Product: The functor
$$\mathcal{D} \times \mathcal{D} \xrightarrow{\ \otimes\ } \mathcal{D}$$
is defined as follows.

Objects: For objects $\underline{r} = (r_1, \ldots, r_m)$ and $\underline{r}' = (r'_1, \ldots, r'_k)$ in $\mathcal{D}$, define the object

$$\underline{r} \otimes \underline{r}' = \Big(\big\{r_j + r'_1\big\}_{j=1}^{m}, \ldots, \big\{r_j + r'_k\big\}_{j=1}^{m}\Big) \tag{4.2.15}$$

with length $|\underline{r}| \cdot |\underline{r}'|$. In this object, the $\big(j + (l-1)m\big)$th entry is $r_j + r'_l$ for each $1 \le j \le m$ and $1 \le l \le k$.

Morphisms: For morphisms $\underline{\sigma} \in \mathcal{D}(\underline{r};\underline{s})$ and $\underline{\sigma}' \in \mathcal{D}(\underline{r}';\underline{s}')$ as in Definition 4.2.7, define the morphism

$$\underline{\sigma} \otimes \underline{\sigma}' = \Big(\sigma \otimes \sigma'; \big\{\sigma_j \oplus \sigma_1'\big\}_{j=1}^m, \ldots, \big\{\sigma_j \oplus \sigma_k'\big\}_{j=1}^m\Big) \in \mathcal{D}\big(\underline{r} \otimes \underline{r}'; \underline{s} \otimes \underline{s}'\big) \tag{4.2.16}$$

with
- $\sigma \otimes \sigma' \in \Sigma_{mk}$ the permutation defined in (2.4.4) and
- each $\sigma_j \oplus \sigma_l' \in \Sigma_{r_j + r_l'}$ a block sum (2.4.2).

In (4.2.16), after $\sigma \otimes \sigma'$, the $\big(j + (l-1)m\big)$th entry is $\sigma_j \oplus \sigma_l'$ for each $1 \le j \le m$ and $1 \le l \le k$.

The Multiplicative Unit: The object $\mathbb{1}$ is the sequence $(0) \in \mathcal{D}$ with length 1 and entry 0.

Associativity and Unity: The natural transformations $\alpha^{\otimes}$, $\lambda^{\otimes}$, and $\rho^{\otimes}$, with components, respectively,

$$(\underline{r} \otimes \underline{r}') \otimes \underline{r}'' \xrightarrow{\alpha^{\otimes}_{\underline{r},\underline{r}',\underline{r}''}} \underline{r} \otimes (\underline{r}' \otimes \underline{r}'')$$
$$(0) \otimes \underline{r} \xrightarrow{\lambda^{\otimes}_{\underline{r}}} \underline{r}$$
$$\underline{r} \otimes (0) \xrightarrow{\rho^{\otimes}_{\underline{r}}} \underline{r}$$

are defined as the identities.

The Multiplicative Symmetry: The natural transformation $\xi^{\otimes}$ has components

$$\underline{r} \otimes \underline{r}' \xrightarrow{\xi^{\otimes}_{\underline{r},\underline{r}'}} \underline{r}' \otimes \underline{r}$$

defined as

$$\xi^{\otimes}_{\underline{r},\underline{r}'} = \Big(\xi^{\otimes}_{m,k}; \big\{\xi^{\oplus}_{r_j,r_1'}\big\}_{j=1}^m, \ldots, \big\{\xi^{\oplus}_{r_j,r_k'}\big\}_{j=1}^m\Big) \tag{4.2.17}$$

with
- $\xi^{\otimes}_{m,k} \in \Sigma_{mk}$ the permutation defined in (2.4.5) and
- each $\xi^{\oplus}_{r_j,r_l'} \in \Sigma_{r_j + r_l'}$ a block permutation (2.4.3).

After $\xi^{\otimes}_{m,k}$, the $\big(j + (l-1)m\big)$th entry is $\xi^{\oplus}_{r_j,r_l'}$ for each $1 \le j \le m$ and $1 \le l \le k$.

This finishes the definition of $\big(\otimes, \mathbb{1}, \alpha^{\otimes}, \lambda^{\otimes}, \rho^{\otimes}, \xi^{\otimes}\big)$ in $\mathcal{D}$. ◇

Explanation 4.2.18. We continue Explanations 4.2.6 and 4.2.11.

(1) The object $\underline{r} \otimes \underline{r}'$ in (4.2.15) is a $k \times m$ matrix, whose rows are indexed by the k intervals in $\underline{r}'$. For each $1 \le l \le k$, the lth row $\{r_j + r_l'\}_{j=1}^m$ is obtained from $\underline{r}$ by concatenating each of its m intervals with the lth interval in $\underline{r}'$. Therefore, for each $1 \le j \le m$, the (l,j)-entry of this matrix is an interval with $r_j + r_l'$ objects.

(2) Consider the morphism $\underline{\sigma} \otimes \underline{\sigma}' \in \mathcal{D}(\underline{r} \otimes \underline{r}'; \underline{s} \otimes \underline{s}')$ in (4.2.16).
- The first entry $\sigma \otimes \sigma' \in \Sigma_{mk}$, which is defined in (2.4.4),
  - permutes the k rows of the matrix $\underline{r} \otimes \underline{r}'$ via $\sigma' \in \Sigma_k$ and
  - permutes the m columns of the matrix via $\sigma \in \Sigma_m$.
- The other entries form a $k \times m$ matrix of permutations. Its (l,j)-entry is the block sum $\sigma_j \oplus \sigma_l' \in \Sigma_{r_j + r_l'}$.

Starting with the $k \times m$ matrix $\underline{r} \otimes \underline{r}'$, first applying the block sums $\sigma_j \oplus \sigma'_l$ to its entries, and then permuting the rows and columns via $\sigma \otimes \sigma'$, the resulting $k \times m$ matrix is $\underline{s} \otimes \underline{s}'$.

(3) Consider the multiplicative symmetry $\xi^{\otimes}_{\underline{r},\underline{r}'} \in \mathcal{D}(\underline{r} \otimes \underline{r}'; \underline{r}' \otimes \underline{r})$ in (4.2.17).
- The first entry $\xi^{\otimes}_{m,k} \in \Sigma_{mk}$, which is defined in (2.4.5), transposes the $k \times m$ matrix $\underline{r} \otimes \underline{r}'$.
- The other entries form a $k \times m$ matrix of permutations. Its (l, j)-entry is the block permutation $\xi^{\oplus}_{r_j, r'_l} \in \Sigma_{r_j + r'_l}$ in (2.4.3).

Starting with the $k \times m$ matrix $\underline{r} \otimes \underline{r}'$, first applying the block permutations $\xi^{\oplus}_{r_j, r'_l}$ to its entries, and then taking the transpose $\xi^{\otimes}_{m,k}$, the resulting $m \times k$ matrix is $\underline{r}' \otimes \underline{r}$. ◇

Lemma 4.2.19. *With the multiplicative structure in Definition 4.2.14, $\mathcal{D}$ is a permutative category.*

Proof. We use the notations in Definitions 4.2.1, 4.2.7, and 4.2.14. We already observed in Lemma 4.2.5 that $\mathcal{D}$ is a groupoid.

To show that $\otimes$ is a functor, first observe that the product in (4.2.16) of two identity morphisms as in (4.2.3) is another identity morphism because

- the block sum preserves identity morphisms and
- $\mathrm{id}_m \otimes \mathrm{id}_k = \mathrm{id}_{mk}$.

The product in (4.2.16) preserves composition as in (4.2.4) because for morphisms

$$\begin{aligned} \underline{\tau} &= (\tau; \tau_1, \ldots, \tau_m) \in \mathcal{D}(\underline{s}; \underline{t}) \quad \text{and} \\ \underline{\tau}' &= (\tau'; \tau'_1, \ldots, \tau'_k) \in \mathcal{D}(\underline{s}'; \underline{t}'), \end{aligned}$$

there are equalities in $\mathcal{D}\big(\underline{r} \otimes \underline{r}'; \underline{t} \otimes \underline{t}'\big)$ as follows.

$$\begin{aligned} &(\underline{\tau} \otimes \underline{\tau}')(\underline{\sigma} \otimes \underline{\sigma}') \\ &= \big(\tau \otimes \tau'; \{\tau_j \oplus \tau'_1\}_{j=1}^m, \ldots, \{\tau_j \oplus \tau'_k\}_{j=1}^m\big)\big(\sigma \otimes \sigma'; \{\sigma_j \oplus \sigma'_1\}_{j=1}^m, \ldots, \{\sigma_j \oplus \sigma'_k\}_{j=1}^m\big) \\ &= \Big((\tau \otimes \tau')(\sigma \otimes \sigma'); \big\{(\tau_{\sigma(j)} \oplus \tau'_{\sigma'(1)})(\sigma_j \oplus \sigma'_1)\big\}_{j=1}^m, \ldots, \big\{(\tau_{\sigma(j)} \oplus \tau'_{\sigma'(k)})(\sigma_j \oplus \sigma'_k)\big\}_{j=1}^m\Big) \\ &= \Big((\tau\sigma) \otimes (\tau'\sigma'); \big\{\tau_{\sigma(j)}\sigma_j \oplus \tau'_{\sigma'(1)}\sigma'_1\big\}_{j=1}^m, \ldots, \big\{\tau_{\sigma(j)}\sigma_j \oplus \tau'_{\sigma'(k)}\sigma'_k\big\}_{j=1}^m\Big) \\ &= \big(\tau\sigma; \tau_{\sigma(1)}\sigma_1, \ldots, \tau_{\sigma(m)}\sigma_m\big) \otimes \big(\tau'\sigma'; \tau'_{\sigma'(1)}\sigma'_1, \ldots, \tau'_{\sigma'(k)}\sigma'_k\big) \\ &= \big(\underline{\tau\sigma}\big) \otimes \big(\underline{\tau}'\underline{\sigma}'\big) \end{aligned}$$

Therefore, $\otimes \colon \mathcal{D} \times \mathcal{D} \longrightarrow \mathcal{D}$ is a functor.

To see that $\alpha^{\otimes}$ is well defined, suppose $\underline{r}'' = (r''_1, \ldots, r''_p)$ is an object in $\mathcal{D}$. Then both $(\underline{r} \otimes \underline{r}') \otimes \underline{r}''$ and $\underline{r} \otimes (\underline{r}' \otimes \underline{r}'')$ are given by the sequence

$$\begin{aligned} &\Big(\big\{r_j + r'_1 + r''_1\big\}_{j=1}^m, \ldots, \big\{r_j + r'_k + r''_1\big\}_{j=1}^m, \\ &\quad \ldots, \big\{r_j + r'_1 + r''_p\big\}_{j=1}^m, \ldots, \big\{r_j + r'_k + r''_p\big\}_{j=1}^m\Big). \end{aligned} \tag{4.2.20}$$

In this sequence, the $\big(j + (l-1)m + (t-1)mk\big)$th entry is $r_j + r'_l + r''_t$ for $1 \leq j \leq m$, $1 \leq l \leq k$, and $1 \leq t \leq p$.

Next suppose

$$\underline{\sigma}'' = (\sigma''; \sigma''_1, \ldots, \sigma''_p) \in \mathcal{D}(\underline{r}''; \underline{s}'')$$

is a morphism. Then there is an equality

$$(\sigma \otimes \sigma') \otimes \sigma'' = \sigma \otimes (\sigma' \otimes \sigma'') \in \Sigma_{mkp}$$

by (2.4.9). So the first entry in $(\underline{\sigma} \otimes \underline{\sigma}') \otimes \underline{\sigma}''$ is equal to the first entry in $\underline{\sigma} \otimes (\underline{\sigma}' \otimes \underline{\sigma}'')$. For the equality of their remaining entries, we

- reuse (4.2.20) by replacing $(+, r_j, r'_l, r''_t)$ with $(\oplus, \sigma_j, \sigma'_l, \sigma''_t)$ for $1 \le j \le m$, $1 \le l \le k$, and $1 \le t \le p$; and
- use the fact that block sums of permutations is strictly associative.

Therefore, we can define $\alpha^{\otimes}$ as the identity.

The identity natural transformations $\lambda^{\otimes}$ and $\rho^{\otimes}$ are well defined by the definitions of the sequence $\mathbb{1} = (0)$ and its identity morphism

$$1_{(0)} = (\mathrm{id}_1; \mathrm{id}_0).$$

The unity axiom (1.2.2) and the pentagon axiom (1.2.3) are satisfied because each morphism involved is an identity. Therefore, $\mathcal{D}$ is a strict monoidal category with the multiplicative structure.

The naturality of $\xi^{\otimes}$ in (4.2.17) means the commutativity of the following diagram for morphisms $\sigma \in \mathcal{D}(\underline{r}; \underline{s})$ and $\sigma' \in \mathcal{D}(\underline{r}'; \underline{s}')$.

$$\begin{array}{ccc}
\underline{r} \otimes \underline{r}' & \xrightarrow{\xi^{\otimes}_{\underline{r},\underline{r}'}} & \underline{r}' \otimes \underline{r} \\
{\scriptstyle \underline{\sigma} \otimes \underline{\sigma}'} \downarrow & & \downarrow {\scriptstyle \underline{\sigma}' \otimes \underline{\sigma}} \\
\underline{s} \otimes \underline{s}' & \xrightarrow{\xi^{\otimes}_{\underline{s},\underline{s}'}} & \underline{s}' \otimes \underline{s}
\end{array}$$

By (4.2.2), (4.2.16), and (4.2.17), the above diagram involves the following four morphisms in $\mathcal{D}$.

$$\begin{aligned}
\underline{\sigma} \otimes \underline{\sigma}' &= \Big(\sigma \otimes \sigma'; \{\sigma_j \oplus \sigma'_1\}_{j=1}^{m}, \ldots, \{\sigma_j \oplus \sigma'_k\}_{j=1}^{m}\Big) \\
\underline{\sigma}' \otimes \underline{\sigma} &= \Big(\sigma' \otimes \sigma; \{\sigma'_l \oplus \sigma_1\}_{l=1}^{k}, \ldots, \{\sigma'_l \oplus \sigma_m\}_{l=1}^{k}\Big) \\
\xi^{\otimes}_{\underline{r},\underline{r}'} &= \Big(\xi^{\otimes}_{m,k}; \big\{\xi^{\oplus}_{r_j,r'_1}\big\}_{j=1}^{m}, \ldots, \big\{\xi^{\oplus}_{r_j,r'_k}\big\}_{j=1}^{m}\Big) \\
\xi^{\otimes}_{\underline{s},\underline{s}'} &= \Big(\xi^{\otimes}_{m,k}; \big\{\xi^{\oplus}_{r_{\sigma^{-1}(j)},r'_{\sigma'^{-1}(1)}}\big\}_{j=1}^{m}, \ldots, \big\{\xi^{\oplus}_{r_{\sigma^{-1}(j)},r'_{\sigma'^{-1}(k)}}\big\}_{j=1}^{m}\Big)
\end{aligned}$$

Using (4.2.4) and interpreting $\xi^{\otimes}_{m,k}$ as taking the transpose of a matrix as in Explanation 4.2.18, there are equalities as follows.

$$\begin{aligned}
&\xi^{\otimes}_{\underline{s},\underline{s}'}(\underline{\sigma} \otimes \underline{\sigma}') \\
&= \Big(\xi^{\otimes}_{m,k}(\sigma \otimes \sigma'); \big\{\xi^{\oplus}_{r_j,r'_1}(\sigma_j \oplus \sigma'_1)\big\}_{j=1}^{m}, \ldots, \big\{\xi^{\oplus}_{r_j,r'_k}(\sigma_j \oplus \sigma'_k)\big\}_{j=1}^{m}\Big) \\
&= \Big((\sigma' \otimes \sigma)\xi^{\otimes}_{m,k}; \big\{(\sigma'_1 \oplus \sigma_j)\xi^{\oplus}_{r_j,r'_1}\big\}_{j=1}^{m}, \ldots, \big\{(\sigma'_k \oplus \sigma_j)\xi^{\oplus}_{r_j,r'_k}\big\}_{j=1}^{m}\Big) \\
&= (\underline{\sigma}' \otimes \underline{\sigma})\xi^{\otimes}_{\underline{r},\underline{r}'}
\end{aligned}$$

Therefore, $\xi^{\otimes}$ is a natural transformation.

Next we check the symmetric monoidal category axioms for $\mathcal{D}$ with the multiplicative structure.

- The symmetry axiom (1.2.20) holds because
  - the permutation $\xi^{\otimes}_{m,k} \in \Sigma_{mk}$ in (2.4.5) has inverse $\xi^{\otimes}_{k,m}$ and
  - the block permutation $\xi^{\oplus}_{r_j,r'_l} \in \Sigma_{r_j+r'_l}$ in (2.4.3) has inverse $\xi^{\oplus}_{r'_l,r_j}$.
- The unit axiom (1.2.21) holds because $\xi^{\otimes}_{\underline{r},(0)}$ is the identity morphism $1_{\underline{r}}$.

It remains to prove the hexagon axiom (1.2.22) for $\mathcal{D}$ with the multiplicative structure. Since $\alpha^{\otimes}$ is the identity, the hexagon axiom means the commutativity of the following diagram for objects $\underline{r}, \underline{r}', \underline{r}'' \in \mathcal{D}$.

(4.2.21)

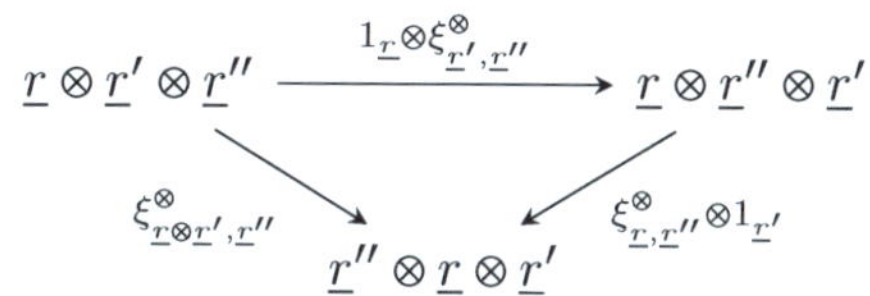

To check that this diagram is commutative, we first compute its three morphisms using (4.2.3), (4.2.16), and (4.2.17) as follows, assuming $|\underline{r}| = m$, $|\underline{r}'| = k$, and $|\underline{r}''| = p$.

(4.2.22)
$$\begin{aligned}
\xi^{\otimes}_{\underline{r}\otimes\underline{r}',\underline{r}''} &= \Big(\xi^{\otimes}_{mk,p}; \big\{\xi^{\oplus}_{r_j+r'_1,r''_1}\big\}_{j=1}^m, \ldots, \big\{\xi^{\oplus}_{r_j+r'_k,r''_1}\big\}_{j=1}^m, \ldots, \\
&\qquad \big\{\xi^{\oplus}_{r_j+r'_1,r''_p}\big\}_{j=1}^m, \ldots, \big\{\xi^{\oplus}_{r_j+r'_k,r''_p}\big\}_{j=1}^m\Big) \\
1_{\underline{r}} \otimes \xi^{\otimes}_{\underline{r}',\underline{r}''} &= \big(\mathrm{id}_m; \mathrm{id}_{r_1}, \ldots, \mathrm{id}_{r_m}\big) \otimes \Big(\xi^{\otimes}_{k,p}; \big\{\xi^{\oplus}_{r'_l,r''_1}\big\}_{l=1}^k, \ldots, \big\{\xi^{\oplus}_{r'_l,r''_p}\big\}_{l=1}^k\Big) \\
&= \Big(\mathrm{id}_m \otimes \xi^{\otimes}_{k,p}; \big\{\mathrm{id}_{r_j} \oplus \xi^{\oplus}_{r'_1,r''_1}\big\}_{j=1}^m, \ldots, \big\{\mathrm{id}_{r_j} \oplus \xi^{\oplus}_{r'_k,r''_1}\big\}_{j=1}^m, \ldots, \\
&\qquad \big\{\mathrm{id}_{r_j} \oplus \xi^{\oplus}_{r'_1,r''_p}\big\}_{j=1}^m, \ldots, \big\{\mathrm{id}_{r_j} \oplus \xi^{\oplus}_{r'_k,r''_p}\big\}_{j=1}^m\Big) \\
\xi^{\otimes}_{\underline{r},\underline{r}''} \otimes 1_{\underline{r}'} &= \Big(\xi^{\otimes}_{m,p}; \big\{\xi^{\oplus}_{r_j,r''_1}\big\}_{j=1}^m, \ldots, \big\{\xi^{\oplus}_{r_j,r''_p}\big\}_{j=1}^m\Big) \otimes \big(\mathrm{id}_k; \mathrm{id}_{r'_1}, \ldots, \mathrm{id}_{r'_k}\big) \\
&= \Big(\xi^{\otimes}_{m,p} \otimes \mathrm{id}_k; \big\{\xi^{\oplus}_{r_j,r''_1} \oplus \mathrm{id}_{r'_1}\big\}_{j=1}^m, \ldots, \big\{\xi^{\oplus}_{r_j,r''_p} \oplus \mathrm{id}_{r'_1}\big\}_{j=1}^m, \ldots, \\
&\qquad \big\{\xi^{\oplus}_{r_j,r''_1} \oplus \mathrm{id}_{r'_k}\big\}_{j=1}^m, \ldots, \big\{\xi^{\oplus}_{r_j,r''_p} \oplus \mathrm{id}_{r'_k}\big\}_{j=1}^m\Big)
\end{aligned}$$

Using (4.2.4) and (4.2.22), the composite in the diagram (4.2.21) is as follows.

$$\begin{aligned}
&\big(\xi^{\otimes}_{\underline{r},\underline{r}''} \otimes 1_{\underline{r}'}\big)\big(1_r \otimes \xi^{\otimes}_{\underline{r}',\underline{r}''}\big) \\
&= \Big(\big(\xi^{\otimes}_{m,p} \otimes \mathrm{id}_k\big)\big(\mathrm{id}_m \otimes \xi^{\otimes}_{k,p}\big); \\
&\quad \Big\{\big(\xi^{\oplus}_{r_j,r''_1} \oplus \mathrm{id}_{r'_1}\big)\big(\mathrm{id}_{r_j} \oplus \xi^{\oplus}_{r'_1,r''_1}\big)\Big\}_{j=1}^m, \ldots, \Big\{\big(\xi^{\oplus}_{r_j,r''_1} \oplus \mathrm{id}_{r'_k}\big)\big(\mathrm{id}_{r_j} \oplus \xi^{\oplus}_{r'_k,r''_1}\big)\Big\}_{j=1}^m, \ldots \\
&\quad \Big\{\big(\xi^{\oplus}_{r_j,r''_p} \oplus \mathrm{id}_{r'_1}\big)\big(\mathrm{id}_{r_j} \oplus \xi^{\oplus}_{r'_1,r''_p}\big)\Big\}_{j=1}^m, \ldots, \Big\{\big(\xi^{\oplus}_{r_j,r''_p} \oplus \mathrm{id}_{r'_k}\big)\big(\mathrm{id}_{r_j} \oplus \xi^{\oplus}_{r'_k,r''_p}\big)\Big\}_{j=1}^m\Big)
\end{aligned}$$

The hexagon axiom for

- the permutation $\xi^{\otimes}_{mk,p}$ in (2.4.10) and
- the block permutation $\xi^{\oplus}_{m+k,p}$ in (4.2.13)

implies the following equalities for $1 \le j \le m$, $1 \le l \le k$, and $1 \le t \le p$.

$$\begin{aligned}
\xi^{\otimes}_{mk,p} &= \big(\xi^{\otimes}_{m,p} \otimes \mathrm{id}_k\big)\big(\mathrm{id}_m \otimes \xi^{\otimes}_{k,p}\big) \\
\xi^{\oplus}_{r_j+r'_l,r''_t} &= \big(\xi^{\oplus}_{r_j,r''_t} \oplus \mathrm{id}_{r'_l}\big)\big(\mathrm{id}_{r_j} \oplus \xi^{\oplus}_{r'_l,r''_t}\big)
\end{aligned}$$

It follows that the composite in the diagram (4.2.21) is equal to $\xi^{\otimes}_{\underline{r}\otimes\underline{r}',\underline{r}''}$ as computed in (4.2.22). This proves the hexagon axiom for $\mathcal{D}$. Therefore, $\mathcal{D}$ is a permutative category with the multiplicative structure. □

The Multiplicative Zeros and Distributivity. Next we define the rest of the left bipermutative category structure for the distortion category. The following definition and lemma use Definitions 4.2.1, 4.2.7, and 4.2.14.

Definition 4.2.23. Define the natural transformations

$$(\lambda^{\bullet}, \rho^{\bullet}, \delta^{l}, \delta^{r})$$

for the distortion category $\mathcal{D}$ as follows.

The Multiplicative Zeros: $\lambda^{\bullet}$ and $\rho^{\bullet}$ are the identity natural transformations

$$\mathbb{0}\otimes\underline{r} \xrightarrow[=]{\lambda^{\bullet}_{\underline{r}}} \mathbb{0} \xleftarrow[=]{\rho^{\bullet}_{\underline{r}}} \underline{r}\otimes\mathbb{0}$$

for objects $\underline{r}\in\mathcal{D}$.

The Left Distributivity: δ^{l} is the identity natural transformation

$$\underline{r}\otimes(\underline{r}'\oplus\underline{r}'') \xrightarrow[=]{\delta^{l}_{\underline{r},\underline{r}',\underline{r}''}} (\underline{r}\otimes\underline{r}')\oplus(\underline{r}\otimes\underline{r}'')$$

for objects $\underline{r},\underline{r}',\underline{r}''\in\mathcal{D}$.

The Right Distributivity: δ^{r} is the natural transformation with components the composites

$$\begin{array}{ccc} (\underline{r}\oplus\underline{r}')\otimes\underline{r}'' & \xrightarrow{\delta^{r}_{\underline{r},\underline{r}',\underline{r}''}} & (\underline{r}\otimes\underline{r}'')\oplus(\underline{r}'\otimes\underline{r}'') \\ {\scriptstyle \xi^{\otimes}_{\underline{r}\oplus\underline{r}',\underline{r}''}}\big\downarrow & & \big\uparrow{\scriptstyle \xi^{\otimes}_{\underline{r}'',\underline{r}}\oplus\xi^{\otimes}_{\underline{r}'',\underline{r}'}} \\ \underline{r}''\otimes(\underline{r}\oplus\underline{r}') & \xrightarrow[=]{\delta^{l}_{\underline{r}'',\underline{r},\underline{r}'}} & (\underline{r}''\otimes\underline{r})\oplus(\underline{r}''\otimes\underline{r}') \end{array} \tag{4.2.24}$$

for objects $\underline{r},\underline{r}',\underline{r}''\in\mathcal{D}$.

This finishes the definition of $(\lambda^{\bullet}, \rho^{\bullet}, \delta^{l}, \delta^{r})$ for $\mathcal{D}$. ◇

Lemma 4.2.25. *The natural transformations $(\lambda^{\bullet}, \rho^{\bullet}, \delta^{l}, \delta^{r})$ in Definition 4.2.23 are well defined.*

Proof. Since $\mathbb{0}$ is the empty sequence $\varnothing\in\mathcal{D}$, which has length 0, by (4.2.15) both $\mathbb{0}\otimes\underline{r}$ and $\underline{r}\otimes\mathbb{0}$ are equal to $\varnothing$. The naturality of $\lambda^{\bullet}$ and $\rho^{\bullet}$ follows from (4.2.16) and the fact that the identity morphism $1_{\varnothing}$ is $(\mathrm{id}_0;)$.

To check that the left distributivity δ^{l} is well defined, first observe that (4.2.8) and (4.2.15) imply the following equalities.

$$\begin{aligned} &\underline{r}\otimes(\underline{r}'\oplus\underline{r}'') \\ &= (r_1,\ldots,r_m)\otimes(r'_1,\ldots,r'_k,r''_1,\ldots,r''_p) \\ &= \left(\{r_j+r'_1\}_{j=1}^{m},\ldots,\{r_j+r'_k\}_{j=1}^{m},\{r_j+r''_1\}_{j=1}^{m},\ldots,\{r_j+r''_p\}_{j=1}^{m}\right) \\ &= (\underline{r}\otimes\underline{r}')\oplus(\underline{r}\otimes\underline{r}'') \end{aligned} \tag{4.2.26}$$

To prove the naturality of δ^{l}, consider morphisms $\sigma\in\mathcal{D}(\underline{r};\underline{s})$, $\sigma'\in\mathcal{D}(\underline{r}';\underline{s}')$, and $\sigma''\in\mathcal{D}(\underline{r}'';\underline{s}'')$. There is an equality of permutations

$$\sigma\otimes(\sigma'\oplus\sigma'') = (\sigma\otimes\sigma')\oplus(\sigma\otimes\sigma'')\in\Sigma_{m(k+p)}$$

because, by (2.4.2) and (2.4.4), both sides are given by the following bijection for $1 \le j \le m$, $1 \le l \le k$, and $1 \le t \le p$.

$$\begin{cases} j + (l-1)m \longmapsto \sigma(j) + \big(\sigma'(l) - 1\big)m \\ j + \big[k + (t-1)\big]m \longmapsto \sigma(j) + \big[k + (\sigma''(t) - 1)\big]m \end{cases}$$

So the first entry in $\underline{\sigma} \otimes (\underline{\sigma}' \oplus \underline{\sigma}'')$ is equal to that in $(\underline{\sigma} \otimes \underline{\sigma}') \oplus (\underline{\sigma} \otimes \underline{\sigma}'')$. To see that their remaining entries are also equal, we use

- (4.2.9), (4.2.16), and
- (4.2.26) by replacing $(+, r_j, r'_l, r''_t)$ with $(\oplus, \sigma_j, \sigma'_l, \sigma''_t)$ for $1 \le j \le m$, $1 \le l \le k$, and $1 \le t \le p$.

This proves that δ^l is a well-defined natural transformation.

Finally, the right distributivity δ^r in (4.2.24) is a well-defined natural transformation by

- the functoriality of $\oplus$ in Lemma 4.2.12,
- the fact that $\xi^{\otimes}$ is a natural transformation in Lemma 4.2.19, and
- the fact that $\delta^l = \mathrm{Id}$ is a natural transformation.

Therefore, $\lambda^\bullet$, $\rho^\bullet$, δ^l, and δ^r in $\mathcal{D}$ are all well defined. □

Next we provide an explicit description of the right distributivity morphism in the distortion category.

Lemma 4.2.27. *For objects $\underline{r}, \underline{r}', \underline{r}'' \in \mathcal{D}$ with lengths $|\underline{r}| = m$, $|\underline{r}'| = k$, and $|\underline{r}''| = p$, the right distributivity morphism*

$$\begin{array}{c} (\underline{r} \oplus \underline{r}') \otimes \underline{r}'' = \Big(\big\{ r_j + r''_1 \big\}_{j=1}^m, \big\{ r'_l + r''_1 \big\}_{l=1}^k, \ldots, \big\{ r_j + r''_p \big\}_{j=1}^m, \big\{ r'_l + r''_p \big\}_{l=1}^k \Big) \\ \Big\downarrow \delta^r_{\underline{r}, \underline{r}', \underline{r}''} \\ (\underline{r} \otimes \underline{r}'') \oplus (\underline{r}' \otimes \underline{r}'') = \Big(\big\{ r_j + r''_1 \big\}_{j=1}^m, \ldots, \big\{ r_j + r''_p \big\}_{j=1}^m, \big\{ r'_l + r''_1 \big\}_{l=1}^k, \ldots, \big\{ r'_l + r''_p \big\}_{l=1}^k \Big) \end{array}$$

in (4.2.24) is given by

$$\Big((\xi^{\otimes}_{p,m} \oplus \xi^{\otimes}_{p,k}) \xi^{\otimes}_{m+k,p}; \big\{ \mathrm{id}_{r_j + r''_1} \big\}_{j=1}^m, \big\{ \mathrm{id}_{r'_l + r''_1} \big\}_{l=1}^k, \ldots, \big\{ \mathrm{id}_{r_j + r''_p} \big\}_{j=1}^m, \big\{ \mathrm{id}_{r'_l + r''_p} \big\}_{l=1}^k \Big).$$

Proof. The domain and the codomain of $\delta^r_{\underline{r}, \underline{r}', \underline{r}''}$ are as stated by (4.2.8) and (4.2.15). Using (4.2.8), (4.2.9), and (4.2.17), we compute the two vertical morphisms in (4.2.24) as follows.

$$\begin{aligned} \xi^{\otimes}_{\underline{r} \oplus \underline{r}', \underline{r}''} &= \Big(\xi^{\otimes}_{m+k,p}; \big\{ \xi^{\oplus}_{r_j, r''_1} \big\}_{j=1}^m, \big\{ \xi^{\oplus}_{r'_l, r''_1} \big\}_{l=1}^k, \ldots, \big\{ \xi^{\oplus}_{r_j, r''_p} \big\}_{j=1}^m, \big\{ \xi^{\oplus}_{r'_l, r''_p} \big\}_{l=1}^k \Big) \\ \xi^{\otimes}_{\underline{r}'', \underline{r}} \oplus \xi^{\otimes}_{\underline{r}'', \underline{r}'} &= \Big(\xi^{\otimes}_{p,m} \oplus \xi^{\otimes}_{p,k}; \big\{ \xi^{\oplus}_{r''_t, r_1} \big\}_{t=1}^p, \ldots, \big\{ \xi^{\oplus}_{r''_t, r_m} \big\}_{t=1}^p, \big\{ \xi^{\oplus}_{r''_t, r'_1} \big\}_{t=1}^p, \ldots, \big\{ \xi^{\oplus}_{r''_t, r'_k} \big\}_{t=1}^p \Big) \end{aligned}$$

The stated formula for $\delta^r_{\underline{r}, \underline{r}', \underline{r}''}$ now follows from (4.2.4) and the fact that the inverse of the block permutation $\xi^{\oplus}_{u,v}$ in (2.4.3) is $\xi^{\oplus}_{v,u}$. □

Explanation 4.2.28. The permutation

$$(\xi^{\otimes}_{p,m} \oplus \xi^{\otimes}_{p,k})\xi^{\otimes}_{m+k,p} \in \Sigma_{(m+k)p}$$

in δ^r in Lemma 4.2.27 is given by the following bijection for $1 \le j \le m$, $1 \le l \le k$, and $1 \le t \le p$.

$$\begin{cases} j + (t-1)(m+k) \longmapsto j + (t-1)m \\ l + m + (t-1)(m+k) \longmapsto l + (t-1)k + mp \end{cases}$$

This permutation rearranges p intervals of $m+k$ objects each to p intervals of m objects each followed by p intervals of k objects each. Except for a small notational difference (with n instead of k), this is the bijection in (2.4.6). A geometric description was given in Explanation 2.4.7. ◇

The Main Result. Recall the notion of a left bipermutative category in Definition 2.5.11. Proposition 2.5.16 states that each left bipermutative category is a tight symmetric bimonoidal category. By Proposition 2.4.8 and Example 2.5.17, Σ is a left bipermutative category. The following result extends that observation to the distortion category.

Theorem 4.2.29. *With the data in Definitions 4.2.7, 4.2.14, and 4.2.23, the distortion category $\mathcal{D}$ is a left bipermutative category.*

Proof. By Lemmas 4.2.12 and 4.2.19, both the additive structure and the multiplicative structure of $\mathcal{D}$ are permutative categories. By Lemma 4.2.25, the natural transformation δ^r and the identity natural transformations $\lambda^{\bullet}$, $\rho^{\bullet}$, and δ^l are well defined.

Next, for an object $\underline{r} \in \mathcal{D}$, both $\varnothing \otimes \underline{r}$ and $\underline{r} \otimes \varnothing$ are equal to $\varnothing$. It follows that

$$\xi^{\otimes}_{\underline{r},\varnothing} = (\mathrm{id}_0;\,) = 1_{\varnothing} : \varnothing \longrightarrow \varnothing.$$

The axiom (2.1.5) holds by the definition of δ^r in (4.2.24). It remains to check the axioms (2.1.6) and (2.1.13).

The axiom (2.1.6) states that the following diagram is commutative for objects $\underline{r}, \underline{r}', \underline{r}'' \in \mathcal{D}$.

$$\begin{array}{ccc} \underline{r} \otimes (\underline{r}' \oplus \underline{r}'') & \xrightarrow[=]{\delta^l_{\underline{r},\underline{r}',\underline{r}''}} & (\underline{r} \otimes \underline{r}') \oplus (\underline{r} \otimes \underline{r}'') \\ {\scriptstyle 1_{\underline{r}} \otimes \xi^{\oplus}_{\underline{r}',\underline{r}''}} \downarrow & & \downarrow {\scriptstyle \xi^{\oplus}_{\underline{r}\otimes\underline{r}', \underline{r}\otimes\underline{r}''}} \\ \underline{r} \otimes (\underline{r}'' \oplus \underline{r}') & \xrightarrow[=]{\delta^l_{\underline{r},\underline{r}'',\underline{r}'}} & (\underline{r} \otimes \underline{r}'') \oplus (\underline{r} \otimes \underline{r}') \end{array}$$

Using (2.4.3), (4.2.3), (4.2.10), and (4.2.15), the previous diagram is commutative by the following equalities.

$$\begin{aligned} & 1_{\underline{r}} \otimes \xi^{\oplus}_{\underline{r}',\underline{r}''} \\ &= \left(\mathrm{id}_m; \mathrm{id}_{r_1}, \ldots, \mathrm{id}_{r_m}\right) \otimes \left(\xi^{\oplus}_{k,p}; \mathrm{id}_{r'_1}, \ldots, \mathrm{id}_{r'_k}, \mathrm{id}_{r''_1}, \ldots, \mathrm{id}_{r''_p}\right) \\ &= \left(\mathrm{id}_m \otimes \xi^{\oplus}_{k,p}; \left\{\mathrm{id}_{r_j} \oplus \mathrm{id}_{r'_1}\right\}_{j=1}^m, \ldots, \left\{\mathrm{id}_{r_j} \oplus \mathrm{id}_{r'_k}\right\}_{j=1}^m, \left\{\mathrm{id}_{r_j} \oplus \mathrm{id}_{r''_1}\right\}_{j=1}^m, \ldots, \left\{\mathrm{id}_{r_j} \oplus \mathrm{id}_{r''_p}\right\}_{j=1}^m\right) \\ &= \left(\xi^{\oplus}_{mk,mp}; \left\{\mathrm{id}_{r_j + r'_1}\right\}_{j=1}^m, \ldots, \left\{\mathrm{id}_{r_j + r'_k}\right\}_{j=1}^m, \left\{\mathrm{id}_{r_j + r''_1}\right\}_{j=1}^m, \ldots, \left\{\mathrm{id}_{r_j + r''_p}\right\}_{j=1}^m\right) \\ &= \xi^{\oplus}_{\underline{r}\otimes\underline{r}', \underline{r}\otimes\underline{r}''} \end{aligned}$$

In the first component, the equality of permutations

$$\mathrm{id}_m \otimes \xi^{\oplus}_{k,p} = \xi^{\oplus}_{mk,mp} \in \Sigma_{m(k+p)}$$

is the axiom (2.1.6) in Σ. We already checked this in (2.4.11) in the proof of Proposition 2.4.8. Moreover, a geometric description was given in Explanation 2.4.15.

The axiom (2.1.13) states that the following diagram is commutative for objects $\underline{a}, \underline{b}, \underline{c}, \underline{d} \in \mathcal{D}$.

$$\begin{array}{ccc}
(\underline{a} \oplus \underline{b}) \otimes (\underline{c} \oplus \underline{d}) & \xrightarrow{\delta^r_{\underline{a},\underline{b},\underline{c}\oplus\underline{d}}} & \underline{a} \otimes (\underline{c} \oplus \underline{d}) \oplus \underline{b} \otimes (\underline{c} \oplus \underline{d}) \\
{\scriptstyle \delta^l_{\underline{a}\oplus\underline{b},\underline{c},\underline{d}}} \downarrow = & & = \downarrow {\scriptstyle \delta^l_{\underline{a},\underline{c},\underline{d}} \oplus \delta^l_{\underline{b},\underline{c},\underline{d}}} \\
(\underline{a} \oplus \underline{b}) \otimes \underline{c} \oplus (\underline{a} \oplus \underline{b}) \otimes \underline{d} & & (\underline{a} \otimes \underline{c}) \oplus (\underline{a} \otimes \underline{d}) \oplus (\underline{b} \otimes \underline{c}) \oplus (\underline{b} \otimes \underline{d}) \\
{\scriptstyle \delta^r_{\underline{a},\underline{b},\underline{c}} \oplus \delta^r_{\underline{a},\underline{b},\underline{d}}} \searrow & & \swarrow {\scriptstyle 1_{\underline{a}\otimes\underline{c}} \oplus \xi^{\oplus}_{\underline{a}\otimes\underline{d},\underline{b}\otimes\underline{c}} \oplus 1_{\underline{b}\otimes\underline{d}}} \\
& (\underline{a} \otimes \underline{c}) \oplus (\underline{b} \otimes \underline{c}) \oplus (\underline{a} \otimes \underline{d}) \oplus (\underline{b} \otimes \underline{d}) &
\end{array} \tag{4.2.30}$$

By (4.2.3), (4.2.4), (4.2.9), (4.2.10), and Lemma 4.2.27, in each of the two composites in (4.2.30), each entry after the first one is an identity permutation. Therefore, it suffices to check the commutativity of the diagram (4.2.30) in the first entry only. In other words, by Lemma 4.2.27, it suffices to check the equality of permutations

$$\begin{aligned}
&\big[(\xi^{\otimes}_{p,m} \oplus \xi^{\otimes}_{p,n})\xi^{\otimes}_{m+n,p}\big] \oplus \big[(\xi^{\otimes}_{q,m} \oplus \xi^{\otimes}_{q,n})\xi^{\otimes}_{m+n,q}\big] \\
&= \big[\mathrm{id}_{mp} \oplus \xi^{\oplus}_{mq,np} \oplus \mathrm{id}_{nq}\big]\big[(\xi^{\otimes}_{p+q,m} \oplus \xi^{\otimes}_{p+q,n})\xi^{\otimes}_{m+n,p+q}\big]
\end{aligned}$$

in $\Sigma_{m+n,p+q}$ if $|\underline{a}| = m$, $|\underline{b}| = n$, $|\underline{c}| = p$, and $|\underline{d}| = q$. This equality is the axiom (2.1.13) in Σ. We already checked this in (2.4.13) in the proof of Proposition 2.4.8. Moreover, a geometric description was given in Explanation 2.4.17. □

Corollary 4.2.31. *The Coherence Theorem 3.9.1 applies to the distortion category.*

Proof. The distortion category is a left bipermutative category, hence in particular a tight symmetric bimonoidal category, by Theorem 4.2.29 and Proposition 2.5.16. As explained in Example 3.9.6, Theorem 3.9.1 applies to $\mathcal{D}$. □

4.3. The Distortion of a Path

In this section, we define the graph morphism from $\mathsf{Gr}(X)$ in Definition 3.1.9 to the distortion category $\mathcal{D}$ that defines the distortion of a path in $\mathsf{Gr}(X)$. Since the distortion category $\mathcal{D}$ is a left bipermutative category by Theorem 4.2.29, it is in particular a tight symmetric bimonoidal category by Proposition 2.5.16. Recall that Convention 3.3.1 still stands in this chapter. In particular, we assume that X is a set with two distinguished elements 0^X and 1^X as in Definition 3.1.6, and X^{fr} is the free $\{\oplus, \otimes\}$-algebra of X in Definition 3.1.2. After the definition of the distortion of a path in $\mathsf{Gr}(X)$, we illustrate it with a series of observations and examples. An important special case of Corollary 4.2.31 is Corollary 4.3.12. It says that any two parallel paths in $\mathsf{Gr}(X)$ whose common domain is regular have the same distortion.

The next definition uses the left bipermutative category structure of the distortion category $\mathcal{D}$ in Definitions 4.2.1, 4.2.7, 4.2.14, and 4.2.23.

Definition 4.3.1. Consider the distortion category $\mathcal{D}$.

- Define the function $\vartheta\colon X \longrightarrow \mathsf{Ob}(\mathcal{D})$ as follows.

$$\vartheta(x) = \begin{cases} (1) & \text{if } x \in X \smallsetminus \{0^X, 1^X\}, \\ \varnothing & \text{if } x = 0^X, \text{ and} \\ (0) & \text{if } x = 1^X. \end{cases} \tag{4.3.2}$$

- Using the same symbol, define the associated graph morphism

$$\mathsf{Gr}(X) \xrightarrow{\ \vartheta\ } \mathcal{D} \tag{4.3.3}$$

  as in Definition 3.1.14, applied to the function in (4.3.2) and the symmetric bimonoidal category $\mathcal{D}$.
- For a path P in $\mathsf{Gr}(X)$, its value $\vartheta P \in \mathcal{D}$ in the sense of (3.1.18) is called the *distortion of* P.

This finishes the definition. ◇

Explanation 4.3.4. By Definitions 4.2.7 and 4.2.14, in the distortion category $\mathcal{D}$, the empty sequence $\varnothing$ is the additive zero $\mathbb{0}$, and the sequence (0) of length 1 is the multiplicative unit $\mathbb{1}$. Therefore, Definition 3.1.14 is indeed applicable to the function ϑ in (4.3.2). Each object $a \in X^{\mathsf{fr}}$ is sent by ϑ to an object in $\mathcal{D}$, which is a finite sequence of nonnegative integers. The distortion of each path in $\mathsf{Gr}(X)$ is a morphism in $\mathcal{D}$ as in Definition 4.2.1, which is in particular a nonempty finite sequence of permutations. ◇

To understand the distortion of a path, first we describe the action of ϑ on monomials and polynomials. Recall from Definition 3.6.3 that a *monomial* in X^{fr} is either an element in X, or a finite product of at least two elements in X with some multiplicative bracketing. A *polynomial* is either a monomial, or a finite sum of at least two monomials with some additive bracketing.

Lemma 4.3.5. *Suppose*

$$a = a_1 \otimes \cdots \otimes a_p \in X^{\mathsf{fr}}$$

for some multiplicative bracketing and some elements $a_j \in X^{\mathsf{fr}}$ *for* $1 \le j \le p$ *with* $p \ge 1$.

(1) $\vartheta(a) = \vartheta(a_1) \otimes \cdots \otimes \vartheta(a_p) \in \mathcal{D}$.

(2) *If* $a_j = 0^X$ *for some* $1 \le j \le p$, *then*

$$\vartheta(a) = \varnothing \in \mathcal{D}.$$

(3) *If* $a_j \in X \smallsetminus \{0^X\}$ *for each* $1 \le j \le p$, *then*

$$\vartheta(a) = (n) \in \mathcal{D},$$

where $n \ge 0$ *is the number of* a_j*'s that belong to* $X \smallsetminus \{0^X, 1^X\}$.

(4) *The image of a monomial under* ϑ *has length either 0 or 1, depending on whether it contains* 0^X *or not.*

Proof. The multiplicative associativity $\alpha^{\otimes}$ in the distortion category $\mathcal{D}$ is the identity by Definition 4.2.14. Therefore, $\vartheta(a) \in \mathcal{D}$ is independent of the choice of a multiplicative bracketing of a. Assertion (1) follows from the definition (3.1.16) of the graph morphism ϑ on objects in X^{fr}.

For assertion (2), by definition $\vartheta(0^X) = \varnothing$. Moreover, $\underline{r} \otimes \varnothing$ and $\varnothing \otimes \underline{r}$ are both $\varnothing \in \mathcal{D}$ for each object $\underline{r} \in \mathcal{D}$. It follows from assertion (1) that $\vartheta(a) = \varnothing$ if $a_j = 0^X$ for some $1 \le j \le p$.

For assertion (3), by definition each

$$\vartheta(a_j) = (\epsilon_j) \in \mathcal{D} \quad \text{with} \quad \epsilon_j = \begin{cases} 0 & \text{if } a_j = 1^X, \text{ and} \\ 1 & \text{if } a_j \in X \smallsetminus \{0^X, 1^X\}. \end{cases}$$

By assertion (1) and (4.2.15), it follows that $\vartheta(a)$ has length 1, and is given by

$$\begin{aligned} \vartheta(a_1) \otimes \cdots \otimes \vartheta(a_p) &= (\epsilon_1) \otimes \cdots \otimes (\epsilon_p) \\ &= \big(\epsilon_1 + \cdots + \epsilon_p\big). \end{aligned}$$

In other words, its only entry is the number of a_j's that belong to $X \smallsetminus \{0^X, 1^X\}$.

Assertion (4) follows from assertions (2) and (3). □

Lemma 4.3.6. *Suppose*

$$a = a^1 \oplus \cdots \oplus a^m \in X^{\mathsf{fr}}$$

for some additive bracketing and some elements $a^i \in X^{\mathsf{fr}}$ for $1 \le i \le m$ with $m \ge 1$. Then:

(1) $\vartheta(a) = \big(\vartheta(a^1), \ldots, \vartheta(a^m)\big) \in \mathcal{D}$.
(2) *If a^i is a monomial for each $1 \le i \le m$, then $\vartheta(a)$ has length $k \ge 0$, where k is the number of a^i's that contain no 0^X.*

Proof. The additive associativity $\alpha^{\oplus}$ in the distortion category $\mathcal{D}$ is the identity by Definition 4.2.7. Therefore, $\vartheta(a) \in \mathcal{D}$ is independent of the choice of an additive bracketing of a. Assertion (1) follows from (3.1.16) and (4.2.8). Assertion (2) follows from assertion (1) and Lemma 4.3.5 (4). □

In the following examples, the symbol $\otimes$ in X^{fr} is omitted.

Example 4.3.7. For objects $u, v, w, x, y, z \in X \smallsetminus \{0^X, 1^X\}$, consider the polynomial

$$a = (uv1^X) \oplus (1^X1^X1^X) \oplus (w1^Xx1^Xy) \oplus (z0^X) \in X^{\mathsf{fr}}$$

with any additive bracketing and any multiplicative bracketing within each monomial. By Lemma 4.3.5 (2) and (3) and Lemma 4.3.6 (1),

$$\begin{aligned} \vartheta(a) &= \big(\vartheta(uv1^X), \vartheta(1^X1^X1^X), \vartheta(w1^Xx1^Xy), \vartheta(z0^X)\big) \\ &= (2, 0, 3) \in \mathcal{D}, \end{aligned}$$

which has length 3. ◇

Example 4.3.8. For objects $r, s, t \in X \smallsetminus \{0^X, 1^X\}$, consider the polynomial

$$b = r \oplus (0^X0^X) \oplus (st) \in X^{\mathsf{fr}}$$

with any additive bracketing. With $a \in X^{\mathsf{fr}}$ the polynomial in Example 4.3.7, there are the following objects in $\mathcal{D}$.

$$\begin{aligned}
\vartheta(b) &= \big(\vartheta(r), \vartheta(0^X 0^X), \vartheta(st)\big)\\
&= (1,2)\\
\vartheta(a \oplus b) &= \big(\vartheta(a), \vartheta(b)\big)\\
&= (2,0,3,1,2)\\
\vartheta(a \otimes b) &= \vartheta(a) \otimes \vartheta(b)\\
&= (2,0,3) \otimes (1,2)\\
&= \big(2+1, 0+1, 3+1, 2+2, 0+2, 3+2\big)\\
&= (3,1,4,4,2,5)
\end{aligned}$$

The second-to-last equality follows from the definition (4.2.15) of $\otimes$ in $\mathcal{D}$. ◇

Next we provide examples of distortions of paths.

Convention 4.3.9. For Lemma 4.3.11, below we interpret the 24 diagrams (2.1.5)–(2.1.28) in Definition 2.1.2 in $\mathsf{Gr}(X)$ as follows.

- We interpret each object there as an element in X^{fr} with $A, B, C, D \in X^{\mathsf{fr}}$, and with $\mathbb{0}$ and $\mathbb{1}$ interpreted as 0^X and 1^X, respectively.
- We interpret each morphism there as the corresponding prime edge in $\mathsf{Gr}(X)$ as in Definition 3.1.8, with one kind of exceptions as stated next.
- If a morphism is the sum of two nonidentity structure morphisms, then we interpret it as a path of length 2 in $\mathsf{Gr}(X)$ consisting of the two corresponding prime edges.

In this way, we interpret each of those 24 diagrams as consisting of two paths in $\mathsf{Gr}(X)$ with a common domain and a common codomain. ◇

Example 4.3.10. The morphism

$$AC \oplus BC \xrightarrow{\xi^{\otimes}_{A,C} \oplus \xi^{\otimes}_{B,C}} CA \oplus CB$$

in (2.1.5) is interpreted as the path

$$AC \oplus BC \xrightarrow{\xi^{\otimes}_{A,C} \oplus 1_{BC}} CA \oplus BC \xrightarrow{1_{CA} \oplus \xi^{\otimes}_{B,C}} CA \oplus CB$$

of length 2 in $\mathsf{Gr}(X)$. Besides (2.1.5), such a sum of two nonidentity structure morphisms happens in (2.1.10)–(2.1.13), (2.1.15), (2.1.16), (2.1.27), and (2.1.28). ◇

Lemma 4.3.11. *Under Convention 4.3.9, in each diagram in* (2.1.5)–(2.1.28), *the two paths in* $\mathsf{Gr}(X)$ *have the same distortion.*

Proof. The assertion means that if we apply the graph morphism $\vartheta\colon \mathsf{Gr}(X) \longrightarrow \mathcal{D}$ in (4.3.3) to each of those 24 diagrams, then the result is a commutative diagram in $\mathcal{D}$. Therefore, the assertion follows from the fact that the distortion category $\mathcal{D}$ is a left bipermutative category, hence in particular a symmetric bimonoidal category, by Theorem 4.2.29 and Proposition 2.5.16. □

Next is Corollary 4.2.31 for the graph morphism $\vartheta\colon \mathsf{Gr}(X) \longrightarrow \mathcal{D}$ in (4.3.3) that defines distortion. In other words, it is the special case of the Coherence Theorem 3.9.1 for the distortion category $\mathcal{D}$ and ϑ. The following result is *not* used in the proof of the Coherence Theorem 4.4.3. Recall the notion of regularity from Definition 3.1.25.

Corollary 4.3.12. *If*

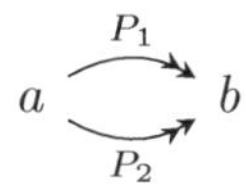

are two paths in $\mathsf{Gr}(X)$ *with* $a \in X^{\mathsf{fr}}$ *regular, then* P_1 *and* P_2 *have the same distortion.*

The next example is an illustration of Corollary 4.3.12.

Example 4.3.13. Consider

- *distinct* elements $a_1, a_2, a_3, b_1, b_2 \in X$;
- polynomials

$$a = (a_1 \oplus a_2) \oplus a_3 \quad \text{and} \quad b = b_1 \oplus b_2 \in X^{\mathsf{fr}};$$

  and
- the following two paths in $\mathsf{Gr}(X)$ in which most of the $\otimes$ symbols are omitted.

(4.3.14)

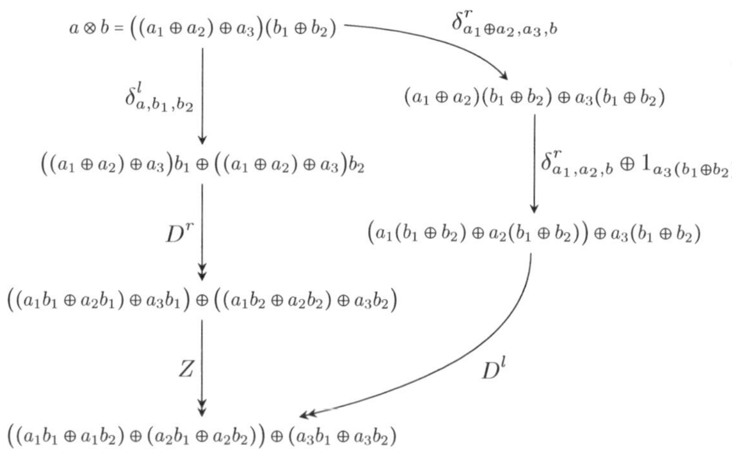

Consider the diagram (4.3.14).

- The path D^l consists of three δ-prime edges as in Definition 3.6.2, each containing one instance of $\delta^l_{a_i, b_1, b_2}$ for $1 \leq i \leq 3$. There are six possibilities of D^l depending on the order of these δ-prime edges. The path D^l may be chosen as any one of them.
- The path D^r consists of four δ-prime edges, each containing one instance of $\delta^r_{a_1 \oplus a_2, a_3, b_j}$ or $\delta^r_{a_1, a_2, b_j}$ for $j = 1, 2$. There are five possibilities of D^r depending on the order of these δ-prime edges, where $\delta^r_{a_1 \oplus a_2, a_3, b_j}$ must occur before $\delta^r_{a_1, a_2, b_j}$ for each $j = 1, 2$. The path D^r may be chosen as any one of them.
- Each prime edge in the path Z contains an instance of $\alpha^{\oplus}$ or $\xi^{\oplus}$. These prime edges move the additive brackets (for $\alpha^{\oplus}$) and permute the six monomials (for $\xi^{\oplus}$). There are infinitely many such paths with the prescribed domain and codomain, and Z may be chosen as any one of them.

Suppose

$$P_1 = \big(Z, D^r, \delta^l_{a,b_1,b_2}\big) \quad \text{and}$$
$$P_2 = \big(D^l, \delta^r_{a_1,a_2,b} \oplus 1_{a_3(b_1 \oplus b_2)}, \delta^r_{a_1 \oplus a_2, a_3, b}\big) \tag{4.3.15}$$

are, respectively, the left path and the right path in (4.3.14). Since the elements a_1, a_2, a_3, b_1, and b_2 are distinct, the common domain $a \otimes b \in X^{\mathsf{fr}}$ of P_1 and P_2 is regular as in Definition 3.1.25. Corollary 4.3.12 implies that the paths P_1 and P_2 have the same distortion. In other words,

$$\vartheta P_1 = \vartheta P_2$$

as morphisms in the distortion category $\mathcal{D}$. ◇

The next two examples are variations of Example 4.3.13.

Example 4.3.16. In Example 4.3.13, suppose instead that the elements

$$a_1, a_2, a_3, b_1, b_2 \in X \smallsetminus \{0^X, 1^X\}$$

are *not* all distinct. This implies that $a \otimes b \in X^{\mathsf{fr}}$ is not regular, so Corollary 4.3.12 does not apply to the two paths in (4.3.14).

However, Corollary 4.3.12 will apply if we first replace the elements as follows.

- Define the set

  $$X' = X \amalg \{a'_1, a'_2, a'_3, b'_1, b'_2\}.$$

  It is obtained from X by
  - adjoining five distinct symbols a'_1, a'_2, a'_3, b'_1, and b'_2 not in X and
  - keeping 0^X and 1^X as the additive zero and the multiplicative unit.
- Next we extend the function $\vartheta\colon X \longrightarrow \mathsf{Ob}(\mathcal{D})$ in (4.3.2) to the function $\vartheta'\colon X' \longrightarrow \mathsf{Ob}(\mathcal{D})$ by defining

  $$\vartheta'(a'_i) = \vartheta'(b'_j) = (1) \quad \text{for} \quad i = 1,2,3 \quad \text{and} \quad j = 1,2.$$

  This is well defined because $a'_i, b'_j \in X' \smallsetminus \{0^X, 1^X\}$.
- Using the associated graph morphism $\vartheta'\colon \mathsf{Gr}(X') \longrightarrow \mathcal{D}$ as in (4.3.3), we consider the prime version of the diagram (4.3.14) in $\mathsf{Gr}(X')$, in which a_i and b_j are replaced by, respectively, a'_i and b'_j for $1 \le i \le 3$ and $1 \le j \le 2$.

The resulting left path P'_1 and right path P'_2 in $\mathsf{Gr}(X')$ satisfy

$$\vartheta P_j = \vartheta' P'_j \in \mathcal{D} \quad \text{for} \quad j = 1,2 \tag{4.3.17}$$

because $a_i, b_j \in X \smallsetminus \{0^X, 1^X\}$ implies

$$\vartheta(a_i) = (1) = \vartheta'(a'_i)$$
$$\vartheta(b_j) = (1) = \vartheta'(b'_j).$$

Moreover, the element

$$a' \otimes b' = \big((a'_1 \oplus a'_2) \oplus a'_3\big) \otimes (b'_1 \oplus b'_2) \in X'^{\mathsf{fr}}$$

is regular because it has the same support (3.1.24) as

$$\big((a'_1 b'_1 \oplus a'_1 b'_2) \oplus (a'_2 b'_1 \oplus a'_2 b'_2)\big) \oplus (a'_3 b'_1 \oplus a'_3 b'_2) \in X'^{\mathsf{fr}},$$

and the elements in $\{a'_1, a'_2, a'_3, b'_1, b'_2\}$ are all distinct. The paths P'_1 and P'_2 have the same distortion by Corollary 4.3.12. By (4.3.17), the paths P_1 and P_2 also have the same distortion. ◇

Example 4.3.18. In Example 4.3.13, as an elementary alternative to applying Corollary 4.3.12, in this example we explicitly compute the distortions of the two paths in the diagram (4.3.14), starting with P_1 in (4.3.15). For the following computation, we assume that a_1, a_2, a_3, b_1, and b_2 belong to $X \smallsetminus \{0^X, 1^X\}$. Lemmas 4.3.5 and 4.3.6 imply that

$$\begin{aligned}\vartheta(a \otimes b) &= \vartheta(a) \otimes \vartheta(b)\\ &= (1,1,1) \otimes (1,1)\\ &= (2,2,2,2,2,2) \in \mathcal{D}.\end{aligned}$$

The common codomain of the two paths is also sent by ϑ to this object in $\mathcal{D}$. If some of the elements in $\{a_1, a_2, a_3, b_1, b_2\}$ are 0^X or 1^X, then the computation below simplifies accordingly, since $\vartheta(0^X) = \varnothing$ and $\vartheta(1^X) = (0)$.

The distortion category $\mathcal{D}$ is a left bipermutative category by Theorem 4.2.29. Its only nonidentity structure isomorphisms are δ^r in (4.2.24), $\xi^\oplus$ in (4.2.10), and $\xi^\otimes$ in (4.2.17), the last of which is not involved in the paths P_1 and P_2. Therefore, we may ignore the additive bracketing, and each prime edge involving either $\alpha^\oplus$ or δ^l has an identity morphism as its distortion.

Lemma 4.2.27 and Explanation 4.2.28 imply that each of the four δ-prime edges in D^r has an identity morphism as its distortion, since $\xi^\otimes_{?,1}$ and $\xi^\otimes_{1,?}$ are both identity permutations. Therefore, the distortion of the path D^r is also an identity morphism.

For the path Z, we simply need to additively permute the middle four monomials to the order in the codomain. We can accomplish this by additively permuting two adjacent monomials at a time. For $1 \le i \le 3$ and $1 \le j \le 2$,

$$\vartheta(a_i b_j) = (1) \otimes (1) = (2) \in \mathcal{D},$$

which has length 1, by Lemma 4.3.5(3). By definition (4.2.10),

$$\xi^\oplus_{(2),(2)} = \big(\xi^\oplus_{1,1}; \mathrm{id}_2, \mathrm{id}_2\big) \colon (2,2) \longrightarrow (2,2) \in \mathcal{D},$$

and $\xi^\oplus_{1,1} \in \Sigma_2$ is the unique nonidentity permutation. We conclude that the distortion of P_1 is the morphism

$$\vartheta P_1 = \vartheta Z = \big(\sigma; \mathrm{id}_2, \ldots, \mathrm{id}_2\big) \colon (2,2,2,2,2,2) \longrightarrow (2,2,2,2,2,2) \in \mathcal{D} \tag{4.3.19}$$

with $\sigma \in \Sigma_6$ the permutation

$$\sigma = \begin{pmatrix} 1 & 2 & 3 & 4 & 5 & 6 \\ 1 & 3 & 5 & 2 & 4 & 6 \end{pmatrix}. \tag{4.3.20}$$

Next, to compute the distortion of the path P_2 in (4.3.15), we apply (4.2.4), Lemma 4.2.27, and Explanation 4.2.28 to its first two prime edges. In the morphism

$$\begin{aligned}&\vartheta\big(\delta^r_{a_1,a_2,b} \oplus 1_{a_3(b_1 \oplus b_2)}\big)\vartheta\big(\delta^r_{a_1 \oplus a_2, a_3, b}\big)\\ &= \big(\delta^r_{(1),(1),(1,1)} \oplus 1_{(2,2)}\big)\delta^r_{(1,1),(1),(1,1)} \in \mathcal{D},\end{aligned} \tag{4.3.21}$$

the first entry is the product permutation

$$\begin{pmatrix} 1 & 2 & 3 & 4 & 5 & 6 \\ 1 & 3 & 2 & 4 & 5 & 6 \end{pmatrix}\begin{pmatrix} 1 & 2 & 3 & 4 & 5 & 6 \\ 1 & 2 & 5 & 3 & 4 & 6 \end{pmatrix} \in \Sigma_6.$$

This is equal to σ in (4.3.20). The other entries of the morphism in (4.3.21) are all identity permutations. So the distortion of the first two prime edges in P_2 is equal to ϑP_1 in (4.3.19). Since the path D^l consists of only prime edges involving

δ^l, whose distortions are identities, ϑD^l is an identity morphism. Therefore, the distortion of the path P_2 is also given by (4.3.19). ◇

4.4. The Second Coherence Theorem

In this section, we prove the main coherence result of this chapter. See Section 4.7 for more discussion of this theorem. We remind the reader that Convention 3.3.1 is still in effect, with C an arbitrary symmetric bimonoidal category. In particular, for a path in $\mathsf{Gr}(X)$:

- Its *value* in C as in (3.1.18) is defined using the graph morphism $\varphi\colon \mathsf{Gr}(X) \longrightarrow \mathsf{C}$ in Definition 3.1.14.
- Its *distortion* is defined in Definition 4.3.1 using the graph morphism $\vartheta\colon \mathsf{Gr}(X) \longrightarrow \mathcal{D}$. The distortion of a path is a morphism in the distortion category $\mathcal{D}$.

Before we prove the Coherence Theorem 4.4.3, we consider the following preliminary version, in which Definitions 3.6.2, 3.8.1, and 3.8.8 are used.

Lemma 4.4.1. *Suppose*

$$a \underset{P_2}{\overset{P_1}{\rightrightarrows}} b$$

are two paths in $\mathsf{Gr}(X)$ *such that the following two statements hold.*

- *P_1 and P_2 have the same distortion and are 1^X-free paths.*
- *a and b contain no 0^X and are δ-reduced and 1^X-reduced.*

Then P_1 and P_2 have the same value in C.

Proof. We reuse parts of the proof of Theorem 3.9.1. First we observe the following.

- By Lemmas 3.6.5 and 3.8.4, the assumptions on a and b mean that each of them is a polynomial as in Definition 3.6.3, in which each monomial is equal to 1^X, or contains no 0^X and 1^X.
- By Explanation 3.8.9, the assumption that each path P_i for $i = 1, 2$ is a 1^X-free path means that each of its prime edges is either an identity, or involves an instance of
  - $\alpha^{\pm\oplus}$, which moves the additive brackets;
  - $\xi^{\pm\oplus}$, which permutes the set of monomials;
  - $\alpha^{\pm\otimes}$, which moves the multiplicative brackets in a monomial; or
  - $\xi^{\pm\otimes}$, which permutes the factors in a monomial.

  Together with the previous remark, it follows that each vertex in each path P_i is a polynomial in which each monomial is equal to 1^X, or contains no 0^X and 1^X.

These remarks, and the naturality of $\alpha^{\oplus}$ and $\xi^{\oplus}$ in each symmetric bimonoidal category, imply that for each $i = 1, 2$, there is a diagram in $\mathsf{Gr}(X)$

$$\begin{array}{ccc} a & \xrightarrow{\quad P_i \quad} & b \\ & {\scriptstyle P_i'}\searrow \quad \nearrow{\scriptstyle P_i''} & \\ & c_i & \end{array}$$

that is commutative in the sense of Definition 3.1.14, in particular in C and $\mathcal{D}$, such that the following two statements hold.

- $P_i': a \twoheadrightarrow c_i$ consists of identities and prime edges involving $\alpha^{\pm\oplus}$ or $\xi^{\pm\oplus}$. In particular, P_i' only moves additive brackets and permutes the set of monomials.
- $P_i'': c_i \twoheadrightarrow b$ consists of identities and prime edges involving $\alpha^{\pm\otimes}$ or $\xi^{\pm\otimes}$. In particular, P_i'' only moves multiplicative brackets and permutes the factors in each monomial.

Our next objective is to prove the following three equalities.

(i) $\vartheta(P_1') = \vartheta(P_2')$.
(ii) $\vartheta(P_1'') = \vartheta(P_2'')$.
(iii) $c_1 = c_2$.

Recall that the only nonidentity structure isomorphisms in the distortion category $\mathcal{D}$ are δ^r in (4.2.24), $\xi^\oplus$ in (4.2.10), and $\xi^\otimes$ in (4.2.17). Suppose that P_1 has distortion

$$\vartheta(P_1) = \big(\sigma; \sigma_1, \ldots, \sigma_m\big) \in \mathcal{D}(\vartheta a; \vartheta b).$$

The above description of P_1' and P_1'', (4.2.4), Lemma 4.3.5 (3), and Lemma 4.3.6 imply the following equalities.

$$\vartheta(P_1') = \big(\sigma; \mathrm{id}_{r_1}, \ldots, \mathrm{id}_{r_m}\big) \in \mathcal{D}(\vartheta a; \vartheta c_1)$$
$$\vartheta(P_1'') = \big(\mathrm{id}_m; \sigma_{\sigma^{-1}(1)}, \ldots, \sigma_{\sigma^{-1}(m)}\big) \in \mathcal{D}(\vartheta c_1; \vartheta b)$$

- m is the common number of monomials in each vertex in P_i' and P_i''.
- For each $1 \le i \le m$, r_i is the number of factors in $X \setminus \{1^X\}$ in the ith monomial in $a \in X^{\mathsf{fr}}$.
- $\sigma \in \Sigma_m$, and each $\sigma_i \in \Sigma_{r_i}$.

The same analysis also applies to P_2. Since $\vartheta(P_1) = \vartheta(P_2)$ by assumption, it follows that there are equalities as follows.

$$\begin{aligned}\vartheta(P_1') &= \vartheta(P_2') = \big(\sigma; \mathrm{id}_{r_1}, \ldots, \mathrm{id}_{r_m}\big)\\ \vartheta(P_1'') &= \vartheta(P_2'') = \big(\mathrm{id}_m; \sigma_{\sigma^{-1}(1)}, \ldots, \sigma_{\sigma^{-1}(m)}\big)\end{aligned} \tag{4.4.2}$$

It follows from (4.2.2) and the first line in (4.4.2) that, ignoring additive brackets, both $P_1': a \twoheadrightarrow c_1$ and $P_2': a \twoheadrightarrow c_2$ permute the set of monomials in a via $\sigma \in \Sigma_m$. Moreover, since $P_1'': c_1 \twoheadrightarrow b$ and $P_2'': c_2 \twoheadrightarrow b$ neither move the additive brackets nor permute the set of monomials, both c_1 and c_2 have the same additive bracketing as b. It follows that

$$c_1 = c_2 \in X^{\mathsf{fr}}.$$

Our final objective is to show that P_1' and P_2' have the same value in C, and similarly for P_1'' and P_2''.

Write c for the element $c_1 = c_2$. Since the paths P_1' and $P_2': a \twoheadrightarrow c$ both permute the set of monomials in a via $\sigma \in \Sigma_m$, the Symmetric Coherence Theorem 1.3.8 for the additive structure

$$(\mathsf{C}, \oplus, \mathbb{0}, \alpha^\oplus, \lambda^\oplus, \rho^\oplus, \xi^\oplus)$$

implies that P_1' and P_2' have the same value in C.

It follows from the second line in (4.4.2) that, ignoring multiplicative brackets, the paths P_1'' and $P_2'': c \twoheadrightarrow b$ both permute the factors in the ith monomial in c via the permutation $\sigma_{\sigma^{-1}(i)}$ for each $1 \le i \le m$. The Symmetric Coherence Theorem 1.3.8 for the multiplicative structure

$$(\mathsf{C}, \otimes, \mathbb{1}, \alpha^\otimes, \lambda^\otimes, \rho^\otimes, \xi^\otimes),$$

when applied to each monomial in c, implies that the paths P_1'' and P_2'' have the same value in C. Therefore, $P_1 = (P_1'', P_1')$ and $P_2 = (P_2'', P_2')$ have the same value in C. □

The following coherence theorem is the main result of this chapter. Recall the concept of a δ-prime edge in Definition 3.6.2.

Theorem 4.4.3 (Laplaza's Second Coherence). *Suppose C is a symmetric bimonoidal category in which the value of each δ-prime edge is a monomorphism. If*

$$a \underset{P_2}{\overset{P_1}{\rightrightarrows}} b$$

are two paths in $\mathsf{Gr}(X)$ with the same distortion, then their values in C are equal.

Proof. We reuse the reduction steps in the proof of Theorem 3.9.1, along with the following remarks.

(1) The distortion category $\mathcal{D}$ is a groupoid by Lemma 4.2.5. Therefore, if two diagrams in $\mathsf{Gr}(X)$

$$\begin{array}{ccc} a & \xrightarrow{P_1} & b \\ {\scriptstyle Q_a}\downarrow & & \downarrow{\scriptstyle Q_b} \\ a' & \xrightarrow{R_1} & b' \end{array} \qquad \text{and} \qquad \begin{array}{ccc} a & \xrightarrow{P_2} & b \\ {\scriptstyle Q_a}\downarrow & & \downarrow{\scriptstyle Q_b} \\ a' & \xrightarrow{R_2} & b' \end{array}$$

are commutative in the sense of Definition 3.1.14, in particular in C and $\mathcal{D}$, such that P_1 and P_2 have the same distortion (that is, $\vartheta P_1 = \vartheta P_2$ in $\mathcal{D}$), then R_1 and R_2 have the same distortion.

(2) As discussed in Explanation 3.9.4 (2), the five reduction steps in the proof of Theorem 3.9.1 can be performed with the regularity condition omitted everywhere. By the previous remark, if the two paths have the same distortion to begin with, then after each of these reduction steps, the two new paths also have the same distortion.

Therefore, by the five reduction steps in the proof of Theorem 3.9.1 with regularity omitted, we reduce to the setting of Lemma 4.4.1, which finishes the proof. □

Remark 4.4.4. In Theorem 4.4.3, the monomorphism assumption on C is needed because, as discussed in Explanation 3.9.4 (1), this condition is used in the reduction step involving the diagram (3.9.2). ◇

Example 4.4.5. As in Examples 3.9.6 through 3.9.8 and 3.9.10, Theorem 4.4.3 is applicable in the following types of categories.

- Tight symmetric bimonoidal categories, that is, those with δ^l and δ^r (2.1.4) natural isomorphisms, not just natural monomorphisms.
- Distributive symmetric monoidal categories in Definition 2.3.1.
- Symmetric monoidal closed categories with finite coproducts in Example 2.3.3.
- The category of modules over a commutative ring in Example 2.3.4.
- Distributive categories in Example 2.3.5.
- Right and left bipermutative categories in Definitions 2.5.2 and 2.5.11.
- The symmetric bimonoidal groupoid Π in Theorem 2.6.2.
- Flat symmetric bimonoidal categories in Definition 3.9.9. ◇

Example 4.4.6. Consider the diagram in Example 3.1.11 with $x, y, z \in X^{\mathsf{fr}}$ arbitrary. Each of the two paths has the identity morphism as its distortion because $\alpha^{\oplus}$, $\rho^{\oplus}$, $\lambda^{\otimes}$, and δ^l are identities in the distortion category $\mathcal{D}$ by Definitions 4.2.7, 4.2.14, and 4.2.23. By Theorem 4.4.3, these two paths have the same value in C, provided that the value of each δ-prime edge is a monomorphism. This is an example where Theorem 4.4.3 applies, but Theorem 3.9.1 does not apply immediately unless the domain $x(y \oplus z) \oplus 0^X$ is regular. $\diamond$

4.5. Coherence of Bimonoidal Categories II

Convention 3.10.1 is in effect throughout this section, so C is a bimonoidal category as in Definition 2.1.2. In this section, we discuss the multiplicatively nonsymmetric analogue of the Coherence Theorem 4.4.3 that applies to bimonoidal categories instead of symmetric bimonoidal categories. As in the symmetric case, the main Coherence Theorem 4.5.8 in this section does *not* require the invertibility of the distributivity morphisms δ^l and δ^r. Instead, it assumes a much weaker monomorphism condition.

Motivation 4.5.1. Just as Theorem 4.4.3 is phrased in terms of the distortion category $\mathcal{D}$ in Definition 4.2.1, Theorem 4.5.8 is phrased in terms of a multiplicatively nonsymmetric analogue of $\mathcal{D}$, which we call the additive distortion category and is denoted by $\mathcal{D}^{\mathsf{ad}}$. As discussed in Section 4.1, in a morphism $\underline{\sigma} = (\sigma; \sigma_1, \ldots, \sigma_m)$ in the distortion category $\mathcal{D}$, we think of each $\sigma_j \in \Sigma_{r_j}$ as permuting the factors in an r_j-fold product

$$a_1^j \otimes \cdots \otimes a_{r_j}^j$$

with some multiplicative bracketing. A bimonoidal category does not have a multiplicative symmetry, so permutations of $\otimes$-factors are no longer allowed. Therefore, in the additive distortion category, each morphism is a permutation that takes one sequence of nonnegative integers to another; see (4.5.3).

We observe in Lemma 4.5.6 that $\mathcal{D}^{\mathsf{ad}}$ is a tight bimonoidal category, and it embeds into the distortion category $\mathcal{D}$. The additive distortion of a path in the nonsymmetric graph $\mathsf{Gr}^{\mathsf{ns}}(X)$ is defined as its value in $\mathcal{D}^{\mathsf{ad}}$. Theorem 4.5.8 states that any two parallel nonsymmetric paths with the same additive distortion also have the same value in C. $\diamond$

Now we define the multiplicatively nonsymmetric analogue of the distortion category.

Definition 4.5.2. Define the *additive distortion category*

$$\big(\mathcal{D}^{\mathsf{ad}}, (\oplus, \mathbb{0}, \alpha^{\oplus}, \lambda^{\oplus}, \rho^{\oplus}, \xi^{\oplus}), (\otimes, \mathbb{1}, \alpha^{\otimes}, \lambda^{\otimes}, \rho^{\otimes}), (\lambda^{\bullet}, \rho^{\bullet}, \delta^l, \delta^r)\big)$$

as follows.

Objects: An object in $\mathcal{D}^{\mathsf{ad}}$ is a finite sequence

$$\underline{r} = (r_1, \ldots, r_m)$$

with $m \geq 0$ and with each r_j for $1 \leq j \leq m$ a nonnegative integer. We call m the *length* of $\underline{r}$, which is denoted by $|\underline{r}|$. The unique sequence with length 0 is denoted by $\varnothing$.

Morphisms: Suppose $\underline{s} = (s_1, \ldots, s_n)$ is an object in $\mathcal{D}^{\mathsf{ad}}$. With $\underline{r}$ as above, the morphism set $\mathcal{D}^{\mathsf{ad}}(\underline{r}; \underline{s})$ consists of permutations

$$\sigma \in \Sigma_m \quad \text{such that} \quad \sigma\underline{r} = \big(r_{\sigma^{-1}(1)}, \ldots, r_{\sigma^{-1}(m)}\big) = \underline{s}. \tag{4.5.3}$$

Other Structures: Except for δ^r, the other structures in $\mathcal{D}^{\mathrm{ad}}$ are defined as in Definitions 4.2.1, 4.2.7, 4.2.14, and 4.2.23 for objects, and by restricting to the first entry for morphisms.

Right Distributivity: δ^r is the natural transformation

$$(\underline{r} \oplus \underline{r}') \otimes \underline{r}'' \xrightarrow{\delta^r_{\underline{r},\underline{r}',\underline{r}''}} (\underline{r} \otimes \underline{r}'') \oplus (\underline{r}' \otimes \underline{r}'')$$

for objects $\underline{r}, \underline{r}', \underline{r}'' \in \mathcal{D}^{\mathrm{ad}}$, that is given by the permutation

$$\delta^r_{\underline{r},\underline{r}',\underline{r}''} = (\xi^{\otimes}_{p,m} \oplus \xi^{\otimes}_{p,k})\xi^{\otimes}_{m+k,p} \in \Sigma_{(m+k)p} \tag{4.5.4}$$

with $|\underline{r}| = m$, $|\underline{r}'| = k$, and $|\underline{r}''| = p$.

This finishes the definition of the additive distortion category. ◇

Explanation 4.5.5. Consider Definition 4.5.2.

- If $\mathcal{D}^{\mathrm{ad}}(\underline{r};\underline{s}) \neq \varnothing$, then $|\underline{r}| = |\underline{s}|$.
- Identity morphisms and composition are those of the symmetric groups.
- $\mathbb{0} = \varnothing$ and $\mathbb{1} = (0)$.
- Consider objects $\underline{r}, \underline{r}' \in \mathcal{D}^{\mathrm{ad}}$.
  - $\underline{r} \oplus \underline{r}'$ is their concatenation as in (4.2.8).
  - $\underline{r} \otimes \underline{r}'$ is as in (4.2.15). Its $(j + (l-1)m)$th entry is $r_j + r'_l$ for $1 \le j \le |\underline{r}|$ and $1 \le l \le |\underline{r}'|$.
- Consider morphisms $\sigma \in \mathcal{D}^{\mathrm{ad}}(\underline{r};\underline{s})$ and $\sigma' \in \mathcal{D}^{\mathrm{ad}}(\underline{r}';\underline{s}')$.
  - $\sigma \oplus \sigma' \in \mathcal{D}^{\mathrm{ad}}(\underline{r} \oplus \underline{r}'; \underline{s} \oplus \underline{s}')$ is their block sum in (2.4.2).
  - $\sigma \otimes \sigma' \in \mathcal{D}^{\mathrm{ad}}(\underline{r} \otimes \underline{r}'; \underline{s} \otimes \underline{s}')$ is the permutation in (2.4.4).
- The additive symmetry is given by

$$\xi^{\oplus}_{\underline{r},\underline{r}'} = \xi^{\oplus}_{|\underline{r}|,|\underline{r}'|} \in \Sigma_{|\underline{r}|+|\underline{r}'|},$$

  which is the block permutation in (2.4.3).
- $\alpha^{\oplus}$, $\lambda^{\oplus}$, $\rho^{\oplus}$, $\alpha^{\otimes}$, $\lambda^{\otimes}$, $\rho^{\otimes}$, $\lambda^{\bullet}$, $\rho^{\bullet}$, and δ^l are identity natural transformations.
- The permutation in (4.5.4) that defines δ^r in $\mathcal{D}^{\mathrm{ad}}$ is the first entry of the corresponding component of δ^r in $\mathcal{D}$. See Lemma 4.2.27 and Explanation 4.2.28.

Note that $\mathcal{D}^{\mathrm{ad}}$ does *not* have a multiplicative symmetry. Instead of using (4.2.24) as in $\mathcal{D}$, we define δ^r in $\mathcal{D}^{\mathrm{ad}}$ by directly specifying the permutation (4.5.4). ◇

Lemma 4.5.6. *The following statements hold.*

(1) *The underlying category of* $\mathcal{D}^{\mathrm{ad}}$ *is a groupoid.*
(2) *There is a faithful embedding*

$$\mathcal{D}^{\mathrm{ad}} \xrightarrow{\iota} \mathcal{D}$$

that is defined as follows.
  - *It is the identity function on objects.*
  - *It sends each morphism* σ *as in* (4.5.3) *to* $(\sigma; \mathrm{id}_{r_1}, \ldots, \mathrm{id}_{r_m})$.

(3) $\mathcal{D}^{\mathrm{ad}}$ *is a tight bimonoidal category as in Definition 2.1.2.*

Proof. For assertion (1), $\mathcal{D}^{\mathsf{ad}}$ is a groupoid because its identity morphisms and composition are defined using, respectively, identity permutations and composition of permutations.

The functoriality of $\iota\colon \mathcal{D}^{\mathsf{ad}} \longrightarrow \mathcal{D}$ follows from (4.2.3) and (4.2.4). It is faithful—that is, injective on morphism sets—by the definition of morphisms in Definition 4.2.1. This proves assertion (2).

For assertion (3), note that we have the following facts:

- $\mathcal{D}^{\mathsf{ad}}$ and $\mathcal{D}$ have the same objects. Structure morphisms in $\mathcal{D}^{\mathsf{ad}}$ are those in $\mathcal{D}$ restricted to the first entry. For δ^r, this follows from Lemma 4.2.27.
- The distortion category $\mathcal{D}$ is a left bipermutative category and, therefore, a tight symmetric bimonoidal category by Proposition 2.5.16 and Theorem 4.2.29. In particular, $\mathcal{D}$ is a tight bimonoidal category.

The faithful embedding ι and these facts imply that $\mathcal{D}^{\mathsf{ad}}$ is also a tight bimonoidal category. □

Recall from Definition 3.10.2 the nonsymmetric graph $\mathsf{Gr}^{\mathsf{ns}}(X)$.

- Its vertex set is the free $\{\oplus, \otimes\}$-algebra X^{fr}.
- Its edges are nonsymmetric prime edges, that is, prime edges that do not involve $\xi^{\pm\otimes}$.

Next we define the multiplicatively nonsymmetric analogue of the distortion of a path.

Definition 4.5.7. Consider the additive distortion category $\mathcal{D}^{\mathsf{ad}}$ in Definition 4.5.2.

- Define the function $\vartheta\colon X \longrightarrow \mathsf{Ob}(\mathcal{D}^{\mathsf{ad}}) = \mathsf{Ob}(\mathcal{D})$ as in (4.3.2), that is,

$$\vartheta(x) = \begin{cases} (1) & \text{if } x \in X \smallsetminus \{0^X, 1^X\}, \\ \varnothing & \text{if } x = 0^X, \text{ and} \\ (0) & \text{if } x = 1^X. \end{cases}$$

- Using the same symbol, define the associated graph morphism

$$\mathsf{Gr}^{\mathsf{ns}}(X) \xrightarrow{\quad \vartheta \quad} \mathcal{D}^{\mathsf{ad}}$$

  as in Definition 3.10.2, applied to the function ϑ and the bimonoidal category $\mathcal{D}^{\mathsf{ad}}$.
- For a path P in $\mathsf{Gr}^{\mathsf{ns}}(X)$, its value $\vartheta P \in \mathcal{D}^{\mathsf{ad}}$ in the sense of Definition 3.10.2 is called the *additive distortion of P*.

This finishes the definition. ◇

We are now ready for the second coherence theorem for bimonoidal categories. It is the multiplicatively nonsymmetric analogue of Theorem 4.4.3.

Theorem 4.5.8 (Bimonoidal Coherence II)**.** *Under Convention 3.10.1, suppose that the value of each δ-prime edge is a monomorphism. If*

$$\begin{array}{c} \xrightarrow{P_1} \\ a \qquad b \\ \xrightarrow[P_2]{} \end{array}$$

are two paths in $\mathsf{Gr}^{\mathsf{ns}}(X)$ with the same additive distortion, then their values in C are equal.

Proof. This proof is obtained from that of Theorem 4.4.3 by the following procedure:

- As in the proof of Theorem 3.10.7, we first adapt the five reduction steps in the proof of Theorem 3.9.1 by
  - removing all instances of $\xi^{\pm\otimes}$;
  - using Definition 3.10.2 instead of their counterparts in Section 3.1; and
  - using Theorem 1.3.3 instead of Theorem 1.3.8 for the monoidal category $(\mathsf{C}, \otimes)$.

  After these reduction steps, we are reduced to the setting of Lemma 4.4.1 with
  - $\xi^{\pm\otimes}$ removed and
  - $\mathsf{Gr}^{\mathsf{ns}}(X)$ and additive distortion instead of, respectively, $\mathsf{Gr}(X)$ and distortion.
- Following the proof of Lemma 4.4.1 with $\xi^{\pm\otimes}$ removed, in place of (4.4.2), we have the equalities

$$\vartheta(P_1') = \vartheta(P_2') = \sigma$$
$$\vartheta(P_1'') = \vartheta(P_2'') = \mathrm{id}_m.$$

  The nonsymmetric paths P_1'' and P_2'' involve only identities and $\alpha^{\pm\otimes}$, so they only move multiplicative brackets within each monomial in c. In the next-to-the-last paragraph, we use Theorem 1.3.3 instead of Theorem 1.3.8 in $(\mathsf{C}, \otimes)$.

With these adjustments, the proof of Theorem 4.4.3 is applicable in the current nonsymmetric setting. □

Example 4.5.9. Theorem 4.5.8 applies to flat, in particular, tight, bimonoidal categories as in Definition 3.9.9. ◇

4.6. Distortion Categories as Grothendieck Constructions

In this section, we describe the distortion category $\mathcal{D}$ and the additive distortion category $\mathcal{D}^{\mathsf{ad}}$ as Grothendieck constructions over the finite ordinal category Σ. We first recall from [**JY21**, 10.1.1] the relevant construction.

Definition 4.6.1. For a category C and a functor $F\colon \mathsf{C}^{\mathsf{op}} \longrightarrow \mathsf{Cat}$, the *Grothendieck construction* $\int_{\mathsf{C}} F$ is the category defined as follows.

- An object in $\int_{\mathsf{C}} F$ is a pair (A, X) consisting of objects $A \in \mathsf{C}$ and $X \in FA$.
- A morphism

$$(f, p)\colon (A, X) \longrightarrow (B, Y) \in \int_{\mathsf{C}} F$$

  consists of
  - a morphism $f\colon A \longrightarrow B$ in C and
  - a morphism $p\colon X \longrightarrow (Ff)(Y)$ in FA.
- The identity morphism of an object (A, X) is the pair $(1_A, 1_X)$ of identity morphisms.

- For another morphism
$$(g,q)\colon (B,Y) \longrightarrow (C,Z),$$
the composition with (f,p) is defined as
$$(g,q)\circ(f,p) = \big(gf, (Ff)(q)\circ p\big)\colon (A,X) \longrightarrow (C,Z).$$

This finishes the definition of $\int_{\mathsf{C}} F$. ◇

Explanation 4.6.2. The notation $\int_{\mathsf{C}} F$ is meant to suggest that it is the integration of the categories FA for $A \in \mathsf{C}$. The composite $(g,q)\circ(f,p)$ may be visualized as follows, where we write f^*, g^*, and $(gf)^*$ for, respectively, Ff, Fg, and $F(gf)$.

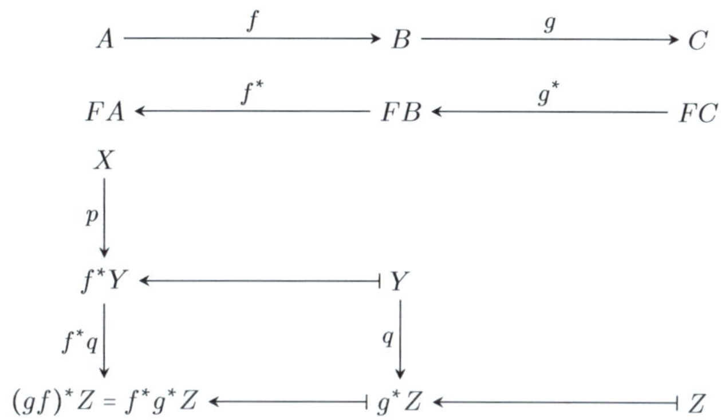

Since $f^*g^* = (gf)^*$ by the functoriality of F, it follows that
$$(f^*q)p\colon X \longrightarrow (gf)^*Z$$
is a well-defined morphism in FA. ◇

The Distortion Category. Recall from Definition 2.4.1 the finite ordinal category Σ, with objects $n \geq 0$ and morphism sets
$$\Sigma(m,n) = \begin{cases} \Sigma_m & \text{if } m = n \text{ and} \\ \varnothing & \text{if } m \neq n. \end{cases}$$

The n-fold Cartesian product of Σ is denoted by $\Sigma^{\times n}$, with $\Sigma^{\times 0} = *$. We will use the following functor to relate the finite ordinal category and the distortion category.

Definition 4.6.3. Define a functor
$$F\colon \Sigma^{\mathsf{op}} \longrightarrow \mathsf{Cat}$$
by the following assignments on objects and morphisms.
$$\begin{aligned} n &\longmapsto \Sigma^{\times n} \\ \big(\Sigma(n,n) = \Sigma_n \ni \sigma\big) &\longmapsto \big(\sigma^{-1} \in \mathsf{Cat}(\Sigma^{\times n}, \Sigma^{\times n})\big) \end{aligned}$$

Here σ^{-1} is the functor that permutes the factors via $\sigma^{-1} \in \Sigma_n$, that is,

(4.6.4) $$\sigma^{-1}(x_1,\ldots,x_n) = \big(x_{\sigma(1)},\ldots,x_{\sigma(n)}\big).$$

This finishes the definition of the functor F. ◇

The following observation provides a conceptual explanation of the distortion category $\mathcal{D}$ in terms of the finite ordinal category Σ.

Proposition 4.6.5. *There is a canonical isomorphism of categories*

$$\mathcal{D} \cong \int_{\Sigma} F$$

with $F: \Sigma^{\mathsf{op}} \longrightarrow \mathsf{Cat}$ *the functor in Definition 4.6.3.*

Proof. This follows from unpacking Definition 4.6.1 of the Grothendieck construction for the functor F. In more detail, an object in $\int_{\Sigma} F$ is a pair $(m, \underline{r})$ consisting of

- an object $m \geq 0$ in Σ and
- an object $\underline{r} \in F(m) = \Sigma^{\times m}$, that is, a sequence $(r_1, \ldots, r_m)$ with each $r_j \geq 0$.

A morphism

$$\underline{\sigma}: (m, \underline{r}) \longrightarrow (n, \underline{s}) \in \int_{\Sigma} F$$

consists of

- a morphism $\sigma \in \Sigma(m, n)$, that is, a permutation $\sigma \in \Sigma_m$ with $m = n$, and
- a morphism

  $$(\sigma_1, \ldots, \sigma_m): \underline{r} \longrightarrow \sigma^{-1}\underline{s} = \big(s_{\sigma(1)}, \ldots, s_{\sigma(m)}\big) \in \Sigma^{\times m},$$

  that is, a permutation

  $$\sigma_j \in \Sigma(r_j, r_j) = \Sigma_{r_j} \quad \text{for each} \quad 1 \leq j \leq m$$

  with $r_j = s_{\sigma(j)}$.

So objects and morphisms in the Grothendieck construction $\int_{\Sigma} F$ are the same as those in the distortion category $\mathcal{D}$ (Definition 4.2.1). Similarly, the identity morphisms and composition in $\int_{\Sigma} F$ are the same as those in $\mathcal{D}$. □

The Additive Distortion Category. There is a similar description for the additive distortion category $\mathcal{D}^{\mathsf{ad}}$ (Definition 4.5.2) that involves the following functor. In the following definition, we regard the set $\mathbb{N}$ of nonnegative integers as a discrete category with only identity morphisms.

Definition 4.6.6. Define a functor

$$F^{\mathsf{ad}}: \Sigma^{\mathsf{op}} \longrightarrow \mathsf{Cat}$$

by the following assignments on objects and morphisms.

$$n \longmapsto \mathbb{N}^{\times n}$$
$$\big(\Sigma(n,n) = \Sigma_n \ni \sigma\big) \longmapsto \big(\sigma^{-1} \in \mathsf{Cat}(\mathbb{N}^{\times n}, \mathbb{N}^{\times n})\big)$$

Here σ^{-1} is the functor that permutes the factors via $\sigma^{-1} \in \Sigma_n$ as in (4.6.4). This finishes the definition of the functor F^{ad}. ◇

Proposition 4.6.7. *There is a canonical isomorphism of categories*

$$\mathcal{D}^{\mathsf{ad}} \cong \int_{\Sigma} F^{\mathsf{ad}}$$

with $F^{\mathsf{ad}}: \Sigma^{\mathsf{op}} \longrightarrow \mathsf{Cat}$ *the functor in Definition 4.6.6.*

Proof. Reuse the proof of Proposition 4.6.5 by replacing the category

$$F(m) = \Sigma^{\times m} \quad \text{with} \quad F^{\mathrm{ad}}(m) = \mathbb{N}^{\times m}.$$

Since $\mathbb{N}$ is a discrete category, the categorical structure of $\int_{\Sigma} F^{\mathrm{ad}}$ agrees with that of $\mathcal{D}^{\mathrm{ad}}$ (Definition 4.5.2). □

The braided version for the braided distortion category is Proposition II.5.5.3. See also Question A.5.6.

4.7. Notes

As we mentioned in the introduction of this chapter, the Coherence Theorem 4.4.3 is due to Laplaza [**Lap72b**], and the proof presented in this chapter partially follows the general outline of Laplaza's original proof. Here we point out the main differences between Laplaza's proof and the one in this chapter. Besides some cosmetic differences in notation and additional detail in this chapter, Laplaza's original proof contains some inaccuracies that we have corrected in this chapter. See Notes 4.7.3 and 4.7.4. In Note 4.7.5, we briefly discuss a possible 2-monad approach.

4.7.1 (Notation)**.** In addition to the differences in notation in Note 3.11.1, the table below shows the correspondence between our notation and the ones in [**Lap72b**].

Definition	Concept	Laplaza [**Lap72b**]
4.2.1	distortion category $\mathcal{D}$	$\underline{D}$
4.2.1	symmetric group Σ_m	P_m
(4.2.10)	block permutation $\xi^{\oplus}_{m,k}$	$t_{m,k}$
(4.2.17)	permutation $\xi^{\otimes}_{m,k}$	$\tau_{m,k}$
4.3.1	distortion ϑP	$\mathrm{dist}(P)$

◇

4.7.2 (Level of Detail)**.** In [**Lap72b**], Laplaza omitted most of the proofs of Lemmas 4.2.5, 4.2.12, 4.2.19, and 4.2.25 and Theorem 4.2.29, which together show that the distortion category is a left bipermutative category. ◇

4.7.3 (Corrections for Laplaza's Parts I–III and Monomorphisms)**.** Laplaza's original proof of Theorem 4.4.3 in [**Lap72b**] is divided into what he called Parts I–V. Parts I–III there correspond to the five reduction steps in the proof of Theorem 3.9.1 without regularity, as used in the proof of Theorem 4.4.3. As we saw in the proof of Theorem 3.9.1, these five reduction steps use essentially all the results in Chapter 3 before Section 3.9. In particular, these include

- Proposition 3.5.32 in reduction step 1 and
- Proposition 3.7.19 in reduction step 4.

These preliminary results correspond to [**Lap72a**, Prop. 5 and 7]. Moreover, the monomorphism assumption in the statement of Theorem 4.4.3, which is used in reduction step 3, is not included in [**Lap72b**]. Therefore, the necessary corrections discussed in Notes 3.11.3 through 3.11.6 still apply. ◇

4.7.4 (Corrections for Laplaza's Parts IV and V)**.** In the proof of Theorem 4.4.3, after the five reduction steps, we finished the proof using Lemma 4.4.1. This is significantly different from the way Laplaza finished his Parts IV and V in [**Lap72b**], which depend on [**Lap72b**, Lemma 3.1]. As we now explain, that lemma is incorrect.

The setting for that lemma is that, in addition to the function $\vartheta\colon X \longrightarrow \mathsf{Ob}(\mathcal{D})$ as in Definition 4.3.1, we are given the following data.

- Another set X' equipped with two distinguished elements $0^{X'}$ and $1^{X'}$.
- A function $f\colon X' \longrightarrow X$ such that $f(0^{X'}) = 0^X$ and $f(1^{X'}) = 1^X$.

The assertion of [**Lap72b**, Lemma 3.1] is that for each path Q in $\mathsf{Gr}(X')$, the distortion of Q is equal to the distortion of $f(Q)$. Here $f(Q)$ is the path in $\mathsf{Gr}(X)$ obtained from Q by applying f to each element in X' that appears in the vertices in Q and in the subscripts of the prime edges. For example, if $a, b, c, d \in X'$, then

$$f\big(\delta^l_{a,b,c} \oplus 1_d\big) = \delta^l_{f(a),f(b),f(c)} \oplus 1_{f(d)}.$$

To see that [**Lap72b**, Lemma 3.1] is incorrect, consider the following data.

(i) The set $X' = X \amalg \{x'\}$ obtained from X by adjoining a new element x', with $0^{X'} = 0^X$ and $1^{X'} = 1^X$.

(ii) The function $f\colon X' \longrightarrow X$ that extends the identity function on X by $f(x') = 1^X$.

(iii) The identity path

$$x' \xrightarrow{\;Q = 1_{x'}\;} x' \quad \text{in} \quad \mathsf{Gr}(X').$$

By Definition 4.3.1, since $x' \in X' \smallsetminus \{0^{X'}, 1^{X'}\}$, the distortion of Q is the identity morphism

$$\vartheta'(1_{x'}) = 1_{\vartheta'(x')} = 1_{(1)} = (\mathrm{id}_1; \mathrm{id}_1) \in \mathcal{D}\big((1);(1)\big).$$

On the other hand, if we first apply f, and then take the distortion, we obtain the identity morphism

$$\vartheta f(1_{x'}) = 1_{\vartheta f(x')} = 1_{\vartheta(1^X)} = 1_{(0)} = (\mathrm{id}_1; \mathrm{id}_0) \in \mathcal{D}\big((0);(0)\big).$$

This is *not* equal to $1_{(1)}$, since even their domains are different objects.

Another simple counterexample is the additive symmetry

$$x' \oplus x' \xrightarrow{\;\xi^{\oplus}_{x',x'}\;} x' \oplus x' \quad \text{in} \quad \mathsf{Gr}(X').$$

Using (4.2.10), its images under ϑ' and ϑf are as follows.

$$\begin{aligned}
\vartheta'(\xi^{\oplus}_{x',x'}) &= \xi^{\oplus}_{(1),(1)} = \big(\xi^{\oplus}_{1,1}; \mathrm{id}_1, \mathrm{id}_1\big) \in \mathcal{D}\big((1) \oplus (1); (1) \oplus (1)\big)\\
\vartheta f(\xi^{\oplus}_{x',x'}) &= \xi^{\oplus}_{(0),(0)} = \big(\xi^{\oplus}_{1,1}; \mathrm{id}_0, \mathrm{id}_0\big) \in \mathcal{D}\big((0) \oplus (0); (0) \oplus (0)\big)
\end{aligned}$$

These are different morphisms in $\mathcal{D}$.

The above setting of (i) and (ii), that is,

$$X' = X \amalg \{x'\}, \qquad f|_X = \mathrm{Id}_X, \quad \text{and} \quad f(x') = 1^X,$$

appeared in the proof of [**Lap72b**, Lemma 3.2], which is used in [**Lap72b**, Part IV, page 233]. Therefore, the setting of (i) and (ii) cannot be ignored. The invalidity of [**Lap72b**, Lemma 3.1] implies that Laplaza's proof of [**Lap72b**, Parts IV and V], which uses that lemma, is also invalid. In our proof of Theorem 4.4.3, we substituted Laplaza's Parts IV and V with Lemma 4.4.1, which completely bypasses all the lemmas in [**Lap72b**, Section 3].

In Example 4.3.16, the assumption that $a_i, b_j \notin \{0^X, 1^X\}$ excludes the situations similar to the counterexamples above. ◇

4.7.5 (2-Monads)**.** By Example 4.4.5, Theorem 4.4.3 applies to *tight* symmetric bimonoidal categories, that is, those with δ^l and δ^r (2.1.4) natural isomorphisms. This special case of Theorem 4.4.3 is also claimed to follow from the 2-monad coherence result in [**Kel74**, pages 371-373]. The discussion in Note 3.11.7 still applies here. ◇

CHAPTER 5

Strictification of Tight Symmetric Bimonoidal Categories

A right bipermutative category C as in Definition 2.5.2 has

- two permutative structures $(\oplus, \mathbb{0}, \xi^{\oplus})$ and $(\otimes, \mathbb{1}, \xi^{\otimes})$ and
- identities for the right distributivity δ^r, the multiplicative zeros $\lambda^{\bullet}$ and $\rho^{\bullet}$, and $\xi^{\otimes}_{-,\mathbb{0}}$,

along with three carefully chosen symmetric bimonoidal category axioms. We observed in Proposition 2.5.7 that each right bipermutative category is a tight symmetric bimonoidal category, where *tight* means that δ^l and δ^r are natural isomorphisms. In this chapter, we show that each tight symmetric bimonoidal category is, in a suitable sense to be defined below, adjoint equivalent to a right bipermutative category. This theorem is originally due to May [**May77**]. The Strictification Theorem 5.4.6 is the bimonoidal analogue of Theorem 1.3.10. Moreover, a multiplicatively nonsymmetric analogue is the Strictification Theorem 5.5.11 for tight bimonoidal categories. A braided version that strictifies tight braided bimonoidal categories is Theorem II.6.3.6. Question A.1.6 is an open question related to strictification of tight symmetric bimonoidal categories.

Organization. To discuss equivalences between symmetric bimonoidal categories, in Section 5.1, we define symmetric bimonoidal functors. Each symmetric bimonoidal functor is a functor with two symmetric monoidal structures, one additive and one multiplicative, that satisfy two compatibility axioms for the multiplicative zeros and the distributivity morphisms. While the axioms are stated for the right multiplicative zeros and the right distributivity morphisms, they are actually equivalent to the left variants. Symmetric bimonoidal functors compose as expected, and they constitute the morphisms in a category $\mathsf{Bi}^{\mathsf{sy}}$ with small symmetric bimonoidal categories as objects. The second half of this section contains examples of symmetric bimonoidal functors and equivalences. This section does not rely on the Coherence Theorems 3.9.1 and 4.4.3, and may be read immediately after Chapter 2.

The proof of the Strictification Theorem 5.4.6 begins in Section 5.2. For a tight symmetric bimonoidal category C, in this section, we construct an associated right bipermutative category A whose objects are formal polynomials in the objects in C. This section contains most of the data of the associated right bipermutative category.

The proof that A is a right bipermutative category is given in Section 5.3. The tightness assumption on C, that is, the invertibility of its distributivity morphisms, is used to define the products of morphisms and the multiplicative symmetry in A. See (5.2.29), (5.2.36), and the explanation following them.

Section 5.4 contains the other half of the proof of the Strictification Theorem 5.4.6. We define two functors between C and A in opposite directions and observe that they (i) constitute an adjoint equivalence and (ii) are both symmetric bimonoidal equivalences. It is in this sense that a given tight symmetric bimonoidal category C is equivalent to a right bipermutative category A. At the end of this section, we record a variant of this strictification result that involves a left bipermutative category; see Theorem 5.4.7.

For monoidal categories, the proof of the Strictification Theorem 1.3.5 relies on the Coherence Theorem 1.3.3. The reason is that the construction of an associated strict monoidal category requires the commutativity of some formally defined diagrams, which are guaranteed by coherence. In a similar manner, for tight symmetric bimonoidal categories, the proof of the Strictification Theorem 5.4.6 also relies heavily on the Coherence Theorem 3.9.1. See, for example, Explanation 5.2.31 and Lemmas 5.3.1, 5.3.4, 5.3.8, and 5.4.4. Each time Theorem 3.9.1 is used to ensure that a certain diagram in C is commutative.

Section 5.5 contains the Strictification Theorems 5.5.11 and 5.5.12 for tight bimonoidal categories. They are the multiplicatively nonsymmetric analogues of Theorems 5.4.6 and 5.4.7. We define the nonsymmetric analogues of symmetric bimonoidal functors and right (left) bipermutative categories, which are called, respectively, bimonoidal functors and right (left) rigid bimonoidal categories. In the nonsymmetric analogue of A in Definition 5.5.5, the left distributivity morphism is defined in terms of δ^l in C, instead of using the diagram (5.3.6) and the multiplicative symmetry. Therefore, the proof of Proposition 5.5.10 that the nonsymmetric analogue of A satisfies the 22 Laplaza axioms for a bimonoidal category is also somewhat different from the symmetric case in Section 5.3.

For a brief history of strictification for tight symmetric bimonoidal categories, the reader is referred to Section 5.6.

Reading Guide. The proof of the Strictification Theorem 5.4.6 is presented with full detail straightly linearly in this chapter. As a possible alternative to reading this chapter linearly, we offer the following suggestion.

(1) In Section 5.1, read Definition 5.1.1 of a symmetric bimonoidal functor.
(2) In Section 5.2, pay special attention to Explanations 5.2.25, 5.2.30, 5.2.31, and 5.2.37. Skip Example 5.2.32 during the first reading.
(3) In Section 5.3, skip all the proofs during the first reading.
(4) In Section 5.4, first read Definitions 5.4.1 and 5.4.2 and the statements of Lemmas 5.4.3 through 5.4.5 and Theorems 5.4.6 and 5.4.7.
(5) With a clear understanding of the structure of the proof of Theorem 5.4.6, go back and read the parts skipped earlier.

As in Chapters 3 and 4, in this chapter we deliberately divided the proof of Theorem 5.4.6 into many lemmas to clarify the overall structure of the proof and to make jumping forward and backward easier. Students are encouraged to regard the many lemmas and their detailed proofs as exercises with full solutions.

Detail. The detailed proof of Theorem 5.4.6 does more than proving this theorem.

(1) We emphasize that Theorem 5.4.6 only applies to *tight* symmetric bimonoidal categories, that is, those with δ^l and δ^r natural isomorphisms. The detailed constructions and proofs in this chapter show precisely how

the invertibility of the distributivity morphisms is used. See Explanations 5.2.25, 5.2.31, and 5.2.37. This is related to an incorrect claim in [**EM06**] that their bipermutative categories, whose factorization morphisms are not invertible in general, are equivalent to right bipermutative categories. The reader is referred to Note 5.6.3 for more discussion related to this issue.

(2) May's proof of Theorem 5.4.6 is given as an outline in [**May77**, 6.3.5], where the Coherence Theorem 3.9.1 is not explicitly mentioned. Our detailed constructions and proofs in this chapter show precisely where Theorem 3.9.1 is used. The reader is referred to Note 5.6.2 for more discussion related to this issue.

(3) The proof of the Strictification Theorem II.6.3.6 of tight braided bimonoidal categories uses modified versions of many of the proofs in this chapter. A detailed treatment here will allow us to be both precise and concise at the same time in the braided case.

5.1. Symmetric Bimonoidal Functors

In this section, we define functors between symmetric bimonoidal categories and observe that there is a category $\mathsf{Bi}^{\mathsf{sy}}$ of small symmetric bimonoidal categories and symmetric bimonoidal functors. Examples of symmetric bimonoidal functors are discussed in the second half of this section.

Recall from Definitions 1.2.11 and 1.2.25 the concept of a symmetric monoidal functor, and from Definition 2.1.2 the concept of a symmetric bimonoidal category.

Definition 5.1.1. Suppose C and D are symmetric bimonoidal categories. A *symmetric bimonoidal functor* from C to D is a tuple

$$\big(F, F^2_\oplus, F^0_\oplus, F^2_\otimes, F^0_\otimes\big) \colon \mathsf{C} \longrightarrow \mathsf{D}$$

consisting of the following data.

- $(F, F^2_\oplus, F^0_\oplus) \colon \mathsf{C} \longrightarrow \mathsf{D}$ is a symmetric monoidal functor from the additive structure of C to the additive structure of D.
- $(F, F^2_\otimes, F^0_\otimes) \colon \mathsf{C} \longrightarrow \mathsf{D}$ is a symmetric monoidal functor from the multiplicative structure of C to the multiplicative structure of D.

These data are required to make the following two diagrams in D commutative for all objects $A, B, C \in \mathsf{C}$.

Multiplicative Zero:

(5.1.2)
$$\begin{array}{ccccc} (FA)\mathbb{0} & & \xrightarrow{\;1_{FA}F^0_\oplus\;} & & (FA)(F\mathbb{0}) \\ \downarrow{\scriptstyle \rho^\bullet_{FA}} & & & & \downarrow{\scriptstyle F^2_\otimes} \\ \mathbb{0} & & & & F(A\mathbb{0}) \\ & \searrow{\scriptstyle F^0_\oplus} & & \swarrow{\scriptstyle F\rho^\bullet_A} & \\ & & F\mathbb{0} & & \end{array}$$

Distributivity:

$$
\begin{array}{ccc}
(FA \oplus FB)(FC) & \xrightarrow{\delta^r} & (FA)(FC) \oplus (FB)(FC) \\
\Big\downarrow {\scriptstyle F^2_{\oplus} 1_{FC}} & & \Big\downarrow {\scriptstyle F^2_{\otimes} \oplus F^2_{\otimes}} \\
F(A \oplus B)(FC) & & F(AC) \oplus F(BC) \\
\Big\downarrow {\scriptstyle F^2_{\otimes}} & & \Big\downarrow {\scriptstyle F^2_{\oplus}} \\
F\big((A \oplus B)C\big) & \xrightarrow{F\delta^r} & F(AC \oplus BC)
\end{array}
\tag{5.1.3}
$$

This finishes the definition of a symmetric bimonoidal functor.

Moreover:

- A symmetric bimonoidal functor as above is sometimes abbreviated to F.
- We write

$$F_{\oplus} = (F, F^2_{\oplus}, F^0_{\oplus}) \quad \text{and} \quad F_{\otimes} = (F, F^2_{\otimes}, F^0_{\otimes})$$

  for the two symmetric monoidal functors, which are called the *additive structure* and the *multiplicative structure*, respectively.
- A symmetric bimonoidal functor is
  - *robust* if $F^2_{\oplus}$, $F^0_{\oplus}$, and $F^0_{\otimes}$ are isomorphisms;
  - *strong* if both $F_{\oplus}$ and $F_{\otimes}$ are strong symmetric monoidal functors, that is, if $F^2_{\oplus}$, $F^0_{\oplus}$, $F^2_{\otimes}$, and $F^0_{\otimes}$ are isomorphisms;
  - *unitary* if it is strong, and if $F^0_{\oplus}$ and $F^0_{\otimes}$ are identities;
  - *strict* if both $F_{\oplus}$ and $F_{\otimes}$ are strict symmetric monoidal functors, that is, if $F^2_{\oplus}$, $F^0_{\oplus}$, $F^2_{\otimes}$, and $F^0_{\otimes}$ are identities; and
  - a *symmetric bimonoidal equivalence* if it is also an equivalence of categories. ◇

The axioms (5.1.2) and (5.1.3) of a symmetric bimonoidal functor are stated in terms of the right multiplicative zeros and the right distributivity morphisms. The following observation says that they can also be stated in terms of the left multiplicative zeros and the left distributivity morphisms.

Proposition 5.1.4. *Suppose*

$$\big(F, F^2_{\oplus}, F^0_{\oplus}, F^2_{\otimes}, F^0_{\otimes}\big) \colon \mathsf{C} \longrightarrow \mathsf{D}$$

consists of the same data as in Definition 5.1.1.

(1) *The multiplicative zero axiom* (5.1.2) *is equivalent to the commutativity of the following diagram for all objects $A \in \mathsf{C}$.*

$$
\begin{array}{ccc}
\mathbb{0}(FA) & \xrightarrow{F^0_{\oplus} 1_{FA}} & (F\mathbb{0})(FA) \\
\Big\downarrow {\scriptstyle \lambda^{\bullet}_{FA}} & & \Big\downarrow {\scriptstyle F^2_{\otimes}} \\
\mathbb{0} & & F(\mathbb{0}A) \\
& {\scriptstyle F^0_{\oplus}} \searrow \quad \swarrow {\scriptstyle F\lambda^{\bullet}_A} & \\
& F\mathbb{0} &
\end{array}
\tag{5.1.5}
$$

(2) *The distributivity axiom* (5.1.3) *is equivalent to the commutativity of the following diagram for all objects* $A, B, C \in \mathsf{C}$.

$$\begin{array}{ccc} (FA)(FB \oplus FC) & \xrightarrow{\delta^l} & (FA)(FB) \oplus (FA)(FC) \\ {\scriptstyle 1_{FA}F^2_\oplus}\downarrow & & \downarrow{\scriptstyle F^2_\otimes \oplus F^2_\otimes} \\ (FA)F(B \oplus C) & & F(AB) \oplus F(AC) \\ {\scriptstyle F^2_\otimes}\downarrow & & \downarrow{\scriptstyle F^2_\oplus} \\ F\big(A(B \oplus C)\big) & \xrightarrow{F\delta^l} & F(AB \oplus AC) \end{array} \qquad (5.1.6)$$

Proof. For the first assertion, consider the following diagram.

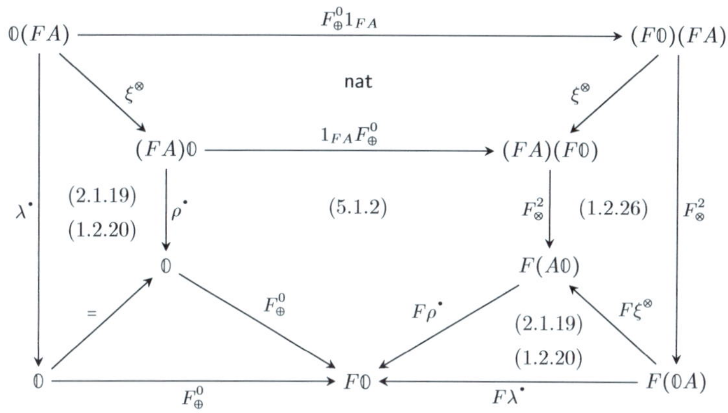

The outer diagram is (5.1.5). Since $\xi^\otimes$ is a natural isomorphism, the outer diagram is commutative if and only if the inside pentagon (5.1.2) is commutative.

For the second assertion, consider the following diagram.

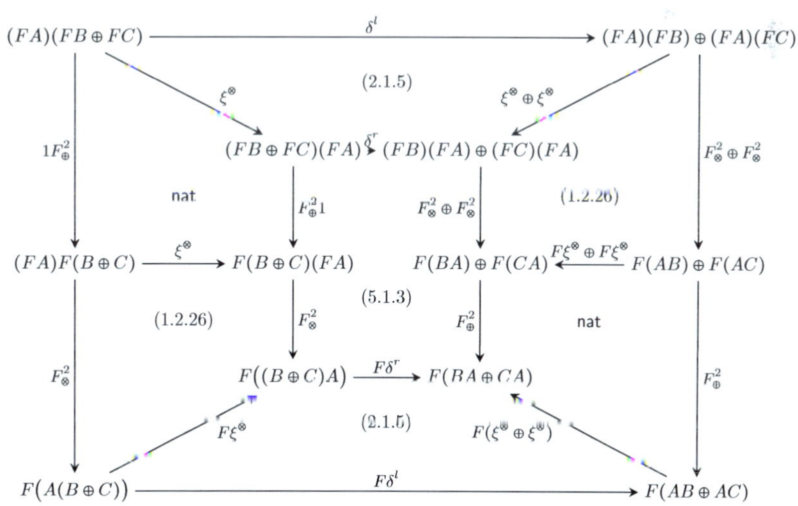

The outer diagram is (5.1.6). Since $\xi^\otimes$ is a natural isomorphism, the outer diagram is commutative if and only if the inside rectangle (5.1.3) is commutative. □

Composition.

Definition 5.1.7. Suppose

$$\mathsf{C} \xrightarrow{F} \mathsf{D} \xrightarrow{G} \mathsf{E}$$

are symmetric bimonoidal functors. The *composite*

$$\big(GF, (GF)^2_\oplus, (GF)^0_\oplus, (GF)^2_\otimes, (GF)^0_\otimes\big) : \mathsf{C} \longrightarrow \mathsf{E}$$

is defined by the composites

$$\big(GF, (GF)^2_\oplus, (GF)^0_\oplus\big) = G_\oplus \circ F_\oplus \quad \text{and}$$
$$\big(GF, (GF)^2_\otimes, (GF)^0_\otimes\big) = G_\otimes \circ F_\otimes$$

of symmetric monoidal functors. ◇

Explanation 5.1.8. In Definition 5.1.7, the underlying functor of the composite is the composite functor $GF : \mathsf{C} \longrightarrow \mathsf{E}$. The additive structure $(GF)_\oplus = G_\oplus \circ F_\oplus$ of the composite is defined by the following composites.

$$(GF)^0_\oplus : \mathbb{0} \xrightarrow{G^0_\oplus} G\mathbb{0} \xrightarrow{G(F^0_\oplus)} GF\mathbb{0}$$

$$(GF)^2_\oplus : GFA \oplus GFB \xrightarrow{G^2_\oplus} G(FA \oplus FB) \xrightarrow{G(F^2_\oplus)} GF(A \oplus B)$$

Two similar composites, with $(\oplus, \mathbb{0})$ replaced by $(\otimes, \mathbb{1})$, define the multiplicative structure $(GF)_\otimes = G_\otimes \circ F_\otimes$ of the composite. ◇

Lemma 5.1.9. *In Definition 5.1.7, $GF : \mathsf{C} \longrightarrow \mathsf{E}$ is a symmetric bimonoidal functor. Moreover, if both F and G are robust (respectively, strong, unitary, or strict), then so is GF.*

Proof. Since the composite of two symmetric monoidal functors is a symmetric monoidal functor, it remains to check the axioms (5.1.2) and (5.1.3) for GF. For (5.1.2), consider the following diagram.

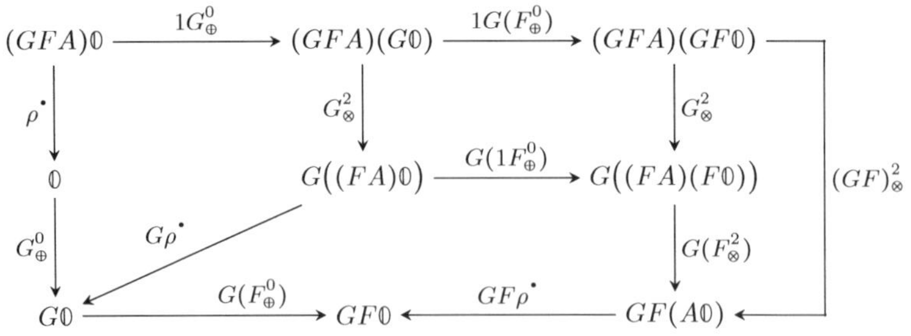

$$\begin{array}{ccccc}
(GFA)\mathbb{0} & \xrightarrow{1G^0_\oplus} & (GFA)(G\mathbb{0}) & \xrightarrow{1G(F^0_\oplus)} & (GFA)(GF\mathbb{0}) \\
\downarrow{\scriptstyle \rho^\bullet} & & \downarrow{\scriptstyle G^2_\otimes} & & \downarrow{\scriptstyle G^2_\otimes} \\
\mathbb{0} & & G((FA)\mathbb{0}) & \xrightarrow{G(1F^0_\oplus)} & G((FA)(F\mathbb{0})) \\
\downarrow{\scriptstyle G^0_\oplus} & \swarrow{\scriptstyle G\rho^\bullet} & & & \downarrow{\scriptstyle G(F^2_\otimes)} \\
G\mathbb{0} & \xrightarrow{G(F^0_\oplus)} & GF\mathbb{0} & \xleftarrow{GF\rho^\bullet} & GF(A\mathbb{0})
\end{array}$$

(with the outer arrow $(GF)^2_\otimes : (GFA)(GF\mathbb{0}) \to GF(A\mathbb{0})$ on the right)

By the definitions of $(GF)^0_\oplus$ and $(GF)^2_\otimes$, the outer diagram is (5.1.2) for GF. The top right rectangle is commutative by the naturality of $G^2_\otimes$. The left and the bottom trapezoids are commutative by the axiom (5.1.2) for F and G.

A similar diagram using

- (5.1.3) for F and G,
- the naturality of $G^2_\oplus$ and $G^2_\otimes$, and
- the definitions of $(GF)^2_\oplus$ and $(GF)^2_\otimes$,

proves the commutativity of (5.1.3) for GF. This shows that GF is a symmetric bimonoidal functor.

The other assertion follows from the following facts:

- If both $F^0_\oplus$ and $G^0_\oplus$ are isomorphisms (respectively, identities), then so is $(GF)^0_\oplus$.
- If both $F^2_\oplus$ and $G^2_\oplus$ are isomorphisms (respectively, identities), then so is $(GF)^2_\oplus$.
- Similar statements for $(F^0_\otimes, G^0_\otimes, (GF)^0_\otimes)$ and $(F^2_\otimes, G^2_\otimes, (GF)^2_\otimes)$ hold.

These statements hold by the definitions of the additive and the multiplicative structures of GF in Explanation 5.1.8. □

A subcategory is called *wide* if it contains all the objects in the larger category.

Proposition 5.1.10. *There is a category* $\mathsf{Bi}^{\mathsf{sy}}$ *defined by the following data.*

- *The objects are small symmetric bimonoidal categories as in Definition 2.1.2.*
- *The morphisms are symmetric bimonoidal functors as in Definition 5.1.1.*
- *Identity morphisms are identity functors with identity monoidal structures.*
- *Composition is as in Definition 5.1.7.*

Moreover, $\mathsf{Bi}^{\mathsf{sy}}$ *has the following wide subcategories:*

- $\mathsf{Bi}^{\mathsf{sy}}_{\mathsf{r}}$ *with robust symmetric bimonoidal functors as morphisms.*
- $\mathsf{Bi}^{\mathsf{sy}}_{\mathsf{sg}}$ *with strong symmetric bimonoidal functors as morphisms.*
- $\mathsf{Bi}^{\mathsf{sy}}_{\mathsf{u}}$ *with unitary symmetric bimonoidal functors as morphisms.*
- $\mathsf{Bi}^{\mathsf{sy}}_{\mathsf{st}}$ *with strict symmetric bimonoidal functors as morphisms.*

Proof. Lemma 5.1.9 shows that composition of symmetric bimonoidal functors is well defined. Associativity and unity of composition follow from those of symmetric monoidal functors. Therefore, $\mathsf{Bi}^{\mathsf{sy}}$ is a category.

The other assertion about wide subcategories holds because each of the four kinds of symmetric bimonoidal functors—robust, strong, unitary, and strict—is closed under composition by Lemma 5.1.9. □

Examples. The rest of this section contains examples of symmetric bimonoidal functors. Recall from Definition 2.3.1 that a distributive symmetric monoidal category is a symmetric monoidal category with finite coproducts over which the monoidal product distributes up to natural isomorphisms. By Proposition 2.3.2, each distributive symmetric monoidal category is a tight symmetric bimonoidal category with $(\oplus, \mathbb{0}) = (\amalg, \varnothing)$.

Proposition 5.1.11. *Suppose* C *and* D *are distributive symmetric monoidal categories, and*

$$(F, F^2_\otimes, F^0_\otimes)\colon (\mathsf{C}, \otimes^{\mathsf{C}}, \mathbb{1}^{\mathsf{C}}) \longrightarrow (\mathsf{D}, \otimes^{\mathsf{D}}, \mathbb{1}^{\mathsf{D}})$$

is a symmetric monoidal functor. Then F *induces a symmetric bimonoidal functor if* C *and* D *are regarded as symmetric bimonoidal categories with additive structures given by coproducts.*

Proof. The additive structure of F is defined by the morphisms

$$\begin{array}{rcl} \varnothing & \xrightarrow{F^0_\oplus} & F\varnothing \\ FA \amalg FB & \xrightarrow{F^2_\oplus} & F(A \amalg B) \end{array} \tag{5.1.12}$$

for objects $A, B \in \mathsf{C}$. The first morphism is the unique morphism from the chosen initial object in D. The second morphism is determined by F applied to the natural morphisms

$$\begin{array}{rcl} A & \longrightarrow & A \amalg B \\ B & \longrightarrow & A \amalg B. \end{array}$$

The universal properties of coproducts and initial objects imply that

$$(F, F^2_\oplus, F^0_\oplus) \colon (\mathsf{C}, \amalg, \varnothing) \longrightarrow (\mathsf{D}, \amalg, \varnothing)$$

is a symmetric monoidal functor. Moreover,

- the multiplicative zero axiom (5.1.2) holds by the universal property of the initial object in D, and
- the distributivity axiom (5.1.3) holds by the naturality of $F^2_\otimes$ and the universal property of coproducts.

Therefore, F is a symmetric bimonoidal functor. □

Example 5.1.13 (Ring Morphisms)**.** Suppose $f \colon R \longrightarrow S$ is a morphism of commutative rings. Then restriction of scalars via f defines a symmetric monoidal functor

$$f^* \colon \mathsf{Mod}(S) \longrightarrow \mathsf{Mod}(R)$$

from the category of S-modules to the category of R-modules. By Example 2.3.4 and Proposition 5.1.11, f^* is a symmetric bimonoidal functor when $\mathsf{Mod}(S)$ and $\mathsf{Mod}(R)$ are regarded as symmetric bimonoidal categories with additive structures given by direct sums. ◇

Example 5.1.14 (Distributive Categories)**.** Suppose $F \colon \mathsf{C} \longrightarrow \mathsf{D}$ is a functor between distributive categories as in Example 2.3.5, such that the natural morphisms

$$\begin{array}{rcl} F* & \xrightarrow{t} & * \\ F(A \times B) & \xrightarrow{p} & FA \times FB \end{array}$$

for objects $A, B \in \mathsf{C}$, which are dual to those in (5.1.12), are isomorphisms. The universal properties of products and terminal objects imply that

$$\big(F, p^{-1}, t^{-1}\big) \colon (\mathsf{C}, \times, *) \longrightarrow (\mathsf{D}, \times, *)$$

is a symmetric monoidal functor. By Example 2.3.5 and Proposition 5.1.11, F is a symmetric bimonoidal functor when C and D are regarded as symmetric bimonoidal categories with $(\oplus, \mathbb{0}) = (\amalg, \varnothing)$. ◇

Recall the small and tight symmetric bimonoidal categories Σ and Σ' in Section 2.4. Their underlying categories are given by finite ordinals and bijections, and they have the same additive structure. In Σ the left distributivity δ^l is the identity, and in Σ' the right distributivity δ^r is the identity. Moreover, Σ' is a right bipermutative category as in Definition 2.5.2, and Σ is a left bipermutative category as in Definition 2.5.11.

Proposition 5.1.15. *There is a symmetric bimonoidal equivalence*

$$\Sigma \xrightarrow{\;F\;} \Sigma'$$

that is defined by the following data.

- *The additive structure* $(F, F^2_\oplus, F^0_\oplus)$ *is the identity symmetric monoidal functor.*
- $F^0_\otimes : 0 \longrightarrow 0$ *is the identity permutation in* Σ_0.
- $F^2_\otimes : mn \longrightarrow mn$ *is the multiplicative symmetry isomorphism* ${\xi'}^{\otimes}_{m,n}$ *in* Σ' *in* (2.4.20).

Proof. To check that the multiplicative structure $F_\otimes = (F, F^2_\otimes, F^0_\otimes)$ is a symmetric monoidal functor, first observe that the unity axioms (1.2.15) hold because each morphism involved is an identity morphism.

Since the multiplicative associativity isomorphisms in both Σ and Σ' are identities, the associativity axiom (1.2.14) becomes the following diagram in Σ' for $m, n, p \geq 0$.

$$\begin{array}{ccc} mnp & \xrightarrow{1_m {\xi'}^{\otimes}_{n,p}} & mpn \\ {\scriptstyle {\xi'}^{\otimes}_{m,n} 1_p}\downarrow & & \downarrow{\scriptstyle {\xi'}^{\otimes}_{m,pn}} \\ nmp & \xrightarrow{{\xi'}^{\otimes}_{nm,p}} & pnm \end{array}$$

This diagram commutes because both composites are given by the bijection

$$k + (j-1)p + (i-1)np \longmapsto i + (j-1)m + (k-1)mn$$

for $1 \leq i \leq m$, $1 \leq j \leq n$, and $1 \leq k \leq p$.

The compatibility of $F^2_\otimes$ with the multiplicative symmetries (1.2.26) holds because each composite is the identity. This follows from the fact that the multiplicative symmetry isomorphism $\xi^{\otimes}_{m,n}$ in Σ in (2.4.5) is the inverse of ${\xi'}^{\otimes}_{m,n}$ in Σ'. So the multiplicative structure $F_\otimes$ is a symmetric monoidal functor.

The multiplicative zero axiom (5.1.2) holds because $\Sigma'(0,0) = \Sigma_0$ contains only the identity morphism of 0.

Since $F^2_\oplus$ and δ^l in Σ are identities, the distributivity axiom, in the equivalent form (5.1.6), is the following diagram in Σ' for $m, n, p \geq 0$.

$$\begin{array}{ccc} m(n+p) & \xrightarrow{\delta^l_{m,n,p}} & mn + mp \\ {\scriptstyle {\xi'}^{\otimes}_{m,n+p}}\downarrow & & \downarrow{\scriptstyle {\xi'}^{\otimes}_{m,n} \oplus {\xi'}^{\otimes}_{m,p}} \\ (n+p)m & \xrightarrow{=} & nm + pm \end{array}$$

This is equal to axiom (2.1.5) in the symmetric bimonoidal category Σ' because δ^r in Σ' is the identity. Therefore, F is a symmetric bimonoidal functor.

Finally, F is a symmetric bimonoidal equivalence because its underlying functor is the identity functor. □

Proposition 5.1.16. *There is a symmetric bimonoidal equivalence*

$$\Sigma' \xrightarrow{\;G\;} \Sigma$$

that is defined by the following data.

- *The additive structure* $(G, G^2_\oplus, G^0_\oplus)$ *is the identity symmetric monoidal functor.*

- $G^0_\otimes \colon 0 \longrightarrow 0$ *is the identity permutation in* Σ_0.
- $G^2_\otimes \colon mn \longrightarrow mn$ *is the multiplicative symmetry isomorphism* $\xi^\otimes_{m,n}$ *in* Σ *in* (2.4.5).

Furthermore, G *and* F *in Proposition 5.1.15 are inverse isomorphisms in* $\mathsf{Bi}^{\mathsf{sy}}$.

Proof. The first assertion, that G is a symmetric bimonoidal equivalence, is proved by a slight modification of the proof of Proposition 5.1.15. The second assertion, that $GF = 1_\Sigma$ and $FG = 1_{\Sigma'}$, follows from the equality

$$\xi^\otimes_{m,n} = \left(\xi'^\otimes_{m,n}\right)^{-1}$$

for all m and n. □

5.2. Associated Right Bipermutative Category: Definitions

As the first step of the strictification theorem, in this section, we give a detailed construction of a right bipermutative category associated to a tight symmetric bimonoidal category. Recall that *tight* means that the distributivity morphisms δ^l and δ^r are natural isomorphisms, not just natural monomorphisms. Most of the data of the associated right bipermutative category are defined in this section. The proof that it actually satisfies the axioms of a right bipermutative category is given in Section 5.3. In Section 5.4, we will show that the associated right bipermutative category is equivalent to the given tight symmetric bimonoidal category.

Convention 5.2.1. For the rest of this section, unless specified otherwise, assume that C is a tight symmetric bimonoidal category as in Definition 2.1.2. ◇

Motivation 5.2.2. The objects of the right bipermutative category A associated to C are formal polynomials as in

$$X = \sum_{i=1}^{r} X^i_1 \cdots X^i_{k_i}$$

with each X^i_j an object in C. In the actual construction, this is defined in (5.2.4), with the ith monomial in (5.2.5). With a suitably defined product, which is (5.2.7) below, the right distributivity is the identity. Similar to the proofs of the Strictification Theorems 1.3.5 and 1.3.10, the proof that A is a right bipermutative category involves using Laplaza's First Coherence Theorem 3.9.1 multiple times. Choosing a suitable convention for additive bracketing and multiplicative bracketing, the object X can be interpreted in C using its tight symmetric bimonoidal structure. This association is then extended to the desired equivalence of symmetric bimonoidal categories. The tightness assumption on C is essential to the constructions of the product of morphisms (5.2.29) and the multiplicative symmetry (5.2.36) in A. ◇

Since the definition of A involves many components, to improve readability we split it into several parts. First we define its objects, the additive zero, the multiplicative unit, the sum, and the product on objects. The following definition only requires a class $\mathsf{Ob}(\mathsf{C})$.

Definition 5.2.3. Define the tuple

$$\left(\mathsf{Ob}(\mathsf{A}), \oplus^{\mathsf{A}}, \mathbb{0}^{\mathsf{A}}, \otimes^{\mathsf{A}}, \mathbb{1}^{\mathsf{A}}\right)$$

as follows.

Objects: $\mathsf{Ob}(\mathsf{A})$ is the class of finite sequences

$$\underline{a} = \{a^1, \ldots, a^r\} \tag{5.2.4}$$

with *additive length* $r \geq 0$, such that for each $1 \leq i \leq r$, a^i is a finite sequence

$$a^i = (a^i_1, \ldots, a^i_{m_i}) \tag{5.2.5}$$

with *multiplicative length* $m_i \geq 0$ and each $a^i_j \in \mathsf{Ob}(\mathsf{C})$. We call

- a^i the *ith monomial* in $\underline{a}$ and
- a^i_j the *jth alphabet* in a^i.

The Additive Zero: Define

$$\mathbb{0}^{\mathsf{A}} = \varnothing \in \mathsf{Ob}(\mathsf{A}),$$

which is the unique object with additive length 0.

The Multiplicative Unit: Define

$$\mathbb{1}^{\mathsf{A}} = \{\varnothing\} \in \mathsf{Ob}(\mathsf{A}),$$

which has additive length 1 and whose only monomial has multiplicative length 0.

The Sum: With $\underline{a}$ as above and $\underline{b} = \{b^1, \ldots, b^s\}$, define their *sum* by concatenation,

$$\underline{a} \oplus^{\mathsf{A}} \underline{b} = \{a^1, \ldots, a^r, b^1, \ldots, b^s\} \in \mathsf{Ob}(\mathsf{A}), \tag{5.2.6}$$

which has additive length $r + s$.

The Product: With $\underline{a}$ and $\underline{b}$ as above, define their *product* by

$$\underline{a} \otimes^{\mathsf{A}} \underline{b} = \{(a^1, b^1), \ldots, (a^1, b^s), \ldots, (a^r, b^1), \ldots, (a^r, b^s)\} \in \mathsf{Ob}(\mathsf{A}), \tag{5.2.7}$$

which has additive length rs. For $1 \leq i \leq r$ and $1 \leq j \leq s$, the $(j+(i-1)s)$th monomial in $\underline{a} \otimes^{\mathsf{A}} \underline{b}$ is the concatenation

$$(a^i, b^j) = (a^i_1, \ldots, a^i_{m_i}, b^j_1, \ldots, b^j_{n_j})$$

if a^i is as in (5.2.5) and

$$b^j = (b^j_1, \ldots, b^j_{n_j}).$$

Its multiplicative length is $m_i + n_j$. ◇

Lemma 5.2.8. *In the context of Definition 5.2.3, the following statements hold.*

(1) $\oplus^{\mathsf{A}}$ *is strictly associative, and* $\mathbb{0}^{\mathsf{A}}$ *is a strict two-sided unit for* $\oplus^{\mathsf{A}}$.
(2) $\otimes^{\mathsf{A}}$ *is strictly associative, and* $\mathbb{1}^{\mathsf{A}}$ *is a strict two-sided unit for* $\otimes^{\mathsf{A}}$.
(3) *The following equalities hold for all objects* $\underline{a}$.

$$\underline{a} \otimes^{\mathsf{A}} \mathbb{0}^{\mathsf{A}} = \mathbb{0}^{\mathsf{A}} = \mathbb{0}^{\mathsf{A}} \otimes^{\mathsf{A}} \underline{a}. \tag{5.2.9}$$

(4) *The right distributive law*

$$(a \oplus^{\mathsf{A}} b) \otimes^{\mathsf{A}} \underline{c} = (\underline{a} \otimes^{\mathsf{A}} \underline{c}) \oplus^{\mathsf{A}} (\underline{b} \otimes^{\mathsf{A}} \underline{c}) \tag{5.2.10}$$

holds for $\underline{a}, \underline{b}, \underline{c} \in \mathsf{Ob}(\mathsf{A})$.

Proof. The first assertion follows from the definitions of $\oplus^{\mathsf{A}}$ as concatenation and $\mathbb{0}^{\mathsf{A}}$ as the object with additive length 0.

For the second assertion, suppose $\underline{c} = \{c^1, \ldots, c^t\}$. Then both $(\underline{a} \otimes^{\mathsf{A}} \underline{b}) \otimes^{\mathsf{A}} \underline{c}$ and $\underline{a} \otimes^{\mathsf{A}} (\underline{b} \otimes^{\mathsf{A}} \underline{c})$ are equal to the object

$$\begin{gathered}\{(a^1,b^1,c^1),\ldots,(a^1,b^1,c^t),\ldots,(a^1,b^s,c^1),\ldots,(a^1,b^s,c^t),\\ \ldots,(a^r,b^1,c^1),\ldots,(a^r,b^1,c^t),\ldots,(a^r,b^s,c^1),\ldots,(a^r,b^s,c^t)\}.\end{gathered}$$

The object $\mathbb{1}^{\mathsf{A}}$ is a strict two-sided unit for $\otimes^{\mathsf{A}}$ because

$$(a^i, \varnothing) = a^i = (\varnothing, a^i).$$

The third assertion holds because the objects $\underline{a} \otimes^{\mathsf{A}} \mathbb{0}^{\mathsf{A}}$ and $\mathbb{0}^{\mathsf{A}} \otimes^{\mathsf{A}} \underline{a}$ both have additive length $r \cdot 0 = 0 \cdot r = 0$.

The right distributive law holds because both sides are equal to the object

$$\begin{gathered}\{(a^1,c^1),\ldots,(a^1,c^t),\ldots,(a^r,c^1),\ldots,(a^r,c^t),\\ (b^1,c^1),\ldots,(b^1,c^t),\ldots,(b^s,c^1),\ldots,(b^s,c^t)\}\end{gathered}$$

in A. □

The morphisms in A are defined by first interpreting the objects in A as objects in C using the following bracketing convention.

Definition 5.2.11. Suppose $(\mathsf{D}, \odot)$ is a monoidal category, and $x_1, x_2, \ldots$ are objects in D.

- Define the *right normalized bracketing* inductively by

$$(x_1 \odot \cdots \odot x_k)_{\mathsf{rt}} = \begin{cases} x_1 & \text{if } k = 1 \text{ and} \\ x_1 \odot (x_2 \odot \cdots \odot x_k)_{\mathsf{rt}} & \text{if } k > 1. \end{cases} \tag{5.2.12}$$

- Define the *left normalized bracketing* inductively by

$$(x_1 \odot \cdots \odot x_k)_{\mathsf{lt}} = \begin{cases} x_1 & \text{if } k = 1 \text{ and} \\ (x_1 \odot \cdots \odot x_{k-1})_{\mathsf{lt}} \odot x_k & \text{if } k > 1. \end{cases} \tag{5.2.13}$$

The subscripts rt and lt stand for *right* and *left*, respectively. ◇

In this chapter, we only use the right normalized bracketing. Left normalized bracketing will be used in Chapter 7; see Definition 7.2.2.

Remark 5.2.14. Definition 5.2.11 requires a lot less than a monoidal category. In particular, with $\odot \in \{\oplus, \otimes\}$, right and left normalized bracketings also make sense for elements in the free $\{\oplus, \otimes\}$-algebra X^{fr} in Definition 3.1.2. ◇

Example 5.2.15. The first few right normalized bracketings are

$$x_1, \quad x_1 \odot x_2, \quad x_1 \odot (x_2 \odot x_3), \quad \text{and} \quad x_1 \odot (x_2 \odot (x_3 \odot x_4)).$$

The first few left normalized bracketings are

$$x_1, \quad x_1 \odot x_2, \quad (x_1 \odot x_2) \odot x_3, \quad \text{and} \quad ((x_1 \odot x_2) \odot x_3) \odot x_4.$$ ◇

We now interpret objects in A as objects in C. The following definition uses the structure $(\oplus, \mathbb{0}, \otimes, \mathbb{1})$ in the assumed symmetric bimonoidal category C.

Definition 5.2.16. Continuing Definition 5.2.3, define a function

$$\mathsf{Ob}(\mathsf{A}) \xrightarrow{\ \pi\ } \mathsf{Ob}(\mathsf{C}) \tag{5.2.17}$$

as follows.

- For a finite sequence $(a_1, \ldots, a_m)$ of objects in C, first define the object

$$\pi(a_1, \ldots, a_m) = \begin{cases} \mathbb{1} \in \mathsf{C} & \text{if } m = 0 \text{ and} \\ (a_1 \otimes \cdots \otimes a_m)_{\mathsf{rt}} \in \mathsf{C} & \text{if } m > 0, \end{cases} \tag{5.2.18}$$

with the right normalized bracketing (5.2.12) defined in $(\mathsf{C}, \otimes)$.

- For an object $\underline{a} = \{a^1, \ldots, a^r\} \in \mathsf{Ob}(\mathsf{A})$ as in (5.2.4), define the object

$$\pi\underline{a} = \begin{cases} \mathbb{0} \in \mathsf{C} & \text{if } r = 0 \text{ and} \\ \left(\pi a^1 \oplus \cdots \oplus \pi a^r\right)_{\mathsf{rt}} \in \mathsf{C} & \text{if } r > 0, \end{cases} \tag{5.2.19}$$

with the right normalized bracketing defined in $(\mathsf{C}, \oplus)$ and each πa^i as in (5.2.18).

This finishes the definition of the function π. ◇

Example 5.2.20. There are objects

$$\begin{gathered} \pi \mathbb{0}^{\mathsf{A}} = \mathbb{0} \\ \pi \mathbb{1}^{\mathsf{A}} = \mathbb{1} \\ \pi\big\{(a_1^1, a_2^1), (\mathbb{1}^{\mathsf{A}}), (a_1^3, a_2^3, a_3^3)\big\} = (a_1^1 \otimes a_2^1) \oplus \left[\mathbb{1} \oplus \left(a_1^3 \otimes (a_2^3 \otimes a_3^3)\right)\right] \\ \pi\left(\underline{a} \oplus^{\mathsf{A}} \underline{b}\right) = \left(\pi a^1 \oplus \cdots \oplus \pi a^r \oplus \pi b^1 \oplus \cdots \oplus \pi b^s\right)_{\mathsf{rt}} \end{gathered}$$

in C for objects $\underline{a}, \underline{b} \in \mathsf{Ob}(\mathsf{A})$ with additive lengths $r, s > 0$. ◇

Definition 5.2.21. Continuing Definition 5.2.16, define the following.

- For objects $\underline{a}, \underline{b} \in \mathsf{Ob}(\mathsf{A})$, define the *morphism* set

$$\mathsf{A}(\underline{a}; \underline{b}) = \mathsf{C}(\pi\underline{a}; \pi\underline{b}) \tag{5.2.22}$$

with π the function in (5.2.17).

- Define the *identity morphism* $1_{\underline{a}} = 1_{\pi\underline{a}} \in \mathsf{A}(\underline{a}; \underline{a})$.
- Define the *composition* in A

$$\mathsf{A}(\underline{b}; \underline{c}) \times \mathsf{A}(\underline{a}; \underline{b}) = \mathsf{C}(\pi\underline{b}; \pi\underline{c}) \times \mathsf{C}(\pi\underline{a}; \pi\underline{b}) \longrightarrow \mathsf{C}(\pi\underline{a}; \pi\underline{c}) = \mathsf{A}(\underline{a}; \underline{c})$$

as the one in C. ◇

Since C is a category, we obtain the following.

Lemma 5.2.23. *Under Definitions 5.2.3 and 5.2.21,* A *is a category.*

To define the sum and the product of morphisms in A, the following notations are used. Recall from Definitions 3.1.9 and 3.1.14 the graph of X and the value in C of a path in $\mathsf{Gr}(X)$.

Definition 5.2.24.

(1) A *Mac Lane coherence isomorphism* in C, which is denoted by $\cong^{\oplus}_{\mathsf{ML}}$, is the value $\varphi P \colon \varphi u \longrightarrow \varphi v$ in C of a path $P \colon u \longrightarrow v$ in $\mathsf{Gr}(X)$ that satisfies the following three conditions.
   (i) P only involves identities, $\alpha^{\pm\oplus}$, $\lambda^{\pm\oplus}$, $\rho^{\pm\oplus}$, and $\xi^{\pm\oplus}$.
   (ii) In addition to the distinguished elements $\{0^X, 1^X\}$, the set X contains a specific element x_m for each monomial m in each object in A that appears in φu.
   (iii) For each monomial m in (ii), the equality

$$\varphi(x_m) = \pi(m) \in \mathsf{C}$$

   holds, with $\pi(m)$ as in (5.2.18).

(2) A *Laplaza coherence isomorphism* in C, which is denoted by $\cong_{\mathsf{Lap}}$, is the value $\varphi P \colon \varphi v \longrightarrow \varphi w$ in C of a path $P \colon v \longrightarrow w$ in $\mathsf{Gr}(X)$ that satisfies the following three conditions.
   (i) P does *not* involve $\xi^{\pm\otimes}$.
   (ii) In addition to the distinguished elements $\{0^X, 1^X\}$, the set X contains a specific element x_a for each alphabet a in each object in A that appears in φv.
   (iii) For each alphabet a in (ii), the equality
$$\varphi(x_a) = a \in \mathsf{C}$$
   holds.

(3) The inverse of a Laplaza coherence isomorphism is denoted by $\cong_{\mathsf{Lap}}^{-1}$. ◇

Explanation 5.2.25. Consider Definition 5.2.24.

(1) A Mac Lane coherence isomorphism is in particular a composite of isomorphisms, each being a $\oplus$ and $\otimes$ of identity morphisms and at most one component of $\alpha^{\pm\oplus}$, $\lambda^{\pm\oplus}$, $\rho^{\pm\oplus}$, and $\xi^{\pm\oplus}$.

(2) A Laplaza coherence isomorphism is in particular a composite of isomorphisms, each being a $\oplus$ and $\otimes$ of identity morphisms and at most one component of $\alpha^{\pm\oplus}$, $\lambda^{\pm\oplus}$, $\rho^{\pm\oplus}$, $\xi^{\pm\oplus}$, $\alpha^{\pm\otimes}$, $\lambda^{\pm\otimes}$, $\rho^{\pm\otimes}$, $\lambda^{\pm\bullet}$, $\rho^{\pm\bullet}$, δ^l, and δ^r. Laplaza coherence isomorphisms exclude $\xi^{\pm\otimes}$ because in each construction below involving $\cong_{\mathsf{Lap}}$, $\xi^{\pm\otimes}$ are never used.

(3) The inverse of a Laplaza coherence isomorphism has the same description as a Laplaza coherence isomorphism, but with δ^l and δ^r replaced by their inverses. This only makes sense if the symmetric bimonoidal category is tight. Moreover, a Mac Lane coherence isomorphism is also a Laplaza coherence isomorphism. ◇

Next we extend the sum and the product in A, defined for objects in (5.2.6) and (5.2.7), to morphisms. The product of two morphisms involves the inverse of a Laplaza coherence isomorphism. This is where we first use the tightness assumption on the given symmetric bimonoidal category C. Recall from Example 3.9.6 that Theorem 3.9.1 applies to each tight symmetric bimonoidal category.

Definition 5.2.26. Continuing Definition 5.2.21, suppose given two morphisms

$$\begin{aligned} &\underline{a} \xrightarrow{f} \underline{b} \in \mathsf{A}(\underline{a};\underline{b}) = \mathsf{C}(\pi\underline{a};\pi\underline{b}) \\ &\underline{c} \xrightarrow{g} \underline{d} \in \mathsf{A}(\underline{c};\underline{d}) = \mathsf{C}(\pi\underline{c};\pi\underline{d}) \end{aligned} \tag{5.2.27}$$

with $\underline{a},\underline{b},\underline{c},\underline{d} \in \mathsf{Ob}(\mathsf{A})$. Their *sum* and *product* are, respectively, the morphisms

$$\begin{aligned} &\underline{a} \oplus^{\mathsf{A}} \underline{c} \xrightarrow{f \oplus^{\mathsf{A}} g} \underline{b} \oplus^{\mathsf{A}} \underline{d} \in \mathsf{A}\big(\underline{a} \oplus^{\mathsf{A}} \underline{c}; \underline{b} \oplus^{\mathsf{A}} \underline{d}\big) = \mathsf{C}\big(\pi(\underline{a} \oplus^{\mathsf{A}} \underline{c}); \pi(\underline{b} \oplus^{\mathsf{A}} \underline{d})\big) \\ &\underline{a} \otimes^{\mathsf{A}} \underline{c} \xrightarrow{f \otimes^{\mathsf{A}} g} \underline{b} \otimes^{\mathsf{A}} \underline{d} \in \mathsf{A}\big(\underline{a} \otimes^{\mathsf{A}} \underline{c}; \underline{b} \otimes^{\mathsf{A}} \underline{d}\big) = \mathsf{C}\big(\pi(\underline{a} \otimes^{\mathsf{A}} \underline{c}); \pi(\underline{b} \otimes^{\mathsf{A}} \underline{d})\big) \end{aligned} \tag{5.2.28}$$

defined as the following composites in C.

$$\begin{array}{ccc} \pi(\underline{a} \oplus^{\mathsf{A}} \underline{c}) & \xrightarrow{f \oplus^{\mathsf{A}} g} & \pi(\underline{b} \oplus^{\mathsf{A}} \underline{d}) \\ {\scriptstyle \cong_{\mathsf{ML}}^{\oplus}} \big\downarrow & & \big\uparrow {\scriptstyle \cong_{\mathsf{ML}}^{\oplus}} \\ \pi\underline{a} \oplus \pi\underline{c} & \xrightarrow{f \oplus g} & \pi\underline{b} \oplus \pi\underline{d} \end{array} \qquad \begin{array}{ccc} \pi(\underline{a} \otimes^{\mathsf{A}} \underline{c}) & \xrightarrow{f \otimes^{\mathsf{A}} g} & \pi(\underline{b} \otimes^{\mathsf{A}} \underline{d}) \\ {\scriptstyle \cong_{\mathsf{Lap}}^{-1}} \big\downarrow & & \big\uparrow {\scriptstyle \cong_{\mathsf{Lap}}} \\ \pi\underline{a} \otimes \pi\underline{c} & \xrightarrow{f \otimes g} & \pi\underline{b} \otimes \pi\underline{d} \end{array} \tag{5.2.29}$$

More detail of these two diagrams are given in Explanations 5.2.30 and 5.2.31 below. ◇

Explanation 5.2.30. Consider the left diagram in (5.2.29) that defines $f \oplus^{\mathsf{A}} g$.

- If $\underline{a} = \mathbb{0}^{\mathsf{A}}$, then the left vertical isomorphism is $\lambda^{-\oplus}_{\pi\underline{c}}$.
- If $\underline{c} = \mathbb{0}^{\mathsf{A}}$, then the left vertical isomorphism is $\rho^{-\oplus}_{\pi\underline{a}}$.
- If $\underline{b} = \mathbb{0}^{\mathsf{A}}$, then the right vertical isomorphism is $\lambda^{\oplus}_{\pi\underline{d}}$.
- If $\underline{d} = \mathbb{0}^{\mathsf{A}}$, then the right vertical isomorphism is $\rho^{\oplus}_{\pi\underline{b}}$.
- If none of $\underline{a}$, $\underline{b}$, $\underline{c}$, and $\underline{d}$ is $\mathbb{0}^{\mathsf{A}}$, then the vertical isomorphisms involve only identity morphisms and $\alpha^{\pm\oplus}$.

The existence and the uniqueness of the vertical isomorphisms are guaranteed by Mac Lane's Coherence Theorem 1.3.3. ◇

Explanation 5.2.31. Consider the right diagram in (5.2.29) that defines $f \otimes^{\mathsf{A}} g$.

- If $\underline{a} = \mathbb{0}^{\mathsf{A}}$ or $\mathbb{1}^{\mathsf{A}}$, then the left vertical isomorphism is $\lambda^{-\bullet}_{\pi\underline{c}}$ or $\lambda^{-\otimes}_{\pi\underline{c}}$, respectively.
- If $\underline{c} = \mathbb{0}^{\mathsf{A}}$ or $\mathbb{1}^{\mathsf{A}}$, then the left vertical isomorphism is $\rho^{-\bullet}_{\pi\underline{a}}$ or $\rho^{-\otimes}_{\pi\underline{a}}$, respectively.
- If $\underline{b} = \mathbb{0}^{\mathsf{A}}$ or $\mathbb{1}^{\mathsf{A}}$, then the right vertical isomorphism is $\lambda^{\bullet}_{\pi\underline{d}}$ or $\lambda^{\otimes}_{\pi\underline{d}}$, respectively.
- If $\underline{d} = \mathbb{0}^{\mathsf{A}}$ or $\mathbb{1}^{\mathsf{A}}$, then the right vertical isomorphism is $\rho^{\bullet}_{\pi\underline{b}}$ or $\rho^{\otimes}_{\pi\underline{b}}$, respectively.
- If none of $\underline{a}$, $\underline{b}$, $\underline{c}$, and $\underline{d}$ is $\mathbb{0}^{\mathsf{A}}$ or $\mathbb{1}^{\mathsf{A}}$, then the following statements hold.
  - $\cong^{-1}_{\mathsf{Lap}}$ involves only identity morphisms, $\alpha^{\pm\oplus}$, $\xi^{\oplus}$, $\alpha^{-\otimes}$, $\lambda^{-\otimes}$, $\rho^{-\otimes}$, δ^{-l}, and δ^{-r}.
  - $\cong_{\mathsf{Lap}}$ involves only identity morphisms, $\alpha^{\pm\oplus}$, $\xi^{\oplus}$, $\alpha^{\otimes}$, $\lambda^{\otimes}$, $\rho^{\otimes}$, δ^{l}, and δ^{r}.

  In $\cong_{\mathsf{Lap}}$, $\lambda^{\otimes}$ and $\rho^{\otimes}$ are involved if and only if, respectively, $\underline{b}$ and $\underline{d}$ contain monomials with multiplicative length 0. Similarly, in $\cong^{-1}_{\mathsf{Lap}}$, $\lambda^{-\otimes}$ and $\rho^{-\otimes}$ are involved if and only if, respectively, $\underline{a}$ and $\underline{c}$ contain monomials with multiplicative length 0.

The existence of the left vertical isomorphism follows from an induction on the sum of the additive lengths of $\underline{a}$ and $\underline{c}$. The existence of the right vertical isomorphism follows similarly from an induction on the sum of the additive lengths of $\underline{b}$ and $\underline{d}$. The uniqueness of the vertical isomorphisms is guaranteed by Theorem 3.9.1. ◇

Example 5.2.32. Here is an example of $\cong_{\mathsf{Lap}}$ in the right diagram in (5.2.29). Suppose

- $\underline{b} = \{(b), (b_1, b_2)\}$ and
- $\underline{d} = \{(d_1, d_2), (d)\}$

are objects in A with b, b_1, b_2, d_1, d_2, $d \in \mathsf{C}$. To save space, we will abbreviate $\otimes$ in C using concatenation, with $\otimes$ taking precedence over $\oplus$. Then there are the following objects.

$$\begin{aligned}
\pi\underline{b} &= b \oplus b_1 b_2 \in \mathsf{C}\\
\pi\underline{d} &= d_1 d_2 \oplus d \in \mathsf{C}\\
\pi\underline{b} \otimes \pi\underline{d} &= \big(b \oplus b_1 b_2\big)\big(d_1 d_2 \oplus d\big)\\
\underline{b} \otimes^{\mathsf{A}} \underline{d} &= \big\{(b, d_1, d_2), (b, d), (b_1, b_2, d_1, d_2), (b_1, b_2, d)\big\} \in \mathsf{A}\\
\pi(\underline{b} \otimes^{\mathsf{A}} \underline{d}) &= b(d_1 d_2) \oplus \big\{bd \oplus \big[b_1(b_2(d_1 d_2)) \oplus b_1(b_2 d)\big]\big\} \in \mathsf{C}
\end{aligned}$$

To go from $\pi\underline{b} \otimes \pi\underline{d}$ to $\pi(\underline{b} \otimes^{\mathsf{A}} \underline{d})$, we regard b, b_1, b_2, d_1, d_2, and d as formal distinct symbols, even if some of them may be the same object in C. Starting from $\pi\underline{b} \otimes \pi\underline{d}$, we distribute using δ^l and δ^r as far as possible. By the constructions of $\otimes^{\mathsf{A}}$ and π, the result matches $\pi(\underline{b} \otimes^{\mathsf{A}} \underline{d})$, except possibly for

- the additive bracketing,
- the order of the monomials, and
- the multiplicative bracketing of each monomial.

To correct these discrepancies, we apply $\alpha^{\pm\oplus}$, $\xi^{\oplus}$, and $\alpha^{\pm\otimes}$.

The following composite in C is one example of a Laplaza coherence isomorphism $\cong_{\mathsf{Lap}}$. Subscripts in the morphisms are omitted to save space.

$$\begin{array}{ccc}
\pi\underline{b} \otimes \pi\underline{d} & \xrightarrow{\ \cong_{\mathsf{Lap}}\ } & \pi(\underline{b} \otimes^{\mathsf{A}} \underline{d})\\
\downarrow \delta^r & & \uparrow 1 \oplus \{1 \oplus [1 \oplus \alpha^{\otimes}]\}\\
b(d_1 d_2 \oplus d) \oplus (b_1 b_2)(d_1 d_2 \oplus d) & & b(d_1 d_2) \oplus \big\{bd \oplus \big[b_1(b_2(d_1 d_2)) \oplus (b_1 b_2)d\big]\big\}\\
\downarrow \delta^l \oplus 1 & & \uparrow 1 \oplus \{1 \oplus [\alpha^{\otimes} \oplus 1]\}\\
[b(d_1 d_2) \oplus bd] \oplus (b_1 b_2)(d_1 d_2 \oplus d) & & b(d_1 d_2) \oplus \big\{bd \oplus \big[(b_1 b_2)(d_1 d_2) \oplus (b_1 b_2)d\big]\big\}\\
\downarrow 1 \oplus \delta^l & & \uparrow \alpha^{\oplus}\\
& [b(d_1 d_2) \oplus bd] \oplus [(b_1 b_2)(d_1 d_2) \oplus (b_1 b_2)d] &
\end{array}$$

Another Laplaza coherence isomorphism

$$\pi\underline{b} \otimes \pi\underline{d} \xrightarrow{\ \cong_{\mathsf{Lap}}\ } \pi(\underline{b} \otimes^{\mathsf{A}} \underline{d})$$

starts with δ^l, followed by two morphisms involving δ^r, then followed by morphisms involving $\xi^{\oplus}$, $\alpha^{\pm\oplus}$, and $\alpha^{\otimes}$. This composite involves an instance of $\xi^{\oplus}$ to swap the middle two terms to match those in $\pi(\underline{b} \otimes^{\mathsf{A}} \underline{d})$. In each case, $\xi^{\otimes}$ is not needed.

To see that these two Laplaza coherence isomorphisms are equal in C, we apply Theorem 3.9.1 as follows. Consider the set

$$X = \big\{0^X, 1^X, b, b_1, b_2, d_1, d_2, d\big\}$$

as in Definition 3.1.6, with b, b_1, b_2, d_1, d_2, and d regarded as formal distinct symbols different from 0^X and 1^X. The element

$$\big(b \oplus (b_1 \otimes b_2)\big) \otimes \big((d_1 \otimes d_2) \oplus d\big) \in X^{\mathsf{fr}}$$

is regular as in Definition 3.1.25 because its support (3.1.24) is

$$bd_1 d_2 \oplus bd \oplus b_1 b_2 d_1 d_2 \oplus b_1 b_2 d \in X^{\mathsf{st}}.$$

Here $\otimes$ is abbreviated to concatenation as above, and the support of an element is denoted by the same symbol.

The graph morphism $\varphi\colon \mathsf{Gr}(X) \longrightarrow \mathsf{C}$ in Definition 3.1.14 is defined as follows:

- $\varphi(0^X) = \mathbb{0}$ and $\varphi(1^X) = \mathbb{1}$.
- φ sends each of the formal symbols b, b_1, b_2, d_1, d_2, and d in X to the corresponding object in C.

Each of the two Laplaza coherence isomorphisms above is the value in C, in the sense of (3.1.18), of a path in $\mathsf{Gr}(X)$ that does not involve $\xi^{\pm\otimes}$. Theorem 3.9.1 implies that these two Laplaza coherence isomorphisms are equal. ◇

The following observation allows us to define the associativity isomorphisms and the unit isomorphisms in A to be identities.

Lemma 5.2.33. *For morphisms $f_i \in \mathsf{A}(\underline{a}_i;\underline{b}_i)$ for $1 \le i \le 3$, the following equalities hold.*

$$\begin{aligned}(f_1 \oplus^{\mathsf{A}} f_2) \oplus^{\mathsf{A}} f_3 &= f_1 \oplus^{\mathsf{A}} (f_2 \oplus^{\mathsf{A}} f_3)\\ (f_1 \otimes^{\mathsf{A}} f_2) \otimes^{\mathsf{A}} f_3 &= f_1 \otimes^{\mathsf{A}} (f_2 \otimes^{\mathsf{A}} f_3)\\ 1_{\mathbb{0}^{\mathsf{A}}} \oplus^{\mathsf{A}} f_1 = f_1 &= f_1 \oplus^{\mathsf{A}} 1_{\mathbb{0}^{\mathsf{A}}}\\ 1_{\mathbb{1}^{\mathsf{A}}} \otimes^{\mathsf{A}} f_1 = f_1 &= f_1 \otimes^{\mathsf{A}} 1_{\mathbb{1}^{\mathsf{A}}}\end{aligned}$$

Proof. In Lemma 5.2.8, we observed that $\oplus^{\mathsf{A}}$ is strictly associative on objects with $\mathbb{0}^{\mathsf{A}}$ as a strict two-sided unit, and similarly for $(\otimes^{\mathsf{A}}, \mathbb{1}^{\mathsf{A}})$.

For the first equality, consider the following diagram in C.

$$\begin{array}{ccc}
\pi\big((\underline{a}_1 \oplus^{\mathsf{A}} \underline{a}_2) \oplus^{\mathsf{A}} \underline{a}_3\big) & \xrightarrow{=} & \pi\big(\underline{a}_1 \oplus^{\mathsf{A}} (\underline{a}_2 \oplus^{\mathsf{A}} \underline{a}_3)\big)\\
{\scriptstyle \cong^{\oplus}_{\mathsf{ML}}}\big\downarrow & & \big\downarrow{\scriptstyle \cong^{\oplus}_{\mathsf{ML}}}\\
(\pi\underline{a}_1 \oplus \pi\underline{a}_2) \oplus \pi\underline{a}_3 & \xrightarrow{\alpha^{\oplus}} & \pi\underline{a}_1 \oplus (\pi\underline{a}_2 \oplus \pi\underline{a}_3)\\
{\scriptstyle (f_1\oplus f_2)\oplus f_3}\big\downarrow & & \big\downarrow{\scriptstyle f_1\oplus(f_2\oplus f_3)}\\
(\pi\underline{b}_1 \oplus \pi\underline{b}_2) \oplus \pi\underline{b}_3 & \xrightarrow{\alpha^{\oplus}} & \pi\underline{b}_1 \oplus (\pi\underline{b}_2 \oplus \pi\underline{b}_3)\\
{\scriptstyle \cong^{\oplus}_{\mathsf{ML}}}\big\downarrow & & \big\downarrow{\scriptstyle \cong^{\oplus}_{\mathsf{ML}}}\\
\pi\big((\underline{b}_1 \oplus^{\mathsf{A}} \underline{b}_2) \oplus^{\mathsf{A}} \underline{b}_3\big) & \xrightarrow{=} & \pi\big(\underline{b}_1 \oplus^{\mathsf{A}} (\underline{b}_2 \oplus^{\mathsf{A}} \underline{b}_3)\big)
\end{array}$$

The left vertical composite defines $(f_1 \oplus^{\mathsf{A}} f_2) \oplus^{\mathsf{A}} f_3$, and the right vertical composite defines $f_1 \oplus^{\mathsf{A}} (f_2 \oplus^{\mathsf{A}} f_3)$. The middle rectangle is commutative by the naturality of $\alpha^{\oplus}$ in C. The top and the bottom rectangles are commutative by Mac Lane's Coherence Theorem 1.3.3. This proves the first equality.

The second equality is proved by modifying the above diagram by replacing

- $(\oplus^{\mathsf{A}}, \oplus, \alpha^{\oplus})$ with $(\otimes^{\mathsf{A}}, \otimes, \alpha^{\otimes})$,
- the top two $\cong^{\oplus}_{\mathsf{ML}}$ with $\cong^{-1}_{\mathsf{Lap}}$, and
- the bottom two $\cong^{\oplus}_{\mathsf{ML}}$ with $\cong_{\mathsf{Lap}}$.

The resulting diagram commutes by the naturality of $\alpha^{\otimes}$ in C and Theorem 3.9.1.

Using Explanation 5.2.30, the equalities

$$1_{\mathbb{0}^{\mathsf{A}}} \oplus^{\mathsf{A}} f_1 = f_1 = f_1 \oplus^{\mathsf{A}} 1_{\mathbb{0}^{\mathsf{A}}}$$

follow from the naturality of $\lambda^{\oplus}$ and $\rho^{\oplus}$ in C.

Using Explanation 5.2.31, the equalities

$$1_{\mathbb{1}^{\mathsf{A}}} \otimes^{\mathsf{A}} f_1 = f_1 = f_1 \otimes^{\mathsf{A}} 1_{\mathbb{1}^{\mathsf{A}}}$$

follow from the naturality of $\lambda^{\otimes}$ and $\rho^{\otimes}$ in C. □

Using the symmetry isomorphisms $\xi^{\oplus}$ and $\xi^{\otimes}$ in C, next we define the symmetry isomorphisms in A. The multiplicative symmetry isomorphism in A requires the inverse of a Laplaza coherence isomorphism and, therefore, uses the tightness assumption on C.

Definition 5.2.34. Continuing Definition 5.2.26, for objects $\underline{a}, \underline{b} \in \mathsf{Ob}(\mathsf{A})$, define the morphisms

$$(5.2.35)\qquad \begin{aligned} \underline{a} \oplus^{\mathsf{A}} \underline{b} &\xrightarrow{\xi^{\oplus\mathsf{A}}_{\underline{a};\underline{b}}} \underline{b} \oplus^{\mathsf{A}} \underline{a} \in \mathsf{A}\big(\underline{a} \oplus^{\mathsf{A}} \underline{b}; \underline{b} \oplus^{\mathsf{A}} \underline{a}\big) = \mathsf{C}\big(\pi(\underline{a} \oplus^{\mathsf{A}} \underline{b}); \pi(\underline{b} \oplus^{\mathsf{A}} \underline{a})\big) \\ \underline{a} \otimes^{\mathsf{A}} \underline{b} &\xrightarrow{\xi^{\otimes\mathsf{A}}_{\underline{a};\underline{b}}} \underline{b} \otimes^{\mathsf{A}} \underline{a} \in \mathsf{A}\big(\underline{a} \otimes^{\mathsf{A}} \underline{b}; \underline{b} \otimes^{\mathsf{A}} \underline{a}\big) = \mathsf{C}\big(\pi(\underline{a} \otimes^{\mathsf{A}} \underline{b}); \pi(\underline{b} \otimes^{\mathsf{A}} \underline{a})\big) \end{aligned}$$

as the following composites in C.

$$(5.2.36)\qquad \begin{array}{ccc} \pi(\underline{a} \oplus^{\mathsf{A}} \underline{b}) & \xrightarrow{\xi^{\oplus\mathsf{A}}_{\underline{a};\underline{b}}} & \pi(\underline{b} \oplus^{\mathsf{A}} \underline{a}) \\ {\scriptstyle \cong^{\oplus}_{\mathsf{ML}}}\big\downarrow & & \big\uparrow{\scriptstyle \cong^{\oplus}_{\mathsf{ML}}} \\ \pi\underline{a} \oplus \pi\underline{b} & \xrightarrow{\xi^{\oplus}_{\pi\underline{a};\pi\underline{b}}} & \pi\underline{b} \oplus \pi\underline{a} \end{array} \qquad \begin{array}{ccc} \pi(\underline{a} \otimes^{\mathsf{A}} \underline{b}) & \xrightarrow{\xi^{\otimes\mathsf{A}}_{\underline{a};\underline{b}}} & \pi(\underline{b} \otimes^{\mathsf{A}} \underline{a}) \\ {\scriptstyle \cong^{-1}_{\mathsf{Lap}}}\big\downarrow & & \big\uparrow{\scriptstyle \cong_{\mathsf{Lap}}} \\ \pi\underline{a} \otimes \pi\underline{b} & \xrightarrow{\xi^{\otimes}_{\pi\underline{a};\pi\underline{b}}} & \pi\underline{b} \otimes \pi\underline{a} \end{array}$$

◇

Explanation 5.2.37. The vertical isomorphisms in (5.2.36) are as in (5.2.29), which are described explicitly in Explanations 5.2.30 and 5.2.31. Moreover, $\xi^{\oplus\mathsf{A}}$ is a Mac Lane coherence isomorphism. ◇

5.3. Associated Right Bipermutative Category: Proofs

We now show in multiple steps that A is a right bipermutative category as in Definition 2.5.2.

Lemma 5.3.1. *Under Definitions 5.2.3, 5.2.21, 5.2.26, and 5.2.34,*

- $(\mathsf{A}, \oplus^{\mathsf{A}}, \mathbb{0}^{\mathsf{A}}, \xi^{\oplus\mathsf{A}})$ *and*
- $(\mathsf{A}, \otimes^{\mathsf{A}}, \mathbb{1}^{\mathsf{A}}, \xi^{\otimes\mathsf{A}})$

are both permutative categories.

Proof. The functoriality of $\oplus^{\mathsf{A}} \colon \mathsf{A} \times \mathsf{A} \longrightarrow \mathsf{A}$ follows from

- its construction in (5.2.29),
- the functoriality of $\oplus$ in C, and
- the uniqueness in Theorem 1.3.8.

We observed in Lemma 5.2.8 that $\oplus^{\mathsf{A}}$ is strictly associative on objects, and $\mathbb{0}^{\mathsf{A}}$ is a strict two-sided unit. Using Lemma 5.2.33, we define

- the associativity isomorphism $\alpha^{\oplus\mathsf{A}}$,
- the left unit isomorphism $\lambda^{\oplus\mathsf{A}}$, and
- the right unit isomorphism $\rho^{\oplus\mathsf{A}}$

in A to be the identity natural transformations. Then

$$(\mathsf{A}, \oplus^{\mathsf{A}}, \mathbb{0}^{\mathsf{A}}, \alpha^{\oplus\mathsf{A}}, \lambda^{\oplus\mathsf{A}}, \rho^{\oplus\mathsf{A}})$$

is a strict monoidal category because the unity axiom (1.2.2) and the pentagon axiom (1.2.3) are satisfied when each edge involved is an identity morphism.

Each component of $\xi^{\oplus\mathsf{A}}$ is by definition an isomorphism. Its naturality follows from the naturality of $\xi^{\oplus}$ in C, the definition of $f \oplus^{\mathsf{A}} g$ in (5.2.29), and the uniqueness in Theorem 1.3.8. The symmetric monoidal category axioms (1.2.20), (1.2.21), and (1.2.22) in $(\mathsf{A}, \oplus^{\mathsf{A}}, \mathbb{0}^{\mathsf{A}}, \xi^{\oplus\mathsf{A}})$ follow from those in $(\mathsf{C}, \oplus)$, the naturality of $\xi^{\oplus}$, and the uniqueness in Theorem 1.3.8. Therefore, $(\mathsf{A}, \oplus^{\mathsf{A}}, \mathbb{0}^{\mathsf{A}}, \xi^{\oplus\mathsf{A}})$ is a permutative category.

The proof for $(\mathsf{A}, \otimes^{\mathsf{A}}, \mathbb{1}^{\mathsf{A}}, \xi^{\otimes\mathsf{A}})$ is essentially the same as above, with $\alpha^{\otimes\mathsf{A}}$, $\lambda^{\otimes\mathsf{A}}$, and $\rho^{\otimes\mathsf{A}}$ defined as the identity natural transformations using Lemma 5.2.33. For the naturality of $\xi^{\otimes\mathsf{A}}$ and the symmetric monoidal category axioms in A, we use the naturality of $\xi^{\otimes}$ in C, the definition of $f \otimes^{\mathsf{A}} g$ in (5.2.29), Theorem 3.9.1, and the symmetric monoidal category axioms in $(\mathsf{C}, \otimes)$. □

Lemma 5.3.2. *Suppose $\underline{a} \in \mathsf{A}$ is an object.*

(1) *The morphisms*

$$\underline{a} \oplus^{\mathsf{A}} \mathbb{0}^{\mathsf{A}} \underset{\xi^{\oplus\mathsf{A}}_{\mathbb{0}^{\mathsf{A}};\underline{a}}}{\overset{\xi^{\oplus\mathsf{A}}_{\underline{a};\mathbb{0}^{\mathsf{A}}}}{\rightleftarrows}} \mathbb{0}^{\mathsf{A}} \oplus^{\mathsf{A}} \underline{a}$$

are both equal to $1_{\underline{a}}$.

(2) *The morphisms*

$$\underline{a} \otimes^{\mathsf{A}} \mathbb{0}^{\mathsf{A}} \underset{\xi^{\otimes\mathsf{A}}_{\mathbb{0}^{\mathsf{A}};\underline{a}}}{\overset{\xi^{\otimes\mathsf{A}}_{\underline{a};\mathbb{0}^{\mathsf{A}}}}{\rightleftarrows}} \mathbb{0}^{\mathsf{A}} \otimes^{\mathsf{A}} \underline{a}$$

are both equal to $1_{\mathbb{0}^{\mathsf{A}}}$.

Proof. For the first assertion, by the left diagram in (5.2.36) with $\underline{b} = \mathbb{0}^{\mathsf{A}}$ and Explanation 5.2.30, $\xi^{\oplus\mathsf{A}}_{\underline{a};\mathbb{0}^{\mathsf{A}}}$ is the following composite in C.

$$\pi\underline{a} \xrightarrow{\rho^{-\oplus}_{\pi\underline{a}}} \pi\underline{a} \oplus \mathbb{0} \xrightarrow{\xi^{\oplus}_{\pi\underline{a};\mathbb{0}}} \mathbb{0} \oplus \pi\underline{a} \xrightarrow{\lambda^{\oplus}_{\pi\underline{a}}} \pi\underline{a}$$

This composite is equal to $1_{\pi\underline{a}} = 1_{\underline{a}}$ by the unit axiom (1.2.21) in the additive structure in C. The other morphism $\xi^{\oplus\mathsf{A}}_{\mathbb{0}^{\mathsf{A}};\underline{a}}$ is also equal to $1_{\underline{a}}$ by the symmetry axiom (1.2.20) in the permutative category $(\mathsf{A}, \oplus^{\mathsf{A}}, \mathbb{0}^{\mathsf{A}}, \xi^{\oplus\mathsf{A}})$.

For the second assertion, by the right diagram in (5.2.36) with $\underline{b} = \mathbb{0}^{\mathsf{A}}$ and Explanation 5.2.31, $\xi^{\otimes\mathsf{A}}_{\underline{a};\mathbb{0}^{\mathsf{A}}}$ is the following composite in C.

$$\mathbb{0} \xrightarrow{\rho^{-\bullet}_{\pi\underline{a}}} \pi\underline{a} \otimes \mathbb{0} \xrightarrow{\xi^{\otimes}_{\pi\underline{a};\mathbb{0}}} \mathbb{0} \otimes \pi\underline{a} \xrightarrow{\lambda^{\bullet}_{\pi\underline{a}}} \mathbb{0}$$

This composite is equal to $1_{\mathbb{0}} = 1_{\mathbb{0}^{\mathsf{A}}}$ by axiom (2.1.19) in C. The other morphism $\xi^{\otimes\mathsf{A}}_{\mathbb{0}^{\mathsf{A}};\underline{a}}$ is also equal to $1_{\mathbb{0}^{\mathsf{A}}}$ by the symmetry axiom (1.2.20) in the permutative category $(\mathsf{A}, \otimes^{\mathsf{A}}, \mathbb{1}^{\mathsf{A}}, \xi^{\otimes\mathsf{A}})$. □

The next observation allows us to define the multiplicative zeros in A to be the identities.

Lemma 5.3.3. *For each morphism $f \in \mathsf{A}(\underline{a};\underline{b})$, the equalities*

$$f \otimes^{\mathsf{A}} 1_{\mathbb{0}^{\mathsf{A}}} = 1_{\mathbb{0}^{\mathsf{A}}} = 1_{\mathbb{0}^{\mathsf{A}}} \otimes^{\mathsf{A}} f$$

hold in $\mathsf{A}(\mathbb{0}^{\mathsf{A}};\mathbb{0}^{\mathsf{A}})$.

Proof. We observed in Lemma 5.2.8 that

$$\underline{a} \otimes^{\mathsf{A}} \mathbb{0}^{\mathsf{A}} = \mathbb{0}^{\mathsf{A}} = \mathbb{0}^{\mathsf{A}} \otimes^{\mathsf{A}} \underline{a}$$

for each object $\underline{a}$ in A. By the right diagram in (5.2.29) with

$$g = 1_{\mathbb{0}^{\mathsf{A}}} = 1_{\pi\mathbb{0}^{\mathsf{A}}} = 1_{\mathbb{0}}$$

and Explanation 5.2.31, $f \otimes^{\mathsf{A}} 1_{\mathbb{0}^{\mathsf{A}}}$ is the following composite in C.

$$\mathbb{0} \xrightarrow{\rho^{\bullet -1}_{\pi\underline{a}}} \pi\underline{a} \otimes \mathbb{0} \xrightarrow{f \otimes 1_{\mathbb{0}}} \pi\underline{b} \otimes \mathbb{0} \xrightarrow{\rho^{\bullet}_{\pi\underline{b}}} \mathbb{0}$$

This composite is equal to $1_{\mathbb{0}}$ by the naturality of $\rho^{\bullet}$ in C. The equality

$$1_{\mathbb{0}^{\mathsf{A}}} = 1_{\mathbb{0}^{\mathsf{A}}} \otimes^{\mathsf{A}} f$$

follows similarly from the naturality of $\lambda^{\bullet}$ in C. □

The next observation allows us to define the right distributivity in A to be the identity.

Lemma 5.3.4. *For morphisms $f_i \in \mathsf{A}(\underline{a}_i; \underline{b}_i)$ for $1 \le i \le 3$, the equality*

$$(f_1 \oplus^{\mathsf{A}} f_2) \otimes^{\mathsf{A}} f_3 = (f_1 \otimes^{\mathsf{A}} f_3) \oplus^{\mathsf{A}} (f_2 \otimes^{\mathsf{A}} f_3)$$

holds in $\mathsf{A}\big((\underline{a}_1 \oplus^{\mathsf{A}} \underline{a}_2) \otimes^{\mathsf{A}} \underline{a}_3; (\underline{b}_1 \oplus^{\mathsf{A}} \underline{b}_2) \otimes^{\mathsf{A}} \underline{b}_3\big)$.

Proof. By (5.2.10), the right distributive law holds strictly for objects in A. For the desired equality, consider the following diagram in C, with $\otimes$ abbreviated to concatenation.

$$\begin{array}{ccc}
\pi\big((\underline{a}_1 \oplus^{\mathsf{A}} \underline{a}_2) \otimes^{\mathsf{A}} \underline{a}_3\big) & \xrightarrow{=} & \pi\big((\underline{a}_1 \otimes^{\mathsf{A}} \underline{a}_3) \oplus^{\mathsf{A}} (\underline{a}_2 \otimes^{\mathsf{A}} \underline{a}_3)\big) \\
\Big\downarrow \cong^{-1}_{\mathsf{Lap}} & & \Big\downarrow \cong^{-1}_{\mathsf{Lap}} \\
(\pi\underline{a}_1 \oplus \pi\underline{a}_2)(\pi\underline{a}_3) & \xrightarrow{\delta^r} & (\pi\underline{a}_1)(\pi\underline{a}_3) \oplus (\pi\underline{a}_2)(\pi\underline{a}_3) \\
\Big\downarrow (f_1 \oplus f_2) f_3 & & \Big\downarrow f_1 f_3 \oplus f_2 f_3 \\
(\pi\underline{b}_1 \oplus \pi\underline{b}_2)(\pi\underline{b}_3) & \xrightarrow{\delta^r} & (\pi\underline{b}_1)(\pi\underline{b}_3) \oplus (\pi\underline{b}_2)(\pi\underline{b}_3) \\
\Big\downarrow \cong_{\mathsf{Lap}} & & \Big\downarrow \cong_{\mathsf{Lap}} \\
\pi\big((\underline{b}_1 \oplus^{\mathsf{A}} \underline{b}_2) \otimes^{\mathsf{A}} \underline{b}_3\big) & \xrightarrow{=} & \pi\big((\underline{b}_1 \otimes^{\mathsf{A}} \underline{b}_3) \oplus^{\mathsf{A}} (\underline{b}_2 \otimes^{\mathsf{A}} \underline{b}_3)\big)
\end{array}$$

By definition (5.2.29), the left vertical composite is

$$(f_1 \oplus^{\mathsf{A}} f_2) \otimes^{\mathsf{A}} f_3,$$

and the right vertical composite is

$$(f_1 \otimes^{\mathsf{A}} f_3) \oplus^{\mathsf{A}} (f_2 \otimes^{\mathsf{A}} f_3).$$

The middle rectangle is commutative by the naturality of δ^r in C. The top and the bottom rectangles are commutative by Theorem 3.9.1 in C. □

Definition 5.3.5. Consider the category A.

- Define the right distributivity morphism

$$(\underline{a} \oplus^{\mathsf{A}} \underline{b}) \otimes^{\mathsf{A}} \underline{c} \xrightarrow{\delta^{r\mathsf{A}}_{\underline{a};\underline{b};\underline{c}}} (\underline{a} \otimes^{\mathsf{A}} \underline{c}) \oplus^{\mathsf{A}} (\underline{b} \otimes^{\mathsf{A}} \underline{c})$$

  for objects $\underline{a}, \underline{b}, \underline{c} \in \mathsf{A}$ as the identity natural transformation. This is well defined by Lemma 5.3.4.

- Define the left distributivity $\delta^{l\mathsf{A}}$ as the composite

$$\begin{array}{ccc} \underline{a}\otimes^{\mathsf{A}}(\underline{b}\oplus^{\mathsf{A}}\underline{c}) & \xrightarrow{\delta^{l\mathsf{A}}_{\underline{a};\underline{b};\underline{c}}} & (\underline{a}\otimes^{\mathsf{A}}\underline{b})\oplus^{\mathsf{A}}(\underline{a}\otimes^{\mathsf{A}}\underline{c}) \\ {\scriptstyle \xi^{\otimes\mathsf{A}}_{\underline{a};\underline{b}\oplus^{\mathsf{A}}\underline{c}}}\big\downarrow & & \big\uparrow{\scriptstyle \xi^{\otimes\mathsf{A}}_{\underline{b};\underline{a}}\oplus^{\mathsf{A}}\xi^{\otimes\mathsf{A}}_{\underline{c};\underline{a}}} \\ (\underline{b}\oplus^{\mathsf{A}}\underline{c})\otimes^{\mathsf{A}}\underline{a} & \xrightarrow[=]{\delta^{r\mathsf{A}}_{\underline{b};\underline{c};\underline{a}}} & (\underline{b}\otimes^{\mathsf{A}}\underline{a})\oplus^{\mathsf{A}}(\underline{c}\otimes^{\mathsf{A}}\underline{a}) \end{array} \tag{5.3.6}$$

for objects $\underline{a},\underline{b},\underline{c}\in\mathsf{A}$. Its naturality follows from the naturality of $\xi^{\otimes\mathsf{A}}$ and the functoriality of $\oplus^{\mathsf{A}}$. ◇

Next we check axioms (2.1.7) and (2.1.13) in A.

Lemma 5.3.7. *Axiom* (2.1.7) *holds in* A.

Proof. Axiom (2.1.7) is the following diagram for objects $\underline{a},\underline{b},\underline{c}\in\mathsf{A}$.

$$\begin{array}{ccc} (\underline{a}\oplus^{\mathsf{A}}\underline{b})\otimes^{\mathsf{A}}\underline{c} & \xrightarrow[=]{\delta^{r\mathsf{A}}} & (\underline{a}\otimes^{\mathsf{A}}\underline{c})\oplus^{\mathsf{A}}(\underline{b}\otimes^{\mathsf{A}}\underline{c}) \\ {\scriptstyle \xi^{\oplus\mathsf{A}}_{\underline{a};\underline{b}}\otimes^{\mathsf{A}}1_{\underline{c}}}\big\downarrow & & \big\downarrow{\scriptstyle \xi^{\oplus\mathsf{A}}_{\underline{a}\otimes^{\mathsf{A}}\underline{c};\underline{b}\otimes^{\mathsf{A}}\underline{c}}} \\ (\underline{b}\oplus^{\mathsf{A}}\underline{a})\otimes^{\mathsf{A}}\underline{c} & \xrightarrow[=]{\delta^{r\mathsf{A}}} & (\underline{b}\otimes^{\mathsf{A}}\underline{c})\oplus^{\mathsf{A}}(\underline{a}\otimes^{\mathsf{A}}\underline{c}) \end{array}$$

By (5.2.29) and (5.2.36), the above diagram is the outer diagram in C below.

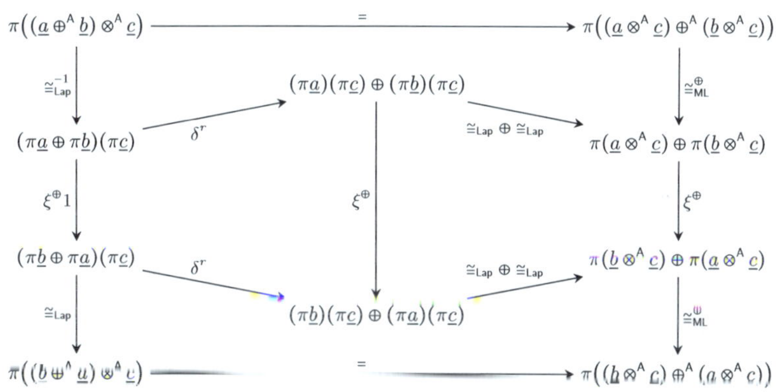

The top and the bottom subdiagrams are commutative by Theorem 3.9.1. The middle left trapezoid is commutative by axiom (2.1.7) in C. The middle right trapezoid is commutative by the naturality of $\xi^{\oplus}$. □

Lemma 5.3.8. *Axiom* (2.1.13) *holds in* A.

Proof. By Lemma 5.3.1 and Definition 5.3.5, the additive associativity $\alpha^{\oplus\mathsf{A}}$ and the right distributivity $\delta^{r\mathsf{A}}$ in A are both identity natural transformations, and the left distributivity $\delta^{l\mathsf{A}}$ is the composite in (5.3.6). By the definitions of $\oplus^{\mathsf{A}}$ in (5.2.29) and $\xi^{\otimes\mathsf{A}}$ in (5.2.36), axiom (2.1.13) in A is the outer diagram in C below for objects

$\underline{a}, \underline{b}, \underline{c}, \underline{d} \in \mathsf{A}$. To save space, we abbreviate $\oplus^{\mathsf{A}}$, $\otimes^{\mathsf{A}}$, and $\pi(?)$ to, respectively, $\oplus$, $\otimes$, and $?'$.

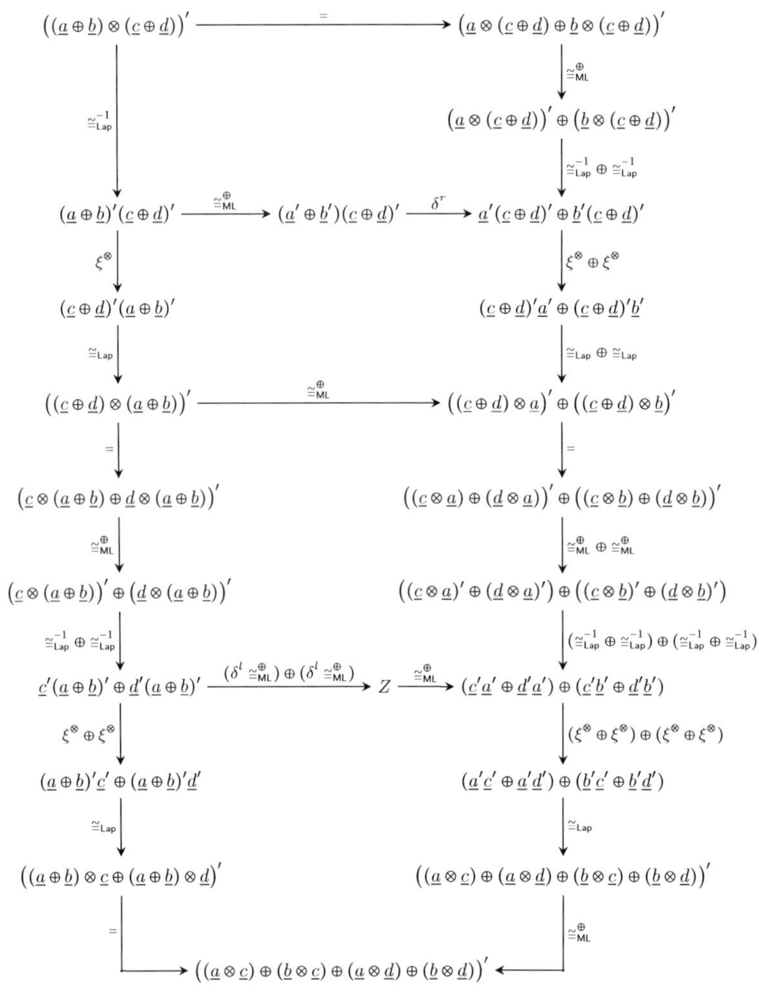

In the above diagram, the object Z is

$$\begin{aligned} Z &= [(\pi\underline{c})(\pi\underline{a}) \oplus (\pi\underline{c})(\pi\underline{b})] \oplus [(\pi\underline{d})(\pi\underline{a}) \oplus (\pi\underline{d})(\pi\underline{b})] \\ &= (\underline{c}'\underline{a}' \oplus \underline{c}'\underline{b}') \oplus (\underline{d}'\underline{a}' \oplus \underline{d}'\underline{b}'). \end{aligned}$$

Each of the four subdiagrams is commutative by Theorem 3.9.1. ☐

Proposition 5.3.9. *Associated to each tight symmetric bimonoidal category* C*, the tuple*

$$\big(\mathsf{A}, (\oplus^{\mathsf{A}}, \mathbb{0}^{\mathsf{A}}, \xi^{\oplus\mathsf{A}}), (\otimes^{\mathsf{A}}, \mathbb{1}^{\mathsf{A}}, \xi^{\otimes\mathsf{A}}), (\lambda^{\bullet\mathsf{A}} = 1, \rho^{\bullet\mathsf{A}} = 1), (\delta^{r\mathsf{A}} = 1, \delta^{l\mathsf{A}})\big)$$

given by Lemmas 5.3.1 and 5.3.3 and Definition 5.3.5 is a right bipermutative category.

Proof. We check the four conditions in Definition 2.5.2.

- The first two conditions say that $\lambda^{\bullet\mathsf{A}}$, $\rho^{\bullet\mathsf{A}}$, and $\delta^{r\mathsf{A}}$ are identity natural transformations. They are true by definition.

- The third condition says that $\xi^{\otimes \mathsf{A}}_{-,0^\mathsf{A}}$ is 1_{0^A}. This holds by Lemma 5.3.2 (2).
- Axioms (2.1.7) and (2.1.13) hold by Lemmas 5.3.7 and 5.3.8, respectively. Axiom (2.1.5) holds by the definition (5.3.6) of $\delta^{l\mathsf{A}}$.

Therefore, A is a right bipermutative category. □

5.4. Strictification

In this section, we finish the proof that each tight symmetric bimonoidal category can be strictified to an equivalent right bipermutative category. There is also a variant involving a left bipermutative category. Convention 5.2.1 is still in effect, so C is a tight symmetric bimonoidal category. Moreover, A denotes the associated right bipermutative category in Proposition 5.3.9. First we define the functors that constitute an equivalence between them.

Definition 5.4.1. Define the functor $\pi\colon \mathsf{A} \longrightarrow \mathsf{C}$ as follows.

- Its assignment on objects is the function in (5.2.17).
- For objects $\underline{a}, \underline{b} \in \mathsf{A}$, its assignment on morphisms

$$\mathsf{A}(\underline{a};\underline{b}) \xrightarrow{\ \pi\ } \mathsf{C}(\pi\underline{a};\pi\underline{b})$$

  is the identity function using (5.2.22).

That π is a functor is part of Definition 5.2.21. ◇

Definition 5.4.2. Define the functor $\iota\colon \mathsf{C} \longrightarrow \mathsf{A}$ as follows.

- Using the notations in (5.2.4) and (5.2.5), for each object $X \in \mathsf{C}$, define the object

$$\iota X = \{(X)\} \in \mathsf{A}.$$

  It has additive length 1, and its only monomial has multiplicative length 1 consisting of X.
- On morphism sets, it is the identity function

$$\mathsf{C}(X;Y) = \mathsf{C}(\pi\iota X;\pi\iota Y) = \mathsf{A}(\iota X;\iota Y).$$

That ι is a functor is part of Definition 5.2.21. ◇

Lemma 5.4.3. *There is an adjoint equivalence*

$$\mathsf{A} \underset{\iota}{\overset{\pi}{\rightleftarrows}} \mathsf{C}$$

with counit $\varepsilon\colon \pi\iota \longrightarrow 1_\mathsf{C}$ *the identity natural transformation.*

Proof. By construction, $\pi\iota = 1_\mathsf{C}$. For objects $\underline{a} \in \mathsf{A}$ and $Y \in \mathsf{C}$, the adjunction $\pi \dashv \iota$ is defined by the natural equalities

$$\mathsf{C}(\pi\underline{a};Y) = \mathsf{C}(\pi\underline{a};\pi\iota Y) = \mathsf{A}(\underline{a};\iota Y).$$

The counit of the adjunction is the identity by construction. The unit of the adjunction associates to each object $\underline{a} \in \mathsf{A}$ the isomorphism

$$\big(\underline{a} \xrightarrow{\ \eta_{\underline{a}}\ } \iota\pi\underline{a}\big) = 1_{\pi\underline{a}} \in \mathsf{A}(\underline{a};\iota\pi\underline{a}) = \mathsf{C}(\pi\underline{a};\pi\iota\pi\underline{a}) = \mathsf{C}(\pi\underline{a};\pi\underline{a}).$$

Therefore, the unit is a natural isomorphism. □

Next we observe that π and ι are symmetric bimonoidal functors. Recall from Definition 5.1.1 that a symmetric bimonoidal functor F is *unitary* if $F^0_\oplus$ and $F^0_\otimes$ are identities, and if $F^2_\oplus$ and $F^2_\otimes$ are natural isomorphisms.

Lemma 5.4.4. *There is a unitary symmetric bimonoidal equivalence* $\mathsf{A} \longrightarrow \mathsf{C}$ *with underlying functor* π.

Proof. Since $\pi \mathbb{0}^{\mathsf{A}} = \mathbb{0}$ and $\pi \mathbb{1}^{\mathsf{A}} = \mathbb{1}$, we define

$$\mathbb{0} \xrightarrow{\pi^0_\oplus} \pi \mathbb{0}^{\mathsf{A}}$$
$$\mathbb{1} \xrightarrow{\pi^0_\otimes} \pi \mathbb{1}^{\mathsf{A}}$$

as the identity morphisms of $\mathbb{0}$ and $\mathbb{1}$ in C, respectively. Next we define

$$\pi \underline{a} \oplus \pi \underline{b} \xrightarrow{\pi^2_\oplus} \pi(\underline{a} \oplus^{\mathsf{A}} \underline{b})$$
$$\pi \underline{a} \otimes \pi \underline{b} \xrightarrow{\pi^2_\otimes} \pi(\underline{a} \otimes^{\mathsf{A}} \underline{b})$$

for objects $\underline{a}, \underline{b} \in \mathsf{A}$ as, respectively, the Mac Lane coherence isomorphism $\cong^{\oplus}_{\mathsf{ML}}$ and the Laplaza coherence isomorphism $\cong_{\mathsf{Lap}}$ in (5.2.29) as the right vertical isomorphisms. Their naturality follows from the definitions of $\oplus^{\mathsf{A}}$ and $\otimes^{\mathsf{A}}$ on morphisms in (5.2.29), and the definition of π as the identity assignment on morphisms.

The triple $(\pi, \pi^2_\oplus, \pi^0_\oplus)$ is a symmetric monoidal functor because each diagram in Definitions 1.2.11 and 1.2.25 is commutative by Theorem 1.3.8. Similarly, the triple $(\pi, \pi^2_\otimes, \pi^0_\otimes)$ is a symmetric monoidal functor because each necessary diagram is commutative by Theorem 3.9.1.

The multiplicative zero axiom (5.1.2) holds for π because

- $\pi^0_\oplus$ and $\rho^{\bullet\mathsf{A}}$ are identity natural transformations, and
- there is an equality

$$\pi^2_\otimes = \rho^\bullet_{\pi \underline{a}} \colon (\pi \underline{a}) \otimes (\pi \mathbb{0}^{\mathsf{A}}) = (\pi \underline{a}) \otimes \mathbb{0} \longrightarrow \mathbb{0} = \pi(\underline{a} \otimes^{\mathsf{A}} \mathbb{0}^{\mathsf{A}}).$$

The distributivity axiom (5.1.3) holds for π by Theorem 3.9.1. Therefore, the tuple

$$(\pi, \pi^2_\oplus, \pi^0_\oplus, \pi^2_\otimes, \pi^0_\otimes)$$

is a unitary symmetric bimonoidal functor. Finally, π is an equivalence of categories by Lemma 5.4.3. □

Recall from Definition 5.1.1 that a symmetric bimonoidal functor F is *strong* if $F^2_\oplus$, $F^0_\oplus$, $F^2_\otimes$, and $F^0_\otimes$ are natural isomorphisms.

Lemma 5.4.5. *There is a strong symmetric bimonoidal equivalence* $\mathsf{C} \longrightarrow \mathsf{A}$ *with underlying functor* ι.

Proof. Define the following morphisms.

$$\left(\mathbb{0}^{\mathsf{A}} = \varnothing \xrightarrow{\iota^0_\oplus} \iota\mathbb{0} = \{(\mathbb{0})\}\right) = 1_{\mathbb{0}} \in \mathsf{A}(\mathbb{0}^{\mathsf{A}}; \iota\mathbb{0}) = \mathsf{C}(\mathbb{0}; \mathbb{0})$$
$$\left(\mathbb{1}^{\mathsf{A}} = \{\varnothing\} \xrightarrow{\iota^0_\otimes} \iota\mathbb{1} = \{(\mathbb{1})\}\right) = 1_{\mathbb{1}} \in \mathsf{A}(\mathbb{1}^{\mathsf{A}}; \iota\mathbb{1}) = \mathsf{C}(\mathbb{1}; \mathbb{1})$$

These are isomorphisms, but not identity morphisms. Define the morphisms

$$\iota X \oplus^{\mathsf{A}} \iota Y = \{(X), (Y)\} \xrightarrow{\iota^2_\oplus} \iota(X \oplus Y) \in \mathsf{A}\big(\iota X \oplus^{\mathsf{A}} \iota Y; \iota(X \oplus Y)\big) = \mathsf{C}(X \oplus Y; X \oplus Y)$$
$$\iota X \otimes^{\mathsf{A}} \iota Y = \{(X, Y)\} \xrightarrow{\iota^2_\otimes} \iota(X \otimes Y) \in \mathsf{A}\big(\iota X \otimes^{\mathsf{A}} \iota Y; \iota(X \otimes Y)\big) = \mathsf{C}(X \otimes Y; X \otimes Y)$$

for objects $X, Y \in \mathsf{C}$ as, respectively, $1_{X \oplus Y}$ and $1_{X \otimes Y}$. These are natural isomorphisms, but not identity morphisms.

Consider the triple $(\iota, \iota^2_\oplus, \iota^0_\oplus)$.

- The associativity axiom (1.2.14) holds because each composite is

$$\begin{aligned} 1_{X\oplus(Y\oplus Z)} &\in \mathsf{A}\big((\iota X \oplus^{\mathsf{A}} \iota Y) \oplus^{\mathsf{A}} \iota Z; \iota(X \oplus (Y \oplus Z))\big) \\ &= \mathsf{C}\big(X \oplus (Y \oplus Z); X \oplus (Y \oplus Z)\big). \end{aligned}$$

- The left unity axiom (1.2.15) holds because each composite is

$$1_X \in \mathsf{A}\big(\mathbb{0}^{\mathsf{A}} \oplus^{\mathsf{A}} \iota X; \iota X\big) = \mathsf{C}(X; X).$$

- The right unity axiom (1.2.15) holds for the analogous reason.
- The axiom (1.2.26) holds because each composite is

$$\xi^{\oplus}_{X,Y} = \mathsf{A}\big(\iota X \oplus^{\mathsf{A}} \iota Y; \iota(Y \oplus X)\big) = \mathsf{C}(X \oplus Y; Y \oplus X).$$

Therefore, $(\iota, \iota^2_\oplus, \iota^0_\oplus)$ is a strong symmetric monoidal functor. The same reasoning, applied to the multiplicative structures in C and A, shows that $(\iota, \iota^2_\otimes, \iota^0_\otimes)$ is a strong symmetric monoidal functor.

The multiplicative zero axiom (5.1.2) holds for ι because each composite is

$$1_{\mathbb{0}} \in \mathsf{A}\big(\iota A \otimes^{\mathsf{A}} \mathbb{0}^{\mathsf{A}}; \iota\mathbb{0}\big) = \mathsf{C}(\mathbb{0}; \mathbb{0}).$$

The distributivity axiom (5.1.3) holds for ι because each composite is

$$\begin{aligned} 1_{AC\oplus BC} &\in \mathsf{A}\big((\iota A \oplus^{\mathsf{A}} \iota B) \otimes^{\mathsf{A}} \iota C; \iota(AC \oplus BC)\big) \\ &= \mathsf{C}(AC \oplus BC; AC \oplus BC). \end{aligned}$$

Therefore, the tuple

$$(\iota, \iota^2_\oplus, \iota^0_\oplus, \iota^2_\otimes, \iota^0_\otimes)$$

is a strong symmetric bimonoidal functor. Finally, ι is an equivalence of categories by Lemma 5.4.3. □

Combining Proposition 5.3.9 and Lemmas 5.4.3 through 5.4.5, we obtain the following strictification result.

Theorem 5.4.6 (Right Bipermutative Strictification). *Suppose C is a tight symmetric bimonoidal category. Then there is an adjoint equivalence*

$$\mathsf{A} \underset{\iota}{\overset{\pi}{\rightleftarrows}} \mathsf{C}$$

with

- *A the right bipermutative category in Proposition 5.3.9;*
- *counit $\varepsilon\colon \pi\iota \longrightarrow 1_{\mathsf{C}}$ the identity natural transformation;*
- *π the unitary symmetric bimonoidal equivalence in Lemma 5.4.4; and*
- *ι the strong symmetric bimonoidal equivalence in Lemma 5.4.5.*

A minor variation of the constructions in this chapter yields the following strictification result to left bipermutative categories as in Definition 2.5.11.

Theorem 5.4.7 (Left Bipermutative Strictification). *Suppose C is a tight symmetric bimonoidal category. Then there is an adjoint equivalence*

$$\mathsf{A}_l \underset{\iota}{\overset{\pi}{\rightleftarrows}} \mathsf{C}$$

with

- *A_l a left bipermutative category;*
- *counit $\varepsilon\colon \pi\iota \longrightarrow 1_{\mathsf{C}}$ the identity natural transformation;*

- *π a unitary symmetric bimonoidal equivalence; and*
- *ι a strong symmetric bimonoidal equivalence.*

Proof. The proof is essentially the same as that of Theorem 5.4.6, with the following modifications.

- In A_l, the objects, the additive zero, the multiplicative unit, and the sum are the same as those in A in Definition 5.2.3.
- The product in A_l is redefined in such a way that the left distributive law holds strictly, that is,

$$\underline{a} \otimes^{\mathsf{A}_l} \underline{b} = \{(a^1,b^1),\ldots,(a^r,b^1),\ldots,(a^1,b^s),\ldots,(a^r,b^s)\} \tag{5.4.8}$$

  instead of (5.2.7).
- Lemma 5.3.4 is replaced by the left distributive law for morphisms.
- Axiom (2.1.5) is used to define δ^r in terms of δ^l, which is the identity in A_l.
- Lemma 5.3.7 is replaced by the assertion that (2.1.6) holds in A_l.
- Proposition 5.3.9 is replaced by the assertion that A_l is a left bipermutative category.

Other structures are defined as in A, and all the proofs require minimal or no changes. □

5.5. Strictification of Tight Bimonoidal Categories

In this section, we prove nonsymmetric analogues of Theorems 5.4.6 and 5.4.7. To state these results, recall from Definition 2.1.2 that a bimonoidal category C has the same definition as a symmetric bimonoidal category, except that

- the multiplicative structure is a monoidal category instead of a symmetric one, and
- the axioms (2.1.5) and (2.1.19) are omitted.

A tight bimonoidal category—that is, one with δ^l and δ^r invertible—is called *right rigid* if

- both the additive and the multiplicative structures are strict monoidal, and
- $\lambda^{\bullet}$, $\rho^{\bullet}$, and δ^r are identity natural transformations.

Similarly, a tight bimonoidal category is *left rigid* if it satisfies these conditions but with $\delta^l = 1$ instead of $\delta^r = 1$. The first main result of this section is Theorem 5.5.11. It states that each tight bimonoidal category is adjoint equivalent, via suitable bimonoidal functors, to a right rigid bimonoidal category. Theorem 5.5.12 is the left rigid analogue.

Here is an outline of this section.

- In Definition 5.5.1, we define bimonoidal functors. These are the nonsymmetric analogues of symmetric bimonoidal functors in Definition 5.1.1. The Strictification Theorems 5.5.11 and 5.5.12 are phrased in terms of these functors.
- In Definition 5.5.5, for a tight bimonoidal category C, we define the object A as in Sections 5.2 and 5.3, except for its left distributivity morphism. Proposition 5.5.10 shows that A is a right rigid bimonoidal category.
- Theorem 5.5.11 shows that C and A are adjoint equivalent via bimonoidal functors, with one unitary and the other one strong.

- With a slight modification of the definition of A and the proofs, Theorem 5.5.12 states that C is adjoint equivalent to a left rigid bimonoidal category via bimonoidal functors, with one unitary and the other one strong.

Bimonoidal Functors.

Definition 5.5.1. Suppose C and D are bimonoidal categories. A *bimonoidal functor* from C to D is a tuple

$$(F, F^2_\oplus, F^0_\oplus, F^2_\otimes, F^0_\otimes) \colon \mathsf{C} \longrightarrow \mathsf{D}$$

consisting of the following data.

-
$$F_\oplus = (F, F^2_\oplus, F^0_\oplus) \colon (\mathsf{C}, \oplus) \longrightarrow (\mathsf{D}, \oplus)$$
is a symmetric monoidal functor from the additive structure of C to the additive structure of D.
-
$$F_\otimes = (F, F^2_\otimes, F^0_\otimes) \colon (\mathsf{C}, \otimes) \longrightarrow (\mathsf{D}, \otimes)$$
is a monoidal functor from the multiplicative structure of C to the multiplicative structure of D.

These data are required to make the diagrams (5.1.2), (5.1.3), (5.1.5), and (5.1.6) commutative. This finishes the definition of a bimonoidal functor.

Moreover, we define the following.

- *Robust*, *strong*, *unitary*, and *strict* bimonoidal functors are defined as in Definition 5.1.1.
- A *bimonoidal equivalence* is a bimonoidal functor whose underlying functor is an equivalence of categories.
- Composites of bimonoidal functors are defined as in Definition 5.1.7. ◇

Explanation 5.5.2. There are two differences between a bimonoidal functor and a *symmetric* bimonoidal functor as in Definition 5.1.1. In a symmetric bimonoidal functor F,

- $F_\otimes$ is a symmetric monoidal functor, and
- F is only assumed to satisfy the multiplicative zero axiom (5.1.2) and the distributivity axiom (5.1.3).

In Proposition 5.1.4, using the multiplicative symmetry $\xi^\otimes$, we showed that (5.1.2) and (5.1.3) are equivalent to, respectively, (5.1.5) and (5.1.6). However, in a bimonoidal category, the multiplicative structure is only assumed to be a monoidal category. So there is no nonsymmetric analogue of Proposition 5.1.4. This is why, in Definition 5.5.1 of a bimonoidal functor, all four axioms are included. ◇

Lemma 5.5.3. *Suppose*

$$\mathsf{C} \xrightarrow{\ F\ } \mathsf{D} \xrightarrow{\ G\ } \mathsf{E}$$

are bimonoidal functors. Then the composite $GF \colon \mathsf{C} \longrightarrow \mathsf{E}$ is also a bimonoidal functor. Moreover, if both F and G are robust (respectively, strong, unitary, or strict), then so is GF.

Proof. The proof of Lemma 5.1.9 shows that the composite GF satisfies the two axioms (5.1.2) and (5.1.3). Two similar diagrams prove the other two axioms (5.1.5) and (5.1.6). The key point is that the multiplicative symmetry $\xi^{\otimes}$ is not involved in these diagrams. □

As in Proposition 5.1.10, Lemma 5.5.3 implies the existence of a category of small bimonoidal categories.

Proposition 5.5.4. *There is a category* Bi *defined by the following data.*

- *The objects are small bimonoidal categories as in Definition 2.1.2.*
- *The morphisms are bimonoidal functors as in Definition 5.5.1.*
- *Identity morphisms are identity functors with identity monoidal structures.*
- *Composition is as in Definition 5.5.1.*

Moreover, Bi *has the following wide subcategories:*

- Bi_{r} *with robust bimonoidal functors as morphisms.*
- $\mathsf{Bi}_{\mathsf{sg}}$ *with strong bimonoidal functors as morphisms.*
- Bi_{u} *with unitary bimonoidal functors as morphisms.*
- $\mathsf{Bi}_{\mathsf{st}}$ *with strict bimonoidal functors as morphisms.*

Strictified Bimonoidal Categories. Next we define the object that will serve as the strictification of a tight bimonoidal category C. In the following definition, A does not have a multiplicative symmetry. Also note that the paths that define Mac Lane coherence isomorphisms $\cong^{\oplus}_{\mathsf{ML}}$ and Laplaza coherence isomorphisms $\cong_{\mathsf{Lap}}$ in Definition 5.2.24 are nonsymmetric paths in the sense of Definition 3.10.2. Therefore, they make sense in a bimonoidal category.

Definition 5.5.5. Suppose C is a tight bimonoidal category as in Definition 2.1.2. Define the data

$$\big(\mathsf{A}, (\oplus^{\mathsf{A}}, \mathbb{0}^{\mathsf{A}}, \xi^{\oplus\mathsf{A}}), (\otimes^{\mathsf{A}}, \mathbb{1}^{\mathsf{A}}), (\lambda^{\bullet\mathsf{A}} = 1, \rho^{\bullet\mathsf{A}} = 1), (\delta^{r\mathsf{A}} = 1, \delta^{l\mathsf{A}})\big)$$

as follows.

- The category A and its structures $\oplus^{\mathsf{A}}$, $\mathbb{0}^{\mathsf{A}}$, $\xi^{\oplus\mathsf{A}}$, $\otimes^{\mathsf{A}}$, $\mathbb{1}^{\mathsf{A}}$, $\lambda^{\bullet\mathsf{A}} = 1$, $\rho^{\bullet\mathsf{A}} = 1$, and $\delta^{r\mathsf{A}} = 1$ are defined as in Definitions 5.2.3, 5.2.16, 5.2.21, 5.2.24, 5.2.26, 5.2.34, 5.3.5, and 5.4.1.
- Using the function π in (5.2.17), $\delta^{l\mathsf{A}}$ is defined as the following composite in C for objects $\underline{a}, \underline{b}, \underline{c} \in \mathsf{A}$.

(5.5.6)
$$\begin{array}{ccc}
\pi\big(\underline{a} \otimes^{\mathsf{A}} (\underline{b} \oplus^{\mathsf{A}} \underline{c})\big) & \xrightarrow{\delta^{l\mathsf{A}}_{\underline{a};\underline{b};\underline{c}}} & \pi\big((\underline{a} \otimes^{\mathsf{A}} \underline{b}) \oplus^{\mathsf{A}} (\underline{a} \otimes^{\mathsf{A}} \underline{c})\big) \\
{\scriptstyle \cong^{-1}_{\mathsf{Lap}}}\big\downarrow & & \big\uparrow{\scriptstyle \cong_{\mathsf{Lap}}} \\
\pi\underline{a} \otimes (\pi\underline{b} \oplus \pi\underline{c}) & \xrightarrow{\delta^{l}_{\pi\underline{a};\pi\underline{b};\pi\underline{c}}} & (\pi\underline{a} \otimes \pi\underline{b}) \oplus (\pi\underline{a} \otimes \pi\underline{c})
\end{array}$$

This finishes the definition of A. ◇

Explanation 5.5.7.

- As in the symmetric case, morphisms in A are defined as morphisms in C, as in the equality (5.2.22), using the function π in (5.2.17). It interprets each object $\underline{a} \in \mathsf{A}$ as the right normalized sum (5.2.19) of the right normalized products (5.2.18) of the alphabets in each monomial in $\underline{a}$.
- In the symmetric case, A has a multiplicative symmetry $\xi^{\otimes\mathsf{A}}$ as in (5.2.36), which is defined using the multiplicative symmetry in the given tight symmetric bimonoidal category. Since C is now a tight bimonoidal category,

which does not have a multiplicative symmetry, A also does not have a multiplicative symmetry.
- For the same reason, instead of using the diagram (5.3.6), in the current nonsymmetric setting, we define the left distributivity morphism $\delta^{l\mathsf{A}}$ directly as in (5.5.6). ◇

We will show shortly that A in Definition 5.5.5 is right rigid in the sense of the next definition.

Definition 5.5.8. A *right rigid bimonoidal category* is a tight bimonoidal category C as in Definition 2.1.2 that satisfies the following three conditions.

- The additive structure $(\mathsf{C}, \oplus)$ is a permutative category.
- The multiplicative structure $(\mathsf{C}, \otimes)$ is a strict monoidal category.
- $\lambda^\bullet$, $\rho^\bullet$, and δ^r are the identity natural transformations.

A *left rigid bimonoidal category* is defined in the same way as a right rigid bimonoidal category, but with $\delta^l = 1$ instead of $\delta^r = 1$. ◇

Example 5.5.9.

- By Proposition 2.5.7, each right bipermutative category is a right rigid bimonoidal category after forgetting the multiplicative symmetry $\xi^\otimes$.
- By Proposition 2.5.16, each left bipermutative category is a left rigid bimonoidal category after forgetting the multiplicative symmetry $\xi^\otimes$. ◇

Next is the nonsymmetric analogue of Proposition 5.3.9.

Proposition 5.5.10. *For each tight bimonoidal category* C, *the data*

$$\big(\mathsf{A}, (\oplus^\mathsf{A}, \mathbb{0}^\mathsf{A}, \xi^{\oplus\mathsf{A}}), (\otimes^\mathsf{A}, \mathbb{1}^\mathsf{A}), (\lambda^{\bullet\mathsf{A}} = 1, \rho^{\bullet\mathsf{A}} = 1), (\delta^{r\mathsf{A}} = 1, \delta^{l\mathsf{A}})\big)$$

in Definition 5.5.5 form a right rigid bimonoidal category.

Proof. The left distributivity morphism $\delta^{l\mathsf{A}}$ defined in (5.5.6) is a natural transformation by

- the naturality of δ^l in C,
- the definitions of $\oplus^\mathsf{A}$ and $\otimes^\mathsf{A}$ on morphisms in (5.2.29),
- Definition 5.4.1 of π as the identity assignment on morphisms, and
- the Coherence Theorem 3.10.7 for bimonoidal categories.

Since C is tight, both δ^l and δ^r are natural isomorphisms in C. This implies that each component of $\delta^{l\mathsf{A}}$ is an isomorphism. Therefore, $\delta^{l\mathsf{A}}$ is a natural isomorphism.

The rest of the proof is obtained from that of Proposition 5.3.9 as follows.

- We use the Coherence Theorem 3.10.7 for bimonoidal categories instead of the Coherence Theorem 3.9.1 for symmetric bimonoidal categories.
- We reuse the proof of Lemma 5.3.1 to infer that
  - the additive structure of A is a permutative category, and
  - the multiplicative structure of A is a strict monoidal category.
- We reuse the proofs of Lemmas 5.3.3 and 5.3.4 to infer that
  - the left multiplicative zero $\lambda^{\bullet\mathsf{A}} = 1$,
  - the right multiplicative zero $\rho^{\bullet\mathsf{A}} = 1$, and
  - the right distributivity morphism $\delta^{r\mathsf{A}} = 1$

  are well-defined natural transformations.

It remains to check that A satisfies the 22 Laplaza axioms in Definition 2.1.2, excluding (2.1.5) and (2.1.19), for a bimonoidal category.

- Each of the following 12 Laplaza axioms holds in A because each edge involved is an identity morphism: (2.1.8), (2.1.11), (2.1.14), (2.1.16)–(2.1.18), (2.1.20)–(2.1.22), (2.1.24), (2.1.26), and (2.1.28).
- Consider the axioms (2.1.15), (2.1.23), (2.1.25), and (2.1.27).
  - (2.1.15) is equivalent to the equality $\delta^{l\mathsf{A}}_{\mathbb{0}^{\mathsf{A}};-;-} = 1$.
  - (2.1.23) is equivalent to the equality $\delta^{l\mathsf{A}}_{-;\mathbb{0}^{\mathsf{A}};-} = 1$.
  - (2.1.25) is equivalent to the equality $\delta^{l\mathsf{A}}_{-;-;\mathbb{0}^{\mathsf{A}}} = 1$.
  - (2.1.27) is equivalent to the equality $\delta^{l\mathsf{A}}_{\mathbb{1}^{\mathsf{A}};-;-} = 1$.

  Each of these four equalities follows from the definition (5.5.6) of $\delta^{l\mathsf{A}}$ and an application of the Coherence Theorem 3.10.7.
- (2.1.7) is proved by reusing the proof of Lemma 5.3.7 and using Theorem 3.10.7 instead of Theorem 3.9.1.

There are five remaining axioms to check in A: (2.1.6), (2.1.9), (2.1.10), (2.1.12), and (2.1.13). Each of these axioms is a consequence of the same axiom in C, along with coherence, naturality, and functoriality properties. To prove these axioms in detail, consider arbitrary objects $\underline{a}, \underline{b}, \underline{c}, \underline{d} \in \mathsf{A}$. To save space in the following diagrams, we abbreviate $\oplus^{\mathsf{A}}$ to $\oplus$, and both $\otimes$ and $\otimes^{\mathsf{A}}$ to concatenation. Moreover, for an object $\underline{e} \in \mathsf{A}$, we write $\pi\underline{e}$ as $\underline{e}'$. For example,

$$\begin{aligned} \pi\big(\underline{a} \otimes^{\mathsf{A}} (\underline{b} \oplus^{\mathsf{A}} \underline{c})\big) &= [\underline{a}(\underline{b} \oplus \underline{c})]', \\ \pi\underline{a} \otimes (\pi\underline{b} \oplus \pi\underline{c}) &= \underline{a}'(\underline{b}' \oplus \underline{c}'), \quad \text{and} \\ \pi\big((\underline{a} \otimes^{\mathsf{A}} \underline{c}) \oplus^{\mathsf{A}} (\underline{a} \otimes^{\mathsf{A}} \underline{b})\big) &= (\underline{ac} \oplus \underline{ab})'. \end{aligned}$$

With these conventions, the axiom (2.1.6) in A states the commutativity of the outer diagram in C below.

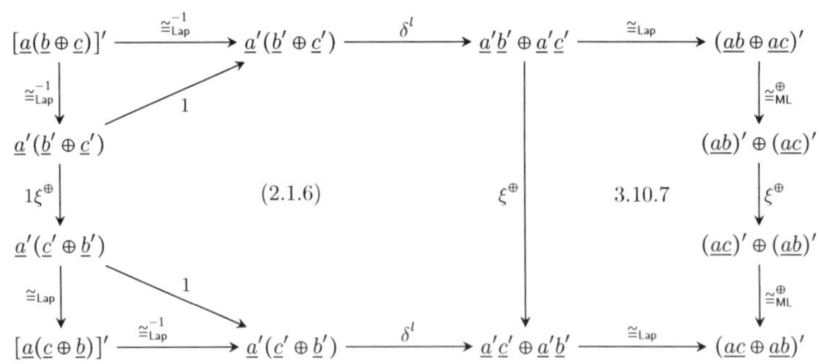

The middle subdiagram is commutative by the axiom (2.1.6) in C. The right subdiagram is commutative by the Coherence Theorem 3.10.7. The two unlabeled triangles are commutative by definition. Similar notations are used in the remaining diagrams in this proof.

Using the strict associativity of $\oplus^{\mathsf{A}}$, the axiom (2.1.9) in A states the commutativity of the outer diagram in C below.

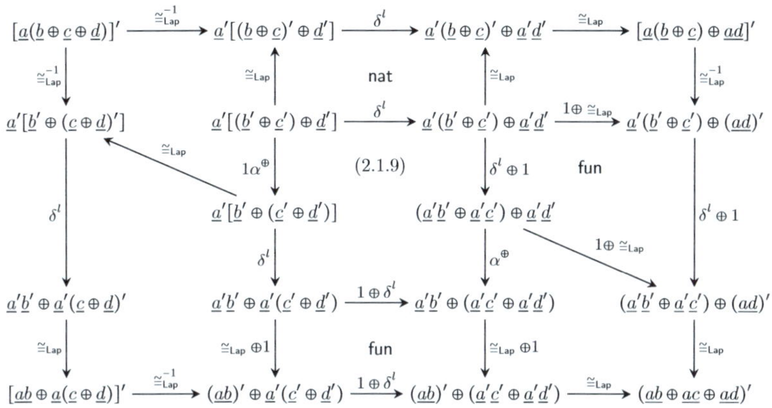

In the middle column, the top two subdiagrams are commutative by the naturality of δ^l and the axiom (2.1.9) in C. The two subdiagrams labeled by fun are commutative by the functoriality of $\oplus$ in C. Each of the other four subdiagrams is commutative by Theorem 3.10.7.

Using the strict associativity of $\otimes^{\mathsf{A}}$, the axiom (2.1.10) in A states the commutativity of the outer diagram in C below.

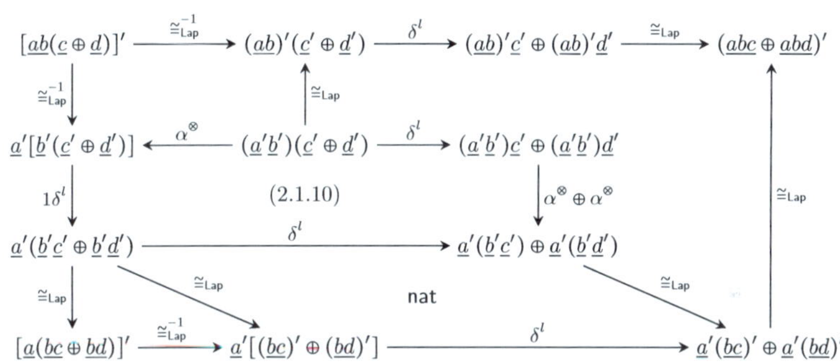

The middle rectangle is commutative by (2.1.10) in C. The bottom parallelogram is commutative by the naturality of δ^l. Each of the other three subdiagrams is commutative by Theorem 3.10.7.

Using the strict associativity of $\otimes^{\mathsf{A}}$ and that $\delta^{r\mathsf{A}} = 1$, the axiom (2.1.12) in A states the commutativity of the outer diagram in C below.

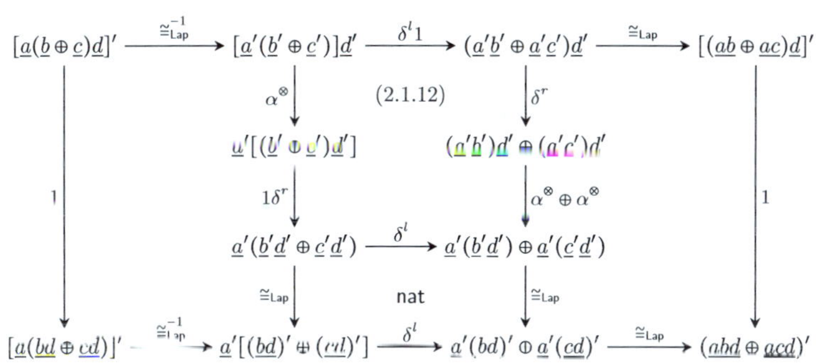

In the middle column, the top and the bottom rectangles are commutative by, respectively, (2.1.12) and the naturality of δ^l in C. Each of the other two rectangles is commutative by Theorem 3.10.7.

When applied to $\underline{a}$, $\underline{b}$, and $\underline{c} \oplus \underline{d}$, $\delta^{r\mathsf{A}} = 1$ is the identity morphism

$$(\underline{a} \oplus \underline{b})(\underline{c} \oplus \underline{d}) \xrightarrow{\delta^{r\mathsf{A}}} \underline{a}(\underline{c} \oplus \underline{d}) \oplus \underline{b}(\underline{c} \oplus \underline{d}).$$

With this identity morphism and $\alpha^{\oplus\mathsf{A}} = 1$ taken into account, the axiom (2.1.13) in A states the commutativity of the outer diagram in C below.

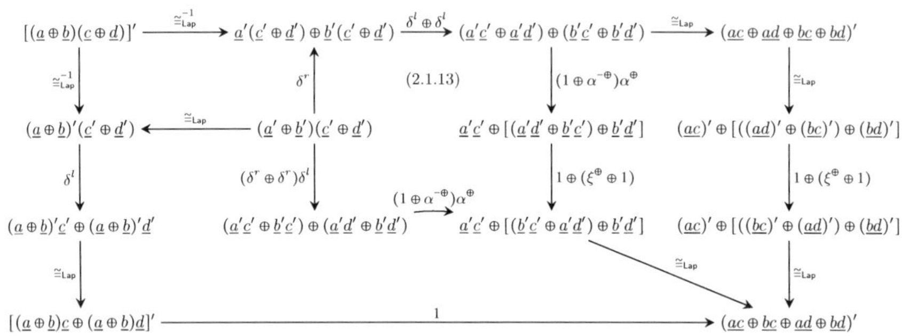

The middle subdiagram is commutative by (2.1.13) in C. Each of the other three subdiagrams is commutative by Theorem 3.10.7. We have proved that A satisfies the 22 Laplaza axioms for a bimonoidal category. □

Strictification Theorems. We are now ready for the main strictification results of this section. The next theorem is the nonsymmetric analogue of Theorem 5.4.6.

Theorem 5.5.11 (Right Rigid Strictification). *Suppose* C *is a tight bimonoidal category. Then there is an adjoint equivalence*

$$\mathsf{A} \underset{\iota}{\overset{\pi}{\rightleftarrows}} \mathsf{C}$$

with

- A *the right rigid bimonoidal category in Proposition 5.5.10;*
- *counit* $\varepsilon\colon \pi\iota \longrightarrow 1_{\mathsf{C}}$ *the identity natural transformation;*
- π *the unitary bimonoidal equivalence as in Definition 5.4.1 and Lemma 5.4.4; and*
- ι *the strong bimonoidal equivalence as in Definition 5.4.2 and Lemma 5.4.5.*

Proof. The proof of Lemma 5.4.3 shows that the pair (π, ι) is an adjoint equivalence, and its counit is the identity.

Reusing the proof of Lemma 5.4.4, to show that π is a unitary bimonoidal equivalence, it remains to prove the axioms (5.1.5) and (5.1.6) for π.

- The axiom (5.1.5) holds for π because
  - $\pi^0_\oplus$ and $\lambda^{\bullet\mathsf{A}}$ are identity natural transformations, and
  - there is an equality

    $$\pi^2_\otimes = \lambda^\bullet_{\pi\underline{a}}\colon (\pi\mathbb{0}^{\mathsf{A}}) \otimes (\pi\underline{a}) = \mathbb{0} \otimes (\pi\underline{a}) \longrightarrow \mathbb{0} = \pi(\mathbb{0}^{\mathsf{A}} \otimes^{\mathsf{A}} \underline{a}).$$

- The axiom (5.1.6) holds by Theorem 3.10.7 and the definition (5.5.6) of $\delta^{l\mathsf{A}}$.

Reusing the proof of Lemma 5.4.5, to show that ι is a strong bimonoidal equivalence, it remains to prove the axioms (5.1.5) and (5.1.6) for ι.

- The axiom (5.1.5) holds for ι because each of the two composites is $1_{\mathbb{0}}$.
- The axiom (5.1.6) holds by Theorem 3.10.7 and the fact that each of the four vertical morphisms is an identity morphism in C.

This finishes the proof. □

As in the symmetric case, the previous theorem has a left rigid analogue. The next theorem is the nonsymmetric analogue of Theorem 5.4.7.

Theorem 5.5.12 (Left Rigid Strictification). *Suppose* C *is a tight bimonoidal category. Then there is an adjoint equivalence*

$$\mathsf{A}_l \underset{\iota}{\overset{\pi}{\rightleftarrows}} \mathsf{C}$$

with

- A_l *a left rigid bimonoidal category as in Definition 5.5.8;*
- *counit* $\varepsilon\colon \pi\iota \longrightarrow 1_{\mathsf{C}}$ *the identity natural transformation;*
- π *a unitary bimonoidal equivalence; and*
- ι *a strong bimonoidal equivalence.*

Proof. As in the symmetric case in Theorem 5.4.7, this proof is obtained from that of Theorem 5.5.11 by redefining the product in A using (5.4.8), so the left distributive law holds strictly.

Moreover, similar to (5.5.6), the right distributivity morphism in A_l is defined as the composite

$$\begin{array}{ccc} \pi\big((\underline{a} \oplus^{\mathsf{A}} \underline{b}) \otimes^{\mathsf{A}} \underline{c}\big) & \xrightarrow{\delta^{r\mathsf{A}_l}_{\underline{a};\underline{b};\underline{c}}} & \pi\big((\underline{a} \otimes^{\mathsf{A}} \underline{c}) \oplus^{\mathsf{A}} (\underline{b} \otimes^{\mathsf{A}} \underline{c})\big) \\ {\scriptstyle \cong^{-1}_{\mathsf{Lap}}}\big\downarrow & & \big\uparrow{\scriptstyle \cong_{\mathsf{Lap}}} \\ (\pi\underline{a} \oplus \pi\underline{b}) \otimes \pi\underline{c} & \xrightarrow{\delta^{r}_{\pi\underline{a};\pi\underline{b};\pi\underline{c}}} & (\pi\underline{a} \otimes \pi\underline{c}) \oplus (\pi\underline{b} \otimes \pi\underline{c}) \end{array}$$

in C for objects $\underline{a}, \underline{b}, \underline{c} \in \mathsf{A}_l$. Other structures are defined as in A in Definition 5.5.5, and all the proofs require minimal or no changes. □

5.6. Notes

5.6.1 (Symmetric Bimonoidal Functors). Our Definition 5.1.1 of a symmetric bimonoidal functor is more general than May's in [**May77**, Def. 6.3.4], which corresponds to our unitary symmetric bimonoidal functors. A *homomorphism between symmetric rig categories* in [**Elg21**, Section 2] corresponds to our strong symmetric bimonoidal functor. ◇

5.6.2 (Strictification Theorem). The proof of the Strictification Theorem 5.4.6 is outlined in [**May77**, Prop. 6.3.5]. We emphasize the necessity of the tightness assumption, that is, the invertibility of the distributivity morphisms. As we saw in (5.2.29) and (5.2.36), the products of morphisms and the symmetry isomorphism in the associated right bipermutative category A use the inverse of a Laplaza coherence isomorphism. By definition, $\cong^{-1}_{\mathsf{Lap}}$ is only defined if the distributivity morphisms δ^l and δ^r are natural isomorphisms.

Since May only gave an outline of the strictification theorem, it may not be apparent that May's proof actually uses the Coherence Theorem 3.9.1 many times. In [**May77**, Prop. 6.3.5], May implicitly used Theorem 3.9.1 when he defined

- the bottom displayed morphism on page 156 there, and
- the second displayed morphism on page 157 there.

These correspond to, respectively, our $f \otimes^{\mathsf{A}} g$ in (5.2.29) and $\xi^{\otimes \mathsf{A}}_{\underline{a};\underline{b}}$ in (5.2.36). Furthermore, showing that A has the desired property of a strictification of a tight symmetric bimonoidal category also uses Theorem 3.9.1 multiple times. See Explanation 5.2.31 and Lemmas 5.3.1, 5.3.4, 5.3.8, and 5.4.4. ◇

5.6.3 (Elmendorf-Mandell Bipermutative Categories)**.** Elmendorf and Mandell in [**EM06**, Def. 3.6] defined their version of bipermutative categories in which the distributivity morphisms

- are not invertible in general, and
- point in the opposite direction as those in (2.1.4).

See Definition II.9.3.2 for the precise definition. Implicitly citing May's result in [**May77**, Prop. 6.3.5], it is stated in [**EM06**, page 178] that their bipermutative categories can be strictified to equivalent right bipermutative categories. As stated in Note 5.6.2, strictification to equivalent right bipermutative categories only applies to *tight* symmetric bimonoidal categories because some constructions and proofs in the strictification require the invertibility of the distributivity morphisms. Therefore, the correct statement is that bipermutative categories in the sense of [**EM06**, Def. 3.6] with invertible distributivity morphisms, which are examples of tight symmetric bimonoidal categories, can be strictified to equivalent right bipermutative categories. This is a special case of our Theorem 5.4.6. See Corollary II.9.3.13. ◇

5.6.4 (Multi-Object Version of Strictification)**.** Guillou proved a multi-object version of Theorem 5.5.11 involving categories weakly enriched in symmetric monoidal categories or in permutative categories in [**Gui10**, Theorem 1.2]. The setting is that of bicategories enriched in the monoidal bicategory of symmetric monoidal categories. Guillou gave two proofs of that theorem, one using (i) the strictification of bicategories to biequivalent 2-categories, and the other using (ii) the Yoneda embedding for bicategories. As we explained in detail in [**JY21**, Chapter 8], both (i) and (ii) depend on the Bicategorical Pasting Theorem 3.6.6 in that book, which is itself a coherence theorem that requires a careful proof. ◇

Part 2

Bicategorical Aspects of Symmetric Bimonoidal Categories

CHAPTER 6

Definitions from Bicategory Theory

Part 2 concerns 2-/bicategorical aspects of symmetric bimonoidal categories.

(1) In Chapter 7, we will define a suitable 2-category $\mathsf{Bi}_{\mathsf{r}}^{\mathsf{fsy}}$ of flat small symmetric bimonoidal categories. The main goal of that chapter is to show that the symmetric bimonoidal category Σ in Proposition 2.4.8 is a lax bicolimit of the 2-functor $\varnothing \longrightarrow \mathsf{Bi}_{\mathsf{r}}^{\mathsf{fsy}}$. In other words, Σ is a bicategorical analogue of an initial object in the 2-category $\mathsf{Bi}_{\mathsf{r}}^{\mathsf{fsy}}$. This theorem confirms a conjecture due to John Baez.

(2) In Chapter 8, we will show that for a tight symmetric bimonoidal category C, the matrix construction $\mathsf{Mat}^{\mathsf{C}}$ is a symmetric monoidal bicategory. It has nonnegative integers as objects, and $n \times m$ matrices with entries objects/morphisms in C as 1-/2-cells $m \longrightarrow n$. Horizontal composition is defined by matrix multiplication with a suitable bracketing for addition. The monoidal composition of the monoidal bicategory is given by matrix tensor product. The braiding of the symmetric monoidal bicategory is induced by permutation matrices in $\mathsf{Mat}^{\mathsf{C}}$. In general, the structure morphisms in $\mathsf{Mat}^{\mathsf{C}}$ are not identities. As an example, the category $\mathsf{2Vect}_{\mathsf{c}}$ of coordinatized 2-vector spaces in Examples 8.4.13 and 8.15.5 is a symmetric monoidal bicategory.

(3) Concepts about 2-/bicategories are also used in
   - Chapters II.7 and II.8, for the braided versions of Baez's Conjecture and the matrix construction $\mathsf{Mat}^{\mathsf{C}}$, and
   - Parts III.1 and III.2, for enriched monoidal categories, multicategories, and the Elmendorf-Mandell K-theory multifunctor.

Without assuming any prior knowledge of 2-dimensional categories, the purpose of this chapter is to recall the relevant concepts that we will use in later chapters. For open questions related to 2-categories and bicategories, see Questions A.1.1, A.1.2, A.1.4, A.1.5, A.2.4, A.4.1, A.5.4, and A.5.5.

Organization. In Section 6.1, we define bicategories and 2-categories, which are due to Bénabou [**Bén67, Bén65**], and provide some examples. In a bicategory, in addition to objects and morphisms between objects, there are also morphisms between morphisms called 2-cells. Just as a category may be regarded as a monoid with multiple objects, a bicategory may be regarded as a monoidal category with multiple objects. A 2-category is a particularly nice bicategory in which the associator and the unitors are identities. The main 2-category in Chapter 7 is the 2-category $\mathsf{Bi}_{\mathsf{r}}^{\mathsf{fsy}}$ in Definition 7.1.8. The main bicategory in Chapter 8 is the matrix bicategory $\mathsf{Mat}^{\mathsf{C}}$, whose horizontal composition is given by the matrix product.

In Section 6.2, we define bicategorical analogues of functors and natural transformations. Due to the presence of 2-cells, there are several versions of functors—lax, pseudo, (strictly) unitary, and strict—between bicategories, depending on the extent to which horizontal composites of 1-cells and identity 1-cells are preserved. A strict functor between 2-categories is called a 2-functor. In Chapter 7, $\varnothing \longrightarrow \mathsf{Bi}_{\mathsf{r}}^{\mathsf{fsy}}$ is an example of a 2-functor. Similarly, transformations between lax functors can be strong or lax, depending on whether the component 2-cells are invertible or not. We also briefly discuss pasting diagrams in bicategories.

In Section 6.3, we define modifications, adjunctions, and adjoint equivalences in bicategories. Modifications are morphisms between lax transformations. Here is an analogy to keep track of the various concepts. If a bicategory is regarded as a topological space, then lax functors, lax transformations, and modifications are analogues of, respectively, continuous maps, homotopies between continuous maps, and homotopies between homotopies between continuous maps. Moreover, the usual concepts of adjunctions and adjoint equivalences also make sense inside a bicategory.

In Section 6.4, we define a monoidal bicategory, which is a one-object tricategory. Since we do not discuss tricategories in this book, we will define a monoidal bicategory with all of its structures explicitly spelled out. It involves the concepts of pseudofunctors, modifications, and adjoint equivalences in a bicategory. A monoidal bicategory has a base bicategory B and a pseudofunctor $\boxtimes\colon \mathsf{B}\times\mathsf{B} \longrightarrow \mathsf{B}$ called the monoidal composition, along with other related structures. The monoidal composition $\boxtimes$ in a monoidal bicategory is an extra level of composition on top of the horizontal composition in the base bicategory. As an analogy, consider complex matrices. There is a base level matrix multiplication AB, which is only defined when the number of columns in A is equal to the number of rows in B. On top of that, the matrix tensor product $A \otimes B$, which is also known as the Kronecker product, is always defined regardless of the sizes of A and B.

In Section 6.5, we define braided, sylleptic, and symmetric monoidal bicategories. Braided and symmetric monoidal bicategories are bicategorical analogues of braided and symmetric monoidal categories. In the bicategorical setting, there is the intermediate concept of a sylleptic monoidal bicategory. The symmetry axiom $\xi_{Y,X}\xi_{X,Y} = 1$ is replaced by an invertible modification ν, called the syllepsis, that connects the braiding-square to the identity, and that satisfies its own coherence axioms. A symmetric monoidal bicategory is a sylleptic monoidal bicategory whose syllepsis satisfies an additional coherence axiom. The main theorem of Chapter 8 is that, for a tight symmetric bimonoidal category C, the matrix bicategory $\mathsf{Mat}^{\mathsf{C}}$ is a symmetric monoidal bicategory. Its monoidal composition is given by the matrix tensor product for matrices in C. The braiding is given by permutation matrices interpreted using $\mathbb{0}$ and $\mathbb{1}$ in C.

In Section 6.6, we define the Gray tensor product for 2-categories and Gray monoids. The Gray tensor product is an interpolation between what we call the box product and the Cartesian product for 2-categories. The 1-category $\mathsf{2Cat}$ of small 2-categories and 2-functors is a symmetric monoidal closed category when equipped with the Gray tensor product. Its monoids are called Gray monoids.

In Section 6.7, we first define permutative Gray monoids and explain them with complete detail. Permutative Gray monoids are to Gray monoids as permutative categories are to strict monoidal categories. The last part of this section contains a brief discussion of strictification related to symmetric monoidal bicategories, permutative Gray monoids, and permutative 2-categories.

References. Most of the topics in this chapter are discussed in much greater detail in the book [**JY21**], where we refer the reader for further discussion, proofs, and examples. Here is a road map between this chapter and the book [**JY21**].

Topics	This chapter	Sections in [**JY21**]
Bicategories	6.1	2.1–2.2
2-categories	6.1	2.3
Pasting diagrams	6.2	Chapter 3
Lax functors	6.2	4.1
Lax transformations	6.2	4.2
Modifications	6.3	4.4
Adjunctions	6.3	6.1–6.2
Monoidal bicategories	6.4	11.1, 11.2, and 12.1
Mates	6.5	6.1
Braided, sylleptic, and symmetric monoidal bicategories	6.5	12.1
The Gray tensor product and Gray monoids	6.6	12.2

The definitions of permutative Gray monoids and permutative 2-categories are from [**GJO17b**]. The articles [**Lac10, Lei∞**] are useful guides for 2-categories and bicategories.

6.1. Bicategories and 2-Categories

In this section, we recall the definitions and some basic examples of bicategories and 2-categories.

Convention 6.1.1. $\mathbb{1}$ denotes the category with one object $*$ and only its identity morphism. For a category C, we identify the categories $\mathsf{C}\times\mathbb{1}$ and $\mathbb{1}\times\mathsf{C}$ with C, and regard the canonical isomorphisms between them as 1_{C}. ◇

A bicategory as defined next is a many-object version of a monoidal category in Definition 1.2.1.

Definition 6.1.2. A *bicategory* is a tuple

$$\big(\mathsf{B}, 1, c, a, \ell, r\big)$$

consisting of the following data.

Objects: B is equipped with a class $\mathsf{Ob}(\mathsf{B})$, whose elements are called *objects* or *0-cells* in B. If $X \in \mathsf{Ob}(\mathsf{B})$, we also write $X \in \mathsf{B}$.

The Hom Categories: For each pair of objects $X, Y \in \mathsf{B}$, B is equipped with a category $\mathsf{B}(X,Y)$, which is called a *hom category*.

- Its objects are called *1-cells* in B. Such a 1-cell f is denoted by either $f\colon X \longrightarrow Y$ or $X \xrightarrow{f} Y$.

- Its morphisms are called *2-cells* in B. For 1-cells $f, f' \in \mathsf{B}(X,Y)$, we display each 2-cell $\alpha : f \longrightarrow f'$ in diagrams as

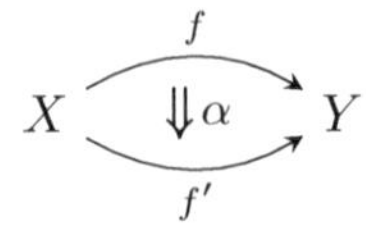

  with a double arrow for the 2-cell.
- Composition and identity morphisms in the category $\mathsf{B}(X,Y)$ are called *vertical composition* and *identity 2-cells*, respectively.
- An isomorphism in $\mathsf{B}(X,Y)$ is called an *invertible 2-cell*, and its inverse is called a *vertical inverse*.
- For a 1-cell f, its identity 2-cell is denoted by 1_f.

Identity 1-Cells: For each object $X \in \mathsf{B}$,

$$\mathbb{1} \xrightarrow{\;1_X\;} \mathsf{B}(X,X)$$

is a functor. We identify the functor 1_X with the 1-cell $1_X(*) \in \mathsf{B}(X,X)$, which is called the *identity 1-cell of* X.

The Horizontal Composition: For each triple of objects $X, Y, Z \in \mathsf{B}$,

$$\mathsf{B}(Y,Z) \times \mathsf{B}(X,Y) \xrightarrow{\;c_{XYZ}\;} \mathsf{B}(X,Z)$$

is a functor, which is called the *horizontal composition*. For 1-cells $f \in \mathsf{B}(X,Y)$ and $g \in \mathsf{B}(Y,Z)$, and 2-cells $\alpha \in \mathsf{B}(X,Y)$ and $\beta \in \mathsf{B}(Y,Z)$, we use the notations

$$\begin{aligned} c_{XYZ}(g,f) &= g \circ f \quad \text{or} \quad gf, \quad \text{and} \\ c_{XYZ}(\beta,\alpha) &= \beta * \alpha. \end{aligned}$$

The Associator: For objects $W, X, Y, Z \in \mathsf{B}$,

$$c_{WXZ}\big(c_{XYZ} \times \mathrm{Id}_{\mathsf{B}(W,X)}\big) \xrightarrow{\;a_{WXYZ}\;} c_{WYZ}\big(\mathrm{Id}_{\mathsf{B}(Y,Z)} \times c_{WXY}\big)$$

is a natural isomorphism, which is called the *associator*, between functors

$$\mathsf{B}(Y,Z) \times \mathsf{B}(X,Y) \times \mathsf{B}(W,X) \longrightarrow \mathsf{B}(W,Z).$$

The Unitors: For each pair of objects $X, Y \in \mathsf{B}$,

$$c_{XYY}\big(1_Y \times \mathrm{Id}_{\mathsf{B}(X,Y)}\big) \xrightarrow{\;\ell_{XY}\;} \mathrm{Id}_{\mathsf{B}(X,Y)} \xleftarrow{\;r_{XY}\;} c_{XXY}\big(\mathrm{Id}_{\mathsf{B}(X,Y)} \times 1_X\big)$$

are natural isomorphisms, which are called the *left unitor* and the *right unitor*, respectively.

The subscripts in c will often be omitted. The subscripts in a, ℓ, and r will often be used to denote their components. The above data are required to satisfy the following two axioms for 1-cells $f \in \mathsf{B}(V,W)$, $g \in \mathsf{B}(W,X)$, $h \in \mathsf{B}(X,Y)$, and $k \in \mathsf{B}(Y,Z)$.

The Unity Axiom: The middle unity diagram

$$
\begin{array}{ccc}
(g1_W)f & \xrightarrow{a_{g,1_W,f}} & g(1_Wf) \\
& {\scriptstyle r_g * 1_f} \searrow \quad \swarrow {\scriptstyle 1_g * \ell_f} & \\
& gf &
\end{array}
\tag{6.1.3}
$$

in $\mathsf{B}(V,X)$ is commutative.

The Pentagon Axiom: The diagram

$$
\begin{array}{ccc}
& (kh)(gf) & \\
{\scriptstyle a_{kh,g,f}} \nearrow & & \searrow {\scriptstyle a_{k,h,gf}} \\
((kh)g)f & & k(h(gf)) \\
{\scriptstyle a_{k,h,g} * 1_f} \downarrow & & \uparrow {\scriptstyle 1_k * a_{h,g,f}} \\
(k(hg))f & \xrightarrow{a_{k,hg,f}} & k((hg)f)
\end{array}
\tag{6.1.4}
$$

in $\mathsf{B}(V,Z)$ is commutative.

This finishes the definition of a bicategory. We sometimes abbreviate a bicategory as above to B. Moreover, a bicategory B is

- *locally small* if each hom category is a small category, and
- *small* if it is locally small and if $\mathsf{Ob}(\mathsf{B})$ is a set. ◇

Explanation 6.1.5. In a bicategory B, we assume that the hom categories $\mathsf{B}(X,Y)$ for objects $X,Y \in \mathsf{B}$ are disjoint. If not, we tacitly replace them with their disjoint union. ◇

Definition 6.1.6. A bicategory B' is called a *subbicategory* of a bicategory B if the following statements hold.

- $\mathsf{Ob}(\mathsf{B}')$ is a subclass of $\mathsf{Ob}(\mathsf{B})$.
- For objects $X,Y \in \mathsf{B}'$, $\mathsf{B}'(X,Y)$ is a subcategory of $\mathsf{B}(X,Y)$.
- The identity 1-cell of X in B' is equal to the identity 1-cell of X in B.
- For objects X,Y,Z in B', the horizontal composition c'_{XYZ} in B' makes the diagram

$$
\begin{array}{ccc}
\mathsf{B}'(Y,Z) \times \mathsf{B}'(X,Y) & \xrightarrow{c'_{XYZ}} & \mathsf{B}'(X,Z) \\
\downarrow & & \downarrow \\
\mathsf{B}(Y,Z) \times \mathsf{B}(X,Y) & \xrightarrow{c_{XYZ}} & \mathsf{B}(X,Z)
\end{array}
$$

  commutative, in which the unnamed arrows are subcategory inclusions.
- Each component of the associator in B' is equal to the corresponding component of the associator in B, and similarly for the left unitors and the right unitors.

This finishes the definition of a subbicategory. Moreover, B' is called a *full* subbicategory of B if for each pair of objects $X,Y \in \mathsf{B}'$, $\mathsf{B}'(X,Y)$ is a full subcategory of $\mathsf{B}(X,Y)$. ◇

Explanation 6.1.7. A subbicategory B' of a bicategory B is full if for each pair of objects $X, Y \in \mathsf{B}'$, and for each pair of 1-cells $f, g \in \mathsf{B}'(X,Y)$, the subcategory inclusion yields an equality

$$\mathsf{B}'(X,Y)(f,g) = \mathsf{B}(X,Y)(f,g)$$

of sets of 2-cells. ◇

Definition 6.1.8. A *2-category* is a bicategory $(\mathsf{B}, 1, c, a, \ell, r)$ in which the associator a, the left unitor ℓ, and the right unitor r are identity natural transformations. ◇

Definition 6.1.9. Suppose B and B' are 2-categories. Then B' is called a *(full) sub-2-category* of B if it is a (full) subbicategory of B in the sense of Definition 6.1.6. ◇

A 2-category can also be characterized explicitly as follows.

Proposition 6.1.10. *A 2-category* B *contains precisely the following data:*

- *A class* $\mathsf{Ob}(\mathsf{B})$ *of objects.*
- *For objects* $X, Y \in \mathsf{B}$, *a class* $\mathsf{B}(X,Y)$ *of 1-cells from* X *to* Y.
- *An identity 1-cell* $1_X \in \mathsf{B}(X,X)$ *for each object* X.
- *For 1-cells* $f, f' \in \mathsf{B}(X,Y)$, *a set* $\mathsf{B}(X,Y)(f,f')$ *of 2-cells from* f *to* f'.
- *An identity 2-cell* $1_f \in \mathsf{B}(X,Y)(f,f)$ *for each 1-cell* $f \in \mathsf{B}(X,Y)$ *and each pair of objects* X, Y.
- *For objects* X *and* Y, *and 1-cells* $f, f', f'' \in \mathsf{B}(X,Y)$, *an assignment*

$$\begin{array}{rcl} \mathsf{B}(X,Y)(f',f'') \times \mathsf{B}(X,Y)(f,f') & \xrightarrow{\;v\;} & \mathsf{B}(X,Y)(f,f'') \\ (\alpha',\alpha) & \longmapsto & \alpha'\alpha, \end{array}$$

  which is called the vertical composition.
- *For objects* X, Y, *and* Z, *an assignment*

$$\begin{array}{rcl} \mathsf{B}(Y,Z) \times \mathsf{B}(X,Y) & \xrightarrow{\;c_1\;} & \mathsf{B}(X,Z) \\ (g,f) & \longmapsto & gf, \end{array}$$

  which is called the horizontal composition of 1-cells.
- *For objects* X, Y, *and* Z, *and 1-cells* $f, f' \in \mathsf{B}(X,Y)$ *and* $g, g' \in \mathsf{B}(Y,Z)$, *an assignment*

$$\begin{array}{rcl} \mathsf{B}(Y,Z)(g,g') \times \mathsf{B}(X,Y)(f,f') & \xrightarrow{\;c_2\;} & \mathsf{B}(X,Z)(gf,g'f') \\ (\beta,\alpha) & \longmapsto & \beta * \alpha, \end{array}$$

  which is called the horizontal composition of 2-cells.

These data are required to satisfy the axioms (i)–(vi) below.

(i) *The vertical composition is associative and unital with respect to the identity 2-cells.*
(ii) *The horizontal composition preserves identity 2-cells and vertical composition.*
(iii) *The horizontal composition of 1-cells is associative, in the sense that for 1-cells* $f \in \mathsf{B}(W,X)$, $g \in \mathsf{B}(X,Y)$, *and* $h \in \mathsf{B}(Y,Z)$, *there is an equality*

$$(hg)f = h(gf) \in \mathsf{B}(W,Z).$$

(iv) *The horizontal composition of 2-cells is associative, in the sense that for 2-cells* $\alpha \in \mathsf{B}(W,X)(f,f')$, $\beta \in \mathsf{B}(X,Y)(g,g')$, *and* $\gamma \in \mathsf{B}(Y,Z)(h,h')$, *there is an equality*

$$(\gamma * \beta) * \alpha = \gamma * (\beta * \alpha)$$

in $\mathsf{B}(W,Z)\big((hg)f, h'(g'f')\big)$.

(v) *The horizontal composition of 1-cells is unital with respect to the identity 1-cells, in the sense that there are equalities*

$$1_Y f = f = f 1_X$$

for each $f \in \mathsf{B}(X,Y)$.

(vi) *The horizontal composition of 2-cells is unital with respect to the identity 2-cells of the identity 1-cells, in the sense that there are equalities*

$$1_{1_Y} * \alpha = \alpha = \alpha * 1_{1_X}$$

for each $\alpha \in \mathsf{B}(X,Y)(f,f')$.

This finishes the list of axioms of a 2-category.

For a 2-category B, its *underlying 1-category* is the category with

- objects $\mathsf{Ob}(\mathsf{B})$,
- morphisms the 1-cells in B,
- identity morphisms the identity 1-cells in B, and
- categorical composition the horizontal composition of 1-cells in B.

By a Cat-*category* we mean a category enriched in the symmetric monoidal category $(\mathsf{Cat}, \times, *)$ of small categories with the Cartesian product as the monoidal product. Discussion of enriched categories can be found in [**Kel05**], [**JY21**, 1.3], and Section III.1.2.

Proposition 6.1.11. *A locally small 2-category is precisely a* Cat-*category.*

Example 6.1.12 (Categories)**.** Each category C yields a 2-category when each morphism set $\mathsf{C}(X,Y)$ is regarded as a category with only identity morphisms. In this 2-category, each 2-cell is an identity 2-cell. In particular, the terminal category $\mathbb{1}$ in Convention 6.1.1 yields a 2-category with one object $*$, one 1-cell 1_*, and one 2-cell 1_{1_*}. ◇

Example 6.1.13 (Small Categories)**.** There is a 2-category Cat defined by the following data.

- It has small categories as objects, functors as 1-cells, and natural transformations as 2-cells.
- Composition of functors are horizontal composition of 1-cells.
- Vertical and horizontal composition of natural transformations are those of 2-cells.
- Identity functors are identity 1-cells.
- Identity natural transformations are identity 2-cells.

From now on, the notation Cat denotes either this 2-category or its underlying 1-category of small categories and functors. ◇

Example 6.1.14 (Monoidal Categories)**.** There is a 2-category MCat with

- small monoidal categories as objects,
- (identity) monoidal functors as (identity) 1-cells,
- monoidal natural transformations as 2-cells, and

- composition of monoidal functors as horizontal composition of 1-cells.

The rest of the structure is defined as in Cat. ◇

Example 6.1.15 (Symmetric Monoidal Categories). SMCat is the 2-category with

- small symmetric monoidal categories as objects,
- (identity) symmetric monoidal functors as (identity) 1-cells,
- monoidal natural transformations as 2-cells, and
- composition of symmetric monoidal functors as horizontal composition of 1-cells.

The rest of the structure is defined as in Cat. ◇

Example 6.1.16 (One-Object Bicategories). Each monoidal category

$$(\mathsf{C}, \otimes, \mathbb{1}, \alpha, \lambda, \rho)$$

yields a one-object bicategory $\Sigma\mathsf{C}$. It has unique hom category $\Sigma\mathsf{C}(*,*) = \mathsf{C}$, identity 1-cell $1_* = \mathbb{1}$, horizontal composition $\otimes$, associator α, left unitor λ, and right unitor ρ. Conversely, for each one-object bicategory, the unique hom category has the structure of a monoidal category as in the previous sentence. ◇

Example 6.1.17 (Bimodules). There is a bicategory Bimod whose objects are unital and associative rings. For two rings R and S, the hom category $\mathsf{Bimod}(R,S)$ has (R,S)-bimodules as objects and (R,S)-bimodule homomorphisms as morphisms.

- For an (R,S)-bimodule M and an (S,T)-bimodule N, the horizontal composition of 1-cells is
$$c_{RST}(N, M) = M \otimes_S N$$
as an (R,T)-bimodule.
- The horizontal composition of 2-cells is given by tensoring of bimodule homomorphisms over a ring.
- Components of the associator, the left unitor, and the right unitor are the usual isomorphisms:
$$M \otimes_S (N \otimes_T P) \cong (M \otimes_S N) \otimes_T P$$
$$M \otimes_S S \cong M$$
$$R \otimes_R M \cong M.$$

Note that the associator seemingly moves brackets from right to left. Moreover, the left unitor has S on the right, and the right unitor has R on the left. This switch of directions is due to the definition of the horizontal composition c_{RST}. ◇

6.2. Lax Functors, Lax Transformations, and Pastings

In this section, we define lax functors between bicategories and lax transformations between lax functors, and briefly discuss pasting diagrams.

Lax Functors. The following definition of a lax functor is modeled after the concept of a monoidal functor in Definition 1.2.11.

Definition 6.2.1. Suppose $(\mathsf{B}, 1, c, a, \ell, r)$ and $(\mathsf{B}', 1', c', a', \ell', r')$ are bicategories as in Definition 6.1.2. A *lax functor*

$$(F, F^2, F^0) \colon \mathsf{B} \longrightarrow \mathsf{B}'$$

from B to B' is a triple consisting of the following data.

Objects: $F \colon \mathsf{Ob}(\mathsf{B}) \longrightarrow \mathsf{Ob}(\mathsf{B}')$ is a function on objects.

The Local Functors: For each pair of objects X, Y in B, it is equipped with a *local functor*

$$F\colon \mathsf{B}(X,Y) \longrightarrow \mathsf{B}'(FX,FY).$$

The Laxity Constraints: For all objects X, Y, Z in B, it is equipped with natural transformations

$$\begin{array}{ccc} \mathsf{B}(Y,Z)\times\mathsf{B}(X,Y) & \xrightarrow{\ c\ } & \mathsf{B}(X,Z) \\ {\scriptstyle F\times F}\downarrow & F^2 \nearrow & \downarrow{\scriptstyle F} \\ \mathsf{B}'(FY,FZ)\times\mathsf{B}'(FX,FY) & \xrightarrow[c']{} & \mathsf{B}'(FX,FZ) \end{array} \qquad \begin{array}{ccc} \mathbb{1} & \xrightarrow{1_X} & \mathsf{B}(X,X) \\ {\scriptstyle 1'_{FX}}\searrow & F^0 \nearrow & \downarrow{\scriptstyle F} \\ & & \mathsf{B}'(FX,FX) \end{array}$$

with component 2-cells

$$Fg \circ Ff \xrightarrow{F^2_{g,f}} F(gf)$$

$$1'_{FX} \xrightarrow{F^0_X} F1_X.$$

We call F^2 the *lax functoriality constraint* and F^0 the *lax unity constraint.*

The above data are required to make the following three diagrams commutative for all 1-cells $f \in \mathsf{B}(W,X)$, $g \in \mathsf{B}(X,Y)$, and $h \in \mathsf{B}(Y,Z)$.

Lax Associativity:

(6.2.2)

$$\begin{array}{ccc} (Fh\circ Fg)\circ Ff & \xrightarrow{\ a'\ } & Fh\circ(Fg\circ Ff) \\ {\scriptstyle F^2_{h,g}*1_{Ff}}\downarrow & & \downarrow{\scriptstyle 1_{Fh}*F^2_{g,f}} \\ F(hg)\circ Ff & & Fh\circ F(gf) \\ {\scriptstyle F^2_{hg,f}}\downarrow & & \downarrow{\scriptstyle F^2_{h,gf}} \\ F((hg)f) & \xrightarrow{\ Fa\ } & F(h(gf)) \end{array}$$

in $B'(FW,FZ)$.

Lax Left and Right Unity:

(6.2.3)

$$\begin{array}{ccc} 1'_{FX}\circ Ff & \xrightarrow{\ \ell'\ } & Ff \\ {\scriptstyle F^0_X*1_{Ff}}\downarrow & & \uparrow{\scriptstyle F\ell} \\ F1_X\circ Ff & \xrightarrow{F^2_{1_X,f}} & F(1_X\circ f) \end{array} \qquad \begin{array}{ccc} Ff\circ 1'_{FW} & \xrightarrow{\ r'\ } & Ff \\ {\scriptstyle 1_{Ff}*F^0_W}\downarrow & & \uparrow{\scriptstyle Fr} \\ Ff\circ F1_W & \xrightarrow{F^2_{f,1_W}} & F(f\circ 1_W) \end{array}$$

in $\mathsf{B}'(FW,FX)$.

This finishes the definition of a lax functor.

Moreover, we define the following.

- A lax functor is *unitary* (respectively, *strictly unitary*) if each lax unity constraint F^0_X is an invertible 2-cell (respectively, identity 2-cell).
- A *pseudofunctor* is a lax functor in which F^2 and F^0 are natural isomorphisms.
- A *strict functor* is a lax functor in which F^2 and F^0 are identity natural transformations.
- A strict functor between two 2-categories is called a *2-functor.* ◇

Explanation 6.2.4. The lax unity constraint F^0 in a lax functor is completely determined by the component 2-cells F^0_X for objects $X \in \mathsf{B}$. Its naturality condition is redundant because the terminal category $\mathbb{1}$ only has the identity morphism of $*$. See [**JY21**, Explanation 4.1.5(3)]. $\diamond$

Proposition 6.2.5. *For 2-categories* A *and* B, *a 2-functor* $F\colon \mathsf{A} \longrightarrow \mathsf{B}$ *consists of precisely the following data.*

- *A function* $F\colon \mathsf{Ob}(\mathsf{A}) \longrightarrow \mathsf{Ob}(\mathsf{B})$ *on objects.*
- *A functor* $F\colon \mathsf{A}(X,Y) \longrightarrow \mathsf{B}(FX,FY)$ *for each pair of objects* X,Y *in* A.

These data are subject to the following two conditions.

(1) F *is a functor between the underlying 1-categories of* A *and* B.
(2) F *preserves horizontal compositions of 2-cells.*

Next we define composites of lax functors.

Definition 6.2.6. Suppose

$$\mathsf{B} \xrightarrow{(F,F^2,F^0)} \mathsf{C} \xrightarrow{(G,G^2,G^0)} \mathsf{D}$$

are lax functors between bicategories. The *composite*

$$\mathsf{B} \xrightarrow{(GF,(GF)^2,(GF)^0)} \mathsf{D}$$

is defined as follows.

Objects: $GF\colon \mathsf{Ob}(\mathsf{B}) \longrightarrow \mathsf{Ob}(\mathsf{D})$ is the composite of the functions F and G on objects.

The Local Functors: For objects X,Y in B, it is equipped with the composite functor

$$\overbrace{\mathsf{B}(X,Y) \xrightarrow{F} \mathsf{C}(FX,FY) \xrightarrow{G} \mathsf{D}(GFX,GFY)}^{GF}.$$

The Lax Unity Constraint: For each object X in B, it is equipped with the vertically composed 2-cell

$$\overbrace{1_{GFX} \xrightarrow{G^0_{FX}} G1_{FX} \xrightarrow{G(F^0_X)} GF1_X}^{(GF)^0_X} \tag{6.2.7}$$

in $\mathsf{D}(GFX,GFX)$.

The Lax Functoriality Constraint: For 1-cells $(g,f) \in \mathsf{B}(Y,Z) \times \mathsf{B}(X,Y)$, it has the vertically composed 2-cell

$$\overbrace{GFg \circ GFf \xrightarrow{G^2_{Fg,Ff}} G(Fg \circ Ff) \xrightarrow{G(F^2_{g,f})} GF(gf)}^{(GF)^2_{g,f}} \tag{6.2.8}$$

in $\mathsf{D}(GFX,GFZ)$.

The finishes the definition of the composite. $\diamond$

Lemma 6.2.9. *Suppose* $F\colon \mathsf{B} \longrightarrow \mathsf{C}$ *and* $G\colon \mathsf{C} \longrightarrow \mathsf{D}$ *are lax functors between bicategories.*

(1) *The composite* $(GF,(GF)^2,(GF)^0)$ *is a lax functor from* B *to* D.

(2) *If both F and G are pseudofunctors (respectively, strict, unitary, or strictly unitary), then so is the composite GF.*

Example 6.2.10 (Identity Strict Functors)**.** Each bicategory B has an identity strict functor $1_{\mathsf{B}}: \mathsf{B} \longrightarrow \mathsf{B}$.

- It is the identity function on the objects in B.
- It is the identity functor on $\mathsf{B}(X,Y)$ for objects X, Y in B.
- For composable 1-cells (g,f), the component $(1_{\mathsf{B}})^2_{g,f}$ is the identity 2-cell $1_{gf} = 1_g * 1_f$.
- The component $(1_{\mathsf{B}})^0_X$ is the identity 2-cell 1_{1_X}.

For 1_{B}, the lax associativity diagram (6.2.2) follows from the naturality of the associator a, and both lax unity diagrams (6.2.3) are commutative by definition. ◇

Theorem 6.2.11. *There is a category* Bicat *with*

- *small bicategories as objects,*
- *lax functors between them as morphisms,*
- *composites of lax functors as in Definition 6.2.6, and*
- *identity strict functors in Example 6.2.10 as identity morphisms.*

Furthermore, Bicat *contains the wide subcategories*

(i) $\mathsf{Bicat}^{\mathsf{u}}$ *with unitary lax functors as morphisms,*
(ii) $\mathsf{Bicat}^{\mathsf{su}}$ *with strictly unitary lax functors as morphisms,*
(iii) $\mathsf{Bicat}^{\mathsf{ps}}$ *with pseudofunctors as morphisms,*
(iv) $\mathsf{Bicat}^{\mathsf{sup}}$ *with strictly unitary pseudofunctors as morphisms, and*
(v) $\mathsf{Bicat}^{\mathsf{st}}$ *with strict functors as morphisms.*

Pastings. Some of the upcoming definitions use the concept of a *pasting diagram* in a bicategory. This is an efficient and visual way to represent iterated vertical composition of 2-cells, each being the horizontal composition of one 2-cell and a finite number of identity 2-cells. This topic is discussed at length in [**JY21**, Ch. 3], where we refer the reader for detailed discussion and many examples. In particular, the Bicategorical Pasting Theorem 3.6.6 there states that each pasting diagram in a bicategory has a unique composite, once a bracketing is chosen for the (co)domain composite 1-cell.

Convention 6.2.12. Pasting diagrams use the left normalized bracketing (5.2.13) for the iterated horizontal composition of (co)domain 1-cells, unless a different bracketing is specified. ◇

Example 6.2.13. Consider the following data in a bicategory B.

- Objects A_1, A_2, A_3, and A_4.
- 1-cells $f: A_1 \longrightarrow A_3$, $g_1: A_1 \longrightarrow A_2$, $g_2: A_2 \longrightarrow A_3$, $g_3: A_2 \longrightarrow A_3$, $g_4: A_3 \longrightarrow A_4$, and $h: A_2 \longrightarrow A_4$.
- 2-cells $\theta_1: f \longrightarrow g_2 g_1$, $\theta: g_2 \longrightarrow g_3$, and $\theta_2: g_4 g_3 \longrightarrow h$.

The pasting diagram

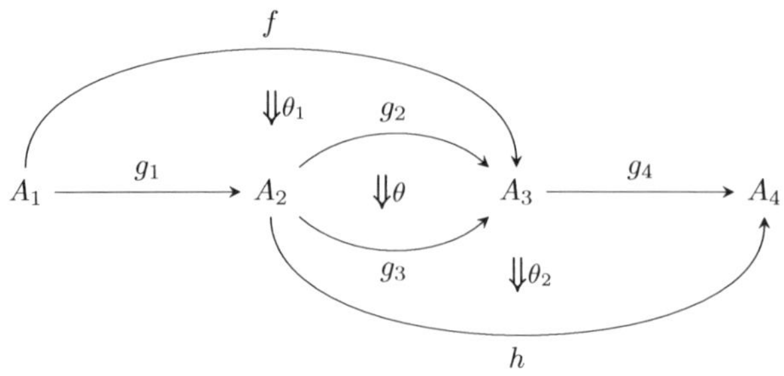

represents the following vertical composite 2-cell.

$$(\theta_2 * 1_{g_1})(a^{-1}_{g_4,g_3,g_1})(1_{g_4} * (\theta * 1_{g_1}))(1_{g_4} * \theta_1): g_4 f \longrightarrow hg_1$$

Note that a component of the inverse of the associator is inserted because the codomain of $1_{g_4} * (\theta * 1_{g_1})$ is the 1-cell $g_4(g_3 g_1)$, while the domain of $\theta_2 * 1_{g_1}$ is the 1-cell $(g_4 g_3)g_1$. ◇

Lax Transformations. The next definition is the bicategorical analogue of a natural transformation in Definition 1.1.7.

Definition 6.2.14. Suppose (F, F^2, F^0) and (G, G^2, G^0) are lax functors $\mathsf{B} \longrightarrow \mathsf{B}'$. A *lax transformation* $\alpha: F \longrightarrow G$ consists of the following data.

Components: It is equipped with a component 1-cell $\alpha_X \in \mathsf{B}'(FX, GX)$ for each object X in B.

The Lax Naturality Constraints: For each pair of objects X, Y in B, it is equipped with a natural transformation

$$\alpha: \alpha_X^* G \longrightarrow (\alpha_Y)_* F: \mathsf{B}(X,Y) \longrightarrow \mathsf{B}'(FX, GY)$$

with a component 2-cell

$$\alpha_f: (Gf)\alpha_X \longrightarrow \alpha_Y(Ff)$$

as in the following diagram for each 1-cell $f \in \mathsf{B}(X,Y)$.

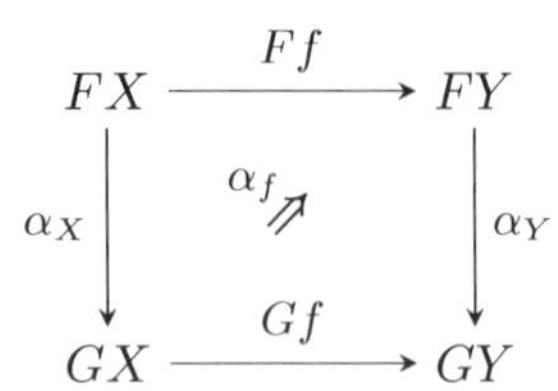

The above data are required to satisfy the following two pasting diagram equalities for all objects X, Y, Z, and 1-cells $f \in \mathsf{B}(X,Y)$ and $g \in \mathsf{B}(Y,Z)$.

Lax Unity:

(6.2.15)

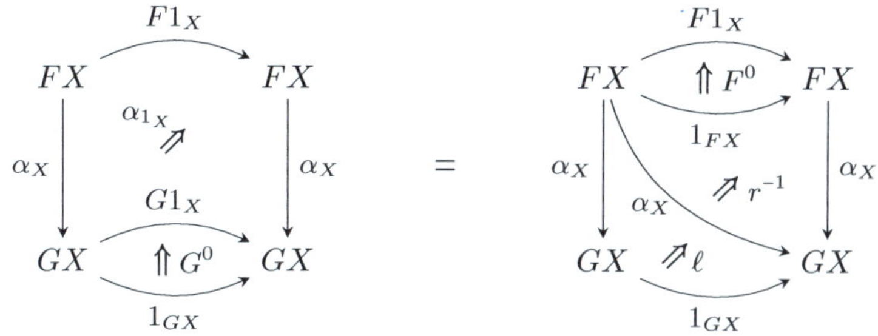

Lax Naturality:

(6.2.16)

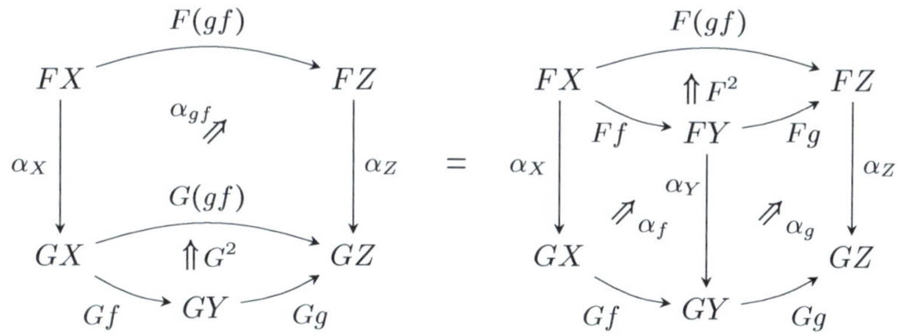

This finishes the definition of a lax transformation.

Moreover, we define the following.

- A *strong transformation* is a lax transformation in which each component α_f is an invertible 2-cell.
- A *strict transformation* is a lax transformation in which each component α_f is an identity 2-cell.
- A *2-natural transformation* is a strict transformation between 2-functors between 2-categories. ◇

Next is the bicategorical analogue of the identity natural transformation of a functor.

Lemma 6.2.17. *Suppose $(F, F^2, F^0): \mathsf{B} \longrightarrow \mathsf{B}'$ is a lax functor between bicategories. Then there is a strong transformation*

$$1_F : F \longrightarrow F$$

defined by the following data.

- *For each object X in B, the component 1-cell $(1_F)_X$ is the identity 1-cell $1_{FX} \in \mathsf{B}(FX, FX)$.*
- *For each 1-cell $f \in \mathsf{B}(X,Y)$, the component 2-cell is the vertical composite*

$$(6.2.18) \qquad \overbrace{(Ff)1_{FX} \xrightarrow{\;r_{Ff}\;} Ff \xrightarrow{\;\ell^{-1}_{Ff}\;} 1_{FY}(Ff)}^{(1_F)_f}$$

in $\mathsf{B}'(FX, FY)$.

Definition 6.2.19. For a lax functor F, the strong transformation $1_F: F \longrightarrow F$ in Lemma 6.2.17 is called the *identity transformation* of F. ◇

Note that if F is a 2-functor, then the identity transformation 1_F is a 2-natural transformation. For a general lax functor F, 1_F is a strong transformation that is not usually strict. Next we define composition of lax transformations.

Definition 6.2.20. Suppose $\alpha: F \longrightarrow G$ and $\beta: G \longrightarrow H$ are lax transformations for lax functors $F, G, H: \mathsf{B} \longrightarrow \mathsf{B}'$. The *horizontal composite* $\beta\alpha: F \longrightarrow H$ is defined with the following data.

Component 1-Cells: For each object X in B, it is equipped with the horizontal composite 1-cell

$$\underbrace{FX \xrightarrow{\alpha_X} GX \xrightarrow{\beta_X} HX}_{(\beta\alpha)_X}$$

in $\mathsf{B}'(FX, HX)$.

Component 2-Cells: For each 1-cell $f \in \mathsf{B}(X, Y)$, $(\beta\alpha)_f$ is the 2-cell

$$\begin{array}{ccc} FX & \xrightarrow{Ff} & FY \\ \downarrow{\scriptstyle \alpha_X} & \overset{\alpha_f}{\Nearrow} & \downarrow{\scriptstyle \alpha_Y} \\ GX & \xrightarrow{Gf} & GY \\ \downarrow{\scriptstyle \beta_X} & \overset{\beta_f}{\Nearrow} & \downarrow{\scriptstyle \beta_Y} \\ HX & \xrightarrow{Hf} & HY \end{array} \qquad (\text{left boundary } \beta_X\alpha_X, \text{ right boundary } \beta_Y\alpha_Y) \tag{6.2.21}$$

whose vertical boundaries are bracketed as indicated. ◇

Lemma 6.2.22. *In Definition 6.2.20, $\beta\alpha: F \longrightarrow H$ is a lax transformation, which is strong if both α and β are strong.*

The 2-Category of 2-Categories.

Proposition 6.2.23. *For 2-functors $F, G : \mathsf{A} \longrightarrow \mathsf{B}$ between 2-categories, a 2-natural transformation $\alpha : F \longrightarrow G$ consists of exactly a component 1-cell $\alpha_X \in \mathsf{B}(FX, GX)$ for each object X in A such that the following two conditions are satisfied.*

1-Cell Naturality: *For each 1-cell $f \in \mathsf{A}(X, Y)$, the two composite 1-cells*

$$\begin{array}{ccc} FX & \xrightarrow{\alpha_X} & GX \\ {\scriptstyle Ff}\downarrow & & \downarrow{\scriptstyle Gf} \\ FY & \xrightarrow{\alpha_Y} & GY \end{array} \tag{6.2.24}$$

in $\mathsf{B}(FX, GY)$ are equal.

2-Cell Naturality: *For each 2-cell $\theta : f \longrightarrow g$ in $\mathsf{A}(X, Y)$, the diagram*

$$\begin{array}{ccc} (Gf)\alpha_X & \xrightarrow{1} & \alpha_Y(Ff) \\ {\scriptstyle G\theta * 1_{\alpha_X}}\downarrow & & \downarrow{\scriptstyle 1_{\alpha_Y} * F\theta} \\ (Gg)\alpha_X & \xrightarrow{1} & \alpha_Y(Fg) \end{array} \tag{6.2.25}$$

in $\mathsf{B}(FX, GY)$ is commutative.

Definition 6.2.26. Denote by $\mathsf{2Cat}$ the 2-category of small 2-categories, 2-functors, and 2-natural transformations.

- Identity 1-cells are the identity strict functors in Example 6.2.10.
- Identity 2-cells are the identity transformations in Lemma 6.2.17.
- Horizontal composition of 2-functors is as in Definition 6.2.6.
- Horizontal and vertical compositions of 2-natural transformations are those of Cat-categories.

A 2-natural transformation $\alpha: F \longrightarrow G$ is a *2-natural isomorphism* if there exists a 2-natural transformation $\beta: G \longrightarrow F$ such that the equalities

$$\beta\alpha = 1_F \quad \text{and} \quad \alpha\beta = 1_G$$

hold. The notation $\mathsf{2Cat}$ is also used for the 1-category of small 2-categories and 2-functors. ◇

Explanation 6.2.27.

(1) For 2-natural transformations $\alpha: F \longrightarrow G$ and $\beta: G \longrightarrow H$ for 2-functors $F, G, H: \mathsf{A} \longrightarrow \mathsf{B}$ between 2-categories, the *vertical composite*

$$\beta\alpha: F \longrightarrow H$$

is defined by the component 1-cells

$$(\beta\alpha)_X = \beta_X \alpha_X \in \mathsf{B}(FX, HX)$$

for objects $X \in \mathsf{A}$.

For a 2-natural transformation $\alpha': F' \longrightarrow G'$ for 2-functors $F', G': \mathsf{B} \longrightarrow \mathsf{C}$ between 2-categories, the *horizontal composite*

$$\alpha' * \alpha: F'F \longrightarrow G'G$$

is defined by the component 1-cells

$$(\alpha' * \alpha)_X = \alpha'_{GX}(F'\alpha_X) \in \mathsf{B}(F'FX, G'GX)$$

for objects $X \in \mathsf{A}$.

(2) If 2-natural transformations are regarded as lax transformations, then the vertical composition of 2-natural transformations in Definition 6.2.26 is the horizontal composition of lax transformations in Definition 6.2.20.

(3) A 2-natural transformation α is a 2-natural isomorphism if and only if each component 1-cell $\alpha_X: FX \longrightarrow GX$ is invertible. In other words, there exists a 1-cell $\beta_X: GX \longrightarrow FX$ such that the equalities

$$\beta_X \alpha_X = 1_{FX} \quad \text{and} \quad \alpha_X \beta_X = 1_{GX}$$

hold. ◇

6.3. Modifications and Adjunctions

In this section, we first define morphisms between lax transformations called modifications. Then we define adjunctions and adjoint equivalences in a bicategory. These concepts will be used in the definition of a monoidal bicategory in Section 6.4.

Definition 6.3.1. Suppose $\alpha, \beta: F \longrightarrow G$ are lax transformations for lax functors $F, G: \mathsf{B} \longrightarrow \mathsf{B}'$. A *modification* $\Gamma: \alpha \longrightarrow \beta$ consists of a component 2-cell

$$\Gamma_X: \alpha_X \longrightarrow \beta_X$$

in $\mathsf{B}'(FX, GX)$ for each object X in B, that satisfies the following *modification axiom* for each 1-cell $f \in \mathsf{B}(X, Y)$.

(6.3.2)

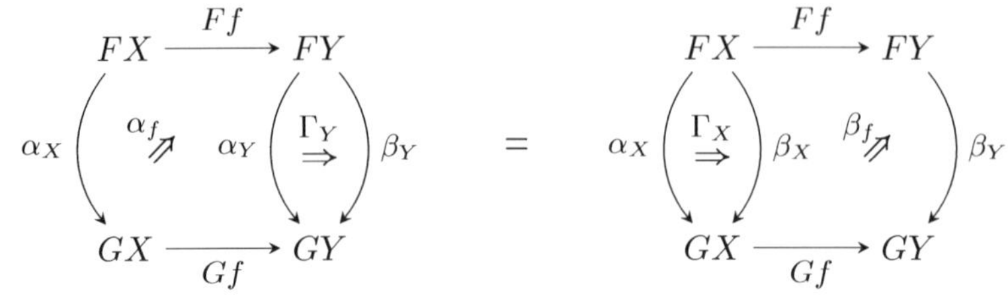

A modification is *invertible* if each component Γ_X is an invertible 2-cell. ◇

Definition 6.3.3. Suppose

- $F, G, H \colon \mathsf{B} \longrightarrow \mathsf{B}'$ are lax functors, and
- $\alpha, \beta, \gamma \colon F \longrightarrow G$ are lax transformations.

Identity Modifications: The *identity modification* of α, denoted by $1_\alpha \colon \alpha \longrightarrow \alpha$, consists of the identity 2-cell

$$(1_\alpha)_X = 1_{\alpha_X} : \alpha_X \longrightarrow \alpha_X$$

in $\mathsf{B}'(FX, GX)$ for each object X in B.

Vertical Composition: Suppose $\Gamma \colon \alpha \longrightarrow \beta$ and $\Omega \colon \beta \longrightarrow \gamma$ are modifications. The *vertical composite*

$$\Omega\Gamma \colon \alpha \longrightarrow \gamma$$

consists of the vertical composite 2-cell

$$\alpha_X \xrightarrow{\Gamma_X} \beta_X \xrightarrow{\Omega_X} \gamma_X, \qquad (\Omega\Gamma)_X \colon \alpha_X \longrightarrow \gamma_X \tag{6.3.4}$$

in $\mathsf{B}'(FX, GX)$ for each object X in B.

Horizontal Composition: With $\Gamma \colon \alpha \longrightarrow \beta$ as above, suppose $\Gamma' \colon \alpha' \longrightarrow \beta'$ is a modification for lax transformations $\alpha', \beta' \colon G \longrightarrow H$. The *horizontal composite*

$$\Gamma' * \Gamma \colon \alpha'\alpha \longrightarrow \beta'\beta$$

consists of the horizontal composite 2-cell

$$(\Gamma' * \Gamma)_X = \Gamma'_X * \Gamma_X \colon (\alpha'\alpha)_X \longrightarrow (\beta'\beta)_X \tag{6.3.5}$$

in $\mathsf{B}'(FX, HX)$ for each object X in B. Here $\alpha'\alpha, \beta'\beta \colon F \longrightarrow H$ are the horizontal composite lax transformations in Definition 6.2.20. ◇

For a small category C and a category D, there is a category with functors $\mathsf{C} \longrightarrow \mathsf{D}$ as objects and natural transformations as morphisms. Next is the bicategorical analogue of that fact.

Definition 6.3.6. Suppose B and B' are bicategories with $\mathsf{Ob}(\mathsf{B})$ a set. Define

$$\mathsf{Bicat}(\mathsf{B}, \mathsf{B}')$$

with the following data.

Objects: The objects in $\mathsf{Bicat}(\mathsf{B},\mathsf{B}')$ are lax functors $\mathsf{B} \longrightarrow \mathsf{B}'$.

The Hom Categories: For lax functors $F, G\colon \mathsf{B} \longrightarrow \mathsf{B}'$, $\mathsf{Bicat}(\mathsf{B},\mathsf{B}')(F,G)$ is the category with
- lax transformations $F \longrightarrow G$ as objects,
- modifications $\Gamma\colon \alpha \longrightarrow \beta$ between such lax transformations as morphisms,
- vertical composition of modifications as composition, and
- identity modifications as identity morphisms.

In other words, in $\mathsf{Bicat}(\mathsf{B},\mathsf{B}')$ 1-cells are lax transformations between lax functors from B to B', and 2-cells are modifications between them.

Identity 1-Cells: For each lax functor $F\colon \mathsf{B} \longrightarrow \mathsf{B}'$, its identity 1-cell is the identity transformation $1_F\colon F \longrightarrow F$ in Definition 6.2.19.

Horizontal Composition: For lax functors $F, G, H\colon \mathsf{B} \longrightarrow \mathsf{B}'$, the horizontal composition

$$\mathsf{Bicat}(\mathsf{B},\mathsf{B}')(G,H) \times \mathsf{Bicat}(\mathsf{B},\mathsf{B}')(F,G) \xrightarrow{\;c\;} \mathsf{Bicat}(\mathsf{B},\mathsf{B}')(F,H)$$

is given by
- the horizontal composition of lax transformations in Definition 6.2.20 for 1-cells and
- the horizontal composition of modifications in Definition 6.3.3 for 2-cells.

The Associator: For
- lax functors $F, G, H, I\colon \mathsf{B} \longrightarrow \mathsf{B}'$ and
- lax transformations $\alpha\colon F \longrightarrow G$, $\beta\colon G \longrightarrow H$, and $\gamma\colon H \longrightarrow I$,

the component

$$a_{\gamma,\beta,\alpha}\colon (\gamma\beta)\alpha \longrightarrow \gamma(\beta\alpha)$$

of the associator a is the modification with, for each object $X \in \mathsf{B}$, the component 2-cell

$$a_{\gamma_X,\beta_X,\alpha_X}\colon (\gamma_X\beta_X)\alpha_X \longrightarrow \gamma_X(\beta_X\alpha_X) \quad \text{in} \quad \mathsf{B}'(FX, IX),$$

which is a component of the associator in B'.

The Unitors: For each lax transformation α as above, the component

$$\ell_\alpha\colon 1_G\alpha \longrightarrow \alpha$$

of the left unitor ℓ is the modification with, for each object $X \in \mathsf{B}$, the component 2-cell

$$\ell_{\alpha_X}\colon 1_{GX}\alpha_X \longrightarrow \alpha_X \quad \text{in} \quad \mathsf{B}'(FX, GX),$$

which is a component of the left unitor in B'. The right unitor r is defined analogously using the right unitor in B'.

This finishes the definition of $\mathsf{Bicat}(\mathsf{B},\mathsf{B}')$. ◇

Theorem 6.3.7. *Suppose* B *and* B' *are bicategories such that* B *has a set of objects. Then* $\mathsf{Bicat}(\mathsf{B},\mathsf{B}')$ *with the structure in Definition 6.3.6 is a bicategory, which is furthermore a 2-category if* B' *is a 2-category.*

Moreover, the bicategory $\mathsf{Bicat}(\mathsf{B},\mathsf{B}')$ *contains a subbicategory* $\mathsf{Bicat}^{\mathsf{ps}}(\mathsf{B},\mathsf{B}')$ *with*
- *pseudofunctors* $\mathsf{B} \longrightarrow \mathsf{B}'$ *as objects,*
- *strong transformations between such pseudofunctors as 1-cells, and*

- *modifications between such strong transformations as 2-cells.*

This is a sub-2-category of $\mathsf{Bicat}(\mathsf{B},\mathsf{B}')$ *if* B' *is a 2-category.*

Motivation 6.3.8. Next we define adjunction and adjoint equivalence in a bicategory. In a category C, an isomorphism is a morphism that is strictly invertible. A similar definition exists in a bicategory B. We call a 1-cell $f\colon X \longrightarrow Y$ in B an *isomorphism* if there exists a 1-cell $g\colon Y \longrightarrow X$ such that the equalities

$$gf = 1_X \quad \text{and} \quad fg = 1_Y$$

of 1-cells hold. However, this notion of sameness is far too strong to hold in practice. A more reasonable concept of sameness should replace the above equalities with invertible 2-cells that are compatible in some sense. The next definition of an adjoint equivalence in a bicategory is modeled after the concept of an adjoint equivalence in Definition 1.1.12. ◇

Definition 6.3.9. An *internal adjunction*, which is also called an *adjunction*, $f \dashv g$ in a bicategory B is a quadruple $(f, g, \eta, \varepsilon)$ consisting of

- 1-cells $f\colon X \longrightarrow Y$ and $g\colon Y \longrightarrow X$ and
- 2-cells $\eta\colon 1_X \longrightarrow gf$ and $\varepsilon\colon fg \longrightarrow 1_Y$.

These data are subject to the following two axioms, in the form of commutative triangles. These are known as the *triangle identities*.

(6.3.10)

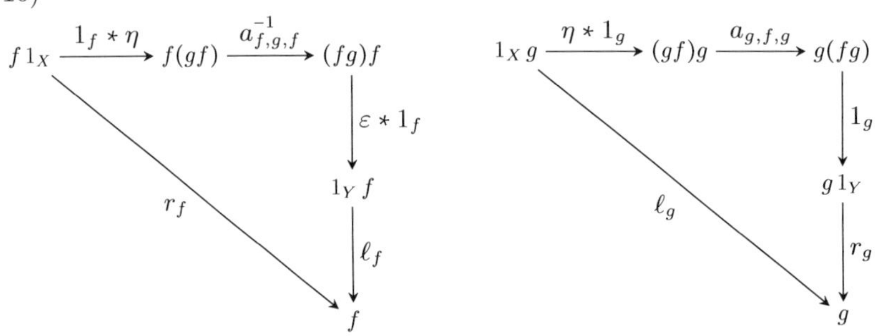

In (6.3.10) the left and the right axioms are called, respectively, the *left triangle identity* and the *right triangle identity*. The 1-cell f is called the *left adjoint*, and g is called the *right adjoint*. The 2-cell η, respectively ε, is called the *unit*, respectively *counit*.

An adjunction $(f, g, \eta, \varepsilon)$ is called an *internal equivalence* or *adjoint equivalence* if η and ε are isomorphisms. We say that f and g are members of an adjoint equivalence in this case, and we let $f^{\bullet}$ denote an adjoint. ◇

6.4. Monoidal Bicategories

In this section, we define monoidal bicategories, which are one-object tricategories. However, this compact definition depends on the concept of a tricategory, which we do not discuss in this book. The following definition spells out explicitly the structure of a monoidal bicategory.

Definition 6.4.1. A *monoidal bicategory* is a tuple

$$\big(\mathsf{B}, \boxtimes, 1_{\boxtimes}, a, \ell, r, \pi, \mu, \lambda, \rho\big)$$

consisting of the following data.

The Base Bicategory: It has a bicategory B, which is called the *base bicategory.* To avoid confusion, we may refer to the associator, the left unitor, and the right unitor in this bicategory B as, respectively, the *base associator*, the *base left unitor*, and the *base right unitor.* The n-fold product bicategory $\mathsf{B} \times \cdots \times \mathsf{B}$ is written as B^n below, and $\mathbb{1}$ is the 2-category with only one object $*$, the identity 1-cell 1_*, and the identity 2-cell 1_{1_*}.

The Monoidal Composition: It has a pseudofunctor

$$\mathsf{B} \times \mathsf{B} \xrightarrow{(\boxtimes, \boxtimes^2, \boxtimes^0)} \mathsf{B},$$

which is called the *monoidal composition.*

The Monoidal Identity: It has a pseudofunctor

$$\mathbb{1} \xrightarrow{(1_\boxtimes, 1_\boxtimes^2, 1_\boxtimes^0)} \mathsf{B},$$

which is called the *monoidal identity.* The object $1_\boxtimes(*) \in \mathsf{B}$ is also denoted by $1_\boxtimes$, which is called the *identity object.*

The Monoidal Associator: It has an adjoint equivalence $(a, a^\bullet, \eta^a, \varepsilon^a)$ with left adjoint

$$\begin{array}{ccc} \mathsf{B}^3 & \xrightarrow{\boxtimes \times 1} & \mathsf{B}^2 \\ {\scriptstyle 1 \times \boxtimes}\downarrow & \overset{a}{\swarrow} & \downarrow{\scriptstyle \boxtimes} \\ \mathsf{B}^2 & \xrightarrow[\boxtimes]{} & \mathsf{B} \end{array}$$

in the bicategory $\mathsf{Bicat}^{\mathsf{ps}}(\mathsf{B}^3, \mathsf{B})$, which is called the *monoidal associator.* Its left and right adjoints have the following component 1-cells for objects $A, B, C \in \mathsf{B}$.

$$(C \boxtimes B) \boxtimes A \underset{a^\bullet_{C,B,A}}{\overset{a_{C,B,A}}{\rightleftarrows}} C \boxtimes (B \boxtimes A) \in \mathsf{B}$$

The Monoidal Unitors: It has adjoint equivalences $(\ell, \ell^\bullet, \eta^\ell, \varepsilon^\ell)$ and $(r, r^\bullet, \eta^r, \varepsilon^r)$ with respective left adjoints

$$\begin{array}{ccccc} & & \mathsf{B}^2 & & \\ & {\scriptstyle 1_\boxtimes \times 1}\nearrow & \Downarrow \ell & \searrow{\scriptstyle \boxtimes} & \\ \mathsf{B} & & \xrightarrow[1]{} & & \mathsf{B} \end{array} \qquad \begin{array}{ccccc} & & \mathsf{B}^2 & & \\ & {\scriptstyle 1 \times 1_\boxtimes}\nearrow & \Downarrow r & \searrow{\scriptstyle \boxtimes} & \\ \mathsf{B} & & \xrightarrow[1]{} & & \mathsf{B} \end{array}$$

in the bicategory $\mathsf{Bicat}^{\mathsf{ps}}(\mathsf{B}, \mathsf{B})$, which are called the *left monoidal unitor* and the *right monoidal unitor*, respectively. Their left and right adjoints have component 1-cells

$$1_\boxtimes \boxtimes A \underset{\ell^\bullet_A}{\overset{\ell_A}{\rightleftarrows}} A \underset{r^\bullet_A}{\overset{r_A}{\leftrightarrows}} A \boxtimes 1_\boxtimes \in \mathsf{B}.$$

The Pentagonator: It has an invertible modification π, which is called the *pentagonator*, with the following component 2-cells for objects $A, B, C, D \in \mathsf{B}$.

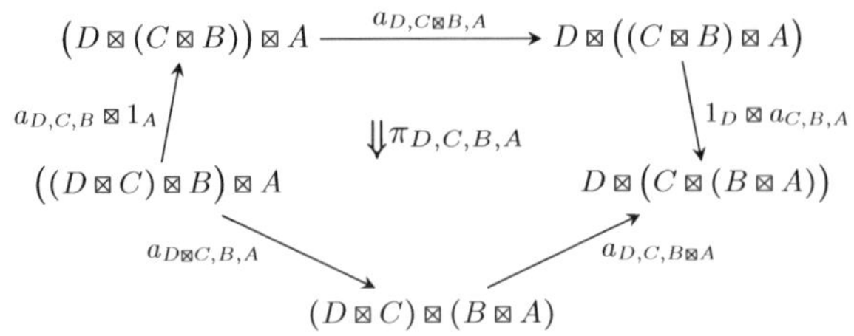

The 2-Unitors: It has invertible modifications μ, λ, and ρ, which are called, respectively, the *middle 2-unitor*, the *left 2-unitor*, and the *right 2-unitor*, with the following component 2-cells in B.

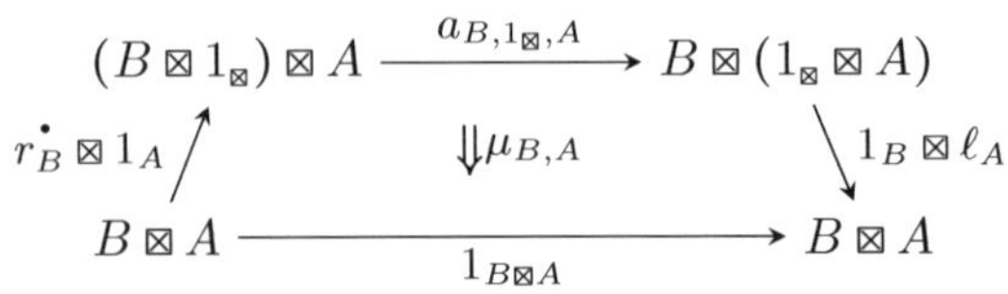

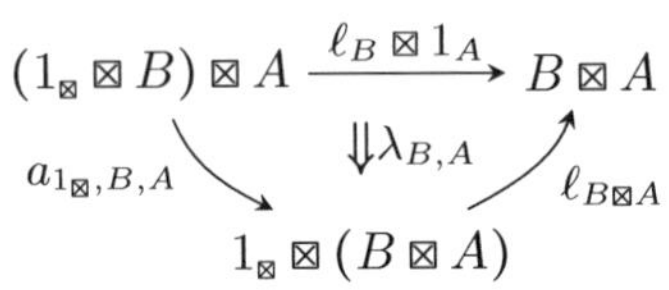

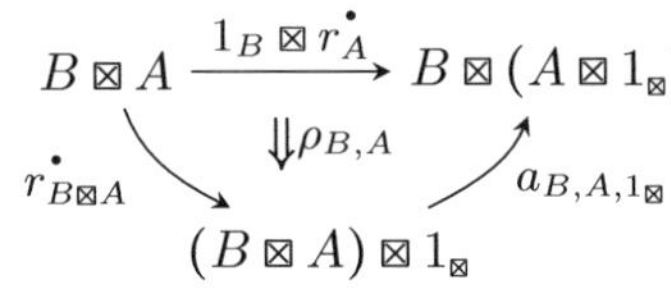

The above data are required to satisfy the following three pasting diagram equalities for objects $A, B, C, D, E \in \mathsf{B}$, with $\boxtimes$ abbreviated to concatenation, and iterates of $\boxtimes$ denoted by parentheses.

The Non-Abelian 4-Cocycle Condition:

(6.4.2)

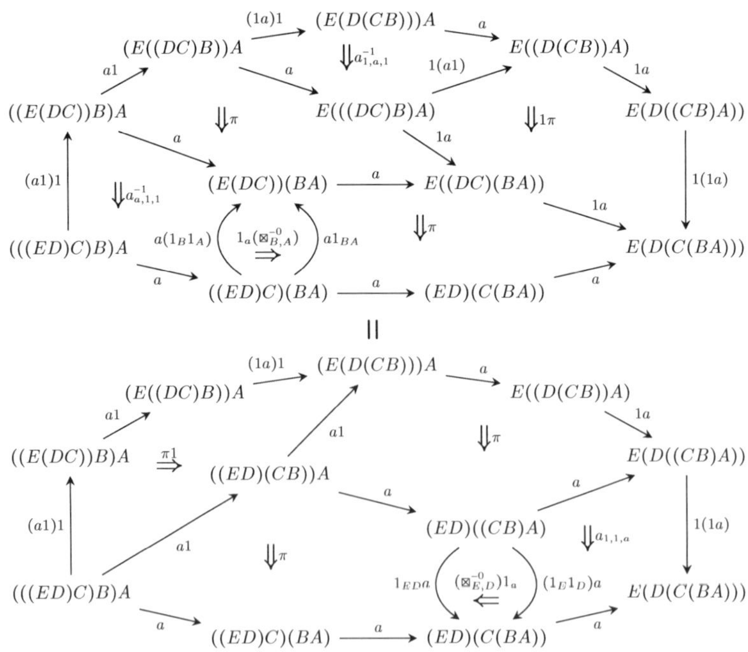

Left Normalization:

(6.4.3)

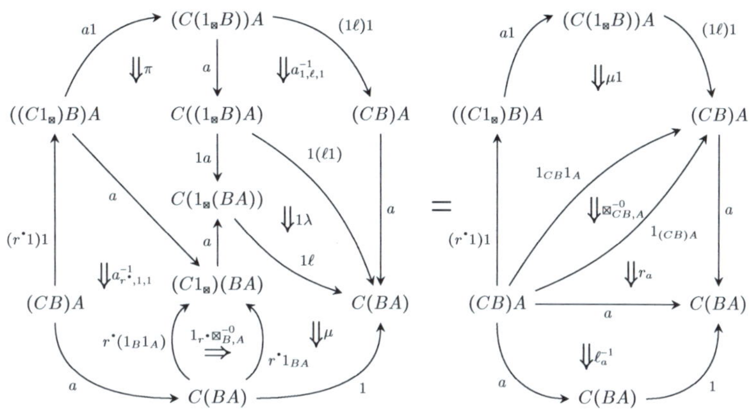

Right Normalization:

(6.4.4)

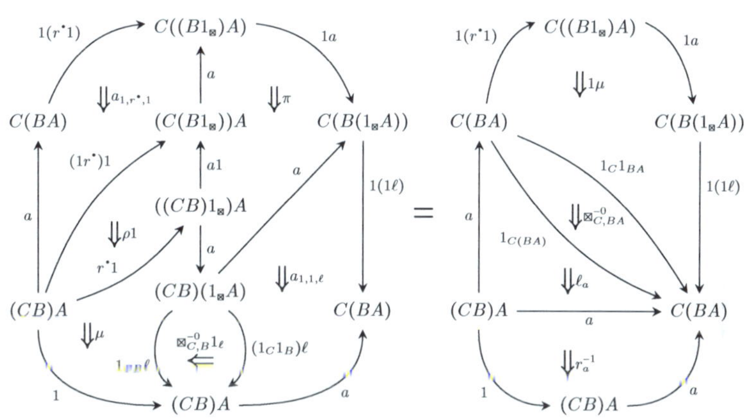

This finishes the definition of a monoidal bicategory. ◇

Remark 6.4.5. The common notation $1_{\boxtimes}\colon \mathbb{1} \longrightarrow \mathsf{B}$ for the identity, which is a pseudofunctor, and for the identity object $1_{\boxtimes} \in \mathsf{B}$ is not to be confused with the identity strong transformation of the pseudofunctor $\boxtimes\colon \mathsf{B} \times \mathsf{B} \longrightarrow \mathsf{B}$. ◇

Explanation 6.4.6 (Data)**.** The monoidal associator and the monoidal unitors in Definition 6.4.1 refer to $\mathsf{Bicat}^{\mathsf{ps}}(\mathsf{B}^n, \mathsf{B})$ in Theorem 6.3.7, which assumes that B has a set of objects. This reference to the bicategory $\mathsf{Bicat}^{\mathsf{ps}}(\mathsf{B}^n, \mathsf{B})$ is simply a matter of convenience to enable us to use the concept of an adjoint equivalence in Definition 6.3.9. For the definition of a monoidal bicategory, this smallness assumption is actually unnecessary.

Indeed, recall

- the composition of pseudofunctors in Definition 6.2.6;
- the identity strong transformation in Lemma 6.2.17;
- the horizontal composition of strong transformations in Definition 6.2.20; and
- the two types of compositions of modifications in Definition 6.3.3.

Using these concepts, in the definition of the monoidal associator $(a, a^{\bullet}, \eta^{a}, \varepsilon^{a})$, we can equivalently require the following conditions.

(i) a and $a^{\bullet}$ are strong transformations as follows.

$$\boxtimes(\boxtimes \times 1) \underset{a^{\bullet}}{\overset{a}{\rightleftarrows}} \boxtimes(1 \times \boxtimes)$$

(ii) η^{a} and ε^{a} are invertible modifications as follows.

$$1_{\boxtimes(\boxtimes \times 1)} \xrightarrow{\eta^{a}} a^{\bullet}a \qquad aa^{\bullet} \xrightarrow{\varepsilon^{a}} 1_{\boxtimes(1 \times \boxtimes)}$$

(iii) The triangle identities (6.3.10) hold. In those two diagrams, the modifications denoted by a, ℓ, and r are defined componentwise in B using, respectively, its associator, left unitor, and right unitor.

Similar remarks apply to the monoidal unitors $(\ell, \ell^{\bullet}, \eta^{\ell}, \varepsilon^{\ell})$ and $(r, r^{\bullet}, \eta^{r}, \varepsilon^{r})$. In summary, with the interpretation above, in the definition of a monoidal bicategory, it is *not* necessary to assume that B has a set of objects. ◇

Explanation 6.4.7 (Axioms). Consider the axioms in Definition 6.4.1.

- The non-abelian 4-cocycle condition (6.4.2) will be abbreviated to NB4 from now on.
- $a_{a,1,1}$, $a_{1,a,1}$, and $a_{1,1,a}$ in NB4 are component 2-cells of the monoidal associator a, which is a strong transformation.
- $\boxtimes^{-0}_{B,A}$ is the inverse of the (B, A) component of $\boxtimes^{0}$.
- In the left normalization axiom (6.4.3) and the right normalization axiom (6.4.4), ℓ_a and r_a are components of, respectively, the left unitor and the right unitor in the base bicategory B.
- The 2-cell 1π in the top side of NB4 is *not* $1 \boxtimes \pi$. To interpret it correctly, first note that π has a component 2-cell

$$\big[(1_D \boxtimes a_{C,B,A})a_{D,C\boxtimes B,A}\big](a_{D,C,B} \boxtimes 1_A) \xrightarrow{\pi_{D,C,B,A}} a_{D,C,B\boxtimes A}a_{D\boxtimes C,B,A}$$

in $\mathsf{B}\big(((DC)B)A, D(C(BA))\big)$. Then 1π is defined as the following vertical composite 2-cell in $\mathsf{B}\big(E(((DC)B)A), E(D(C(BA)))\big)$, with $\boxtimes^{-2}$ the inverse of $\boxtimes^2$.

$$
\begin{array}{ll}
\ulcorner & \big[\big(1_E \boxtimes (1_D \boxtimes a_{C,B,A})\big)\big(1_E \boxtimes a_{D,C\boxtimes B,A}\big)\big]\big[1_E \boxtimes (a_{D,C,B} \boxtimes 1_A)\big] \\
\Big| & \Big\downarrow \boxtimes^2 * 1 \\
\Big| & \big[(1_E 1_E) \boxtimes \big((1_D \boxtimes a_{C,B,A}) a_{D,C\boxtimes B,A}\big)\big]\big[1_E \boxtimes (a_{D,C,B} \boxtimes 1_A)\big] \\
\Big| & \Big\downarrow \boxtimes^2 \\
\Big| & \big[(1_E 1_E) 1_E\big] \boxtimes \big[\big((1_D \boxtimes a_{C,B,A}) a_{D,C\boxtimes B,A}\big)(a_{D,C,B} \boxtimes 1_A)\big] \\
\Big| & \Big\downarrow (\ell_{1_E} * 1) \boxtimes \pi_{D,C,B,A} \\
\Big| 1\pi & (1_E 1_E) \boxtimes \big(a_{D,C,B\boxtimes A} a_{D\boxtimes C,B,A}\big) \\
\Big| & \Big\downarrow \boxtimes^{-2} \\
\llcorner\!\longrightarrow & \big(1_E \boxtimes a_{D,C,B\boxtimes A}\big)\big(1_E \boxtimes a_{D\boxtimes C,B,A}\big)
\end{array}
\tag{6.4.8}
$$

- The 2-cells
  - $\pi 1$ in the bottom side of NB4 (6.4.2),
  - 1λ and $\mu 1$ in the left normalization axiom (6.4.3), and
  - $\rho 1$ and 1μ in the right normalization axiom (6.4.4)

  are interpreted in a similar way to (6.4.8). ◇

6.5. Symmetric Monoidal Bicategories

In this section, we define braided, sylleptic, and symmetric analogues of monoidal bicategories.

- Braided monoidal bicategories are in Definition 6.5.3.
- Sylleptic monoidal bicategories are in Definition 6.5.7.
- Symmetric monoidal bicategories are in Definition 6.5.9.

Motivation 6.5.1. A braided monoidal category in Definition II.1.3.15 has an underlying monoidal category and a natural isomorphism $\xi \colon X \otimes Y \longrightarrow Y \otimes X$, which is called the braiding, that satisfies two hexagon axioms. In a similar manner, a braided monoidal bicategory has

- an underlying monoidal bicategory;
- an adjoint equivalence β, which is called the braiding, that relates the monoidal composition $\boxtimes$ and its opposite; and
- two invertible modifications, which are called the left and the right hexagonators, that fill the diagrams in the hexagon axioms with invertible 2-cells.

Moreover, there are four coherence axioms for the braiding and the hexagonators.

In the 1-categorical setting, a symmetric monoidal category is a braided monoidal category whose braiding satisfies the symmetry axiom, $\xi_{Y,X}\xi_{X,Y} = 1$. In the monoidal bicategorical setting, the symmetry axiom is replaced by an invertible modification ν, which is called the syllepsis, that satisfies two coherence axioms. This intermediate structure is called a sylleptic monoidal bicategory. A symmetric monoidal bicategory is a sylleptic monoidal bicategory whose syllepsis satisfies an additional axiom. ◇

The definition of a braided monoidal bicategory requires the theory of *mates* discussed in detail in [**JY21**, Section 6.1]. Briefly, a mate of a 2-cell θ is obtained by replacing some of the boundary 1-cells in θ by their adjoints as in Definition 6.3.9, by pasting with the (co)units, their inverses, and the left and the right unitors. The following preliminary observation contains examples of mates of the pentagonator π that are used in the definition of a braided monoidal bicategory. The symbols $\boxtimes$ for the monoidal composition among the objects are omitted to save space.

Lemma 6.5.2. *In a monoidal bicategory* B*, the pentagonator* π *induces invertible 2-cells* π_n *for* $1 \leq n \leq 10$*, with component 2-cells as displayed below.*

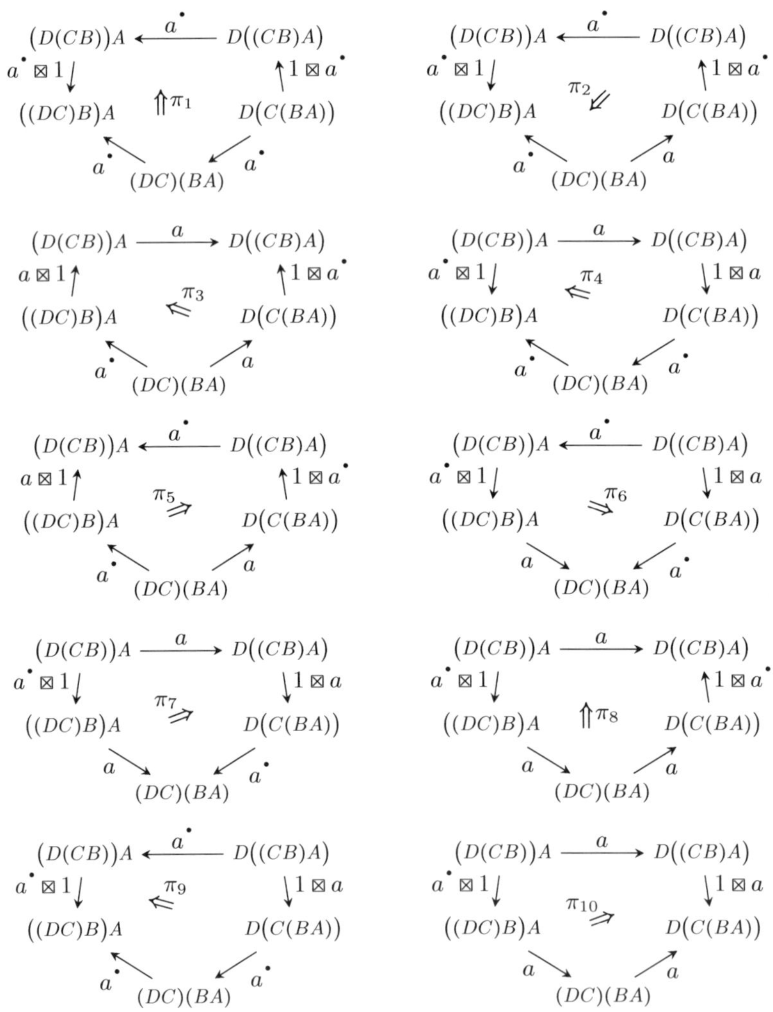

Definition 6.5.3. A *braided monoidal bicategory* is a quadruple

$$(\mathsf{B}, \beta, R_{-|--}, R_{--|-})$$

consisting of the following data.

(1) B is a monoidal bicategory $\big(\mathsf{B}, \boxtimes, 1_{\boxtimes}, a, \ell, r, \pi, \mu, \lambda, \rho\big)$ as in Definition 6.4.1.

(2) $(\beta, \beta^{\bullet}, \eta^{\beta}, \varepsilon^{\beta})$ is an adjoint equivalence as in

$$\begin{array}{ccccc} \mathsf{B}^2 & & \xrightarrow{\quad\boxtimes\quad} & & \mathsf{B} \\ & \underset{\tau}{\searrow} & \Downarrow\beta & \underset{\boxtimes}{\nearrow} & \\ & & \mathsf{B}^2 & & \end{array}$$

in $\mathsf{Bicat}^{\mathsf{ps}}(\mathsf{B}^2, \mathsf{B})$, which is called the *braiding*, in which τ switches the two arguments. For objects $A, B \in \mathsf{B}$, the component 1-cells of β and $\beta^{\bullet}$ are

$$A \boxtimes B \underset{\beta^{\bullet}_{A,B}}{\overset{\beta_{A,B}}{\rightleftarrows}} B \boxtimes A.$$

(3) $R_{-|--}$ is an invertible modification, which is called the *left hexagonator*. Its components are invertible 2-cells

$$\begin{array}{ccc} (B \boxtimes A) \boxtimes C & \xrightarrow{a_{B,A,C}} & B \boxtimes (A \boxtimes C) \\ {\scriptstyle \beta_{A,B} \boxtimes 1_C} \uparrow & & \downarrow {\scriptstyle 1_B \boxtimes \beta_{A,C}} \\ (A \boxtimes B) \boxtimes C & \Downarrow R_{A|B,C} & B \boxtimes (C \boxtimes A) \\ {\scriptstyle a_{A,B,C}} \downarrow & & \uparrow {\scriptstyle a_{B,C,A}} \\ A \boxtimes (B \boxtimes C) & \xrightarrow{\beta_{A,B\boxtimes C}} & (B \boxtimes C) \boxtimes A \end{array}$$

in $\mathsf{B}\big((A \boxtimes B) \boxtimes C, B \boxtimes (C \boxtimes A)\big)$ with the left normalized bracketings in the (co)domain, for objects $A, B, C \in \mathsf{B}$.

(4) $R_{--|-}$ is an invertible modification, which is called the *right hexagonator*. Its components are invertible 2-cells

$$\begin{array}{ccc} A \boxtimes (C \boxtimes B) & \xrightarrow{a^{\bullet}_{A,C,B}} & (A \boxtimes C) \boxtimes B \\ {\scriptstyle 1_A \boxtimes \beta_{B,C}} \uparrow & & \downarrow {\scriptstyle \beta_{A,C} \boxtimes 1_B} \\ A \boxtimes (B \boxtimes C) & \Downarrow R_{A,B|C} & (C \boxtimes A) \boxtimes B \\ {\scriptstyle a^{\bullet}_{A,B,C}} \downarrow & & \uparrow {\scriptstyle a^{\bullet}_{C,A,B}} \\ (A \boxtimes B) \boxtimes C & \xrightarrow{\beta_{A\boxtimes B,C}} & C \boxtimes (A \boxtimes B) \end{array}$$

in $\mathsf{B}\big(A \boxtimes (B \boxtimes C), (C \boxtimes A) \boxtimes B\big)$ with the left normalized bracketings in the (co)domain.

The above data are required to satisfy the following four pasting diagram equalities for objects $A, B, C, D \in \mathsf{B}$, with $\boxtimes$ abbreviated to concatenation, and iterates of $\boxtimes$ denoted by parentheses. The 2-cells π_n in Lemma 6.5.2 are used in the axioms below.

The (3,1)-Crossing Axiom:

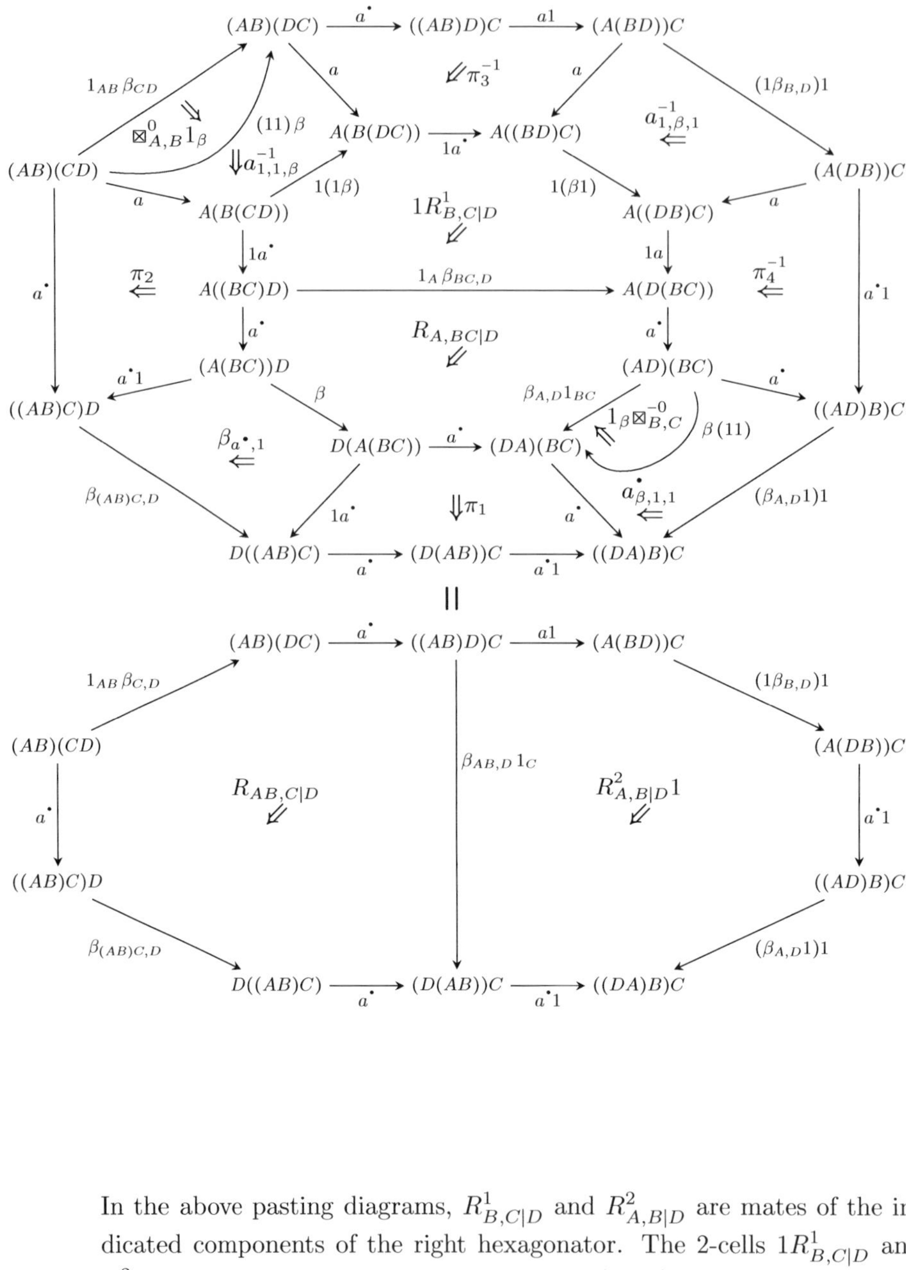

In the above pasting diagrams, $R^1_{B,C|D}$ and $R^2_{A,B|D}$ are mates of the indicated components of the right hexagonator. The 2-cells $1R^1_{B,C|D}$ and $R^2_{A,B|D}1$ are interpreted in a similar way to (6.4.8).

The (1,3)-Crossing Axiom:

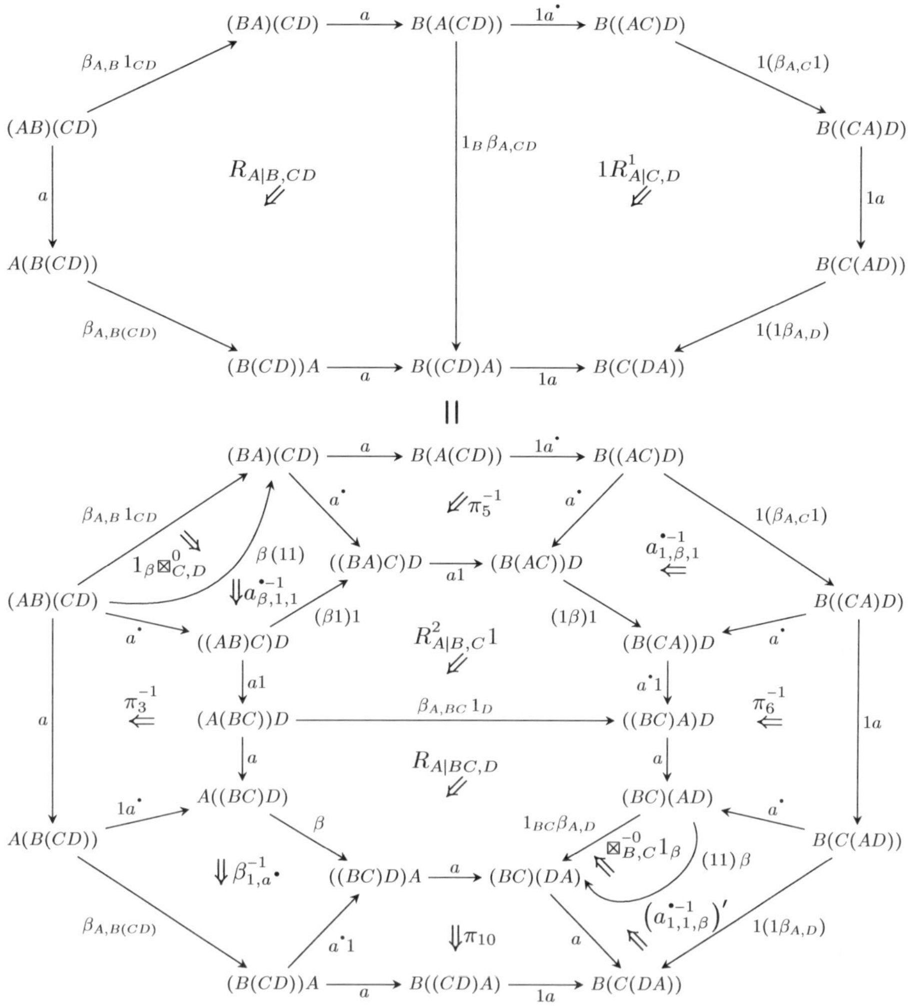

In the above pasting diagrams, $R^1_{A|C,D}$ and $R^2_{A|B,C}$ are mates of the indicated components of the left hexagonator. The 2-cells $1R^1_{A|C,D}$ and $R^2_{A|B,C}1$ are interpreted in a similar way to (6.4.8). The 2-cell $\left(a^{\bullet -1}_{1,1,\beta}\right)'$ is a mate of $a^{\bullet -1}_{1,1,\beta}$.

The (2,2)-Crossing Axiom:

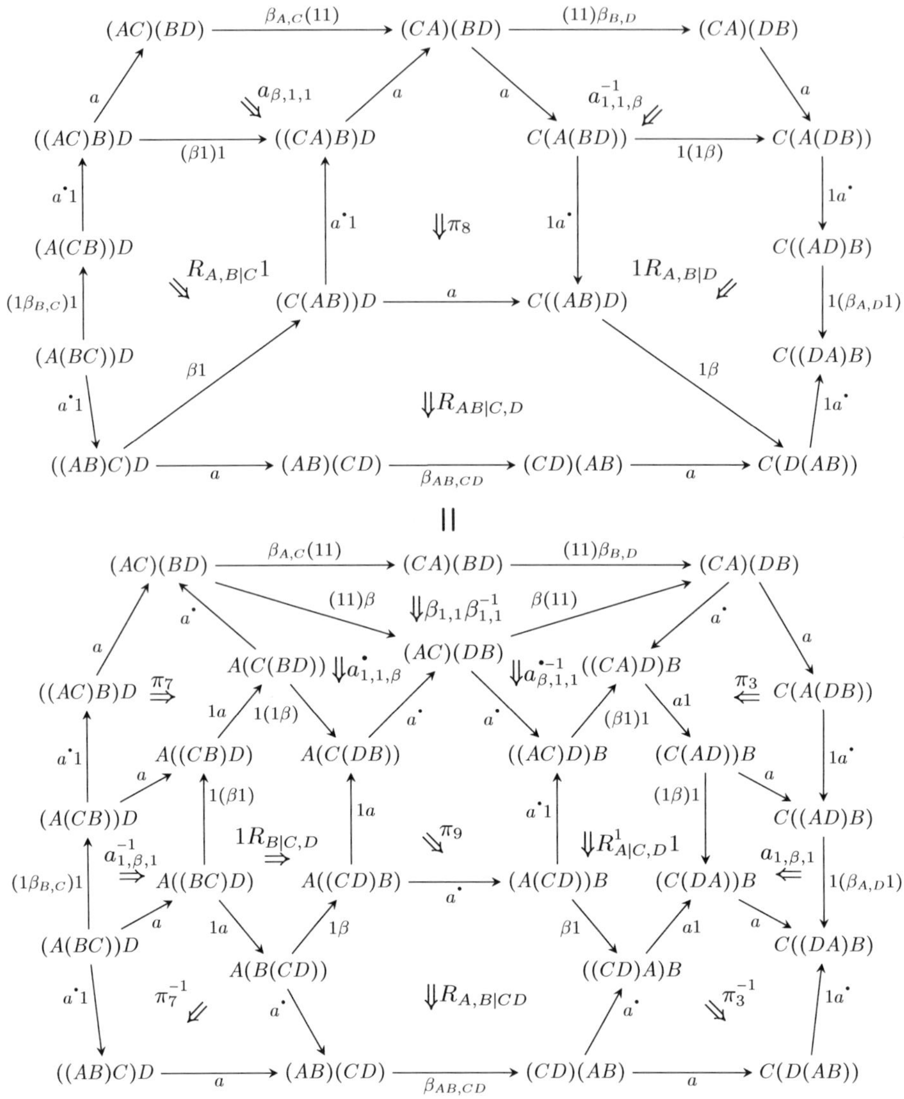

In the above pasting diagrams, $R^1_{A|C,D}$ is a mate of the indicated component of the left hexagonator. The 2-cells $R_{A,B|C}1$, $1R_{A,B|D}$, $1R_{B|C,D}$, and $R^1_{A|C,D}1$ are interpreted in a similar way to (6.4.8).

The Yang-Baxter Axiom:

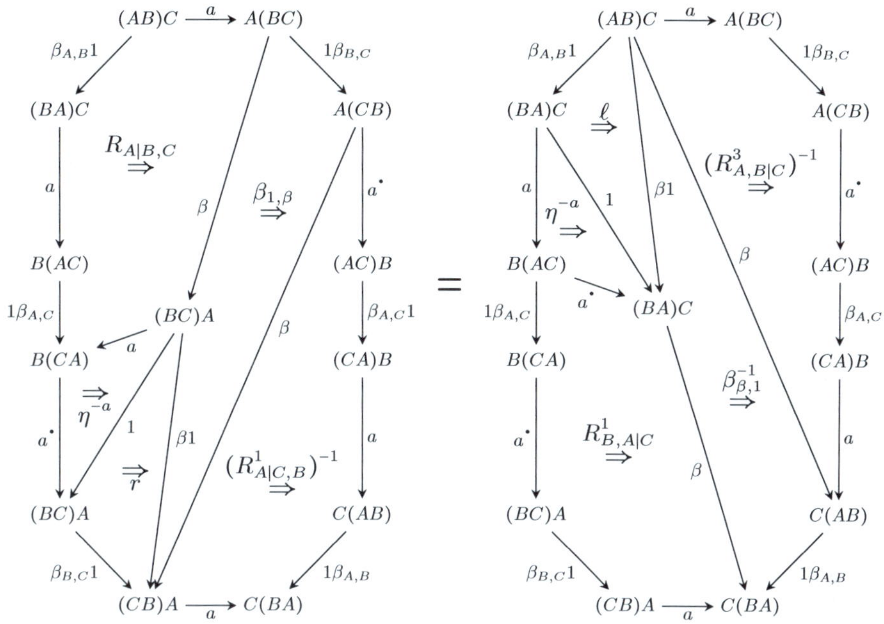

- ℓ and r are, respectively, the left and the right unitors in the base bicategory B.
- $\eta^a : 1 \longrightarrow a^\bullet a$ is the unit of the adjoint equivalence $(a, a^\bullet, \eta^a, \varepsilon^a)$, whose inverse is denoted by η^{-a}.
- $R^1_{A|C,B}$ is a mate of the indicated component of the left hexagonator.
- $R^1_{B,A|C}$ and $R^3_{A,B|C}$ are mates of the indicated components of the right hexagonator.

This finishes the definition of a braided monoidal bicategory. ◇

Explanation 6.5.4 (Mates). In the (3,1)-crossing axiom, $R^1_{B,C|D}$ is the composite of the following pasting diagram in B, with all the ⊠ symbols omitted to save space.

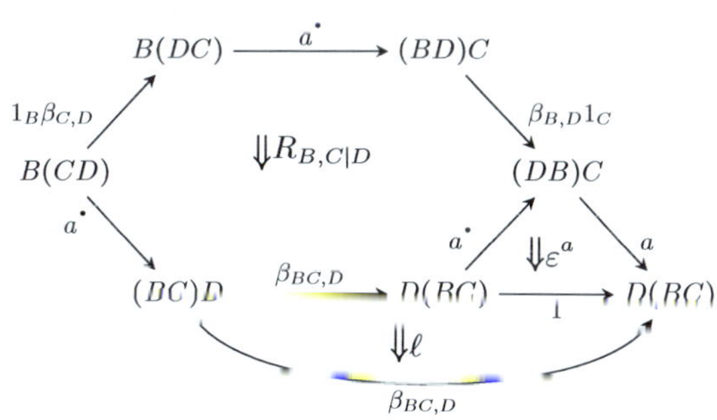

Other mates of components of the left or the right hexagonators are defined in a similar way. ◇

Explanation 6.5.5 (Visualization)**.** Consider Definition 6.5.3.

The Braiding: The braiding β may be visualized as the generating braid  in the braid group B_2. However, in this case, the braiding does not admit a strict inverse. Instead, it is the left adjoint of an adjoint equivalence with right adjoint $\beta^{\bullet}$.

The Hexagonators: The left hexagonator $R_{-|--}$ may be visualized as the braid  with the last two strings crossing over the first string. The domain of $R_{-|--}$ corresponds to first crossing the second string over the first string, followed by crossing the last string over the first string. The codomain corresponds to crossing the last two strings over the first string in one step.

The right hexagonator $R_{--|-}$ admits a similar interpretation using the braid . The domain corresponds to crossing one string over two strings to its left, one string at a time. The codomain corresponds to crossing one string over two strings in one step.

The Crossing Axioms: The (3,1)-crossing axiom may be visualized using the left-most picture below.

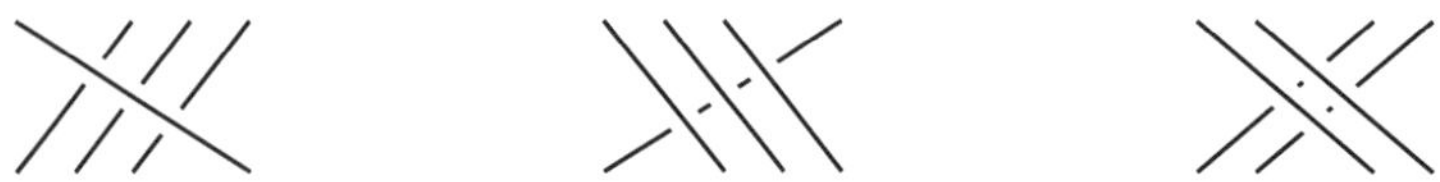

(3,1)-crossing (1,3)-crossing (2,2)-crossing

The common domain of the two pasting diagrams in the (3,1)-crossing axiom corresponds to crossing one string over three strings, one string at a time. The common codomain corresponds to crossing one string over three strings in one step. The two pasting diagrams correspond to two ways to transform from the common domain to the common codomain using the structures in a braided monoidal bicategory. The (1,3)-crossing axiom and the (2,2)-crossing axiom admit similar interpretations, using the middle and the right pictures above.

The Yang-Baxter Axiom: The Yang-Baxter axiom may be visualized using the following pictures.

domain codomain

The common domain of the two pasting diagrams in the Yang-Baxter axiom corresponds to the left picture above, while the common codomain corresponds to the right picture. The two pasting diagrams correspond to two ways to transform from the domain to the codomain using the structures in a braided monoidal bicategory. ◇

Explanation 6.5.6 (Braiding and Size)**.** Similar to Explanation 6.4.6, the assumption that the tuple $(\beta, \beta^{\bullet}, \eta^{\beta}, \varepsilon^{\beta})$ be an adjoint equivalence is equivalent to the following three statements.

(i) β and $\beta^\bullet$ are strong transformations as follows.

$$\boxtimes \underset{\beta^\bullet}{\overset{\beta}{\rightleftarrows}} \boxtimes\tau$$

(ii) η^β and ε^β are invertible modifications as follows.

$$1_\boxtimes \xrightarrow{\eta^\beta} \beta^\bullet\beta$$
$$\beta\beta^\bullet \xrightarrow{\varepsilon^\beta} 1_{\boxtimes\tau}$$

(iii) The triangle identities (6.3.10) hold. In those two diagrams, the modifications denoted by a, ℓ, and r are defined componentwise in B using, respectively, its associator, left unitor, and right unitor.

In particular, in the definition of a braided monoidal bicategory, it is *not* necessary to assume that B has a set of objects. ◇

Definition 6.5.7. A *sylleptic monoidal bicategory* is a quintuple

$$(\mathsf{B}, \beta, R_{-|--}, R_{--|-}, \nu)$$

consisting of the following data.

(1) $(\mathsf{B}, \beta, R_{-|--}, R_{--|-})$ is a braided monoidal bicategory.
(2) ν is an invertible modification, which is called the *syllepsis*. Its components are invertible 2-cells

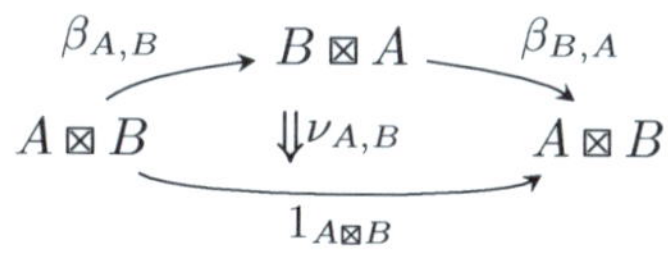

in $\mathsf{B}(A \boxtimes B, A \boxtimes B)$.

The following two pasting diagram equalities are required to hold for objects $A, B, C \in \mathsf{B}$, with the same conventions as in the axioms in Definition 6.5.3.

The (2,1)-Syllepsis Axiom:

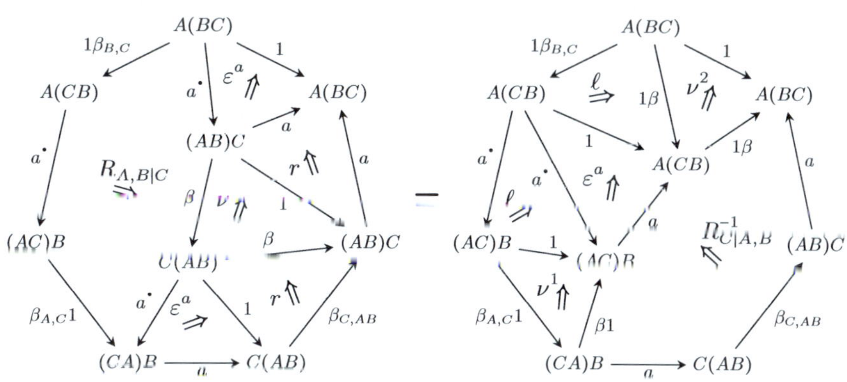

The (1,2)-Syllepsis Axiom:

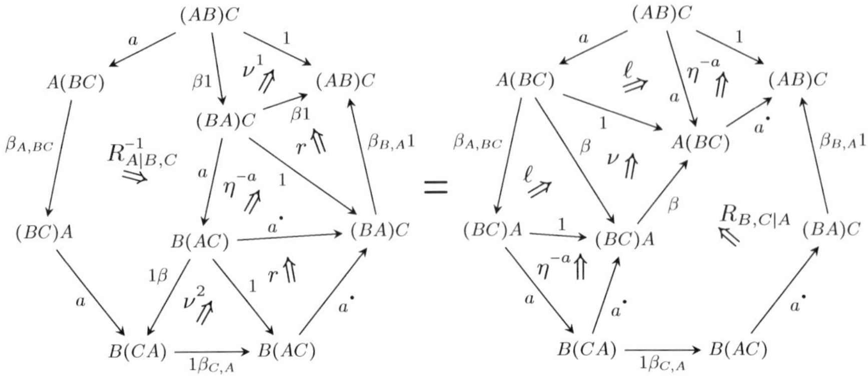

The 2-cells ν^1 and ν^2 are induced by the syllepsis ν, and are interpreted in a similar way to (6.4.8) using $\boxtimes^2$, $\boxtimes^{-0}$, and ℓ. This finishes the definition of a sylleptic monoidal bicategory. ◇

Explanation 6.5.8 (Visualization)**.** Consider Definition 6.5.7.

The Braiding: Due to the existence of the syllepsis, the braiding β in a sylleptic monoidal bicategory may be visualized as the virtual crossing $\times$.

The Syllepsis: It is the isomorphism ⧖ $\cong$ || that straightens its domain to the identity.

The Axioms: In the (2,1)-syllepsis axiom, the common domain is the left picture below

(2,1)-syllepsis (1,2)-syllepsis

with common codomain the identity |||. The two pasting diagrams correspond to two ways to transform from the domain to the codomain using the structures in a sylleptic monoidal bicategory. The (1,2)-syllepsis axiom admits a similar interpretation with common domain the right picture above. ◇

Definition 6.5.9. A *symmetric monoidal bicategory* is a sylleptic monoidal bicategory as in Definition 6.5.7 that satisfies the *triple braid axiom*

(6.5.10)

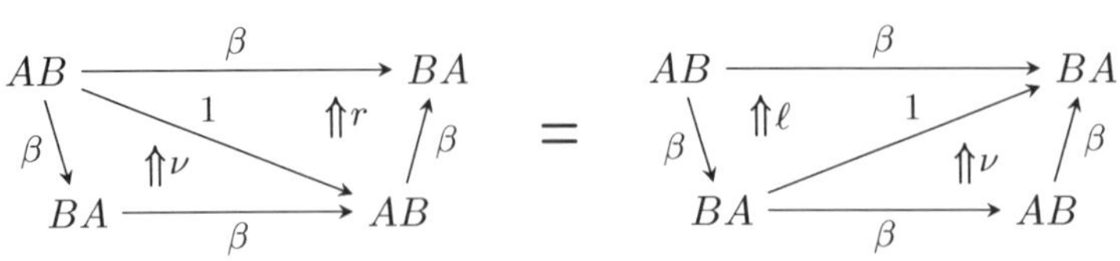

for objects A and B in B. ◇

Explanation 6.5.11 (Visualization). The triple braid axiom (6.5.10) may be visualized as the commutativity of the following diagram.

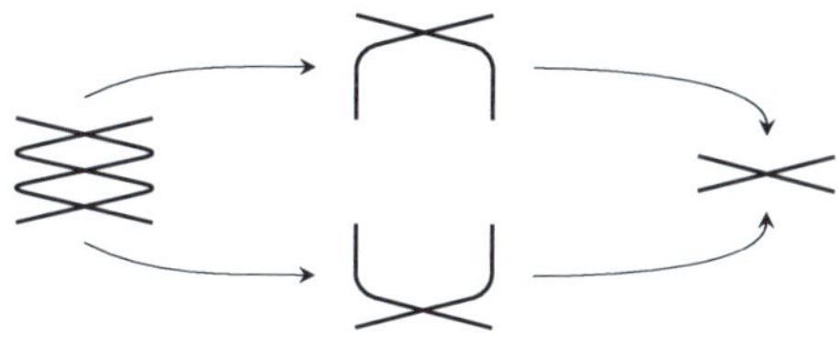

In other words, given three consecutive virtual crossings, straightening the first two virtual crossings is the same as straightening the last two virtual crossings. ◇

6.6. The Gray Tensor Product

In this section, we define the Gray tensor product for 2-categories and the corresponding Gray monoids. The Gray tensor product is a weakening of the Cartesian product for 2-categories. The 1-category 2Cat of small 2-categories and 2-functors becomes a symmetric monoidal closed category with the Gray tensor product, and with the internal hom given by strict functors, strong transformations, and modifications of 2-categories. From now on, whenever a 2-category is regarded as a Cat-category via Proposition 6.1.11, we automatically assume that it is locally small without explicitly stating it. Similarly, whenever a 2-category is regarded as an object in the 1-category 2Cat, it is automatically assumed to be small. Whenever necessary, we use Grothendieck's *Axiom of Universes*—that every set belongs to some universe—to move to a bigger universe. More discussion of universes may be found in [**JY21**, Section 1.1].

We begin by defining a simpler product that will be used in the definition of the Gray tensor product.

The Box Product.

Definition 6.6.1. Suppose C and D are 2-categories. The *box product* $\mathsf{C} \,\square\, \mathsf{D}$ is the Cat-enriched pushout induced by the inclusions $\mathsf{Ob}\,\mathsf{C} \longrightarrow \mathsf{C}$ and $\mathsf{Ob}\,\mathsf{D} \longrightarrow \mathsf{D}$ in the following diagram.

$$\begin{array}{ccc} \mathsf{Ob}\,\mathsf{C} \times \mathsf{Ob}\,\mathsf{D} & \longrightarrow & \mathsf{C} \times (\mathsf{Ob}\,\mathsf{D}) \\ \downarrow & & \downarrow \\ (\mathsf{Ob}\,\mathsf{C}) \times \mathsf{D} & \longrightarrow & \mathsf{C} \,\square\, \mathsf{D} \end{array} \tag{6.6.2}$$

This finishes the definition of the box product. ◇

Explanation 6.6.3.

(1) The 1-cells of $\mathsf{C} \times \mathsf{Ob}\,\mathsf{D}$ are given by $(f, 1_Y)$ for a 1-cell $f \in \mathsf{C}(X, X')$ and an object $Y \in \mathsf{D}$, and the 2-cells of $\mathsf{C} \times \mathsf{Ob}\,\mathsf{D}$ are given by $(\alpha, 1_{1_Y})$ for a 2-cell $\alpha \in \mathsf{C}(X, X')(f_1, f_2)$ and an object $Y \in \mathsf{D}$. We denote their images in $\mathsf{C} \,\square\, \mathsf{D}$ as $f \,\square\, Y$ and $\alpha \,\square\, Y$, respectively, and do likewise for 1-cells and 2-cells in $(\mathsf{Ob}\,\mathsf{C}) \times \mathsf{D}$.

(2) Unpacking Definition 6.6.1, we can describe $\mathsf{C} \,\square\, \mathsf{D}$ as follows.

Objects: The objects are pairs (X, Y), written $X \,\square\, Y$, with $X \in \mathsf{C}$ and $Y \in \mathsf{D}$.

1-Cells: The 1-cells are generated under composition by pairs consisting of a 1-cell and an object, which are called *basic 1-cells* and written as

- $f \,\square\, Y : X \,\square\, Y \longrightarrow X' \,\square\, Y$, for $f \in \mathsf{C}(X, X')$ and $Y \in \mathsf{D}$; and
- $X \,\square\, g : X \,\square\, Y \longrightarrow X \,\square\, Y'$, for $g \in \mathsf{D}(Y, Y')$ and $X \in \mathsf{C}$.

Because the arrows in (6.6.2) are 2-functors, these 1-cells are subject to the following conditions.

(i) For $X \in \mathsf{C}$ and $Y \in \mathsf{D}$, we have
$$1_X \,\square\, Y = 1_{X \square Y} = X \,\square\, 1_Y.$$

(ii) For $f \in \mathsf{C}(X, X')$, $f' \in \mathsf{C}(X', X'')$, and $Y \in \mathsf{D}$, we have
$$(f' \,\square\, Y)(f \,\square\, Y) = (f'f) \,\square\, Y.$$

(iii) For $g \in \mathsf{D}(Y, Y')$, $g' \in \mathsf{D}(Y', Y'')$, and $X \in \mathsf{C}$, we have
$$(X \,\square\, g')(X \,\square\, g) = X \,\square\, (g'g).$$

2-Cells: The 2-cells are generated under horizontal and vertical composition by pairs consisting of a 2-cell and an object, which are called *basic 2-cells* and written as

- $\alpha \,\square\, Y : f_1 \,\square\, Y \longrightarrow f_2 \,\square\, Y$ for $\alpha \in \mathsf{C}(X, X')(f_1, f_2)$ and $Y \in \mathsf{D}$; and
- $X \,\square\, \beta : X \,\square\, g_1 \longrightarrow X \,\square\, g_2$ for $\beta \in \mathsf{D}(Y, Y')(g_1, g_2)$ and $X \in \mathsf{C}$.

Because the arrows in (6.6.2) are 2-functors, these 2-cells are subject to the following conditions.

(i) For $f \in \mathsf{C}(X, X')$ and $g \in \mathsf{D}(Y, Y')$, we have
$$1_f \,\square\, Y = 1_{f \square Y} \quad \text{and} \quad X \,\square\, 1_g = 1_{X \square g}.$$

(ii) For $\alpha \in \mathsf{C}(X, X')(f_1, f_2)$, $\alpha' \in \mathsf{C}(X', X'')(f_1', f_2')$, and $Y \in \mathsf{D}$, we have
$$(\alpha' \,\square\, Y) * (\alpha \,\square\, Y) = (\alpha' * \alpha) \,\square\, Y.$$

(iii) For $\beta \in \mathsf{D}(Y, Y')(g_1, g_2)$, $\beta' \in \mathsf{D}(Y', Y'')(g_1', g_2')$, and $X \in \mathsf{C}$, we have
$$(X \,\square\, \beta') * (X \,\square\, \beta) = X \,\square\, (\beta' * \beta).$$

(iv) For $\alpha \in \mathsf{C}(X, X')(f_1, f_2)$ and $\alpha' \in \mathsf{C}(X, X')(f_2, f_3)$, we have
$$(\alpha' \,\square\, Y)(\alpha \,\square\, Y) = (\alpha'\alpha) \,\square\, Y.$$

(v) For $\beta \in \mathsf{D}(Y, Y')(g_1, g_2)$ and $\beta' \in \mathsf{D}(Y, Y')(g_2, g_3)$, we have
$$(X \,\square\, \beta')(X \,\square\, \beta) = X \,\square\, (\beta'\beta).$$

This concludes the unpacking of Definition 6.6.1.

(3) By the universal property of the pushout, there is a 2-functor
$$j : \mathsf{C} \,\square\, \mathsf{D} \longrightarrow \mathsf{C} \times \mathsf{D}$$
that is bijective on objects. It sends a 1-cell $f \,\square\, Y$ to $f \times 1_Y$, a 2-cell $\alpha \,\square\, Y$ to $\alpha \times 1_{1_Y}$, and similarly for $X \,\square\, g$ or $X \,\square\, \beta$. The composites
$$(f \,\square\, Y')(X \,\square\, g) \quad \text{and} \quad (X' \,\square\, g)(f \,\square\, Y)$$
are distinct in $\mathsf{C} \,\square\, \mathsf{D}$, but both are mapped by j to $f \times g$. This observation is a basis for Motivation 6.6.4 below. ◇

Defining the Gray Tensor Product.

Motivation 6.6.4. We now turn to the definition of the Gray tensor product. This will have the same 0-cells and 1-cells as the box product, but additional 2-cells. Recall that the two composites in the square below are unrelated in $\mathsf{C} \square \mathsf{D}$, and their images in $\mathsf{C} \times \mathsf{D}$ are equal.

$$\begin{array}{ccc} X \square Y & \xrightarrow{f \square Y} & X' \square Y \\ {\scriptstyle X \square g}\downarrow & & \downarrow{\scriptstyle X' \square g} \\ X \square Y' & \xrightarrow{f \square Y'} & X' \square Y' \end{array}$$

In the Gray tensor product, the corresponding square is filled by a generally nontrivial isomorphism $\Sigma_{f,g}$. In this way, the Gray tensor product is an intermediary between $\mathsf{C} \square \mathsf{D}$ and $\mathsf{C} \times \mathsf{D}$. ◇

Definition 6.6.5. For two 2-categories C and D, the *Gray tensor product* $\mathsf{C} \circledast \mathsf{D}$ is a 2-category defined as follows. The objects and 1-cells of $\mathsf{C} \circledast \mathsf{D}$ are the same as those of $\mathsf{C} \square \mathsf{D}$, now denoted with $\circledast$ instead of $\square$.

The 2-cells are defined in two stages as follows. The *proto-2-cells* are generated under horizontal composition by

- the basic 2-cells of $\mathsf{C} \square \mathsf{D}$, which are now denoted $\alpha \circledast Y$ and $X \circledast \beta$; and
- the *transition 2-cells*

$$\Sigma_{f,g} \colon (f \circledast Y')(X \circledast g) \longrightarrow (X' \circledast g)(f \circledast Y) \quad \text{and}$$

$$\Sigma_{f,g}^{-1} \colon (X' \circledast g)(f \circledast Y) \longrightarrow (f \circledast Y')(X \circledast g)$$

for each pair of nonidentity 1-cells $f \in \mathsf{C}(X, X')$ and $g \in \mathsf{D}(Y, Y')$. If either f or g is an identity 1-cell, then $\Sigma_{f,g}$ is the respective identity 2-cell.

This horizontal composition is required to be associative and unital, satisfying the relations induced by $\square$, that is,

- $(\alpha' \circledast Y) * (\alpha \circledast Y) = (\alpha' * \alpha) \circledast Y$ and
- $(X \circledast \beta') * (X \circledast \beta) = X \circledast (\beta' * \beta)$

for horizontally composable 2-cells α and α' in C, respectively β and β' in D.

The 2-cells of $\mathsf{C} \circledast \mathsf{D}$ are equivalence classes of vertical composites of proto-2-cells, where the equivalence relation is the smallest one that includes the following.

(1) The vertical composites $\Sigma_{f,g}\Sigma_{f,g}^{-1}$ and $\Sigma_{f,g}^{-1}\Sigma_{f,g}$ are equivalent to the respective identities.
(2) The basic 2-cells from $\mathsf{C} \square \mathsf{D}$ satisfy the vertical composition relations induced by $\square$, namely,
   - $(\alpha_2 \circledast Y)(\alpha_1 \circledast Y) \sim (\alpha_2\alpha_1) \circledast Y$ and
   - $(X \circledast \beta_2)(X \circledast \beta_1) \sim X \circledast (\beta_2\beta_1)$.
(3) For $f \in \mathsf{C}(X, X')$, $f' \in \mathsf{C}(X', X'')$, and $g \in \mathsf{D}(Y, Y')$, we have

$$\big(\Sigma_{f',g} * (1_f \circledast Y)\big)\big((1_{f'} \circledast Y') * \Sigma_{f,g}\big) \sim \Sigma_{f'f,g}.$$

(4) For $g \in \mathsf{D}(Y, Y')$, $g' \in \mathsf{D}(Y', Y'')$, and $f \in \mathsf{C}(X, X')$, we have

$$\big((X' \circledast 1_{g'}) * \Sigma_{f,g}\big)\big(\Sigma_{f,g'} * (X \circledast 1_g)\big) \sim \Sigma_{f,g'g}.$$

(5) For f, f', g, and g' as above, we have

$$\big((X'' \circledast 1_{g'}) * (1_{f'} \circledast Y') * \Sigma_{f,g}\big)\big(\Sigma_{f',g'} * (1_f \circledast Y') * (X \circledast 1_g)\big) \sim$$

$$\big(\Sigma_{f',g'} * (X' \circledast 1_g) * (1_f \circledast Y)\big)\big((1_{f'} \circledast Y'') * (X' \circledast 1_{g'}) * \Sigma_{f,g}\big).$$

(6) For $\alpha \in \mathsf{C}(X,X')(f_1,f_2)$ and $\beta \in \mathsf{D}(Y,Y')(g_1,g_2)$, we have

$$\big((X' \circledast \beta) * (\alpha \circledast Y)\big)\Sigma_{f_1,g_1} \sim \Sigma_{f_2,g_2}\big((\alpha \circledast Y') * (X \circledast \beta)\big).$$

(7) The equivalence relation is closed under vertical composition.

(8) For any horizontally composable proto-2-cells λ and λ', we have

$$(1 * \lambda)(\lambda' * 1) \sim (\lambda' * \lambda) \sim (\lambda' * 1)(1 * \lambda).$$

Each 2-cell Λ is represented by a vertical composite of proto-2-cells, $\lambda_1 \cdots \lambda_n$, and thus the vertical composition of 2-cells is defined by concatenation. The horizontal composition of 2-cells is defined by

$$(\lambda_1'\lambda_2') * (\lambda_1\lambda_2) = (\lambda_1' * \lambda_1)(\lambda_2' * \lambda_2) \tag{6.6.6}$$

for appropriately composable proto-2-cells λ_1, λ_2, λ_1', and λ_2'. This extends to define horizontal composition of general vertical composites

$$(\lambda_1' \cdots \lambda_n') * (\lambda_1 \cdots \lambda_m)$$

with proto-2-cells

- λ_i in $(\mathsf{C} \circledast \mathsf{D})(X \circledast Y, X' \circledast Y')$ and
- λ_i' in $(\mathsf{C} \circledast \mathsf{D})(X' \circledast Y', X'' \circledast Y'')$

by inserting appropriate identity 2-cells so that $m = n$, and then by induction on (6.6.6).

Condition (8) implies that this definition is independent of how identities are inserted and satisfies the middle four exchange property, that is, that the horizontal composition preserves vertical composition of 2-cells. Preservation of units follows from the corresponding properties of $\Box$, together with conditions (3) and (4) with f' and g' being identities. This finishes the definition of the Gray tensor product $\mathsf{C} \circledast \mathsf{D}$. ◇

Explanation 6.6.7 (The Transition 2-Cells)**.** Since $\Sigma_{f,g}$ is an identity 2-cell whenever f or g is an identity 1-cell, conditions (3), (4), and (5) are equivalent to the requirement that all possible composites formed from the following pasting diagram are equal to $\Sigma_{f'f,g'g}$ for all f,f',g, and g'.

(6.6.8)

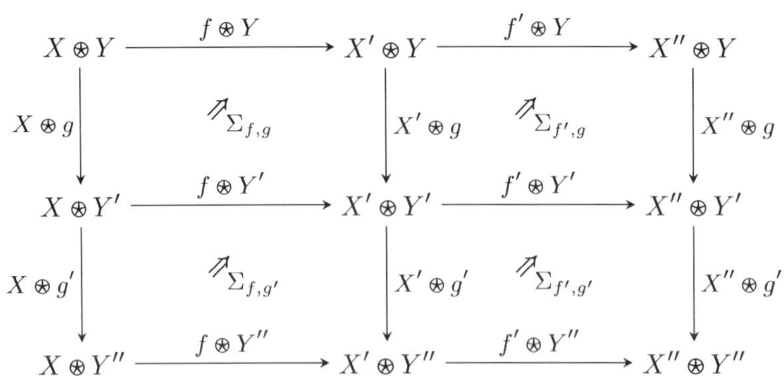

In particular, condition (5) means that the two ways of forming a vertical composite from $\Sigma_{f',g'} * \Sigma_{f,g}$ are equal. Furthermore, condition (6) means that the composites of the following pasting diagrams are equal in $\mathsf{C} \otimes \mathsf{D}$.

(6.6.9)

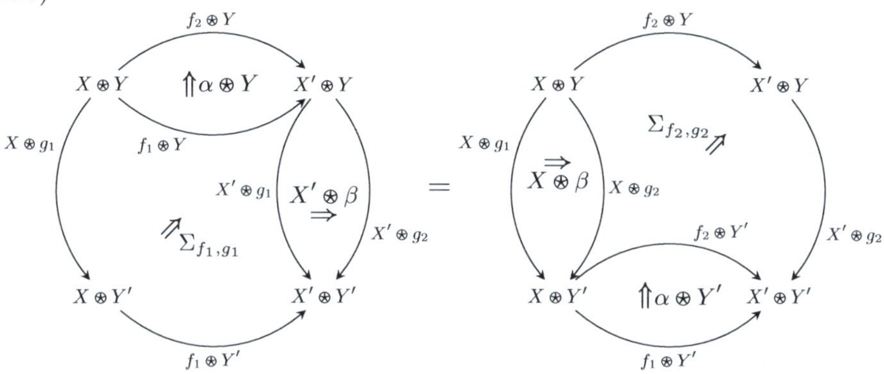

◇

Proposition 6.6.10. *For 2-categories* C *and* D*, the Gray tensor product* $\mathsf{C} \otimes \mathsf{D}$ *is a 2-category.*

Notation 6.6.11. For bicategories C and D, we let $\mathsf{Hom}(\mathsf{C}, \mathsf{D})$ denote the full subbicategory of $\mathsf{Bicat}^{\mathsf{ps}}(\mathsf{C}, \mathsf{D})$ consisting of strict functors, strong transformations, and modifications. Recall from Theorem 6.3.7 that $\mathsf{Hom}(\mathsf{C}, \mathsf{D})$ is a 2-category whenever D is a 2-category. ◇

Recall from Definition 1.2.28 the concept of a symmetric monoidal closed category.

Theorem 6.6.12. *There is a symmetric monoidal closed category*

$$\mathsf{Gray} = \big(\mathsf{2Cat}, \otimes, \mathbb{1}, a, \lambda, \rho, \xi, \mathsf{Hom}\big)$$

with the following data.

- *The underlying category is* $\mathsf{2Cat}$*, the 1-category of small 2-categories and 2-functors.*
- *The monoidal product is the Gray tensor product* $\otimes$*.*
- *The unit object is the terminal 2-category* $\mathbb{1}$*.*
- *The associativity isomorphism* a*, the left unit isomorphism* λ*, and the right unit isomorphism* ρ *are induced by those of the Cartesian product.*
- *For 2-categories* C *and* D*, the* (C, D)*-component*

$$\mathsf{C} \otimes \mathsf{D} \xrightarrow{\xi_{\mathsf{C},\mathsf{D}}} \mathsf{D} \otimes \mathsf{C}$$

of the symmetry isomorphism ξ *is defined on generating cells as follows for objects* $X \in \mathsf{C}$ *and* $Y \in \mathsf{D}$*, 1-cells* $f \in \mathsf{C}$ *and* $g \in \mathsf{D}$*, and 2-cells* $\alpha \in \mathsf{C}$ *and* $\beta \in \mathsf{D}$*.*

$$\xi(X \otimes Y) = Y \otimes X$$
$$\xi(f \otimes Y) = Y \otimes f \qquad \xi(X \otimes g) = g \otimes X$$
$$\xi(\alpha \otimes Y) = Y \otimes \alpha \qquad \xi(X \otimes \beta) = \beta \otimes X$$
$$\xi \Sigma_{f,g} = \Sigma_{g,f}^{-1}$$

- *The internal hom is* Hom *in Notation 6.6.11.*

Gray Monoids. Recall from Definition 1.2.8 that a monoid in a monoidal category consists of an object together with multiplication and unit morphisms satisfying axioms for associativity and unity.

Definition 6.6.13. A *Gray monoid* is a monoid $(\mathsf{C}, \odot, I)$ in Gray. ◇

Explanation 6.6.14 (Monoids in Gray)**.** Rewriting Definition 1.2.8 in this context, a Gray monoid is a triple $(\mathsf{C}, \odot, I)$ consisting of a 2-category C and 2-functors

$$\begin{aligned} \odot \colon \mathsf{C} \otimes \mathsf{C} &\longrightarrow \mathsf{C} \\ I \colon \mathbb{1} &\longrightarrow \mathsf{C} \end{aligned}$$

such that the following diagrams of 2-categories and 2-functors commute.

$$\begin{array}{ccc} (\mathsf{C}\otimes\mathsf{C})\otimes\mathsf{C} & \xrightarrow{a} & \mathsf{C}\otimes(\mathsf{C}\otimes\mathsf{C}) \\ & & \downarrow{\scriptstyle 1\otimes\odot} \\ \downarrow{\scriptstyle \odot\otimes 1} & & \mathsf{C}\otimes\mathsf{C} \\ & & \downarrow{\scriptstyle \odot} \\ \mathsf{C}\otimes\mathsf{C} & \xrightarrow{\odot} & \mathsf{C} \end{array} \qquad \begin{array}{ccccc} \mathbb{1}\otimes\mathsf{C} & \xrightarrow{I\otimes 1} & \mathsf{C}\otimes\mathsf{C} & \xleftarrow{1\otimes I} & \mathsf{C}\otimes\mathbb{1} \\ & {\scriptstyle\lambda}\searrow & \downarrow{\scriptstyle\odot} & \swarrow{\scriptstyle\rho} & \\ & & \mathsf{C} & & \end{array} \tag{6.6.15}$$

◇

Explanation 6.6.16 (Data and Axioms for Gray Monoids)**.** We have an even more explicit list of data and axioms by unpacking the definition of the Gray tensor product $\otimes$. A Gray monoid

$$(\mathsf{C}, \odot, I)$$

consists of a 2-category C together with the following data.

The Unit: A distinguished object I, which is called the *Gray unit.*

Objects: For each pair of objects W and X, an object $W \odot X$.

1-Cells: For each object W and 1-cell $f \colon X \longrightarrow X'$, 1-cells

$$\begin{aligned} W \odot f \colon W \odot X &\longrightarrow W \odot X' \quad \text{and} \\ f \odot W \colon X \odot W &\longrightarrow X' \odot W. \end{aligned}$$

2-Cells: For each object W and 2-cell $\alpha \colon f_1 \longrightarrow f_2$ in $\mathsf{C}(X, X')$, *basic 2-cells*

$$\begin{aligned} W \odot \alpha \colon W \odot f_1 &\longrightarrow W \odot f_2 \quad \text{and} \\ \alpha \odot W \colon f_1 \odot W &\longrightarrow f_2 \odot W. \end{aligned}$$

For each 1-cell $f \colon X \longrightarrow X'$ and 1-cell $g \colon Y \longrightarrow Y'$, a 2-cell isomorphism

$$\Sigma_{f,g} \colon (f \odot Y')(X \odot g) \xrightarrow{\cong} (X' \odot g)(f \odot Y), \tag{6.6.17}$$

which is called a *transition 2-cell.*

These data are subject to the following axioms.

(1) For each object W, the assignments on cells

$$\begin{aligned} W \odot - \colon \mathsf{C} &\longrightarrow \mathsf{C} \quad \text{and} \\ - \odot W \colon \mathsf{C} &\longrightarrow \mathsf{C} \end{aligned}$$

are 2-functors. By Proposition 6.2.5, 2-functoriality of $W \odot -$ is equivalent to the following five equalities for each object X along with appropriately composable 1-cells f and g, and 2-cells α, α', and γ.

$$W \odot 1_X = 1_{W \odot X}$$
$$W \odot 1_f = 1_{W \odot f}$$
$$W \odot (gf) = (W \odot g)(W \odot f)$$
$$W \odot (\alpha\gamma) = (W \odot \alpha)(W \odot \gamma)$$
$$W \odot (\alpha * \alpha') = (W \odot \alpha) * (W \odot \alpha')$$

2-functoriality of $- \odot W$ is equivalent to the following five equalities, provided that the compositions are defined.

$$1_X \odot W = 1_{X \odot W}$$
$$1_f \odot W = 1_{f \odot W}$$
$$(gf) \odot W = (g \odot W)(f \odot W)$$
$$(\alpha\gamma) \odot W = (\alpha \odot W)(\gamma \odot W)$$
$$(\alpha * \alpha') \odot W = (\alpha \odot W) * (\alpha' \odot W)$$

(2) The Gray unit I is strict. That is, for each object X, 1-cell f, and 2-cell α, we have the following equalities.

$$I \odot X = X = X \odot I$$
$$I \odot f = f = f \odot I$$
$$I \odot \alpha = \alpha = \alpha \odot I$$

(3) The product $\odot$ is strictly associative. That is, for objects Z, W, and X, we have

$$(Z \odot W) \odot X = Z \odot (W \odot X).$$

For each 1-cell f and 2-cell α, we have the following equalities.

$$(Z \odot W) \odot f = Z \odot (W \odot f) \qquad (Z \odot W) \odot \alpha = Z \odot (W \odot \alpha)$$
$$(Z \odot f) \odot W = Z \odot (f \odot W) \qquad (Z \odot \alpha) \odot W = Z \odot (\alpha \odot W)$$
$$(f \odot Z) \odot W = f \odot (Z \odot W) \qquad (\alpha \odot Z) \odot W = \alpha \odot (Z \odot W)$$

(4) For 1-cells $f\colon X \longrightarrow X'$, $g\colon Y \longrightarrow Y'$, and $h\colon Z \longrightarrow Z'$, we have the following equalities.

$$\Sigma_{f,g} \odot Z = \Sigma_{f,g \odot Z}$$
$$\Sigma_{f \odot Y,h} = \Sigma_{f,Y \odot h}$$
$$\Sigma_{X \odot g,h} = X \odot \Sigma_{g,h}$$

(5) For $f \in \mathsf{C}(X,X')$, $f' \in \mathsf{C}(X',X'')$, $g \in \mathsf{C}(Y,Y')$, and $g' \in \mathsf{C}(Y',Y'')$, we have the following two equalities of pasting diagrams.

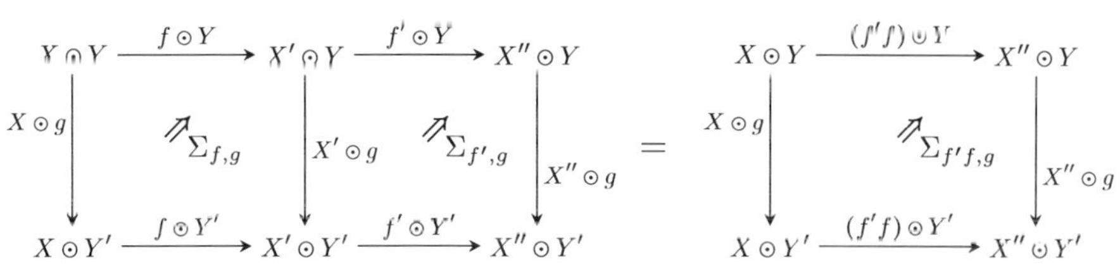

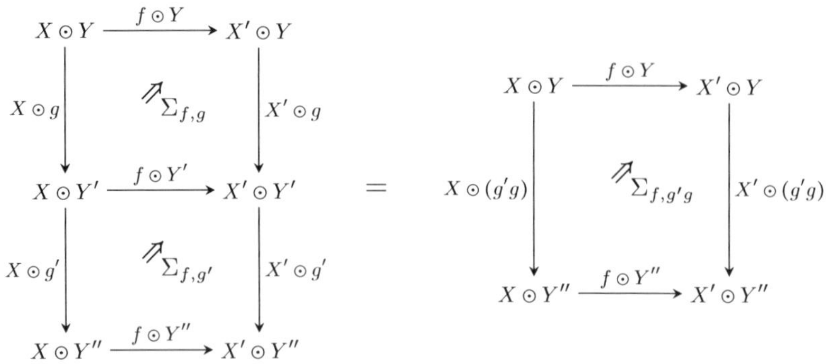

(6) For $\alpha \in \mathsf{C}(X, X')(f_1, f_2)$ and $\beta \in \mathsf{D}(Y, Y')(g_1, g_2)$, we have the following equality of pasting diagrams.

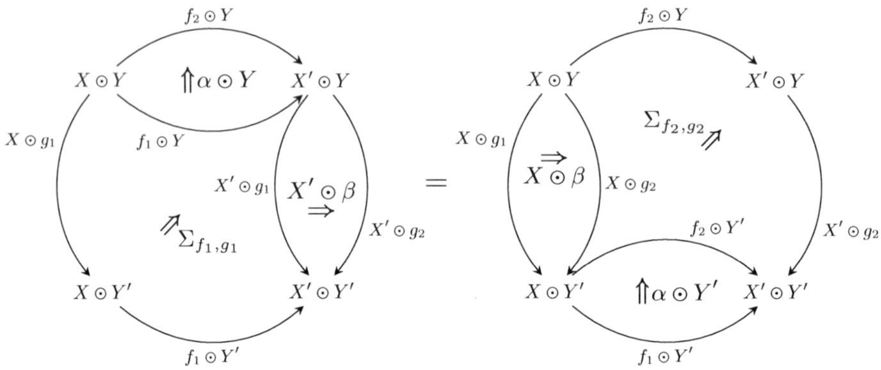

This finishes the list of axioms of a Gray monoid. In particular, condition (5) together with the invertibility of $\Sigma_{f,g}$ implies that $\Sigma_{1,g}$ and $\Sigma_{f,1}$ are identity 2-cells. ◇

6.7. Permutative Gray Monoids and 2-Categories

In this section, we recall from [**GJO17b**, Def. 3.28 and 3.45] the concepts of

- permutative Gray monoids in Definition 6.7.1 and
- permutative 2-categories in Definition 6.7.16.

This section ends with a brief discussion of strictification related to these structures.

Permutative Gray Monoids. Recall from Definition 1.2.18 that a permutative category is a symmetric monoidal category whose underlying monoidal category is strict. Permutative Gray monoids are their analogues in the context of 2-categories with the Gray tensor product $\otimes$.

Definition 6.7.1. A *permutative Gray monoid* is a quadruple

$$(\mathsf{C}, \odot, I, \beta)$$

consisting of the following data.

- $(\mathsf{C}, \odot, I)$ is a Gray monoid as in Definition 6.6.13.
- $\beta \colon \odot \longrightarrow \odot\xi$ is a 2-natural isomorphism, which is called the *Gray symmetry*, as in

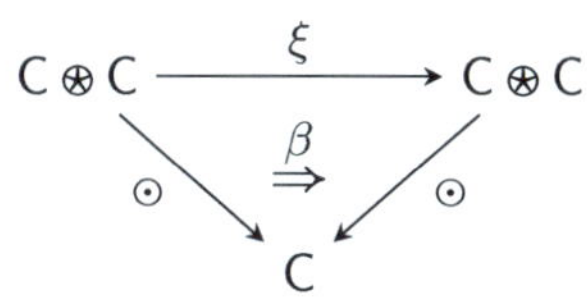

  with ξ the symmetry isomorphism in Gray in Theorem 6.6.12.

These data are subject to the following three pasting diagram axioms in 2Cat, in which each unlabeled region is strictly commutative.

The Symmetry Axiom: The following pasting diagram is equal to $1_{\odot}$.

(6.7.2)

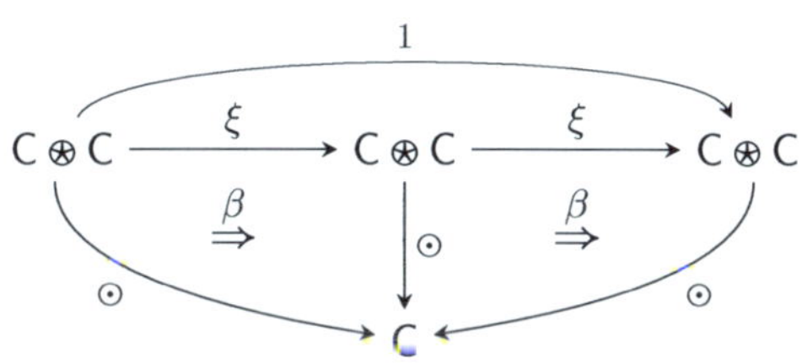

The Unit Axiom: The following pasting diagram is equal to 1_{λ}, with λ and ρ the left and the right unit isomorphisms in Gray.

(6.7.3)

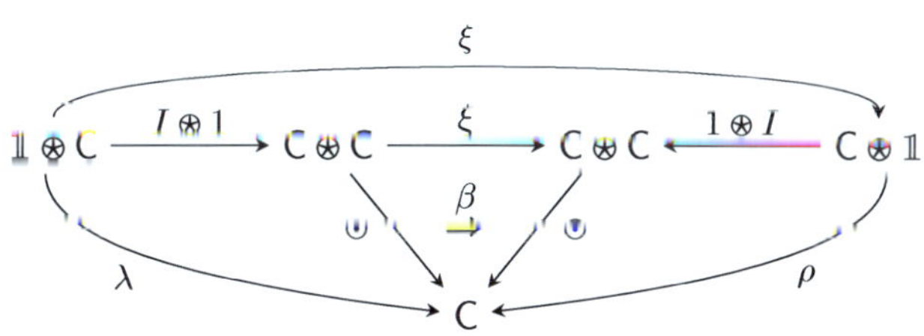

The Hexagon Axiom:

(6.7.4)

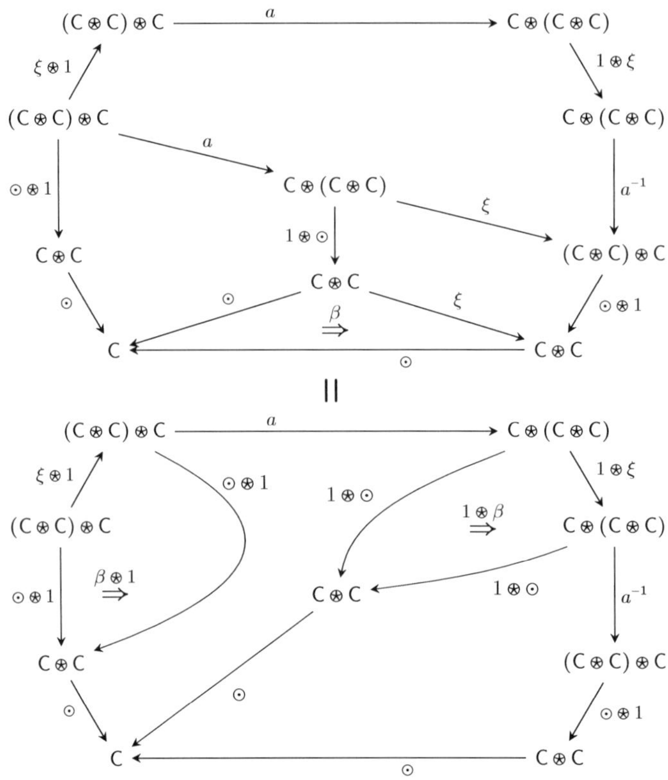

This finishes the definition of a permutative Gray monoid. ◇

Explanation 6.7.5 (Data). The Gray symmetry $\beta\colon \odot \longrightarrow \odot\xi$ in a permutative Gray monoid is a 2-natural isomorphism. By Proposition 6.2.23 and Explanation 6.6.16, it consists of a component 1-cell

$$X \odot Y \xrightarrow{\beta_{X,Y}} Y \odot X \ \in \mathsf{C} \tag{6.7.6}$$

for each pair of objects $X, Y \in \mathsf{C}$ such that the following invertibility and naturality conditions are satisfied.

Invertibility: Each $\beta_{X,Y}$ is invertible. In other words, there exists a 1-cell

$$Y \odot X \xrightarrow{\beta^{-1}_{X,Y}} X \odot Y \ \in \mathsf{C}$$

such that

$$\beta^{-1}_{X,Y}\beta_{X,Y} = 1_{X\odot Y} \quad \text{and} \quad \beta_{X,Y}\beta^{-1}_{X,Y} = 1_{Y\odot X}. \tag{6.7.7}$$

In fact, this inverse is given by

$$\beta^{-1}_{X,Y} = \beta_{Y,X}$$

by the symmetry axiom as interpreted in (6.7.12) below.

1-Cell Naturality: For each object W and 1-cell $f: X \longrightarrow X'$ in C, the following diagram in C is commutative.

(6.7.8)
$$\begin{array}{ccccc} X \odot W & \xrightarrow{\beta_{X,W}} & W \odot X & \xrightarrow{\beta_{W,X}} & X \odot W \\ {\scriptstyle f \odot W}\downarrow & & \downarrow{\scriptstyle W \odot f} & & \downarrow{\scriptstyle f \odot W} \\ X' \odot W & \xrightarrow{\beta_{X',W}} & W \odot X' & \xrightarrow{\beta_{W,X'}} & X' \odot W \end{array}$$

In fact, the commutativity of the right square follows from that of the left square and the symmetry axiom (6.7.12). Therefore, only the left square is necessary.

2-Cell Naturality: There are two types of 2-cell naturality conditions, one for basic 2-cells and one for transition 2-cells.

(1) For each object W and 2-cell $\alpha: f_1 \longrightarrow f_2 \in \mathsf{C}(X, X')$, the following equality holds in $\mathsf{C}(X \odot W, W \odot X')$.

(6.7.9)
$$(W \odot \alpha) * 1_{\beta_{X,W}} = 1_{\beta_{X',W}} * (\alpha \odot W)$$

This is equivalent to the following pasting diagram equality, with the unlabeled subregions commutative by (6.7.8).

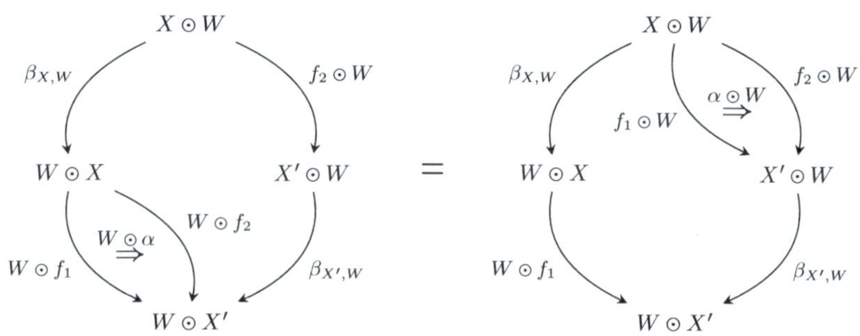

(2) For each pair of 1-cells $f: X \longrightarrow X'$ and $g: Y \longrightarrow Y'$, the following equality holds in $\mathsf{C}(X \odot Y, Y' \odot X')$.

(6.7.10)
$$\Sigma_{g,f}^{-1} * 1_{\beta_{X,Y}} = 1_{\beta_{X',Y'}} * \Sigma_{f,g}.$$

This is equivalent to the following pasting diagram equality, with the unlabeled subregions commutative by (6.7.8).

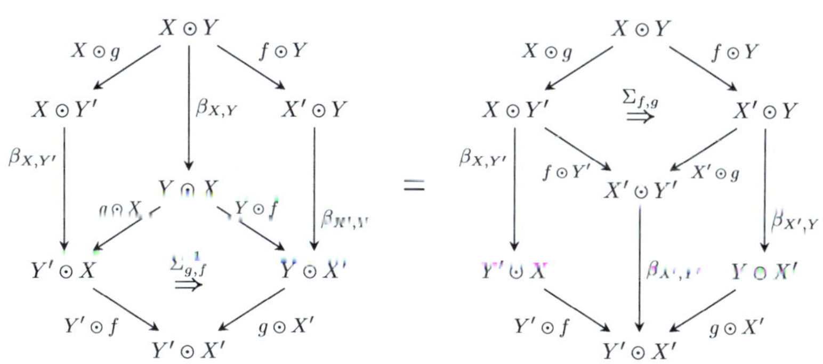

This completes the conditions for $\beta: \odot \longrightarrow \odot\xi$ to be a 2-natural isomorphism. $\diamond$

Explanation 6.7.11 (Axioms)**.** Consider the axioms of a permutative Gray monoid in Definition 6.7.1.

The Symmetry Axiom: In (6.7.2), the equality $\xi\xi = 1$ follows from the symmetry axiom (1.2.20) in Gray. In terms of component 1-cells, the symmetry axiom means the commutativity of the diagram

$$\begin{array}{ccc} X \odot Y & \xrightarrow{1_{X \odot Y}} & X \odot Y \\ & {\scriptstyle \beta_{X,Y}} \searrow \quad \nearrow {\scriptstyle \beta_{Y,X}} & \\ & Y \odot X & \end{array} \tag{6.7.12}$$

for objects $X, Y \in \mathsf{C}$. This is equivalent to the invertibility condition (6.7.7).

The Unit Axiom: Consider (6.7.3).

- The top subregion is commutative by the naturality of ξ.
- The left and the right subregions are commutative by the unity axiom of the Gray monoid $(\mathsf{C}, \odot, I)$, that is, the right diagram in (6.6.15).
- The equality $\rho\xi = \lambda$ follows from the symmetry axiom (1.2.20) and the unit axiom (1.2.21) in Gray.

Since λ and ρ in Gray are induced by those of the Cartesian product, in terms of component 1-cells the unit axiom means the commutativity of the diagram

$$\begin{array}{ccc} X & \xrightarrow{1_X} & X \\ \Vert & & \Vert \\ I \odot X & \xrightarrow{\beta_{I,X}} & X \odot I \end{array} \tag{6.7.13}$$

for each object $X \in \mathsf{C}$, with I the Gray unit of C. It can be shown, using an argument similar to the proof of Proposition II.1.3.21, which is the corresponding fact for braided monoidal categories, that the unit axiom follows from the symmetry axiom and the hexagon axiom.

The Hexagon Axiom: Consider (6.7.4).

- In the top pasting diagram, the top hexagon is commutative by the symmetry axiom (1.2.20) and the hexagon axiom (1.2.22) in Gray. The quadrilateral is commutative by the naturality of ξ.
- The other three unlabeled subregions in (6.7.4) are commutative by the associativity axiom of the Gray monoid $(\mathsf{C}, \odot, I)$, that is, the left diagram in (6.6.15).

In terms of component 1-cells, the hexagon axiom means that the diagram

$$\begin{array}{ccc} (Y \odot X) \odot Z & \xrightarrow{1} & Y \odot (X \odot Z) \\ {\scriptstyle \beta_{X,Y} \odot Z} \nearrow & & \searrow {\scriptstyle Y \odot \beta_{X,Z}} \\ (X \odot Y) \odot Z & & Y \odot (Z \odot X) \\ {\scriptstyle 1} \searrow & & \nearrow {\scriptstyle 1} \\ X \odot (Y \odot Z) & \xrightarrow{\beta_{X, Y \odot Z}} & (Y \odot Z) \odot X \end{array} \tag{6.7.14}$$

is commutative for objects $X, Y, Z \in \mathsf{C}$.

This finishes the explanation of the axioms of a permutative Gray monoid. $\diamond$

Permutative 2-Categories. The 1-category $\mathsf{2Cat}$ has another symmetric monoidal structure other than Gray in Theorem 6.6.12.

Definition 6.7.15. Denote by $(\mathsf{2Cat}, \times)$ the symmetric monoidal category with the following data.

- The underlying category is $\mathsf{2Cat}$.
- The monoidal product is the Cartesian product $\times$.
- The monoidal unit is the terminal 2-category $\mathbb{1}$.
- The associativity isomorphism a is the identity.
- The left and the right unit isomorphisms λ and ρ are defined by dropping the $\mathbb{1}$ argument.
- The symmetry isomorphism ξ is given by swapping the two arguments.

This finishes the definition of $(\mathsf{2Cat}, \times)$. ◇

Definition 6.7.16. A *permutative 2-category* is a quadruple

$$(\mathsf{C}, \boxdot, I, \beta)$$

consisting of the following data.

- $(\mathsf{C}, \boxdot, I)$ is a monoid in $(\mathsf{2Cat}, \times)$ with
  - C a small 2-category,
  - multiplication 2-functor $- \boxdot -: \mathsf{C} \times \mathsf{C} \longrightarrow \mathsf{C}$, and
  - unit 2-functor $I: \mathbb{1} \longrightarrow \mathsf{C}$.
- $\beta: \boxdot \longrightarrow \boxdot\,\xi$ is a 2-natural isomorphism as follows.

$$\begin{array}{ccc} \mathsf{C} \times \mathsf{C} & \xrightarrow{\quad \xi \quad} & \mathsf{C} \times \mathsf{C} \\ & {\scriptstyle\boxdot} \searrow \quad \overset{\beta}{\Rightarrow} \quad \swarrow {\scriptstyle\boxdot} & \\ & \mathsf{C} & \end{array}$$

These data are subject to the same three axioms as for permutative Gray monoids in Definition 6.7.1 with $(\otimes, \odot)$ replaced by $(\times, \boxdot)$. This finishes the definition of a permutative 2-category. ◇

Next is the analogue of Explanations 6.6.14 and 6.6.16 with the Cartesian product instead of the Gray tensor product.

Explanation 6.7.17 (Monoid). A monoid

$$(\mathsf{C}, \boxdot, I)$$

in $(\mathsf{2Cat}, \times)$ consists of a small 2-category C together with the following data.

The Unit: A distinguished object I.

Objects: For each pair of objects W and X, an object $W \boxdot X$.

1-Cells: For each pair of 1-cells $f: X \longrightarrow X'$ and $g: Y \longrightarrow Y'$, a 1-cell

$$f \boxdot g: X \boxdot Y \longrightarrow X' \boxdot Y'.$$

2-Cells: For each pair of 2-cells $\alpha: f_1 \longrightarrow f_2$ in $\mathsf{C}(X, X')$ and $\gamma: g_1 \longrightarrow g_2$ in $\mathsf{C}(Y, Y')$, a 2-cell

$$\alpha \boxdot \gamma: f_1 \boxdot g_1 \longrightarrow f_2 \boxdot g_2.$$

These data are subject to the following 2-functoriality, unity, and associativity conditions.

The 2-Functoriality of the Multiplication:

$$-\boxdot-\colon \mathsf{C}\times\mathsf{C}\longrightarrow\mathsf{C}$$

is a 2-functor. This is equivalent to the following list of equalities whenever they are defined for objects W and X, 1-cells f, f', g, and g', and 2-cells α, α', $\tilde{\alpha}$, γ, γ', and $\tilde{\gamma}$.

$$\begin{aligned} 1_W\boxdot 1_X &= 1_{W\boxdot X}\\ 1_f\boxdot 1_g &= 1_{f\boxdot g}\\ (f'f)\boxdot(g'g) &= (f'\boxdot g')(f\boxdot g)\\ (\alpha'\alpha)\boxdot(\gamma'\gamma) &= (\alpha'\boxdot\gamma')(\alpha\boxdot\gamma)\\ (\tilde{\alpha}*\alpha)\boxdot(\tilde{\gamma}*\gamma) &= (\tilde{\alpha}\boxdot\tilde{\gamma})*(\alpha\boxdot\gamma)\end{aligned}\tag{6.7.18}$$

Unity: For each object X, 1-cell f, and 2-cell α, the following equalities hold.

$$\begin{aligned} I\boxdot X = X &= X\boxdot I\\ 1_I\boxdot f = f &= f\boxdot 1_I\\ 1_{1_I}\boxdot\alpha = \alpha &= \alpha\boxdot 1_{1_I}\end{aligned}\tag{6.7.19}$$

Associativity: For objects X, Y, and Z, 1-cells f, g, and h, and 2-cells α, γ, and θ, the following equalities hold.

$$\begin{aligned}(X\boxdot Y)\boxdot Z &= X\boxdot(Y\boxdot Z)\\ (f\boxdot g)\boxdot h &= f\boxdot(g\boxdot h)\\ (\alpha\boxdot\gamma)\boxdot\theta &= \alpha\boxdot(\gamma\boxdot\theta)\end{aligned}\tag{6.7.20}$$

This finishes the explicit description of a monoid in $(\mathsf{2Cat},\times)$. ◇

Next is the analogue of Explanations 6.7.5 and 6.7.11 with the Cartesian product instead of the Gray tensor product.

Explanation 6.7.21 (Permutative 2-Category)**.** For a monoid $(\mathsf{C},\boxdot,I)$ in $(\mathsf{2Cat},\times)$ as in Explanation 6.7.17, consider the 2-natural isomorphism $\beta\colon\boxdot\longrightarrow\boxdot\,\xi$ of a permutative 2-category in Definition 6.7.16. It consists of a component 1-cell

$$X\boxdot Y\xrightarrow{\beta_{X,Y}} Y\boxdot X\ \in\mathsf{C}$$

for each pair of objects $X,Y\in\mathsf{C}$ such that the following invertibility and naturality conditions are satisfied.

Invertibility: The following diagram is commutative for objects $X,Y\in\mathsf{C}$.

$$\begin{array}{ccc} X\boxdot Y & \xrightarrow{1_{X\boxdot Y}} & X\boxdot Y\\ & {\scriptstyle\beta_{X,Y}}\searrow\quad\nearrow{\scriptstyle\beta_{Y,X}} & \\ & Y\boxdot X & \end{array}\tag{6.7.22}$$

1-Cell Naturality: For 1-cells $f\colon X\longrightarrow X'$ and $g\colon Y\longrightarrow Y'$ in C, the following diagram in $\mathsf{C}(X\boxdot Y, Y'\boxdot X')$ is commutative.

$$\begin{array}{ccc} X\boxdot Y & \xrightarrow{\beta_{X,Y}} & Y\boxdot X\\ {\scriptstyle f\boxdot g}\downarrow & & \downarrow{\scriptstyle g\boxdot f}\\ X'\boxdot Y' & \xrightarrow{\beta_{X',Y'}} & Y'\boxdot X'\end{array}\tag{6.7.23}$$

2-Cell Naturality: For 2-cells

- $\alpha: f_1 \longrightarrow f_2$ in $\mathsf{C}(X, X')$ and
- $\gamma: g_1 \longrightarrow g_2$ in $\mathsf{C}(Y, Y')$,

the following equality holds in $\mathsf{C}(X \boxdot Y, Y' \boxdot X')$.

$$(\gamma \boxdot \alpha) * 1_{\beta_{X,Y}} = 1_{\beta_{X',Y'}} * (\alpha \boxdot \gamma) \tag{6.7.24}$$

This is equivalent to the following pasting diagram equality, with the unlabeled subregions commutative by (6.7.23).

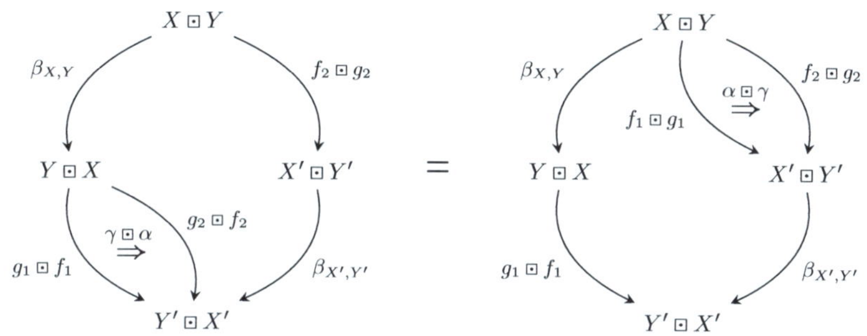

This completes the conditions for $\beta: \boxdot \longrightarrow \boxdot\,\xi$ to be a 2-natural isomorphism.

The three axioms of a permutative 2-category are equivalent to, respectively, the commutative diagrams (6.7.12), (6.7.13), and (6.7.14) with $\odot$ replaced by $\boxdot$. Note that the symmetry axiom is the same as the invertibility condition (6.7.22). $\diamond$

Strictification. We close this chapter with a brief discussion of strictification related to symmetric monoidal bicategories, permutative Gray monoids, and permutative 2-categories. Since we will not use these results in this book, we will not provide detailed definitions and statements here. The reader is referred to the cited references for detail.

A 2-functor $F: \mathsf{C} \longrightarrow \mathsf{D}$ between permutative Gray monoids is a *permutative Gray functor* if it strictly preserves the Gray unit, the multiplication $\odot$, and the Gray symmetry β. There is a 1-category PGray, which is denoted by *PermGrayMon* in [**GJO17b**], of permutative Gray monoids and permutative Gray functors.

A *quasi-strict symmetric monoidal 2-category* [**SP**∞, Def. 2.28] is a symmetric monoidal bicategory B as in Definition 6.5.9 that satisfies the following conditions.

- The underlying monoidal bicategory of B is a Gray monoid as in Definition 6.6.13.
- The braided monoidal bicategory structure of B as in Definition 6.5.3 is a *semistrict braided monoidal 2-category* in the sense of [**Cra98**, Def. 2.2].
- The left hexagonator $R_{-|--}$, the right hexagonator $R_{--|-}$, and the syllepsis ν are all identities.
- The component 2-cells of the braiding β in Definition 6.5.3 of the forms $\beta_{f,1}$ and $\beta_{1,g}$ are identities.
- Transition 2-cells (6.6.17) of the forms $\Sigma_{f,\beta}$ and $\Sigma_{\beta,g}$ are identities for 1-cells f and g, and for component 1-cells of the braiding β.

A *quasi-strict symmetric monoidal 2-functor* $F: \mathsf{C} \longrightarrow \mathsf{D}$ between quasi-strict symmetric monoidal 2-categories is a 2-functor of the underlying 2-categories that strictly preserves all the structures, with naturality constraints either the identity (when this makes sense) or the unique coherence isomorphisms in D. There is a 1-category $\mathsf{2Cat}^{\mathsf{qst}}$, which is denoted by $qsSM2Cat$ in [**GJO17b**], of quasi-strict symmetric monoidal 2-categories and quasi-strict symmetric monoidal 2-functors.

By [**GJO17b**, Theorem 3.43], there is an isomorphism

$$\mathsf{PGray} \cong \mathsf{2Cat}^{\mathsf{qst}}$$

between the 1-categories of permutative Gray monoids and the 1-category of quasi-strict symmetric monoidal 2-categories. Moreover, by [**SP**∞, Theorem 2.97] every symmetric monoidal bicategory is connected to a quasi-strict symmetric monoidal 2-category via a zigzag of symmetric monoidal strict functors. Therefore, every symmetric monoidal bicategory is symmetric monoidal biequivalent to a permutative Gray monoid.

On the other hand, *not* every permutative Gray monoid is symmetric monoidal biequivalent to a permutative 2-category; see [**SP**∞, Ex. 2.30]. However, every permutative Gray monoid is *weakly equivalent* to a permutative 2-category in a homotopical sense; see [**GJO17b**, Theorem 1.1]. Therefore, permutative 2-categories model all weak homotopy types, but not all categorical equivalence types, of permutative Gray monoids.

CHAPTER 7

Baez's Conjecture

The purpose of this chapter is to prove a conjecture due to John Baez on the existence of an initial object, in a suitably weakened sense, in a suitable 2-category of symmetric bimonoidal categories. We will prove two versions of Baez's Conjecture. Denote by $\mathsf{Bi}_{\mathsf{r}}^{\mathsf{fsy}}$ the 2-category in Definition 7.1.8 with

- flat small symmetric bimonoidal categories as objects,
- robust symmetric bimonoidal functors as 1-cells, and
- bimonoidal natural transformations as 2-cells.

Denote by $\varnothing$ the empty 2-category with no objects, no 1-cells, and no 2-cells. The first version of Baez's Conjecture is Theorem 7.8.1. It states that the left bipermutative category Σ in Proposition 2.4.8 is a lax bicolimit of the unique 2-functor

$$\varnothing \longrightarrow \mathsf{Bi}_{\mathsf{r}}^{\mathsf{fsy}}.$$

The second version is Theorem 7.8.3. It states that the right bipermutative category Σ' in Proposition 2.4.23 is also a lax bicolimit of the unique 2-functor $\varnothing \longrightarrow \mathsf{Bi}_{\mathsf{r}}^{\mathsf{fsy}}$. For an open question related to Baez's Conjecture, see Question A.2.6.

Motivation. To motivate Baez's Conjecture, recall that a *rig* is a ring without additive inverses. Among rigs, $\mathbb{N}$, consisting of the natural numbers with its usual addition and multiplication, is an initial object. Indeed, given a rig R, a rig map $\mathbb{N} \longrightarrow R$ must send 0 and 1 in $\mathbb{N}$ to those in R. The image of each $n \in \mathbb{N}$ for $n \geq 2$ must be the sum

$$1 + \cdots + 1 \in R$$

of n copies of $1 \in R$, since the rig map preserves addition.

Since symmetric bimonoidal categories are categorical analogues of multiplicatively commutative rigs, it is natural to ask whether there is an analogue of $\mathbb{N}$ that is initial among symmetric bimonoidal categories. Using our terminology, the following conjecture due to John Baez predicts the existence of a symmetric bimonoidal analogue of $\mathbb{N}$.

> **Baez's Conjecture** [**Bae18**]. The groupoid of finite sets and bijections is an initial object, in a suitably weakened sense, in a 2-category of symmetric bimonoidal categories.

In this chapter, we will prove two precise versions of this conjecture.

Restrictions. Let us discuss some conditions that we need to impose on the statement of Baez's Conjecture.

Smallness: The 2-category Cat has *small* categories as objects. The smallness condition ensures that for each pair of functors $F, G\colon \mathsf{C} \longrightarrow \mathsf{D}$ with C and D small categories, there is a *set* of natural transformations $F \longrightarrow G$.

Similarly, to discuss a 2-category of symmetric bimonoidal categories, as we will see in Proposition 7.1.7, we need to restrict to small symmetric bimonoidal categories.

Finite Ordinals: The groupoid of finite sets and bijections is not small, since every infinite set has infinitely many finite subsets. Therefore, for Baez's Conjecture we should consider a small version of this groupoid, that is, a groupoid of representatives of finite sets and bijections. There are two such candidates:

- the left bipermutative category Σ in Proposition 2.4.8 and
- the right bipermutative category Σ' in Proposition 2.4.23.

Both of them have objects $n \geq 0$, and the morphisms are given by the symmetric groups. The main difference between them is that the object $m \otimes n = mn$ is interpreted as an $n \times m$ matrix in Σ and as an $m \times n$ matrix in Σ'. See Explanations 2.4.7 and 2.4.22. We will show that each of Σ and Σ' satisfies Baez's Conjecture.

Flatness: The Coherence Theorems 3.9.1 and 4.4.3, which have a monomorphism assumption, are used in the proof of Baez's Conjecture. Therefore, we will restrict to small symmetric bimonoidal categories that are *flat* in the sense of Definition 3.9.9. The flatness assumption means that if we start with a component of either distributivity morphism δ^l or δ^r, and take iterated sums and products with a finite number of identity morphisms, then the result is still a monomorphism. For example, tight symmetric bimonoidal categories—that is, those with δ^l and δ^r natural isomorphisms—are flat.

Robustness: In the proof of Baez's Conjecture, we need the symmetric bimonoidal functors in question to have some invertible structure morphisms. To be specific, we want our symmetric bimonoidal functors to be *robust* in the sense of Definition 5.1.1. This means that the following structure morphisms are isomorphisms:

- the additive monoidal constraint

$$G^2_{\oplus} : GA \oplus GB \longrightarrow G(A \oplus B),$$

- the additive zero constraint

$$G^0_{\oplus} : \mathbb{0} \longrightarrow G(0),$$

and

- the multiplicative unit constraint

$$G^0_{\otimes} : \mathbb{1} \longrightarrow G(1).$$

A strong symmetric bimonoidal functor is automatically robust. However, Baez's Conjecture does *not* need the invertibility of the multiplicative monoidal constraint

$$G^2_{\otimes} : GA \otimes GB \longrightarrow G(A \otimes B).$$

Lax Bicolimit: There is the issue of which kind of weak colimits we should use. For a general 2-functor $F : \mathsf{A} \longrightarrow \mathsf{B}$ between 2-categories with $\mathsf{Ob}(\mathsf{A})$ a set, there are five different kinds of colimits for F [**JY21**, Chapter 5]:

- *lax bicolimits* involving lax transformations between cones and equivalences of categories;
- *lax colimits* involving lax transformations between cones and isomorphisms of categories;

- *pseudo bicolimits* involving strong transformations between cones and equivalences of categories;
- *pseudo colimits* involving strong transformations between cones and isomorphisms of categories; and
- *2-colimits* in terms of 2-natural transformations between cones and isomorphisms of categories.

Since the structure morphisms $G^2_\oplus$, $G^0_\oplus$, and $G^0_\otimes$ are isomorphisms instead of identity morphisms, we should use bicolimits instead of colimits. Moreover, the domain being the empty 2-category $\varnothing$ implies that lax bicolimits and pseudo bicolimits are the same in the current setting. Therefore, Baez's Conjecture involves lax bicolimits.

Precise Formulation. With the discussion above in mind, the precise version of Baez's Conjecture, Theorem 7.8.1, states that the left bipermutative category Σ is a lax bicolimit of the 2-functor $\varnothing \longrightarrow \mathsf{Bi}^{\mathsf{fsy}}_{\mathsf{r}}$. More concretely, this means that for each flat small symmetric bimonoidal category C, the unique functor

$$\mathsf{Bi}^{\mathsf{fsy}}_{\mathsf{r}}(\Sigma, \mathsf{C}) \xrightarrow{\;T\;} \mathbb{1} \tag{7.0.1}$$

to the terminal category $\mathbb{1}$ is an equivalence of categories, that is, fully faithful and essentially surjective. The second version of Baez's Conjecture, Theorem 7.8.3, states the same thing for the right bipermutative category Σ'.

While the statement of Baez's Conjecture requires the smallness assumption for the 2-category $\mathsf{Bi}^{\mathsf{fsy}}_{\mathsf{r}}$, its proof only uses smallness at the end to make sure that C is an object in $\mathsf{Bi}^{\mathsf{fsy}}_{\mathsf{r}}$. See Remark 7.8.2.

The validity of Baez's Conjecture may seem obvious at first glance because of the definitions (7.2.3) and (7.2.4). We stress that this is not the case, since it relies on nontrivial coherence properties of symmetric bimonoidal categories and symmetric bimonoidal functors. In fact, its proof involves both Coherence Theorems 3.9.1 and 4.4.3 for symmetric bimonoidal categories, although we do not need to use the Strictification Theorem 5.4.6. See Note 7.9.1. The proof in this chapter also requires a version of Epstein's Coherence Theorem 1.3.12 for symmetric bimonoidal functors in Theorem 7.5.8.

Outline. In Section 7.1, the category $\mathsf{Bi}^{\mathsf{sy}}$ of small symmetric bimonoidal categories and symmetric bimonoidal functors in Proposition 5.1.10 is extended to a 2-category. The 2-cells in this 2-category are bimonoidal natural transformations, which are monoidal natural transformations with respect to both the additive structures and the multiplicative structures. The wide subcategories $\mathsf{Bi}^{\mathsf{sy}}_{\mathsf{r}}$, $\mathsf{Bi}^{\mathsf{sy}}_{\mathsf{sg}}$, $\mathsf{Bi}^{\mathsf{sy}}_{\mathsf{u}}$, and $\mathsf{Bi}^{\mathsf{sy}}_{\mathsf{st}}$ with, respectively, robust, strong, unitary, and strict symmetric bimonoidal functors extend to sub-2-categories of the 2-category $\mathsf{Bi}^{\mathsf{sy}}$. Furthermore, the sub-2-category $\mathsf{Bi}^{\mathsf{fsy}}_{\mathsf{r}}$ of $\mathsf{Bi}^{\mathsf{sy}}_{\mathsf{r}}$ has *flat* small symmetric bimonoidal categories as objects.

Now we provide an outline of the proof of Baez's Conjecture, which is divided into two parts:

(1) the essential surjectivity of the functor T in (7.0.1) and
(2) the fully faithfulness of the functor T.

The first part begins in Section 7.2. For each symmetric bimonoidal category C, in this section, we construct a strong symmetric monoidal functor

$$F_\oplus = (F, F^2_\oplus, F^0_\oplus) \colon \Sigma \longrightarrow \mathsf{C}$$

between the additive structures. The flatness of C is not needed in this section because the Coherence Theorems 3.9.1 and 4.4.3 are not used yet. The functor F sends

- $0 \in \Sigma$ to $\mathbb{0} \in \mathsf{C}$,
- $1 \in \Sigma$ to $\mathbb{1} \in \mathsf{C}$, and
- $n \in \Sigma$ for $n \geq 2$ to a sum of n copies of $\mathbb{1} \in \mathsf{C}$ with some preselected convention for additive bracketing.

Mac Lane's Coherence Theorems 1.3.3 and 1.3.8 are used (i) to define the assignment of F on morphisms and the additive monoidal constraint $F^2_\oplus$, and (ii) to prove that F is a symmetric monoidal functor.

In Section 7.3, the strong symmetric monoidal functor $F_\oplus$ is extended to a symmetric monoidal functor

$$F_\otimes = (F, F^2_\otimes, F^0_\otimes) \colon \Sigma \longrightarrow \mathsf{C}$$

between the multiplicative structures, where C is now assumed to be flat. The definition of the multiplicative monoidal constraint $F^2_\otimes$ involves both $\oplus$ and $\otimes$. The Coherence Theorem 4.4.3 is used in the proofs of (i) the naturality of $F^2_\otimes$ and (ii) the associativity axiom (1.2.14) of $F_\otimes$ in Lemmas 7.3.15, 7.3.24, and 7.3.25. The Coherence Theorem 3.9.1 is used in the proof of Lemma 7.3.28 to show that F satisfies the axiom (1.2.26) of a symmetric monoidal functor.

In Section 7.4, we show that

$$(F, F^2_\otimes, F^0_\otimes, F^2_\oplus, F^0_\oplus) \colon \Sigma \longrightarrow \mathsf{C}$$

is a robust symmetric bimonoidal functor. The proof of the distributivity axiom (5.1.3), in the equivalent form (5.1.6), for F requires the Coherence Theorem 3.9.1. See Lemma 7.4.3, which is used at the end of the proof of Proposition 7.4.4. The canonical robust symmetric bimonoidal functor F implies that the functor T in (7.0.1) is essentially surjective.

The second part of the proof of Baez's Conjecture begins in Section 7.5. To show that the functor T in (7.0.1) is fully faithful, we will need to use a coherence property of symmetric bimonoidal functors. In this section, we formulate and prove this coherence property; see Theorem 7.5.8. When we actually apply this result in the proof of Baez's Conjecture, the domain of the symmetric bimonoidal functor is Σ. Theorem 7.5.8 is stated and proved in the much more general case of symmetric bimonoidal functors $G \colon \mathsf{C} \longrightarrow \mathsf{D}$ with C and D flat. The Coherence Theorem 4.4.3 is used to make sure that the diagram (7.5.2) is well defined; see Explanation 7.5.5. Moreover, Epstein's Coherence Theorem 1.3.12 is used near the end of the proof of Theorem 7.5.8.

In Section 7.6, we show that for any symmetric bimonoidal category C and any two robust symmetric bimonoidal functors $G, H \colon \Sigma \longrightarrow \mathsf{C}$, there is at most one bimonoidal natural transformation $G \longrightarrow H$, which must be a bimonoidal natural isomorphism. See Lemma 7.6.3. The main ingredient of this observation is Lemma 7.6.2, where we give an explicit description of the components of such a bimonoidal natural transformation. It is at this point that we need the *robust* condition on symmetric bimonoidal functors. The statement of Lemma 7.6.2 states precisely where we need the invertibility of the structure morphisms $G^2_\oplus$, $G^0_\oplus$, and $G^0_\otimes$.

In Section 7.7, we observe that the robust symmetric bimonoidal functor $F\colon \Sigma \longrightarrow \mathsf{C}$ with C flat is initial in a suitable sense. More precisely, in Lemma 7.7.9 we show that for each robust symmetric bimonoidal functor $G\colon \Sigma \longrightarrow \mathsf{C}$, there exists a unique bimonoidal natural transformation $\theta^G\colon F \longrightarrow G$, which is furthermore an isomorphism. Epstein's Coherence Theorem 1.3.12 is used in Lemmas 7.7.3 and 7.7.4 to prove (i) the naturality of θ^G and (ii) that θ^G is a monoidal natural transformation for the additive structures. Theorem 7.5.8 is used in Lemma 7.7.6 to prove that θ^G is a monoidal natural transformation for the multiplicative structures.

In Section 7.8, we collect the results from earlier sections to finish the proof of Baez's Conjecture, which is Theorem 7.8.1. Another version using Σ' is Theorem 7.8.3. Elgueta [**Elg21**] has a more restricted version of Baez's Conjecture for bimonoidal categories without multiplicative symmetries. Its relationship to this chapter is discussed in Section 7.9.

Reading Guide. As in Chapters 3, 4, and 5, we offer the following possible alternative to reading this chapter strictly linearly.

(1) Read Definitions 7.2.2 and 7.3.12 of $F\colon \Sigma \longrightarrow \mathsf{C}$.
(2) Read Definition 7.7.1 of $\theta^G\colon F \longrightarrow G$.
(3) Read Section 7.8, which contains both versions of Baez's Conjecture.
(4) Go back and read the parts skipped earlier.

7.1. The 2-Category of Symmetric Bimonoidal Categories

In this section, we define the 2-category of small symmetric bimonoidal categories and some of its sub-2-categories. The 2-category for Baez's Conjecture is in Definition 7.1.8. Proposition 5.1.10 contains the 1-categories of small symmetric bimonoidal categories and different subclasses of symmetric bimonoidal functors. Therefore, our task is to define the 2-cells, which are suitable natural transformations. For monoidal functors $F, G\colon \mathsf{C} \longrightarrow \mathsf{D}$, recall from Definition 1.2.16 that a monoidal natural transformation $\theta\colon F \longrightarrow G$ is a natural transformation that is also compatible with the structure morphisms (F^2, F^0) of F and (G^2, G^0) of G.

Convention 7.1.1. For the rest of this chapter, unless otherwise specified, C and D are arbitrary symmetric bimonoidal categories as in Definition 2.1.2. Sometimes they are required to be small or flat (as in Definition 3.9.9) as specified. To clarify that some structure belongs to a particular category, we decorate it with the name of the category. For example, $\mathbb{0}^{\mathsf{C}}$ and $\mathbb{1}^{\mathsf{C}}$ are, respectively, the additive zero and the multiplicative unit in C. ◇

Recall from Definition 5.1.1 that a symmetric bimonoidal functor $F\colon \mathsf{C} \longrightarrow \mathsf{D}$ is a functor equipped with

- an additive symmetric monoidal functor structure

$$F_\oplus = (F, F^2_\oplus, F^0_\oplus)\colon (\mathsf{C}, \oplus, \mathbb{0}, \alpha^\oplus, \lambda^\oplus, \rho^\oplus, \xi^\oplus) \longrightarrow (\mathsf{D}, \oplus, \mathbb{0}, \alpha^\oplus, \lambda^\oplus, \rho^\oplus, \xi^\oplus)$$

  and
- a multiplicative symmetric monoidal functor structure

$$F_\otimes = (F, F^2_\otimes, F^0_\otimes)\colon (\mathsf{C}, \otimes, \mathbb{1}, \alpha^\otimes, \lambda^\otimes, \rho^\otimes, \xi^\otimes) \longrightarrow (\mathsf{D}, \otimes, \mathbb{1}, \alpha^\otimes, \lambda^\otimes, \rho^\otimes, \xi^\otimes)$$

that are compatible with the multiplicative zeros (5.1.2) and the distributivity morphisms (5.1.3).

Definition 7.1.2. Suppose $F, G\colon \mathsf{C} \longrightarrow \mathsf{D}$ are two symmetric bimonoidal functors.

(1) A *bimonoidal natural transformation* $\theta\colon F \longrightarrow G$ is a natural transformation of the underlying functors such that both
- $\theta\colon F_\oplus \longrightarrow G_\oplus$ and
- $\theta\colon F_\otimes \longrightarrow G_\otimes$

are monoidal natural transformations.

(2) A *bimonoidal natural isomorphism* is an invertible bimonoidal natural transformation. $\diamond$

Explanation 7.1.3. In Definition 7.1.2, the assumption that

$$F_\oplus = (F, F^2_\oplus, F^0_\oplus) \xrightarrow{\ \theta\ } (G, G^2_\oplus, G^0_\oplus) = G_\oplus$$

is a monoidal natural transformation means that the diagrams

$$\begin{array}{ccc} FX \oplus FY & \xrightarrow{\theta_X \oplus \theta_Y} & GX \oplus GY \\ {\scriptstyle F^2_\oplus}\downarrow & & \downarrow{\scriptstyle G^2_\oplus} \\ F(X \oplus Y) & \xrightarrow{\theta_{X\oplus Y}} & G(X \oplus Y) \end{array} \qquad \begin{array}{ccc} \mathbb{0}^{\mathsf{D}} & \xrightarrow{F^0_\oplus} & F\mathbb{0}^{\mathsf{C}} \\ \| & & \downarrow{\scriptstyle \theta_{\mathbb{0}^{\mathsf{C}}}} \\ \mathbb{0}^{\mathsf{D}} & \xrightarrow{G^0_\oplus} & G\mathbb{0}^{\mathsf{C}} \end{array} \tag{7.1.4}$$

are commutative in D for all objects $X, Y \in \mathsf{C}$. Similarly, the assumption that

$$F_\otimes = (F, F^2_\otimes, F^0_\otimes) \xrightarrow{\ \theta\ } (G, G^2_\otimes, G^0_\otimes) = G_\otimes$$

is a monoidal natural transformation means that the diagrams

$$\begin{array}{ccc} FX \otimes FY & \xrightarrow{\theta_X \otimes \theta_Y} & GX \otimes GY \\ {\scriptstyle F^2_\otimes}\downarrow & & \downarrow{\scriptstyle G^2_\otimes} \\ F(X \otimes Y) & \xrightarrow{\theta_{X\otimes Y}} & G(X \otimes Y) \end{array} \qquad \begin{array}{ccc} \mathbb{1}^{\mathsf{D}} & \xrightarrow{F^0_\otimes} & F\mathbb{1}^{\mathsf{C}} \\ \| & & \downarrow{\scriptstyle \theta_{\mathbb{1}^{\mathsf{C}}}} \\ \mathbb{1}^{\mathsf{D}} & \xrightarrow{G^0_\otimes} & G\mathbb{1}^{\mathsf{C}} \end{array} \tag{7.1.5}$$

are commutative in D. Moreover, a bimonoidal natural transformation $\theta\colon F \longrightarrow G$ is a bimonoidal natural isomorphism if and only if each component of θ is an isomorphism in D. $\diamond$

Recall from Proposition 5.1.10 that $\mathsf{Bi}^{\mathsf{sy}}$ is the 1-category of *small* symmetric bimonoidal categories and symmetric bimonoidal functors. Also recall its wide subcategories $\mathsf{Bi}^{\mathsf{sy}}_{\mathsf{r}}$, $\mathsf{Bi}^{\mathsf{sy}}_{\mathsf{sg}}$, $\mathsf{Bi}^{\mathsf{sy}}_{\mathsf{u}}$, and $\mathsf{Bi}^{\mathsf{sy}}_{\mathsf{st}}$ with, respectively, robust, strong, unitary, and strict symmetric bimonoidal functors. Now we extend these 1-categories to 2-categories as follows.

Definition 7.1.6. Define the 2-categorical data for $\mathsf{Bi}^{\mathsf{sy}}$ as follows.

Objects: Objects are small symmetric bimonoidal categories.

1-Cells: The 1-cells in $\mathsf{Bi}^{\mathsf{sy}}(\mathsf{C}, \mathsf{D})$ are the symmetric bimonoidal functors $\mathsf{C} \longrightarrow \mathsf{D}$.

Identity 1-Cells: $1_{\mathsf{C}} \in \mathsf{Bi}^{\mathsf{sy}}(\mathsf{C}, \mathsf{C})$ is the identity symmetric bimonoidal functor.

1-Cell Composition: Horizontal composition of 1-cells is as in Definition 5.1.7.

2-Cells: The 2-cells in $\mathsf{Bi}^{\mathsf{sy}}(\mathsf{C}, \mathsf{D})(F, G)$ are the bimonoidal natural transformations $F \longrightarrow G$ in Definition 7.1.2.

Identity 2-Cells: For each 1-cell $F\colon \mathsf{C} \longrightarrow \mathsf{D}$, the identity 2-cell

$$1_F \in \mathsf{Bi}^{\mathsf{sy}}(\mathsf{C}, \mathsf{D})(F, F)$$

is the identity natural transformation of F.

2-Cell Compositions: Vertical and horizontal compositions of 2-cells are those of natural transformations in Definition 1.1.8.

This finishes the definition of the 2-categorical data for $\mathsf{Bi}^{\mathsf{sy}}$.

Moreover, similar definitions define the 2-categorical data for

- $\mathsf{Bi}^{\mathsf{sy}}_{\mathsf{r}}$ with robust symmetric bimonoidal functors as 1-cells,
- $\mathsf{Bi}^{\mathsf{sy}}_{\mathsf{sg}}$ with strong symmetric bimonoidal functors as 1-cells,
- $\mathsf{Bi}^{\mathsf{sy}}_{\mathsf{u}}$ with unitary symmetric bimonoidal functors as 1-cells, and
- $\mathsf{Bi}^{\mathsf{sy}}_{\mathsf{st}}$ with strict symmetric bimonoidal functors as 1-cells. ◇

Proposition 7.1.7. *With the data in Definition 7.1.6,* $\mathsf{Bi}^{\mathsf{sy}}$ *is a 2-category that contains the following full sub-2-categories:*

$$\mathsf{Bi}^{\mathsf{sy}}_{\mathsf{st}} \subset \mathsf{Bi}^{\mathsf{sy}}_{\mathsf{u}} \subset \mathsf{Bi}^{\mathsf{sy}}_{\mathsf{sg}} \subset \mathsf{Bi}^{\mathsf{sy}}_{\mathsf{r}} \subset \mathsf{Bi}^{\mathsf{sy}}.$$

Proof. The smallness assumption on symmetric bimonoidal categories ensures that for each pair of 1-cells $F, G\colon \mathsf{C} \longrightarrow \mathsf{D}$, there is a *set* of 2-cells in $\mathsf{Bi}^{\mathsf{sy}}(\mathsf{C},\mathsf{D})(F,G)$. By Proposition 5.1.10, the objects, 1-cells, identity 1-cells, and their horizontal composition in $\mathsf{Bi}^{\mathsf{sy}}$ constitute a 1-category. Vertical and horizontal compositions of 2-cells are well defined because monoidal natural transformations are closed under these compositions. Since symmetric monoidal functors and monoidal natural transformations satisfy axioms (i)–(vi) in Proposition 6.1.10, so do the 1-cells and 2-cells in $\mathsf{Bi}^{\mathsf{sy}}$. Therefore, $\mathsf{Bi}^{\mathsf{sy}}$ is a 2-category.

The existence of the four sub-2-categories follows from the fact that, in each case, the 1-cells are closed under horizontal composition by Lemma 5.1.9. □

For Baez's Conjecture, in addition to restricting the 1-cells to the robust ones, we also need to restrict the objects as follows. Recall from Definition 3.9.9 that a symmetric bimonoidal category is *flat* if each iterated sum and product of a component of either δ^l or δ^r (2.1.4) with a finite number of identity morphisms is a monomorphism. For example,

- flatness ensures that the morphisms in (3.9.5) are monomorphisms, and
- tight symmetric bimonoidal categories—that is, those with δ^l and δ^r natural isomorphisms—are flat.

The Coherence Theorems 3.9.1 and 4.4.3 apply to flat symmetric bimonoidal categories because the value of each δ-prime edge is of the type in the definition of flatness. Also recall from Definition 5.1.1 that a *robust* symmetric bimonoidal functor F has $F^2_\oplus$, $F^0_\oplus$, and $F^0_\otimes$ isomorphisms.

Definition 7.1.8. Denote by $\mathsf{Bi}^{\mathsf{fsy}}_{\mathsf{r}}$ the full sub-2-category of $\mathsf{Bi}^{\mathsf{sy}}$ with

- flat small symmetric bimonoidal categories as objects and
- robust symmetric bimonoidal functors as 1-cells. ◇

The rest of this section contains examples. By Proposition 2.3.2, each distributive symmetric monoidal category, whose monoidal product is denoted by $\otimes$, yields a tight symmetric bimonoidal category with additive structure given by coproducts, that is, $(\oplus, \mathbb{0}) = (\amalg, \varnothing)$. Moreover, by Proposition 5.1.11, each symmetric monoidal functor $F\colon \mathsf{C} \longrightarrow \mathsf{D}$ between distributive symmetric monoidal categories canonically extends to a symmetric bimonoidal functor with the additive structure $(F^2_\oplus, F^0_\oplus)$ in (5.1.12). The next observation does not require the relevant categories to be small.

Proposition 7.1.9. *Suppose $\theta\colon F \longrightarrow G$ is a monoidal natural transformation between symmetric monoidal functors $F, G\colon \mathsf{C} \longrightarrow \mathsf{D}$ with C and D distributive symmetric monoidal categories. Then θ is a bimonoidal natural transformation when F and G are regarded as symmetric bimonoidal functors as in Proposition 5.1.11.*

Proof. By assumption, $\theta: F_\otimes \longrightarrow G_\otimes$ is a monoidal natural transformation.

For $\theta: F_\oplus \longrightarrow G_\oplus$, we check the commutativity of the diagrams in (7.1.4).

(1) The right diagram in (7.1.4) is commutative by the universal property of the initial object in D.

(2) The left diagram in (7.1.4) is commutative by
- the universal property of coproducts in D,
- the definitions of $F_\oplus^2$ and $G_\oplus^2$ in (5.1.12), and
- the naturality of θ.

Therefore, $\theta: F_\oplus \longrightarrow G_\oplus$ is a monoidal natural transformation. □

Example 7.1.10 (Distributive Categories). Suppose $F, G: \mathsf{C} \longrightarrow D$ are functors between distributive categories as in Example 2.3.5, such that the natural morphisms

$$F* \xrightarrow{\;t\;} *$$

$$F(A \times B) \xrightarrow{\;p\;} FA \times FB$$

for objects $A, B \in \mathsf{C}$, which are dual to those in (5.1.12), are isomorphisms, and similarly for G. As we mentioned in Example 5.1.14,

$$F_\otimes = \left(F, p^{-1}, t^{-1}\right),\ G_\otimes = \left(G, p^{-1}, t^{-1}\right): (\mathsf{C}, \times, *) \longrightarrow (\mathsf{D}, \times, *)$$

are symmetric monoidal functors. If $\theta: F \longrightarrow G$ is a natural transformation, then it is also a monoidal natural transformation $\theta: F_\otimes \longrightarrow G_\otimes$ for the following reasons.

(1) The right diagram in (7.1.5) is commutative by
- the universal property of the terminal object $*$ in D and
- the invertibility of t for both F and G.

(2) The left diagram in (7.1.5) is commutative by
- the universal property of products in D,
- the invertibility of p for both F and G, and
- the naturality of θ.

By Proposition 7.1.9, θ is a bimonoidal natural transformation when F and G are regarded as symmetric bimonoidal functors as in Proposition 5.1.11. ◇

7.2. The Additive Structure

This section contains the first step of the proof of Baez's Conjecture. For each symmetric bimonoidal category C, we show that there exists a canonical strong symmetric monoidal functor $F_\oplus: \Sigma \longrightarrow \mathsf{C}$ between the additive structures of Σ and C. Here Σ is the tight symmetric bimonoidal category in Definition 2.4.1 and Proposition 2.4.8. It is furthermore a left bipermutative category in the sense of Definition 2.5.11. The flatness assumption on C is not needed until we define the multiplicative structure $F_\otimes$ in Section 7.3. The smallness of C is not needed until Section 7.8.

Definition 7.2.1. For each integer $n \geq 0$, define the following object in C.

$$\overline{n} = \begin{cases} \mathbb{0} & \text{if } n = 0, \\ \mathbb{1} & \text{if } n = 1, \text{ and} \\ (\mathbb{1} \oplus \cdots \oplus \mathbb{1})_{\mathsf{lt}} & \text{if } n > 1. \end{cases}$$

In the last case, $\overline{n}$ is the left normalized sum (5.2.13) of n copies of the multiplicative unit $\mathbb{1}$ in C. ◇

In the next definition and the rest of this chapter, a *Mac Lane coherence isomorphism* means a component of a permuted canonical map as in Definition 1.3.6, applied to the additive structure

$$(\mathsf{C}, \oplus, \mathbb{0}, \alpha^{\oplus}, \lambda^{\oplus}, \rho^{\oplus}, \xi^{\oplus}),$$

which is often abbreviated to $(\mathsf{C}, \oplus, \mathbb{0})$ or $(\mathsf{C}, \oplus)$. This is an adaptation of Definition 5.2.24 to the current context.

Definition 7.2.2. Using the additive structures in Σ and C, define the data

$$F_{\oplus} = (F, F^2_{\oplus}, F^0_{\oplus}) \colon \Sigma \longrightarrow \mathsf{C}$$

of a symmetric monoidal functor as follows.

The Functor: The functor $F \colon \Sigma \longrightarrow \mathsf{C}$ is defined as follows.

Objects: For each $n \geq 0$, define

$$F(n) = \overline{n} \in \mathsf{C} \tag{7.2.3}$$

with $\overline{n}$ as in Definition 7.2.1.

Morphisms: For each morphism $\sigma \in \Sigma(n,n) = \Sigma_n$, define the morphism

$$F(n) = \overline{n} \xrightarrow{F(\sigma)} \overline{n} = F(n) \in \mathsf{C} \tag{7.2.4}$$

as the unique Mac Lane coherence isomorphism in the symmetric monoidal category $(\mathsf{C}, \oplus, \mathbb{0})$ that permutes the n copies of $\mathbb{1}$ in $\overline{n}$ as $\sigma \in \Sigma_n$ permutes n letters.

The Additive Zero Constraint: The morphism

$$\mathbb{0} \xrightarrow{F^0_{\oplus}} F(0) = \mathbb{0} \in \mathsf{C} \tag{7.2.5}$$

is the identity morphism $1_{\mathbb{0}}$.

The Additive Monoidal Constraint: For $m, n \geq 0$, define the morphism

$$\overline{m} \oplus \overline{n} = F(m) \oplus F(n) \xrightarrow{F^2_{\oplus}} F(m+n) = \overline{m+n} \in \mathsf{C} \tag{7.2.6}$$

as the unique Mac Lane coherence isomorphism in $(\mathsf{C}, \oplus, \mathbb{0})$ that does *not* involve $\xi^{\pm\oplus}$.

This finishes the definition of $F_{\oplus}$. ◇

Explanation 7.2.7. Consider the definition (7.2.6) of $F^2_{\oplus}$.

(1) If $m = n = 0$, then

$$\mathbb{0} \oplus \mathbb{0} \xrightarrow{F^2_{\oplus}} \mathbb{0}$$

is the left additive zero $\lambda^{\oplus}_{\mathbb{0}}$, which is also equal to the right additive zero $\rho^{\oplus}_{\mathbb{0}}$ by (1.2.6).

(2) If $m = 0$ and $n > 0$, then

$$\mathbb{0} \oplus \overline{n} \xrightarrow{F^2_{\oplus}} \overline{n}$$

is the left additive zero $\lambda^{\oplus}_{\overline{n}}$.

(3) If $m > 0$ and $n = 0$, then

$$\overline{m} \oplus \mathbb{0} \xrightarrow{F^2_{\oplus}} \overline{m}$$

is the right additive zero $\rho^{\oplus}_{\overline{m}}$.

(4) If $m, n > 0$, then

$$\left(\bigoplus_{i=1}^{m} \mathbb{1}\right)_{\mathsf{lt}} \oplus \left(\bigoplus_{j=1}^{n} \mathbb{1}\right)_{\mathsf{lt}} = \overline{m} \oplus \overline{n} \xrightarrow{F_{\oplus}^{2}} \overline{m+n} = \left(\bigoplus_{i=1}^{m+n} \mathbb{1}\right)_{\mathsf{lt}} \tag{7.2.8}$$

- is the identity morphism $1_{\overline{m+1}}$ if $n = 1$, and
- involves $\alpha^{-\oplus}_{\overline{m},\overline{k},\mathbb{1}}$ if $n > 1$, with $\alpha^{-\oplus}$ the inverse of $\alpha^{\oplus}$ and $1 \leq k \leq n-1$.

For example, in the last case, if $n = 3$, then $F_{\oplus}^{2}$ is the following composite in C.

$$\begin{array}{l} \overline{m} \oplus \big((\mathbb{1} \oplus \mathbb{1}) \oplus \mathbb{1}\big) \\ \quad \big\downarrow \alpha^{-\oplus}_{\overline{m},\mathbb{1}\oplus\mathbb{1},\mathbb{1}} \\ \big(\overline{m} \oplus (\mathbb{1} \oplus \mathbb{1})\big) \oplus \mathbb{1} \xrightarrow{\alpha^{-\oplus}_{\overline{m},\mathbb{1},\mathbb{1}} \oplus 1_{\mathbb{1}}} \big((\overline{m} \oplus \mathbb{1}) \oplus \mathbb{1}\big) \oplus \mathbb{1} = \overline{m+3} \end{array}$$

We stress that $F_{\oplus}^{2}$ does not involve the additive symmetry $\xi^{\oplus}$ and its inverse. ◇

Lemma 7.2.9. *For each symmetric bimonoidal category* C,

$$F_{\oplus} \colon \Sigma \longrightarrow \mathsf{C}$$

in Definition 7.2.2 is a strong symmetric monoidal functor.

Proof. The functoriality of $F \colon \Sigma \longrightarrow \mathsf{C}$ follows from

- the definitions of identity morphisms and composition in Σ as those of permutations and
- the uniqueness in Theorem 1.3.8 for the symmetric monoidal category $(\mathsf{C}, \oplus, \mathbb{0})$.

By definition, $F_{\oplus}^{0}$ is the identity morphism $1_{\mathbb{0}}$, and $F_{\oplus}^{2}$ is componentwise an isomorphism.

The uniqueness in Theorem 1.3.8 for $(\mathsf{C}, \oplus, \mathbb{0})$ implies the following statements.

- $F_{\oplus}^{2}$ is a natural isomorphism.
- The associativity axiom (1.2.14) holds for $(F, F_{\oplus}^{2})$.
- The compatibility axiom (1.2.26) holds for $(F, F_{\oplus}^{2})$ with the additive symmetries $\xi^{\oplus}$ in Σ and C.

The unity axiom (1.2.15) for $F_{\oplus}$ follows from

- the definition $F_{\oplus}^{0} = 1_{\mathbb{0}}$,
- that $\lambda^{\oplus}$ and $\rho^{\oplus}$ in Σ are identities, and
- the first three cases in Explanation 7.2.7.

Therefore, $F_{\oplus}$ is a strong symmetric monoidal functor between the additive structures. □

7.3. The Multiplicative Structure

The purpose of this section is to extend the strong symmetric monoidal functor $F_{\oplus} \colon \Sigma \longrightarrow \mathsf{C}$ in Lemma 7.2.9 to a symmetric monoidal functor $F_{\otimes} \colon \Sigma \longrightarrow \mathsf{C}$ between the multiplicative structures of Σ and C.

Convention 7.3.1. For the rest of this chapter, unless otherwise specified, C denotes an arbitrary *flat* symmetric bimonoidal category as in Definition 3.9.9. ◇

Flatness implies that the monomorphism condition in the Coherence Theorems 3.9.1 and 4.4.3 are satisfied. Here is a brief outline of this section.

- The multiplicative structure $F_\otimes$ is in Definition 7.3.12, after the preliminary Definition 7.3.3 and Lemmas 7.3.6 and 7.3.7.
- We verify that $F^2_\otimes$ is well defined in Lemma 7.3.15.
- We prove a factorization of $F^2_\otimes$ in Lemma 7.3.21.
- The associativity axiom (1.2.14) and the unity axioms (1.2.15) for a monoidal functor are verified in Lemmas 7.3.25 and 7.3.27, respectively.
- We show that $F_\otimes$ is a symmetric monoidal functor in Lemma 7.3.28.

Recall from (5.2.13) that the subscript lt means left normalized bracketing.

Motivation 7.3.2. The definition of the multiplicative monoidal constraint $F^2_\otimes$ is more complicated than that of $F^2_\oplus$ because it involves a mixture of $\oplus$ and $\otimes$. Here we motivate its definition.

By definition, $F^2_\otimes$ involves a structure morphism

$$\begin{array}{ccc} F(m)\otimes F(n) = \overline{m}\otimes\overline{n} & & \overline{mn} = F(mn) \\ \| & & \| \\ \Big(\bigoplus_{i=1}^{m}\mathbb{1}\Big)_{\mathsf{lt}}\otimes\Big(\bigoplus_{j=1}^{n}\mathbb{1}\Big)_{\mathsf{lt}} & \xrightarrow{F^2_\otimes} & \Big(\bigoplus_{j=1}^{n}\bigoplus_{i=1}^{m}\mathbb{1}\Big)_{\mathsf{lt}} \end{array}$$

in C for $m,n > 0$, in addition to the cases where either m or n is 0. A natural way to define this morphism is to first distribute, using δ^l and δ^r, the product in the domain all the way down to a sum of mn copies of $\mathbb{1}\otimes\mathbb{1}$. Applying the multiplicative unit

$$\mathbb{1}\otimes\mathbb{1} \xrightarrow{\lambda^\otimes_{\mathbb{1}}} \mathbb{1}$$

in each summand results in the desired codomain $\overline{mn}$.

In the first step of distributing the domain, we need to specify

- an order of the mn copies of $\mathbb{1}\otimes\mathbb{1}$ and
- an additive bracketing of their sum.

For each such choice, there exist multiple ways to distributive the domain to the sum of mn copies of $\mathbb{1}\otimes\mathbb{1}$. In the second step, in addition to applying copies of $\lambda^\otimes_{\mathbb{1}}$, we also need to make sure that the additive bracketings match. One can certainly choose a specific path that achieves these two steps.

However, in order to show that F is a symmetric bimonoidal functor and to prove Baez's Conjecture using F, we need to verify the commutativity of a number of diagrams. This involves applying the Coherence Theorems 3.9.1 and 4.4.3. Therefore, we define $F^2_\otimes$ in terms of a coherence property, rather than as a specific path, that will make it easier to apply those theorems. As we will see in Definitions 7.3.3 and 7.3.12, $F^2_\otimes$ uses the *distortion* of a path in Definition 4.3.1, which is a crucial ingredient of Theorem 4.4.3. This in turn uses

- the graph in Definition 3.1.9,
- the distortion category $\mathcal{D}$ in Section 4.2, and
- the left normalized bracketing (5.2.13) applied to elements in the free $\{\oplus,\otimes\}$-algebra as in Remark 5.2.14.

The reader may want to review those definitions before proceeding. ◇

We will return to the definition of $F^2_\otimes$ after some preliminary definitions and lemmas.

Definition 7.3.3. Suppose $m, n \geq 1$.

(1) Define the set
$$B = \{0^B, 1^B, b_1, \ldots, b_n\}$$
with $n+2$ elements.

(2) With each 1^B_i denoting a copy of 1^B, define
$$\Big(\bigoplus_{i=1}^{m} 1^B_i\Big)_{\mathsf{lt}} \otimes \Big(\bigoplus_{j=1}^{n} b_j\Big)_{\mathsf{lt}} \xrightarrow{\ \ P\ \ } \Big(\bigoplus_{j=1}^{n}\bigoplus_{i=1}^{m} b_j\Big)_{\mathsf{lt}} \tag{7.3.4}$$
as any path in $\mathsf{Gr}(B)$ whose distortion is the identity.

(3) Define the function $\varphi\colon B \longrightarrow \mathsf{Ob}(\mathsf{C})$ by
$$\varphi(x) = \begin{cases} \mathbb{0} & \text{if } x = 0^B \text{ and} \\ \mathbb{1} & \text{if } x \in \{1^B, b_1, \ldots, b_n\}. \end{cases} \tag{7.3.5}$$
The value of the path P in C is defined as in Definition 3.1.14 via the associated graph morphism $\varphi\colon \mathsf{Gr}(B) \longrightarrow \mathsf{C}$. ◇

First we check that the path P in (7.3.4) is well defined. Recall from Definition 4.2.1 that each object in the distortion category $\mathcal{D}$ is a sequence of nonnegative integers. The next lemma uses the graph morphism
$$\mathsf{Gr}(B) \xrightarrow{\ \ \vartheta\ \ } \mathcal{D}$$
in Definition 4.3.1, applied to the set B in Definition 7.3.3 (1).

Lemma 7.3.6. *The domain and the codomain of P in (7.3.4) are both sent by the graph morphism ϑ to the object*
$$(1, \ldots, 1) \in \mathcal{D}$$
with length mn and each entry equal to 1.

Proof. The definitions (3.1.16), (4.2.8), and (4.3.2) imply the following equalities of objects in $\mathcal{D}$.
$$\vartheta\Big(\bigoplus_{i=1}^{m} 1^B_i\Big)_{\mathsf{lt}} = (\overbrace{0, \ldots, 0}^{m})$$
$$\vartheta\Big(\bigoplus_{j=1}^{n} b_j\Big)_{\mathsf{lt}} = (\overbrace{1, \ldots, 1}^{n})$$
These equalities and (4.2.15) imply that the domain of P is sent by ϑ to the object
$$\begin{aligned}(\overbrace{0, \ldots, 0}^{m}) \otimes (\overbrace{1, \ldots, 1}^{n}) &= (\overbrace{0+1, \ldots, 0+1}^{mn}) \\ &= (\overbrace{1, \ldots, 1}^{mn}) \\ &= \vartheta\Big(\bigoplus_{j=1}^{n}\bigoplus_{i=1}^{m} b_j\Big)_{\mathsf{lt}}.\end{aligned}$$
This proves the assertion. □

We also need to make sure that such a path P actually exists.

Lemma 7.3.7. *There exists a path P as in (7.3.4) whose distortion is the identity.*

Proof. Consider the path

$$P = (P_4, P_3, P_2, P_1) \tag{7.3.8}$$

in $\mathsf{Gr}(B)$ defined below.

$$\begin{array}{ccc}
\Big(\bigoplus_{i=1}^{m} 1_i^B\Big)_{\mathrm{lt}} \otimes \Big(\bigoplus_{j=1}^{n} b_j\Big)_{\mathrm{lt}} & \xrightarrow{\quad P \quad} & \Big(\bigoplus_{j=1}^{n}\bigoplus_{i=1}^{m} b_j\Big)_{\mathrm{lt}} \\
{\scriptstyle P_1}\downarrow & & \uparrow{\scriptstyle P_4} \\
\Big(\bigoplus_{j=1}^{n}\Big(\bigoplus_{i=1}^{m} 1_i^B\Big)_{\mathrm{lt}} \otimes b_j\Big)_{\mathrm{lt}} & & \Big(\bigoplus_{j=1}^{n}\Big(\bigoplus_{i=1}^{m} b_j\Big)_{\mathrm{lt}}\Big)_{\mathrm{lt}} \\
{\scriptstyle P_2}\searrow & & \nearrow{\scriptstyle P_3} \\
 & \Big(\bigoplus_{j=1}^{n}\Big(\bigoplus_{i=1}^{m} (1_i^B \otimes b_j)\Big)_{\mathrm{lt}}\Big)_{\mathrm{lt}} &
\end{array}$$

(1) The path P_1 is the identity if $n = 1$. If $n > 1$, then P_1 has $n-1$ prime edges, each containing an instance of the left distributivity

$$\begin{array}{c}
\Big(\bigoplus_{i=1}^{m} 1_i^B\Big)_{\mathrm{lt}} \otimes \Big(\bigoplus_{l=1}^{n-j+1} b_l\Big)_{\mathrm{lt}} \\
\Big\downarrow \delta^l_{(\oplus_{i=1}^m 1_i^B)_{\mathrm{lt}},\,(\oplus_{l=1}^{n-j} b_l)_{\mathrm{lt}},\, b_{n-j+1}} \\
\Big[\Big(\bigoplus_{i=1}^{m} 1_i^B\Big)_{\mathrm{lt}} \otimes \Big(\bigoplus_{l=1}^{n-j} b_l\Big)_{\mathrm{lt}}\Big] \oplus \Big[\Big(\bigoplus_{i=1}^{m} 1_i^B\Big)_{\mathrm{lt}} \otimes b_{n-j+1}\Big]
\end{array}$$

for $1 \le j \le n-1$.

(2) The path P_2 is the identity if $m = 1$. If $m > 1$, then P_2 has $(m-1)n$ prime edges, each containing an instance of the right distributivity

$$\begin{array}{c}
\Big(\bigoplus_{k=1}^{m-i+1} 1_k^B\Big)_{\mathrm{lt}} \otimes b_j \\
\Big\downarrow \delta^r_{(\oplus_{k=1}^{m-i} 1_k^B)_{\mathrm{lt}},\, 1^B_{m-i+1},\, b_j} \\
\Big[\Big(\bigoplus_{k=1}^{m-i} 1_k^B\Big)_{\mathrm{lt}} \otimes b_j\Big] \oplus \big(1^B_{m-i+1} \otimes b_j\big)
\end{array} \tag{7.3.9}$$

for $1 \le j \le n$ and $1 \le i \le m-1$.

(3) The path P_3 has mn prime edges, each containing an instance of the left multiplicative unit

$$1_i^B \otimes b_j \xrightarrow{\ \lambda^{\otimes}_{b_j}\ } b_j$$

for $1 \le j \le n$ and $1 \le i \le m$.

(4) The path P_4 is the identity if either $m = 1$ or $n = 1$. If $m, n > 1$, then P_4 has $(m-1)(n-1)$ prime edges, each containing an instance of $\alpha^{-\oplus}$.

Since this path P has the desired domain and codomain, it remains to show that its distortion is the identity. We will show that each of P_1, P_2, P_3, and P_4 has identity distortion.

The distortion category $\mathcal{D}$ is a left bipermutative category by Theorem 4.2.29. Its only nonidentity structure isomorphisms are δ^r in (4.2.24), $\xi^{\oplus}$ in (4.2.10), and $\xi^{\otimes}$ in (4.2.17). Therefore, the paths P_1, P_3, and P_4 have identity distortions.

To see that the path P_2 has identity distortion, recall that P_2 is made up of prime edges containing δ^r in (7.3.9). By Definition 3.1.14 and Lemma 4.2.27, it suffices to show that the first entry of each distortion $\vartheta\delta^r$ is the identity permutation. By Definition 3.1.14, (4.2.8), and (4.3.2), the distortion of δ^r in (7.3.9) is the following morphism in $\mathcal{D}$.

$$\begin{aligned}\vartheta\delta^r_{(\oplus_{k=1}^{m-i}1^B_k)_{\mathrm{lt}},\, 1^B_{m-i+1},\, b_j} &= \delta^r_{\vartheta(\oplus_{k=1}^{m-i}1^B_k)_{\mathrm{lt}},\, \vartheta 1^B_{m-i+1},\, \vartheta b_j}\\ &= \delta^r_{(0,\ldots,0),(0),(1)}\end{aligned}$$

Here $(0,\ldots,0)$ has length $m-i$. By Lemma 4.2.27, the first entry of this δ^r in $\mathcal{D}$ is the permutation

$$(\xi^{\otimes}_{1,m-i} \oplus \xi^{\otimes}_{1,1})\xi^{\otimes}_{m-i+1,1} \in \Sigma_{m-i+1}. \tag{7.3.10}$$

This is the identity permutation because both $\xi^{\otimes}_{?,1}$ and $\xi^{\otimes}_{1,?}$ are identity permutations by definition (2.4.5). □

Example 7.3.11. For the case $m = n = 2$, the path P in (7.3.8) has the following 8 prime edges.

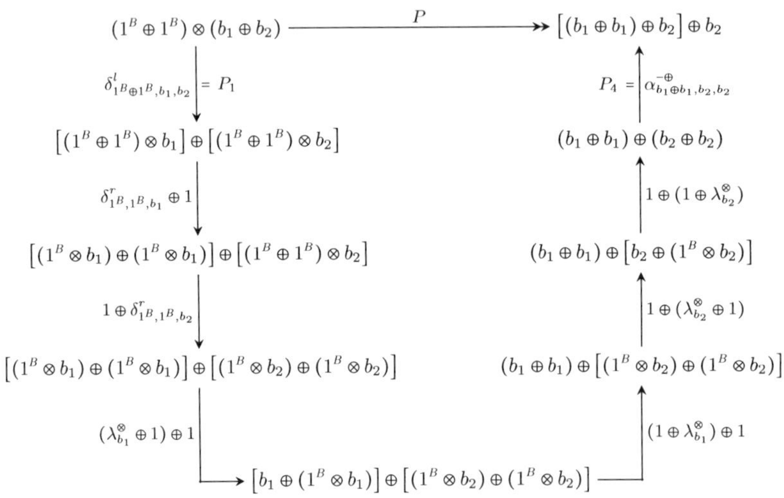

The path P_2 has two prime edges, and the path P_3 has four prime edges. ◇

Now we define the multiplicative structure of F.

Definition 7.3.12. For a flat symmetric bimonoidal category C, extend the functor F in Definition 7.2.2 to the data

$$F_{\otimes} = (F, F^2_{\otimes}, F^0_{\otimes}) \colon (\Sigma, \otimes, 1) \longrightarrow (\mathsf{C}, \otimes, \mathbb{1})$$

of a symmetric monoidal functor as follows.

The Multiplicative Unit Constraint: The morphism

$$\mathbb{1} \xrightarrow{F^0_{\otimes}} F(1) = \mathbb{1} \in \mathsf{C} \tag{7.3.13}$$

is the identity morphism $1_{\mathbb{1}}$.

The Multiplicative Monoidal Constraint: For $m,n \geq 0$, define the morphism

$$\overline{m} \otimes \overline{n} = F(m) \otimes F(n) \xrightarrow{F^2_\otimes} F(mn) = \overline{mn} \in \mathsf{C} \tag{7.3.14}$$

as follows.

(1) If $m = 0$, then $F^2_\otimes$ is the left multiplicative zero

$$\mathbb{0} \otimes \overline{n} \xrightarrow{\lambda^\bullet_{\overline{n}}} \mathbb{0}.$$

(2) If $n = 0$, then $F^2_\otimes$ is the right multiplicative zero

$$\overline{m} \otimes \mathbb{0} \xrightarrow{\rho^\bullet_{\overline{m}}} \mathbb{0}.$$

If $m = n = 0$, then $\lambda^\bullet_{\mathbb{0}} = \rho^\bullet_{\mathbb{0}}$ by the axiom (2.1.14) in C.

(3) If $m,n > 0$, then

$$\Big(\bigoplus_{i=1}^m \mathbb{1}\Big)_{\mathsf{lt}} \otimes \Big(\bigoplus_{j=1}^n \mathbb{1}\Big)_{\mathsf{lt}} \xrightarrow{F^2_\otimes} \Big(\bigoplus_{j=1}^n \bigoplus_{i=1}^m \mathbb{1}\Big)_{\mathsf{lt}}$$

is the value in C of any path P as in (7.3.4).

This finishes the definition of $F_\otimes$. ◇

Lemma 7.3.15. *For a flat symmetric bimonoidal category C, $F^2_\otimes$ in (7.3.14) is well defined and is a natural transformation.*

Proof. To see that $F^2_\otimes$ is well defined in the case $m,n > 0$, recall from Lemma 7.3.7 that such a path P in $\mathsf{Gr}(B)$ with identity distortion exists. In fact, we constructed an explicit path (7.3.8) with identity distortion. Suppose P' is any other path in $\mathsf{Gr}(B)$ with the same (co)domain as P and with identity distortion. Since C is assumed to be flat, the Coherence Theorem 4.4.3 implies that P and P' have the same value in C. Moreover, the value of P in C has the stated (co)domain of $F^2_\otimes$ by the definition (7.3.5) of the function $\varphi\colon B \longrightarrow \mathsf{Ob}(\mathsf{C})$. Therefore, $F^2_\otimes$ is well defined.

To see that $F^2_\otimes$ is a natural transformation, first note that the cases with either $m = 0$ or $n = 0$ follow from the naturality of $\lambda^\bullet$ and $\rho^\bullet$.

For $m,n > 0$, by definition (7.2.4) to check the naturality of $F^2_\otimes$, we need to show the commutativity of the following diagram in C for permutations $\sigma \in \Sigma_m$ and $\tau \in \Sigma_n$.

$$\begin{array}{ccc} \Big(\bigoplus_{i=1}^m \mathbb{1}\Big)_{\mathsf{lt}} \otimes \Big(\bigoplus_{j=1}^n \mathbb{1}\Big)_{\mathsf{lt}} & \xrightarrow{F^2_\otimes} & \Big(\bigoplus_{j=1}^n \bigoplus_{i=1}^m \mathbb{1}\Big)_{\mathsf{lt}} \\ {\scriptstyle \cong^\sigma_{\mathsf{ML}} \otimes \cong^\tau_{\mathsf{ML}}} \downarrow & & \downarrow {\scriptstyle \cong^{\sigma\otimes\tau}_{\mathsf{ML}}} \\ \Big(\bigoplus_{i=1}^m \mathbb{1}\Big)_{\mathsf{lt}} \otimes \Big(\bigoplus_{j=1}^n \mathbb{1}\Big)_{\mathsf{lt}} & \xrightarrow{F^2_\otimes} & \Big(\bigoplus_{j=1}^n \bigoplus_{i=1}^m \mathbb{1}\Big)_{\mathsf{lt}} \end{array} \tag{7.3.16}$$

Consider (7.3.16).

- $\cong^\sigma_{\mathsf{ML}}\colon \overline{m} \longrightarrow \overline{m}$ is the unique Mac Lane coherence isomorphism in the symmetric monoidal category $(\mathsf{C}, \oplus, \mathbb{0})$ that additively permutes the m copies of $\mathbb{1}$ as $\sigma \in \Sigma_m$ permutes m letters.
- $\cong^\tau_{\mathsf{ML}}\colon \overline{n} \longrightarrow \overline{n}$ is interpreted in the same way using $\tau \in \Sigma_n$.

- $\cong_{\mathsf{ML}}^{\sigma\otimes\tau} : \overline{mn} \longrightarrow \overline{mn}$ is interpreted in the same way using the permutation $\sigma \otimes \tau \in \Sigma_{mn}$ in (2.4.4). From Explanation 2.4.7, if $\sigma \otimes \tau$ is regarded as acting on the entries of an $n \times m$ matrix, then τ permutes the n rows, and σ permutes the m columns.

To show the commutativity of the diagram (7.3.16), we first realize its four morphisms as the values in C of the four corresponding sides in the following diagram in $\mathsf{Gr}(B)$, with each b_j^i a copy of $b_j \in B$.

$$\begin{array}{ccc}
\Big(\bigoplus_{i=1}^{m} 1_i^B\Big)_{\mathsf{lt}} \otimes \Big(\bigoplus_{j=1}^{n} b_j\Big)_{\mathsf{lt}} & \xrightarrow{\quad P \quad} & \Big(\bigoplus_{j=1}^{n}\bigoplus_{i=1}^{m} b_j^i\Big)_{\mathsf{lt}} \\
\Big\downarrow P_{\sigma\otimes 1} & & \\
\Big(\bigoplus_{i=1}^{m} 1_{\sigma^{-1}(i)}^B\Big)_{\mathsf{lt}} \otimes \Big(\bigoplus_{j=1}^{n} b_j\Big)_{\mathsf{lt}} & & \Big\downarrow P_{\sigma\otimes\tau} \\
\Big\downarrow P_{1\otimes\tau} & & \\
\Big(\bigoplus_{i=1}^{m} 1_{\sigma^{-1}(i)}^B\Big)_{\mathsf{lt}} \otimes \Big(\bigoplus_{j=1}^{n} b_{\tau^{-1}(j)}\Big)_{\mathsf{lt}} & \xrightarrow{\quad P' \quad} & \Big(\bigoplus_{j=1}^{n}\bigoplus_{i=1}^{m} b_{\tau^{-1}(j)}^{\sigma^{-1}(i)}\Big)_{\mathsf{lt}}
\end{array} \tag{7.3.17}$$

Consider (7.3.17).

- Both paths P and P' have identity distortion. They exist by Lemma 7.3.7.
- $P_{\sigma\otimes 1}$ is a path in which each prime edge has an instance of $\alpha^{\pm\oplus}$ or $\xi^{\pm\oplus}$ applied to the 1_i^B's. They are permuted as $\sigma \in \Sigma_m$ permutes m letters. Such a path exists because each symmetric group is generated by adjacent transpositions that swap two consecutive letters. The same remark applies in the next two items.
- $P_{1\otimes\tau}$ is a path in which each prime edge has an instance of $\alpha^{\pm\oplus}$ or $\xi^{\pm\oplus}$ applied to the b_j's. They are permuted as $\tau \in \Sigma_n$ permutes n letters.
- $P_{\sigma\otimes\tau}$ is a path in which each prime edge has an instance of $\alpha^{\pm\oplus}$ or $\xi^{\pm\oplus}$, such that the b_j^i's are permuted by $\sigma \otimes \tau \in \Sigma_{mn}$ as indicated by the subscript and superscript.

Now we check that the four morphisms in the diagram (7.3.16) are the values in C of the four corresponding paths in (7.3.17).

- The values in C of the paths P and P', which have identity distortion, are both $F_{\otimes}^2$ by the latter's definition.
- The value in C of the left vertical path $(P_{1\otimes\tau}, P_{\sigma\otimes 1})$ is $\cong_{\mathsf{ML}}^{\sigma} \otimes \cong_{\mathsf{ML}}^{\tau}$ by
  - the definitions of the paths $P_{\sigma\otimes 1}$ and $P_{1\otimes\tau}$,
  - the functoriality of $\otimes$ in C, and
  - the uniqueness part in the Symmetric Coherence Theorem 1.3.8.
- Similarly, the value in C of the right vertical path $P_{\sigma\otimes\tau}$ is $\cong_{\mathsf{ML}}^{\sigma\otimes\tau}$ by Theorem 1.3.8.

Therefore, the values in C of the four sides in the diagram (7.3.17) are the four corresponding morphisms in the diagram (7.3.16).

The Coherence Theorem 4.4.3 is applicable to C because it is assumed to be flat. Together with the discussion in the previous paragraph, to show that the diagram (7.3.16) is commutative, it suffices to show that the diagram (7.3.17) is commutative in the distortion category $\mathcal{D}$. Furthermore, since P and P' have identity distortion, it suffices to show that the two vertical paths in (7.3.17) have the same distortion.

By Lemma 7.3.6, all five vertices in (7.3.17) are sent by the graph morphism $\vartheta\colon \mathsf{Gr}(B) \longrightarrow \mathcal{D}$ in Definitions 4.3.1 and 7.3.3 to the same object $(1,\ldots,1) \in \mathcal{D}$. Each entry of this object is 1, and the symmetric group Σ_1 is the trivial group. Therefore, it remains to show that the distortions

$$\vartheta(P_{1\otimes\tau}, P_{\sigma\otimes 1}) \quad \text{and} \quad \vartheta P_{\sigma\otimes\tau}$$

have the same first entry. By (4.2.16),

- the first entry of $\vartheta P_{\sigma\otimes 1}$ is $\sigma \otimes \mathrm{id}_n$, and
- the first entry of $\vartheta P_{1\otimes\tau}$ is $\mathrm{id}_m \otimes \tau$.

By (4.2.4), the first entry of $\vartheta(P_{1\otimes\tau}, P_{\sigma\otimes 1})$ is the permutation

$$\begin{aligned}(\mathrm{id}_m \otimes \tau)(\sigma \otimes \mathrm{id}_n) &= (\mathrm{id}_m\sigma) \otimes (\tau\mathrm{id}_n)\\ &= \sigma \otimes \tau \in \Sigma_{mn}.\end{aligned}$$

Since this is equal to the first entry of $\vartheta P_{\sigma\otimes\tau}$, the two vertical paths in (7.3.17) have the same distortion. □

To prove the rest of the properties of $F^2_\otimes$, we will factor $F^2_\otimes$ using the following definition, which should be compared to Definition 7.3.3.

Definition 7.3.18. Suppose C is a flat symmetric bimonoidal category, and $m, n > 0$.

- Define the set
$$B' = \{0^{B'}, 1^{B'}, a_1, \ldots, a_m, b_1, \ldots, b_n\}$$
with $m + n + 2$ elements.
- Define
$$\Big(\bigoplus_{i=1}^m a_i\Big)_{\mathsf{lt}} \otimes \Big(\bigoplus_{j=1}^n b_j\Big)_{\mathsf{lt}} \xrightarrow{\;Q\;} \Big(\bigoplus_{j=1}^n \bigoplus_{i=1}^m (a_i \otimes b_j)\Big)_{\mathsf{lt}} \tag{7.3.19}$$
as any path in $\mathsf{Gr}(B')$ with identity distortion.
- Define the function $\varphi\colon B' \longrightarrow \mathsf{Ob}(\mathsf{C})$ by
$$\varphi(x) = \begin{cases} \mathbb{0} & \text{if } x = 0^{B'} \text{ and}\\ \mathbb{1} & \text{if } x \in B' \smallsetminus \{0^{B'}\}.\end{cases} \tag{7.3.20}$$
- Define $F'_\otimes$ as the value in C of any path Q in (7.3.19), defined via the associated graph morphism $\varphi\colon \mathsf{Gr}(B') \longrightarrow \mathsf{C}$ in Definition 3.1.14.
- Define $F''_\otimes$ as the left normalized sum of mn copies of the left multiplicative unit $\lambda^\otimes_{\mathbb{1}}\colon \mathbb{1} \otimes \mathbb{1} \longrightarrow \mathbb{1}$. ◇

Lemma 7.3.21. *For a flat symmetric bimonoidal category* C *and* $m, n > 0$*, the morphism*

$$F^2_\otimes\colon \overline{m} \otimes \overline{n} \longrightarrow \overline{mn}$$

in (7.3.14) factors as follows, where $F'_\otimes$ *and* $F''_\otimes$ *are as in Definition 7.3.18.*

$$\begin{array}{ccc}
\Big(\bigoplus_{i=1}^m \mathbb{1}\Big)_{\mathsf{lt}} \otimes \Big(\bigoplus_{j=1}^n \mathbb{1}\Big)_{\mathsf{lt}} & \xrightarrow{\;F^2_\otimes\;} & \Big(\bigoplus_{j=1}^n \bigoplus_{i=1}^m \mathbb{1}\Big)_{\mathsf{lt}}\\
{\scriptstyle F'_\otimes}\Big\downarrow & & \Big\uparrow{\scriptstyle F''_\otimes}\\
 & \Big(\bigoplus_{j=1}^n \bigoplus_{i=1}^m (\mathbb{1} \otimes \mathbb{1})\Big)_{\mathsf{lt}} &
\end{array}$$

Proof. The existence of a path Q with identity distortion as in (7.3.19) is proved by the construction (7.3.8) with only (P_4, P_2, P_1) and with a_i replacing 1_i^B for $1 \le i \le m$. The Coherence Theorem 4.4.3 implies that $F'_\otimes$ is well defined. In other words, if Q' is any other path with the same (co)domain as Q and with identity distortion, then Q and Q' have the same value in C. In particular, we may choose the path Q such that each of its prime edges contains an instance of δ^l, δ^r, or $\alpha^{-\oplus}$.

The morphism $F^2_\otimes$ factors as $F''_\otimes F'_\otimes$ by (i) the construction of the path (7.3.8), whose value in C is $F^2_\otimes$, and (ii) the naturality of $\alpha^\oplus$ in C. □

The next definition and lemma will be used to show that $F^2_\otimes$ satisfies the associativity axiom (1.2.14).

Definition 7.3.22. Suppose $m, n, p > 0$.

- Define the set

$$B'' = \{0^{B''}, 1^{B''}, a_1, \ldots, a_m, b_1, \ldots, b_n, c_1, \ldots, c_p\}$$

with $m + n + p + 2$ elements.
- Define the following diagram in $\mathsf{Gr}(B'')$.

(7.3.23)
$$\begin{array}{ccc}
\Big[\big(\bigoplus_{i=1}^{m} a_i\big)_{\mathsf{lt}} \otimes \big(\bigoplus_{j=1}^{n} b_j\big)_{\mathsf{lt}}\Big] \otimes \big(\bigoplus_{k=1}^{p} c_k\big)_{\mathsf{lt}} & \xrightarrow{\alpha^\otimes} & \big(\bigoplus_{i=1}^{m} a_i\big)_{\mathsf{lt}} \otimes \Big[\big(\bigoplus_{j=1}^{n} b_j\big)_{\mathsf{lt}} \otimes \big(\bigoplus_{k=1}^{p} c_k\big)_{\mathsf{lt}}\Big] \\
\downarrow{\scriptstyle Q_1 \otimes 1} & & \downarrow{\scriptstyle 1 \otimes Q_3} \\
\big(\bigoplus_{j=1}^{n}\bigoplus_{i=1}^{m} (a_i \otimes b_j)\big)_{\mathsf{lt}} \otimes \big(\bigoplus_{k=1}^{p} c_k\big)_{\mathsf{lt}} & & \big(\bigoplus_{i=1}^{m} a_i\big)_{\mathsf{lt}} \otimes \big(\bigoplus_{k=1}^{p}\bigoplus_{j=1}^{n} (b_j \otimes c_k)\big)_{\mathsf{lt}} \\
\downarrow{\scriptstyle Q_2} & & \downarrow{\scriptstyle Q_4} \\
\big(\bigoplus_{k=1}^{p}\bigoplus_{j=1}^{n}\bigoplus_{i=1}^{m} (a_i \otimes b_j) \otimes c_k\big)_{\mathsf{lt}} & \xrightarrow{R} & \big(\bigoplus_{k=1}^{p}\bigoplus_{j=1}^{n}\bigoplus_{i=1}^{m} a_i \otimes (b_j \otimes c_k)\big)_{\mathsf{lt}}
\end{array}$$

  - Up to a change of symbols, each of the paths Q_1, Q_2, Q_3, and Q_4 is the path Q in (7.3.19).
  - $Q_1 \otimes 1$ and $1 \otimes Q_3$ are defined in Notation 3.3.10.
  - The path R has mnp prime edges, each containing one instance of the multiplicative associativity

$$(a_i \otimes b_j) \otimes c_k \xrightarrow{\alpha^\otimes_{a_i, b_j, c_k}} a_i \otimes (b_j \otimes c_k)$$

for $1 \le i \le m$, $1 \le j \le n$, and $1 \le k \le p$. ◇

The following lemma uses Definition 3.1.14.

Lemma 7.3.24. *The diagram (7.3.23) is commutative in each flat symmetric bimonoidal category C with respect to each function $\varphi \colon B'' \longrightarrow \mathsf{Ob}(\mathsf{C})$ that satisfies*

$$\varphi(0^{B''}) = \mathbb{0} \quad \textit{and} \quad \varphi(1^{B''}) = \mathbb{1}.$$

Proof. First we make two remarks.

- By Definition 7.3.18, Q has identity distortion. Therefore, each of the paths $Q_1 \otimes 1$, Q_2, $1 \otimes Q_3$, and Q_4 has identity distortion.
- $\alpha^{\otimes}$ in the distortion category is the identity by Definition 4.2.14. Therefore, the path R has identity distortion.

It follows that both the left-bottom path and the top-right path in (7.3.23) have identity distortion. Since C is assumed to be flat, the Coherence Theorem 4.4.3 implies that these two paths have the same value in C with respect to any choice of the function φ satisfying the stated conditions.

Alternatively, we may also use the Coherence Theorem 3.9.1 by noting that the lower right vertex in (7.3.23) is regular in the sense of Definition 3.1.25. Therefore, the upper left vertex is also regular by Lemma 3.1.29. Theorem 3.9.1 now implies that the two paths have the same value in C. □

Lemma 7.3.25. *For a flat symmetric bimonoidal category* C, $F_{\otimes}$ *in Definition 7.3.12 satisfies the associativity axiom* (1.2.14).

Proof. The axiom (1.2.14) for $F_{\otimes}$ means the commutativity of the diagram

$$\begin{array}{ccc} (\overline{m}\otimes\overline{n})\otimes\overline{p} & \xrightarrow{\alpha^{\otimes}} & \overline{m}\otimes(\overline{n}\otimes\overline{p}) \\ {\scriptstyle F^2_{\otimes}\otimes 1}\big\downarrow & & \big\downarrow{\scriptstyle 1\otimes F^2_{\otimes}} \\ \overline{mn}\otimes\overline{p} & & \overline{m}\otimes\overline{np} \\ {\scriptstyle F^2_{\otimes}}\big\downarrow & & \big\downarrow{\scriptstyle F^2_{\otimes}} \\ \overline{mnp} & \xrightarrow{1} & \overline{mnp} \end{array} \tag{7.3.26}$$

in C for $m,n,p \geq 0$. There are four cases.

(1) If $m = 0$, then (7.3.26) is commutative by the axiom (2.1.22) and the naturality of $\lambda^{\bullet}$ in C.
(2) If $n = 0$, then (7.3.26) is commutative by the axiom (2.1.21) in C.
(3) If $p = 0$, then (7.3.26) is commutative by the axiom (2.1.20) and the naturality of $\rho^{\bullet}$ in C.

For the case $m,n,p > 0$, first we make some remarks.

(i) The diagram

$$\begin{array}{ccc} (\mathbb{1}\otimes\mathbb{1})\otimes\mathbb{1} & \xrightarrow{\alpha^{\otimes}_{\mathbb{1},\mathbb{1},\mathbb{1}}} & \mathbb{1}\otimes(\mathbb{1}\otimes\mathbb{1}) \\ {\scriptstyle \lambda^{\otimes}_{\mathbb{1}}\otimes 1_{\mathbb{1}}}\big\downarrow & & \big\downarrow{\scriptstyle 1_{\mathbb{1}}\otimes\lambda^{\otimes}_{\mathbb{1}}} \\ \mathbb{1}\otimes\mathbb{1} & \xrightarrow{1_{\mathbb{1}\otimes\mathbb{1}}} & \mathbb{1}\otimes\mathbb{1} \\ {\scriptstyle \lambda^{\otimes}_{\mathbb{1}}}\big\downarrow & & \big\downarrow{\scriptstyle \lambda^{\otimes}_{\mathbb{1}}} \\ \mathbb{1} & \xrightarrow{1_{\mathbb{1}}} & \mathbb{1} \end{array}$$

in C is commutative. Indeed, the top square is commutative by the unity axiom (1.2.2) and the equality $\lambda^{\otimes}_{\mathbb{1}} = \rho^{\otimes}_{\mathbb{1}}$ in (1.2.6). The bottom square is commutative by definition.

(ii) $F^2_{\otimes}$ factors as $F''_{\otimes}F'_{\otimes}$ by Lemma 7.3.21, where $F'_{\otimes}$ (respectively, $F''_{\otimes}$) only involves δ^l, δ^r, and $\alpha^{-\oplus}$ (respectively, $\lambda^{\otimes}_{\mathbb{1}}$).
(iii) $\alpha^{\oplus}$, δ^l, and δ^r are natural transformations in C.

These remarks imply that, to prove the commutativity of the diagram (7.3.26) when $m, n, p > 0$, it suffices to show that the diagram (7.3.23) is commutative in C with respect to the function $\varphi\colon B'' \longrightarrow \mathsf{Ob}(\mathsf{C})$ defined by

$$\varphi(x) = \begin{cases} \mathbb{0} & \text{if } x = 0^{B''} \text{ and} \\ \mathbb{1} & \text{if } x \in B'' \smallsetminus \{0^{B''}\}. \end{cases}$$

Therefore, Lemma 7.3.24 finishes the proof. □

Lemma 7.3.27. *In the context of Definition 7.3.12, $F_\otimes$ satisfies the unity axioms* (1.2.15).

Proof. Recall that $F^0_\otimes$ is the identity morphism $1_{\mathbb{1}}$, and $\lambda^\otimes$ in Σ is the identity by Definition 2.4.1. Therefore, the left unity axiom in (1.2.15) for $F_\otimes$ means that the two morphisms

$$\mathbb{1} \otimes \overline{n} \underset{F^2_\otimes}{\overset{\lambda^\otimes_{\overline{n}}}{\rightrightarrows}} \overline{n}$$

in C are equal for each $n \geq 0$. There are two cases.

(1) If $n = 0$, then this $F^2_\otimes$ is $\rho^\bullet_{\mathbb{1}}$, which is equal to $\lambda^\otimes_{\mathbb{0}}$ by the axiom (2.1.18) in C.

(2) If $n > 0$, then the equality $\lambda^\otimes_{\overline{n}} = F^2_\otimes$ follows from
- the definition of $F^2_\otimes$ (7.3.14) in the case $m, n > 0$ and
- the fact that $\lambda^\otimes$ in the distortion category is the identity by Definition 4.2.14.

This proves the left unity axiom in (1.2.15).

The right unity axiom in (1.2.15) is proved in the same way using
- the axiom (2.1.17) instead of (2.1.18) and
- the fact that $\rho^\otimes$ in the distortion category is the identity.

Therefore, $F_\otimes$ satisfies the unity axioms (1.2.15). □

Lemma 7.3.28. *For a flat symmetric bimonoidal category C, $F_\otimes$ in Definition 7.3.12 is a symmetric monoidal functor.*

Proof. By Lemmas 7.3.15, 7.3.25, and 7.3.27, $F_\otimes$ is a monoidal functor. It remains to check the axiom (1.2.26) of a symmetric monoidal functor. This axiom means that the diagram

(7.3.29)
$$\begin{array}{ccc} \overline{m} \otimes \overline{n} & \xrightarrow{\xi^\otimes_{\overline{m},\overline{n}}} & \overline{n} \otimes \overline{m} \\ {\scriptstyle F^2_\otimes}\downarrow & & \downarrow{\scriptstyle F^2_\otimes} \\ \overline{mn} & \xrightarrow{F\xi^\otimes_{m,n}} & \overline{nm} \end{array}$$

in C is commutative for $m, n \geq 0$. In the bottom horizontal arrow, $\xi^\otimes_{m,n} \in \Sigma_{mn}$ is the permutation in (2.4.5). It may be interpreted as taking the transpose of an $n \times m$ matrix as in Explanation 2.4.16. There are three cases.

(1) If $m = 0$, then (7.3.29) becomes the following diagram.

$$\begin{array}{ccc} \mathbb{0} \otimes \overline{n} & \xrightarrow{\xi^\otimes_{\mathbb{0},\overline{n}}} & \overline{n} \otimes \mathbb{0} \\ {\scriptstyle \lambda^\bullet_{\overline{n}}}\downarrow & & \downarrow{\scriptstyle \rho^\bullet_{\overline{n}}} \\ \mathbb{0} & \xrightarrow{1} & \mathbb{0} \end{array}$$

This diagram is commutative by

- the symmetry axiom $\xi^{\otimes}\xi^{\otimes} = 1$ in (1.2.20) and
- the axiom $\rho^{\bullet} = \lambda^{\bullet}\xi^{\otimes}_{-,\mathbb{0}}$ in (2.1.19).

(2) Similarly, if $n = 0$, then (7.3.29) is commutative by the axiom (2.1.19).

For the case $m, n > 0$, by Lemma 7.3.21, the diagram (7.3.29) is the outer diagram below.

(7.3.30)
$$\begin{array}{ccc}
\Big(\bigoplus_{i=1}^{m} \mathbb{1}\Big)_{\mathsf{lt}} \otimes \Big(\bigoplus_{j=1}^{n} \mathbb{1}\Big)_{\mathsf{lt}} & \xrightarrow{\xi^{\otimes}_{\overline{m},\overline{n}}} & \Big(\bigoplus_{j=1}^{n} \mathbb{1}\Big)_{\mathsf{lt}} \otimes \Big(\bigoplus_{i=1}^{m} \mathbb{1}\Big)_{\mathsf{lt}} \\
{\scriptstyle F'_{\otimes}}\downarrow & & \downarrow{\scriptstyle F'_{\otimes}} \\
\Big(\bigoplus_{j=1}^{n}\bigoplus_{i=1}^{m} (\mathbb{1} \otimes \mathbb{1})\Big)_{\mathsf{lt}} & \xrightarrow{\xi^{\otimes}_{m,n}} & \Big(\bigoplus_{i=1}^{m}\bigoplus_{j=1}^{n} (\mathbb{1} \otimes \mathbb{1})\Big)_{\mathsf{lt}} \\
{\scriptstyle (\oplus_j \oplus_i \lambda^{\otimes}_{\mathbb{1}})_{\mathsf{lt}}}\downarrow & & \downarrow{\scriptstyle (\oplus_i \oplus_j \lambda^{\otimes}_{\mathbb{1}})_{\mathsf{lt}}} \\
\Big(\bigoplus_{j=1}^{n}\bigoplus_{i=1}^{m} \mathbb{1}\Big)_{\mathsf{lt}} & \xrightarrow{\xi^{\otimes}_{m,n}} & \Big(\bigoplus_{i=1}^{m}\bigoplus_{j=1}^{n} \mathbb{1}\Big)_{\mathsf{lt}}
\end{array}$$

Consider (7.3.30).

- The upper left $F'_{\otimes}$ is the value in C of the path Q in (7.3.19) with identity distortion.
- The upper right $F'_{\otimes}$ is the same, but with the roles of the a_i's and of the b_j's switched.
- Each of the middle and the bottom horizontal arrows $\xi^{\otimes}_{m,n}$ is the unique Mac Lane coherence isomorphism in $(\mathsf{C}, \oplus, \mathbb{0})$ that additively permutes the mn summands as the transpose of an $n \times m$ matrix.

The bottom square in (7.3.30) is commutative by the naturality of Mac Lane coherence isomorphisms. It remains to show that the top square in (7.3.30) is commutative.

By Proposition II.1.3.26, $\xi^{\otimes}_{\mathbb{1},\mathbb{1}} = 1_{\mathbb{1}\otimes\mathbb{1}}$. It follows that the top square in (7.3.30) is the value in C of the following diagram in $\mathsf{Gr}(B')$ with B' as in Definition 7.3.18.

(7.3.31)
$$\begin{array}{ccc}
\Big(\bigoplus_{i=1}^{m} a_i\Big)_{\mathsf{lt}} \otimes \Big(\bigoplus_{j=1}^{n} b_j\Big)_{\mathsf{lt}} & \xrightarrow{\xi^{\otimes}} & \Big(\bigoplus_{j=1}^{n} b_j\Big)_{\mathsf{lt}} \otimes \Big(\bigoplus_{i=1}^{m} a_i\Big)_{\mathsf{lt}} \\
{\scriptstyle Q}\downarrow & & \downarrow{\scriptstyle Q'} \\
\Big(\bigoplus_{j=1}^{n}\bigoplus_{i=1}^{m} (a_i \otimes b_j)\Big)_{\mathsf{lt}} & \overset{R}{\twoheadrightarrow} & \Big(\bigoplus_{i=1}^{m}\bigoplus_{j=1}^{n} (b_j \otimes a_i)\Big)_{\mathsf{lt}}
\end{array}$$

Consider (7.3.31).

- Q is the path in (7.3.19).
- Q' is the same but with the roles of the a_i's and of the b_j's switched.
- Each prime edge in the path R contains an instance of one of the following:
  - $\alpha^{\pm\oplus}$ or $\xi^{\oplus}$, which additively permute the mn summands as the transpose of an $n \times m$ matrix, or
  - $\xi^{\otimes}_{a_i,b_j} \colon a_i \otimes b_j \longrightarrow b_j \otimes a_i$ for $1 \leq i \leq m$ and $1 \leq j \leq n$.
- The lower right vertex is regular in the sense of Definition 3.1.25. By Lemma 3.1.29, the upper left vertex is also regular.

Since C is assumed to be flat, the Coherence Theorem 3.9.1 implies that the diagram (7.3.31) is commutative in C. So the top square in (7.3.30) is commutative. □

7.4. Weakly Initial Symmetric Bimonoidal Category

The purpose of this section is to prove half of Baez's Conjecture. For each flat symmetric bimonoidal category C as in Definition 3.9.9, we prove that the data

$$(F, F^2_\oplus, F^0_\oplus, F^2_\otimes, F^0_\otimes) \colon \Sigma \longrightarrow \mathsf{C}$$

in Definitions 7.2.2 and 7.3.12 constitute a robust symmetric bimonoidal functor as in Definition 5.1.1. The next definition and lemma will be used to prove the distributivity axiom (5.1.3) for F. Recall from (5.2.13) that the subscript lt means left normalized bracketing.

Definition 7.4.1. Suppose $m, n, p > 0$.

- Define the set
$$B' = \left\{0^{B'}, 1^{B'}, a_1, \ldots, a_m, b_1, \ldots, b_{n+p}\right\}$$
with $m + n + p + 2$ elements. This is the set B' in Definition 7.3.18, but with $n + p$ in place of n.
- Define the following diagram in $\mathsf{Gr}(B')$.

$$\begin{array}{ccc}
\Big(\bigoplus_{i=1}^m a_i\Big)_{\mathsf{lt}} \otimes \Big[\Big(\bigoplus_{j=1}^n b_j\Big)_{\mathsf{lt}} \oplus \Big(\bigoplus_{k=1}^p b_{n+k}\Big)_{\mathsf{lt}}\Big] & \xrightarrow{\delta^l} & \Big[\Big(\bigoplus_{i=1}^m a_i\Big)_{\mathsf{lt}} \otimes \Big(\bigoplus_{j=1}^n b_j\Big)_{\mathsf{lt}}\Big] \oplus \Big[\Big(\bigoplus_{i=1}^m a_i\Big)_{\mathsf{lt}} \otimes \Big(\bigoplus_{k=1}^p b_{n+k}\Big)_{\mathsf{lt}}\Big] \\
\Big\downarrow{\scriptstyle 1 \otimes R_1} & & \Big\downarrow{\scriptstyle Q_2 \oplus 1} \\
\Big(\bigoplus_{i=1}^m a_i\Big)_{\mathsf{lt}} \otimes \Big(\bigoplus_{j=1}^{n+p} b_j\Big)_{\mathsf{lt}} & & \Big(\bigoplus_{j=1}^n \bigoplus_{i=1}^m (a_i \otimes b_j)\Big)_{\mathsf{lt}} \oplus \Big[\Big(\bigoplus_{i=1}^m a_i\Big)_{\mathsf{lt}} \otimes \Big(\bigoplus_{k=1}^p b_{n+k}\Big)_{\mathsf{lt}}\Big] \\
\Big\downarrow{\scriptstyle Q_1} & & \Big\downarrow{\scriptstyle 1 \oplus Q_3} \\
\Big(\bigoplus_{j=1}^{n+p} \bigoplus_{i=1}^m (a_i \otimes b_j)\Big)_{\mathsf{lt}} & \xleftarrow{R_2} & \Big(\bigoplus_{j=1}^n \bigoplus_{i=1}^m (a_i \otimes b_j)\Big)_{\mathsf{lt}} \oplus \Big(\bigoplus_{k=1}^p \bigoplus_{i=1}^m (a_i \otimes b_{n+k})\Big)_{\mathsf{lt}}
\end{array} \tag{7.4.2}$$

  - Up to a change of symbols, each of the paths Q_1, Q_2, and Q_3 is the path Q in (7.3.19).
  - Each prime edge in each of the paths R_1 and R_2 contains an instance of $\alpha^{-\oplus}$.
  - The paths $Q_2 \oplus 1$, $1 \oplus Q_3$, and $1 \otimes R_1$ are as in Notation 3.3.10. ◇

Lemma 7.4.3. *The diagram* (7.4.2) *is commutative in each flat symmetric bimonoidal category* C *with respect to each function* $\varphi \colon B' \longrightarrow \mathsf{Ob}(\mathsf{C})$ *that satisfies*

$$\varphi(0^{B'}) = \mathbb{0} \quad \textit{and} \quad \varphi(1^{B'}) = \mathbb{1}.$$

Proof. The existence of the paths Q_1, Q_2, and Q_3 is proved by the construction (7.3.8) with only (P_4, P_2, P_1) and with a_i replacing 1^B_i for $1 \leq i \leq m$. Each of the paths R_1 and R_2 exists because, in each case, the domain and the codomain agree up to additive bracketings.

The lower left vertex in (7.4.2) is regular in the sense of Definition 3.1.25. Therefore, the upper left vertex is also regular by Lemma 3.1.29. Since C is assumed to be flat, the Coherence Theorem 3.9.1 implies that the two paths have the same value in C. □

Next is the main result of this section that establishes the first half of Baez's Conjecture.

Proposition 7.4.4. *For each flat symmetric bimonoidal category* C, *the data*

$$(F, F^2_\oplus, F^0_\oplus, F^2_\otimes, F^0_\otimes) : \Sigma \longrightarrow \mathsf{C}$$

in Definitions 7.2.2 and 7.3.12 constitute a robust symmetric bimonoidal functor.

Proof. First recall the following facts.

- Lemma 7.2.9 shows that $F_\oplus = (F, F^2_\oplus, F^0_\oplus)$ is a strong symmetric monoidal functor.
- Lemma 7.3.28 shows that $F_\otimes = (F, F^2_\otimes, F^0_\otimes)$ is a symmetric monoidal functor.
- $F^0_\otimes$ is the identity morphism $1_{\mathbb{1}}$ by definition.

Therefore, it remains to check the two axioms in Definition 5.1.1.

The multiplicative zero axiom (5.1.2) holds because

- $F^0_\oplus : \mathbb{0} \longrightarrow \mathbb{0}$ is the identity morphism $1_{\mathbb{0}}$, and
- the right multiplicative zero $\rho^\bullet$ in Σ is the identity.

So each of the two composites in (5.1.2) is the same component of $\rho^\bullet$.

By Proposition 5.1.4, the distributivity axiom (5.1.3) is equivalent to the commutativity of the diagram (5.1.6). Since δ^l in Σ is the identity by Definition 2.4.1, the diagram (5.1.6) is the following diagram in C for $m, n, p \geq 0$.

$$\begin{array}{ccc}
\overline{m} \otimes (\overline{n} \oplus \overline{p}) & \xrightarrow{\delta^l_{\overline{m},\overline{n},\overline{p}}} & (\overline{m} \otimes \overline{n}) \oplus (\overline{m} \otimes \overline{p}) \\
{\scriptstyle 1 \otimes F^2_\oplus} \downarrow & & \downarrow {\scriptstyle F^2_\otimes \oplus F^2_\otimes} \\
\overline{m} \otimes \overline{n+p} & & \overline{mn} \oplus \overline{mp} \\
{\scriptstyle F^2_\otimes} \downarrow & & \downarrow {\scriptstyle F^2_\oplus} \\
\overline{m(n+p)} & \xrightarrow[=]{F\delta^l} & \overline{mn + mp}
\end{array} \tag{7.4.5}$$

To prove the commutativity of the diagram (7.4.5), consider the following four cases.

(1) If $m = 0$, then (7.4.5) is commutative by
- the axiom (2.1.15) in C and
- the naturality of $\lambda^\bullet$.

(2) If $n = 0$, then (7.4.5) factors as follows.

$$\begin{array}{ccccc}
\overline{m} \otimes (\mathbb{0} \oplus \overline{p}) & & \xrightarrow{\delta^l_{\overline{m},\mathbb{0},\overline{p}}} & & (\overline{m} \otimes \mathbb{0}) \oplus (\overline{m} \otimes \overline{p}) \\
{\scriptstyle 1 \otimes \lambda^\oplus_{\overline{p}}} \downarrow & & \swarrow {\scriptstyle \rho^\bullet_{\overline{m}} \oplus 1} & & \downarrow {\scriptstyle \rho^\bullet_{\overline{m}} \oplus F^2_\otimes} \\
\overline{m} \otimes \overline{p} & \xleftarrow{\lambda^\oplus_{\overline{m} \otimes \overline{p}}} & \mathbb{0} \oplus (\overline{m} \otimes \overline{p}) & \xrightarrow{1 \oplus F^2_\otimes} & \mathbb{0} \oplus \overline{mp} \\
{\scriptstyle F^2_\otimes} \downarrow & & & & \downarrow {\scriptstyle \lambda^\oplus_{\overline{mp}}} \\
\overline{mp} & & \xrightarrow{=} & & \overline{mp}
\end{array}$$

- The upper left trapezoid is commutative by the axiom (2.1.23) in C.
- The upper right triangle is commutative by the functoriality of $\oplus$.
- The bottom rectangle is commutative by the naturality of $\lambda^\oplus$.

(3) If $p = 0$, then (7.4.5) factors as follows.

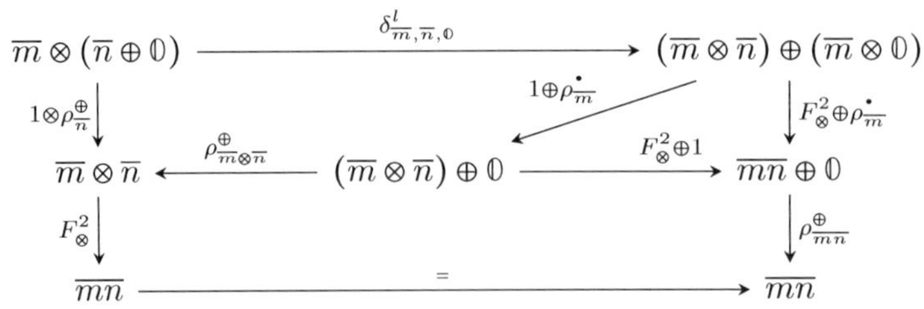

- The upper left trapezoid is commutative by the axiom (2.1.25) in C.
- The upper right triangle is commutative by the functoriality of $\oplus$.
- The bottom rectangle is commutative by the naturality of $\rho^{\oplus}$.

For the case $m, n, p > 0$, we use

- the description of $F^2_{\oplus}$ in Explanation 7.2.7 and
- the factorization of $F^2_{\otimes}$ in Lemma 7.3.21.

They imply that the diagram (7.4.5) is the outer diagram below.

(7.4.6)

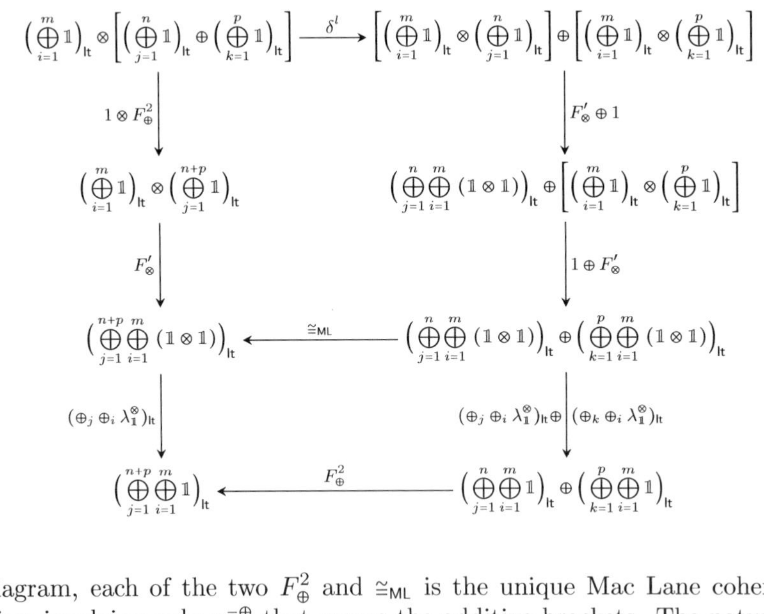

In this diagram, each of the two $F^2_{\oplus}$ and $\cong_{\mathsf{ML}}$ is the unique Mac Lane coherence isomorphism involving only $\alpha^{-\oplus}$ that moves the additive brackets. The naturality of Mac Lane coherence isomorphisms implies that the bottom rectangle in (7.4.6) is commutative.

It remains to show that the top square in (7.4.6) is commutative. In the context of Definition 7.4.1, consider the function $\varphi \colon B' \longrightarrow \mathsf{Ob}(\mathsf{C})$ defined as in (7.3.20) and the associated graph morphism $\varphi \colon \mathsf{Gr}(B') \longrightarrow \mathsf{C}$. Then the top square in (7.4.6) is the value in C of the diagram (7.4.2) in $\mathsf{Gr}(B')$. By Lemma 7.4.3, the top square in (7.4.6) is commutative. $\square$

7.5. Coherence of Symmetric Bimonoidal Functors

The purpose of this section is to prove a coherence property of symmetric bimonoidal functors. In Section 7.7, we will use this coherence property to prove that the canonical robust symmetric bimonoidal functor $F\colon \Sigma \longrightarrow \mathsf{C}$ in Proposition 7.4.4 is initial in a suitable sense. See Lemma 7.7.9 for the precise statement.

The desired coherence property is about the diagram (7.5.2) defined below. It is used in the proof of Lemma 7.7.6. Recall from (5.2.13) that the subscript lt means left normalized bracketing.

Definition 7.5.1. Suppose given the following data.

- C and D are flat symmetric bimonoidal categories, and

$$(G, G^2_\oplus, G^0_\oplus, G^2_\otimes, G^0_\otimes)\colon \mathsf{C} \longrightarrow \mathsf{D}$$

  is a symmetric bimonoidal functor.
- $A_i, B_j \in \mathsf{C}$ are objects for $1 \le i \le m$ and $1 \le j \le n$ with $m, n \ge 1$.

Define the following diagram in D.

$$\begin{array}{ccc}
\Big(\bigoplus_{i=1}^m GA_i\Big)_{\mathsf{lt}} \otimes \Big(\bigoplus_{j=1}^n GB_j\Big)_{\mathsf{lt}} & \xrightarrow{\;p_{m,n}\;} & \Big(\bigoplus_{j=1}^n \bigoplus_{i=1}^m (GA_i \otimes GB_j)\Big)_{\mathsf{lt}} \\
\Big\downarrow {\scriptstyle G_\oplus \otimes G_\oplus} & & \Big\downarrow {\scriptstyle (\oplus_j \oplus_i G^2_\otimes)_{\mathsf{lt}}} \\
G\Big(\bigoplus_{i=1}^m A_i\Big)_{\mathsf{lt}} \otimes G\Big(\bigoplus_{j=1}^n B_j\Big)_{\mathsf{lt}} & & \Big(\bigoplus_{j=1}^n \bigoplus_{i=1}^m G(A_i \otimes B_j)\Big)_{\mathsf{lt}} \\
\Big\downarrow {\scriptstyle G^2_\otimes} & & \Big\downarrow {\scriptstyle G_\oplus} \\
G\Big[\Big(\bigoplus_{i=1}^m A_i\Big)_{\mathsf{lt}} \otimes \Big(\bigoplus_{j=1}^n B_j\Big)_{\mathsf{lt}}\Big] & \xrightarrow{\;Gq_{m,n}\;} & G\Big(\bigoplus_{j=1}^n \bigoplus_{i=1}^m (A_i \otimes B_j)\Big)_{\mathsf{lt}}
\end{array} \tag{7.5.2}$$

Consider (7.5.2).

(1) Each of the three morphisms $G_\oplus$ is a composite of morphisms, each being the sum of identity morphisms and at most one component of the additive monoidal constraint $G^2_\oplus$.

(2) The morphism

$$\Big(\bigoplus_{i=1}^m GA_i\Big)_{\mathsf{lt}} \otimes \Big(\bigoplus_{j=1}^n GB_j\Big)_{\mathsf{lt}} \xrightarrow{\;p_{m,n}\;} \Big(\bigoplus_{j=1}^n \bigoplus_{i=1}^m (GA_i \otimes GB_j)\Big)_{\mathsf{lt}} \tag{7.5.3}$$

is the value in D of the path Q in (7.3.19), taken with respect to the function $\varphi^p\colon B' \longrightarrow \mathsf{Ob}(\mathsf{D})$ defined by

$$\varphi^p(x) = \begin{cases} 0^{\mathsf{D}} & \text{if } x = 0^{B'}, \\ 1^{\mathsf{D}} & \text{if } x = 1^{B'}, \\ GA_i & \text{if } x = a_i \text{ for } 1 \le i \le m, \text{ and} \\ GB_j & \text{if } x = b_j \text{ for } 1 \le j \le n. \end{cases}$$

(3) The morphism

$$\Big(\bigoplus_{i=1}^{m} A_i\Big)_{\mathsf{lt}} \otimes \Big(\bigoplus_{j=1}^{n} B_j\Big)_{\mathsf{lt}} \xrightarrow{q_{m,n}} \Big(\bigoplus_{j=1}^{n}\bigoplus_{i=1}^{m} (A_i \otimes B_j)\Big)_{\mathsf{lt}} \tag{7.5.4}$$

is the value in C of the path Q in (7.3.19), taken with respect to the function $\varphi^q \colon B' \longrightarrow \mathsf{Ob}(\mathsf{C})$ defined by

$$\varphi^q(x) = \begin{cases} \mathbb{0}^{\mathsf{C}} & \text{if } x = 0^{B'}, \\ \mathbb{1}^{\mathsf{C}} & \text{if } x = 1^{B'}, \\ A_i & \text{if } x = a_i \text{ for } 1 \le i \le m, \text{ and} \\ B_j & \text{if } x = b_j \text{ for } 1 \le j \le n. \end{cases}$$

The finishes the definition of the diagram (7.5.2). ◇

Explanation 7.5.5. Consider Definition 7.5.1.

(1) Consider the morphism

$$\Big(\bigoplus_{i=1}^{m} GA_i\Big)_{\mathsf{lt}} \xrightarrow{G_\oplus} G\Big(\bigoplus_{i=1}^{m} A_i\Big)_{\mathsf{lt}} \in \mathsf{D}$$

in the diagram (7.5.2).

- If $m = 1$, then $G_\oplus$ is the identity morphism 1_{GA_1}.
- Inductively, if $m \ge 2$, then $G_\oplus$ is the composite below.

$$\begin{array}{ccc} \Big(\bigoplus_{i=1}^{m} GA_i\Big)_{\mathsf{lt}} & & G\Big(\bigoplus_{i=1}^{m} A_i\Big)_{\mathsf{lt}} \\ \Big\| & & \Big\uparrow G^2_\oplus \\ \Big(\bigoplus_{i=1}^{m-1} GA_i\Big)_{\mathsf{lt}} \oplus GA_m & \xrightarrow{G_\oplus \oplus 1} & G\Big(\bigoplus_{i=1}^{m-1} A_i\Big)_{\mathsf{lt}} \oplus GA_m \end{array} \tag{7.5.6}$$

For example, if $m = 2$, then $G_\oplus$ is the additive monoidal constraint

$$GA_1 \oplus GA_2 \xrightarrow{G^2_\oplus} G(A_1 \oplus A_2).$$

The flatness of C and D are *not* needed to define the morphism $G_\oplus$. Moreover, if $G^2_\oplus$ is an isomorphism, then so is $G_\oplus$.

(2) We use the symbol $G_\oplus$ for both the additive structure $(G, G^2_\oplus, G^0_\oplus)$ of G and the morphism in the diagram (7.5.2). This should not cause any confusion, since the morphism $G_\oplus$ is entirely made up of the additive monoidal constraint $G^2_\oplus$ and identity morphisms.

(3) In the definitions of $p_{m,n}$ and $q_{m,n}$, the existence of the path Q is explained in the first paragraph of the proof of Lemma 7.3.21. The Coherence Theorem 4.4.3, which is applicable by the flatness assumption of C and D, ensures that the value of Q in each of C and D is unique regardless of how Q is chosen. ◇

The objective of this section is to show that the diagram (7.5.2) is commutative. First we consider the following preliminary case.

Lemma 7.5.7. *If $n = 1$, then the diagram (7.5.2) is commutative.*

Proof. We proceed by induction on $m \geq 1$. If $m = 1$, then all three $G_\oplus$'s, q, and g are identity morphisms. So both composites in the diagram (7.5.2) are the morphism

$$GA_1 \otimes GB_1 \xrightarrow{G^2_\otimes} G(A_1 \otimes B_1) \in \mathsf{D}.$$

Inductively, suppose $m \geq 2$. To save space below, we write B_1 as B, $G?$ as $?^G$, and $\otimes$ as concatenation, with $\otimes$ taking precedence over $\oplus$. For example, the upper left corner in (7.5.2) is written as

$$\Big(\bigoplus_{i=1}^m A_i^G\Big)_{\mathsf{lt}}\Big(\bigoplus_{j=1}^n B_j^G\Big)_{\mathsf{lt}}.$$

With these abbreviations, the diagram (7.5.2) with $n = 1$ is the outer diagram below.

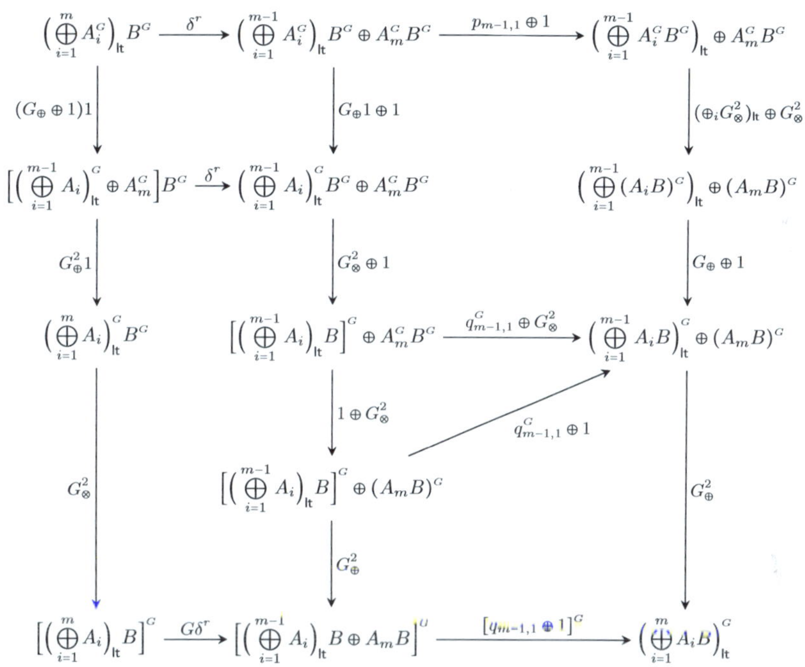

Consider the previous diagram.

- The upper left rectangle is commutative by the naturality of δ^r.
- The lower left rectangle is commutative by the distributivity axiom (5.1.3) for G and the functoriality of $\oplus$.
- The upper right square is commutative by the induction hypothesis and the functoriality of $\oplus$.
- The middle right triangle is commutative by the functoriality of $\oplus$.
- The lower right trapezoid is commutative by the naturality of $G^2_\oplus$.

This finishes the induction step. □

Theorem 7.5.8. *The diagram* (7.5.2) *is commutative for* $m, n \geq 1$.

Proof. We proceed by induction on $n \geq 1$. The case $n = 1$ is proved in Lemma 7.5.7.

Inductively, suppose $n \geq 2$. Using the notation in the proof of Lemma 7.5.7, we factor the diagram (7.5.2) into six subdiagrams as follows.

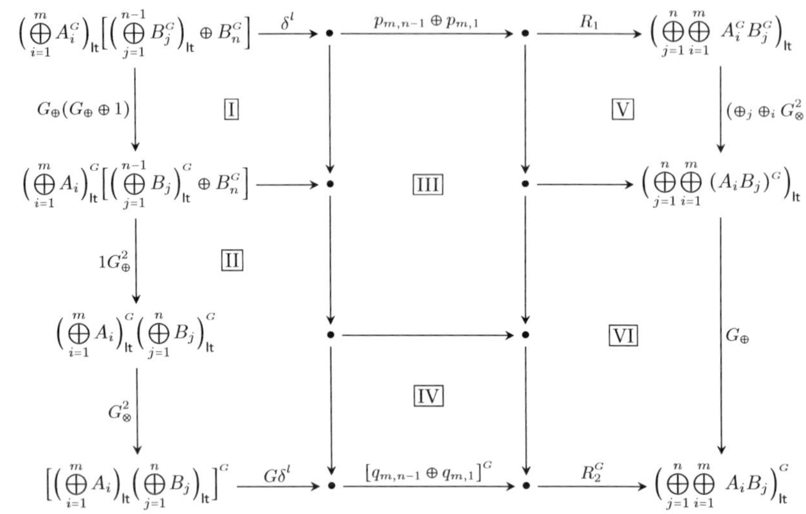

The subdiagrams $\boxed{\text{I}}$ and $\boxed{\text{II}}$ are as follows.

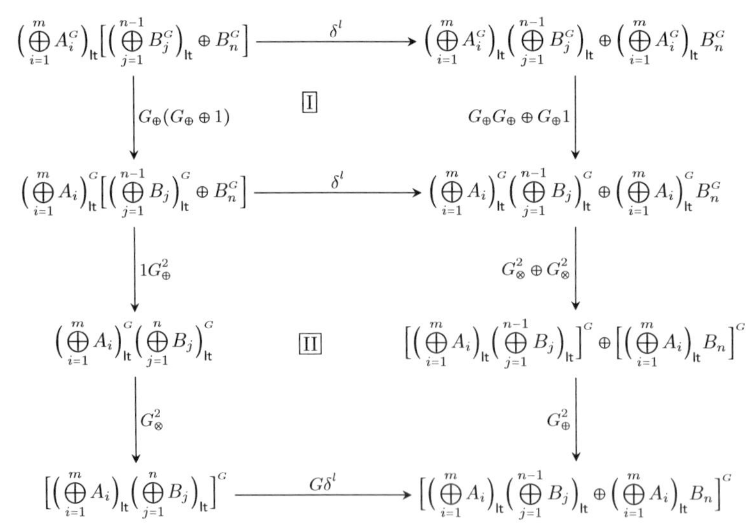

The subdiagram $\boxed{\text{I}}$ is commutative by the naturality of δ^l. The subdiagram $\boxed{\text{II}}$ is commutative by (5.1.6), which is equivalent to the distributivity axiom (5.1.3) for G.

The subdiagrams $\boxed{\text{III}}$ and $\boxed{\text{IV}}$ are as follows.

$$
\begin{array}{ccc}
\Big(\bigoplus_{i=1}^{m} A_i^G\Big)_{\mathsf{lt}}\Big(\bigoplus_{j=1}^{n-1} B_j^G\Big)_{\mathsf{lt}} \oplus \Big(\bigoplus_{i=1}^{m} A_i^G\Big)_{\mathsf{lt}} B_n^G & \xrightarrow{\;p_{m,n-1}\,\oplus\, p_{m,1}\;} & \Big(\bigoplus_{j=1}^{n-1}\bigoplus_{i=1}^{m} A_i^G B_j^G\Big)_{\mathsf{lt}} \oplus \Big(\bigoplus_{i=1}^{m} A_i^G B_n^G\Big)_{\mathsf{lt}} \\
{\scriptstyle G_\oplus G_\oplus \,\oplus\, G_\oplus 1}\Big\downarrow & & \Big\downarrow{\scriptstyle (\oplus_j \oplus_i G_\otimes^2)_{\mathsf{lt}} \,\oplus\, (\oplus_i G_\otimes^2)_{\mathsf{lt}}} \\
\Big(\bigoplus_{i=1}^{m} A_i\Big)^G_{\mathsf{lt}}\Big(\bigoplus_{j=1}^{n-1} B_j\Big)^G_{\mathsf{lt}} \oplus \Big(\bigoplus_{i=1}^{m} A_i\Big)^G_{\mathsf{lt}} B_n^G & \boxed{\text{III}} & \Big(\bigoplus_{j=1}^{n-1}\bigoplus_{i=1}^{m} (A_iB_j)^G\Big)_{\mathsf{lt}} \oplus \Big(\bigoplus_{i=1}^{m} (A_iB_n)^G\Big)_{\mathsf{lt}} \\
{\scriptstyle G_\otimes^2 \,\oplus\, G_\otimes^2}\Big\downarrow & & \Big\downarrow{\scriptstyle G_\oplus \,\oplus\, G_\oplus} \\
\Big[\Big(\bigoplus_{i=1}^{m} A_i\Big)_{\mathsf{lt}}\Big(\bigoplus_{j=1}^{n-1} B_j\Big)_{\mathsf{lt}}\Big]^G \oplus \Big[\Big(\bigoplus_{i=1}^{m} A_i\Big)_{\mathsf{lt}} B_n\Big]^G & \xrightarrow{\;q^G_{m,n-1}\,\oplus\, q^G_{m,1}\;} & \Big(\bigoplus_{j=1}^{n-1}\bigoplus_{i=1}^{m} A_iB_j\Big)^G_{\mathsf{lt}} \oplus \Big(\bigoplus_{i=1}^{m} A_iB_n\Big)^G_{\mathsf{lt}} \\
{\scriptstyle G_\oplus^2}\Big\downarrow & \boxed{\text{IV}} & \Big\downarrow{\scriptstyle G_\oplus^2} \\
\Big[\Big(\bigoplus_{i=1}^{m} A_i\Big)_{\mathsf{lt}}\Big(\bigoplus_{j=1}^{n-1} B_j\Big)_{\mathsf{lt}} \oplus \Big(\bigoplus_{i=1}^{m} A_i\Big)_{\mathsf{lt}} B_n\Big]^G & \xrightarrow{\;[q_{m,n-1}\,\oplus\, q_{m,1}]^G\;} & \Big[\Big(\bigoplus_{j=1}^{n-1}\bigoplus_{i=1}^{m} A_iB_j\Big)_{\mathsf{lt}} \oplus \Big(\bigoplus_{i=1}^{m} A_iB_n\Big)_{\mathsf{lt}}\Big]^G
\end{array}
$$

The subdiagram $\boxed{\text{III}}$ is commutative by

- the induction hypothesis for the summands involving $p_{m,n-1}$ and $q^G_{m,n-1}$;
- the case $n = 1$ in Lemma 7.5.7 for the summands involving $p_{m,1}$ and $q^G_{m,1}$; and
- the functoriality of $\oplus$.

The subdiagram $\boxed{\text{IV}}$ is commutative by the naturality of $G^2_\oplus$.

The subdiagrams $\boxed{\text{V}}$ and $\boxed{\text{VI}}$ are as follows.

$$
\begin{array}{ccc}
\Big(\bigoplus_{j=1}^{n-1}\bigoplus_{i=1}^{m} A_i^G B_j^G\Big)_{\mathsf{lt}} \oplus \Big(\bigoplus_{i=1}^{m} A_i^G B_n^G\Big)_{\mathsf{lt}} & \xrightarrow{\;R_1\;} & \Big(\bigoplus_{j=1}^{n}\bigoplus_{i=1}^{m} A_i^G B_j^G\Big)_{\mathsf{lt}} \\
{\scriptstyle (\oplus_j \oplus_i G_\otimes^2)_{\mathsf{lt}} \,\oplus\, (\oplus_i G_\otimes^2)_{\mathsf{lt}}}\Big\downarrow & \boxed{\text{V}} & \Big\downarrow{\scriptstyle (\oplus_j \oplus_i G_\otimes^2)_{\mathsf{lt}}} \\
\Big(\bigoplus_{j=1}^{n-1}\bigoplus_{i=1}^{m} (A_iB_j)^G\Big)_{\mathsf{lt}} \oplus \Big(\bigoplus_{i=1}^{m} (A_iB_n)^G\Big)_{\mathsf{lt}} & \xrightarrow{\;R\;} & \Big(\bigoplus_{j=1}^{n}\bigoplus_{i=1}^{m} (A_iB_j)^G\Big)_{\mathsf{lt}} \\
{\scriptstyle G_\oplus \,\oplus\, G_\oplus}\Big\downarrow & & \Big\downarrow \\
\Big(\bigoplus_{j=1}^{n-1}\bigoplus_{i=1}^{m} A_iB_j\Big)^G_{\mathsf{lt}} \oplus \Big(\bigoplus_{i=1}^{m} A_iB_n\Big)^G_{\mathsf{lt}} & \boxed{\text{VI}} & \Big\downarrow{\scriptstyle G_\oplus} \\
{\scriptstyle G_\oplus^2}\Big\downarrow & & \Big\downarrow \\
\Big[\Big(\bigoplus_{j=1}^{n-1}\bigoplus_{i=1}^{m} A_iB_j\Big)_{\mathsf{lt}} \oplus \Big(\bigoplus_{i=1}^{m} A_iB_n\Big)_{\mathsf{lt}}\Big]^G & \xrightarrow{\;R_2^G\;} & \Big(\bigoplus_{j=1}^{n}\bigoplus_{i=1}^{m} A_iB_j\Big)^G_{\mathsf{lt}}
\end{array}
$$

Each of R_1, R, and R_2 is a composite of morphisms, each being the sum of identity morphisms and at most one component of $\alpha^{-\oplus}$. Each of them exists because in each case the domain and the codomain agree up to additive bracketings. The subdiagram $\boxed{\text{V}}$ is commutative by the naturality of $\alpha^{\oplus}$. The subdiagram $\boxed{\text{VI}}$ is commutative by Epstein's Coherence Theorem 1.3.12 for the additive structure $(G, G^2_{\oplus})$. This finishes the induction step. $\square$

7.6. Uniqueness of 2-Cells

In this section, we show that for each symmetric bimonoidal category C and robust symmetric bimonoidal functors $G, H \colon \Sigma \longrightarrow \mathsf{C}$, there is at most one bimonoidal natural transformation $G \longrightarrow H$. Moreover, such a bimonoidal natural transformation must be an isomorphism. The main part is Lemma 7.6.2, where we describe explicitly the components of such a bimonoidal natural transformation.

Notation 7.6.1. Recall from Definition 5.1.1 that a robust symmetric bimonoidal functor G has invertible structure morphisms $G^2_{\oplus}$, $G^0_{\oplus}$, and $G^0_{\otimes}$.

- We write the inverses of $G^2_{\oplus}$, $G^0_{\oplus}$, and $G^0_{\otimes}$ as, respectively, $G^{-2}_{\oplus}$, $G^{-0}_{\oplus}$, and $G^{-0}_{\otimes}$.
- The inverse of the morphism $G_{\oplus}$ in Definition 7.5.1 and Explanation 7.5.5 is denoted by $G^{-1}_{\oplus}$. $\diamond$

Recall from Definition 2.4.18 that the objects in Σ are $n \geq 0$.

Lemma 7.6.2. *Suppose given the following data.*

- C *is a symmetric bimonoidal category as in Definition 2.1.2.*
- $G, H \colon \Sigma \longrightarrow \mathsf{C}$ *are symmetric bimonoidal functors with G robust.*
- $\pi \colon G \longrightarrow H$ *is a bimonoidal natural transformation as in Definition 7.1.2.*

Then the following statements hold.

(1) π_0 *is the following composite in* C.

$$\pi_0 \colon G(0) \xrightarrow{G^{-0}_{\oplus}} \mathbb{0} \xrightarrow{H^0_{\oplus}} H(0)$$

(2) π_1 *is the following composite in* C.

$$\pi_1 \colon G(1) \xrightarrow{G^{-0}_{\otimes}} \mathbb{1} \xrightarrow{H^0_{\otimes}} H(1)$$

(3) *For each $n \geq 2$, π_n is the following composite in* C.

$$\begin{array}{ccc} G(n) & \xrightarrow{\quad \pi_n \quad} & H(n) \\ {\scriptstyle G^{-1}_{\oplus}} \big\downarrow & & \big\uparrow {\scriptstyle H_{\oplus}} \\ \Big(\bigoplus_{i=1}^{n} G(1)\Big)_{\mathsf{lt}} & \xrightarrow{(\oplus_i \pi_1)_{\mathsf{lt}}} & \Big(\bigoplus_{i=1}^{n} H(1)\Big)_{\mathsf{lt}} \end{array}$$

Proof. The first two assertions for π_0 and π_1 follow from

- the right diagrams in (7.1.4) and (7.1.5), respectively, for the monoidal natural transformations $\pi \colon G_{\oplus} \longrightarrow H_{\oplus}$ and $\pi \colon G_{\otimes} \longrightarrow H_{\otimes}$; and
- the invertibility of $G^0_{\oplus}$ and $G^0_{\otimes}$.

The last assertion about π_n is proved by induction on $n \geq 2$. The case $n = 2$ follows from

- the left diagram in (7.1.4),
- the invertibility of $G^2_{\oplus}$, and
- the fact that $2 = 1 \oplus 1$ in Σ.

Inductively, suppose $n > 2$. In Σ, $n = (n-1) \oplus 1$. Consider the following diagram in C.

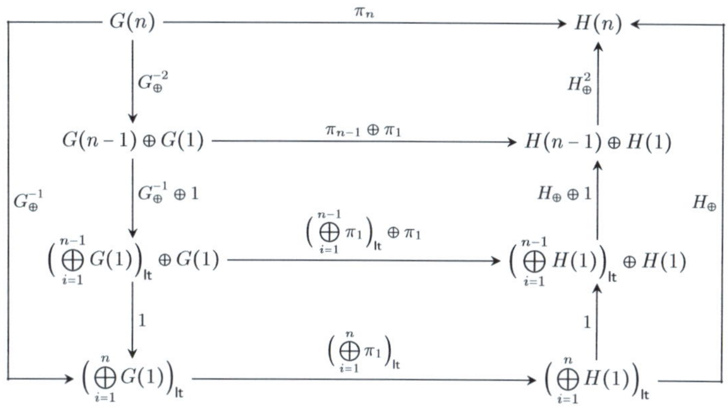

Consider the previous diagram.

- The top rectangle is commutative by the left diagram in (7.1.4) and the invertibility of $G^2_{\oplus}$.
- The middle rectangle is commutative by the induction hypothesis for the summands involving π_{n-1} and the functoriality of $\oplus$.
- The bottom rectangle is commutative by the definition (5.2.13) of the left normalized bracketing.
- By (7.5.6), the left and the right vertical composites are, respectively, $G^{-1}_{\oplus}$ and $H_{\oplus}$.

This finishes the induction step. □

Lemma 7.6.3. *In the context of Lemma 7.6.2, the following statements hold.*

(1) *π is the only bimonoidal natural transformation $G \longrightarrow H$.*
(2) *If H is also robust, then π is a bimonoidal natural isomorphism.*

Proof. The first assertion follows from Lemma 7.6.2 because each component of π is uniquely expressed in terms of the structure morphisms of G and H. The second assertion holds because, if H is also robust, then each component of π is a composite of isomorphisms. □

7.7. Initial 1-Cell

The purpose of this section is to establish the second half of Baez's Conjecture. In Lemma 7.7.9, we observe that the robust symmetric bimonoidal functor $F\colon \Sigma \longrightarrow \mathsf{C}$ in Definitions 7.2.2 and 7.3.12 and Proposition 7.4.4 is initial in a suitable sense. To prove the initial property of F, first we define the components of the expected unique bimonoidal natural transformation.

Definition 7.7.1. Suppose

- C is a flat symmetric bimonoidal category and
- $G\colon \Sigma \longrightarrow \mathsf{C}$ is a symmetric bimonoidal functor.

Define $\theta^G\colon F \longrightarrow G$ with the component morphisms

$$F(n) = \overline{n} \xrightarrow{\theta_n^G} G(n) \in \mathsf{C}$$

for $n \geq 0$ as follows.

- θ_0^G is the additive zero constraint

$$F(0) = \mathbb{0} \xrightarrow{G_\oplus^0} G(0).$$

- θ_1^G is the multiplicative unit constraint

$$F(1) = \mathbb{1} \xrightarrow{G_\otimes^0} G(1).$$

- For each $n \geq 2$, θ_n^G is the following composite in C, with $G_\oplus$ the morphism in (7.5.6).

$$\begin{array}{ccc} F(n) = \overline{n} & \xrightarrow{\quad\theta_n^G\quad} & G(n) \\ \Big\| & & \Big\uparrow{\scriptstyle G_\oplus} \\ \Big(\bigoplus_{j=1}^{n} \mathbb{1}\Big)_{\mathsf{lt}} & \xrightarrow{(\oplus_j G_\otimes^0)_{\mathsf{lt}}} & \Big(\bigoplus_{j=1}^{n} G(1)\Big)_{\mathsf{lt}} \end{array}$$

This finishes the definition of θ^G. ◇

Explanation 7.7.2. The components of θ^G in Definition 7.7.1 are the results of applying the three diagrams in Lemma 7.6.2 to F and G because of the following statements.

(1) $F_\oplus^0$ is the identity morphism $1_\mathbb{0}$ by (7.2.5).
(2) $F_\otimes^0$ is the identity morphism $1_\mathbb{1}$ by (7.3.13).
(3) For each $n \geq 2$, the morphism

$$\Big(\bigoplus_{j=1}^{n} F(1)\Big)_{\mathsf{lt}} = \overline{n} \xrightarrow{F_\oplus} \overline{n} = F(n)$$

is the identity morphism $1_{\overline{n}}$ by (7.2.8) and Explanation 7.5.5. ◇

We now show that θ^G is a bimonoidal natural transformation. To clarify the argument, we separate the proof into several lemmas. Recall from Definition 2.4.1 that morphisms in Σ are permutations.

Lemma 7.7.3. *$\theta^G\colon F \longrightarrow G$ in Definition 7.7.1 is a natural transformation.*

Proof. The naturality of θ^G with respect to morphisms in $\Sigma(n,n) = \Sigma_n$ for $n = 0, 1$ holds because Σ_0 and Σ_1 are the trivial group.

Suppose $\sigma \in \Sigma_n$ is a permutation for some $n \geq 2$. The naturality of θ^G with respect to σ is the outer diagram below.

$$\begin{array}{ccc}
F(n) = \Big(\bigoplus_{j=1}^{n} \mathbb{1}\Big)_{\mathsf{lt}} & \xrightarrow{F(\sigma)} & \Big(\bigoplus_{j=1}^{n} \mathbb{1}\Big)_{\mathsf{lt}} = F(n) \\
{\scriptstyle (\oplus_j G^0_{\otimes})_{\mathsf{lt}}} \big\downarrow & & \big\downarrow {\scriptstyle (\oplus_j G^0_{\otimes})_{\mathsf{lt}}} \\
\Big(\bigoplus_{j=1}^{n} G(1)\Big)_{\mathsf{lt}} & \xrightarrow{\sigma} & \Big(\bigoplus_{j=1}^{n} G(1)\Big)_{\mathsf{lt}} \\
{\scriptstyle G_{\oplus}} \big\downarrow & & \big\downarrow {\scriptstyle G_{\oplus}} \\
G(n) & \xrightarrow{G(\sigma)} & G(n)
\end{array}$$

Consider the previous diagram.

- In the top square, each of $F(\sigma)$ and σ is the unique Mac Lane coherence isomorphism in $(\mathsf{C}, \oplus)$ that additively permutes the n copies of $\mathbb{1}$, or of $G(1)$, as $\sigma \in \Sigma_n$ permutes n letters. The top square is commutative by the naturality of $\alpha^{\oplus}$ and $\xi^{\oplus}$ in C.
- In the bottom square, since

$$n = \Big(\bigoplus_{j=1}^{n} 1\Big)_{\mathsf{lt}} \in \Sigma,$$

  by (7.5.6), each of the two composites is the same component of a G-coherent map in the sense of Definition 1.3.11. Therefore, the bottom square is commutative by Epstein's Coherence Theorem 1.3.12 for the symmetric monoidal functor $(G, G^2_{\oplus}) \colon \Sigma \longrightarrow \mathsf{C}$.

Therefore, θ^G is a natural transformation. $\square$

Lemma 7.7.4. *In Definition 7.7.1,*

$$\theta^G \colon (F, F^2_{\oplus}, F^0_{\oplus}) \longrightarrow (G, G^2_{\oplus}, G^0_{\oplus})$$

is a monoidal natural transformation.

Proof. By Lemma 7.7.3, it remains to check the two axioms in Definition 1.2.16 for the additive structures. The right diagram in (1.2.17) is commutative by the definition of $\theta^G_0 = G^0_{\oplus}$, since $F^0_{\oplus} \colon \mathbb{0} \longrightarrow F(0)$ is the identity morphism $1_{\mathbb{0}}$.

The left diagram in (1.2.17) is the following diagram in C for $m, n \geq 0$.

$$\begin{array}{ccc}
F(m) \oplus F(n) = m \oplus n & \xrightarrow{\theta^G_m \oplus \theta^G_n} & G(m) \oplus G(n) \\
{\scriptstyle F^2_{\oplus}} \big\downarrow & & \big\downarrow {\scriptstyle G^2_{\oplus}} \\
F(m+n) = \overline{m+n} & \xrightarrow{\theta^G_{m+n}} & G(m+n)
\end{array} \tag{7.7.5}$$

To prove its commutativity, we consider all the possible cases.

If $m = 0$, then (7.7.5) is the outer diagram below.

$$\begin{array}{ccccc}
\mathbb{0} \oplus \overline{n} & & \xrightarrow{G^0_\oplus \oplus \theta^G_n} & & G(0) \oplus G(n) \\
 & \searrow^{1 \oplus \theta^G_n} & & \nearrow^{G^0_\oplus \oplus 1} & \\
\Big\downarrow \lambda^\oplus_{\overline{n}} & & \mathbb{0} \oplus G(n) & & \Big\downarrow G^2_\oplus \\
 & & & \searrow^{\lambda^\oplus_{G(n)}} & \\
\overline{n} & & \xrightarrow{\theta^G_n} & & G(n)
\end{array}$$

Starting from the top and going counterclockwise, the three triangles in the diagram above are commutative by, respectively,

- the functoriality of $\oplus$,
- the naturality of $\lambda^\oplus$ in C, and
- the left unity axiom (1.2.15) for $(G, G^2_\oplus, G^0_\oplus)$ and the fact that $\lambda^\oplus$ is the identity in Σ.

If $n = 0$, then (7.7.5) decomposes analogously. The resulting diagram is commutative by the functoriality of $\oplus$, the naturality of $\rho^\oplus$ in C, the right unity axiom (1.2.15) for $(G, G^2_\oplus, G^0_\oplus)$, and the fact that $\rho^\oplus$ is the identity in Σ.

If $m, n \geq 1$, then (7.7.5) is the outer diagram below.

$$\begin{array}{ccccc}
\Big(\bigoplus_{i=1}^m \mathbb{1}\Big)_{\mathsf{lt}} \oplus \Big(\bigoplus_{j=1}^n \mathbb{1}\Big)_{\mathsf{lt}} & \xrightarrow{(\oplus_i G^0_\otimes)_{\mathsf{lt}} \oplus (\oplus_j G^0_\otimes)_{\mathsf{lt}}} & \Big(\bigoplus_{i=1}^m G(1)\Big)_{\mathsf{lt}} \oplus \Big(\bigoplus_{j=1}^n G(1)\Big)_{\mathsf{lt}} & \xrightarrow{G_\oplus \oplus G_\oplus} & G(m) \oplus G(n) \\
\Big\downarrow F^2_\oplus & & \Big\downarrow \cong_{\mathsf{ML}} & & \Big\downarrow G^2_\oplus \\
\Big(\bigoplus_{k=1}^{m+n} \mathbb{1}\Big)_{\mathsf{lt}} & \xrightarrow{(\oplus_k G^0_\otimes)_{\mathsf{lt}}} & \Big(\bigoplus_{k=1}^{m+n} G(1)\Big)_{\mathsf{lt}} & \xrightarrow{G_\oplus} & G(m+n)
\end{array}$$

Consider the previous diagram.

- Both $F^2_\oplus$ and $\cong_{\mathsf{ML}}$ are the unique Mac Lane coherence isomorphism in $(\mathsf{C}, \oplus)$ that involve only $\alpha^{-\oplus}$ and move the additive brackets in the same way. The left square is commutative by the naturality of $\alpha^\oplus$.
- In the right square, if $m = 1$, then the top left $G_\oplus$ is the identity morphism. If $n = 1$, then the top right $G_\oplus$ is the identity morphism. This square is commutative by Epstein's Coherence Theorem 1.3.12 for the symmetric monoidal functor $(G, G^2_\oplus): \Sigma \longrightarrow \mathsf{C}$.

This finishes the proof of the commutativity of (7.7.5). □

Lemma 7.7.6. *In Definition 7.7.1,*

$$\theta^G : (F, F^2_\otimes, F^0_\otimes) \longrightarrow (G, G^2_\otimes, G^0_\otimes)$$

is a monoidal natural transformation.

Proof. By Lemma 7.7.3, it remains to check the two axioms in Definition 1.2.16 for the multiplicative structures. The right diagram in (1.2.17) is commutative by the definition of $\theta^G_1 = G^0_\otimes$, since $F^0_\otimes : \mathbb{1} \longrightarrow F(1)$ is the identity morphism $1_{\mathbb{1}}$.

The left diagram in (1.2.17) is the following diagram in C for $m, n \geq 0$.

$$\begin{array}{ccc} F(m) \otimes F(n) = \overline{m} \otimes \overline{n} & \xrightarrow{\theta^G_m \otimes \theta^G_n} & G(m) \otimes G(n) \\ {\scriptstyle F^2_\otimes}\downarrow & & \downarrow{\scriptstyle G^2_\otimes} \\ F(mn) = \overline{mn} & \xrightarrow{\theta^G_{mn}} & G(mn) \end{array} \tag{7.7.7}$$

To prove its commutativity, we consider all the possible cases.

If $m = 0$, then (7.7.7) is the outer diagram below.

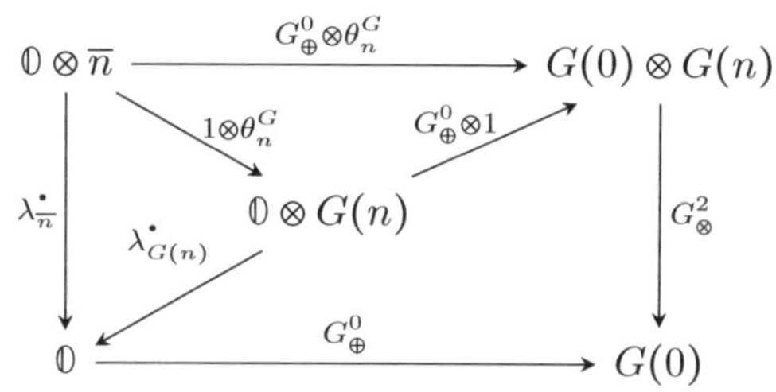

Starting from the top and going counterclockwise, the three triangles in the diagram above are commutative by, respectively,

- the functoriality of $\otimes$,
- the naturality of $\lambda^\bullet$ in C, and
- (5.1.5), which is equivalent to the multiplicative zero axiom (5.1.2) for G, and the fact that $\lambda^\bullet$ is the identity in Σ.

If $n = 0$, then (7.7.7) decomposes analogously. The resulting diagram is commutative by the functoriality of $\otimes$, the naturality of $\rho^\bullet$ in C, the axiom (5.1.2) for G, and the fact that $\rho^\bullet$ is the identity in Σ.

If $m = n = 1$, then (7.7.7) is the outer diagram below.

(7.7.8)

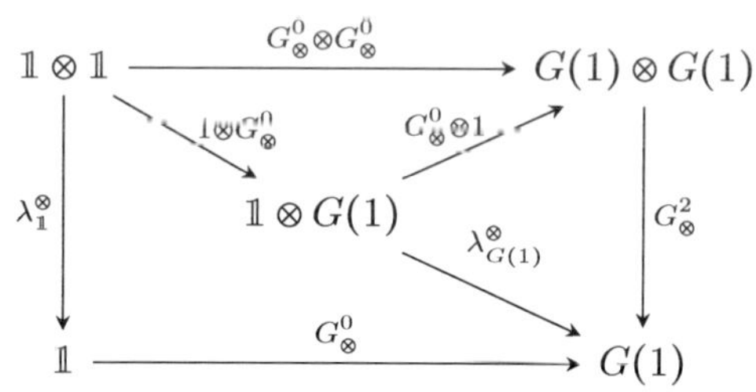

Starting from the top and going counterclockwise, the three triangles in the diagram above are commutative by, respectively,

- the functoriality of $\otimes$,
- the naturality of $\lambda^\otimes$ in C, and
- the left unity axiom (1.2.15) for $(G, G^2_\otimes, G^0_\otimes)$ and the fact that $\lambda^\otimes$ is the identity in Σ.

If $m, n \geq 1$, then (7.7.7) is the outer diagram below.

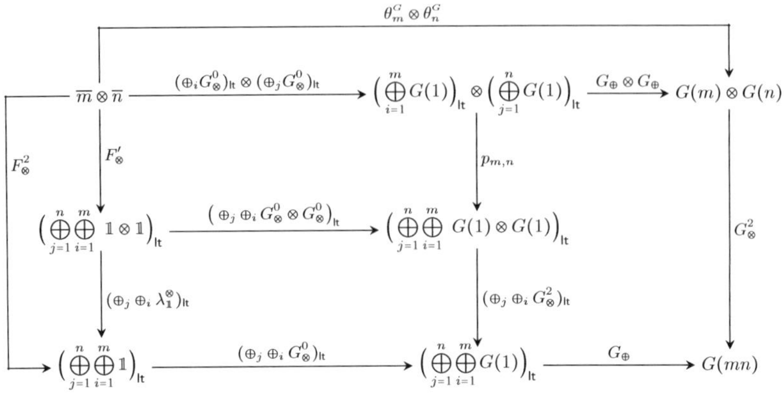

Consider the previous diagram.

- The top rectangle is commutative by the functoriality of $\otimes$ and the definition of θ_n^G for $n \geq 1$. In the top right horizontal morphism, the left $G_\oplus$ is the identity morphism if $m = 1$, and the right $G_\oplus$ is the identity morphism if $n = 1$.
- The left rectangle is the factorization of $F_\otimes^2$ in Lemma 7.3.21.
- The morphism $p_{m,n}$ is the one in (7.5.3) with

$$A_i = \mathbb{1} = B_j \quad \text{for} \quad 1 \leq i \leq m \quad \text{and} \quad 1 \leq j \leq n.$$

  Both $F'_\otimes$ and $p_{m,n}$ are defined using the path Q in (7.3.19), which may be chosen as consisting of δ^l, δ^r, and $\alpha^{-\oplus}$. The upper left square is commutative by the naturality of δ^l, δ^r, and $\alpha^\oplus$ in C.
- The lower left square is commutative by the case $m = n = 1$ in (7.7.8), applied to each $1 \leq i \leq m$ and $1 \leq j \leq n$.
- The right rectangle is the diagram (7.5.2) for $G\colon \Sigma \longrightarrow \mathsf{C}$ with each A_i and each B_j equal to $\mathbb{1}$. This rectangle is commutative by Theorem 7.5.8. Note that in this case, the morphism $q_{m,n}$ in (7.5.4), which is defined by the path Q in (7.3.19), is the identity morphism. Indeed, the only nonidentity structure morphisms in Σ are δ^r, $\xi^\oplus$, and $\xi^\otimes$, the last two of which are not in Q. Each δ^r in $q_{m,n}$ is the identity permutation, as we showed in (7.3.10).

This finishes the proof of the commutativity of (7.7.7). $\square$

Lemma 7.7.9. *Suppose $G\colon \Sigma \longrightarrow \mathsf{C}$ is a robust symmetric bimonoidal functor with C a flat symmetric bimonoidal category. Then there exists a unique bimonoidal natural transformation*

$$\theta\colon F \longrightarrow G.$$

Moreover, θ is as in Definition 7.7.1, and is a bimonoidal natural isomorphism.

Proof. In Lemmas 7.7.3, 7.7.4, and 7.7.6, we showed that $\theta^G\colon F \longrightarrow G$ in Definition 7.7.1 is a bimonoidal natural transformation. The uniqueness and the invertibility of θ^G are established in Lemma 7.6.3. $\square$

7.8. Bi-Initial Symmetric Bimonoidal Category

In this section, we assemble the results so far in this chapter to prove Baez's Conjecture. In fact, we prove two versions, one with Σ and the other with Σ'.

Recall the following.

- Σ is the small tight symmetric bimonoidal category in Proposition 2.4.8.
- $\varnothing$ is the empty 2-category, with no objects, no 1-cells, and no 2-cells.
- $\mathsf{Bi}_{\mathsf{r}}^{\mathsf{fsy}}$ is the 2-category in Definition 7.1.8 with
  - flat small symmetric bimonoidal categories in Definition 3.9.9 as objects,
  - robust symmetric bimonoidal functors in Definition 5.1.1 as 1-cells, and
  - bimonoidal natural transformations in Definition 7.1.2 as 2-cells.

In the following assertion, the precise definition of a *lax bicolimit* can be found in [**JY21**, Section 5.2]. Since we only need it for a special case, we will unpack the following statement in the proof in terms of 1-categorical concepts.

Theorem 7.8.1 (Baez's Conjecture). *Σ is a lax bicolimit of the 2-functor* $\varnothing \longrightarrow \mathsf{Bi}_{\mathsf{r}}^{\mathsf{fsy}}$.

Proof. Since $\varnothing$ is the empty 2-category, the assertion means that for each flat small symmetric bimonoidal category C, the unique functor

$$\mathsf{Bi}_{\mathsf{r}}^{\mathsf{fsy}}(\Sigma, \mathsf{C}) \xrightarrow{\ T\ } \mathbb{1}$$

to the terminal category is an equivalence of categories, that is, fully faithful and essentially surjective.

Since $\mathbb{1}$ is the terminal category, the essential surjectivity of T means the existence of a robust symmetric bimonoidal functor $\Sigma \longrightarrow \mathsf{C}$. Even without the smallness assumption on C, this is true by Proposition 7.4.4, where we constructed a canonical robust symmetric bimonoidal functor $F\colon \Sigma \longrightarrow \mathsf{C}$.

The fully faithfulness of T means that for each pair of robust symmetric bimonoidal functors $G, H\colon \Sigma \longrightarrow \mathsf{C}$, there exists a unique bimonoidal natural transformation $G \longrightarrow H$. Even without the smallness assumption on C, such a bimonoidal natural transformation is given by the vertical composite

$$G \xrightarrow{\ (\theta^G)^{-1}\ } F \xrightarrow{\ \theta^H\ } H$$

with θ^G and θ^H from Lemma 7.7.9. Its uniqueness follows from Lemma 7.6.3. The proof of Baez's Conjecture is now complete. $\square$

Remark 7.8.2. We actually proved something stronger than the statement of Theorem 7.8.1. In the proof of Baez's Conjecture, the smallness of C is only used at the very end to make sure that it is an object in the 2-category $\mathsf{Bi}_{\mathsf{r}}^{\mathsf{fsy}}$. The construction of F starting in Section 7.2 and the proofs of its properties through Lemma 7.7.9 do not require any smallness condition. ◊

Recall the small tight symmetric bimonoidal category Σ' in Definition 2.4.18 and Proposition 2.4.23. It is, furthermore, a right bipermutative category in the sense of Definition 2.5.2. In Proposition 5.1.16, we observed that Σ and Σ' are canonically isomorphic by exhibiting a pair of inverse isomorphisms. Therefore, Theorem 7.8.1 also holds for Σ'.

Theorem 7.8.3 (Baez's Conjecture, Version 2). *Σ' is a lax bicolimit of the 2-functor $\varnothing \longrightarrow \mathsf{Bi}_r^{\mathsf{fsy}}$. In other words, for each flat small symmetric bimonoidal category C, the unique functor*

$$\mathsf{Bi}_r^{\mathsf{fsy}}(\Sigma', \mathsf{C}) \longrightarrow \mathbb{1}$$

is an equivalence of categories.

7.9. Notes

7.9.1 (Nonsymmetric Version). In [**Elg21**, Theorem 4.3.3], Elgueta proved a version of Baez's Conjecture that is different from our Theorems 7.8.1 and 7.8.3 in several aspects. Here we discuss these differences.

Objects. Elgueta's version is about a 2-category **RigCat** whose objects are *rig categories*. A rig category is different from a symmetric bimonoidal category in two ways.

- A rig category does not have a multiplicative symmetry $\xi^{\otimes}$, so the multiplicative structure is a monoidal category, instead of a symmetric monoidal category. The axioms (2.1.5) (relating δ^l and δ^r) and (2.1.19) (symmetry of the multiplicative zeros) involving $\xi^{\otimes}$ are omitted.
- In a rig category, the distributivity morphisms δ^l and δ^r are natural isomorphisms, instead of monomorphisms as in (2.1.4).

Elgueta did not explicitly state that his rig categories should be small. However, as discussed in the introduction of this chapter, to form the 2-category **RigCat**, its objects should be small rig categories.

In our versions of Baez's Conjecture, the objects in the 2-category $\mathsf{Bi}_r^{\mathsf{fsy}}$ are flat small symmetric bimonoidal categories. In particular, δ^l and δ^r in (2.1.4) are natural monomorphisms. Instead of asking for their invertibility, we require the much weaker condition of flatness in Definition 3.9.9.

1-Cells. For 1-cells, Elgueta uses what he calls *homomorphisms*. These are functors whose additive structure is a strong symmetric monoidal functor, and whose multiplicative structure is a strong monoidal functor.

In contrast, our 1-cells in $\mathsf{Bi}_r^{\mathsf{fsy}}$ are robust symmetric bimonoidal functors in Definition 5.1.1.

- The additive structure is a strong symmetric monoidal functor, just like Elgueta's homomorphism.
- The multiplicative structure is a *symmetric* monoidal functor whose multiplicative unit constraint

  $$G^0_{\otimes} \colon \mathbb{1} \longrightarrow G(1)$$

  is an isomorphism. However, the multiplicative monoidal constraint $G^2_{\otimes}$ is not required to be invertible. In particular, in Lemma 7.3.28, we need to check the axiom (1.2.26) of a symmetric monoidal functor for $F_{\otimes}$, which is the diagram (7.3.29).

2-Cells. Elgueta's 2-cells are basically the same as our bimonoidal natural transformations in Definition 7.1.2, which he calls *rig transformations*. The only difference is that rig transformations go between his homomorphisms.

Proof Strategy. Elgueta's proof [**Elg21**, start of Section 4] involves replacing his rig category with an equivalent semistrict rig category, using a strictification theorem due to Guillou [**Gui10**]. See Note 5.6.4. This is closely related to the restriction on the objects of **RigCat**. As we pointed out multiple times in Chapter 5, in particular in Note 5.6.2, the Strictification Theorem 5.4.6 only applies to *tight* symmetric bimonoidal categories, that is, those with δ^l and δ^r natural isomorphisms. The same is true for the multiplicatively nonsymmetric version in [**Gui10**, Def. 2.1]. Therefore, using strictification means that one must insist on the invertibility of the distributivity morphisms.

In contrast, our Theorems 7.8.1 and 7.8.3 do *not* use the Strictification Theorem 5.4.6. Instead, we work directly with a flat small symmetric bimonoidal category in the entire proof. Here are some advantages of this approach:

(1) It allows us to be very specific about where and how the Coherence Theorems 3.9.1 and 4.4.3 and Epstein's Coherence Theorem 1.3.12 are used.
(2) We can make precise all the relevant structure morphisms with respect to the actual symmetric bimonoidal category in question, instead of an equivalent one. See Definitions 7.2.2, 7.3.12, and 7.7.1.
(3) As we pointed out above, we only need our small symmetric bimonoidal categories to be flat, which is a much weaker assumption than the invertibility of the distributivity morphisms. $\diamond$

7.9.2 (Nonsymmetric Version via Sheet Diagrams)**.** The nonsymmetric version of Baez's Conjecture [**Elg21**] discussed in Note 7.9.1 also follows from the sheet diagram results in [**CDH∞**, Section 8]. Both papers [**CDH∞, Elg21**] use the strictification theorem in [**Gui10**]. For strictification of tight bimonoidal categories, our Theorems 5.5.11 and 5.5.12 may also be used. $\diamond$

CHAPTER 8

Symmetric Monoidal Bicategorification

The main theorem of this chapter, Theorem 8.15.4, says that a matrix construction

$$\mathsf{C} \longmapsto \mathsf{Mat}^{\mathsf{C}}$$

sends each tight symmetric bimonoidal category C as in Definition 2.1.2 to a symmetric monoidal bicategory $\mathsf{Mat}^{\mathsf{C}}$ as in Definition 6.5.9. This process is called *symmetric monoidal bicategorification* because it starts with a 1-categorical structure, namely a tight symmetric bimonoidal category C, and produces a symmetric monoidal bicategory $\mathsf{Mat}^{\mathsf{C}}$. There is also a braided analogue in Theorem II.8.4.7.

The main tools for the constructions and proofs in this chapter are the Coherence Theorems 3.9.1 and 3.10.7 and the definitions in Section 3.1. In fact, the many constructions and proofs in this chapter provide a good illustration of the utility of the graph theoretic framework introduced in Section 3.1. For open questions related to the matrix construction, see Appendix A. In the rest of this introduction, we discuss aspects of Theorem 8.15.4 without going into too much detail and provide an outline and a reading guide for this chapter.

Motivation. To motivate the Bicategorification Theorem 8.15.4, consider a rig $(R, +, \times, 0_R, 1_R)$, that is, a ring without additive inverses, such as the set of natural numbers $\mathbb{N}$ with its usual addition and multiplication. Matrices with entries in R have two natural types of products.

- For an $n \times m$ matrix $A = (A_{ji})$ and a $p \times n$ matrix $B = (B_{kj})$, there is the usual matrix product BA, which is a $p \times m$ matrix, whose (k, i)-entry is the sum
$$(BA)_{ki} = \sum_{j=1}^{n} B_{kj} A_{ji}.$$
- For an $n \times m$ matrix A and a $q \times p$ matrix $B = (B_{lk})$, the matrix tensor product, which is also called the Kronecker product,
$$A \otimes B = \begin{bmatrix} A_{11}B & \cdots & A_{1m}B \\ \vdots & \ddots & \vdots \\ A_{n1}B & \cdots & A_{nm}B \end{bmatrix}$$
is the $nq \times mp$ block matrix obtained from A by replacing each entry A_{ji} with the scalar product $A_{ji}B = (A_{ji}B_{lk})_{l,k}$.

These products satisfy some well-known properties, including the following associativity properties, where the first one assumes that the matrix products are defined.

$$\begin{gathered} (AB)C = A(BC) \\ (A \otimes B) \otimes C = A \otimes (B \otimes C) \end{gathered} \tag{8.0.1}$$

Moreover, if R is a multiplicatively commutative rig, such as $\mathbb{N}$, then the *mixed-product property*

$$(B \otimes B')(A \otimes A') = BA \otimes B'A' \tag{8.0.2}$$

holds whenever the matrix products BA and $B'A'$ are defined. These properties imply that for a commutative rig R, there is a permutative category

$$(\mathsf{Mat}^R, \otimes, 1, \xi)$$

as in Definition 1.2.18 with the following data.

- The objects in Mat^R are nonnegative integers.
- The morphisms $m \longrightarrow n$ are $n \times m$ matrices with entries in R.
- The categorical composition sends (B, A) to the matrix product BA.
- The identity morphism on n is the $n \times n$ identity matrix 1_n with the unit 1_R along the diagonal and 0_R in every other entry.
- The monoidal unit is the integer 1.
- The monoidal product $\otimes\colon (\mathsf{Mat}^R)^2 \longrightarrow \mathsf{Mat}^R$ is defined by
  - $m \otimes n = mn$ on objects and
  - the matrix tensor product on morphisms.
- The symmetry isomorphism $\xi_{m,n}\colon mn \longrightarrow nm$ is the permutation matrix obtained from the identity matrix 1_{mn} by permuting its columns using the permutation in (2.4.5).

For example, the mixed-product property (8.0.2) says that $\otimes$ preserves categorical composition. The three symmetric monoidal category axioms for Mat^R are checked in the proof of Proposition 2.4.8. In particular, the hexagon axiom (1.2.22) for Mat^R corresponds to the permutation in (2.4.10).

Symmetric bimonoidal categories are categorical analogues of commutative rigs. For a symmetric bimonoidal category C as in Definition 2.1.2, matrices whose entries are all objects, or all morphisms, in C make sense. Using the sum $\oplus$ and the product $\otimes$ in C, and fixing a convention for iterated sums, the matrix product makes sense for matrices with objects or morphisms in C. Similarly, using $\otimes$ in C, the matrix tensor product makes sense for matrices of objects or morphisms in C. To avoid confusion with the product $\otimes$ in C, for matrices A and B of objects or morphisms in C, their matrix tensor product is denoted by $A \boxtimes B$. The existence of the permutative category Mat^R for a commutative rig R leads to a natural question.

> For a symmetric bimonoidal category C, is there a symmetric monoidal structure on the collection $\mathsf{Mat}^{\mathsf{C}}$ of matrices whose entries are objects or morphisms in C?

A commutative rig R is a *set* with an addition and a commutative multiplication. The matrix construction Mat^R produces a permutative category, that is, a *category* with a multiplication that is symmetric in a categorical sense. Heuristically, we should expect an analogous level-shifting if we start with a symmetric bimonoidal category C. In other words, we expect $\mathsf{Mat}^{\mathsf{C}}$ to be a *bicategory* with a multiplication that is symmetric in a bicategorical sense. Since the two monoidal structures in C are not strict in general, the associativity properties (8.0.1) and the mixed-product property (8.0.2) should hold up to natural isomorphisms. The goal of this chapter is to answer the above question positively: if C is a tight symmetric bimonoidal category, then $\mathsf{Mat}^{\mathsf{C}}$ is a symmetric monoidal bicategory as in

Definition 6.5.9. In general, all the structure morphisms in the symmetric monoidal bicategory $\mathsf{Mat}^{\mathsf{C}}$ are not identities.

Formulation. For each tight symmetric bimonoidal category C, the matrix construction $\mathsf{Mat}^{\mathsf{C}}$ is defined by the following data.

- The objects are nonnegative integers.
- A 1-cell $m \longrightarrow n$ is an $n \times m$ matrix $A = (A_{ji})$ with each entry A_{ji} an object in C.
- For 1-cells $A = (A_{ji}), B = (B_{ji}) \colon m \longrightarrow n$, a 2-cell $f = (f_{ji}) \colon A \longrightarrow B$ is an $n \times m$ matrix with each (j,i)-entry a morphism $f_{ji} \colon A_{ji} \longrightarrow B_{ji}$ in C.
- The horizontal composition
$$\mathsf{Mat}^{\mathsf{C}}_{n,p} \times \mathsf{Mat}^{\mathsf{C}}_{m,n} \longrightarrow \mathsf{Mat}^{\mathsf{C}}_{m,p}$$
is given by the usual matrix product (that is, multiplication of a $p \times n$ matrix with an $n \times m$ matrix to yield a $p \times m$ matrix), defined via $\oplus$ and $\otimes$ in C, applied to matrices of objects or morphisms in C, with the left normalized bracketing for iterated sums.
- The vertical composition of 2-cells is given by entrywise composition of morphisms in C.
- The structure morphisms in C induce canonical associator a, left unitor ℓ, and right unitor r in $\mathsf{Mat}^{\mathsf{C}}$, making it into a bicategory, called the *matrix bicategory*. See Theorem 8.4.12, which holds more generally for tight bimonoidal categories because the multiplicative symmetry $\xi^{\otimes}$ is not needed up to this point.
- For the monoidal bicategory structure, the monoidal composition
$$\mathsf{Mat}^{\mathsf{C}} \times \mathsf{Mat}^{\mathsf{C}} \xrightarrow{\boxtimes} \mathsf{Mat}^{\mathsf{C}}$$
is given by
  - $m \boxtimes n = mn$ on objects and
  - the matrix tensor product on 1-cells and 2-cells, defined via $\otimes$ in C.
- The structure morphisms in C induce the monoidal associator $a^{\boxtimes}$, the left monoidal unitor $\ell^{\boxtimes}$, the right monoidal unitor $r^{\boxtimes}$, the pentagonator π, the middle 2-unitor μ, the left 2-unitor $\lambda^{\boxtimes}$, and the right 2-unitor $\rho^{\boxtimes}$. With these structures, $\mathsf{Mat}^{\mathsf{C}}$ is a monoidal bicategory; see Theorem 8.12.9.
- This monoidal bicategory is equipped with a braiding β induced by permutation matrices, which are defined via the additive zero $\mathbb{0}$ and the multiplicative unit $\mathbb{1}$ in C. The structure morphisms in C induce canonical left and right hexagonators, $R_{-|--}$ and $R_{--|-}$, and a syllepsis ν.

Theorem 8.15.4 states that, with these structures, $\mathsf{Mat}^{\mathsf{C}}$ is a symmetric monoidal bicategory.

Connections. The constructions and results in this chapter have interesting connections with algebraic K-theory and linear algebra.

For each object A in a bicategory B, the hom category $\mathsf{B}(A,A)$ inherits the structure of a monoidal category. Its monoidal product is a restriction of the horizontal composition in B. For the matrix bicategory $\mathsf{Mat}^{\mathsf{C}}$, this means that each $\mathsf{Mat}^{\mathsf{C}}_{n,n}$ is a monoidal category; see Corollary 8.4.14. Its objects are $n \times n$ matrices with entries in C, and its morphisms are $n \times n$ matrices of morphisms in C between corresponding objects. The monoidal product is given by the horizontal composition in $\mathsf{Mat}^{\mathsf{C}}$, that is, the matrix product. The existence of the monoidal category $\mathsf{Mat}^{\mathsf{C}}_{n,n}$ is stated in [**BDR04**, Prop. 3.3] without proof. It is an important ingredient in the algebraic K-theory of C as defined in [**BDR04**, Def. 3.12].

In the braided monoidal bicategory structure of $\mathsf{Mat}^{\mathsf{C}}$, the braiding β is a strong transformation. Its component 1-cells $\beta_{m,n}$ (8.13.23) are permutation matrices in $\mathsf{Mat}^{\mathsf{C}}$. In the usual setting of linear algebra, a column permutation of a matrix A is equal to the matrix product AP for a suitable permutation matrix P. In the setting of $\mathsf{Mat}^{\mathsf{C}}$, this equality is replaced by a natural isomorphism (8.13.13), and similarly for row permutations as in (8.13.16). The component 2-cells $\beta_{A,B}$ of the braiding involves connecting $(B \boxtimes A)\beta_{m,n}$ and $\beta_{m',n'}(A \boxtimes B)$. This in turn involves connecting a column permutation of $B \boxtimes A$ and a row permutation of $A \boxtimes B$. These column and row permutations are generalizations to $\mathsf{Mat}^{\mathsf{C}}$ of examples of the Khatri-Rao product and the face product, which are both related to the Tracy-Singh product in linear algebra. See Explanation 8.13.21.

Non-Strictness of Structures. The symmetric monoidal bicategory $\mathsf{Mat}^{\mathsf{C}}$ is a genuine symmetric monoidal bicategory in the following sense.

(i) The left unitor ℓ, the right unitor r, and especially the associator a are not identities in general. This means that in general $\mathsf{Mat}^{\mathsf{C}}$ is a genuine bicategory and not a 2-category. In fact, the associator a, with entries in (8.3.5), involves $\alpha^{\pm\oplus}$, $\xi^{\oplus}$, $\alpha^{\otimes}$, δ^{r}, and δ^{-l}. It is rare for a symmetric bimonoidal category, even a right or a left bipermutative category as in Definitions 2.5.2 and 2.5.11, to have all of these structure morphisms equal to the identity.

(ii) The monoidal composition $\boxtimes$ is a pseudofunctor, but not a strict functor in general. Its lax functoriality constraint $\boxtimes^2$, with entries in (8.6.20), involves $\alpha^{\oplus}$, $\alpha^{\pm\otimes}$, $\xi^{\otimes}$, δ^{-r}, and δ^{-l} in most cases. Once again, it is rare to have all of these structure morphisms equal to the identity at the same time.

(iii) The symmetric monoidal bicategorical data $a^{\boxtimes}$, $\ell^{\boxtimes}$, $r^{\boxtimes}$, π, μ, $\lambda^{\boxtimes}$, $\rho^{\boxtimes}$, β, $R_{-|--}$, $R_{--|-}$, and ν are also far from strict in general.

For example, suppose C is the tight symmetric bimonoidal category $\mathsf{Vect}^{\mathbb{C}}$ in Example 2.1.32, with finite dimensional complex vector spaces as objects and $\mathbb{C}$-linear maps as morphisms. Then $\mathsf{Mat}^{\mathsf{C}}$ is a genuine symmetric monoidal bicategory as described above. The 1-cells in $\mathsf{Mat}^{\mathsf{C}}$ are called coordinatized 2-vector spaces. See Examples 8.4.13 and 8.15.5.

Symmetry and Tightness. Let us discuss why the multiplicative symmetry $\xi^{\otimes}$ and the tightness assumption in C are needed to construct the symmetric monoidal bicategory $\mathsf{Mat}^{\mathsf{C}}$.

(i) The lax functoriality constraint

$$(B \boxtimes B')(A \boxtimes A') \xrightarrow{\boxtimes^2_{(B,B'),(A,A')}} BA \boxtimes B'A' \in \mathsf{Mat}^{\mathsf{C}}_{mm',pp'}$$

of the monoidal composition $\boxtimes$ is the analogue of the mixed-product property (8.0.2) in $\mathsf{Mat}^{\mathsf{C}}$. The positions of the middle two matrices B' and A are switched. In a typical entry, this switch uses the multiplicative symmetry $\xi^{\otimes}$ in (8.6.17).

(ii) The component 2-cells of the braiding β (8.13.24) involves an entrywise application of $\xi^{\otimes}$.

(iii) Recall that a symmetric bimonoidal category is *tight* if the distributivity morphisms δ^r and δ^l are natural isomorphisms, instead of natural monomorphisms. The Coherence Theorems 3.9.1 and 4.4.3 do not require tightness, but only a condition about monomorphisms. For instance, flatness as in Definition 3.9.9 is enough for the two coherence theorems. However, as we mentioned above, in $\mathsf{Mat}^{\mathsf{C}}$,
   - the associator a involves δ^{-l}, and
   - the lax functoriality constraint $\boxtimes^2$ of the monoidal composition involves both δ^{-r} and δ^{-l}.

   Therefore, the distributivity morphisms in C must be invertible for these constructions.

Constructions and Proofs. To define the many parts of the symmetric monoidal bicategory $\mathsf{Mat}^{\mathsf{C}}$, in each case, we write down the matrices that we want to connect. Then we use the most straightforward combination of structure morphisms in C to connect those matrices. For example, for the left unitor this step is performed in (8.2.3) and (8.2.4). The difficult part is in showing that the structures so defined have the required properties for $\mathsf{Mat}^{\mathsf{C}}$ to be a symmetric monoidal bicategory. We accomplish this by consistently reducing each such proof to an instance of the Coherence Theorem 3.9.1 or the simpler case in Proposition 3.5.33. These coherence results are formulated in terms of the graph theoretic framework in Section 3.1, which we use throughout this chapter.

There is an additional subtlety in a number of proofs in this chapter that arises from the following two facts.

- The associator a in $\mathsf{Mat}^{\mathsf{C}}$, with entries in (8.3.5), involves both δ^r and δ^{-l}.
- The lax functoriality constraint $\boxtimes^2$, with entries in (8.6.20), of the monoidal composition $\boxtimes$ in $\mathsf{Mat}^{\mathsf{C}}$ involves δ^{-r} and δ^{-l} in most cases.

The Coherence Theorem 3.9.1 and the simpler case in Proposition 3.5.33 are stated in terms of paths in $\mathsf{Gr}(X)$ in Definition 3.1.9. By Definition 3.1.6, the elementary edges δ^l and δ^r do *not* have formal inverses, so paths cannot have formal inverses of δ^l and δ^r. As a result, in some of the proofs that involve the associator a or the lax functoriality constraint $\boxtimes^2$, we will need to subdivide the diagram in question with additional morphisms in such a way that the coherence results can be applied in each subdiagram.

Relatively simple examples of this subdivision process are the following subdivided diagrams:

- (8.4.5) for the unity axiom in $\mathsf{Mat}^{\mathsf{C}}$ and
- the diagrams in cases (6)–(8) in the proof of Lemma 8.4.8 for the pentagon axiom.

More involved and typical examples are the following subdivided diagrams:

- (8.4.10) for the general case of the pentagon axiom in $\mathsf{Mat}^{\mathsf{C}}$;
- (8.7.26) for the lax associativity axiom for the pseudofunctor $\boxtimes$;
- (8.8.28) for the lax naturality axiom for the monoidal associator $a^{\boxtimes}$;
- (8.8.44) for the modification axiom for the unit $\eta^a\colon 1_{\boxtimes(\boxtimes\times 1)} \longrightarrow a^{\boxtimes\bullet}a^{\boxtimes}$; and
- (8.10.8) for the modification axiom for the pentagonator π.

Notation Guide. In Sections 8.1 through 8.5,

$$\big(\mathsf{C},(\oplus,\mathbb{0},\alpha^{\oplus},\lambda^{\oplus},\rho^{\oplus},\xi^{\oplus}),(\otimes,\mathbb{1},\alpha^{\otimes},\lambda^{\otimes},\rho^{\otimes}),(\lambda^{\bullet},\rho^{\bullet}),(\delta^{l},\delta^{r})\big)$$

denotes a bimonoidal category as in Definition 2.1.2. In the remaining sections, C is a symmetric bimonoidal category with multiplicative symmetry $\xi^{\otimes}$. If we want C to be *flat* as in Definition 3.9.9, or *tight*—that is, to have invertible distributivity morphisms δ^l and δ^r—we will state so explicitly, sometimes at the beginning of a section. In the absence of clarifying parentheses, $\otimes$ takes precedence over $\oplus$, and $\otimes$ is sometimes omitted to save space.

For the bicategory $\mathsf{Mat}^{\mathsf{C}}$, we sometimes add the adjective *base* before the associator and the unitors. To avoid confusion with the two monoidal products $\{\oplus,\otimes\}$ in C, the monoidal composition in $\mathsf{Mat}^{\mathsf{C}}$ is denoted by $\boxtimes$. Similarly, to avoid confusion with the base associator and the base unitors, the monoidal associator $a^{\boxtimes}$ and the monoidal unitors $\ell^{\boxtimes}$ and $r^{\boxtimes}$ are decorated with the superscript $\boxtimes$. Moreover, the left and the right 2-unitors $\lambda^{\boxtimes}$ an $\rho^{\boxtimes}$ are also decorated with $\boxtimes$ to avoid confusion with the various λ and ρ in C.

Outline. In Section 8.1, we define the hom categories $\mathsf{Mat}^{\mathsf{C}}_{m,n}$ and the matrix product, which will form the horizontal composition. We also discuss coherence properties of the matrix product when one of the matrices is a 0 matrix. The left unitor ℓ and the right unitor r in $\mathsf{Mat}^{\mathsf{C}}$ are defined in Section 8.2. The associator a is defined in Section 8.3. The unity axiom (6.1.3) and the pentagon axiom (6.1.4) are verified in Section 8.4. By the end of this section, we will have proved that $\mathsf{Mat}^{\mathsf{C}}$ is a bicategory.

In the next few sections, we construct a monoidal bicategory structure on the matrix bicategory $\mathsf{Mat}^{\mathsf{C}}$. The monoidal identity is constructed in Section 8.5. The matrix tensor product and the monoidal composition $\boxtimes$ are defined in Section 8.6. The lax associativity axiom (6.2.2) and the lax unity axiom (6.2.3) for the pseudofunctor $\boxtimes$ are verified in Section 8.7. The monoidal associator $a^{\boxtimes}$ is constructed in Section 8.8. The left monoidal unitor $\ell^{\boxtimes}$ and the right monoidal unitor $r^{\boxtimes}$ are constructed in Section 8.9. The pentagonator is defined in Section 8.10. The middle 2-unitor μ, the left 2-unitor $\lambda^{\boxtimes}$, and the right 2-unitor $\rho^{\boxtimes}$ are defined in Section 8.11. The fact that $\mathsf{Mat}^{\mathsf{C}}$ is a monoidal bicategory is stated and verified in Section 8.12.

Permutation matrices, row and column permutations, and the braiding β are defined in Section 8.13. In Section 8.14, we define the left and the right hexagonators, and verify that $\mathsf{Mat}^{\mathsf{C}}$ is a braided monoidal bicategory. In Section 8.15, we define the syllepsis ν, and conclude the proof of the Bicategorification Theorem 8.15.4.

Conventions. The following conventions are in effect throughout this chapter.

- A *Mac Lane coherence isomorphism*, which is denoted by $\cong^{\oplus}_{\mathsf{ML}}$, means the value in C of a path that only involves identities, $\alpha^{\pm\oplus}$, $\lambda^{\pm\oplus}$, $\rho^{\pm\oplus}$, and $\xi^{\pm\oplus}$.
- A *Laplaza coherence isomorphism*, which is denoted by $\cong_{\mathsf{Lap}}$, means the value in C of a path that does *not* involve $\xi^{\pm\otimes}$.

These are adaptations of the notions in Definition 5.2.24 to the current context.

Reading Guide. Instead of reading this chapter linearly, the reader may first read the key definitions and statements of results listed below to obtain a general idea of the structure of $\mathsf{Mat}^{\mathsf{C}}$.

(1) The base bicategory.
- Definition 8.1.1: The hom categories $\mathsf{Mat}^{\mathsf{C}}_{m,n}$.
- Definition 8.1.3: The matrix product and the identity matrices.
- Lemmas 8.2.1 and 8.2.7: The base unitors ℓ and r.
- Lemma 8.3.1: The base associator a.
- Definition 8.4.11: The matrix bicategory $\mathsf{Mat}^{\mathsf{C}}$.

(2) The monoidal bicategory.
- Definition 8.5.1: The monoidal identity $1_{\boxtimes}\colon \mathbb{1} \longrightarrow \mathsf{Mat}^{\mathsf{C}}$.
- Definition 8.6.1: The matrix tensor product $A \boxtimes B$.
- Lemma 8.6.8: The lax unity constraint $\boxtimes^0$.
- Definition 8.6.19: The monoidal composition $\boxtimes$.
- Definitions 8.8.33 and 8.8.37: The monoidal associator $(a^{\boxtimes}, a^{\boxtimes\bullet}, \eta^a, \varepsilon^a)$.
- Definitions 8.9.1, 8.9.5, and 8.9.8: The left monoidal unitor.
- Definitions 8.9.14, 8.9.17, and 8.9.20: The right monoidal unitor.
- Definition 8.10.1: The pentagonator π.
- Definitions 8.11.1, 8.11.6, and 8.11.11: The 2-unitors μ, $\lambda^{\boxtimes}$, and $\rho^{\boxtimes}$.

(3) The symmetric monoidal bicategory.
- Definition 8.13.3: Row/column permutations and permutation matrices.
- Lemmas 8.13.12 and 8.13.15: Taking matrix products with permutation matrices and row/column permutations.
- Definitions 8.13.22, 8.13.30, 8.13.35, and 8.13.38: The braiding.
- Lemma 8.14.4 and Definition 8.14.9: The left hexagonator $R_{-|--}$.
- Lemma 8.14.16 and Definition 8.14.21: The right hexagonator $R_{--|-}$.
- Definition 8.15.1: The syllepsis ν.
- Theorem 8.15.4: The Bicategorification Theorem.
- Example 8.15.5: Applying the Bicategorification Theorem.

Before reading the proofs in the rest of this chapter, the reader may want to review Section 3.1. As in Chapters 3, 4, 5, and 7, the reader may consider the many detailed constructions and proofs in this chapter as exercises with full solutions.

8.1. Matrix Construction

To construct the bicategory $\mathsf{Mat}^{\mathsf{C}}$ of matrices for a tight bimonoidal category C, in this section, we first define its hom categories and the associated matrix product. The latter will eventually be used to define the horizontal composition in the bicategory $\mathsf{Mat}^{\mathsf{C}}$. In the second half of this section, we discuss coherence properties of multiplying by a 0 matrix. These coherence properties will be used in Lemma 8.3.1 when we discuss associativity properties of the matrix product. Coherence properties of the matrix product with respect to the identity matrices and associativity will be discussed in Sections 8.2 and 8.3.

The Matrix Product.

Definition 8.1.1. Suppose C is a category. For integers $m, n \geq 0$, define a category $\mathsf{Mat}^{\mathsf{C}}_{m,n}$ as follows.

Objects: An object in $\mathsf{Mat}^{\mathsf{C}}_{m,n}$ is an $n \times m$ matrix

$$A = (A_{ji})_{1 \leq j \leq n,\, 1 \leq i \leq m}$$

with each A_{ji} an object in C. We call A an $n \times m$ *matrix in* C. When n and m are understood, we will also write A as (A_{ji}).

Morphisms: A morphism

$$f\colon A = (A_{ji}) \longrightarrow (A'_{ji}) = A'$$

is an $n \times m$ matrix

$$f = (f_{ji})_{1 \le j \le n,\, 1 \le i \le m}$$

with each $f_{ji}\colon A_{ji} \longrightarrow A'_{ji}$ a morphism in C.

Identity Morphisms: For an object $A = (A_{ji})$, its identity morphism is the $n \times m$ matrix

$$1_A = \left(1_{A_{ji}}\right)$$

of identity morphisms in C.

Composition: If $f\colon A \longrightarrow A'$ is a morphism as above and if $f'\colon A' \longrightarrow A''$ is another morphism, then their composite $f'f\colon A \longrightarrow A''$ is defined entrywise in C as

$$(f'f)_{ji} = f'_{ji} f_{ji}\colon A_{ji} \longrightarrow A''_{ji}.$$

This finishes the definition of the category $\mathsf{Mat}^{\mathsf{C}}_{m,n}$.

Moreover, we define the following.

- An object in $\mathsf{Mat}^{\mathsf{C}}_{m,n}$ is also called a *matrix.*
- Consider an $n \times m$ matrix $A = (A_{ji})$.
  - $A_{ji} \in \mathsf{C}$ is called the (j,i)*-entry.*
  - For each $1 \le j \le n$, the j*th row* is the $1 \times m$ matrix $(A_{ji})_{1 \le i \le m}$.
  - For each $1 \le i \le m$, the i*th column* is the $n \times 1$ matrix $(A_{ji})_{1 \le j \le n}$.
- If C has a distinguished object $\mathbb{0}$, then the *0 matrix* $\mathbb{0}_{m,n} \in \mathsf{Mat}^{\mathsf{C}}_{m,n}$ is the matrix with each entry $\mathbb{0} \in \mathsf{C}$ if $m, n > 0$. If either m or n is 0, then $\mathbb{0}_{m,n}$ denotes the empty matrix.
- An $n \times n$ matrix is also called a *square matrix.*
- Consider an $n \times n$ square matrix $A = (A_{ji})$.
  - A_{jj} is said to be *on the diagonal*, and is called the j*th diagonal entry* for $1 \le j \le n$.
  - A_{ji} with $j \ne i$ is said to be *off the diagonal*, and is called an *off-diagonal entry.*
  - A_{ji} with $j > i$ is said to be *below the diagonal*, and is called a *below-diagonal entry.*
  - A_{ji} with $j < i$ is said to be *above the diagonal*, and is called an *above-diagonal entry.*

Similar terminology applies to morphisms in $\mathsf{Mat}^{\mathsf{C}}_{m,n}$. ◇

Explanation 8.1.2. Consider Definition 8.1.1.

- $\mathsf{Mat}^{\mathsf{C}}_{m,n}$ is indeed a category because everything is defined entrywise in C.
- If either m or n is 0, then $\mathsf{Mat}^{\mathsf{C}}_{m,n}$ is the terminal category with one object—namely, the unique empty matrix—and its identity morphism.
- If $m = n = 1$, then $\mathsf{Mat}^{\mathsf{C}}_{1,1}$ is canonically identified with C. ◇

We now extend Definition 8.1.1 to include the matrix product and identity matrices, which will be used to define, respectively, the horizontal composition and the identity 1-cells in the bicategory $\mathsf{Mat}^{\mathsf{C}}$. Recall from (5.2.13) that the subscript lt means left normalized bracketing.

Definition 8.1.3. Suppose C is a bimonoidal category, and $m, n, p \geq 0$.

- For
  - an $n \times m$ matrix $A = (A_{ji}) \in \mathsf{Mat}^{\mathsf{C}}_{m,n}$ and
  - a $p \times n$ matrix $B = (B_{kj}) \in \mathsf{Mat}^{\mathsf{C}}_{n,p}$,

  define their *matrix product* $BA \in \mathsf{Mat}^{\mathsf{C}}_{m,p}$ whose (k,i)-entry, for $1 \leq i \leq m$ and $1 \leq k \leq p$, is the following object in C:

$$(BA)_{ki} = \begin{cases} \Big(\bigoplus_{j=1}^{n} (B_{kj} \otimes A_{ji})\Big)_{\mathsf{lt}} & \text{if } n \geq 1 \text{ and} \\ \mathbb{0} & \text{if } n = 0. \end{cases} \tag{8.1.4}$$

  If either m or p is 0, then BA is the empty matrix.
- For morphisms
  - $f = (f_{ji}) \in \mathsf{Mat}^{\mathsf{C}}_{m,n}(A, A')$ and
  - $g = (g_{kj}) \in \mathsf{Mat}^{\mathsf{C}}_{n,p}(B, B')$,

  define their *matrix product*

$$g \star f \in \mathsf{Mat}^{\mathsf{C}}_{m,p}(BA, B'A')$$

  as the $p \times m$ matrix whose (k,i)-entry, for $1 \leq i \leq m$ and $1 \leq k \leq p$, is the following morphism in C:

$$(g \star f)_{ki} = \begin{cases} \Big(\bigoplus_{j=1}^{n} (g_{kj} \otimes f_{ji})\Big)_{\mathsf{lt}} : (BA)_{ki} \longrightarrow (B'A')_{ki} & \text{if } n \geq 1 \text{ and} \\ 1_{\mathbb{0}} : \mathbb{0} \longrightarrow \mathbb{0} & \text{if } n = 0. \end{cases} \tag{8.1.5}$$

  If either m or p is 0, then $g \star f$ is the identity morphism of the empty matrix.
- The $n \times n$ *identity matrix* is the square matrix $\mathbb{1}^n \in \mathsf{Mat}^{\mathsf{C}}_{n,n}$ with entries

$$\mathbb{1}^n_{ji} = \begin{cases} \mathbb{1} & \text{if } i = j \text{ and} \\ \mathbb{0} & \text{if } i \neq j. \end{cases} \tag{8.1.6}$$

  If $n = 0$, then $\mathbb{1}^0$ is the unique empty matrix. ◇

Explanation 8.1.7. In Definition 8.1.3, suppose $m, p > 0$ and $n = 0$.

- Then $A \in \mathsf{Mat}^{\mathsf{C}}_{m,n}$ and $B \in \mathsf{Mat}^{\mathsf{C}}_{n,p}$ are both empty matrices. However, in the matrix product (8.1.4), $BA \in \mathsf{Mat}^{\mathsf{C}}_{m,p}$ is the 0 matrix $\mathbb{0}_{m,p}$ in which each entry is the additive zero $\mathbb{0} \in \mathsf{C}$.
- $f \in \mathsf{Mat}^{\mathsf{C}}_{m,n}(A, A')$ and $g \in \mathsf{Mat}^{\mathsf{C}}_{n,p}(B, B')$ are both identity morphisms of the empty matrix. In the matrix product (8.1.5), $g \star f$ is the identity morphism of the 0 matrix $\mathbb{0}_{m,p}$. ◇

Lemma 8.1.8. *Suppose C is a bimonoidal category, and $m, n, p \geq 0$. Then the matrix product*

$$\mathsf{Mat}^{\mathsf{C}}_{n,p} \times \mathsf{Mat}^{\mathsf{C}}_{m,n} \longrightarrow \mathsf{Mat}^{\mathsf{C}}_{m,p}$$

in Definition 8.1.3 is a functor.

Proof. If one of m, n, or p is 0, then functoriality is true by definition.

If $m, n, p > 0$, then the matrix product preserves identity morphisms, that is,

$$1_B \star 1_A = 1_{BA}.$$

This is true because each of 1_A, 1_B, and 1_{BA} is entrywise an identity morphism, and $\oplus$ and $\otimes$ preserves identity morphisms.

For morphisms $f' = (f'_{ji}) \in \mathsf{Mat}^{\mathsf{C}}_{m,n}(A', A'')$ and $g' = (g'_{kj}) \in \mathsf{Mat}^{\mathsf{C}}_{n,p}(B', B'')$, the functoriality of $\oplus$ and $\otimes$ implies the following equalities for $1 \le i \le m$ and $1 \le k \le p$.

$$\begin{aligned}
\big[(g' \star f')(g \star f)\big]_{ki} &= \Big(\bigoplus_{j=1}^{n} (g'_{kj} \otimes f'_{ji})\Big)_{\mathsf{lt}} \Big(\bigoplus_{j=1}^{n} (g_{kj} \otimes f_{ji})\Big)_{\mathsf{lt}} \\
&= \Big(\bigoplus_{j=1}^{n} (g'_{kj} \otimes f'_{ji})(g_{kj} \otimes f_{ji})\Big)_{\mathsf{lt}} \\
&= \Big(\bigoplus_{j=1}^{n} (g'_{kj} g_{kj}) \otimes (f'_{ji} f_{ji})\Big)_{\mathsf{lt}} \\
&= \big[(g'g) \star (f'f)\big]_{ki}
\end{aligned} \tag{8.1.9}$$

Therefore, the matrix product preserves composition. □

Multiplying with Zero Matrices. To show that $\mathsf{Mat}^{\mathsf{C}}$ is a bicategory and a symmetric monoidal bicategory later, we will need to understand the coherence properties of the matrix product (8.1.4) with respect to the 0 matrices, the identity matrices, and iterated products. As a warm up exercise, we first consider the matrix product with a 0 matrix. The other coherence properties will be discussed in subsequent sections. Recall that $\mathbb{0}_{m,n}$ denotes the $n \times m$ matrix with each entry the additive zero $\mathbb{0} \in \mathsf{C}$.

Lemma 8.1.10. *Suppose* C *is a bimonoidal category, and* $m, n, p \ge 0$. *Then there is a natural isomorphism*

$$\mathbb{0}_{n,p} A \xrightarrow[\cong]{\zeta^{\ell}_A} \mathbb{0}_{m,p} \in \mathsf{Mat}^{\mathsf{C}}_{m,p} \tag{8.1.11}$$

for $A \in \mathsf{Mat}^{\mathsf{C}}_{m,n}$.

Proof. There are three cases.

- If either m or p is 0, then both $\mathbb{0}_{n,p}A$ and $\mathbb{0}_{m,p}$ are the empty matrix.
- If $m, p > 0$ and $n = 0$, then $\mathbb{0}_{n,p}$ and A are the empty matrix, but $\mathbb{0}_{n,p}A = \mathbb{0}_{m,p}$.

In each of these two cases, ζ^{ℓ}_A is the identity morphism.

For $m, n, p > 0$, suppose $A = (A_{ji})$. For $1 \le k \le p$ and $1 \le i \le m$, $\mathbb{0}_{n,p}A$ has (k,i)-entry

$$(\mathbb{0}_{n,p}A)_{ki} = \Big(\bigoplus_{j=1}^{n} (\mathbb{0} \otimes A_{ji})\Big)_{\mathsf{lt}}. \tag{8.1.12}$$

This object is isomorphic to the additive zero $\mathbb{0}$ via the following composite in C.

$$
\begin{array}{ccc}
(\mathbb{0}_{n,p}A)_{ki} & \xrightarrow{\quad(\zeta^{\ell}_{A})_{ki}\quad} & \mathbb{0} \\
{\scriptstyle (\lambda^{\bullet}_{A_{1i}}\oplus\cdots\oplus\lambda^{\bullet}_{A_{ni}})_{\mathsf{lt}}}\big\downarrow & & \| \\
\Big(\bigoplus_{j=1}^{n}\mathbb{0}\Big)_{\mathsf{lt}} & \xrightarrow{\quad Z\quad} & \mathbb{0}
\end{array}
\tag{8.1.13}
$$

- If $n = 1$, then $Z = 1_{\mathbb{0}}$.
- If $n > 1$, then Z is a composite of $n - 1$ morphisms, each being the sum of identity morphisms and one $\lambda^{\oplus}_{\mathbb{0}} \colon \mathbb{0} \oplus \mathbb{0} \longrightarrow \mathbb{0}$.

The naturality of ζ^{ℓ}_{A} with respect to A follows from the naturality of $\lambda^{\bullet}$ and the functoriality of $\oplus$ in C. □

Next we interpret the natural isomorphism ζ^{ℓ} in Lemma 8.1.10 in terms of paths and coherence.

Example 8.1.14 (Left Zero via Paths). In (8.1.13), $(\zeta^{\ell}_{A})_{ki}$ is the unique value in C of any 0^X-reduction

$$\Big(\bigoplus_{j=1}^{n}(0^X \otimes x_j)\Big)_{\mathsf{lt}} \xrightarrow{\;Z^{\ell}\;} \!\!\!\rightarrow 0^X \tag{8.1.15}$$

in the sense of (3.1.18) and Definition 3.3.4. Here we take the set

$$X^{\ell} = \{0^X, 1^X, x_1, \ldots, x_n\} \tag{8.1.16}$$

and the function $\varphi^{\ell} \colon X^{\ell} \longrightarrow \mathsf{Ob}(\mathsf{C})$ defined by

$$\varphi^{\ell}(x) = \begin{cases} \mathbb{0} & \text{if } x = 0^X, \\ \mathbb{1} & \text{if } x = 1^X, \text{ and} \\ A_{ji} & \text{if } x = x_j \text{ for } 1 \leq j \leq n. \end{cases} \tag{8.1.17}$$

The existence of a 0^X-reduction Z^{ℓ} in (8.1.15) with the given (co)domain follows from Lemmas 3.3.6 and 3.3.11. The uniqueness of its value in C follows from Lemma 3.3.12. ◇

Next is the case with the 0 matrix on the right side.

Lemma 8.1.18. *Suppose C is a bimonoidal category, and $m, n, q \geq 0$. Then there is a natural isomorphism*

$$A\mathbb{0}_{q,m} \xrightarrow[\cong]{\;\zeta^{r}_{A}\;} \mathbb{0}_{q,n} \in \mathsf{Mat}^{\mathsf{C}}_{q,n} \tag{8.1.19}$$

for $A \in \mathsf{Mat}^{\mathsf{C}}_{m,n}$.

Proof. There are three cases.

- If either q or n is 0, then both $A\mathbb{0}_{q,m}$ and $\mathbb{0}_{q,n}$ are the empty matrix.
- If $q, n > 0$ and $m = 0$, then both A and $\mathbb{0}_{q,m}$ are the empty matrix, but $A\mathbb{0}_{q,m} = \mathbb{0}_{q,n}$.

In each of these two cases, ζ^{r}_{A} is the identity morphism.

For $m,n,q > 0$, suppose $A = (A_{ji})$. For $1 \leq j \leq n$ and $1 \leq l \leq q$, $A\mathbb{0}_{q,m}$ has (j,l)-entry

$$(A\mathbb{0}_{q,m})_{jl} = \Big(\bigoplus_{i=1}^{m} (A_{ji} \otimes \mathbb{0})\Big)_{\mathsf{lt}}. \tag{8.1.20}$$

This object is isomorphic to the additive zero $\mathbb{0}$ via the following composite in C.

$$\begin{array}{ccc} (A\mathbb{0}_{q,m})_{jl} & \xrightarrow{(\zeta^r_A)_{jl}} & \mathbb{0} \\ {\scriptstyle (\rho^\bullet_{A_{j1}} \oplus \cdots \oplus \rho^\bullet_{A_{jm}})_{\mathsf{lt}}} \downarrow & & \| \\ \Big(\bigoplus_{i=1}^{m} \mathbb{0}\Big)_{\mathsf{lt}} & \xrightarrow{Z} & \mathbb{0} \end{array} \tag{8.1.21}$$

- If $m = 1$, then $Z = 1_{\mathbb{0}}$.
- If $m > 1$, then Z is a composite of $m-1$ morphisms, each being the sum of identity morphisms and one $\lambda^{\oplus}_{\mathbb{0}} \colon \mathbb{0} \oplus \mathbb{0} \longrightarrow \mathbb{0}$.

The naturality of ζ^r_A with respect to A follows from the naturality of $\rho^\bullet$ and the functoriality of $\oplus$ in C. □

The following example interprets the natural isomorphism ζ^r in Lemma 8.1.18 in terms of paths and coherence.

Example 8.1.22 (Right Zero via Paths)**.** The isomorphism $(\zeta^r_A)_{jl}$ in (8.1.21) is the unique value in C of any 0^X-reduction

$$\Big(\bigoplus_{i=1}^{m} (x_i \otimes 0^X)\Big)_{\mathsf{lt}} \xrightarrow{Z^r} 0^X \tag{8.1.23}$$

in the sense of (3.1.18) and Definition 3.3.4. Here we take the set

$$X^r = \{0^X, 1^X, x_1, \ldots, x_m\} \tag{8.1.24}$$

and the function $\varphi^r \colon X^r \longrightarrow \mathsf{Ob}(\mathsf{C})$ defined by

$$\varphi^r(x) = \begin{cases} \mathbb{0} & \text{if } x = 0^X, \\ \mathbb{1} & \text{if } x = 1^X, \text{ and} \\ A_{ji} & \text{if } x = x_i \text{ for } 1 \leq i \leq m. \end{cases} \tag{8.1.25}$$

The existence and the uniqueness of its value in C of a 0^X-reduction Z^r in (8.1.23) with the given (co)domain follow from Lemmas 3.3.6, 3.3.11, and 3.3.12. ◇

8.2. The Base Unitors

In this section, we discuss coherence properties of the matrix product (8.1.4) with respect to the identity matrices (8.1.6). These coherence properties will be used to define the left unitor and the right unitor in $\mathsf{Mat}^{\mathsf{C}}$. Recall from Definition 3.9.9 the notion of a *flat* bimonoidal category.

The Base Left Unitor.

Lemma 8.2.1. *Suppose* C *is a flat bimonoidal category, and* $m,n \geq 0$. *Then there is a natural isomorphism*

$$\mathbb{1}^n A \xrightarrow[\cong]{\ell_A} A \tag{8.2.2}$$

for $A = (A_{ji}) \in \mathsf{Mat}^{\mathsf{C}}_{m,n}$.

Proof. If either m or n is 0, then both $\mathbb{1}^n A$ and A are the empty matrix, and ℓ_A is the identity morphism of the empty matrix.

For $m,n > 0$, $1 \le k \le n$, and $1 \le i \le m$, $\mathbb{1}^n A$ has (k,i)-entry

$$\begin{aligned}(\mathbb{1}^n A)_{ki} &= \Big(\bigoplus_{j=1}^{n} (\mathbb{1}^n_{kj} \otimes A_{ji})\Big)_{\mathrm{lt}} \\ &= \Big(\underbrace{(\mathbb{0}\otimes A_{1i})\oplus\cdots\oplus(\mathbb{0}\otimes A_{k-1,i})}_{\text{empty if } k=1}\oplus(\mathbb{1}\otimes A_{ki})\oplus\underbrace{(\mathbb{0}\otimes A_{k+1,i})\oplus\cdots\oplus(\mathbb{0}\otimes A_{ni})}_{\text{empty if } k=n}\Big)_{\mathrm{lt}}.\end{aligned} \tag{8.2.3}$$

This object is isomorphic to A_{ki} via the following composite in C.

$$\begin{array}{ccc} (\mathbb{1}^n A)_{ki} & \xrightarrow{(\ell_A)_{ki}} & A_{ki} \\ \Big\downarrow{\scriptstyle (\lambda^{\bullet}\oplus\cdots\oplus\lambda^{\bullet}\oplus\lambda^{\otimes}\oplus\lambda^{\bullet}\oplus\cdots\oplus\lambda^{\bullet})_{\mathrm{lt}}} & & \Big\| \\ \Big(\underbrace{\mathbb{0}\oplus\cdots\oplus\mathbb{0}\oplus}_{k-1 \text{ copies of } \mathbb{0}} A_{ki} \underbrace{\oplus\mathbb{0}\oplus\cdots\oplus\mathbb{0}}_{n-k \text{ copies of } \mathbb{0}}\Big)_{\mathrm{lt}} & \xrightarrow{Z} & A_{ki} \end{array} \tag{8.2.4}$$

- If $n = 1$, then Z is the identity morphism.
- If $n > 1$, then Z is a composite of $n-1$ morphisms, each being the sum of identity morphisms and one of $\lambda^{\oplus}_{\mathbb{0}}$, $\lambda^{\oplus}_{A_{ki}}$, or $\rho^{\oplus}_{A_{ki}}$.

The naturality of ℓ_A with respect to A follows from the naturality of $\lambda^{\bullet}$, $\lambda^{\otimes}$, $\lambda^{\oplus}$, and $\rho^{\oplus}$, and the functoriality of $\oplus$ in C. □

Next we interpret the natural isomorphism ℓ in Lemma 8.2.1 in terms of paths and coherence.

Example 8.2.5 (The Left Unitor via Paths). In (8.2.4), $(\ell_A)_{ki}$ is the unique value in C of any path

$$\begin{array}{c} \Big(\underbrace{(0^X\otimes x_1)\oplus\cdots\oplus(0^X\otimes x_{k-1})}_{\text{empty if } k=1}\oplus(1^X\otimes x_k)\oplus\underbrace{(0^X\otimes x_{k+1})\oplus\cdots\oplus(0^X\otimes x_n)}_{\text{empty if } k=n}\Big)_{\mathrm{lt}} \\ \Big\downarrow{\scriptstyle P_\ell} \\ x_k \end{array} \tag{8.2.6}$$

in the sense of Definition 3.1.9 and (3.1.18). Here we take the set

$$X^\ell = \{0^X, 1^X, x_1, \ldots, x_n\}$$

in (8.1.16) and the function $\varphi^\ell \colon X^\ell \longrightarrow \mathsf{Ob}(\mathsf{C})$ in (8.1.17). In such a path P_ℓ in $\mathsf{Gr}(X^\ell)$,

- the domain is obtained from the last line in (8.2.3) by replacing $(\mathbb{0}, \mathbb{1}, A_{ji})$ with $(0^X, 1^X, x_j)$ for $1 \le j \le n$, and
- the codomain is x_k.

For example, if $k = 2$ and $n = 3$, then one such P_ℓ is the following path.

$$\begin{array}{ccc}
\big[(0^X \otimes x_1) \oplus (1^X \otimes x_2)\big] \oplus (0^X \otimes x_3) & \xrightarrow{\;P_\ell\;} & x_2 \\
{\scriptstyle [\lambda^{\bullet}_{x_1} \oplus 1] \oplus 1} \downarrow & & \uparrow {\scriptstyle \rho^{\oplus}_{x_2}} \\
\big[0^X \oplus (1^X \otimes x_2)\big] \oplus (0^X \otimes x_3) & & x_2 \oplus 0^X \\
{\scriptstyle 1 \oplus \lambda^{\bullet}_{x_3}} \downarrow & & \uparrow {\scriptstyle \lambda^{\oplus}_{x_2} \oplus 1} \\
\big[0^X \oplus (1^X \otimes x_2)\big] \oplus 0^X & \xrightarrow{\;[1 \oplus \lambda^{\otimes}_{x_2}] \oplus 1\;} & \big[0^X \oplus x_2\big] \oplus 0^X
\end{array}$$

There are other paths in $\mathsf{Gr}(X^\ell)$ with the same (co)domain as P_ℓ. For instance, after the first two prime edges, we can also use the prime edges involving $\lambda^{\oplus}_{1^X \otimes x_2}$ and $\rho^{\oplus}_{1^X \otimes x_2}$ before $\lambda^{\otimes}_{x_2}$. As we will explain below, all such paths yield the same morphism in C, so it does not matter which one we choose.

In general, one such path P_ℓ whose value in C gives the morphisms in (8.2.4) is as follows.

- Each of the first $n - 1$ prime edges in P_ℓ contains an instance of the left multiplicative zero
$$0^X \otimes x_j \xrightarrow{\lambda^{\bullet}_{x_j}} 0^X \quad \text{for} \quad 1 \leq j \neq k \leq n.$$
- The nth prime edge in P_ℓ contains an instance of the left multiplicative unit
$$1^X \otimes x_k \xrightarrow{\lambda^{\otimes}_{x_k}} x_k.$$
- Each of the last $n - 1$ prime edges in P_ℓ contains an instance of the left additive zero $\lambda^{\oplus}$ or the right additive zero $\rho^{\oplus}$. Together they remove the $n - 1$ copies of 0^X.

This construction ensures the existence of at least one path P_ℓ with the given (co)domain.

There are infinitely many different paths in $\mathsf{Gr}(X^\ell)$ with the same (co)domain as P_ℓ. The canonical isomorphism $(\ell_A)_{ki}$ in (8.2.4) is the *unique* value in C of any path in $\mathsf{Gr}(X^\ell)$ with the given (co)domain. Indeed, since the codomain x_k is nonsymmetric regular in the sense of Definition 3.10.2, so is the domain by Lemma 3.10.3. The uniqueness of the value in C of any such path follows from the Bimonoidal Coherence Theorem 3.10.7, which is applicable by the flatness assumption on C. ◇

The Base Right Unitor. Next is the case with the identity matrix on the right. It will be used to define the right unitor in $\mathsf{Mat}^{\mathsf{C}}$.

Lemma 8.2.7. *Suppose C is a flat bimonoidal category, and $m, n \geq 0$. Then there is a natural isomorphism*

$$A\mathbb{1}^m \xrightarrow[\cong]{r_A} A \tag{8.2.8}$$

for $A = (A_{ji}) \in \mathsf{Mat}^{\mathsf{C}}_{m,n}$.

Proof. If either m or n is 0, then both $A\mathbb{1}^m$ and A are the empty matrix, and r_A is the identity morphism of the empty matrix.

For $m, n > 0$, $1 \le j \le n$, and $1 \le h \le m$, $A\mathbb{1}^m$ has (j,h)-entry

$$
\begin{aligned}
&(A\mathbb{1}^m)_{jh} \\
&= \Big(\bigoplus_{i=1}^{m} \big(A_{ji} \otimes \mathbb{1}^m_{ih}\big)\Big)_{\mathsf{lt}} \\
&= \Big(\underbrace{(A_{j1}\otimes \mathbb{0}) \oplus \cdots \oplus (A_{j,h-1}\otimes \mathbb{0})}_{\text{empty if } h=1} \oplus (A_{jh}\otimes \mathbb{1}) \oplus \underbrace{(A_{j,h+1}\otimes \mathbb{0}) \oplus \cdots \oplus (A_{jm}\otimes \mathbb{0})}_{\text{empty if } h=m}\Big)_{\mathsf{lt}}.
\end{aligned}
\tag{8.2.9}
$$

This object is isomorphic to A_{jh} via the following composite in C.

$$
\begin{array}{ccc}
(A\mathbb{1}^m)_{jh} & \xrightarrow{(r_A)_{jh}} & A_{jh} \\
\Big\downarrow{\scriptstyle (\rho^{\bullet}\oplus\cdots\oplus\rho^{\bullet}\oplus\rho^{\otimes}\oplus\rho^{\bullet}\oplus\cdots\oplus\rho^{\bullet})_{\mathsf{lt}}} & & \\
\Big(\underbrace{\mathbb{0}\oplus\cdots\oplus\mathbb{0}\oplus}_{h-1 \text{ copies of } \mathbb{0}} A_{jh} \underbrace{\oplus\mathbb{0}\oplus\cdots\oplus\mathbb{0}}_{m-h \text{ copies of } \mathbb{0}}\Big)_{\mathsf{lt}} & \xrightarrow{Z} & A_{jh}
\end{array}
\tag{8.2.10}
$$

- If $m = 1$, then Z is the identity morphism.
- If $m > 1$, then Z is a composite of $m - 1$ morphisms, each being the sum of identity morphisms and one of $\lambda^{\oplus}_{\mathbb{0}}$, $\lambda^{\oplus}_{A_{jh}}$, or $\rho^{\oplus}_{A_{jh}}$.

The naturality of r_A with respect to A follows from the naturality of $\rho^{\bullet}$, $\rho^{\otimes}$, $\lambda^{\oplus}$, and $\rho^{\oplus}$, and the functoriality of $\oplus$ in C. □

The following example interprets the natural isomorphism r in Lemma 8.2.7 in terms of paths and coherence.

Example 8.2.11 (The Right Unitor via Paths). The isomorphism $(r_A)_{jh}$ in (8.2.10) is the unique value in C of any path P_r as follows.

$$
\begin{array}{c}
\Big(\underbrace{(x_1\otimes 0^X) \oplus \cdots \oplus (x_{h-1}\otimes 0^X)}_{\text{empty if } h=1} \oplus (x_h\otimes 1^X) \oplus \underbrace{(x_{h+1}\otimes 0^X) \oplus \cdots \oplus (x_m\otimes 0^X)}_{\text{empty if } h=m}\Big)_{\mathsf{lt}} \\
\Big\downarrow{\scriptstyle P_r} \\
x_h
\end{array}
\tag{8.2.12}
$$

Here we take the set

$$X^r = \{0^X, 1^X, x_1, \ldots, x_m\}$$

in (8.1.24) and the function $\varphi^r : X^r \longrightarrow \mathsf{Ob}(\mathsf{C})$ in (8.1.25). In such a path P_r in $\mathsf{Gr}(X^r)$,

- the domain is obtained from the last line in (8.2.9) by replacing $(\mathbb{0}, \mathbb{1}, A_{ji})$ with $(0^X, 1^X, x_i)$ for $1 \le i \le m$, and
- the codomain is x_h.

One such path P_r whose value in C gives the morphisms in (8.2.10) is as follows.

- Each of the first $m-1$ prime edges in P_r contains an instance of the right multiplicative zero

$$x_i \otimes 0^X \xrightarrow{\rho^{\bullet}_{x_i}} 0^X \quad \text{for} \quad 1 \le i \ne h \le m.$$

- The mth prime edge in P_r contains an instance of the right multiplicative unit

$$x_h \otimes 1^X \xrightarrow{\rho^{\otimes}_{x_h}} x_h.$$

- Each of the last $m-1$ prime edges in P_r contains an instance of the left additive zero $\lambda^{\oplus}$ or the right additive zero $\rho^{\oplus}$. Together they remove the $m-1$ copies of 0^X.

This construction ensures the existence of at least one path P_r with the given (co)domain.

There are infinitely many different paths in $\mathsf{Gr}(X^r)$ with the same (co)domain as P_r. The canonical isomorphism $(r_A)_{jh}$ in (8.2.10) is the *unique* value in C of any path in $\mathsf{Gr}(X^r)$ with the given (co)domain. Indeed, since the codomain x_h is nonsymmetric regular in the sense of Definition 3.10.2, so is the domain by Lemma 3.10.3. The uniqueness of the value in C of any such path follows from the Bimonoidal Coherence Theorem 3.10.7, which is applicable by the flatness assumption on C. ◇

8.3. The Base Associator

In this section, we discuss coherence properties of the matrix product (8.1.4) with respect to associativity. The next lemma will be used to define the associator in $\mathsf{Mat}^{\mathsf{C}}$. Recall from Definition 2.1.2 that a *tight* bimonoidal category is one in which the distributivity morphisms δ^l and δ^r are natural isomorphisms, not just monomorphisms. Also recall from the introduction of this chapter our conventions for *Mac Lane coherence isomorphisms* and *Laplaza coherence isomorphisms*.

Lemma 8.3.1. *Suppose C is a tight bimonoidal category. For $m,n,p,q \geq 0$, there is a natural isomorphism*

$$(CB)A \xrightarrow[\cong]{a_{C,B,A}} C(BA) \in \mathsf{Mat}^{\mathsf{C}}_{m,q} \tag{8.3.2}$$

for $(A,B,C) \in \mathsf{Mat}^{\mathsf{C}}_{m,n} \times \mathsf{Mat}^{\mathsf{C}}_{n,p} \times \mathsf{Mat}^{\mathsf{C}}_{p,q}$.

Proof. There are five cases.

(1) If either m or q is 0, then both $(CB)A$ and $C(BA)$ are the empty matrix, and $a_{C,B,A}$ is the identity morphism of the empty matrix.
(2) If $m,q > 0$ and $n = p = 0$, then both $(CB)A$ and $C(BA)$ are the 0 matrix $\mathbb{0}_{m,q}$, and $a_{C,B,A}$ is the identity morphism.
(3) If $m,p,q > 0$ and $n = 0$, then $a_{C,B,A}$ is

$$(CB)A = \mathbb{0}_{m,q} \xrightarrow[\cong]{(\zeta^r_C)^{-1}} C\mathbb{0}_{m,p} = C(BA)$$

with ζ^r the natural isomorphism in (8.1.19).
(4) If $m,n,q > 0$ and $p = 0$, then $a_{C,B,A}$ is

$$(CB)A = \mathbb{0}_{n,q}A \xrightarrow[\cong]{\zeta^\ell_A} \mathbb{0}_{m,q} = C(BA)$$

with ζ^ℓ the natural isomorphism in (8.1.11).

In each of the above cases, a is natural.

For the remaining case with $m,n,p,q > 0$, consider the following data.

- $A = (A_{ji}) \in \mathsf{Mat}^{\mathsf{C}}_{m,n}$ is an $n \times m$ matrix.
- $B = (B_{kj}) \in \mathsf{Mat}^{\mathsf{C}}_{n,p}$ is a $p \times n$ matrix.
- $C = (C_{lk}) \in \mathsf{Mat}^{\mathsf{C}}_{p,q}$ is a $q \times p$ matrix.

We first compute the entries of the matrix products $(CB)A$ and $C(BA) \in \mathsf{Mat}^{\mathsf{C}}_{m,q}$. For $1 \le i \le m$ and $1 \le l \le q$, the (l,i)-entry of $(CB)A$ is the following object in C.

$$\begin{aligned}[(CB)A]_{li} &= \Big(\bigoplus_{j=1}^{n} (CB)_{lj} \otimes A_{ji}\Big)_{\mathsf{lt}} \\ &= \Big(\bigoplus_{j=1}^{n}\Big(\bigoplus_{k=1}^{p} C_{lk} \otimes B_{kj}\Big)_{\mathsf{lt}} \otimes A_{ji}\Big)_{\mathsf{lt}}\end{aligned} \tag{8.3.3}$$

Similarly, the (l,i)-entry of $C(BA)$ is the following object in C.

$$\begin{aligned}[C(BA)]_{li} &= \Big(\bigoplus_{k=1}^{p} C_{lk} \otimes (BA)_{ki}\Big)_{\mathsf{lt}} \\ &= \Big(\bigoplus_{k=1}^{p} C_{lk} \otimes \Big(\bigoplus_{j=1}^{n} B_{kj} \otimes A_{ji}\Big)_{\mathsf{lt}}\Big)_{\mathsf{lt}}\end{aligned} \tag{8.3.4}$$

The (l,i)-entry of $a_{C,B,A}$ is the following composite of isomorphisms in C.

$$\begin{array}{ccc} [(CB)A]_{li} & \xrightarrow{(a_{C,B,A})_{li}} & [C(BA)]_{li} \\ \| & & \| \\ \Big(\bigoplus_{j=1}^{n}\Big(\bigoplus_{k=1}^{p} C_{lk} \otimes B_{kj}\Big)_{\mathsf{lt}} \otimes A_{ji}\Big)_{\mathsf{lt}} & & \Big(\bigoplus_{k=1}^{p} C_{lk} \otimes \Big(\bigoplus_{j=1}^{n} B_{kj} \otimes A_{ji}\Big)_{\mathsf{lt}}\Big)_{\mathsf{lt}} \\ \downarrow a^1 & & \uparrow a^4 \\ \Big(\bigoplus_{j=1}^{n}\Big(\bigoplus_{k=1}^{p} (C_{lk} \otimes B_{kj}) \otimes A_{ji}\Big)_{\mathsf{lt}}\Big)_{\mathsf{lt}} & & \Big(\bigoplus_{k=1}^{p}\Big(\bigoplus_{j=1}^{n} C_{lk} \otimes (B_{kj} \otimes A_{ji})\Big)_{\mathsf{lt}}\Big)_{\mathsf{lt}} \\ \downarrow a^2 & & \uparrow a^3 \\ & \Big(\bigoplus_{j=1}^{n}\Big(\bigoplus_{k=1}^{p} C_{lk} \otimes (B_{kj} \otimes A_{ji})\Big)_{\mathsf{lt}}\Big)_{\mathsf{lt}} & \end{array} \tag{8.3.5}$$

Consider the diagram (8.3.5).

(1) a^1 is the identity morphism if $p = 1$. If $p > 1$, then a^1 is a composite of $n(p-1)$ morphisms, each being the sum of identity morphisms and one component of the right distributivity morphism

$$\begin{array}{c} \Big[\Big(\bigoplus_{k=1}^{p'-1} C_{lk} \otimes B_{kj}\Big)_{\mathsf{lt}} \oplus (C_{lp'} \otimes B_{p'j})\Big] \otimes A_{ji} \\ \Big\downarrow \delta^r_{\left(\oplus_{k=1}^{p'-1} C_{lk}\otimes B_{kj}\right)_{\mathsf{lt}},\, (C_{lp'}\otimes B_{p'j}),\, A_{ji}} \\ \Big[\Big(\bigoplus_{k=1}^{p'-1} C_{lk} \otimes B_{kj}\Big)_{\mathsf{lt}} \otimes A_{ji}\Big] \oplus \Big[(C_{lp'} \otimes B_{p'j}) \otimes A_{ji}\Big] \end{array} \tag{8.3.6}$$

for $1 \le j \le n$ and $2 \le p' \le p$. This is an isomorphism because C is tight.

(2) a^2 is a composite of np morphisms, each being the sum of identity morphisms and one component of the multiplicative associativity

$$(C_{lk} \otimes B_{kj}) \otimes A_{ji} \xrightarrow{\alpha^{\otimes}_{C_{lk}, B_{kj}, A_{ji}}} C_{lk} \otimes (B_{kj} \otimes A_{ji}) \tag{8.3.7}$$

for $1 \le j \le n$ and $1 \le k \le p$. Equivalently, by the functoriality of $\oplus$, a^2 is the sum

$$a^2 = \Big(\bigoplus_{j=1}^{n}\Big(\bigoplus_{k=1}^{p} \alpha^{\otimes}_{C_{lk},B_{kj},A_{ji}}\Big)_{\mathsf{lt}}\Big)_{\mathsf{lt}}.$$

(3) a^3 is a Mac Lane coherence isomorphism $\cong^{\oplus}_{\mathsf{ML}}$ in $(\mathsf{C},\oplus)$ that involves only $\alpha^{\pm\oplus}$ and $\xi^{\oplus}$. It regards the np objects $C_{lk}\otimes(B_{kj}\otimes A_{ji})$ as formal variables. Its existence and uniqueness follow from Theorem 1.3.8.

(4) a^4 is the identity morphism if $n = 1$. If $n > 1$, then a^4 is a composite of $p(n-1)$ morphisms, each being the sum of identity morphisms and one instance of

$$\begin{array}{c} \Big[C_{lk}\otimes\Big(\bigoplus_{j=1}^{n'-1} B_{kj}\otimes A_{ji}\Big)_{\mathsf{lt}}\Big]\oplus\Big[C_{lk}\otimes(B_{kn'}\otimes A_{n'i})\Big] \\ \Big\downarrow (\delta^l)^{-1}_{C_{lk},\left(\oplus_{j=1}^{n'-1}B_{kj}\otimes A_{ji}\right)_{\mathsf{lt}},(B_{kn'}\otimes A_{n'i})} \\ C_{lk}\otimes\Big(\bigoplus_{j=1}^{n'} B_{kj}\otimes A_{ji}\Big)_{\mathsf{lt}} \end{array} \tag{8.3.8}$$

for $1 \le k \le p$ and $2 \le n' \le n$. The tightness assumption on C is used here to make sure that the left distributivity morphism δ^l is invertible.

The naturality of $a_{C,B,A}$ follows from the naturality of $\alpha^{\oplus}$, $\xi^{\oplus}$, $\alpha^{\otimes}$, δ^l, and δ^r, and the functoriality of $\oplus$ in C. □

Next we interpret the natural isomorphism a in Lemma 8.3.1 in terms of paths and coherence.

Example 8.3.9 (The Associator via Paths)**.** To realize $(a_{C,B,A})_{li}$ in (8.3.5) as the values in C of paths, for $1 \le i \le m$ and $1 \le l \le q$, we consider the set

$$X^a = \{0^X, 1^X, a_{ji}, b_{kj}, c_{lk}\}_{1\le j\le n,\, 1\le k\le p} \tag{8.3.10}$$

with $n + np + p + 2$ elements and the function $\varphi^a \colon X^a \longrightarrow \mathsf{Ob}(\mathsf{C})$ defined as

$$\varphi^a(x) = \begin{cases} \mathbb{0} & \text{if } x = 0^X, \\ \mathbb{1} & \text{if } x = 1^X, \\ A_{ji} & \text{if } x = a_{ji} \text{ for } 1 \le j \le n, \\ B_{kj} & \text{if } x = b_{kj} \text{ for } 1 \le j \le n \text{ and } 1 \le k \le p, \text{ and} \\ C_{lk} & \text{if } x = c_{lk} \text{ for } 1 \le k \le p. \end{cases} \tag{8.3.11}$$

Define the following four paths in $\mathsf{Gr}(X^a)$, in which the direction of P_a^4 is not an error.

$$(8.3.12)\qquad \begin{array}{ccc}
\Big(\bigoplus_{j=1}^{n}\Big(\bigoplus_{k=1}^{p} c_{lk}\otimes b_{kj}\Big)_{\mathsf{lt}}\otimes a_{ji}\Big)_{\mathsf{lt}} & & \Big(\bigoplus_{k=1}^{p} c_{lk}\otimes\Big(\bigoplus_{j=1}^{n} b_{kj}\otimes a_{ji}\Big)_{\mathsf{lt}}\Big)_{\mathsf{lt}} \\
\Big\downarrow P_a^1 & & \Big\downarrow P_a^4 \\
\Big(\bigoplus_{j=1}^{n}\Big(\bigoplus_{k=1}^{p} (c_{lk}\otimes b_{kj})\otimes a_{ji}\Big)_{\mathsf{lt}}\Big)_{\mathsf{lt}} & & \Big(\bigoplus_{k=1}^{p}\Big(\bigoplus_{j=1}^{n} c_{lk}\otimes (b_{kj}\otimes a_{ji})\Big)_{\mathsf{lt}}\Big)_{\mathsf{lt}} \\
\Big\downarrow P_a^2 & & \Big\uparrow P_a^3 \\
 & \Big(\bigoplus_{j=1}^{n}\Big(\bigoplus_{k=1}^{p} c_{lk}\otimes (b_{kj}\otimes a_{ji})\Big)_{\mathsf{lt}}\Big)_{\mathsf{lt}} &
\end{array}$$

Consider (8.3.12).

(1) For each $1 \leq h \leq 3$, the path P_a^h is obtained from the corresponding isomorphism a^h in (8.3.5) by replacing the objects (A_{ji}, B_{kj}, C_{lk}) in C with the elements (a_{ji}, b_{kj}, c_{lk}) in X^a.
- Each prime edge in P_a^1 is an identity or involves an instance of δ^r.
- Each prime edge in P_a^2 is an identity or involves an instance of $\alpha^{\otimes}$.
- Each prime edge in P_a^3 is an identity or involves an instance of $\alpha^{\pm\oplus}$ or $\xi^{\oplus}$.

The value of the path P_a^h in C is the isomorphism a^h in (8.3.5). Since $\xi^{\pm\otimes}$ are not involved, the path (P_a^3, P_a^2, P_a^1) in $\mathsf{Gr}(X^a)$ realizes the composite $a^3a^2a^1$ in C as a Laplaza coherence isomorphism $\cong_{\mathsf{Lap}}$.

(2) The path P_a^4 is obtained from the isomorphism a^4 in (8.3.5) by
- replacing the objects (A_{ji}, B_{kj}, C_{lk}) in C with the elements (a_{ji}, b_{kj}, c_{lk}) in X^a and
- reversing the direction of each $(\delta^l)^{-1}$ in (8.3.8) and replacing it with δ^l in $\mathsf{Gr}(X^a)$.

The reason that P_a^4 is defined in this way is that the elementary edges δ^l and δ^r do *not* have formal inverses; see Definitions 3.1.6 and 3.1.10. The value of P_a^4 in C is $(a^4)^{-1}$. Since $\xi^{\pm\otimes}$ are not involved, the value of P_a^4 in C is a Laplaza coherence isomorphism. Therefore, a^4 is an inverse Laplaza coherence isomorphism $\cong_{\mathsf{Lap}}^{-1}$.

In summary, using the paths P_a^h in $\mathsf{Gr}(X^a)$ for $1 \leq h \leq 4$ in (8.3.12), the isomorphism $(a_{C,B,A})_{li}$ in (8.3.5) is the following composite of a Laplaza coherence isomorphism followed by an inverse Laplaza coherence isomorphism.

$$(8.3.13)\qquad \begin{array}{ccc}
[(CB)A]_{li} & \xrightarrow{\quad (a_{C,B,A})_{li}\quad} & [C(BA)]_{li} \\
\| & & \| \\
\Big(\bigoplus_{j=1}^{n}\Big(\bigoplus_{k=1}^{p} C_{lk}\otimes B_{kj}\Big)_{\mathsf{lt}}\otimes A_{ji}\Big)_{\mathsf{lt}} & & \Big(\bigoplus_{k=1}^{p} C_{lk}\otimes\Big(\bigoplus_{j=1}^{n} B_{kj}\otimes A_{ji}\Big)_{\mathsf{lt}}\Big)_{\mathsf{lt}} \\
a^3a^2a^1 \Big\downarrow \cong_{\mathsf{Lap}} & & \cong_{\mathsf{Lap}}^{-1} \Big\uparrow a^4 \\
 & \Big(\bigoplus_{k=1}^{p}\Big(\bigoplus_{j=1}^{n} C_{lk}\otimes (B_{kj}\otimes A_{ji})\Big)_{\mathsf{lt}}\Big)_{\mathsf{lt}} &
\end{array}$$

The decomposition (8.3.13) has the following uniqueness property. Suppose given a solid arrow diagram in C of the following form.

(8.3.14)

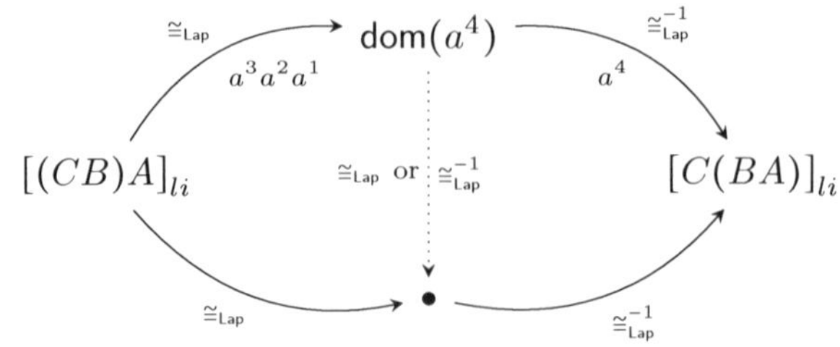

- The top composite is the factorization of $(a_{C,B,A})_{li}$ in (8.3.13).
- The bottom composite is some Laplaza coherence isomorphism followed by some inverse Laplaza coherence isomorphism, defined using the same set X^a in (8.3.10) and the same function φ^a in (8.3.11).

Suppose, furthermore, that there exists a dotted arrow as indicated that is either a Laplaza coherence isomorphism or an inverse Laplaza coherence isomorphism, defined with the same set X^a and function φ^a. Then the bottom composite in (8.3.14) is also equal to $(a_{C,B,A})_{li}$.

Indeed, first note that the codomain of P^1_a is nonsymmetric regular in the sense of Definition 3.10.2. Lemma 3.10.3 then implies that each vertex in (8.3.12) is nonsymmetric regular. By the tightness assumption on C, the Bimonoidal Coherence Theorem 3.9.1 implies that each of the two subdiagrams in (8.3.14) is commutative in C. This uniqueness property and similar technique will be used in later sections to prove (monoidal) bicategory axioms related to a. ◇

Example 8.3.15 (Bipermutative Categories)**.** Suppose C is a right or left bipermutative category as in Definitions 2.5.2 and 2.5.11. Then $\mathbb{1}$ is a strict multiplicative unit, and $\mathbb{0}$ is a strict additive zero and a strict multiplicative zero. It follows from (8.1.13), (8.1.21), (8.2.4), and (8.2.10) that there are equalities as follows for $A \in \mathsf{Mat}^{\mathsf{C}}_{m,n}$.

$$\begin{aligned}
\mathbb{0}_{n,p}A &= \mathbb{0}_{m,p} \in \mathsf{Mat}^{\mathsf{C}}_{m,p}\\
A\mathbb{0}_{q,m} &= \mathbb{0}_{q,n} \in \mathsf{Mat}^{\mathsf{C}}_{q,n}\\
\mathbb{1}^n A = A &= A\mathbb{1}^m \in \mathsf{Mat}^{\mathsf{C}}_{m,n}
\end{aligned}$$

Moreover, in the context of Lemma 8.3.1, the following statements hold.

- If one of m, n, p, q is 0 (the first four cases), then $a_{C,B,A}$ is the identity morphism.
- If $m, n, p, q > 0$, then $a_{C,B,A}$ in (8.3.5) is *not* the identity in general because it involves both δ^r in (8.3.6) and δ^l in (8.3.8).

This means that even for a right or a left bipermutative category C, the bicategory $\mathsf{Mat}^{\mathsf{C}}$ in Definition 8.4.11 below is *not* in general a 2-category in the sense of Definition 6.1.8. ◇

8.4. The Matrix Bicategory

In this section, we assemble the constructions in the previous sections to construct the matrix bicategory $\mathsf{Mat}^{\mathsf{C}}$ for a tight bimonoidal category C; see Theorem 8.4.12. As an immediate consequence, we obtain a monoidal category $\mathsf{Mat}^{\mathsf{C}}_{n,n}$ with objects $n \times n$ matrices in C and with the monoidal product given by the matrix product; see Corollary 8.4.14. This result on $\mathsf{Mat}^{\mathsf{C}}_{n,n}$ is stated in [**BDR04**, Prop. 3.3], with the additional assumption that C is symmetric.

Motivation 8.4.1. Recall from Definition 6.1.2 the notion of a bicategory, which includes the unity axiom (6.1.3) and the pentagon axiom (6.1.4). Each of these axioms asserts that some diagram in some category $\mathsf{Mat}^{\mathsf{C}}_{m,n}$ is commutative. Since morphisms in each $\mathsf{Mat}^{\mathsf{C}}_{m,n}$ are defined entrywise in C, checking those two axioms amounts to checking some commutative diagrams in C. In each case, we will realize the relevant diagram in C using paths in $\mathsf{Gr}(X)$ in the sense of Definition 3.1.14 for a suitable set X of formal variables. The Bimonoidal Coherence Theorem 3.10.7 then implies the commutativity of the diagram in C. ◇

The Unity Axiom. The unity axiom in $\mathsf{Mat}^{\mathsf{C}}$ is proved in the next lemma. Recall the matrix product and the identity matrices in Definition 8.1.3 and Explanation 8.1.7 and the natural isomorphisms ℓ, r, and a in, respectively, Lemmas 8.2.1, 8.2.7, and 8.3.1.

Lemma 8.4.2. *For each tight bimonoidal category* C, *the diagram*

$$\begin{array}{ccc} (B\mathbb{1}^n)A & \xrightarrow{\;a_{B,\mathbb{1}^n,A}\;} & B(\mathbb{1}^n A) \\ & {\scriptstyle r_B \star 1_A} \searrow \quad \swarrow {\scriptstyle 1_B \star \ell_A} & \\ & BA & \end{array} \tag{8.4.3}$$

in $\mathsf{Mat}^{\mathsf{C}}_{m,p}$ *is commutative for matrices* $A \in \mathsf{Mat}^{\mathsf{C}}_{m,n}$ *and* $B \in \mathsf{Mat}^{\mathsf{C}}_{n,p}$.

Proof. If one of m, n, or p is 0, then all three morphisms in (8.4.3) are identity morphisms.

Suppose $m,n,p > 0$. It suffices to show that the (k,i)-entries of the two composites in (8.4.3) are equal for $1 \le i \le m$ and $1 \le k \le p$. Using (8.1.4), (8.1.5), (8.3.3), and (8.3.4), the (k,i)-entry of the diagram (8.4.3) is the following diagram in C.

$$\begin{array}{ccccc} [(B\mathbb{1}^n)A]_{ki} & & \xrightarrow{\;(a_{B,\mathbb{1}^n,A})_{ki}\;} & & [B(\mathbb{1}^n A)]_{ki} \\ \| & & & & \| \\ \Big(\bigoplus_{j=1}^n \Big(\bigoplus_{l=1}^n B_{kl}\otimes \mathbb{1}^n_{lj}\Big)_{\mathsf{lt}} \otimes A_{ji}\Big)_{\mathsf{lt}} & & & & \Big(\bigoplus_{l=1}^n B_{kl}\otimes \Big(\bigoplus_{j=1}^n \mathbb{1}^n_{lj}\otimes A_{ji}\Big)_{\mathsf{lt}}\Big)_{\mathsf{lt}} \\ \Big\downarrow {\scriptstyle \big(\bigoplus_{j=1}^n (r_B)_{kj}\otimes 1_{A_{ji}}\big)_{\mathsf{lt}}} & & & & \Big\downarrow {\scriptstyle \big(\bigoplus_{l=1}^n 1_{B_{kl}}\otimes (\ell_A)_{li}\big)_{\mathsf{lt}}} \\ \Big(\bigoplus_{j=1}^n B_{kj}\otimes A_{ji}\Big)_{\mathsf{lt}} & \xrightarrow{\;=\;} & (BA)_{ki} & \xleftarrow{\;=\;} & \Big(\bigoplus_{l=1}^n B_{kl}\otimes A_{li}\Big)_{\mathsf{lt}} \end{array} \tag{8.4.4}$$

To show that this diagram is commutative, we first realize its edges as the values in C of paths, in the sense of (3.1.18), defined as follows.

Consider the set

$$X = \{0^X, 1^X, a_1, \ldots, a_n, b_1, \ldots, b_n\}$$

and the function $\varphi\colon X \to \mathsf{Ob}(\mathsf{C})$ defined by

$$\varphi(x) = \begin{cases} 0 & \text{if } x = 0^X, \\ \mathbb{1} & \text{if } x = 1^X, \\ A_{ji} & \text{if } x = a_j \text{ for } 1 \le j \le n, \text{ and} \\ B_{kj} & \text{if } x = b_j \text{ for } 1 \le j \le n. \end{cases}$$

There is a diagram

(8.4.5)
$$\begin{array}{ccc}
 & \Big(\bigoplus_{l=1}^{n}\Big(\bigoplus_{j=1}^{n} b_l \otimes (\delta_{lj}^X \otimes a_j)\Big)_{\mathsf{lt}}\Big)_{\mathsf{lt}} & \\
\overset{(P_a^3, P_a^2, P_a^1)}{\nearrow} & & \overset{P_a^4}{\nwarrow} \\
\Big(\bigoplus_{j=1}^{n}\Big(\bigoplus_{l=1}^{n} b_l \otimes \delta_{lj}^X\Big)_{\mathsf{lt}} \otimes a_j\Big)_{\mathsf{lt}} & \Big\downarrow M & \Big(\bigoplus_{l=1}^{n} b_l \otimes \Big(\bigoplus_{j=1}^{n} \delta_{lj}^X \otimes a_j\Big)_{\mathsf{lt}}\Big)_{\mathsf{lt}} \\
\underset{R}{\searrow} & & \underset{L}{\swarrow} \\
 & \Big(\bigoplus_{j=1}^{n} b_j \otimes a_j\Big)_{\mathsf{lt}} &
\end{array}$$

in $\mathsf{Gr}(X)$ defined as follows.

(1) The element $\delta_{lj}^X \in X$ is defined as

(8.4.6)
$$\delta_{lj}^X = \begin{cases} 0^X & \text{if } l \neq j \text{ and} \\ 1^X & \text{if } l = j. \end{cases}$$

(2) The component $(a_{B,\mathbb{1}^n,A})_{ki}$ in (8.4.4) decomposes as $a^4a^3a^2a^1$ as in (8.3.5). The paths P_a^h for $1 \le h \le 4$ in (8.4.5) are the ones in (8.3.12) with the symbols $(k,p,l,c_{lk},b_{kj},a_{ji})$ replaced by $(l,n,k,b_l,\delta_{lj}^X,a_j)$. As we explained in Example 8.3.9,
- the path (P_a^3, P_a^2, P_a^1) has value $a^3a^2a^1$ in C, and
- the path P_a^4 has value $(a^4)^{-1}$ in C.

(3) For $1 \le j \le n$, the component $(r_B)_{kj}$ in (8.4.4) is the isomorphism in (8.2.10) with the symbols (A,m,j,h) replaced by (B,n,k,j). By Example 8.2.11, each $(r_B)_{kj}$ is the value in C of a path P_r^j as in (8.2.12), defined with the subset $\{0^X, 1^X, b_1, \ldots, b_n\}$. The path R in (8.4.5) is defined using
- the paths P_r^j for $1 \le j \le n$ and
- the constructions in Notation 3.3.10

such that the value of R in C is the left edge in (8.4.4).

(4) For $1 \le l \le n$, the component $(\ell_A)_{li}$ in (8.4.4) is the isomorphism in (8.2.4) with the symbol k replaced by l. By Example 8.2.5, each $(\ell_A)_{li}$ is the value in C of a path P_ℓ^l as in (8.2.6), defined with the subset $\{0^X, 1^X, a_1, \ldots, a_n\}$. The path L in (8.4.5) is defined using the paths
- P_ℓ^l for $1 \le l \le n$ and
- the constructions in Notation 3.3.10

such that the value of L in C is the right edge in (8.4.4).

(5) Since δ_{lj}^X in (8.4.6) is either 0^X or 1^X, we define M in (8.4.5) as the path with the following prime edges.
- For $1 \le l \neq j \le n$, M has prime edges containing

$$b_l \otimes (0^X \otimes a_j) \xrightarrow{1_{b_l} \otimes \lambda_{a_j}^{\bullet}} b_l \otimes 0^X \xrightarrow{\rho_{b_l}^{\bullet}} 0^X.$$

- For $1 \le l = j \le n$, M has prime edges containing

$$b_j \otimes (1^X \otimes a_j) \xrightarrow{1_{b_j} \otimes \lambda_{a_j}^{\otimes}} b_j \otimes a_j.$$

- After the above prime edges, M has prime edges containing $\lambda^{\oplus}$ or $\rho^{\oplus}$ that remove the $n(n-1)$ copies of 0^X additively.

The bottom middle vertex in (8.4.5) is nonsymmetric regular in the sense of Definition 3.10.2. By Lemma 3.10.3, the domain of R and the domain of L are also nonsymmetric regular. Using the tightness assumption on C, Theorem 3.10.7 implies that the diagram (8.4.5) is commutative in C. This, in turn, implies that the diagram (8.4.4) is commutative. □

The Pentagon Axiom. Next we prove the pentagon axiom (6.1.4) in $\mathsf{Mat}^{\mathsf{C}}$. To clarify the cases, we separate the proof into two lemmas. For each tight bimonoidal category C, consider the diagram

(8.4.7)

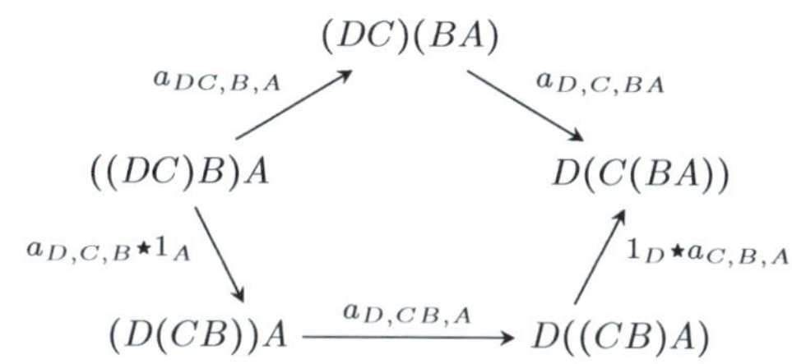

in $\mathsf{Mat}^{\mathsf{C}}_{m,t}$ for

$$(A,B,C,D) \in \mathsf{Mat}^{\mathsf{C}}_{m,n} \times \mathsf{Mat}^{\mathsf{C}}_{n,p} \times \mathsf{Mat}^{\mathsf{C}}_{p,q} \times \mathsf{Mat}^{\mathsf{C}}_{q,t}.$$

The following proof involves the natural isomorphisms ζ^ℓ in (8.1.11) and ζ^r in (8.1.19).

Lemma 8.4.8. *If at least one of m, n, p, q, and t is 0, then the pentagon (8.4.7) is commutative.*

Proof. We consider all the possible cases.

(1) If either m or t is 0, then each edge in (8.4.7) is the identity morphism of the empty matrix.
(2) If $m, t > 0$ and $n = p = q = 0$, then each edge in (8.4.7) is the identity morphism of the 0 matrix $\mathbb{0}_{m,t}$.
(3) If $m, n, t > 0$ and $p = q = 0$, then the pentagon (8.4.7) is the following diagram.

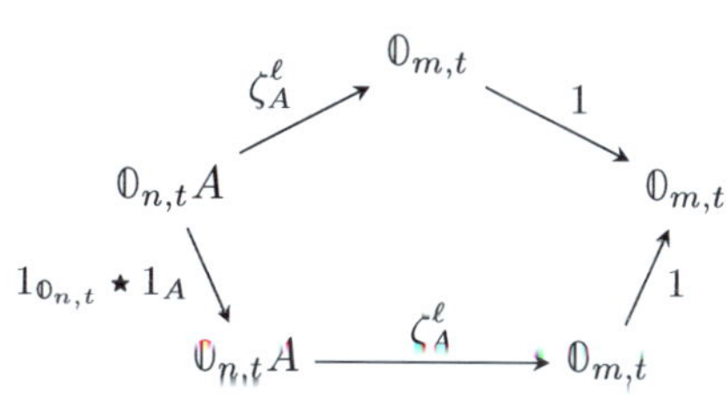

Since

$$1_{\mathbb{0}_{n,t}} \star 1_A = 1_{\mathbb{0}_{n,t}A}$$

by Lemma 8.1.8, the above pentagon is commutative.

(4) If $m, q, t > 0$ and $n = p = 0$, then the pentagon (8.4.7) is the following diagram.

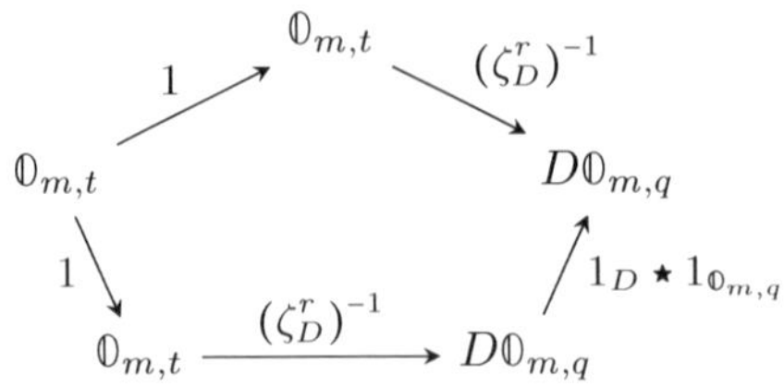

As in the previous case, this pentagon is commutative by Lemma 8.1.8.

(5) If $m, p, t > 0$ and $n = q = 0$, then the pentagon (8.4.7) is the following diagram.

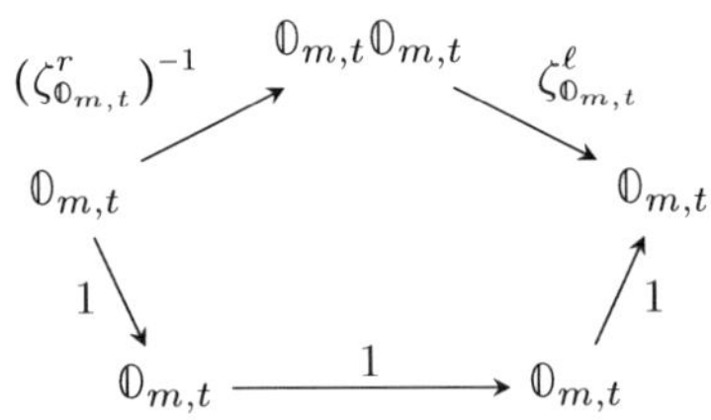

Since

$$\lambda^\bullet_{\mathbb{0}} = \rho^\bullet_{\mathbb{0}} \colon \mathbb{0} \otimes \mathbb{0} \longrightarrow \mathbb{0}$$

by the axiom (2.1.14) in C, each entry of $\zeta^\ell_{\mathbb{0}_{m,t}}$ in (8.1.13) is equal to the corresponding entry of $\zeta^r_{\mathbb{0}_{m,t}}$ in (8.1.21). Therefore, the above pentagon is commutative.

(6) If $m, p, q, t > 0$ and $n = 0$, then the pentagon (8.4.7) is the left diagram below.

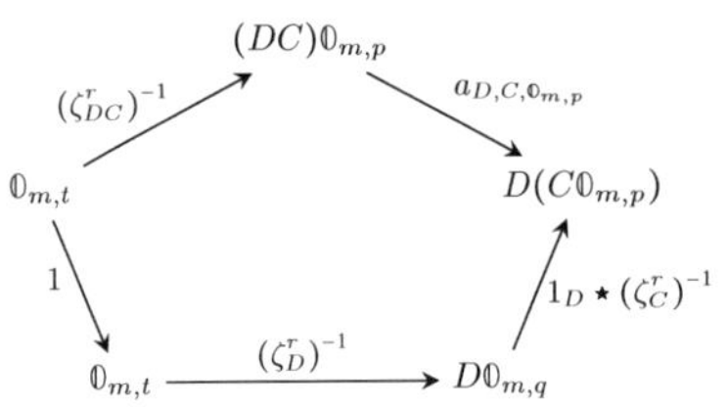

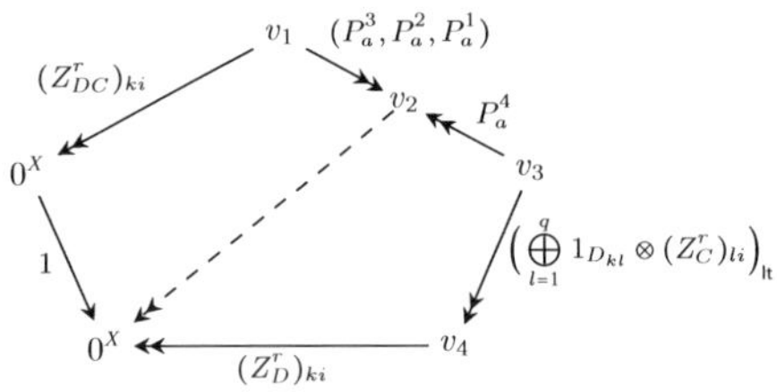

It suffices to prove the commutativity of the left pentagon in the (k,i)-entry for $1 \le i \le m$ and $1 \le k \le t$. As in Example 8.1.22, the (k,i)-entry of ζ^r_{DC} is the unique value in C of a 0^X-reduction $(Z^r_{DC})_{ki}$. Similarly, $(Z^r_D)_{ki}$ and $(Z^r_C)_{li}$ are 0^X-reductions whose values in C are, respectively, $(\zeta^r_D)_{ki}$ and $(\zeta^r_C)_{li}$.

As in Example 8.3.9, the entry $(a_{D,C,\mathbb{0}_{m,p}})_{ki}$ is the value in C of a zigzag of paths $\big(P^4_a, (P^3_a, P^2_a, P^1_a)\big)$. With the appropriate inverses taken into account, the (k,i)-entry of the left pentagon is the value in C of the right pentagon for some elements v_1, v_2, v_3, and v_4. By Lemma 3.10.3, each of the four vertices $v_?$ has the same support as 0^X. By Lemmas 3.3.6 and 3.3.11, v_2 has a 0^X-reduction denoted by the dashed arrow. By Theorem 3.10.7, the following two statements hold.

- The two paths from v_1 to the lower left 0^X have the same value in C.
- The two paths from v_3 to the lower left 0^X have the same value in C.

Therefore, the (k,i)-entry of the left pentagon is commutative.

(7) If $m,n,q,t > 0$ and $p = 0$, then the pentagon (8.4.7) is the diagram below.

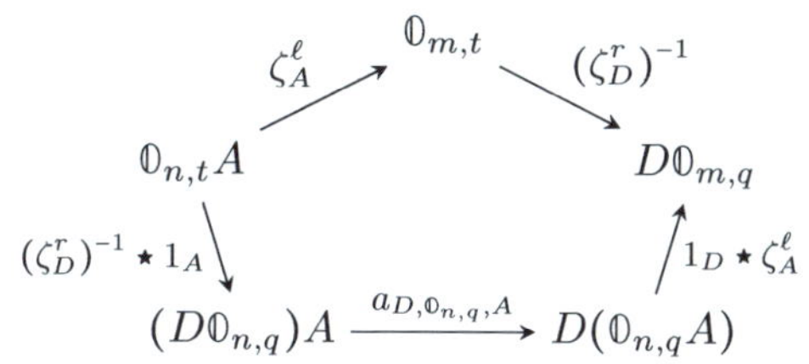

As in case (6), it suffices to show the commutativity of the pentagon above in the (k,i)-entry, which is the value in C of the pentagon below with the appropriate inverses taken into account.

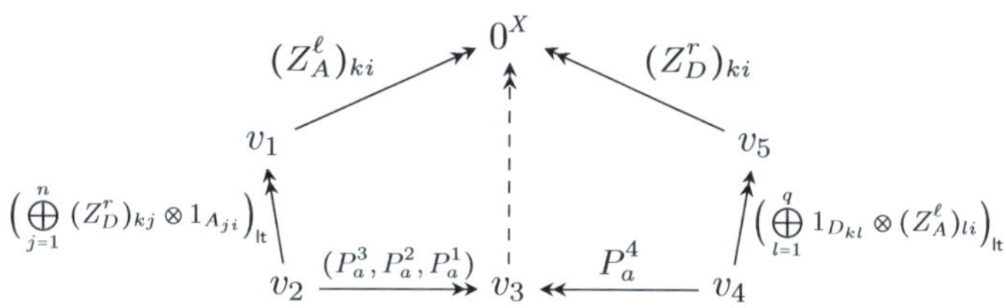

Here $(Z_A^\ell)_{ki}$ is any 0^X-reduction as in (8.1.15) whose value in C is the (k,i)-entry of ζ_A^ℓ, and similarly for $(Z_A^\ell)_{li}$. Because of the top 0^X, each of the other five vertices has the same support as 0^X, and v_3 has a 0^X-reduction denoted by the dashed arrow. For each $h \in \{2,4\}$, the two paths from v_h to the top 0^X have the same value in C by Theorem 3.10.7. Therefore, the (k,i)-entry of the pentagon is commutative.

(8) If $m,n,p,t > 0$ and $q = 0$, then the pentagon (8.4.7) is the diagram below.

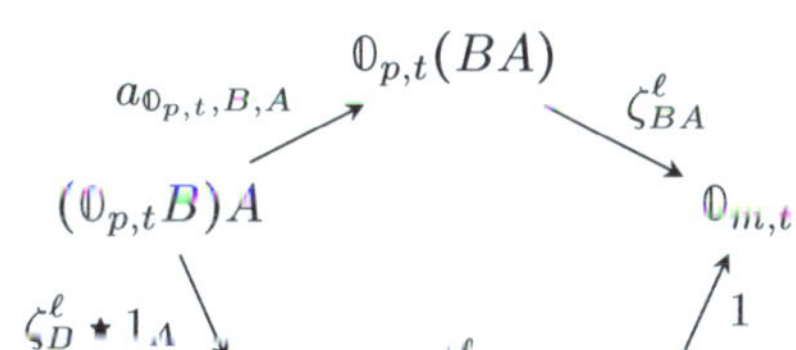

As in cases (6) and (7), it suffices to show the commutativity of the pentagon above in the (k,i)-entry, which is the value in C of the pentagon below with the appropriate inverses taken into account.

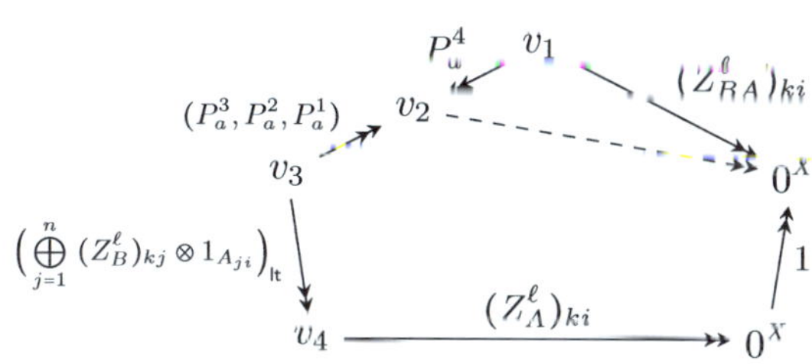

Because of the right 0^X, each of the four vertices $v_?$ has the same support as 0^X, and v_2 has a 0^X-reduction denoted by the dashed arrow. For each $h \in \{1,3\}$, the two paths from v_h to the right 0^X have the same value in C by Theorem 3.10.7. Therefore, the (k,i)-entry of the pentagon is commutative.

All the cases have been proved. □

Lemma 8.4.9. *If $m,n,p,q,t > 0$, then the pentagon (8.4.7) is commutative.*

Proof. It suffices to show that the (t',m')-entry of the pentagon (8.4.7) is commutative for $1 \le m' \le m$ and $1 \le t' \le t$. First we realize the (t',m')-entry of the pentagon (8.4.7) as the value in C of a diagram in $\mathsf{Gr}(X)$ with the following data.

- The set of formal variables is
$$X = \left\{0^X, 1^X, a_{jm'}, b_{kj}, c_{lk}, d_{t'l}\right\}$$
in which $1 \le j \le n$, $1 \le k \le p$, and $1 \le l \le q$.
- The function $\varphi\colon X \longrightarrow \mathsf{Ob}(\mathsf{C})$ is defined by
$$\varphi(x) = \begin{cases} \mathbb{0} & \text{if } x = 0^X, \\ \mathbb{1} & \text{if } x = 1^X, \\ A_{jm'} & \text{if } x = a_{jm'} \text{ for } 1 \le j \le n, \\ B_{kj} & \text{if } x = b_{kj} \text{ for } 1 \le k \le p \text{ and } 1 \le j \le n, \\ C_{lk} & \text{if } x = c_{lk} \text{ for } 1 \le l \le q \text{ and } 1 \le k \le p, \text{ and} \\ D_{t'l} & \text{if } x = d_{t'l} \text{ for } 1 \le l \le q. \end{cases}$$

By Example 8.3.9, the (t',m')-entry of the pentagon (8.4.7) is the value in C of the outer pentagon in $\mathsf{Gr}(X)$ below with the appropriate inverses taken into account.

(8.4.10)

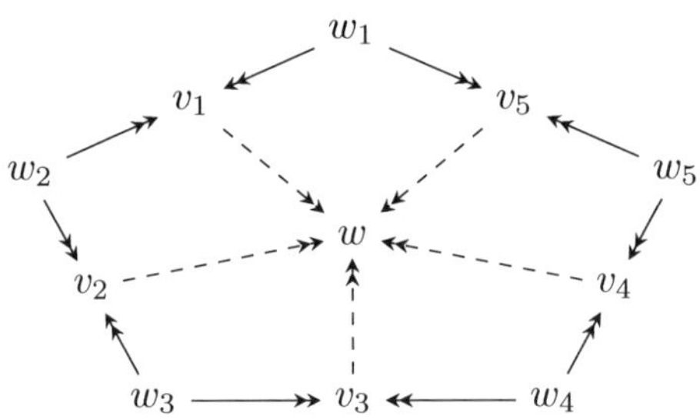

The corner vertices in (8.4.10) and their images under φ are as follows.

$$w_1 = \Big[\bigoplus_{k=1}^{p}\Big(\bigoplus_{l=1}^{q} d_{t'l} \otimes c_{lk}\Big)_{\mathsf{lt}} \otimes \Big(\bigoplus_{j=1}^{n} b_{kj} \otimes a_{jm'}\Big)_{\mathsf{lt}}\Big]_{\mathsf{lt}} \qquad \varphi w_1 = \big[(DC)(BA)\big]_{t'm'}$$

$$w_2 = \Big[\bigoplus_{j=1}^{n}\Big\{\bigoplus_{k=1}^{p}\Big(\bigoplus_{l=1}^{q} d_{t'l} \otimes c_{lk}\Big)_{\mathsf{lt}} \otimes b_{kj}\Big\}_{\mathsf{lt}} \otimes a_{jm'}\Big]_{\mathsf{lt}} \qquad \varphi w_2 = \big[((DC)B)A\big]_{t'm'}$$

$$w_3 = \Big[\bigoplus_{j=1}^{n}\Big\{\bigoplus_{l=1}^{q} d_{t'l} \otimes \Big(\bigoplus_{k=1}^{p} c_{lk} \otimes b_{kj}\Big)_{\mathsf{lt}}\Big\}_{\mathsf{lt}} \otimes a_{jm'}\Big]_{\mathsf{lt}} \qquad \varphi w_3 = \big[(D(CB))A\big]_{t'm'}$$

$$w_4 = \Big[\bigoplus_{l=1}^{q} d_{t'l} \otimes \Big\{\bigoplus_{j=1}^{n}\Big(\bigoplus_{k=1}^{p} c_{lk} \otimes b_{kj}\Big)_{\mathsf{lt}} \otimes a_{jm'}\Big\}_{\mathsf{lt}}\Big]_{\mathsf{lt}} \qquad \varphi w_4 = \big[D((CB)A)\big]_{t'm'}$$

$$w_5 = \Big[\bigoplus_{l=1}^{q} d_{t'l} \otimes \Big\{\bigoplus_{k=1}^{p} c_{lk} \otimes \Big(\bigoplus_{j=1}^{n} b_{kj} \otimes a_{jm'}\Big)_{\mathsf{lt}}\Big\}_{\mathsf{lt}}\Big]_{\mathsf{lt}} \qquad \varphi w_5 = \big[D(C(BA))\big]_{t'm'}$$

The intermediate vertices in (8.4.10) are defined as follows.

$$v_1 = \Big[\bigoplus_{k=1}^{p} \Big\{\bigoplus_{j=1}^{n} \Big(\bigoplus_{l=1}^{q} d_{t'l} \otimes c_{lk}\Big)_{\mathsf{lt}} \otimes \big(b_{kj} \otimes a_{jm'}\big)\Big\}_{\mathsf{lt}}\Big]_{\mathsf{lt}}$$

$$v_2 = \Big[\bigoplus_{j=1}^{n} \Big\{\bigoplus_{l=1}^{q} \Big(\bigoplus_{k=1}^{p} d_{t'l} \otimes \big(c_{lk} \otimes b_{kj}\big)\Big)_{\mathsf{lt}}\Big\}_{\mathsf{lt}} \otimes a_{jm'}\Big]_{\mathsf{lt}}$$

$$v_3 = \Big[\bigoplus_{l=1}^{q} \Big\{\bigoplus_{j=1}^{n} d_{t'l} \otimes \Big[\Big(\bigoplus_{k=1}^{p} c_{lk} \otimes b_{kj}\Big)_{\mathsf{lt}} \otimes a_{jm'}\Big]\Big\}_{\mathsf{lt}}\Big]_{\mathsf{lt}}$$

$$v_4 = \Big[\bigoplus_{l=1}^{q} d_{t'l} \otimes \Big\{\bigoplus_{k=1}^{p} \Big(\bigoplus_{j=1}^{n} c_{lk} \otimes \big(b_{kj} \otimes a_{jm'}\big)\Big)_{\mathsf{lt}}\Big\}_{\mathsf{lt}}\Big]_{\mathsf{lt}}$$

$$v_5 = \Big[\bigoplus_{l=1}^{q} \Big\{\bigoplus_{k=1}^{p} d_{t'l} \otimes \Big[c_{lk} \otimes \Big(\bigoplus_{j=1}^{n} b_{kj} \otimes a_{jm'}\Big)_{\mathsf{lt}}\Big]\Big\}_{\mathsf{lt}}\Big]_{\mathsf{lt}}$$

The middle vertex in (8.4.10) is defined as follows.

$$w = \Big[\bigoplus_{l=1}^{q} \Big\{\bigoplus_{k=1}^{p} \Big(\bigoplus_{j=1}^{n} d_{t'l} \otimes \big[c_{lk} \otimes \big(b_{kj} \otimes a_{jm'}\big)\big]\Big)_{\mathsf{lt}}\Big\}_{\mathsf{lt}}\Big]_{\mathsf{lt}}$$

For example, the (t', m')-entry

$$\varphi w_2 = \big[((DC)B)A\big]_{t'm'} \xrightarrow{(a_{DC,B,A})_{t'm'}} \big[(DC)(BA)\big]_{t'm'} = \varphi w_1$$

is the value in C of the zigzag of paths

$$w_2 \xrightarrow{(P_a^3, P_a^2, P_a^1)} v_1 \xleftarrow{P_a^4} w_1$$

as in (8.3.12), up to a change of symbols.

To show the commutativity of the outer pentagon in (8.4.10) in C, we join the intermediate vertices $v_?$ at the center as follows. The middle vertex w is not the domain of any distributivity morphism δ^l and δ^r. This fact implies that for each $1 \le s \le 5$, there is a path $v_s \longrightarrow w$ that is denoted by a dashed arrow in (8.4.10) and is defined as follows.

(1) Consider the path $v_1 \longrightarrow w$.

(i) For $1 \le k \le p$ and $1 \le j \le n$, it contains the following instances of δ^r,

$$\begin{array}{c}\Big[\Big(\bigoplus_{l=1}^{q'-1} d_{t'l} \otimes c_{lk}\Big)_{\mathsf{lt}} \oplus \big(d_{t'q'} \otimes c_{q'k}\big)\Big] \otimes \big(b_{kj} \otimes a_{jm'}\big) \\ \Big\downarrow \delta^r_{\left(\oplus_{l=1}^{q'-1} d_{t'l} \otimes c_{lk}\right)_{\mathsf{lt}},\, (d_{t'q'} \otimes c_{q'k}),\, (b_{kj} \otimes a_{jm'})} \\ \Big[\Big(\bigoplus_{l=1}^{q'-1} d_{t'l} \otimes c_{lk}\Big)_{\mathsf{lt}} \otimes \big(b_{kj} \otimes a_{jm'}\big)\Big] \oplus \Big[\big(d_{t'q'} \otimes c_{q'k}\big) \otimes \big(b_{kj} \otimes a_{jm'}\big)\Big]\end{array}$$

for $2 \le q' \le q$ if $q > 1$, or the identity if $q = 1$.

(ii) It contains the following npq instances of $\alpha^{\otimes}$.

$$\big(d_{t'l} \otimes c_{lk}\big) \otimes \big(b_{kj} \otimes a_{jm'}\big) \xrightarrow{\alpha^{\otimes}_{d_{t'l}, c_{lk}, (b_{kj} \otimes a_{jm'})}} d_{t'l} \otimes \big[c_{lk} \otimes \big(b_{kj} \otimes a_{jm'}\big)\big]$$

(iii) It contains $\alpha^{\pm\oplus}$ and $\xi^{\oplus}$ that move brackets and permute additively.

(2) Consider the path $v_2 \twoheadrightarrow w$.

(i) For $1 \le j \le n$, it contains the following instances of δ^r,

$$\begin{array}{c}
\Big[\Big\{\bigoplus_{l=1}^{q'-1}\Big(\bigoplus_{k=1}^{p} d_{t'l} \otimes (c_{lk} \otimes b_{kj})\Big)_{\mathsf{lt}}\Big\}_{\mathsf{lt}} \oplus \Big(\bigoplus_{k=1}^{p} d_{t'q'} \otimes (c_{q'k} \otimes b_{kj})\Big)_{\mathsf{lt}}\Big] \otimes a_{jm'} \\
\Big\downarrow \delta^r_{\{\cdots\}_{\mathsf{lt}},\, (\cdots)_{\mathsf{lt}},\, a_{jm'}} \\
\Big[\Big\{\bigoplus_{l=1}^{q'-1}\Big(\bigoplus_{k=1}^{p} d_{t'l} \otimes (c_{lk} \otimes b_{kj})\Big)_{\mathsf{lt}}\Big\}_{\mathsf{lt}} \otimes a_{jm'}\Big] \oplus \Big[\Big(\bigoplus_{k=1}^{p} d_{t'q'} \otimes (c_{q'k} \otimes b_{kj})\Big)_{\mathsf{lt}} \otimes a_{jm'}\Big]
\end{array}$$

for $2 \le q' \le q$ if $q > 1$, or the identity if $q = 1$.

(ii) For $1 \le j \le n$ and $1 \le l \le q$, it contains the following instances of δ^r,

$$\begin{array}{c}
\Big[\Big(\bigoplus_{k=1}^{p'-1} d_{t'l} \otimes (c_{lk} \otimes b_{kj})\Big)_{\mathsf{lt}} \oplus \big(d_{t'l} \otimes (c_{lp'} \otimes b_{p'j})\big)\Big] \otimes a_{jm'} \\
\Big\downarrow \delta^r_{(\cdots)_{\mathsf{lt}},\, (\cdots),\, a_{jm'}} \\
\Big[\Big(\bigoplus_{k=1}^{p'-1} d_{t'l} \otimes (c_{lk} \otimes b_{kj})\Big)_{\mathsf{lt}} \otimes a_{jm'}\Big] \oplus \Big[\big(d_{t'l} \otimes (c_{lp'} \otimes b_{p'j})\big) \otimes a_{jm'}\Big]
\end{array}$$

for $2 \le p' \le p$ if $p > 1$, or the identity if $p = 1$.

(iii) It contains the following $2npq$ instances of $\alpha^{\otimes}$.

$$\big(d_{t'l} \otimes (c_{lk} \otimes b_{kj})\big) \otimes a_{jm'} \xrightarrow{\alpha^{\otimes}} d_{t'l} \otimes \big((c_{lk} \otimes b_{kj}) \otimes a_{jm'}\big)$$

$$(c_{lk} \otimes b_{kj}) \otimes a_{jm'} \xrightarrow{\alpha^{\otimes}} c_{lk} \otimes (b_{kj} \otimes a_{jm'})$$

(iv) It contains $\alpha^{\pm\oplus}$ and $\xi^{\oplus}$ that move brackets and permute additively.

(3) Consider the path $v_3 \twoheadrightarrow w$.

(i) For $1 \le l \le q$ and $1 \le j \le n$, it contains the following instances of δ^r,

$$\begin{array}{c}
\Big[\Big(\bigoplus_{k=1}^{p'-1} c_{lk} \otimes b_{kj}\Big)_{\mathsf{lt}} \oplus (c_{lp'} \otimes b_{p'j})\Big] \otimes a_{jm'} \\
\Big\downarrow \delta^r_{(\cdots)_{\mathsf{lt}},\, (c_{lp'} \otimes b_{p'j}),\, a_{jm'}} \\
\Big[\Big(\bigoplus_{k=1}^{p'-1} c_{lk} \otimes b_{kj}\Big)_{\mathsf{lt}} \otimes a_{jm'}\Big] \oplus \Big[(c_{lp'} \otimes b_{p'j}) \otimes a_{jm'}\Big]
\end{array}$$

for $2 \le p' \le p$ if $p > 1$, or the identity if $p = 1$.

(ii) For $1 \le l \le q$ and $1 \le j \le n$, it contains the following instances of δ^l,

$$\begin{array}{c} d_{t'l} \otimes \Big[\Big(\bigoplus_{k=1}^{p'-1} (c_{lk} \otimes b_{kj}) \otimes a_{jm'} \Big)_{\mathrm{lt}} \oplus \big((c_{lp'} \otimes b_{p'j}) \otimes a_{jm'} \big) \Big] \\ \Big\downarrow \delta^l_{d_{t'l},\, (\cdots)_{\mathrm{lt}},\, ((c_{lp'} \otimes b_{p'j}) \otimes a_{jm'})} \\ \Big[d_{t'l} \otimes \Big(\bigoplus_{k=1}^{p'-1} (c_{lk} \otimes b_{kj}) \otimes a_{jm'} \Big)_{\mathrm{lt}} \Big] \oplus \Big[d_{t'l} \otimes \big((c_{lp'} \otimes b_{p'j}) \otimes a_{jm'} \big) \Big] \end{array}$$

for $2 \le p' \le p$ if $p > 1$, or the identity if $p = 1$.

(iii) It contains the following npq instances of $\alpha^{\otimes}$.

$$(c_{lk} \otimes b_{kj}) \otimes a_{jm'} \xrightarrow{\alpha^{\otimes}} c_{lk} \otimes (b_{kj} \otimes a_{jm'})$$

(iv) It contains $\alpha^{\pm\oplus}$ and $\xi^{\oplus}$ that move brackets and permute additively.

(4) Consider the path $v_4 \twoheadrightarrow w$.

(i) For $1 \le l \le q$, it contains the following instances of δ^l,

$$\begin{array}{c} d_{t'l} \otimes \Big[\Big\{ \bigoplus_{k=1}^{p'-1} \Big(\bigoplus_{j=1}^{n} c_{lk} \otimes (b_{kj} \otimes a_{jm'}) \Big)_{\mathrm{lt}} \Big\}_{\mathrm{lt}} \oplus \Big(\bigoplus_{j=1}^{n} c_{lp'} \otimes (b_{p'j} \otimes a_{jm'}) \Big)_{\mathrm{lt}} \Big] \\ \Big\downarrow \delta^l_{d_{t'l},\, \{\cdots\}_{\mathrm{lt}},\, (\cdots)_{\mathrm{lt}}} \\ \Big[d_{t'l} \otimes \Big\{ \bigoplus_{k=1}^{p'-1} \Big(\bigoplus_{j=1}^{n} c_{lk} \otimes (b_{kj} \otimes a_{jm'}) \Big)_{\mathrm{lt}} \Big\}_{\mathrm{lt}} \Big] \oplus \Big[d_{t'l} \otimes \Big(\bigoplus_{j=1}^{n} c_{lp'} \otimes (b_{p'j} \otimes a_{jm'}) \Big)_{\mathrm{lt}} \Big] \end{array}$$

for $2 \le p' \le p$ if $p > 1$, or the identity if $p = 1$.

(ii) For $1 \le l \le q$ and $1 \le k \le p$, it contains the following instances of δ^l,

$$\begin{array}{c} d_{t'l} \otimes \Big[\Big(\bigoplus_{j=1}^{n'-1} c_{lk} \otimes (b_{kj} \otimes a_{jm'}) \Big)_{\mathrm{lt}} \oplus \big(c_{lk} \otimes (b_{kn'} \otimes a_{n'm'}) \big) \Big] \\ \Big\downarrow \delta^l_{d_{t'l},\, (\cdots)_{\mathrm{lt}},\, (\cdots)} \\ \Big[d_{t'l} \otimes \Big(\bigoplus_{j=1}^{n'-1} c_{lk} \otimes (b_{kj} \otimes a_{jm'}) \Big)_{\mathrm{lt}} \Big] \oplus \Big[d_{t'l} \otimes \big(c_{lk} \otimes (b_{kn'} \otimes a_{n'm'}) \big) \Big] \end{array}$$

for $2 \le n' \le n$ if $n > 1$, or the identity if $n = 1$.

(5) Consider the path $v_5 \twoheadrightarrow w$.

(i) For $1 \le l \le q$ and $1 \le k \le p$, it contains the following instances of δ^l,

$$\begin{array}{c} c_{lk} \otimes \Big[\Big(\bigoplus_{j=1}^{n'-1} b_{kj} \otimes a_{jm'} \Big)_{\mathrm{lt}} \oplus (b_{kn'} \otimes a_{n'm'}) \Big] \\ \Big\downarrow \delta^l_{c_{lk},\, (\cdots)_{\mathrm{lt}},\, (b_{kn'} \otimes a_{n'm'})} \\ \Big[c_{lk} \otimes \Big(\bigoplus_{j=1}^{n'-1} b_{kj} \otimes a_{jm'} \Big)_{\mathrm{lt}} \Big] \oplus \Big[c_{lk} \otimes (b_{kn'} \otimes a_{n'm'}) \Big] \end{array}$$

for $2 \le n' \le n$ if $n > 1$, or the identity if $n = 1$.

(ii) For $1 \leq l \leq q$ and $1 \leq k \leq p$, it contains the following instances of δ^l,

$$\begin{array}{c} d_{t'l} \otimes \Big[\Big(\bigoplus_{j=1}^{n'-1} c_{lk} \otimes (b_{kj} \otimes a_{jm'}) \Big)_{\mathsf{lt}} \oplus \big(c_{lk} \otimes (b_{kn'} \otimes a_{n'm'}) \big) \Big] \\ \Big\downarrow \delta^l_{d_{t'l}, (\cdots)_{\mathsf{lt}}, (c_{lk} \otimes (b_{kn'} \otimes a_{n'm'}))} \\ \Big[d_{t'l} \otimes \Big(\bigoplus_{j=1}^{n'-1} c_{lk} \otimes (b_{kj} \otimes a_{jm'}) \Big)_{\mathsf{lt}} \Big] \oplus \Big[d_{t'l} \otimes \big(c_{lk} \otimes (b_{kn'} \otimes a_{n'm'}) \big) \Big] \end{array}$$

for $2 \leq n' \leq n$ if $n > 1$, or the identity if $n = 1$.

The vertex w is nonsymmetric regular in the sense of Definition 3.10.2. By Lemma 3.10.3, each vertex w_s for $1 \leq s \leq 5$ is also nonsymmetric regular. By the tightness assumption on C, Theorem 3.10.7 implies that each of the five subdiagrams in (8.4.10) is commutative in C. Therefore, the (t', m')-entry of the pentagon (8.4.7) is commutative. □

The Matrix Bicategory. Recall the concept of a bicategory in Definition 6.1.2.

Definition 8.4.11. Suppose C is a tight bimonoidal category. Define the *matrix bicategory*

$$\big(\mathsf{Mat}^{\mathsf{C}}, \mathbb{1}, c, a, \ell, r\big)$$

as consisting of the following data.

Objects: The objects in $\mathsf{Mat}^{\mathsf{C}}$ are nonnegative integers $n \geq 0$.

The Hom Categories: For $m, n \geq 0$, the hom category $\mathsf{Mat}^{\mathsf{C}}(m, n)$ is the category $\mathsf{Mat}^{\mathsf{C}}_{m,n}$ in Definition 8.1.1.

The Identity 1-Cells: For each object $n \geq 0$, its identity 1-cell 1_n is the $n \times n$ identity matrix $\mathbb{1}^n \in \mathsf{Mat}^{\mathsf{C}}_{n,n}$ defined entrywise in (8.1.6).

The Horizontal Composition: For objects $m, n, p \geq 0$, the horizontal composition

$$\mathsf{Mat}^{\mathsf{C}}_{n,p} \times \mathsf{Mat}^{\mathsf{C}}_{m,n} \xrightarrow{\ c\ } \mathsf{Mat}^{\mathsf{C}}_{m,p}$$

is given by the matrix product in Definition 8.1.3.

The Associator: For objects $m, n, p, q \geq 0$, the associator is the natural isomorphism a in Lemma 8.3.1.

The Unitors: For objects $m, n \geq 0$, the left unitor ℓ and the right unitor r are the natural isomorphisms in, respectively, Lemmas 8.2.1 and 8.2.7.

This finishes the definition of $\mathsf{Mat}^{\mathsf{C}}$. ◇

Theorem 8.4.12. *In Definition 8.4.11,* $\mathsf{Mat}^{\mathsf{C}}$ *is a bicategory.*

Proof. The horizontal composition, which is given by the matrix product, is a functor by Lemma 8.1.8. Lemmas 8.2.1, 8.2.7, and 8.3.1 establish the naturality of ℓ, r, and a, respectively. The unity axiom (6.1.3) is proved in Lemma 8.4.2. The pentagon axiom (6.1.4) is proved in Lemmas 8.4.8 and 8.4.9. □

Example 8.4.13 (Coordinatized 2-Vector Spaces)**.** Suppose C is the tight symmetric bimonoidal category $\mathsf{Vect}^{\mathbb{C}}$ of finite dimensional complex vector spaces in

Example 2.1.32. Then the matrix bicategory $\mathsf{Mat}^{\mathsf{C}}$ is the bicategory of *coordinatized 2-vector spaces* in [**JY21**, Ex. 2.1.28], denoted by $\mathsf{2Vect}_{\mathsf{c}}$ there. This is one incarnation of 2-vector spaces [**KV94**, Def. 5.2]. We will revisit 2-vector spaces in Example 8.15.5. ◇

If A is an object in a bicategory B, then the hom category $\mathsf{B}(A, A)$ becomes a monoidal category with

- monoidal unit the identity 1-cell 1_A,
- monoidal product the horizontal composition

$$\mathsf{B}(A, A) \times \mathsf{B}(A, A) \longrightarrow \mathsf{B}(A, A),$$

  and
- associativity isomorphism and left/right unit isomorphisms the restrictions of, respectively, the associator and the left/right unitors of B.

The unity axiom (1.2.2) and the pentagon axiom (1.2.3) for the monoidal category $\mathsf{B}(A, A)$ are special cases of the corresponding axioms (6.1.3) and (6.1.4) for B. When applied to the matrix bicategory $\mathsf{Mat}^{\mathsf{C}}$ in Theorem 8.4.12, we obtain the following result, which is stated in [**BDR04**, Prop. 3.3] without proof, under the additional assumption that C is symmetric.

Corollary 8.4.14. *For each tight bimonoidal category* C *and each* $n \geq 1$, *the category* $\mathsf{Mat}^{\mathsf{C}}_{n,n}$ *in Definition 8.1.1 is a monoidal category with*

- *monoidal unit the identity matrix* $\mathbb{1}^n$ *in* (8.1.6),
- *monoidal product the matrix product in* (8.1.4) *and* (8.1.5),
- *left unit isomorphism in* (8.2.2),
- *right unit isomorphism in* (8.2.8), *and*
- *associativity isomorphism in* (8.3.2).

In the rest of this chapter, we will extend the bicategory $\mathsf{Mat}^{\mathsf{C}}$ in Theorem 8.4.12 to a symmetric monoidal bicategory.

8.5. The Monoidal Identity

For each tight bimonoidal category C as in Definition 2.1.2, we observed in Theorem 8.4.12 that $\mathsf{Mat}^{\mathsf{C}}$ in Definition 8.4.11 is a bicategory in the sense of Definition 6.1.2. The main goal for the rest of this chapter is to show that, if C is also symmetric, then $\mathsf{Mat}^{\mathsf{C}}$ has the structure of a symmetric monoidal bicategory as in Definition 6.5.9. For the next several sections, we construct a monoidal bicategory structure as in Definition 6.4.1 on $\mathsf{Mat}^{\mathsf{C}}$. In this section, we construct its monoidal identity, which is a pseudofunctor in the sense of Definition 6.2.1.

Definition 8.5.1. For each tight bimonoidal category C, define the data of a lax functor

$$\mathbb{1} \xrightarrow{(1_\boxtimes, 1^2_\boxtimes, 1^0_\boxtimes)} \mathsf{Mat}^{\mathsf{C}}$$

as follows.

Object: The identity object $1_\boxtimes(*)$ is the integer $1 \in \mathsf{Mat}^{\mathsf{C}}$.

1-Cell: The identity 1-cell $1_* \in \mathbb{1}(*, *)$ is sent by $1_\boxtimes$ to the 1×1 identity matrix

$$\mathbb{1}^1 = (\mathbb{1}) \in \mathsf{Mat}^{\mathsf{C}}_{1,1}$$

whose only entry is the multiplicative unit $\mathbb{1} \in \mathsf{C}$.

2-Cell: The identity 2-cell $1_{1_*} \in \mathbb{1}(1_*, 1_*)$ is sent by $1_\boxtimes$ to the 1×1 matrix
$$(1_\mathbb{1}) \in \mathsf{Mat}^{\mathsf{C}}_{1,1}(\mathbb{1}^1, \mathbb{1}^1)$$
whose only entry is the identity morphism $1_\mathbb{1} \colon \mathbb{1} \longrightarrow \mathbb{1}$ in C.

The Lax Unity Constraint: Define
$$1_{1_\boxtimes(*)} = 1_1 = \mathbb{1}^1 = (\mathbb{1}) \xrightarrow{\;1^0_\boxtimes\;} (\mathbb{1}) = \mathbb{1}^1 = 1_\boxtimes(1_*)$$
as the identity 2-cell $(1_\mathbb{1})$.

The Lax Functoriality Constraint: Using the matrix product (8.1.4), define
$$1_\boxtimes(1_*)1_\boxtimes(1_*) = (\mathbb{1})(\mathbb{1}) = (\mathbb{1} \otimes \mathbb{1}) \xrightarrow{\;1^2_\boxtimes\;} (\mathbb{1}) = 1_\boxtimes(1_*) = 1_\boxtimes(1_* 1_*)$$
as the 1×1 matrix whose only entry is the left multiplicative unit
$$\lambda^\otimes_\mathbb{1} \colon \mathbb{1} \otimes \mathbb{1} \longrightarrow \mathbb{1} \in \mathsf{C}.$$

This finishes the definition of the tuple $(1_\boxtimes, 1^2_\boxtimes, 1^0_\boxtimes)$. ◇

Lemma 8.5.2. *The tuple $(1_\boxtimes, 1^2_\boxtimes, 1^0_\boxtimes)$ in Definition 8.5.1 is a strictly unitary pseudofunctor.*

Proof. The assignment
$$\mathbb{1}(*, *) \xrightarrow{\;1_\boxtimes\;} \mathsf{Mat}^{\mathsf{C}}_{1,1}$$
sends the identity 2-cell 1_{1_*} to the identity 2-cell $(1_\mathbb{1})$, which is also equal to the vertical composition
$$(1_\mathbb{1})(1_\mathbb{1}) = (1_\mathbb{1} 1_\mathbb{1}).$$
Therefore, $1_\boxtimes$ is a functor.

The lax functoriality constraint $1^2_\boxtimes$ is natural because the only 2-cell in $\mathbb{1}$ is the identity 2-cell of 1_*. To show that $1_\boxtimes$ is a lax functor, it remains to check the two axioms (6.2.2) and (6.2.3).

Since
$$1_\boxtimes(1_*) = \mathbb{1}^1 = (\mathbb{1}),$$
the lax associativity axiom (6.2.2) in the current setting is the left diagram below in $\mathsf{Mat}^{\mathsf{C}}_{1,1}$.

$$(8.5.3)\qquad \begin{array}{ccc} (\mathbb{1}^1\mathbb{1}^1)\mathbb{1}^1 & \xrightarrow{\;a\;} & \mathbb{1}^1(\mathbb{1}^1\mathbb{1}^1) \\ {\scriptstyle 1^2_\boxtimes * 1}\downarrow & & \downarrow{\scriptstyle 1 * 1^2_\boxtimes} \\ \mathbb{1}^1\mathbb{1}^1 & & \mathbb{1}^1\mathbb{1}^1 \\ {\scriptstyle 1^2_\boxtimes}\downarrow & & \downarrow{\scriptstyle 1^2_\boxtimes} \\ \mathbb{1}^1 & \xrightarrow{\;1_\boxtimes(a)\;} & \mathbb{1}^1 \end{array} \qquad \begin{array}{ccc} (\mathbb{1} \otimes \mathbb{1}) \otimes \mathbb{1} & \xrightarrow{\;\alpha^\otimes_{\mathbb{1},\mathbb{1},\mathbb{1}}\;} & \mathbb{1} \otimes (\mathbb{1} \otimes \mathbb{1}) \\ {\scriptstyle \lambda^\otimes_\mathbb{1} \otimes 1_\mathbb{1}}\downarrow & & \downarrow{\scriptstyle 1_\mathbb{1} \otimes \lambda^\otimes_\mathbb{1}} \\ \mathbb{1} \otimes \mathbb{1} & \xrightarrow{\;1\;} & \mathbb{1} \otimes \mathbb{1} \\ {\scriptstyle \lambda^\otimes_\mathbb{1}}\downarrow & & \downarrow{\scriptstyle \lambda^\otimes_\mathbb{1}} \\ \mathbb{1} & \xrightarrow{\;1\;} & \mathbb{1} \end{array}$$

The associator in the matrix bicategory is defined in Lemma 8.3.1. Using its explicit construction in (8.3.5), the only entry of a in (8.5.3) is the multiplicative associativity $\alpha^\otimes_{\mathbb{1},\mathbb{1},\mathbb{1}}$, which corresponds to (8.3.7). Moreover, the associator in $\mathbb{1}$ is the identity. Therefore, in (8.5.3) the commutativity of the left diagram is equivalent to the commutativity of the outer diagram in C on the right.

In the right diagram in (8.5.3), the following statements hold.

- The bottom square is commutative by definition.
- The top square is commutative by
  - the unity axiom (1.2.2) in $(\mathsf{C}, \otimes, \mathbb{1})$ and
  - the equality $\lambda^{\otimes}_{\mathbb{1}} = \rho^{\otimes}_{\mathbb{1}}$ in (1.2.6), applied to the top left vertical morphism.

This proves the lax associativity axiom (6.2.2).

In the current setting, the lax unity axiom (6.2.3) consists of the following two diagrams in $\mathsf{Mat}^{\mathsf{C}}_{1,1}$.

$$
\begin{array}{ccc}
\mathbb{1}^1\mathbb{1}^1 & \xrightarrow{\ \ell\ } & \mathbb{1}^1\\
{\scriptstyle 1^0_{\boxtimes} * 1}\big\downarrow & & \big\uparrow{\scriptstyle 1_{\boxtimes}(\ell)}\\
\mathbb{1}^1\mathbb{1}^1 & \xrightarrow{\ 1^2_{\boxtimes}\ } & \mathbb{1}^1
\end{array}
\qquad\qquad
\begin{array}{ccc}
\mathbb{1}^1\mathbb{1}^1 & \xrightarrow{\ r\ } & \mathbb{1}^1\\
{\scriptstyle 1^0_{\boxtimes} * 1}\big\downarrow & & \big\uparrow{\scriptstyle 1_{\boxtimes}(r)}\\
\mathbb{1}^1\mathbb{1}^1 & \xrightarrow{\ 1^2_{\boxtimes}\ } & \mathbb{1}^1
\end{array}
\tag{8.5.4}
$$

Both ℓ and r in $\mathbb{1}$ are the identity. The unitors ℓ and r in $\mathsf{Mat}^{\mathsf{C}}$ are defined in Lemmas 8.2.1 and 8.2.7, respectively. Using their explicit constructions in (8.2.4) and (8.2.10), the two diagrams in (8.5.4) are commutative if and only if the following two diagrams in C are commutative.

$$
\begin{array}{ccc}
\mathbb{1}\otimes\mathbb{1} & \xrightarrow{\ \lambda^{\otimes}_{\mathbb{1}}\ } & \mathbb{1}\\
{\scriptstyle 1}\big\downarrow & & \big\uparrow{\scriptstyle 1}\\
\mathbb{1}\otimes\mathbb{1} & \xrightarrow{\ \lambda^{\otimes}_{\mathbb{1}}\ } & \mathbb{1}
\end{array}
\qquad\qquad
\begin{array}{ccc}
\mathbb{1}\otimes\mathbb{1} & \xrightarrow{\ \rho^{\otimes}_{\mathbb{1}}\ } & \mathbb{1}\\
{\scriptstyle 1}\big\downarrow & & \big\uparrow{\scriptstyle 1}\\
\mathbb{1}\otimes\mathbb{1} & \xrightarrow{\ \lambda^{\otimes}_{\mathbb{1}}\ } & \mathbb{1}
\end{array}
$$

The left square is commutative by definition. The right square is commutative by the equality $\lambda^{\otimes}_{\mathbb{1}} = \rho^{\otimes}_{\mathbb{1}}$ in (1.2.6). This proves the lax unity axiom (6.2.3), so $1_{\boxtimes}$ is a lax functor.

Since $1^0_{\boxtimes}$ is the identity 2-cell $(1_{\mathbb{1}})$ and since $1^2_{\boxtimes} = (\lambda^{\otimes}_{\mathbb{1}})$ is an isomorphism, $1_{\boxtimes}$ is a strictly unitary pseudofunctor. □

8.6. The Monoidal Composition

For the rest of this chapter, C denotes a tight *symmetric* bimonoidal category as in Definition 2.1.2. We are in the process of constructing a monoidal bicategory structure on the matrix bicategory $\mathsf{Mat}^{\mathsf{C}}$. In Section 8.5, we constructed its monoidal identity. In this section, we construct the monoidal composition

$$(\boxtimes, \boxtimes^2, \boxtimes^0) \colon \mathsf{Mat}^{\mathsf{C}} \times \mathsf{Mat}^{\mathsf{C}} \longrightarrow \mathsf{Mat}^{\mathsf{C}}.$$

Here is an outline of this section.

- We define the matrix tensor product for matrices in C in Definition 8.6.1 and check that it is a functor in Lemma 8.6.7.
- Using the matrix tensor product, the lax unity constraint $\boxtimes^0$ is constructed in Lemma 8.6.8.
- The lax functoriality constraint $\boxtimes^2$ is constructed in several steps in Lemmas 8.6.12, 8.6.13, 8.6.16, and 8.6.21.
- The data $(\boxtimes, \boxtimes^2, \boxtimes^0)$ are assembled in Definition 8.6.19.

To see that the triple $(\boxtimes, \boxtimes^2, \boxtimes^0)$ is a pseudofunctor in the sense of Definition 6.2.1, we will check the lax associativity axiom (6.2.2) and the lax unity axiom (6.2.3) in Section 8.7. The multiplicative symmetry $\xi^{\otimes}$ is first used in Lemma 8.6.16.

The Matrix Tensor Product. The construction uses the following generalizations of the scalar product and the tensor product of complex matrices.

Definition 8.6.1. Suppose $A = (A_{ji}) \in \mathsf{Mat}^{\mathsf{C}}_{m,n}$ and $B = (B_{lk}) \in \mathsf{Mat}^{\mathsf{C}}_{p,q}$ for some $m, n, p, q \geq 0$, and C is a tight symmetric bimonoidal category.

- For each object $C \in \mathsf{C}$, define the *scalar product*

$$C \boxtimes A = \big(C \otimes A_{ji}\big) \in \mathsf{Mat}^{\mathsf{C}}_{m,n} \tag{8.6.2}$$

  as the $n \times m$ matrix obtained from A by replacing each entry A_{ji} by the product $C \otimes A_{ji}$.
- Define the *matrix tensor product*

$$A \boxtimes B = \big(A_{ji} \boxtimes B\big)_{1\leq j\leq n,\, 1\leq i\leq m} \in \mathsf{Mat}^{\mathsf{C}}_{mp,nq} \tag{8.6.3}$$

  as the $nq \times mp$ matrix obtained from A by replacing each entry A_{ji} by the scalar product $A_{ji} \boxtimes B$. In other words, it has entries

$$(A \boxtimes B)_{(j-1)q+l,\, (i-1)p+k} = A_{ji} \otimes B_{lk} \in \mathsf{C} \tag{8.6.4}$$

  for $1 \leq i \leq m$, $1 \leq j \leq n$, $1 \leq k \leq p$, and $1 \leq l \leq q$.

The same notation and terminology apply if C, A, and B are morphisms in, respectively, C, $\mathsf{Mat}^{\mathsf{C}}_{m,n}$, and $\mathsf{Mat}^{\mathsf{C}}_{p,q}$. ◇

Explanation 8.6.5. Suppose $A = (A_{ji}) \in \mathsf{Mat}^{\mathsf{C}}_{m,n}$ is displayed using the following matrix notation.

$$A = \begin{bmatrix} A_{11} & \cdots & A_{1m} \\ \vdots & \ddots & \vdots \\ A_{n1} & \cdots & A_{nm} \end{bmatrix}$$

Then the scalar product in (8.6.2) and the matrix tensor product in (8.6.3) are displayed as follows.

$$C \boxtimes A = \begin{bmatrix} C \otimes A_{11} & \cdots & C \otimes A_{1m} \\ \vdots & \ddots & \vdots \\ C \otimes A_{n1} & \cdots & C \otimes A_{nm} \end{bmatrix} \qquad A \boxtimes B = \begin{bmatrix} A_{11} \boxtimes B & \cdots & A_{1m} \boxtimes B \\ \vdots & \ddots & \vdots \\ A_{n1} \boxtimes B & \cdots & A_{nm} \boxtimes B \end{bmatrix}$$

If either m or n is 0, then $C \boxtimes A \in \mathsf{Mat}^{\mathsf{C}}_{m,n}$ is the empty matrix. If one of m, n, p, or q is 0, then $A \boxtimes B \in \mathsf{Mat}^{\mathsf{C}}_{mp,nq}$ is the empty matrix. ◇

Example 8.6.6. Suppose R is a commutative ring, regarded as a tight symmetric bimonoidal category with its additive and multiplicative structures, and with only identity morphisms. Then the scalar product (8.6.2) and the matrix tensor product (8.6.3) in Mat^R are the usual ones for matrices with entries in R. For example, if $R = \mathbb{C}$, then the matrix tensor product in $\mathsf{Mat}^{\mathbb{C}}$ is the usual tensor product of complex matrices. ◇

Lemma 8.6.7. *For $m, n, p, q \geq 0$, the matrix tensor product*

$$\mathsf{Mat}^{\mathsf{C}}_{m,n} \times \mathsf{Mat}^{\mathsf{C}}_{p,q} \xrightarrow{\ \boxtimes\ } \mathsf{Mat}^{\mathsf{C}}_{mp,nq}$$

in (8.6.3) is a functor.

Proof. Suppose $f\colon A \longrightarrow A' \in \mathsf{Mat}^{\mathsf{C}}_{m,n}$ and $g\colon B \longrightarrow B' \in \mathsf{Mat}^{\mathsf{C}}_{p,q}$ are morphisms. Then for $1 \le j \le n$, $1 \le i \le m$, $1 \le l \le q$, and $1 \le k \le p$,

$$(f \boxtimes g)_{(j-1)q+l,\,(i-1)p+k} = f_{ji} \otimes g_{lk}\colon A_{ji} \otimes B_{lk} \longrightarrow A'_{ji} \otimes B'_{lk} \in \mathsf{C}.$$

Therefore,

$$f \boxtimes g\colon A \boxtimes B \longrightarrow A' \boxtimes B' \in \mathsf{Mat}^{\mathsf{C}}_{mp,nq}$$

is a well-defined morphism. That $\boxtimes$ preserves identity morphisms and composition follows from (i) Definition 8.1.1 that these notions are defined entrywise in C and (ii) the functoriality of $\otimes$. □

The Lax Unity Constraint. Recall from (8.1.6) the identity matrix $\mathbb{1}^m \in \mathsf{Mat}^{\mathsf{C}}_{m,m}$ with m copies of the multiplicative unit $\mathbb{1}$ along the diagonal and the additive zero $\mathbb{0}$ in every other entry. The next lemma will be used to define the lax unity constraint of the monoidal composition in $\mathsf{Mat}^{\mathsf{C}}$.

Lemma 8.6.8. *For $m, p \ge 0$, there is a canonical isomorphism*

$$\mathbb{1}^{mp} \xrightarrow[\cong]{\boxtimes^0_{(m,p)}} \mathbb{1}^m \boxtimes \mathbb{1}^p \in \mathsf{Mat}^{\mathsf{C}}_{mp,mp} \tag{8.6.9}$$

with each entry $\lambda_{\mathbb{1}}^{-\otimes}$, $\rho_{\mathbb{1}}^{-\bullet}$, $\lambda_{\mathbb{1}}^{-\bullet}$, or $\lambda_{\mathbb{0}}^{-\bullet}$ if $m, p > 0$.

Proof. If either m or p is 0, then $\mathbb{1}^{mp}$ and $\mathbb{1}^m \boxtimes \mathbb{1}^p$ are both the empty matrix. In this case, $\boxtimes^0_{(m,p)}$ is the identity morphism.

For $m, p > 0$, $\mathbb{1}^m \boxtimes \mathbb{1}^p$ is computed as follows, with each $\mathbb{1}_i$ a copy of $\mathbb{1}$.

$$\begin{aligned}
\mathbb{1}^m \boxtimes \mathbb{1}^p &= \begin{bmatrix} \mathbb{1}_1 \boxtimes \begin{bmatrix} \mathbb{1}_1 & \cdots & \mathbb{0} \\ \vdots & \ddots & \vdots \\ \mathbb{0} & \cdots & \mathbb{1}_p \end{bmatrix} & \cdots & \mathbb{0} \boxtimes \begin{bmatrix} \mathbb{1}_1 & \cdots & \mathbb{0} \\ \vdots & \ddots & \vdots \\ \mathbb{0} & \cdots & \mathbb{1}_p \end{bmatrix} \\ \vdots & \ddots & \vdots \\ \mathbb{0} \boxtimes \begin{bmatrix} \mathbb{1}_1 & \cdots & \mathbb{0} \\ \vdots & \ddots & \vdots \\ \mathbb{0} & \cdots & \mathbb{1}_p \end{bmatrix} & \cdots & \mathbb{1}_m \boxtimes \begin{bmatrix} \mathbb{1}_1 & \cdots & \mathbb{0} \\ \vdots & \ddots & \vdots \\ \mathbb{0} & \cdots & \mathbb{1}_p \end{bmatrix} \end{bmatrix} \\
&= \begin{bmatrix} \mathbb{1}_1 \otimes \mathbb{1}_1 &&&&& \\ & \ddots &&&& \\ && \mathbb{1}_1 \otimes \mathbb{1}_p &&& \\ &&& \ddots && \\ &&&& \mathbb{1}_m \otimes \mathbb{1}_1 & \\ &&&&& \ddots \\ &&&&&& \mathbb{1}_m \otimes \mathbb{1}_p \end{bmatrix}
\end{aligned} \tag{8.6.10}$$

In other words, $\mathbb{1}^m \boxtimes \mathbb{1}^p$ is an $mp \times mp$ matrix with

- each diagonal entry $\mathbb{1} \otimes \mathbb{1}$; and
- each off-diagonal entry
  - $\mathbb{1} \otimes \mathbb{0}$ (= each off-diagonal entry in each of the m diagonal blocks),
  - $\mathbb{0} \otimes \mathbb{1}$ (= each diagonal entry in each off-diagonal block), or
  - $\mathbb{0} \otimes \mathbb{0}$ (= each off-diagonal entry in each off-diagonal block).

The desired isomorphism $\boxtimes^0_{(m,p)}$ in (8.6.9) has

- each diagonal entry $\lambda_{\mathbb{1}}^{-\otimes}\colon \mathbb{1} \longrightarrow \mathbb{1} \otimes \mathbb{1}$ and
- each off-diagonal entry

$$\mathbb{0} \xrightarrow{\rho_{\mathbb{1}}^{-\bullet}} \mathbb{1} \otimes \mathbb{0}, \qquad \mathbb{0} \xrightarrow{\lambda_{\mathbb{1}}^{-\bullet}} \mathbb{0} \otimes \mathbb{1}, \qquad \text{or} \qquad \mathbb{0} \xrightarrow{\lambda_{\mathbb{0}}^{-\bullet}} \mathbb{0} \otimes \mathbb{0}.$$

Here $\lambda^{-\otimes}$, $\lambda^{-\bullet}$, and $\rho^{-\bullet}$ are the inverses of, respectively, the left multiplicative unit $\lambda^{\otimes}$, the left multiplicative zero $\lambda^{\bullet}$, and the right multiplicative zero $\rho^{\bullet}$ in C. □

The Lax Functoriality Constraint. To define the lax functoriality constraint $\boxtimes^2$ of the monoidal composition in $\mathsf{Mat}^{\mathsf{C}}$, we first compute explicitly the entries of its (co)domain. Suppose A, B, A', and B' are arbitrary 1-cells in $\mathsf{Mat}^{\mathsf{C}}$ as follows.

$$\begin{array}{c} m \xrightarrow{A} n \xrightarrow{B} p \\ m' \xrightarrow{A'} n' \xrightarrow{B'} p' \end{array} \tag{8.6.11}$$

In other words, they are matrices in C as follows.

- $(B,A) = \big((B_{kj}),(A_{ji})\big) \in \mathsf{Mat}^{\mathsf{C}}_{n,p} \times \mathsf{Mat}^{\mathsf{C}}_{m,n}$
- $(B',A') = \big((B'_{k'j'}),(A'_{j'i'})\big) \in \mathsf{Mat}^{\mathsf{C}}_{n',p'} \times \mathsf{Mat}^{\mathsf{C}}_{m',n'}$

Using the matrix product (8.1.4) and the matrix tensor product (8.6.3), there are the following matrices in C.

- $\big(B \boxtimes B', A \boxtimes A'\big) \in \mathsf{Mat}^{\mathsf{C}}_{nn',pp'} \times \mathsf{Mat}^{\mathsf{C}}_{mm',nn'}$
- $(B \boxtimes B')(A \boxtimes A') \in \mathsf{Mat}^{\mathsf{C}}_{mm',pp'}$
- $(BA, B'A') \in \mathsf{Mat}^{\mathsf{C}}_{m,p} \times \mathsf{Mat}^{\mathsf{C}}_{m',p'}$
- $BA \boxtimes B'A' \in \mathsf{Mat}^{\mathsf{C}}_{mm',pp'}$

The next lemma describes the entries in $(B \boxtimes B')(A \boxtimes A')$, which is the domain of $\boxtimes^2_{(B,B'),(A,A')}$. Recall the left normalized bracketing in (5.2.13).

Lemma 8.6.12. *In the setting of* (8.6.11), *for* $1 \le i \le m$, $1 \le i' \le m'$, $1 \le k \le p$, *and* $1 \le k' \le p'$, *the following equality of objects in* C *holds.*

$$\begin{aligned} &\big[(B \boxtimes B')(A \boxtimes A')\big]_{(k-1)p'+k',\,(i-1)m'+i'} \\ &= \begin{cases} \mathbb{0} & \text{if } n \text{ or } n' \text{ is } 0, \text{ and} \\ \Big[\bigoplus_{j=1}^{n}\bigoplus_{j'=1}^{n'} \big[(B_{kj} \otimes B'_{k'j'}) \otimes (A_{ji} \otimes A'_{j'i'})\big]\Big]_{\mathsf{lt}} & \text{if } n,n' > 0. \end{cases} \end{aligned}$$

Proof. If one of m, m', p, or p' is 0, then $(B \boxtimes B')(A \boxtimes A')$ is the empty matrix, in which case there is nothing to prove. So we assume that $m, m', p, p' > 0$.

If either n or n' is 0, then $nn' = 0$, and $(B \boxtimes B')(A \boxtimes A')$ is the 0 matrix $\mathbb{0}_{mm',pp'}$ by definition (8.1.4). So we furthermore assume that $n, n' > 0$. By the definition (8.1.4) of the matrix product, the entry of $(B \boxtimes B')(A \boxtimes A')$ in the statement of the lemma is the unique entry of the matrix product of

- the $[(k-1)p' + k']$th row in $B \boxtimes B'$ and
- the $[(i-1)m' + i']$th column in $A \boxtimes A'$.

In other words, it is the unique entry of the following matrix product, in which the $\otimes$ symbol in each entry is abbreviated to concatenation.

$$\Big[\overbrace{B_{k1}B'_{k'1} \cdots B_{k1}B'_{k'n'}}^{n' \text{ entries}} \cdots \overbrace{B_{kn}B'_{k'1} \cdots B_{kn}B'_{k'n'}}^{n' \text{ entries}}\Big] \begin{bmatrix} A_{1i}A'_{1i'} \\ \vdots \\ A_{1i}A'_{n'i'} \\ \vdots \\ A_{ni}A'_{1i'} \\ \vdots \\ A_{ni}A'_{n'i'} \end{bmatrix}$$

The unique entry of this matrix product is the left normalized sum stated in the lemma. □

The next lemma describes the entries in $BA \boxtimes B'A'$, which is the codomain of $\boxtimes^2_{(B,B'),(A,A')}$.

Lemma 8.6.13. *In the setting of* (8.6.11), *for* $1 \le i \le m$, $1 \le i' \le m'$, $1 \le k \le p$, *and* $1 \le k' \le p'$, *the following equality of objects in* C *holds.*

$$\begin{aligned}&\big(BA \boxtimes B'A'\big)_{(k-1)p'+k',\,(i-1)m'+i'}\\ &= \begin{cases} \mathbb{0} \otimes \mathbb{0} & \text{if } n = n' = 0,\\ \mathbb{0} \otimes \left[\bigoplus_{j'=1}^{n'} \big(B'_{k'j'} \otimes A'_{j'i'}\big)\right]_{\mathsf{lt}} & \text{if } n = 0 \text{ and } n' > 0,\\ \left[\bigoplus_{j=1}^{n} \big(B_{kj} \otimes A_{ji}\big)\right]_{\mathsf{lt}} \otimes \mathbb{0} & \text{if } n > 0 \text{ and } n' = 0, \text{ and}\\ \left[\bigoplus_{j=1}^{n} \big(B_{kj} \otimes A_{ji}\big)\right]_{\mathsf{lt}} \otimes \left[\bigoplus_{j'=1}^{n'} \big(B'_{k'j'} \otimes A'_{j'i'}\big)\right]_{\mathsf{lt}} & \text{if } n, n' > 0.\end{cases}\end{aligned}$$

Proof. If one of m, m', p, or p' is 0, then $BA \boxtimes B'A'$ is the empty matrix, in which case there is nothing to prove. So we assume that $m, m', p, p' > 0$.

By the definition (8.6.4) of the matrix tensor product, the entry of $BA \boxtimes B'A'$ in the lemma is the product

$$(BA)_{ki} \otimes (B'A')_{k'i'}.$$

By definition (8.1.4) this product is as stated in the statement of the lemma. Note that if $n = 0$, then BA is the 0 matrix $\mathbb{0}_{m,p}$. If $n' = 0$, then $B'A'$ is the 0 matrix $\mathbb{0}_{m',p'}$. □

To define the lax functoriality constraint $\boxtimes^2_{(B,B'),(A,A')}$, we connect the objects in Lemmas 8.6.12 and 8.6.13 using paths in the sense of Definition 3.1.14 with the following setting.

Definition 8.6.14. In the setting of (8.6.11), suppose $m, n, p, m', n', p' > 0$.

- Define the set of formal variables

$$X^{\boxtimes} = \{0^X, 1^X, a_{ji}, b_{kj}, a'_{j'i'}, b'_{k'j'}\}$$

with $1 \le i \le m$, $1 \le j \le n$, $1 \le k \le p$, $1 \le i' \le m'$, $1 \le j' \le n'$, and $1 \le k' \le p'$.
- Define the function $\varphi^{\boxtimes}\colon X^{\boxtimes} \longrightarrow \mathsf{Ob}(\mathsf{C})$ as follows.

$$\varphi^{\boxtimes}(x) = \begin{cases} \mathbb{0} & \text{if } x = 0^X.\\ \mathbb{1} & \text{if } x = 1^X.\\ A_{ji} & \text{if } x = a_{ji}.\\ B_{kj} & \text{if } x = b_{kj}.\\ A'_{j'i'} & \text{if } x = a'_{j'i'}.\\ B'_{k'j'} & \text{if } x = b'_{k'j'}.\end{cases}$$

Paths in $\mathsf{Gr}(X^{\boxtimes})$ take values in C via the function $\varphi^{\boxtimes}$. ◇

Motivation 8.6.15. It is possible to define an explicit morphism in C from the object in Lemma 8.6.12 to the object in Lemma 8.6.13. If $m, n, p, m', n', p' > 0$, then this involves

- permuting $B'_{k'j'}$ and A_{ji} in each object
$$\big(B_{kj} \otimes B'_{k'j'}\big) \otimes \big(A_{ji} \otimes A'_{j'i'}\big)$$
via $\alpha^{\pm\otimes}$ and $\xi^{\otimes}$ and
- factoring via the inverses of δ^l and δ^r.

However, to prove that $\boxtimes^2$ has the desired properties of the lax functoriality constraint, we will need to use the Coherence Theorem 3.9.1. So we need to realize such a morphism in C using paths in $\mathsf{Gr}(X^{\boxtimes})$.

Recall from Definition 3.1.6 that the elementary edges δ^l and δ^r do not have formal inverses in $\mathsf{Gr}(X^{\boxtimes})$. Therefore, in general, it is not possible to have a single path in $\mathsf{Gr}(X^{\boxtimes})$ whose value in C has the object in Lemma 8.6.12 as its domain and the object in Lemma 8.6.13 as its codomain. Instead, they are connected by a zigzag of paths in $\mathsf{Gr}(X^{\boxtimes})$ as follows. ◇

Lemma 8.6.16. *In the setting of Definition 8.6.14, the following statements hold.*

(1) *There exist paths*

$$\begin{array}{c}
\Big[\bigoplus_{j=1}^{n}\bigoplus_{j'=1}^{n'}\big[(b_{kj}\otimes b'_{k'j'})\otimes(a_{ji}\otimes a'_{j'i'})\big]\Big]_{\mathsf{lt}} \\
\Big\downarrow P \\
\Big[\bigoplus_{j=1}^{n}\Big[\bigoplus_{j'=1}^{n'}(b_{kj}\otimes a_{ji})\otimes(b'_{k'j'}\otimes a'_{j'i'})\Big]_{\mathsf{lt}}\Big]_{\mathsf{lt}} \\
\Big\uparrow Q \\
\Big[\bigoplus_{j=1}^{n}(b_{kj}\otimes a_{ji})\Big]_{\mathsf{lt}}\otimes\Big[\bigoplus_{j'=1}^{n'}(b'_{k'j'}\otimes a'_{j'i'})\Big]_{\mathsf{lt}}
\end{array}$$

in $\mathsf{Gr}(X^{\boxtimes})$.

(2) *All the paths in* $\mathsf{Gr}(X^{\boxtimes})$ *with the same (co)domain as P, respectively Q, have the same value in* C.

Proof. For the existence of the path P, first we use prime edges involving $\alpha^{\pm\otimes}$ and $\xi^{\otimes}$ to permute each monomial

$$\big(b_{kj}\otimes b'_{k'j'}\big)\otimes\big(a_{ji}\otimes a'_{j'i'}\big) \quad \text{to} \quad \big(b_{kj}\otimes a_{ji}\big)\otimes\big(b'_{k'j'}\otimes a'_{j'i'}\big). \tag{8.6.17}$$

After these prime edges, we use $\alpha^{\oplus}$ to move the additive brackets to match with the additive bracketing of the codomain of P.

The path $Q = (Q_2, Q_1)$ is the following concatenated path in $\mathsf{Gr}(X^{\boxtimes})$.

$$\begin{array}{c}
\Big[\bigoplus_{j=1}^{n}\Big[\bigoplus_{j'=1}^{n'}(b_{kj}\otimes a_{ji})\otimes(b'_{k'j'}\otimes a'_{j'i'})\Big]_{\mathsf{lt}}\Big]_{\mathsf{lt}} \\
\Big\uparrow Q_2 \\
\Big[\bigoplus_{j=1}^{n}(b_{kj}\otimes a_{ji})\otimes\Big[\bigoplus_{j'=1}^{n'}(b'_{k'j'}\otimes a'_{j'i'})\Big]_{\mathsf{lt}}\Big]_{\mathsf{lt}} \\
\Big\uparrow Q_1 \\
\Big[\bigoplus_{j=1}^{n}(b_{kj}\otimes a_{ji})\Big]_{\mathsf{lt}}\otimes\Big[\bigoplus_{j'=1}^{n'}(b'_{k'j'}\otimes a'_{j'i'})\Big]_{\mathsf{lt}}
\end{array}$$

- Q_1 is the identity if $n = 1$. If $n > 1$, then each prime edge in Q_1 contains an instance of δ^r.
- Q_2 is the identity if $n' = 1$. If $n' > 1$, then each prime edge in Q_2 contains an instance of δ^l.

This proves the existence of the paths P and Q.

For the second assertion, observe that the domain of P is regular in the sense of Definition 3.1.25. By Lemma 3.1.29, the domain of Q is also regular. Since C is assumed to be a tight symmetric bimonoidal category, the Coherence Theorem 3.9.1 implies the uniqueness of the value of the path P, respectively Q, in C provided that it has the stated (co)domain. □

Remark 8.6.18. One of the reasons that we need C to be a tight *symmetric* bimonoidal category is the permutation in (8.6.17). It uses the multiplicative symmetric monoidal structure $(\mathsf{C}, \otimes, \alpha^{\otimes}, \xi^{\otimes})$. ◇

We are now ready to define the monoidal composition in $\mathsf{Mat}^{\mathsf{C}}$ in the sense of Definition 6.4.1.

Definition 8.6.19. For a tight symmetric bimonoidal category C, define the data of a lax functor

$$\mathsf{Mat}^{\mathsf{C}} \times \mathsf{Mat}^{\mathsf{C}} \xrightarrow{(\boxtimes, \boxtimes^2, \boxtimes^0)} \mathsf{Mat}^{\mathsf{C}}$$

as follows.

Objects: For each pair of objects $(m,p) \in \mathsf{Mat}^{\mathsf{C}} \times \mathsf{Mat}^{\mathsf{C}}$, define the object

$$\boxtimes(m,p) = m \boxtimes p = mp \in \mathsf{Mat}^{\mathsf{C}}.$$

The Local Functors: For $m,n,p,q \geq 0$, the local functor

$$\big(\mathsf{Mat}^{\mathsf{C}} \times \mathsf{Mat}^{\mathsf{C}}\big)\big((m,p),(n,q)\big) = \mathsf{Mat}^{\mathsf{C}}_{m,n} \times \mathsf{Mat}^{\mathsf{C}}_{p,q} \xrightarrow{\boxtimes} \mathsf{Mat}^{\mathsf{C}}_{mp,nq}$$

is defined as the matrix tensor product (8.6.3) for matrices in C and for their morphisms.

The Lax Unity Constraint: For each pair of objects $(m,p) \in \mathsf{Mat}^{\mathsf{C}} \times \mathsf{Mat}^{\mathsf{C}}$, define the component 2-cell

$$1_{m \boxtimes p} = \mathbb{1}^{mp} \xrightarrow{\boxtimes^0_{(m,p)}} \mathbb{1}^m \boxtimes \mathbb{1}^p = \boxtimes(1_{(m,p)}) \in \mathsf{Mat}^{\mathsf{C}}_{mp,mp}$$

as the canonical isomorphism in (8.6.9).

The Lax Functoriality Constraint: In the setting of (8.6.11), the component 2-cell

$$(B \boxtimes B')(A \boxtimes A') \xrightarrow{\boxtimes^2_{(B,B'),(A,A')}} BA \boxtimes B'A' \in \mathsf{Mat}^{\mathsf{C}}_{mm',pp'}$$

is the identity morphism of the empty matrix if $m, m', p,$ or p' is 0. If $m, m', p, p' > 0$, then its $\big((k-1)p' + k', (i-1)m' + i'\big)$-entry is

$$\begin{cases} \lambda_0^{-\bullet} & \text{if } n = n' = 0, \\ \lambda^{-\bullet}_{(B'A')_{k'i'}} & \text{if } n = 0 \text{ and } n' > 0, \\ \rho^{-\bullet}_{(BA)_{ki}} & \text{if } n > 0 \text{ and } n' = 0, \text{ and} \\ (\varphi^{\boxtimes}Q)^{-1}(\varphi^{\boxtimes}P) & \text{if } n, n' > 0. \end{cases} \tag{8.6.20}$$

In (8.6.20), the following statements hold.

- $1 \leq i \leq m$, $1 \leq i' \leq m'$, $1 \leq k \leq p$, and $1 \leq k' \leq p'$.
- $\lambda^{-\bullet}$ and $\rho^{-\bullet}$ are the inverses of, respectively, $\lambda^{\bullet}$ and $\rho^{\bullet}$ in C.
- P and Q are the paths in Lemma 8.6.16, with $\varphi^{\boxtimes}P$ and $\varphi^{\boxtimes}Q$ their values in C as in Definitions 3.1.14 and 8.6.14.

This finishes the definition of the triple $(\boxtimes, \boxtimes^2, \boxtimes^0)$. ◇

First we check the naturality of $\boxtimes^2$.

Lemma 8.6.21. *In Definition 8.6.19, $\boxtimes^2$ is a natural isomorphism.*

Proof. Each entry of $\boxtimes^2_{(B,B'),(A,A')}$ is a well-defined isomorphism in C by Lemmas 8.6.12, 8.6.13, and 8.6.16. If m, m', p, or p' is 0, then $\mathsf{Mat}^{\mathsf{C}}_{mm',pp'}$ is the terminal category. The naturality of $\boxtimes^2$ holds in this case. So we assume that $m, m', p, p' > 0$.

The 2-cells in $\mathsf{Mat}^{\mathsf{C}}$ are entrywise morphisms in C.

- If either n or n' is 0, then the naturality of $\boxtimes^2$ follows from the naturality of $\lambda^{\bullet}$ and $\rho^{\bullet}$ in C.
- If $n, n' > 0$, then by the proof of Lemma 8.6.16, each entry of $\boxtimes^2$ is a composite of isomorphisms, each being the iterated sum and product of identity morphisms and one component of $\alpha^{\oplus}$, $\alpha^{\pm\otimes}$, $\xi^{\otimes}$, δ^{-l}, or δ^{-r}. The naturality of $\boxtimes^2$ follows from the naturality of these structure morphisms in C.

Moreover, $\boxtimes^2$ is a natural isomorphism because each of its components is an isomorphism or a composite of isomorphisms. □

8.7. The Pseudofunctoriality of the Monoidal Composition

For a tight symmetric bimonoidal category C, we are in the process of constructing a monoidal bicategory structure, in the sense of Definition 6.4.1, on the matrix bicategory $\mathsf{Mat}^{\mathsf{C}}$. In Section 8.6, we defined the data $(\boxtimes, \boxtimes^2, \boxtimes^0)$ that are supposed to constitute the monoidal composition in $\mathsf{Mat}^{\mathsf{C}}$. In this section, we check that the triple $(\boxtimes, \boxtimes^2, \boxtimes^0)$ is indeed a pseudofunctor in the sense of Definition 6.2.1. We check

- the lax associativity axiom (6.2.2) in Lemmas 8.7.3, 8.7.4, 8.7.13, and 8.7.23, and
- the lax unity axiom (6.2.3) in Lemma 8.7.27.

The Lax Associativity Axiom. To check the lax associativity axiom (6.2.2) for the triple $(\boxtimes, \boxtimes^2, \boxtimes^0)$, consider arbitrary composable 1-cells in $\mathsf{Mat}^{\mathsf{C}}$ as follows.

$$\begin{array}{c} m \xrightarrow{\ A\ } n \xrightarrow{\ B\ } p \xrightarrow{\ C\ } q \\ m' \xrightarrow{\ A'\ } n' \xrightarrow{\ B'\ } p' \xrightarrow{\ C'\ } q' \end{array} \tag{8.7.1}$$

Here $A = (A_{ji})$, $A' = (A'_{j'i'})$, $B = (B_{kj})$, and $B' = (B'_{k'j'})$ are as in (8.6.11), and

$$(C, C') = \big((C_{lk}), (C'_{l'k'})\big) \in \mathsf{Mat}^{\mathsf{C}}_{p,q} \times \mathsf{Mat}^{\mathsf{C}}_{p',q'}.$$

The lax associativity axiom (6.2.2) for these 1-cells is the commutativity of the following diagram in $\mathsf{Mat}^{\mathsf{C}}_{mm',qq'}$, with a the associator in Lemma 8.3.1.

$$\begin{array}{ccc} \big[(C \boxtimes C')(B \boxtimes B')\big](A \boxtimes A') & \xrightarrow{\;a\;} & (C \boxtimes C')\big[(B \boxtimes B')(A \boxtimes A')\big] \\ {\scriptstyle \boxtimes^2 \star 1}\downarrow & & \downarrow{\scriptstyle 1 \star \boxtimes^2} \\ (CB \boxtimes C'B')(A \boxtimes A') & & (C \boxtimes C')(BA \boxtimes B'A') \\ {\scriptstyle \boxtimes^2}\downarrow & & \downarrow{\scriptstyle \boxtimes^2} \\ (CB)A \boxtimes (C'B')A' & \xrightarrow{\;a \boxtimes a\;} & C(BA) \boxtimes C'(B'A') \end{array} \tag{8.7.2}$$

Since morphisms in $\mathsf{Mat}^{\mathsf{C}}_{mm',qq'}$ are entrywise morphisms in C, it suffices to prove the commutativity of (8.7.2) in a typical entry. We use (8.6.20) to interpret the entries of $\boxtimes^2$. In some diagrams, some subscripts are omitted to save space. When we restrict to a typical entry, the indices are

$$1 \le i \le m, \quad 1 \le l \le q, \quad 1 \le i' \le m', \quad \text{and} \quad 1 \le l' \le q'.$$

To clarify the many cases, we split the proof of the commutativity of (8.7.2) into several lemmas.

Lemma 8.7.3. *In the setting of* (8.7.1), *suppose that either*

(1) *at least one of* $m, m', q,$ *and* q' *is* 0, *or*
(2) $m, m', q, q' > 0$, *and at most one of* $n, n', p,$ *and* p' *is* > 0.

Then the diagram (8.7.2) *is commutative.*

Proof. If at least one of $m, m', q,$ and q' is 0, then $\mathsf{Mat}^{\mathsf{C}}_{mm',qq'}$ is the terminal category. The diagram (8.7.2) is commutative in this case.

For the other case, suppose that $m, m', q, q' > 0$ and that at most one of $n, n', p,$ and p' is > 0. There are 5 subcases.

(1) If $n = n' = p = p' = 0$, then (8.7.2) is the following diagram.

$$\begin{array}{ccc} \mathbb{0}_{mm',qq'} & \xrightarrow{\;1\;} & \mathbb{0}_{mm',qq'} \\ {\scriptstyle 1}\downarrow & & \downarrow{\scriptstyle 1} \\ \mathbb{0}_{mm',qq'} & & \mathbb{0}_{mm',qq'} \\ {\scriptstyle \boxtimes^{?}}\downarrow & & \downarrow{\scriptstyle \boxtimes^{?}} \\ \mathbb{0}_{m,q} \boxtimes \mathbb{0}_{m',q'} & \xrightarrow{\;1 \boxtimes 1\;} & \mathbb{0}_{m,q} \boxtimes \mathbb{0}_{m',q'} \end{array}$$

This is commutative because $1 \boxtimes 1 = 1$ by Lemma 8.6.7.

(2) If $n > 0$ and $n' = p = p' = 0$, then (8.7.2) is the left diagram below, with ζ^ℓ the natural isomorphism in (8.1.11).

$$\begin{array}{ccc} \mathbb{0}_{mm',qq'} & \xrightarrow{\;1\;} & \mathbb{0}_{mm',qq'} \\ {\scriptstyle 1}\downarrow & & \downarrow{\scriptstyle 1} \\ \mathbb{0}_{mm',qq'} & & \mathbb{0}_{mm',qq'} \\ {\scriptstyle \boxtimes^2}\downarrow & & \downarrow{\scriptstyle \boxtimes^2} \\ \mathbb{0}_{n,q}A \boxtimes \mathbb{0}_{m',q'} & \xrightarrow{\;\zeta^\ell_A \boxtimes 1\;} & \mathbb{0}_{m,q} \boxtimes \mathbb{0}_{m',q'} \end{array} \qquad \begin{array}{ccc} \mathbb{0} & \xrightarrow{\;1\;} & \mathbb{0} \\ {\scriptstyle 1}\downarrow & & \downarrow{\scriptstyle 1} \\ \mathbb{0} & & \mathbb{0} \\ {\scriptstyle \rho^{-\bullet}_{(\mathbb{0}_{n,q}A)_{li}}}\downarrow & & \downarrow{\scriptstyle \lambda^{-\bullet}_{\mathbb{0}} = \rho^{-\bullet}_{\mathbb{0}}} \\ (\mathbb{0}_{n,q}A)_{li} \otimes \mathbb{0} & \xrightarrow{\;(\zeta^\ell_A)_{li} \otimes 1\;} & \mathbb{0} \otimes \mathbb{0} \end{array}$$

A typical entry of the left diagram above is the right diagram in C. The right diagram is commutative by

- the equality $\lambda_{\mathbb{0}}^{\bullet} = \rho_{\mathbb{0}}^{\bullet}$, which is the axiom (2.1.14) in C, and
- the naturality of $\rho^{\bullet}$.

(3) If $n' > 0$ and $n = p = p' = 0$, then (8.7.2) is the left diagram below.

$$\begin{array}{ccc}
\mathbb{0}_{mm',qq'} & \xrightarrow{\ 1\ } & \mathbb{0}_{mm',qq'} \\
{\scriptstyle 1}\downarrow & & \downarrow{\scriptstyle 1} \\
\mathbb{0}_{mm',qq'} & & \mathbb{0}_{mm',qq'} \\
{\scriptstyle \boxtimes^2}\downarrow & & \downarrow{\scriptstyle \boxtimes^2} \\
\mathbb{0}_{m,q} \boxtimes \mathbb{0}_{n',q'} A' & \xrightarrow{1 \boxtimes \zeta^{\ell}_{A'}} & \mathbb{0}_{m,q} \boxtimes \mathbb{0}_{m',q'}
\end{array}
\qquad
\begin{array}{ccc}
\mathbb{0} & \xrightarrow{\ 1\ } & \mathbb{0} \\
{\scriptstyle 1}\downarrow & & \downarrow{\scriptstyle 1} \\
\mathbb{0} & & \mathbb{0} \\
{\scriptstyle \lambda^{-\bullet}_{(\mathbb{0}_{n',q'}A')_{l'i'}}}\downarrow & & \downarrow{\scriptstyle \lambda^{-\bullet}_{\mathbb{0}}} \\
\mathbb{0} \otimes (\mathbb{0}_{n',q'} A')_{l'i'} & \xrightarrow{1 \otimes (\zeta^{\ell}_{A'})_{l'i'}} & \mathbb{0} \otimes \mathbb{0}
\end{array}$$

A typical entry of the left diagram above is the right diagram in C, which is commutative by the naturality of $\lambda^{\bullet}$.

(4) If $p > 0$ and $n = n' = p' = 0$, then (8.7.2) is the left diagram below, with ζ^r the natural isomorphism in (8.1.19).

$$\begin{array}{ccc}
\mathbb{0}_{mm',qq'} & \xrightarrow{\ 1\ } & \mathbb{0}_{mm',qq'} \\
{\scriptstyle 1}\downarrow & & \downarrow{\scriptstyle 1} \\
\mathbb{0}_{mm',qq'} & & \mathbb{0}_{mm',qq'} \\
{\scriptstyle \boxtimes^2}\downarrow & & \downarrow{\scriptstyle \boxtimes^2} \\
\mathbb{0}_{m,q} \boxtimes \mathbb{0}_{m',q'} & \xrightarrow{(\zeta^{r}_{C})^{-1} \boxtimes 1} & C\mathbb{0}_{m,p} \boxtimes \mathbb{0}_{m',q'}
\end{array}
\qquad
\begin{array}{ccc}
\mathbb{0} & \xrightarrow{\ 1\ } & \mathbb{0} \\
{\scriptstyle 1}\downarrow & & \downarrow{\scriptstyle 1} \\
\mathbb{0} & & \mathbb{0} \\
{\scriptstyle \lambda^{-\bullet}_{\mathbb{0}} = \rho^{-\bullet}_{\mathbb{0}}}\downarrow & & \downarrow{\scriptstyle \rho^{-\bullet}_{(C\mathbb{0}_{m,p})_{li}}} \\
\mathbb{0} \otimes \mathbb{0} & \xrightarrow{(\zeta^{r}_{C})^{-1}_{li} \otimes 1} & (C\mathbb{0}_{m,p})_{li} \otimes \mathbb{0}
\end{array}$$

As in case (2), a typical entry of the left diagram above is the right diagram in C. It is commutative by the equality $\lambda_{\mathbb{0}}^{\bullet} = \rho_{\mathbb{0}}^{\bullet}$ in (2.1.14) and the naturality of $\rho^{\bullet}$.

(5) If $p' > 0$ and $n = n' = p = 0$, then (8.7.2) is the left diagram below.

$$\begin{array}{ccc}
\mathbb{0}_{mm',qq'} & \xrightarrow{\ 1\ } & \mathbb{0}_{mm',qq'} \\
{\scriptstyle 1}\downarrow & & \downarrow{\scriptstyle 1} \\
\mathbb{0}_{mm',qq'} & & \mathbb{0}_{mm',qq'} \\
{\scriptstyle \boxtimes^2}\downarrow & & \downarrow{\scriptstyle \boxtimes^2} \\
\mathbb{0}_{m,q} \boxtimes \mathbb{0}_{m',q'} & \xrightarrow{1 \boxtimes (\zeta^{r}_{C'})^{-1}} & \mathbb{0}_{m,q} \boxtimes C'\mathbb{0}_{m',p'}
\end{array}
\qquad
\begin{array}{ccc}
\mathbb{0} & \xrightarrow{\ 1\ } & \mathbb{0} \\
{\scriptstyle 1}\downarrow & & \downarrow{\scriptstyle 1} \\
\mathbb{0} & & \mathbb{0} \\
{\scriptstyle \lambda^{-\bullet}_{\mathbb{0}}}\downarrow & & \downarrow{\scriptstyle \lambda^{-\bullet}_{(C'\mathbb{0}_{m',p'})_{l'i'}}} \\
\mathbb{0} \otimes \mathbb{0} & \xrightarrow{1 \otimes (\zeta^{r}_{C'})^{-1}_{l'i'}} & \mathbb{0} \otimes (C'\mathbb{0}_{m',p'})_{l'i'}
\end{array}$$

A typical entry of the left diagram above is the right diagram in C, which is commutative by the naturality of $\lambda^{\bullet}$.

This finishes the proofs of all the subcases. □

Lemma 8.7.4. *In the setting of* (8.7.1), *suppose that*

- $m, m', q, q' > 0$ *and*
- *precisely two of* $n, n', p,$ *and* p' *are 0.*

Then the diagram (8.7.2) *is commutative.*

Proof. There are six cases.

(1) If $n, n' > 0$ and $p = p' = 0$, then (8.7.2) is the diagram below.

$$\begin{array}{ccc} \mathbb{0}_{nn',qq'}(A \boxtimes A') & \xrightarrow{\ \zeta^{\ell}_{A\boxtimes A'}\ } & \mathbb{0}_{mm',qq'} \\ {\scriptstyle \boxtimes^2 \star 1}\big\downarrow & & \big\downarrow{\scriptstyle 1} \\ (\mathbb{0}_{n,q} \boxtimes \mathbb{0}_{n',q'})(A \boxtimes A') & & \mathbb{0}_{mm',qq'} \\ {\scriptstyle \boxtimes^2}\big\downarrow & & \big\downarrow{\scriptstyle \boxtimes^2} \\ \mathbb{0}_{n,q}A \boxtimes \mathbb{0}_{n',q'}A' & \xrightarrow{\ \zeta^{\ell}_{A}\boxtimes\zeta^{\ell}_{A'}\ } & \mathbb{0}_{m,q} \boxtimes \mathbb{0}_{m',q'} \end{array} \tag{8.7.5}$$

Since $\boxtimes^2 1 = \boxtimes^2$, the $\big((l-1)q' + l', (i-1)m' + i'\big)$-entry of (8.7.5) is the diagram in C below.

$$\begin{array}{ccc} \Big[\bigoplus_{j=1}^{n}\bigoplus_{j'=1}^{n'} \mathbb{0} \otimes (A_{ji} \otimes A'_{j'i'})\Big]_{\mathsf{lt}} & \xrightarrow{\ (\lambda^{\bullet}, \lambda^{\oplus})\ } & \mathbb{0} \\ {\scriptstyle [\oplus_j \oplus_{j'} (\lambda_{\mathbb{0}}^{-\bullet} \otimes 1)]_{\mathsf{lt}}}\big\downarrow & & \\ \Big[\bigoplus_{j=1}^{n}\bigoplus_{j'=1}^{n'} (\mathbb{0} \otimes \mathbb{0}) \otimes (A_{ji} \otimes A'_{j'i'})\Big]_{\mathsf{lt}} & & \\ {\scriptstyle (\alpha^{\oplus}, \alpha^{\pm\otimes}, \xi^{\otimes})}\big\downarrow & & \big\downarrow{\scriptstyle \lambda_{\mathbb{0}}^{-\bullet}} \\ \Big[\bigoplus_{j=1}^{n}\Big(\bigoplus_{j'=1}^{n'} (\mathbb{0} \otimes A_{ji}) \otimes (\mathbb{0} \otimes A'_{j'i'})\Big)_{\mathsf{lt}}\Big]_{\mathsf{lt}} & & \\ {\scriptstyle (\delta^{-r}, \delta^{-l})}\big\downarrow & & \\ \Big[\bigoplus_{j=1}^{n} (\mathbb{0} \otimes A_{ji})\Big]_{\mathsf{lt}} \otimes \Big[\bigoplus_{j'=1}^{n'} (\mathbb{0} \otimes A'_{j'i'})\Big]_{\mathsf{lt}} & \xrightarrow{\ (\lambda^{\bullet}, \lambda^{\oplus})\ } & \mathbb{0} \otimes \mathbb{0} \end{array} \tag{8.7.6}$$

Consider the diagram (8.7.6).

- By (8.1.13), each entry of each component of ζ^{ℓ} is a composite of morphisms, each being a sum of identity morphisms and one component of $\lambda^{\bullet}$ or $\lambda^{\oplus}$. The notation $(\lambda^{\bullet}, \lambda^{\oplus})$ denotes a composite of morphisms, each being an iterated sum and product of identity morphisms and at most one component of $\lambda^{\bullet}$ or $\lambda^{\oplus}$.
- The morphisms $(\alpha^{\oplus}, \alpha^{\pm\otimes}, \xi^{\otimes})$ and $(\delta^{-r}, \delta^{-l})$ are interpreted similarly, with δ^{-r} and δ^{-l} the inverses of δ^{r} and δ^{l}, respectively. Each of them is a composite of morphisms, each being an iterated sum and product of identity morphisms and at most one component of one of the indicated structure morphisms in C.
- In interpreting the lower left $\boxtimes^2$ in (8.7.5), we use the paths P and Q in the proof of Lemma 8.6.16.

To prove the commutativity of (8.7.6), we realize each morphism or its inverse as the value in C of a path in $\mathsf{Gr}(X)$ using the set of formal variables

$$X = \{0^X, 1^X, a_{ji}, a'_{j'i'}\}_{1 \le j \le n,\ 1 \le j' \le n'}$$

and the function $\varphi\colon X \longrightarrow \mathsf{Ob}(\mathsf{C})$ defined by

$$\varphi(x) = \begin{cases} \mathbb{0} & \text{if } x = 0^X, \\ \mathbb{1} & \text{if } x = 1^X, \\ A_{ji} & \text{if } x = a_{ji}, \text{ and} \\ A'_{j'i'} & \text{if } x = a'_{j'i'}. \end{cases}$$

By Notation 3.3.10, Example 8.1.14, and Lemma 8.6.16, there is a diagram in $\mathsf{Gr}(X)$ as follows.

$$\tag{8.7.7} \begin{array}{ccc} \Big[\bigoplus_{j=1}^{n}\bigoplus_{j'=1}^{n'} 0^X \otimes \big(a_{ji} \otimes a'_{j'i'}\big)\Big]_{\mathsf{lt}} & \xleftarrow{\;L_2\;} & 0^X \\ \big\downarrow L_3 & & \\ \Big[\bigoplus_{j=1}^{n}\bigoplus_{j'=1}^{n'} (0^X \otimes 0^X) \otimes \big(a_{ji} \otimes a'_{j'i'}\big)\Big]_{\mathsf{lt}} & & \\ \big\downarrow P & & \big\uparrow \lambda^{\bullet}_{0^X} \\ \Big[\bigoplus_{j=1}^{n}\Big(\bigoplus_{j'=1}^{n'} (0^X \otimes a_{ji}) \otimes (0^X \otimes a'_{j'i'})\Big)_{\mathsf{lt}}\Big]_{\mathsf{lt}} & & \\ \big\uparrow Q & & \\ \Big[\bigoplus_{j=1}^{n} (0^X \otimes a_{ji})\Big]_{\mathsf{lt}} \otimes \Big[\bigoplus_{j'=1}^{n'} (0^X \otimes a'_{j'i'})\Big]_{\mathsf{lt}} & \xrightarrow{\;L_1\;} & 0^X \otimes 0^X \end{array}$$

Consider the diagram (8.7.7).

- The paths P and Q are those in Lemma 8.6.16, with each b_{kj} and each $b'_{k'j'}$ replaced by 0^X.
  - Each prime edge in P is an identity or has an instance of $\alpha^{\oplus}$, $\alpha^{\pm\otimes}$, or $\xi^{\otimes}$. Its value in C is the morphism $(\alpha^{\oplus}, \alpha^{\pm\otimes}, \xi^{\otimes})$ in (8.7.6).
  - Each prime edge in Q is an identity or has an instance of δ^l or δ^r. Its value in C is the *inverse* of the morphism $(\delta^{-r}, \delta^{-l})$ in (8.7.6).
- Each prime edge in L_1 is an identity or has an instance of $\lambda^{\bullet}$ or $\lambda^{\oplus}$. Its value in C is the bottom horizontal morphism in (8.7.6).
- Each prime edge in L_2 is an identity or has an instance of $\lambda^{-\bullet}$ or $\lambda^{-\oplus}$. Its value in C is the *inverse* of the top horizontal morphism in (8.7.6).
- Each prime edge in L_3 is an identity or has an instance of $\lambda^{-\bullet}$. Its value in C is the top left vertical morphism in (8.7.6).

Since the upper right vertex in (8.7.7) is 0^X, by Lemma 3.1.29, the lower left vertex has the same support as 0^X. By Proposition 3.5.33, the two paths

$$Q \quad \text{and} \quad (P, L_3, L_2, \lambda^{\bullet}_{0^X}, L_1)$$

have the same value in C. Since all the structure morphisms involved are invertible (because C is assumed to be tight), this implies that (8.7.6) is commutative.

(2) If $p, p' > 0$ and $n = n' = 0$, then (8.7.2) is the diagram below, with ζ^{-r} the inverse of ζ^r in (8.1.19).

$$\begin{CD}
\mathbb{0}_{mm',qq'} @>{\zeta^{-r}_{C\boxtimes C'}}>> (C\boxtimes C')\mathbb{0}_{mm',pp'} \\
@V{1}VV @VV{1\star\boxtimes^2}V \\
\mathbb{0}_{mm',qq'} @. (C\boxtimes C')(\mathbb{0}_{m,p}\boxtimes\mathbb{0}_{m',p'}) \\
@V{\boxtimes^2}VV @VV{\boxtimes^2}V \\
\mathbb{0}_{m,q}\boxtimes\mathbb{0}_{m',q'} @>{\zeta^{-r}_C\boxtimes\zeta^{-r}_{C'}}>> C\mathbb{0}_{m,p}\boxtimes C'\mathbb{0}_{m',p'}
\end{CD} \tag{8.7.8}$$

Composing the upper left 1 away and using the notation in (8.7.6), the $\big((l-1)q'+l', (i-1)m'+i'\big)$-entry of (8.7.8) is the diagram in C below.

$$\begin{CD}
\mathbb{0} @>{(\lambda^{-\oplus},\rho^{-\bullet})}>> \Big[\bigoplus_{k=1}^{p}\bigoplus_{k'=1}^{p'}(C_{lk}\otimes C'_{l'k'})\otimes\mathbb{0}\Big]_{\mathsf{lt}} \\
@| @VV{\big[\oplus_k\oplus_{k'}(1\otimes\lambda_{\mathbb{0}}^{-\bullet})\big]_{\mathsf{lt}}}V \\
@. \Big[\bigoplus_{k=1}^{p}\bigoplus_{k'=1}^{p'}(C_{lk}\otimes C'_{l'k'})\otimes(\mathbb{0}\otimes\mathbb{0})\Big]_{\mathsf{lt}} \\
@V{\lambda_{\mathbb{0}}^{-\bullet}}VV @VV{(\alpha^{\oplus},\alpha^{\pm\otimes},\xi^{\otimes})}V \\
@. \Big[\bigoplus_{k=1}^{p}\Big(\bigoplus_{k'=1}^{p'}(C_{lk}\otimes\mathbb{0})\otimes(C'_{l'k'}\otimes\mathbb{0})\Big)_{\mathsf{lt}}\Big]_{\mathsf{lt}} \\
@| @VV{(\delta^{-r},\delta^{-l})}V \\
\mathbb{0}\otimes\mathbb{0} @>{(\lambda^{-\oplus},\rho^{-\bullet})}>> \Big[\bigoplus_{k=1}^{p}(C_{lk}\otimes\mathbb{0})\Big]_{\mathsf{lt}}\otimes\Big[\bigoplus_{k'=1}^{p'}(C'_{l'k'}\otimes\mathbb{0})\Big]_{\mathsf{lt}}
\end{CD} \tag{8.7.9}$$

To prove the commutativity of (8.7.9), we realize each morphism or its inverse as the value in C of a path in $\mathsf{Gr}(X)$ using the set of formal variables

$$X = \{0^X, 1^X, c_{lk}, c'_{l'k'}\}_{1\le k\le p,\, 1\le k'\le p'}$$

and the function $\varphi\colon X \longrightarrow \mathsf{Ob}(\mathsf{C})$ defined by

$$\varphi(x) = \begin{cases} \mathbb{0} & \text{if } x = 0^X, \\ \mathbb{1} & \text{if } x = 1^X, \\ C_{lk} & \text{if } x = c_{lk}, \text{ and} \\ C'_{l'k'} & \text{if } x = c'_{l'k'}. \end{cases}$$

Similar to (8.7.7), by Notation 3.3.10, Example 8.1.22, and Lemma 8.6.16, there is a diagram in $\mathsf{Gr}(X)$ as follows in which each path has value in C the corresponding edge in (8.7.9) or its inverse.

(8.7.10)

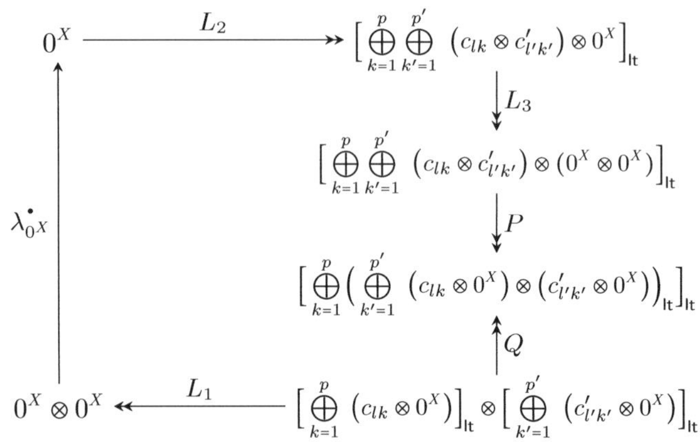

Consider the diagram (8.7.10).

- The paths P and Q are those in Lemma 8.6.16 up to a change of symbols.
  - The value of P in C is the morphism $(\alpha^{\oplus}, \alpha^{\pm\otimes}, \xi^{\otimes})$ in (8.7.9).
  - The value of Q in C is the *inverse* of the morphism $(\delta^{-r}, \delta^{-l})$ in (8.7.9).
- Each prime edge in L_1 is an identity or has an instance of $\lambda^{\oplus}$ or $\rho^{\bullet}$. Its value in C is the *inverse* of the bottom horizontal morphism in (8.7.9).
- Each prime edge in L_2 is an identity or has an instance of $\lambda^{-\oplus}$ or $\rho^{-\bullet}$. Its value in C is the top horizontal morphism in (8.7.9).
- Each prime edge in L_3 is an identity or has an instance of $\lambda^{-\bullet}$. Its value in C is the top right vertical morphism in (8.7.9).

Since the upper left vertex in (8.7.10) is 0^X, by Lemma 3.1.29, the lower right vertex has the same support as 0^X. By Proposition 3.5.33, the two paths

$$Q \quad \text{and} \quad (P, L_3, L_2, \lambda^{\bullet}_{0^X}, L_1)$$

have the same value in C. This implies that (8.7.9) is commutative.

(3) If $n, p' > 0$ and $n' = p = 0$, then (8.7.2) is the diagram below.

(8.7.11)

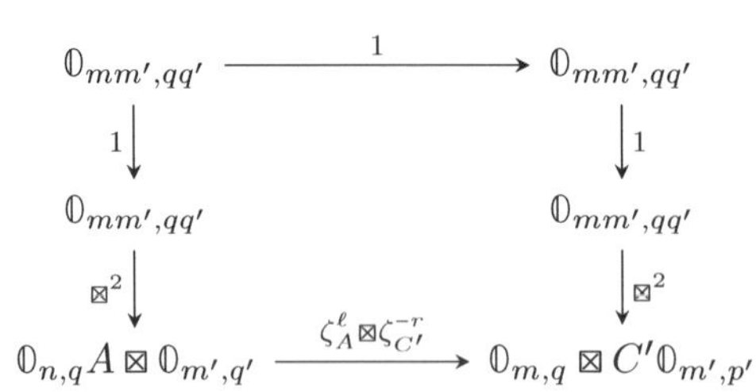

First composing away the three 1's, the $\big((l-1)q'+l',(i-1)m'+i'\big)$-entry of (8.7.11) becomes the outer diagram in C below.

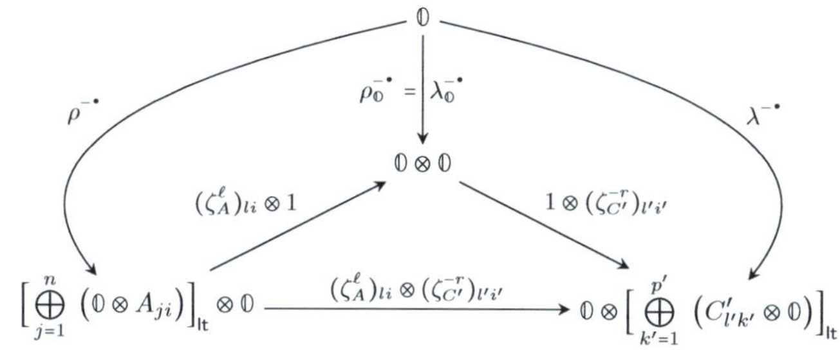

- The equality $\lambda_{\mathbb{0}}^{-\bullet} = \rho_{\mathbb{0}}^{-\bullet}$ follows from the axiom (2.1.14).
- The bottom subdiagram is commutative by the functoriality of $\otimes$.
- The left and the right subdiagrams are commutative by the naturality of $\rho^{\bullet}$ and $\lambda^{\bullet}$, respectively.

(4) If $n',p>0$ and $n=p'=0$, then (8.7.2) is the diagram below.

(8.7.12)

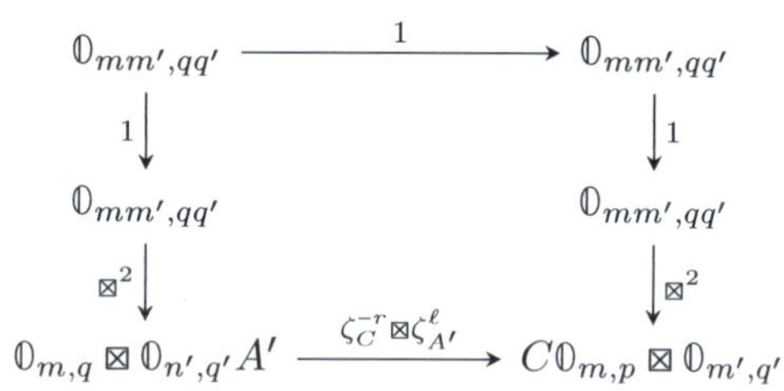

First composing away the three 1's, the $\big((l-1)q'+l',(i-1)m'+i'\big)$-entry of (8.7.12) becomes the outer diagram in C below.

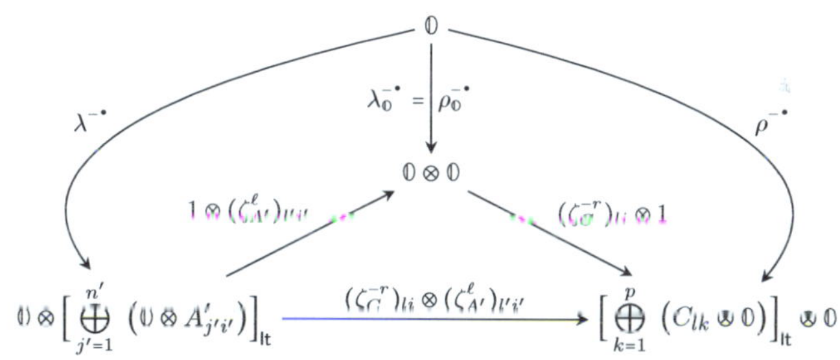

As in case (3), this diagram is commutative by the functoriality of $\otimes$ and the naturality of $\lambda^{\bullet}$ and $\rho^{\bullet}$.

(5) If $n,p>0$ and $n'=p'=0$, then (8.7.2) is the left diagram below.

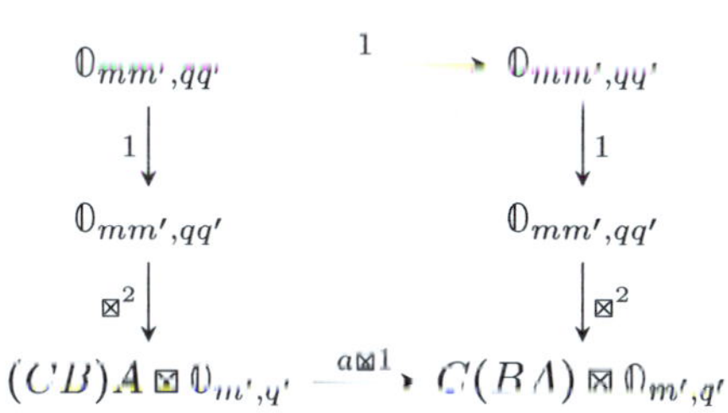

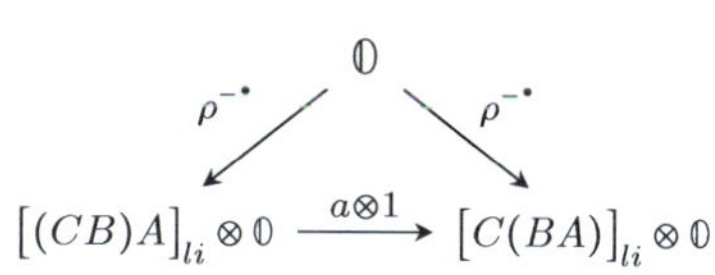

First composing the three 1's away, the $\big((l-1)q'+l',(i-1)m'+i'\big)$-entry of the left diagram above is the right diagram in C. It is commutative by the naturality of $\rho^{\bullet}$.

(6) If $n',p'>0$ and $n=p=0$, then (8.7.2) is the left diagram below.

$$\begin{array}{ccc}
\mathbb{0}_{mm',qq'} & \xrightarrow{\;1\;} & \mathbb{0}_{mm',qq'} \\
{\scriptstyle 1}\downarrow & & \downarrow{\scriptstyle 1} \\
\mathbb{0}_{mm',qq'} & & \mathbb{0}_{mm',qq'} \\
{\scriptstyle \boxtimes^2}\downarrow & & \downarrow{\scriptstyle \boxtimes^2} \\
\mathbb{0}_{m,q}\boxtimes(C'B')A' & \xrightarrow{1\boxtimes a} & \mathbb{0}_{m,q}\boxtimes C'(B'A')
\end{array}
\qquad
\begin{array}{ccc}
 & \mathbb{0} & \\
{\scriptstyle \lambda^{-\bullet}}\swarrow & & \searrow{\scriptstyle \lambda^{-\bullet}} \\
\mathbb{0}\otimes\big[(C'B')A'\big]_{l'i'} & \xrightarrow{1\otimes a} & \mathbb{0}\otimes\big[C'(B'A')\big]_{l'i'}
\end{array}$$

First composing the three 1's away, the $\big((l-1)q'+l',(i-1)m'+i'\big)$-entry of the left diagram above is the right diagram in C. It is commutative by the naturality of $\lambda^{\bullet}$.

This finishes the proofs of all six cases. □

Lemma 8.7.13. *In the setting of* (8.7.1), *suppose that*

- *$m,m',q,q'>0$ and*
- *precisely one of $n,n',p,$ and p' is 0.*

Then the diagram (8.7.2) *is commutative.*

Proof. There are four cases.

(1) If $n=0$ and $n',p,p'>0$, then (8.7.2) is the following diagram.

$$\begin{array}{ccc}
\mathbb{0}_{mm',qq'} & \xrightarrow{\;\zeta^{-r}_{C\boxtimes C'}\;} & (C\boxtimes C')\mathbb{0}_{mm',pp'} \\
{\scriptstyle 1}\downarrow & & \downarrow{\scriptstyle 1\star\boxtimes^2} \\
\mathbb{0}_{mm',qq'} & & (C\boxtimes C')(\mathbb{0}_{m,p}\boxtimes B'A') \\
{\scriptstyle \boxtimes^2}\downarrow & & \downarrow{\scriptstyle \boxtimes^2} \\
\mathbb{0}_{m,q}\boxtimes(C'B')A' & \xrightarrow{\zeta^{-r}_{C}\boxtimes a_{C',B',A'}} & C\mathbb{0}_{m,p}\boxtimes C'(B'A')
\end{array}
\tag{8.7.14}$$

Composing the top left 1 away, the $\big((l-1)q'+l',(i-1)m'+i'\big)$-entry of (8.7.14) is the diagram in C below, with the notation in (8.7.6) and with $\otimes$ abbreviated to concatenation.

(8.7.15)

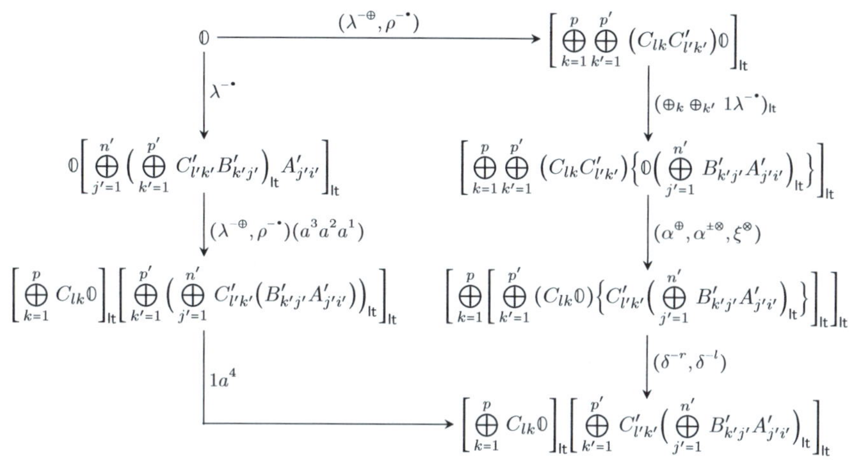

In the diagram (8.7.15), the factorization

$$(a_{C',B',A'})_{l'i'} = a^4a^3a^2a^1$$

is from (8.3.5) up to a change of symbols.

- a^1 is built from identity morphisms and δ^r.
- a^2 is built from identity morphisms and $\alpha^{\otimes}$.
- a^3 is built from identity morphisms, $\alpha^{\pm\oplus}$, and $\xi^{\oplus}$.
- a^4 is built from identity morphisms and δ^{-l}.

To prove that (8.7.15) is commutative, we realize each morphism or its inverse as the value in C of a path in $\mathsf{Gr}(X)$ using the set of formal variables

$$X = \left\{0^X, 1^X, c_{lk}, a'_{j'i'}, b'_{k'j'}, c'_{l'k'}\right\}_{1\le k\le p,\,1\le j'\le n',\,1\le k'\le p'}$$

and the function $\varphi\colon X \longrightarrow \mathsf{Ob}(\mathsf{C})$ defined as follows.

$$\varphi(x) = \begin{cases} \mathbb{0} & \text{if } x = 0^X, \\ \mathbb{1} & \text{if } x = 1^X, \\ C_{lk} & \text{if } x = c_{lk}, \\ A'_{j'i'} & \text{if } x = a'_{j'i'}, \\ B'_{k'j'} & \text{if } x = b'_{k'j'}, \\ C'_{l'k'} & \text{if } x = c'_{l'k'}. \end{cases}$$

By Notation 3.3.10 and Example 8.3.9, there is a diagram in $\mathsf{Gr}(X)$ as follows, in which the value in C of each path is the corresponding edge in (8.7.15) or its inverse. We continue to abbreviate $\otimes$ to concatenation.

(8.7.16)

$$\begin{array}{ccc}
0^X & \xleftarrow{\ \ L_2\ \ } & \Big[\bigoplus_{k=1}^{p}\bigoplus_{k'=1}^{p'}(c_{lk}c'_{l'k'})0^X\Big]_{\mathsf{lt}} \\
\Big\downarrow \lambda^{-\bullet} & & \Big\uparrow L_1 \\
0^X\Big[\bigoplus_{j'=1}^{n'}\Big(\bigoplus_{k'=1}^{p'}c'_{l'k'}b'_{k'j'}\Big)_{\mathsf{lt}}a'_{j'i'}\Big]_{\mathsf{lt}} & & \Big[\bigoplus_{k=1}^{p}\bigoplus_{k'=1}^{p'}(c_{lk}c'_{l'k'})\Big\{0^X\Big(\bigoplus_{j'=1}^{n'}b'_{k'j'}a'_{j'i'}\Big)_{\mathsf{lt}}\Big\}\Big]_{\mathsf{lt}} \\
\Big\downarrow L_3 & & \Big\uparrow P^{-1} \\
\Big[\bigoplus_{k=1}^{p}c_{lk}0^X\Big]_{\mathsf{lt}}\Big[\bigoplus_{k'=1}^{p'}\Big(\bigoplus_{j'=1}^{n'}c'_{l'k'}(b'_{k'j'}a'_{j'i'})\Big)_{\mathsf{lt}}\Big]_{\mathsf{lt}} & & \Big[\bigoplus_{k=1}^{p}\Big[\bigoplus_{k'=1}^{p'}(c_{lk}0^X)\Big\{c'_{l'k'}\Big(\bigoplus_{j'=1}^{n'}b'_{k'j'}a'_{j'i'}\Big)_{\mathsf{lt}}\Big\}\Big]_{\mathsf{lt}}\Big]_{\mathsf{lt}} \\
\Big\uparrow L & & \Big\uparrow Q \\
 & \xleftarrow{\qquad\qquad} & \Big[\bigoplus_{k=1}^{p}c_{lk}0^X\Big]_{\mathsf{lt}}\Big[\bigoplus_{k'=1}^{p'}c'_{l'k'}\Big(\bigoplus_{j'=1}^{n'}b'_{k'j'}a'_{j'i'}\Big)_{\mathsf{lt}}\Big]_{\mathsf{lt}}
\end{array}$$

Consider the diagram (8.7.16).

- Each prime edge in L is an identity or contains δ^l.
- P and Q are the paths in Lemma 8.6.16 up to a change of symbols, and P^{-1} is the formal inverse of P in the sense of Definition 3.1.10.
  - Each prime edge in Q is an identity or contains δ^r or δ^l.
  - Each prime edge in P^{-1} is an identity or contains $\alpha^{-\oplus}$, $\alpha^{\pm\otimes}$, or $\xi^{-\otimes}$.
- Each prime edge in L_1 is an identity or contains $\lambda^{\bullet}$.
- Each prime edge in L_2 is an identity or contains $\lambda^{\oplus}$ or $\rho^{\bullet}$.
- Each prime edge in L_3 is an identity or contains $\alpha^{\pm\oplus}$, $\xi^{\oplus}$, $\lambda^{-\oplus}$, $\alpha^{\otimes}$, $\rho^{-\bullet}$, or δ^r.

Since the upper left vertex in (8.7.16) is 0^X, by Lemma 3.1.29, the lower right vertex has the same support as 0^X. By Proposition 3.5.33, the two paths

$$L \quad \text{and} \quad (L_3, \lambda^{-\bullet}, L_2, L_1, P^{-1}, Q)$$

have the same value in C. This implies that (8.7.15) is commutative.

(2) If $n' = 0$ and $n, p, p' > 0$, then (8.7.2) is the following diagram.

$$\begin{array}{ccc}
\mathbb{0}_{mm',qq'} & \xrightarrow{\ \zeta^{-r}_{C\boxtimes C'}\ } & (C\boxtimes C')\mathbb{0}_{mm',pp'} \\
\Big\downarrow 1 & & \Big\downarrow 1\star\boxtimes^2 \\
\mathbb{0}_{mm',qq'} & & (C\boxtimes C')(BA\boxtimes\mathbb{0}_{m',p'}) \\
\Big\downarrow \boxtimes^2 & & \Big\downarrow \boxtimes^2 \\
(CB)A\boxtimes\mathbb{0}_{m',q'} & \xrightarrow{\ a_{C,B,A}\boxtimes\zeta^{-r}_{C'}\ } & C(BA)\boxtimes C'\mathbb{0}_{m',p'}
\end{array} \tag{8.7.17}$$

With an argument almost identical to that of case (1), each entry of (8.7.17) is a diagram in C similar to (8.7.15). Its commutativity is proved by realizing each edge or its inverse as the value in C of a path in $\mathsf{Gr}(X)$, similar to (8.7.16), for some set X of formal variables and function $\varphi\colon X \longrightarrow \mathsf{Ob}(\mathsf{C})$. Since one of its vertices is 0^X, this diagram is commutative in C by Lemma 3.1.29 and Proposition 3.5.33.

(3) If $p = 0$ and $n, n', p' > 0$, then (8.7.2) is the following diagram.

(8.7.18)
$$\begin{array}{ccc}
\mathbb{0}_{nn',qq'}(A \boxtimes A') & \xrightarrow{\zeta^{\ell}_{A\boxtimes A'}} & \mathbb{0}_{mm',qq'} \\
\big\downarrow{\scriptstyle \boxtimes^2 \star 1} & & \big\downarrow{\scriptstyle 1} \\
(\mathbb{0}_{n,q} \boxtimes C'B')(A \boxtimes A') & & \mathbb{0}_{mm',qq'} \\
\big\downarrow{\scriptstyle \boxtimes^2} & & \big\downarrow{\scriptstyle \boxtimes^2} \\
(\mathbb{0}_{n,q}A) \boxtimes (C'B')A' & \xrightarrow{\zeta^{\ell}_{A}\boxtimes a_{C',B',A'}} & \mathbb{0}_{m,q} \boxtimes C'(B'A')
\end{array}$$

Composing the top right 1 away, the $\big((l-1)q' + l', (i-1)m' + i'\big)$-entry of (8.7.18) is the diagram in C below, with the notation in (8.7.6) and with $\otimes$ abbreviated to concatenation.

(8.7.19)
$$\begin{array}{ccc}
\Big[\bigoplus_{j=1}^{n}\bigoplus_{j'=1}^{n'} \mathbb{0}(A_{ji}A'_{j'i'})\Big]_{\mathsf{lt}} & \xrightarrow{(\lambda^{\oplus},\lambda^{\bullet})} & \mathbb{0} \\
\big\downarrow{\scriptstyle (\oplus_j \oplus_{j'} \lambda^{-\bullet}1)_{\mathsf{lt}}} & & \big\downarrow{\scriptstyle \lambda^{-\bullet}} \\
\Big[\bigoplus_{j=1}^{n}\bigoplus_{j'=1}^{n'} \Big[\mathbb{0}\Big(\bigoplus_{k'=1}^{p'} C'_{l'k'}B'_{k'j'}\Big)_{\mathsf{lt}}\Big](A_{ji}A'_{j'i'})\Big]_{\mathsf{lt}} & & \mathbb{0}\Big[\bigoplus_{k'=1}^{p'} C'_{l'k'}\Big(\bigoplus_{j'=1}^{n'} B'_{k'j'}A'_{j'i'}\Big)_{\mathsf{lt}}\Big]_{\mathsf{lt}} \\
\big\downarrow{\scriptstyle (\alpha^{\oplus},\alpha^{\pm\otimes},\xi^{\otimes})} & & \big\uparrow{\scriptstyle 1a^4} \\
\Big[\bigoplus_{j=1}^{n}\Big[\bigoplus_{j'=1}^{n'} (\mathbb{0}A_{ji})\Big\{\Big(\bigoplus_{k'=1}^{p'} C'_{l'k'}B'_{k'j'}\Big)_{\mathsf{lt}} A'_{j'i'}\Big\}\Big]_{\mathsf{lt}}\Big]_{\mathsf{lt}} & & \mathbb{0}\Big[\bigoplus_{k'=1}^{p'}\Big(\bigoplus_{j'=1}^{n'} C'_{l'k'}(B'_{k'j'}A'_{j'i'})\Big)_{\mathsf{lt}}\Big]_{\mathsf{lt}} \\
\big\downarrow{\scriptstyle (\delta^{-r},\delta^{-l})} & & \big\uparrow{\scriptstyle (\lambda^{\oplus},\lambda^{\bullet})(a^3a^2a^1)} \\
\Big[\bigoplus_{j=1}^{n} \mathbb{0}A_{ji}\Big]_{\mathsf{lt}}\Big[\bigoplus_{j'=1}^{n'}\Big(\bigoplus_{k'=1}^{p'} C'_{l'k'}B'_{k'j'}\Big)_{\mathsf{lt}} A'_{j'i'}\Big]_{\mathsf{lt}} & \longrightarrow & \lrcorner
\end{array}$$

As in (8.7.15), the factorization

$$(a_{C',B',A'})_{l'i'} = a^4a^3a^2a^1$$

in (8.7.19) is from (8.3.5) up to a change of symbols.

To prove that (8.7.19) is commutative, we realize each morphism or its inverse as the value in C of a path in $\mathsf{Gr}(X)$ using the set of formal variables

$$X = \big\{0^X, 1^X, a_{ji}, a'_{j'i'}, b'_{k'j'}, c'_{l'k'}\big\}_{1\le j\le n,\, 1\le j'\le n',\, 1\le k'\le p'}$$

and the function $\varphi\colon X \longrightarrow \mathsf{Ob}(\mathsf{C})$ defined as follows.

$$\varphi(x) = \begin{cases} \mathbb{0} & \text{if } x = 0^X. \\ \mathbb{1} & \text{if } x = 1^X. \\ A_{ji} & \text{if } x = a_{ji}. \\ A'_{j'i'} & \text{if } x = a'_{j'i'}. \\ B'_{k'j'} & \text{if } x = b'_{k'j'}. \\ C'_{l'k'} & \text{if } x = c'_{l'k'}. \end{cases}$$

By Notation 3.3.10 and Example 8.3.9, there is a diagram in $\mathsf{Gr}(X)$ as follows, in which the value in C of each path is the corresponding edge in (8.7.19) or its inverse. We continue to abbreviate $\otimes$ to concatenation.

(8.7.20)
$$\begin{array}{ccc}
\Big[\bigoplus_{j=1}^{n}\bigoplus_{j'=1}^{n'} 0^X(a_{ji}a'_{j'i'})\Big]_{\mathsf{lt}} & \xrightarrow{\;L_2\;} & 0^X \\
\uparrow{\scriptstyle L_1} & & \downarrow{\scriptstyle \lambda^{-\bullet}} \\
\Big[\bigoplus_{j=1}^{n}\bigoplus_{j'=1}^{n'} \Big[0^X\Big(\bigoplus_{k'=1}^{p'} c'_{l'k'}b'_{k'j'}\Big)_{\mathsf{lt}}\Big](a_{ji}a'_{j'i'})\Big]_{\mathsf{lt}} & & 0^X\Big[\bigoplus_{k'=1}^{p'} c'_{l'k'}\Big(\bigoplus_{j'=1}^{n'} b'_{k'j'}a'_{j'i'}\Big)_{\mathsf{lt}}\Big]_{\mathsf{lt}} \\
\uparrow{\scriptstyle P^{-1}} & & \downarrow{\scriptstyle L_3} \\
\Big[\bigoplus_{j=1}^{n}\Big[\bigoplus_{j'=1}^{n'} (0^X a_{ji})\Big\{\Big(\bigoplus_{k'=1}^{p'} c'_{l'k'}b'_{k'j'}\Big)_{\mathsf{lt}} a'_{j'i'}\Big\}\Big]_{\mathsf{lt}}\Big]_{\mathsf{lt}} & & 0^X\Big[\bigoplus_{k'=1}^{p'}\Big(\bigoplus_{j'=1}^{n'} c'_{l'k'}(b'_{k'j'}a'_{j'i'})\Big)_{\mathsf{lt}}\Big]_{\mathsf{lt}} \\
\uparrow{\scriptstyle Q} & & \uparrow{\scriptstyle L} \\
\Big[\bigoplus_{j=1}^{n} 0^X a_{ji}\Big]_{\mathsf{lt}}\Big[\bigoplus_{j'=1}^{n'}\Big(\bigoplus_{k'=1}^{p'} c'_{l'k'}b'_{k'j'}\Big)_{\mathsf{lt}} a'_{j'i'}\Big]_{\mathsf{lt}} & \longrightarrow &
\end{array}$$

Consider the diagram (8.7.20).

- Each prime edge in L is an identity or contains $\alpha^{\pm\oplus}$, $\xi^{\oplus}$, $\lambda^{\oplus}$, $\alpha^{\otimes}$, $\lambda^{\bullet}$, or δ^{r}.
- The paths Q, P^{-1}, and L_1 are interpreted as in (8.7.16).
- Each prime edge in L_2 is an identity or contains $\lambda^{\oplus}$ or $\lambda^{\bullet}$.
- Each prime edge in L_3 is an identity or contains δ^{l}.

Since the upper right vertex in (8.7.20) is 0^X, by Lemma 3.1.29, the lower left vertex has the same support as 0^X. By Proposition 3.5.33, the two paths

$$L \quad \text{and} \quad (L_3, \lambda^{-\bullet}, L_2, L_1, P^{-1}, Q)$$

have the same value in C. This implies that (8.7.19) is commutative.

(4) If $p' = 0$ and $n, n', p > 0$, then (8.7.2) is the following diagram.

(8.7.21)
$$\begin{array}{ccc}
\mathbb{0}_{nn',qq'}(A \boxtimes A') & \xrightarrow{\;\zeta^{\ell}_{A\boxtimes A'}\;} & \mathbb{0}_{mm',qq'} \\
{\scriptstyle \boxtimes^2 \star 1}\downarrow & & \downarrow{\scriptstyle 1} \\
(CB \boxtimes \mathbb{0}_{n',q'})(A \boxtimes A') & & \mathbb{0}_{mm',qq'} \\
{\scriptstyle \boxtimes^2}\downarrow & & \downarrow{\scriptstyle \boxtimes^2} \\
(CB)A \boxtimes (\mathbb{0}_{n',q'}A') & \xrightarrow{\;a_{C,B,A}\boxtimes \zeta^{\ell}_{A'}\;} & C(BA) \boxtimes \mathbb{0}_{m',q'}
\end{array}$$

With an argument almost identical to that of case (3), each entry of (8.7.21) is a diagram in C similar to (8.7.19). Its commutativity is proved by realizing each edge or its inverse as the value in C of a path in $\mathsf{Gr}(X)$, similar to (8.7.20), for some set X of formal variables and function $\varphi: X \longrightarrow \mathsf{Ob}(\mathsf{C})$. Since one of its vertices is 0^X, this diagram is commutative in C by Lemma 3.1.29 and Proposition 3.5.33.

This finishes the proofs of all four cases. □

The remaining case of the lax associativity axiom (6.2.2) for the triple $(\boxtimes, \boxtimes^2, \boxtimes)$ is the most involved. We first compute the entries of the six objects in (8.7.2).

Lemma 8.7.22. *In the setting of* (8.7.1), *suppose that* $m, n, p, q, m', n', p', q' > 0$. *Then the* $\big((l-1)q' + l', (i-1)m' + i'\big)$*-entries of the six objects in* (8.7.2) *are as follows.*

(1)

$$\begin{aligned}&\Big[\big[(C \boxtimes C')(B \boxtimes B')\big](A \boxtimes A')\Big]_{(l-1)q'+l',(i-1)m'+i'}\\ &= \left[\bigoplus_{j=1}^{n}\bigoplus_{j'=1}^{n'}\left[\bigoplus_{k=1}^{p}\bigoplus_{k'=1}^{p'}(C_{lk} \otimes C'_{l'k'}) \otimes (B_{kj} \otimes B'_{k'j'})\right]_{\mathsf{lt}} \otimes (A_{ji} \otimes A'_{j'i'})\right]_{\mathsf{lt}}\end{aligned}$$

(2)

$$\begin{aligned}&\big[(CB \boxtimes C'B')(A \boxtimes A')\big]_{(l-1)q'+l',(i-1)m'+i'}\\ &= \left[\bigoplus_{j=1}^{n}\bigoplus_{j'=1}^{n'}\left\{\Big(\bigoplus_{k=1}^{p} C_{lk} \otimes B_{kj}\Big)_{\mathsf{lt}} \otimes \Big(\bigoplus_{k'=1}^{p'} C'_{l'k'} \otimes B'_{k'j'}\Big)_{\mathsf{lt}}\right\} \otimes (A_{ji} \otimes A'_{j'i'})\right]_{\mathsf{lt}}\end{aligned}$$

(3)

$$\begin{aligned}&\big[(CB)A \boxtimes (C'B')A'\big]_{(l-1)q'+l',(i-1)m'+i'}\\ &= \left[\bigoplus_{j=1}^{n}\Big(\bigoplus_{k=1}^{p} C_{lk} \otimes B_{kj}\Big)_{\mathsf{lt}} \otimes A_{ji}\right]_{\mathsf{lt}} \otimes \left[\bigoplus_{j'=1}^{n'}\Big(\bigoplus_{k'=1}^{p'} C'_{l'k'} \otimes B'_{k'j'}\Big)_{\mathsf{lt}} \otimes A'_{j'i'}\right]_{\mathsf{lt}}\end{aligned}$$

(4)

$$\begin{aligned}&\Big[(C \boxtimes C')\big[(B \boxtimes B')(A \boxtimes A')\big]\Big]_{(l-1)q'+l',(i-1)m'+i'}\\ &= \left[\bigoplus_{k=1}^{p}\bigoplus_{k'=1}^{p'}(C_{lk} \otimes C'_{l'k'}) \otimes \left[\bigoplus_{j=1}^{n}\bigoplus_{j'=1}^{n'}(B_{kj} \otimes B'_{k'j'}) \otimes (A_{ji} \otimes A'_{j'i'})\right]_{\mathsf{lt}}\right]_{\mathsf{lt}}\end{aligned}$$

(5)

$$\begin{aligned}&\big[(C \boxtimes C')(BA \boxtimes B'A')\big]_{(l-1)q'+l',(i-1)m'+i'}\\ &= \left[\bigoplus_{k=1}^{p}\bigoplus_{k'=1}^{p'}(C_{lk} \otimes C'_{l'k'}) \otimes \left\{\Big(\bigoplus_{j=1}^{n} B_{kj} \otimes A_{ji}\Big)_{\mathsf{lt}} \otimes \Big(\bigoplus_{j'=1}^{n'} B'_{k'j'} \otimes A'_{j'i'}\Big)_{\mathsf{lt}}\right\}\right]_{\mathsf{lt}}\end{aligned}$$

(6)

$$\begin{aligned}&\big[C(BA) \boxtimes C'(B'A')\big]_{(l-1)q'+l',(i-1)m'+i'}\\ &= \left[\bigoplus_{k=1}^{p} C_{lk} \otimes \Big(\bigoplus_{j=1}^{n} B_{kj} \otimes A_{ji}\Big)_{\mathsf{lt}}\right]_{\mathsf{lt}} \otimes \left[\bigoplus_{k'=1}^{p'} C'_{l'k'} \otimes \Big(\bigoplus_{j'=1}^{n'} B'_{k'j'} \otimes A'_{j'i'}\Big)_{\mathsf{lt}}\right]_{\mathsf{lt}}\end{aligned}$$

Proof. Each of these equalities is proved by applying the definitions of the matrix product (8.1.4) and of the matrix tensor product (8.6.4). For example, the first equality is proved as follows.

$$\begin{aligned}
&\Big[\big[(C \boxtimes C')(B \boxtimes B')\big](A \boxtimes A')\Big]_{(l-1)q'+l',(i-1)m'+i'}\\
&= \left[\bigoplus_{j=1}^{n}\bigoplus_{j'=1}^{n'} \big[(C \boxtimes C')(B \boxtimes B')\big]_{(l-1)q'+l',(j-1)n'+j'} \otimes (A \boxtimes A')_{(j-1)n'+j',(i-1)m'+i'}\right]_{\mathrm{lt}}\\
&= \left[\bigoplus_{j=1}^{n}\bigoplus_{j'=1}^{n'} \left[\bigoplus_{k=1}^{p}\bigoplus_{k'=1}^{p'} (C \boxtimes C')_{(l-1)q'+l',(k-1)p'+k'} \otimes (B \boxtimes B')_{(k-1)p'+k',(j-1)n'+j'}\right]_{\mathrm{lt}}\right.\\
&\qquad\qquad\qquad\qquad \left. \otimes\, (A \boxtimes A')_{(j-1)n'+j',(i-1)m'+i'}\right]_{\mathrm{lt}}
\end{aligned}$$

Both of these equalities follow from the definition (8.1.4) of the matrix product. The desired expression is obtained by applying (8.6.4) to each of the three entries of the matrix tensor products. The other five equalities in the lemma are proved in the same way. □

Next we prove the remaining case of the lax associativity axiom for $(\boxtimes, \boxtimes^2, \boxtimes^0)$.

Lemma 8.7.23. *In the setting of (8.7.1), if $m, n, p, q, m', n', p', q' > 0$, then the diagram (8.7.2) is commutative.*

Proof. The $\big((l-1)q' + l', (i-1)m' + i'\big)$-entry of (8.7.2) is the following diagram in C, with (1)–(6) denoting the six objects in Lemma 8.7.22.

(8.7.24)
$$\begin{array}{ccc}
(1) & \xrightarrow{\;a_{C\boxtimes C', B\boxtimes B', A\boxtimes A'}\;} & (4)\\
{\scriptstyle \boxtimes^2 \star 1}\downarrow & & \downarrow{\scriptstyle 1 \star \boxtimes^2}\\
(2) & & (5)\\
{\scriptstyle \boxtimes^2}\downarrow & & \downarrow{\scriptstyle \boxtimes^2}\\
(3) & \xrightarrow{\;a_{C,B,A} \boxtimes a_{C',B',A'}\;} & (6)
\end{array}$$

With

- $\otimes$ abbreviated to concatenation,
- the notation in (8.7.6),
- the factorization of the associator in (8.3.5), and
- the factorization of $\boxtimes^2$ represented by the zigzag of paths in Lemma 8.6.16,

the diagram (8.7.24) factors as follows.

(8.7.25)

$$(1) \xrightarrow{(\alpha^{\pm\oplus},\xi^{\oplus},\alpha^{\otimes},\delta^{r})} \Big[\bigoplus_{k=1}^{p}\bigoplus_{k'=1}^{p'}\Big[\bigoplus_{j=1}^{n}\bigoplus_{j'=1}^{n'}(C_{lk}C'_{l'k'})\big((B_{kj}B'_{k'j'})(A_{ji}A'_{j'i'})\big)\Big]_{\mathsf{lt}}\Big]_{\mathsf{lt}} \xrightarrow{(\delta^{-l})} (4)$$

$$(4) \xrightarrow{(\alpha^{\oplus},\alpha^{\pm\otimes},\xi^{\otimes})} \Big[\bigoplus_{k=1}^{p}\bigoplus_{k'=1}^{p'}(C_{lk}C'_{l'k'})\Big[\bigoplus_{j=1}^{n}\Big\{\bigoplus_{j'=1}^{n'}(B_{kj}A_{ji})(B'_{k'j'}A'_{j'i'})\Big\}_{\mathsf{lt}}\Big]_{\mathsf{lt}}\Big]_{\mathsf{lt}} \xrightarrow{(\delta^{-r},\delta^{-l})} (5)$$

$$(5) \xrightarrow{(\alpha^{\oplus},\alpha^{\pm\otimes},\xi^{\otimes})} \Big[\bigoplus_{k=1}^{p}\Big[\bigoplus_{k'=1}^{p'}\Big\{C_{lk}\Big(\bigoplus_{j=1}^{n}B_{kj}A_{ji}\Big)_{\mathsf{lt}}\Big\}\Big\{C'_{l'k'}\Big(\bigoplus_{j'=1}^{n'}B'_{k'j'}A'_{j'i'}\Big)_{\mathsf{lt}}\Big\}\Big]_{\mathsf{lt}}\Big]_{\mathsf{lt}} \xrightarrow{(\delta^{-r},\delta^{-l})} (6)$$

$$(1) \xrightarrow{(\alpha^{\oplus},\alpha^{\pm\otimes},\xi^{\otimes})} \Big[\bigoplus_{j=1}^{n}\bigoplus_{j'=1}^{n'}\Big[\bigoplus_{k=1}^{p}\Big\{\bigoplus_{k'=1}^{p'}(C_{lk}B_{kj})(C'_{l'k'}B'_{k'j'})\Big\}_{\mathsf{lt}}\Big]_{\mathsf{lt}}(A_{ji}A'_{j'i'})\Big]_{\mathsf{lt}} \xrightarrow{(\delta^{-r},\delta^{-l})} (2)$$

$$(2) \xrightarrow{(\alpha^{\oplus},\alpha^{\pm\otimes},\xi^{\otimes})} \Big[\bigoplus_{j=1}^{n}\Big[\bigoplus_{j'=1}^{n'}\Big\{\Big(\bigoplus_{k=1}^{p}C_{lk}B_{kj}\Big)A_{ji}\Big\}\Big\{\Big(\bigoplus_{k'=1}^{p'}C'_{l'k'}B'_{k'j'}\Big)_{\mathsf{lt}}A'_{j'i'}\Big\}\Big]_{\mathsf{lt}}\Big]_{\mathsf{lt}} \xrightarrow{(\delta^{-r},\delta^{-l})} (3)$$

$$(3) \xrightarrow{(\alpha^{\pm\oplus},\xi^{\oplus},\alpha^{\otimes},\delta^{r})} \Big[\bigoplus_{k=1}^{p}\Big(\bigoplus_{j=1}^{n}C_{lk}(B_{kj}A_{ji})\Big)_{\mathsf{lt}}\Big]_{\mathsf{lt}}\Big[\bigoplus_{k'=1}^{p'}\Big(\bigoplus_{j'=1}^{n'}C'_{l'k'}(B'_{k'j'}A'_{j'i'})\Big)_{\mathsf{lt}}\Big]_{\mathsf{lt}} \xrightarrow{(\delta^{-l})} (6)$$

For example, (δ^{-l}) denotes a composite of morphisms, each being an iterated sum and product of identity morphisms and one component of δ^{-l}.

To prove that (8.7.25) is commutative, we realize each morphism or its inverse as the value in C of a path in $\mathsf{Gr}(X)$ using the set of formal variables

$$X = \left\{0^X, 1^X, a_{ji}, b_{kj}, c_{lk}, a'_{j'i'}, b'_{k'j'}, c'_{l'k'}\right\}_{1\le j\le n,\, 1\le k\le p,\, 1\le j'\le n',\, 1\le k'\le p'}$$

and the function $\varphi\colon X \longrightarrow \mathsf{Ob}(\mathsf{C})$ defined as follows.

$$\varphi(x) = \begin{cases} \mathbb{0} & \text{if } x = 0^X.\\ \mathbb{1} & \text{if } x = 1^X.\\ A_{ji} & \text{if } x = a_{ji}.\\ B_{kj} & \text{if } x = b_{kj}.\\ C_{lk} & \text{if } x = c_{lk}.\\ A'_{j'i'} & \text{if } x = a'_{j'i'}.\\ B'_{k'j'} & \text{if } x = b'_{k'j'}.\\ C'_{l'k'} & \text{if } x = c'_{l'k'}.\end{cases}$$

There is a diagram in $\mathsf{Gr}(X)$ as follows, in which the value in C of each path in the boundary is the corresponding edge in (8.7.25) or its inverse. We continue to abbreviate $\otimes$ to concatenation. Moreover, to save space, we omit the subscripts in the formal variables, so a, b, c, a', b', and c' mean, respectively, a_{ji}, b_{kj}, c_{lk}, $a'_{j'i'}$, $b'_{k'j'}$, and $c'_{l'k'}$.

(8.7.26)

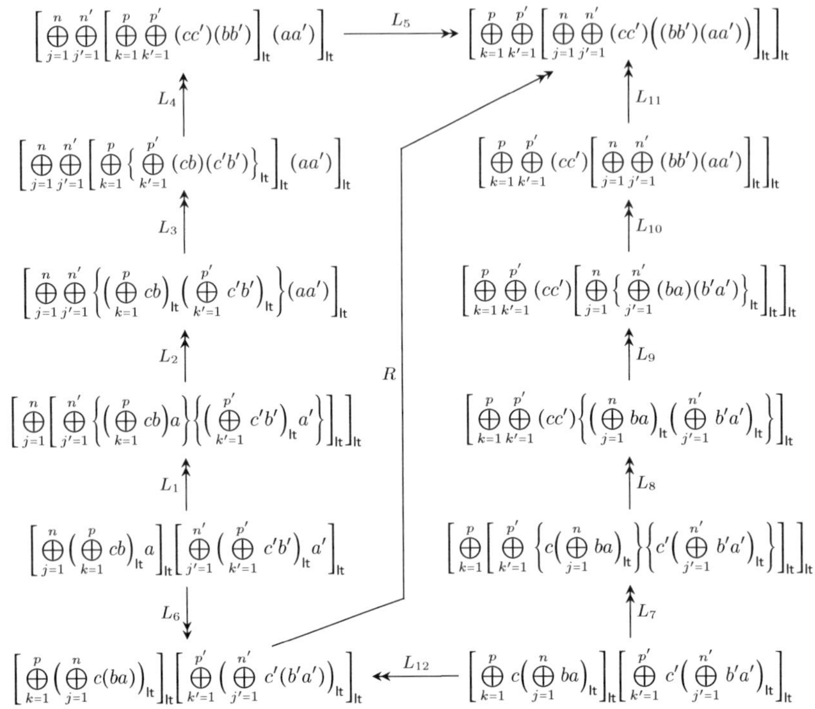

Consider the diagram (8.7.26).

- Each prime edge in L_1, L_3, L_7, and L_9 is an identity or contains δ^r or δ^l.
- Each prime edge in L_2, L_4, L_8, and L_{10} is an identity or contains $\alpha^{-\oplus}$, $\alpha^{\pm\otimes}$, or $\xi^{-\otimes}$.
- Each prime edge in L_5 and L_6 is an identity or contains $\alpha^{\pm\oplus}$, $\xi^{\oplus}$, $\alpha^{\otimes}$, or δ^r.
- Each prime edge in L_{11} and L_{12} is an identity or contains δ^l.
- Each prime edge in the path R is an identity or contains
  - δ^r or δ^l to bring the four sums to the front;
  - $\alpha^{\pm\otimes}$ or $\xi^{\otimes}$ to permute the monomial

$$\big[c_{lk}(b_{kj}a_{ji})\big]\big[c'_{l'k'}(b'_{k'j'}a'_{j'i'})\big] \quad \text{to} \quad (c_{lk}c'_{l'k'})\big((b_{kj}b'_{k'j'})(a_{ji}a'_{j'i'})\big);$$

  or
  - $\alpha^{\pm\oplus}$ or $\xi^{\oplus}$ to move additive brackets and to permute additively to match with the additive bracketing of the codomain of R.

In (8.7.26), the upper right vertex is regular in the sense of Definition 3.1.25, so each vertex is regular by Lemma 3.1.29. Since C is assumed to be tight, the Coherence Theorem 3.9.1 implies that each of the two subdiagrams is commutative in C. This implies that (8.7.25) is commutative. □

The Lax Unity Axiom. To finish the proof that $(\boxtimes, \boxtimes^2, \boxtimes^0)$ is a pseudofunctor, next we check the lax unity axiom. Recall the identity matrix in (8.1.6).

Lemma 8.7.27. *The data* $(\boxtimes, \boxtimes^2, \boxtimes^0)$ *in Definition 8.6.19 satisfy the lax unity axiom* (6.2.3).

Proof. We check the lax left unity axiom; the proof for the lax right unity axiom is almost identical.

The lax left unity axiom (6.2.3) states the commutativity of the diagram

$$\begin{array}{ccc} \mathbb{1}^{nq}(A \boxtimes B) & \xrightarrow{\ \ell_{A\boxtimes B}\ } & A \boxtimes B \\ {\scriptstyle \boxtimes^0_{(n,q)} \star 1}\big\downarrow & & \big\uparrow{\scriptstyle \ell_A \boxtimes \ell_B} \\ (\mathbb{1}^n \boxtimes \mathbb{1}^q)(A \boxtimes B) & \xrightarrow{\ \boxtimes^2_{(\mathbb{1}^n,\mathbb{1}^q),(A,B)}\ } & (\mathbb{1}^n A) \boxtimes (\mathbb{1}^q B) \end{array} \tag{8.7.28}$$

in $\mathsf{Mat}^{\mathsf{C}}_{mp,nq}$ for $(A, B) \in \mathsf{Mat}^{\mathsf{C}}_{m,n} \times \mathsf{Mat}^{\mathsf{C}}_{p,q}$, with ℓ the natural isomorphism in (8.2.2). If one of m, n, p, or q is 0, then $\mathsf{Mat}^{\mathsf{C}}_{mp,nq}$ is the terminal category, and (8.7.28) is commutative. For the rest of this proof, we assume that $m, n, p, q > 0$.

For $1 \le i \le m$, $1 \le j \le n$, $1 \le k \le p$, and $1 \le l \le q$, it suffices to show that the (J, I)-entry of (8.7.28) is commutative in C with

$$J = (j-1)q + l \quad \text{and} \quad I = (i-1)p + k.$$

By definitions (8.1.4), (8.1.5), (8.2.4), (8.6.4), (8.6.9), and (8.6.20), the (J, I)-entry of (8.7.28) is the following diagram in C. We abbreviate $\otimes$ to concatenation and use the notation in (8.7.6). Moreover, $J' = (j'-1)q + l'$.

$$\begin{array}{ccc} \Big[\bigoplus_{j'=1}^{n}\bigoplus_{l'=1}^{q} \mathbb{1}^{nq}_{(J,J')}(A_{j'i}B_{l'k})\Big]_{\mathsf{lt}} & \xrightarrow{\ (\lambda^{\oplus},\rho^{\oplus},\lambda^{\otimes},\lambda^{\bullet})\ } & A_{ji}B_{lk} \\ {\scriptstyle (\lambda^{-\otimes},\lambda^{-\bullet},\rho^{-\bullet})}\big\downarrow & & \big\uparrow{\scriptstyle (\lambda^{\oplus},\rho^{\oplus},\lambda^{\otimes},\lambda^{\bullet})} \\ \Big[\bigoplus_{j'=1}^{n}\bigoplus_{l'=1}^{q} (\mathbb{1}^n_{jj'}\mathbb{1}^q_{ll'})(A_{j'i}B_{l'k})\Big]_{\mathsf{lt}} & & \Big[\bigoplus_{j'=1}^{n} \mathbb{1}^n_{jj'}A_{j'i}\Big]_{\mathsf{lt}}\Big[\bigoplus_{l'=1}^{q} \mathbb{1}^q_{ll'}B_{l'k}\Big]_{\mathsf{lt}} \\ {\scriptstyle (\alpha^{\oplus},\alpha^{\pm\otimes},\xi^{\otimes})}\big\downarrow & & \big\uparrow{\scriptstyle (\delta^{-r},\delta^{-l})} \\ & \Big[\bigoplus_{j'=1}^{n}\Big\{\bigoplus_{l'=1}^{q} (\mathbb{1}^n_{jj'}A_{j'i})(\mathbb{1}^q_{ll'}B_{l'k})\Big\}_{\mathsf{lt}}\Big]_{\mathsf{lt}} & \end{array} \tag{8.7.29}$$

For example, $(\lambda^{\oplus}, \rho^{\oplus}, \lambda^{\otimes}, \lambda^{\bullet})$ denotes a composite of morphisms, each being an iterated sum and product of identity morphisms and one component of $\lambda^{\oplus}$, $\rho^{\oplus}$, $\lambda^{\otimes}$, or $\lambda^{\bullet}$.

To prove that (8.7.29) is commutative, we realize each morphism or its inverse as the value in C of a path in $\mathsf{Gr}(X)$ using the set of formal variables

$$X = \Big\{0^X, 1^X, a_{j'i}, b_{l'k}\Big\}_{1 \le j' \le n,\ 1 \le l' \le q}$$

and the function $\varphi\colon X \longrightarrow \mathsf{Ob}(\mathsf{C})$ defined as follows.

$$\varphi(x) = \begin{cases} \mathbb{0} & \text{if } x = 0^X. \\ \mathbb{1} & \text{if } x = 1^X. \\ A_{j'i} & \text{if } x = a_{j'i}. \\ B_{l'k} & \text{if } x = b_{l'k}. \end{cases}$$

Below we continue to abbreviate $\otimes$ to concatenation. The element $\delta^X_{rs} \in \{0^X, 1^X\}$ is defined in (8.4.6). There is a diagram in $\mathsf{Gr}(X)$ as follows, in which the value in C of each path is the corresponding edge in (8.7.29) or its inverse.

$$\begin{array}{ccc}
\left[\bigoplus_{j'=1}^{n}\bigoplus_{l'=1}^{q} \delta^X_{(J,J')}\big(a_{j'i}b_{l'k}\big)\right]_{\mathsf{lt}} & \xleftarrow{\;L_2\;} & a_{ji}b_{lk} \\
\Big\downarrow L_3 & & \Big\uparrow L_1 \\
\left[\bigoplus_{j'=1}^{n}\bigoplus_{l'=1}^{q} \big(\delta^X_{jj'}\delta^X_{ll'}\big)\big(a_{j'i}b_{l'k}\big)\right]_{\mathsf{lt}} & & \left[\bigoplus_{j'=1}^{n} \delta^X_{jj'}a_{j'i}\right]_{\mathsf{lt}}\left[\bigoplus_{l'=1}^{q} \delta^X_{ll'}b_{l'k}\right]_{\mathsf{lt}} \\
\Big\downarrow L_4 & & \Big\downarrow L \\
& \left[\bigoplus_{j'=1}^{n}\left\{\bigoplus_{l'=1}^{q} \big(\delta^X_{jj'}a_{j'i}\big)\big(\delta^X_{ll'}b_{l'k}\big)\right\}_{\mathsf{lt}}\right]_{\mathsf{lt}} &
\end{array} \tag{8.7.30}$$

Consider the diagram (8.7.30).

- Each prime edge in L is an identity or contains δ^r or δ^l.
- Each prime edge in L_1 is an identity or contains $\lambda^{\oplus}$, $\rho^{\oplus}$, $\lambda^{\otimes}$, or $\lambda^{\bullet}$.
- Each prime edge in L_2 is an identity or contains $\lambda^{-\oplus}$, $\rho^{-\oplus}$, $\lambda^{-\otimes}$, or $\lambda^{-\bullet}$.
- Each prime edge in L_3 is an identity or contains $\lambda^{-\otimes}$, $\lambda^{-\bullet}$, or $\rho^{-\bullet}$.
- Each prime edge in L_4 is an identity or contains $\alpha^{\oplus}$, $\alpha^{\pm\otimes}$, or $\xi^{\otimes}$.

The vertex $a_{ji}b_{lk}$ in (8.7.30) is regular in the sense of Definition 3.1.25. By Lemma 3.1.29, the middle right vertex is also regular. Since C is assumed to be tight, the Coherence Theorem 3.9.1 implies that the two paths

$$L \quad \text{and} \quad (L_4, L_3, L_2, L_1)$$

have the same value in C. This implies that (8.7.29) is commutative. □

Lemma 8.7.31. *The data $(\boxtimes, \boxtimes^2, \boxtimes^0)$ in Definition 8.6.19 is a pseudofunctor.*

Proof. We already proved all the necessary statements.

- Lemma 8.6.7 shows that $\boxtimes$ defines local functors.
- Each component of $\boxtimes^0$ is an isomorphism by Lemma 8.6.8.
- $\boxtimes^2$ is a natural isomorphism by Lemma 8.6.21.
- Lemmas 8.7.3, 8.7.4, 8.7.13, and 8.7.23 prove the lax associativity axiom (6.2.2).
- Lemma 8.7.27 proves the lax unity axiom (6.2.3).

Therefore, $(\boxtimes, \boxtimes^2, \boxtimes^0)$ is a pseudofunctor. □

8.8. The Monoidal Associator

Throughout this section, C denotes an arbitrary tight symmetric bimonoidal category as in Definition 2.1.2. We are in the process of constructing a monoidal bicategory structure on the matrix bicategory $\mathsf{Mat}^{\mathsf{C}}$. In this section, we define the monoidal associator, in the sense of Definition 6.4.1, for the monoidal composition $\boxtimes$ on $\mathsf{Mat}^{\mathsf{C}}$. In Lemma 8.7.31, we proved that the triple

$$(\boxtimes, \boxtimes^2, \boxtimes^0)\colon \mathsf{Mat}^{\mathsf{C}} \times \mathsf{Mat}^{\mathsf{C}} \longrightarrow \mathsf{Mat}^{\mathsf{C}}$$

in Definition 8.6.19 is a pseudofunctor. This induces two pseudofunctors

$$\mathsf{Mat}^{\mathsf{C}} \times \mathsf{Mat}^{\mathsf{C}} \times \mathsf{Mat}^{\mathsf{C}} \overset{\boxtimes(\boxtimes\times 1)}{\underset{\boxtimes(1\times\boxtimes)}{\rightrightarrows}} \mathsf{Mat}^{\mathsf{C}}.$$

As discussed in Explanation 6.4.6, the monoidal associator $(a^{\boxtimes}, a^{\boxtimes\bullet}, \eta^a, \varepsilon^a)$ consists of the following data.

(i) $a^{\boxtimes}$ and $a^{\boxtimes\bullet}$ are strong transformations as in Definition 6.2.14 as follows.

$$\boxtimes(\boxtimes \times 1) \underset{a^{\boxtimes\bullet}}{\overset{a^{\boxtimes}}{\rightleftarrows}} \boxtimes(1 \times \boxtimes)$$

These strong transformations are decorated with $\boxtimes$ to avoid confusion with the base associator (8.3.2) in $\mathsf{Mat}^{\mathsf{C}}$.

(ii) η^a and ε^a are invertible modifications as in Definition 6.3.1 as follows.

$$1_{\boxtimes(\boxtimes\times 1)} \xrightarrow{\eta^a} a^{\boxtimes\bullet} a^{\boxtimes} \qquad a^{\boxtimes} a^{\boxtimes\bullet} \xrightarrow{\varepsilon^a} 1_{\boxtimes(1\times\boxtimes)}$$

Moreover, these data are required to satisfy the triangle identities (6.3.10).

Here is an outline of this section.

- The left adjoint $a^{\boxtimes}$ is defined in Definition 8.8.1. Lemmas 8.8.5, 8.8.11, 8.8.17, and 8.8.26 show that $a^{\boxtimes}$ is a strong transformation.
- The right adjoint $a^{\boxtimes\bullet}$ is defined in Definition 8.8.33.
- The modifications η^a and ε^a are defined in Definition 8.8.37 and verified in Lemma 8.8.39.
- The triangle identities (6.3.10) are proved in Lemma 8.8.45.

The Left Adjoint of the Monoidal Associator.

Definition 8.8.1. With respect to the pseudofunctor $(\boxtimes, \boxtimes^2, \boxtimes^0)$ in Definition 8.6.19 and Lemma 8.7.31, define the data of a lax transformation

$$\boxtimes(\boxtimes \times 1) \xrightarrow{a^{\boxtimes}} \boxtimes(1 \times \boxtimes)$$

as follows.

Component 1-Cells: For each triple of objects $(m,n,p) \in (\mathsf{Mat}^{\mathsf{C}})^3$, define

$$mnp = \big((m \boxtimes n) \boxtimes p\big) \xrightarrow{a^{\boxtimes}_{m,n,p}} \big(m \boxtimes (n \boxtimes p)\big) = mnp \tag{8.8.2}$$

as the identity matrix $\mathbb{1}^{mnp} \in \mathsf{Mat}^{\mathsf{C}}_{mnp,mnp}$ in (8.1.6).

Component 2-Cells: For each triple of 1-cells

$$\big(A = (A_{i'i}), B = (B_{j'j}), C = (C_{k'k})\big) \in \mathsf{Mat}^{\mathsf{C}}_{m,m'} \times \mathsf{Mat}^{\mathsf{C}}_{n,n'} \times \mathsf{Mat}^{\mathsf{C}}_{p,p'},$$

define the component 2-cell

$$a^{\boxtimes}_{A,B,C} \in \mathsf{Mat}^{\mathsf{C}}_{mnp,m'n'p'}\Big([A \boxtimes (B \boxtimes C)]a^{\boxtimes}_{m,n,p}; a^{\boxtimes}_{m',n',p'}[(A \boxtimes B) \boxtimes C]\Big)$$

as the following vertical composite.

(8.8.3)

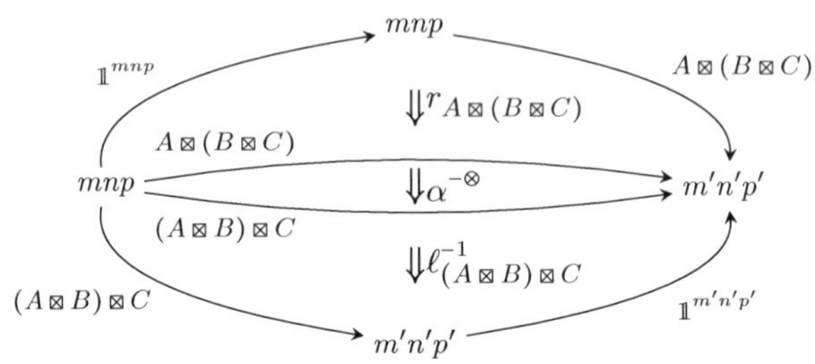

- ℓ is the base left unitor in $\mathsf{Mat}^{\mathsf{C}}$ in (8.2.2).
- r is the base right unitor in $\mathsf{Mat}^{\mathsf{C}}$ in (8.2.8).
- $\alpha^{-\otimes}$ is the 2-cell with $((i'-1)n'p' + (j'-1)p' + k', (i-1)np + (j-1)p + k)$-entry the structure morphism in C,

$$A_{i'i} \otimes \big(B_{j'j} \otimes C_{k'k}\big) \xrightarrow{\alpha^{-\otimes}_{A_{i'i},B_{j'j},C_{k'k}}} \big(A_{i'i} \otimes B_{j'j}\big) \otimes C_{k'k}$$

  for $1 \le i \le m$, $1 \le j \le n$, $1 \le k \le p$, $1 \le i' \le m'$, $1 \le j' \le n'$, and $1 \le k' \le p'$.

This finishes the definition of $a^{\boxtimes}$. ◇

Explanation 8.8.4. By (8.2.4) and (8.2.10), each entry of the 2-cell $a^{\boxtimes}_{A,B,C}$ in (8.8.3) is a composite of isomorphisms, each being an iterated sum of identity morphisms and one component of $\lambda^{\pm\oplus}$, $\rho^{\pm\oplus}$, $\alpha^{-\otimes}$, $\lambda^{-\otimes}$, $\rho^{\otimes}$, $\lambda^{-\bullet}$, or $\rho^{\bullet}$ in C. ◇

Lemma 8.8.5. *For each pair of objects (m,n,p) and (m',n',p') in $(\mathsf{Mat}^{\mathsf{C}})^3$, $a^{\boxtimes}$ in (8.8.3) is a natural isomorphism.*

Proof. The naturality of $a^{\boxtimes}$ means that, for each triple of 2-cells

$$(f,g,h) \in \mathsf{Mat}^{\mathsf{C}}_{m,m'}(A,A') \times \mathsf{Mat}^{\mathsf{C}}_{n,n'}(B,B') \times \mathsf{Mat}^{\mathsf{C}}_{p,p'}(C,C'),$$

the diagram

(8.8.6)
$$\begin{array}{ccc} [A \boxtimes (B \boxtimes C)]\, \mathbb{1}^{mnp} & \xrightarrow{a^{\boxtimes}_{A,B,C}} & \mathbb{1}^{m'n'p'}\, [(A \boxtimes B) \boxtimes C] \\ {\scriptstyle [f\boxtimes(g\boxtimes h)]\star 1}\big\downarrow & & \big\downarrow{\scriptstyle 1\star[(f\boxtimes g)\boxtimes h]} \\ [A' \boxtimes (B' \boxtimes C')]\, \mathbb{1}^{mnp} & \xrightarrow{a^{\boxtimes}_{A',B',C'}} & \mathbb{1}^{m'n'p'}\, [(A' \boxtimes B') \boxtimes C'] \end{array}$$

in $\mathsf{Mat}^{\mathsf{C}}_{mnp,m'n'p'}$ is commutative. Since 2-cells in $\mathsf{Mat}^{\mathsf{C}}$ are entrywise morphisms in C, it suffices to prove the commutativity of (8.8.6) in each entry. Restricted to each entry, the commutativity of (8.8.6) follows from the naturality of the structure morphisms $\lambda^{\oplus}$, $\rho^{\oplus}$, $\alpha^{\otimes}$, $\lambda^{\otimes}$, $\rho^{\otimes}$, $\lambda^{\bullet}$, and $\rho^{\bullet}$ in C. Moreover, since these are natural isomorphisms in C, $a^{\boxtimes}$ is a natural isomorphism. □

The Lax Unity of the Left Adjoint. Before we show that $a^{\boxtimes}$ satisfies the lax unity axiom (6.2.15), we first describe the entries of the relevant 1-cells. The next few lemmas follow from the definitions of the matrix product (8.1.4) and of the matrix tensor product (8.6.4), applied to the identity matrices (8.1.6). The next lemma describes the entries in the domain 1-cell of the lax unity axiom.

Lemma 8.8.7. *For $m,n,p > 0$, each entry of the matrix product*

$$\mathbb{1}^{mnp}\mathbb{1}^{mnp} \in \mathsf{Mat}^{\mathsf{C}}_{mnp,mnp}$$

is a left normalized sum of mnp objects as follows.

(1) *For $1 \le s \le mnp$, the sth diagonal entry is*

$$\Bigg[\overbrace{\Big[\bigoplus_{i=1}^{s-1} \mathbb{0}\otimes\mathbb{0}\Big]}^{\textit{empty if } s=1} \oplus (\mathbb{1}\otimes\mathbb{1}) \oplus \overbrace{\Big[\bigoplus_{i=s+1}^{mnp} \mathbb{0}\otimes\mathbb{0}\Big]}^{\textit{empty if } s=mnp}\Bigg]_{\mathsf{lt}}.$$

(2) *The (s,t)-entry with $s < t$ is*

$$\Bigg[\overbrace{\Big[\bigoplus_{i=1}^{s-1} \mathbb{0}\otimes\mathbb{0}\Big]}^{\textit{empty if } s=1} \oplus (\mathbb{1}\otimes\mathbb{0}) \oplus \overbrace{\Big[\bigoplus_{i=s+1}^{t-1} \mathbb{0}\otimes\mathbb{0}\Big]}^{\textit{empty if } s=t-1} \oplus (\mathbb{0}\otimes\mathbb{1}) \oplus \overbrace{\Big[\bigoplus_{i=t+1}^{mnp} \mathbb{0}\otimes\mathbb{0}\Big]}^{\textit{empty if } t=mnp}\Bigg]_{\mathsf{lt}}.$$

(3) *The (s,t)-entry with $s > t$ is*

$$\Bigg[\overbrace{\Big[\bigoplus_{i=1}^{t-1} \mathbb{0}\otimes\mathbb{0}\Big]}^{\textit{empty if } t=1} \oplus (\mathbb{0}\otimes\mathbb{1}) \oplus \overbrace{\Big[\bigoplus_{i=t+1}^{s-1} \mathbb{0}\otimes\mathbb{0}\Big]}^{\textit{empty if } t=s-1} \oplus (\mathbb{1}\otimes\mathbb{0}) \oplus \overbrace{\Big[\bigoplus_{i=s+1}^{mnp} \mathbb{0}\otimes\mathbb{0}\Big]}^{\textit{empty if } s=mnp}\Bigg]_{\mathsf{lt}}.$$

The lax unity axiom also involves the following matrices.

Lemma 8.8.8. *Suppose $m,n,p > 0$.*

(1) *For the matrix*

$$\mathbb{1}^m \boxtimes (\mathbb{1}^n \boxtimes \mathbb{1}^p) \in \mathsf{Mat}^{\mathsf{C}}_{mnp,mnp},$$

the following two statements hold.

- *Each diagonal entry is $\mathbb{1}\otimes(\mathbb{1}\otimes\mathbb{1})$.*
- *Each off-diagonal entry is $\mathbb{1}\otimes(\mathbb{1}\otimes\mathbb{0})$, $\mathbb{1}\otimes(\mathbb{0}\otimes\mathbb{1})$, $\mathbb{1}\otimes(\mathbb{0}\otimes\mathbb{0})$, $\mathbb{0}\otimes(\mathbb{1}\otimes\mathbb{1})$, $\mathbb{0}\otimes(\mathbb{1}\otimes\mathbb{0})$, $\mathbb{0}\otimes(\mathbb{0}\otimes\mathbb{1})$, or $\mathbb{0}\otimes(\mathbb{0}\otimes\mathbb{0})$.*

(2) *For the matrix*

$$(\mathbb{1}^m \boxtimes \mathbb{1}^n) \boxtimes \mathbb{1}^p \in \mathsf{Mat}^{\mathsf{C}}_{mnp,mnp},$$

the following two statements hold.

- *Each diagonal entry is $(\mathbb{1}\otimes\mathbb{1})\otimes\mathbb{1}$.*
- *Each off-diagonal entry is $(\mathbb{1}\otimes\mathbb{1})\otimes\mathbb{0}$, $(\mathbb{1}\otimes\mathbb{0})\otimes\mathbb{1}$, $(\mathbb{1}\otimes\mathbb{0})\otimes\mathbb{0}$, $(\mathbb{0}\otimes\mathbb{1})\otimes\mathbb{1}$, $(\mathbb{0}\otimes\mathbb{1})\otimes\mathbb{0}$, $(\mathbb{0}\otimes\mathbb{0})\otimes\mathbb{1}$, or $(\mathbb{0}\otimes\mathbb{0})\otimes\mathbb{0}$.*

The next lemma describes an intermediate 1-cell in the left-hand side of the lax unity axiom.

Lemma 8.8.9. *For $m,n,p > 0$, each entry of the matrix product*

$$\big[\mathbb{1}^m \boxtimes (\mathbb{1}^n \boxtimes \mathbb{1}^p)\big]\mathbb{1}^{mnp} \in \mathsf{Mat}^{\mathsf{C}}_{mnp,mnp}$$

is a left normalized sum of mnp objects as follows, with $z_{j,i}$ denoting the (j,i)-entry in $\mathbb{1}^m \boxtimes (\mathbb{1}^n \boxtimes \mathbb{1}^p)$.

(1) *For $1 \leq s \leq mnp$, the sth diagonal entry is*

$$\Bigg[\overbrace{\Big[\bigoplus_{i=1}^{s-1} z_{s,i} \otimes \mathbb{0}\Big]}^{\textit{empty if } s=1} \oplus \Big[\big(\mathbb{1} \otimes (\mathbb{1} \otimes \mathbb{1})\big) \otimes \mathbb{1}\Big] \oplus \overbrace{\Big[\bigoplus_{i=s+1}^{mnp} z_{s,i} \otimes \mathbb{0}\Big]}^{\textit{empty if } s=mnp}\Bigg]_{\mathsf{lt}}.$$

(2) *The (s,t)-entry with $s < t$ is*

$$\Bigg[\overbrace{\Big[\bigoplus_{i=1}^{s-1} z_{s,i} \otimes \mathbb{0}\Big]}^{\textit{empty if } s=1} \oplus \Big[\big(\mathbb{1} \otimes (\mathbb{1} \otimes \mathbb{1})\big) \otimes \mathbb{0}\Big] \oplus \overbrace{\Big[\bigoplus_{i=s+1}^{t-1} z_{s,i} \otimes \mathbb{0}\Big]}^{\textit{empty if } s=t-1} \oplus (z_{s,t} \otimes \mathbb{1}) \oplus \overbrace{\Big[\bigoplus_{i=t+1}^{mnp} z_{s,i} \otimes \mathbb{0}\Big]}^{\textit{empty if } t=mnp}\Bigg]_{\mathsf{lt}}.$$

(3) *The (s,t)-entry with $s > t$ is*

$$\Bigg[\overbrace{\Big[\bigoplus_{i=1}^{t-1} z_{s,i} \otimes \mathbb{0}\Big]}^{\textit{empty if } t=1} \oplus (z_{s,t} \otimes \mathbb{1}) \oplus \overbrace{\Big[\bigoplus_{i=t+1}^{s-1} z_{s,i} \otimes \mathbb{0}\Big]}^{\textit{empty if } t=s-1} \oplus \Big[\big(\mathbb{1} \otimes (\mathbb{1} \otimes \mathbb{1})\big) \otimes \mathbb{0}\Big] \oplus \overbrace{\Big[\bigoplus_{i=s+1}^{mnp} z_{s,i} \otimes \mathbb{0}\Big]}^{\textit{empty if } s=mnp}\Bigg]_{\mathsf{lt}}.$$

The next lemma describes the entries in the codomain 1-cell of the lax unity axiom.

Lemma 8.8.10. *For $m,n,p > 0$, each entry of the matrix product*

$$\mathbb{1}^{mnp}\big[(\mathbb{1}^m \boxtimes \mathbb{1}^n) \boxtimes \mathbb{1}^p\big] \in \mathsf{Mat}^{\mathsf{C}}_{mnp,mnp}$$

is a left normalized sum of mnp objects as follows, with $y_{i,j}$ denoting the (i,j)-entry in $(\mathbb{1}^m \boxtimes \mathbb{1}^n) \boxtimes \mathbb{1}^p$.

(1) *For $1 \leq s \leq mnp$, the sth diagonal entry is*

$$\Bigg[\overbrace{\Big[\bigoplus_{i=1}^{s-1} \mathbb{0} \otimes y_{i,s}\Big]}^{\textit{empty if } s=1} \oplus \Big[\mathbb{1} \otimes \big((\mathbb{1} \otimes \mathbb{1}) \otimes \mathbb{1}\big)\Big] \oplus \overbrace{\Big[\bigoplus_{i=s+1}^{mnp} \mathbb{0} \otimes y_{i,s}\Big]}^{\textit{empty if } s=mnp}\Bigg]_{\mathsf{lt}}.$$

(2) *The (s,t)-entry with $s < t$ is*

$$\Bigg[\overbrace{\Big[\bigoplus_{i=1}^{s-1} \mathbb{0} \otimes y_{i,t}\Big]}^{\textit{empty if } s=1} \oplus (\mathbb{1} \otimes y_{s,t}) \oplus \overbrace{\Big[\bigoplus_{i=s+1}^{t-1} \mathbb{0} \otimes y_{i,t}\Big]}^{\textit{empty if } s=t-1} \oplus \Big[\mathbb{0} \otimes \big((\mathbb{1} \otimes \mathbb{1}) \otimes \mathbb{1}\big)\Big] \oplus \overbrace{\Big[\bigoplus_{i=t+1}^{mnp} \mathbb{0} \otimes y_{i,t}\Big]}^{\textit{empty if } t=mnp}\Bigg]_{\mathsf{lt}}.$$

(3) *The (s,t)-entry with $s > t$ is*

$$\Bigg[\overbrace{\Big[\bigoplus_{i=1}^{t-1} \mathbb{0} \otimes y_{i,t}\Big]}^{\textit{empty if } t=1} \oplus \Big[\mathbb{0} \otimes \big((\mathbb{1} \otimes \mathbb{1}) \otimes \mathbb{1}\big)\Big] \oplus \overbrace{\Big[\bigoplus_{i=t+1}^{s-1} \mathbb{0} \otimes y_{i,t}\Big]}^{\textit{empty if } t=s-1} \oplus (\mathbb{1} \otimes y_{s,t}) \oplus \overbrace{\Big[\bigoplus_{i=s+1}^{mnp} \mathbb{0} \otimes y_{i,t}\Big]}^{\textit{empty if } s=mnp}\Bigg]_{\mathsf{lt}}.$$

To see that $a^{\boxtimes}$ is a strong transformation, next we check the axioms in Definition 6.2.14, starting with the lax unity axiom.

Lemma 8.8.11. *$a^{\boxtimes}$ in Definition 8.8.1 satisfies the lax unity axiom (6.2.15).*

Proof. By (6.2.7) and (8.8.2), the lax unity axiom for $a^{\boxtimes}$ states the following equality of pasting diagrams in $\mathsf{Mat}^{\mathsf{C}}_{mnp,mnp}$ for $m,n,p \geq 0$.

(8.8.12)

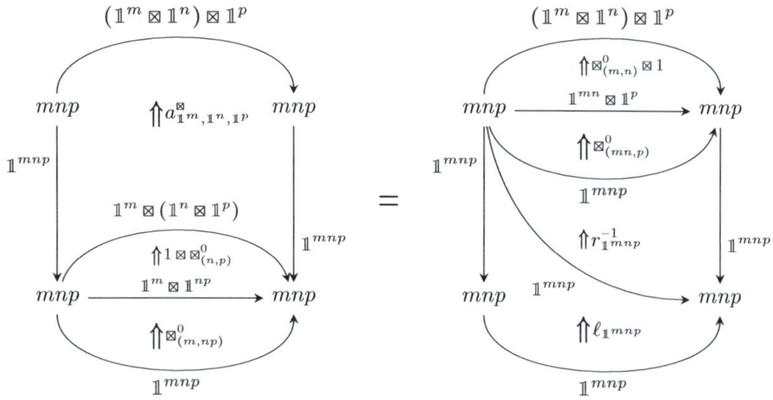

If m, n, or p is 0, then $\mathsf{Mat}^{\mathsf{C}}_{mnp,mnp}$ is the terminal category, and the equality in (8.8.12) holds. For the rest of this proof, we assume that $m,n,p > 0$.

Since 2-cells in $\mathsf{Mat}^{\mathsf{C}}$ are entrywise morphisms in C, it suffices to prove the equality in (8.8.12) in a typical entry. The entries of the 2-cells in (8.8.12) are interpreted using the definitions of

- the lax unity constraint $\boxtimes^0$ in Lemma 8.6.8,
- the component 2-cell $a^{\boxtimes}_{A,B,C}$ in (8.8.3),
- the base left unitor ℓ in (8.2.4), and
- the base right unitor r in (8.2.10).

The entries of the 1-cells in (8.8.12) are described in Lemmas 8.8.7 through 8.8.10. We will realize each 2-cell entry in (8.8.12) using paths in $\mathsf{Gr}(X)$ with

- the set $X = \{0^X, 1^X\}$ and
- the function $\varphi\colon X \longrightarrow \mathsf{Ob}(\mathsf{C})$ defined by $\varphi(0^X) = \mathbb{0}$ and $\varphi(1^X) = \mathbb{1}$.

The rest of this proof is split into two cases, one for the diagonal entries and one for the off-diagonal entries.

(1) For each $1 \leq s \leq mnp$, there exists a diagram in $\mathsf{Gr}(X)$ as in (8.8.13) below, whose left path (L_2, L_1), respectively right path (R_3, R_2, R_1), has value in C the sth diagonal entry of the left, respectively right, pasting diagram in (8.8.12). We abbreviate $\otimes$ to concatenation.

(8.8.13)

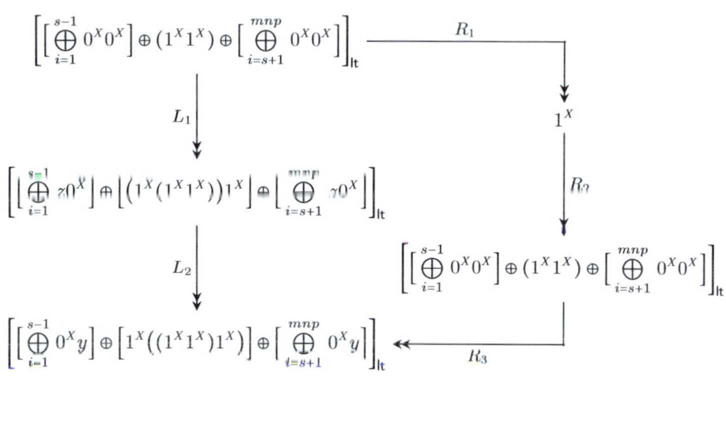

Consider the diagram (8.8.13).

- In the upper left vertex, as in Lemma 8.8.7, the sequence of symbols

$$\Big[\bigoplus_{i=1}^{s-1} 0^X 0^X\Big]\oplus$$

is empty if $s = 1$, and the sum after $(1^X 1^X)$ is empty if $s = mnp$. Similar interpretations apply to the sums in the other vertices.
- Each instance of z is an element $\delta_1^X(\delta_2^X\delta_3^X) \in X^{\mathsf{fr}}$ with each $\delta_j^X \in X = \{0^X, 1^X\}$, at least one of which is 0^X.
- Each instance of y is an element $(\delta_1^X\delta_2^X)\delta_3^X \in X^{\mathsf{fr}}$ with each $\delta_j^X \in X = \{0^X, 1^X\}$, at least one of which is 0^X.
- The value of L_1 in C is the morphism

$$\varphi L_1 = \Big(\big[(1 \boxtimes \boxtimes^0_{(n,p)}) \boxtimes^0_{(m,np)}\big] \star 1_{\mathbb{1}^{mnp}}\Big)_{s,s}.$$

Each prime edge in L_1 is an identity or contains $\lambda^{-\otimes}$, $\lambda^{-\bullet}$, or $\rho^{-\bullet}$.
- The value of L_2 in C is the morphism

$$\varphi L_2 = \big(a^{\boxtimes}_{\mathbb{1}^m,\mathbb{1}^n,\mathbb{1}^p}\big)_{s,s}.$$

Each prime edge in L_2 is an identity or contains $\lambda^{\pm\oplus}$, $\rho^{\pm\oplus}$, $\alpha^{-\otimes}$, $\lambda^{-\otimes}$, $\rho^{\otimes}$, $\lambda^{-\bullet}$, or $\rho^{\bullet}$.
- The value of R_1 in C is the morphism

$$\varphi R_1 = (\ell_{\mathbb{1}^{mnp}})_{s,s}.$$

Each prime edge in R_1 is an identity or contains $\lambda^{\oplus}$, $\rho^{\oplus}$, $\lambda^{\otimes}$, or $\lambda^{\bullet}$.
- The value of R_2 in C is the morphism

$$\varphi R_2 = \big(r^{-1}_{\mathbb{1}^{mnp}}\big)_{s,s}.$$

Each prime edge in R_2 is an identity or contains $\lambda^{-\oplus}$, $\rho^{-\oplus}$, $\rho^{-\otimes}$, or $\rho^{-\bullet}$.
- The value of R_3 in C is the morphism

$$\varphi R_3 = \Big(1_{\mathbb{1}^{mnp}} \star \big[(\boxtimes^0_{(m,n)} \boxtimes 1) \boxtimes^0_{(mn,p)}\big]\Big)_{s,s}.$$

Each prime edge in R_3 is an identity or contains $\lambda^{-\otimes}$, $\lambda^{-\bullet}$, or $\rho^{-\bullet}$.

Since the vertex 1^X in (8.8.13) is regular as in Definition 3.1.25, the common domain of L_1 and R_1 is also regular by Lemma 3.1.29. Since C is assumed to be tight, the Coherence Theorem 3.9.1 implies that the two paths (L_2, L_1) and (R_3, R_2, R_1) have the same value in C. This implies that the sth diagonal entries of the two pasting diagrams in (8.8.12) are equal.

(2) For $1 \leq s < t \leq mnp$, there exists a diagram in $\mathsf{Gr}(X)$ as follows.

(8.8.14)
$$\begin{array}{ccc}
\Big[\big[\bigoplus_{i=1}^{s-1} 0^X0^X\big] \oplus (1^X0^X) \oplus \big[\bigoplus_{i=s+1}^{t-1} 0^X0^X\big] \oplus (0^X1^X) \oplus \big[\bigoplus_{i=t+1}^{mnp} 0^X0^X\big]\Big]_{\mathsf{lt}} & \xrightarrow{R_1} & 0^X \\
\Big\downarrow L_1 & & \Big\downarrow R_2 \\
\Big[\big[\bigoplus_{i=1}^{s-1} z0^X\big] \oplus \big[(1^X(1^X1^X))0^X\big] \oplus \big[\bigoplus_{i=s+1}^{t-1} z0^X\big] \oplus (z1^X) \oplus \big[\bigoplus_{i=t+1}^{mnp} z0^X\big]\Big]_{\mathsf{lt}} & & \Big[\big[\bigoplus_{i=1}^{s-1} 0^X0^X\big] \oplus (1^X0^X) \oplus \big[\bigoplus_{i=s+1}^{t-1} 0^X0^X\big] \oplus (0^X1^X) \oplus \big[\bigoplus_{i=t+1}^{mnp} 0^X0^X\big]\Big]_{\mathsf{lt}} \\
\Big\downarrow L_2 & & \Big\downarrow R_3 \\
\Big[\big[\bigoplus_{i=1}^{s-1} 0^Xy\big] \oplus (1^Xy) \oplus \big[\bigoplus_{i=s+1}^{t-1} 0^Xy\big] \oplus \big[0^X((1^X1^X)1^X)\big] \oplus \big[\bigoplus_{i=t+1}^{mnp} 0^Xy\big]\Big]_{\mathsf{lt}} & \longleftarrow &
\end{array}$$

Consider the diagram (8.8.14).

- As in Lemma 8.8.7, the sums
$$\bigoplus_{i=1}^{s-1}, \quad \bigoplus_{i=s+1}^{t-1}, \quad \text{and} \quad \bigoplus_{i=t+1}^{mnp}$$
are empty if, respectively, $s = 1$, $s = t - 1$, and $t = mnp$.
- The elements $y, z \in X^{\mathsf{fr}}$ are interpreted as in case (1), and similarly for the paths L_1, L_2, R_1, R_2, and R_3 with the (s,t)-entry instead of the (s,s)-entry.

Since one vertex in (8.8.14) is 0^X, the common domain of L_1 and R_1 has the same support as 0^X by Lemma 3.1.29. Proposition 3.5.33 implies that the two paths (L_2, L_1) and (R_3, R_2, R_1) have the same value in C. This implies that the (s,t)-entries of the two pasting diagrams in (8.8.12) are equal.

The case with $s > t$ is proved in the same way, using the $s > t$ cases in Lemmas 8.8.7 through 8.8.10 to interpret the 1-cells.

This finishes the proof that $a^{\boxtimes}$ satisfies the lax unity axiom (6.2.15). □

The Lax Naturality of the Left Adjoint. The lax naturality axiom (6.2.16) for $a^{\boxtimes}$ states that, for 1-cells

(8.8.15)
$$\begin{array}{ccc}
m & m' & m'' \\
A\big\downarrow & A'\big\downarrow & A''\big\downarrow \\
n & n' & n'' \\
B\big\downarrow & B'\big\downarrow & B''\big\downarrow \\
p & p' & p''
\end{array}$$

in $\mathsf{Mat}^{\mathsf{C}}$, the following equality of pasting diagrams in $\mathsf{Mat}^{\mathsf{C}}_{mm'm'',pp'p''}$ holds.

(8.8.16)

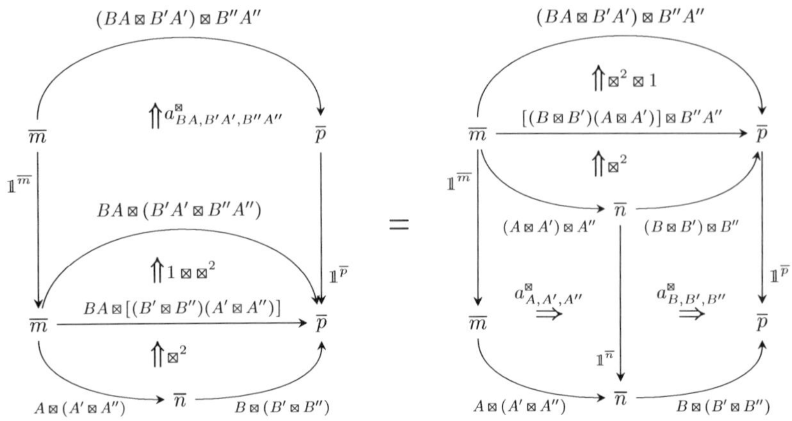

Consider (8.8.16).

(1) $\overline{m} = mm'm''$, $\overline{n} = nn'n''$, and $\overline{p} = pp'p''$.
(2) (6.2.8) is used to obtain the following decompositions of lax functoriality constraints for the composites, with $\boxtimes^2$ as in Definition 8.6.19.

$$\big(\boxtimes(\boxtimes \times 1)\big)^2 = \big(\boxtimes^2 \boxtimes 1\big)\boxtimes^2$$
$$\big(\boxtimes(1 \times \boxtimes)\big)^2 = \big(1 \boxtimes \boxtimes^2\big)\boxtimes^2$$

(3) In the left pasting diagram, by (8.8.3),

$$a^{\boxtimes}_{BA,B'A',B''A''} = \ell^{-1}\alpha^{-\otimes}r$$

with
- ℓ the base left unitor in (8.2.2),
- $\alpha^{-\otimes}$ entrywise a component of $\alpha^{-\otimes}$ in C, and
- r the base right unitor in (8.2.8).

(4) In the right pasting diagram, there are two instances of the base associator a in (8.3.2) and one instance of a^{-1} that are not explicitly displayed. They are necessary because horizontal composition of 1-cells in $\mathsf{Mat}^{\mathsf{C}}$ is not strictly associative, but only up to the natural isomorphism a.

To clarify the argument of the various cases, we split the proof of the lax naturality axiom for $a^{\boxtimes}$ into two lemmas.

Lemma 8.8.17. *In the setting of* (8.8.15), *suppose that either*

(1) *at least one of* m, m', m'', p, p', *or* p'' *is 0, or*
(2) $m, m', m'', p, p', p'' > 0$, *and at least one of* n, n', *or* n'' *is 0.*

Then the equality in (8.8.16) *holds.*

Proof. If at least one of m, m', m'', p, p', or p'' is 0, then $\mathsf{Mat}^{\mathsf{C}}_{\overline{m},\overline{p}}$ is the terminal category, and the equality in (8.8.16) holds.

For the second case, suppose that $m, m', m'', p, p', p'' > 0$, and that at least one of n, n', or n'' is 0. Since $\overline{n} = 0$, the two pasting diagrams in (8.8.16) become the following ones, where $\varnothing$ denotes the empty matrix.

(8.8.18)

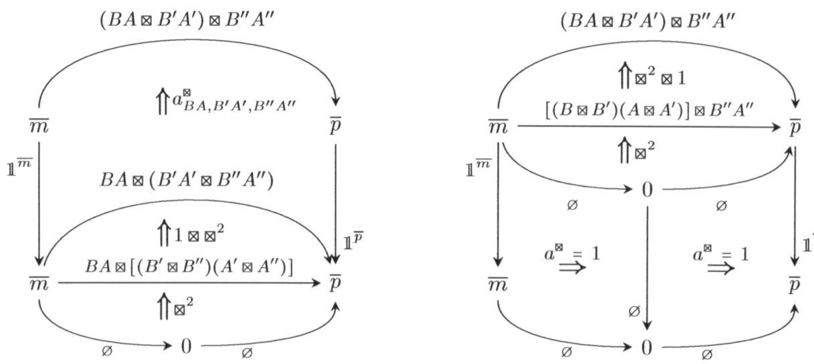

Consider (8.8.18).

- By Convention 6.2.12 and (8.1.4), each domain 1-cell is the matrix product

$$(\varnothing\varnothing)\mathbb{1}^{\overline{m}} = \mathbb{0}_{\overline{m},\overline{p}}\mathbb{1}^{\overline{m}} \in \mathsf{Mat}^{\mathsf{C}}_{\overline{m},\overline{p}}.$$

- If $n = 0$, then BA and $(B \boxtimes B')(A \boxtimes A')$ are 0 matrices.
- If $n' = 0$, then $B'A'$, $(B' \boxtimes B'')(A' \boxtimes A'')$, and $(B \boxtimes B')(A \boxtimes A')$ are 0 matrices.
- If $n'' = 0$, then $B''A''$ and $(B' \boxtimes B'')(A' \boxtimes A'')$ are 0 matrices.
- Each of the two 2-cells

$$\varnothing\varnothing = \mathbb{0}_{\overline{m},\overline{p}} \xrightarrow{\boxtimes^2} BA \boxtimes \big[(B' \boxtimes B'')(A' \boxtimes A'')\big]$$
$$\varnothing\varnothing = \mathbb{0}_{\overline{m},\overline{p}} \xrightarrow{\boxtimes^2} \big[(B \boxtimes B')(A \boxtimes A')\big] \boxtimes B''A''$$

  is entrywise a component of $\lambda^{-\bullet}$ or $\rho^{-\bullet}$.
- Each of the two 2-cells

(8.8.19)
$$\begin{aligned} (B' \boxtimes B'')(A' \boxtimes A'') &\xrightarrow{\boxtimes^2} B'A' \boxtimes B''A'' \\ (B \boxtimes B')(A \boxtimes A') &\xrightarrow{\boxtimes^2} BA \boxtimes B'A' \end{aligned}$$

  is as in (8.6.20).

The left and the right pasting diagrams in (8.8.18) are, respectively, the left and the right composites below in $\mathsf{Mat}^{\mathsf{C}}_{\overline{m},\overline{p}}$, with ζ^{ℓ} and ζ^{r} the natural isomorphisms in (8.1.11) and (8.1.19).

(8.8.20)
$$
\begin{array}{ccc}
 & \varnothing(\varnothing\mathbb{1}^{\overline{m}}) = \mathbb{0}_{\overline{m},\overline{p}} & \\
 \nearrow{\scriptstyle a\,=\,\zeta^{\ell}_{\mathbb{1}^{\overline{m}}}} & & \searrow{\scriptstyle 1\,\star\, a^{\boxtimes}\,=\,1} \\
(\varnothing\varnothing)\mathbb{1}^{\overline{m}} = \mathbb{0}_{\overline{m},\overline{p}}\mathbb{1}^{\overline{m}} & & \varnothing(\varnothing\varnothing) \\
{\scriptstyle \boxtimes^2 \star 1}\downarrow & & \downarrow{\scriptstyle a^{-1}\,=\,1} \\
\big[BA\boxtimes[(B'\boxtimes B'')(A'\boxtimes A'')]\big]\mathbb{1}^{\overline{m}} & & (\varnothing\varnothing)\varnothing \\
{\scriptstyle [1\boxtimes\boxtimes^2]\star 1}\downarrow & & \downarrow{\scriptstyle a^{\boxtimes}\star 1\,=\,1} \\
\big[BA\boxtimes(B'A'\boxtimes B''A'')\big]\mathbb{1}^{\overline{m}} & & (\mathbb{1}^{\overline{p}}\varnothing)\varnothing = \mathbb{0}_{\overline{m},\overline{p}} \\
{\scriptstyle r}\downarrow & & \downarrow{\scriptstyle a\,=\,\zeta^{-r}_{\mathbb{1}^{\overline{p}}}} \\
BA\boxtimes(B'A'\boxtimes B''A'') & & \mathbb{1}^{\overline{p}}(\varnothing\varnothing) = \mathbb{1}^{\overline{p}}\mathbb{0}_{\overline{m},\overline{p}} \\
{\scriptstyle \alpha^{-\otimes}}\downarrow & & \downarrow{\scriptstyle 1\star\boxtimes^2} \\
(BA\boxtimes B'A')\boxtimes B''A'' & & \mathbb{1}^{\overline{p}}\big[[(B\boxtimes B')(A\boxtimes A')]\boxtimes B''A''\big] \\
\searrow{\scriptstyle \ell^{-1}} & & \swarrow{\scriptstyle 1\star[\boxtimes^2\boxtimes 1]} \\
 & \mathbb{1}^{\overline{p}}\big[(BA\boxtimes B'A')\boxtimes B''A''\big] &
\end{array}
$$

Using the description above of the morphisms involved, similar to the proofs in Section 8.7, the commutativity of (8.8.20) in each entry is proved by realizing each edge or its inverse as the value in C of a path in $\mathsf{Gr}(X)$ for some set X of formal variables and some function $\varphi\colon X \longrightarrow \mathsf{Ob}(\mathsf{C})$. The detailed argument is given below.

(1) Suppose $nn' > 0$ and $n'' = 0$. First composing the three 1's away, the diagram (8.8.20) becomes the following diagram in $\mathsf{Mat}^{\mathsf{C}}_{\overline{m},\overline{p}}$.

(8.8.21)
$$
\begin{array}{ccc}
\mathbb{0}_{\overline{m},\overline{p}}\mathbb{1}^{\overline{m}} & \xrightarrow{\quad\zeta^{\ell}_{\mathbb{1}^{\overline{m}}}\quad} & \mathbb{0}_{\overline{m},\overline{p}} \\
{\scriptstyle \boxtimes^2\star 1}\downarrow & & \downarrow{\scriptstyle \zeta^{-r}_{\mathbb{1}^{\overline{p}}}} \\
\big[BA\boxtimes\mathbb{0}_{m'm'',p'p''}\big]\mathbb{1}^{\overline{m}} & & \mathbb{1}^{\overline{p}}\mathbb{0}_{\overline{m},\overline{p}} \\
{\scriptstyle [1\boxtimes\boxtimes^2]\star 1}\downarrow & & \downarrow{\scriptstyle 1\star\boxtimes^2} \\
\big[BA\boxtimes(B'A'\boxtimes\mathbb{0}_{m'',p''})\big]\mathbb{1}^{\overline{m}} & & \mathbb{1}^{\overline{p}}\big[[(B\boxtimes B')(A\boxtimes A')]\boxtimes\mathbb{0}_{m'',p''}\big] \\
{\scriptstyle r}\downarrow & & \\
BA\boxtimes(B'A'\boxtimes\mathbb{0}_{m'',p''}) & & \downarrow{\scriptstyle 1\star[\boxtimes^2\boxtimes 1]} \\
{\scriptstyle \alpha^{-\otimes}}\downarrow & & \\
(BA\boxtimes B'A')\boxtimes\mathbb{0}_{m'',p''} & \xrightarrow{\quad\ell^{-1}\quad} & \mathbb{1}^{\overline{p}}\big[(BA\boxtimes B'A')\boxtimes\mathbb{0}_{m'',p''}\big]
\end{array}
$$

For indices

(8.8.22)
$$
\begin{array}{lll}
1 \le i, e \le m & 1 \le i', e' \le m' & 1 \le i'', e'' \le m'' \\
1 \le k, c \le p & 1 \le k', c' \le p' & 1 \le k'', c'' \le p''
\end{array}
$$

we use the following abbreviations.

$$
\begin{aligned}
K &= (k-1)p'p'' + (k'-1)p'' + k'' \\
I &= (i-1)m'm'' + (i'-1)m'' + i'' \\
E &= (e-1)m'm'' + (e'-1)m'' + e'' \\
C &= (c-1)p'p'' + (c'-1)p'' + c'' \\
\bigoplus_{\overline{e}=1}^{\overline{m}} &= \bigoplus_{e=1}^{m}\bigoplus_{e'=1}^{m'}\bigoplus_{e''=1}^{m''} \qquad \bigoplus_{\overline{c}=1}^{\overline{p}} = \bigoplus_{c=1}^{p}\bigoplus_{c'=1}^{p'}\bigoplus_{c''=1}^{p''} \\
\bigoplus_{j,j'=1,1}^{n,n'} &= \bigoplus_{j=1}^{n}\bigoplus_{j'=1}^{n'}
\end{aligned}
\tag{8.8.23}
$$

Using the notation in (8.7.6) for morphisms, the (K,I)-entry of (8.8.21) is the following diagram in C, with $\otimes$ abbreviated to concatenation.

(8.8.24)

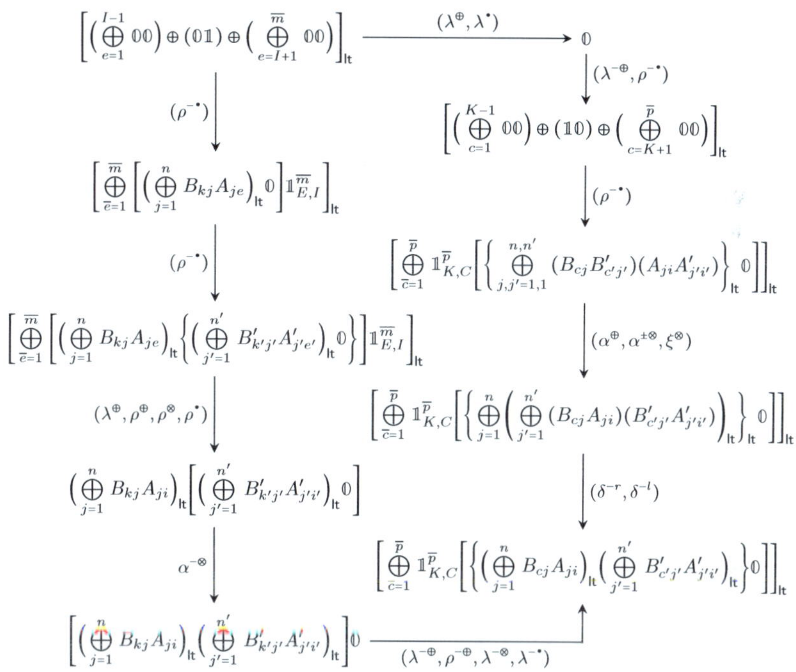

In the top left object, the symbols

$$\Big(\bigoplus_{e=1}^{I-1} \mathbb{0}\mathbb{0}\Big)\oplus \quad \text{and} \quad \oplus\Big(\bigoplus_{e=I+1}^{\overline{m}} \mathbb{0}\mathbb{0}\Big)$$

are empty if, respectively, $I = 1$ and $I = \overline{m}$. Similar interpretations apply to the sums in the second-to-top object in the right column.

To show that (8.8.24) is commutative, we use the set of formal variables

$$X = \left\{0^X, 1^X, a_{je}, b_{cj}, a'_{j'e'}, b'_{c'j'}\right\}_{1\le e\le m,\, 1\le j\le n,\, 1\le c\le p,\, 1\le e'\le m',\, 1\le j'\le n',\, 1\le c'\le p'}$$

and the function $\varphi\colon X \longrightarrow \mathsf{Ob}(\mathsf{C})$ defined as follows.

$$\varphi(x) = \begin{cases} \mathbb{0} & \text{if } x = 0^X. \\ \mathbb{1} & \text{if } x = 1^X. \\ A_{je} & \text{if } x = a_{je}. \\ B_{cj} & \text{if } x = b_{cj}. \\ A'_{j'e'} & \text{if } x = a'_{j'e'}. \\ B'_{c'j'} & \text{if } x = b'_{c'j'}. \end{cases}$$

There is a diagram in $\mathsf{Gr}(X)$ as follows. Each path has value in C the corresponding edge in (8.8.24) or its inverse, and $\delta^X_{\bullet\bullet}$ is as in (8.4.6).

$$\begin{array}{ccc}
\Big[\big(\bigoplus_{e=1}^{I-1} 0^X0^X\big) \oplus (0^X1^X) \oplus \big(\bigoplus_{e=I+1}^{\overline{m}} 0^X0^X\big)\Big]_{\mathsf{lt}} & \xrightarrow{(\lambda^{\oplus},\lambda^{\bullet})} & 0^X \\
\uparrow{\scriptstyle (\rho^{\bullet})} & & \downarrow{\scriptstyle (\lambda^{-\oplus},\rho^{-\bullet})} \\
 & & \Big[\big(\bigoplus_{c=1}^{K-1} 0^X0^X\big) \oplus (1^X0^X) \oplus \big(\bigoplus_{c=K+1}^{\overline{p}} 0^X0^X\big)\Big]_{\mathsf{lt}} \\
\Big[\bigoplus_{\overline{e}=1}^{\overline{m}} \Big[\big(\bigoplus_{j=1}^{n} b_{kj}a_{je}\big)_{\mathsf{lt}} 0^X\Big]\delta^X_{E,I}\Big]_{\mathsf{lt}} & & \downarrow{\scriptstyle (\rho^{-\bullet})} \\
\uparrow{\scriptstyle (\rho^{\bullet})} & & \Big[\bigoplus_{\overline{c}=1}^{\overline{p}} \delta^X_{K,C}\Big[\Big\{\bigoplus_{j,j'=1,1}^{n,n'} (b_{cj}b'_{c'j'})(a_{ji}a'_{j'i'})\Big\}_{\mathsf{lt}} 0^X\Big]\Big]_{\mathsf{lt}} \\
\Big[\bigoplus_{\overline{e}=1}^{\overline{m}} \Big[\big(\bigoplus_{j=1}^{n} b_{kj}a_{je}\big)_{\mathsf{lt}} \Big\{\big(\bigoplus_{j'=1}^{n'} b'_{k'j'}a'_{j'e'}\big)_{\mathsf{lt}} 0^X\Big\}\Big]\delta^X_{E,I}\Big]_{\mathsf{lt}} & & \downarrow{\scriptstyle (\alpha^{\oplus},\alpha^{\pm\otimes},\xi^{\otimes})} \\
\uparrow{\scriptstyle (\lambda^{-\oplus},\rho^{-\oplus},\rho^{-\otimes},\rho^{-\bullet})} & & \Big[\bigoplus_{\overline{c}=1}^{\overline{p}} \delta^X_{K,C}\Big[\Big\{\bigoplus_{j=1}^{n}\Big(\bigoplus_{j'=1}^{n'} (b_{cj}a_{ji})(b'_{c'j'}a'_{j'i'})\Big)_{\mathsf{lt}}\Big\}_{\mathsf{lt}} 0^X\Big]\Big]_{\mathsf{lt}} \\
\big(\bigoplus_{j=1}^{n} b_{kj}a_{ji}\big)_{\mathsf{lt}}\Big[\big(\bigoplus_{j'=1}^{n'} b'_{k'j'}a'_{j'i'}\big)_{\mathsf{lt}} 0^X\Big] & & \uparrow{\scriptstyle (\delta^r,\delta^l)} \\
\uparrow{\scriptstyle \alpha^{\otimes}} & & \Big[\bigoplus_{\overline{c}=1}^{\overline{p}} \delta^X_{K,C}\Big[\Big\{\big(\bigoplus_{j=1}^{n} b_{cj}a_{ji}\big)_{\mathsf{lt}}\big(\bigoplus_{j'=1}^{n'} b'_{c'j'}a'_{j'i'}\big)_{\mathsf{lt}}\Big\}0^X\Big]\Big]_{\mathsf{lt}} \\
\Big[\big(\bigoplus_{j=1}^{n} b_{kj}a_{ji}\big)_{\mathsf{lt}}\big(\bigoplus_{j'=1}^{n'} b'_{k'j'}a'_{j'i'}\big)_{\mathsf{lt}}\Big]0^X & \xleftarrow{(\lambda^{\oplus},\rho^{\oplus},\lambda^{\otimes},\lambda^{\bullet})} &
\end{array} \tag{8.8.25}$$

In the lower right path (δ^r, δ^l), each prime edge is an identity or contains δ^r or δ^l. Similar notation is used in the other paths.

Since the upper right vertex in (8.8.25) is 0^X, the lower right vertex has the same support as 0^X by Lemma 3.1.29. Proposition 3.5.33 implies that the long concatenated path and (δ^r, δ^l) in (8.8.25) have the same value in C. This implies that (8.8.24) is commutative.

(2) Suppose $n = 0$ and $n'n'' > 0$. This case is similar to case (1). Each entry of the diagram (8.8.20) is the value in C of some diagram in $\mathsf{Gr}(X)$, similar to (8.8.25), for some set X of formal variables and some function $\varphi: X \longrightarrow \mathsf{Ob}(\mathsf{C})$. There is one path of the form (δ^r, δ^l) that arises from the 2-cell

$$(B' \boxtimes B'')(A' \boxtimes A'') \xrightarrow{\boxtimes^2} B'A' \boxtimes B''A''$$

in (8.8.20) as part of the last case of (8.6.20).

(3) The other cases have either (i) $n' = 0$, or (ii) $n' > 0$ and $n = n'' = 0$. Each of these cases is similar to case (1). In each case, each entry of (8.8.20) is the value in C of some diagram in $\mathsf{Gr}(X)$, similar to (8.8.25), for some set

X of formal variables and some function $\varphi\colon X \longrightarrow \mathsf{Ob}(\mathsf{C})$. However, the elementary edges δ^r and δ^l do *not* appear in this diagram because each entry of each of the two 2-cells in (8.8.19) is a component of $\lambda^{-\bullet}$ or $\rho^{-\bullet}$.

This finishes the proof of the lemma. □

Lemma 8.8.26. *In the setting of* (8.8.15), *suppose that* $m, m', m'', n, n', n'', p, p', p'' > 0$. *Then the equality in* (8.8.16) *holds.*

Proof. The two pasting diagrams in (8.8.16) are the two composites below.

(8.8.27)
$$\begin{array}{ccc}
 & (B \boxtimes (B' \boxtimes B''))\big[(A \boxtimes (A' \boxtimes A''))\mathbb{1}^{\overline{m}}\big] & \\
\big[(B \boxtimes (B' \boxtimes B''))(A \boxtimes (A' \boxtimes A''))\big]\mathbb{1}^{\overline{m}} \ \xrightarrow{\ a\ } \text{(up to top)} & & \text{(from top)}\ \downarrow 1 \star a^{\boxtimes} \\
 & & (B \boxtimes (B' \boxtimes B''))\big[\mathbb{1}^{\overline{n}}((A \boxtimes A') \boxtimes A'')\big] \\
\downarrow \boxtimes^2 \star 1 & & \downarrow a^{-1} \\
\big[BA \boxtimes (B' \boxtimes B'')(A' \boxtimes A'')\big]\mathbb{1}^{\overline{m}} & & \big[(B \boxtimes (B' \boxtimes B''))\mathbb{1}^{\overline{n}}\big]((A \boxtimes A') \boxtimes A'') \\
\downarrow [1 \boxtimes \boxtimes^2] \star 1 & & \downarrow a^{\boxtimes} \star 1 \\
\big[BA \boxtimes (B'A' \boxtimes B''A'')\big]\mathbb{1}^{\overline{m}} & & \big[\mathbb{1}^{\overline{p}}((B \boxtimes B') \boxtimes B'')\big]((A \boxtimes A') \boxtimes A'') \\
\downarrow a^{\boxtimes} & & \downarrow a \\
\mathbb{1}^{\overline{p}}\big[(BA \boxtimes B'A') \boxtimes B''A''\big] & & \mathbb{1}^{\overline{p}}\big[((B \boxtimes B') \boxtimes B'')((A \boxtimes A') \boxtimes A'')\big] \\
\uparrow 1 \star [\boxtimes^2 \boxtimes 1] & & \downarrow 1 \star \boxtimes^2 \\
 & \mathbb{1}^{\overline{p}}\big[(B \boxtimes B')(A \boxtimes A') \boxtimes B''A''\big] &
\end{array}$$

Consider the diagram (8.8.27).

- Each instance of a is entrywise $a^4a^3a^2a^1$ as in (8.3.5). It involves δ^r (8.3.6) in a^1 and δ^{-l} (8.3.8) in a^4, but neither δ^{-r} nor δ^l.
- Each instance of $a^{\boxtimes} = \ell^{-1}\alpha^{-\otimes}r$ is as in (8.8.3). It has neither $\delta^{\pm r}$ nor $\delta^{\pm l}$.
- Each instance of $\boxtimes^2$ is entrywise as in the last case in (8.6.20). As explained in Lemma 8.6.16, it involves δ^{-r} and δ^{-l}, but neither δ^r nor δ^l. In particular, each entry of $(\boxtimes^2)^{-1}$ involves δ^r and δ^l, but neither δ^{-r} nor δ^{-l}.

As a brief outline, similar to Section 8.7, we prove the commutativity of each entry of the diagram (8.8.27) by

- expressing its morphisms using paths in $\mathsf{Gr}(X)$ for some set X of formal variables and some function $\varphi\colon X \longrightarrow \mathsf{Ob}(\mathsf{C})$ and
- applying the Coherence Theorem 3.9.1.

Due to the presence of $\delta^{\pm r}$ and $\delta^{\pm l}$ in some of the morphisms as discussed in the previous paragraph, we need to consider the *inverses* of a^4 and $\boxtimes^2$. This direction reversal creates a zigzag of morphisms. To show that the resulting diagram is commutative, we need to subdivide it in such a way that the additional morphisms are also paths in $\mathsf{Gr}(X)$ with the same set X and function φ. The detailed argument is given below.

Using the notation in (8.8.23), consider the (K, I)-entry of the diagram (8.8.27). Denote by

- R_1 the codomain of r in $a^{\boxtimes} = \ell^{-1}\alpha^{-\otimes}r$ along the left side;
- R_2 the codomain of $1 \star r$ in $1 \star a^{\boxtimes}$ in the upper right;

- R_3 the codomain of $r \star 1$ in $a^{\boxtimes} \star 1$ in the middle right;
- D_1 the domain of a^4 in $a = a^4 a^3 a^2 a^1$ in the upper left;
- D_2 the codomain of $(a^4)^{-1}$ in a^{-1} in the upper right; and
- D_3 the domain of a^4 in a in the lower right.

Using these notations and (8.8.23), the (K,I)-entry of the diagram (8.8.27) is equivalent to the outer diagram in C below.

(8.8.28)

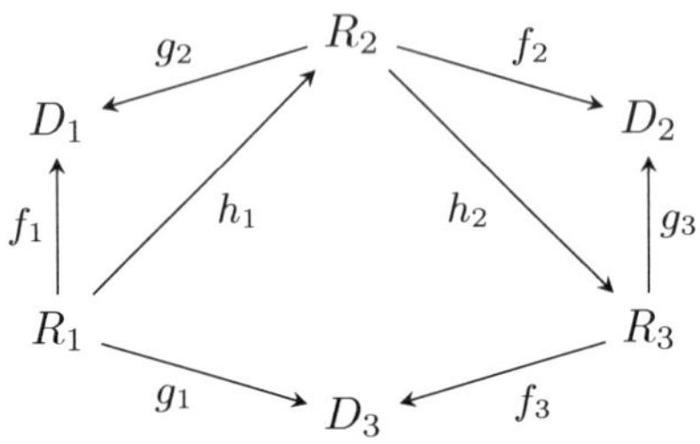

The objects and morphisms in the diagram (8.8.28) are described explicitly below.

- In addition to (8.8.23), we use the following indices and abbreviations.

(8.8.29)
$$\begin{gathered}
1 \le j, l \le n \qquad 1 \le j', l' \le n' \qquad 1 \le j'', l'' \le n'' \\
J = (j-1)n'n'' + (j'-1)n'' + j'' \\
L = (l-1)n'n'' + (l'-1)n'' + l'' \\
\bigoplus_{\overline{j}=1}^{\overline{n}} = \bigoplus_{j=1}^{n} \bigoplus_{j'=1}^{n'} \bigoplus_{j''=1}^{n''} \qquad \bigoplus_{\overline{l}=1}^{\overline{n}} = \bigoplus_{l=1}^{n} \bigoplus_{l'=1}^{n'} \bigoplus_{l''=1}^{n''}
\end{gathered}$$

- Abbreviating $\otimes$ to concatenation, the objects R_1, R_2, and R_3 are as follows.

$$\begin{aligned}
R_1 &= \big[BA \boxtimes (B'A' \boxtimes B''A'')\big]_{K,I} \\
&= (BA)_{ki}\big[(B'A')_{k'i'}(B''A'')_{k''i''}\big] \\
&= \Big[\bigoplus_{j=1}^{n} B_{kj}A_{ji}\Big]_{\mathsf{lt}} \Big[\Big\{\bigoplus_{j'=1}^{n'} B'_{k'j'}A'_{j'i'}\Big\}_{\mathsf{lt}} \Big\{\bigoplus_{j''=1}^{n''} B''_{k''j''}A''_{j''i''}\Big\}_{\mathsf{lt}}\Big] \\
R_2 &= \Big[\big(B \boxtimes (B' \boxtimes B'')\big)\big(A \boxtimes (A' \boxtimes A'')\big)\Big]_{K,I} \\
&= \Big[\bigoplus_{\overline{j}=1}^{\overline{n}} \big(B \boxtimes (B' \boxtimes B'')\big)_{K,J} \big(A \boxtimes (A' \boxtimes A'')\big)_{J,I}\Big]_{\mathsf{lt}} \\
&= \Big[\bigoplus_{\overline{j}=1}^{\overline{n}} \big(B_{kj}(B'_{k'j'}B''_{k''j''})\big)\big(A_{ji}(A'_{j'i'}A''_{j''i''})\big)\Big]_{\mathsf{lt}} \\
R_3 &= \Big[\big(B \boxtimes (B' \boxtimes B'')\big)\big((A \boxtimes A') \boxtimes A''\big)\Big]_{K,I} \\
&= \Big[\bigoplus_{\overline{j}=1}^{\overline{n}} \big(B \boxtimes (B' \boxtimes B'')\big)_{K,J} \big((A \boxtimes A') \boxtimes A''\big)_{J,I}\Big]_{\mathsf{lt}} \\
&= \Big[\bigoplus_{\overline{j}=1}^{\overline{n}} \big(B_{kj}(B'_{k'j'}B''_{k''j''})\big)\big((A_{ji}A'_{j'i'})A''_{j''i''}\big)\Big]_{\mathsf{lt}}
\end{aligned}$$

- The objects D_1, D_2, and D_3 are as follows.

$$
\begin{aligned}
D_1 &= \Bigg[\bigoplus_{\overline{j}=1}^{\overline{n}} \Bigg[\bigoplus_{\overline{e}=1}^{\overline{m}} \big(B \boxtimes (B' \boxtimes B'')\big)_{K,J} \Big\{\big(A \boxtimes (A' \boxtimes A'')\big)_{J,E} \mathbb{1}^{\overline{m}}_{E,I}\Big\}\Bigg]_{\mathsf{lt}}\Bigg]_{\mathsf{lt}} \\
&= \Bigg[\bigoplus_{\overline{j}=1}^{\overline{n}} \Bigg[\bigoplus_{\overline{e}=1}^{\overline{m}} \big(B_{kj}(B'_{k'j'}B''_{k''j''})\big) \Big\{\big(A_{je}(A'_{j'e'}A''_{j''e''})\big) \mathbb{1}^{\overline{m}}_{E,I}\Big\}\Bigg]_{\mathsf{lt}}\Bigg]_{\mathsf{lt}} \\
D_2 &= \Bigg[\bigoplus_{\overline{j}=1}^{\overline{n}} \Bigg[\bigoplus_{\overline{l}=1}^{\overline{n}} \big(B \boxtimes (B' \boxtimes B'')\big)_{K,J} \Big\{\mathbb{1}^{\overline{n}}_{J,L}\big((A \boxtimes A') \boxtimes A''\big)_{L,I}\Big\}\Bigg]_{\mathsf{lt}}\Bigg]_{\mathsf{lt}} \\
&= \Bigg[\bigoplus_{\overline{j}=1}^{\overline{n}} \Bigg[\bigoplus_{\overline{l}=1}^{\overline{n}} \big(B_{kj}(B'_{k'j'}B''_{k''j''})\big) \Big\{\mathbb{1}^{\overline{n}}_{J,L}\big((A_{li}A'_{l'i'})A''_{l''i''}\big)\Big\}\Bigg]_{\mathsf{lt}}\Bigg]_{\mathsf{lt}} \\
D_3 &= \Bigg[\bigoplus_{\overline{c}=1}^{\overline{p}} \Bigg[\bigoplus_{\overline{j}=1}^{\overline{n}} \mathbb{1}^{\overline{p}}_{K,C} \Big\{\big((B \boxtimes B') \boxtimes B''\big)_{C,J}\big((A \boxtimes A') \boxtimes A''\big)_{J,I}\Big\}\Bigg]_{\mathsf{lt}}\Bigg]_{\mathsf{lt}} \\
&= \Bigg[\bigoplus_{\overline{c}=1}^{\overline{p}} \Bigg[\bigoplus_{\overline{j}=1}^{\overline{n}} \mathbb{1}^{\overline{p}}_{K,C} \Big\{\big((B_{cj}B'_{c'j'})B''_{c''j''}\big)\big((A_{ji}A'_{j'i'})A''_{j''i''}\big)\Big\}\Bigg]_{\mathsf{lt}}\Bigg]_{\mathsf{lt}}
\end{aligned}
$$

- With $\boxtimes^{-2} = (\boxtimes^2)^{-1}$ and $a^{-4} = (a^4)^{-1}$, the morphisms in the boundary are as follows.

$$
\begin{aligned}
f_1 &= \big(a^3a^2a^1\big)\Big[(\boxtimes^{-2} \star 1)\big([1 \boxtimes \boxtimes^{-2}] \star 1\big)r^{-1}\Big]_{K,I} \\
f_2 &= a^{-4}\Big[(1 \star \ell^{-1})(1 \star \alpha^{-\otimes})\Big]_{K,I} \\
f_3 &= \big(a^3a^2a^1\big)\Big[(\ell^{-1} \star 1)(\alpha^{-\otimes} \star 1)\Big]_{K,I} \\
g_1 &= a^{-4}\Big[(1 \star \boxtimes^{-2})\big(1 \star [\boxtimes^{-2} \boxtimes 1]\big)\ell^{-1}\alpha^{-\otimes}\Big]_{K,I} \\
g_2 &= a^{-4}\big[1 \star r^{-1}\big]_{K,I} \\
g_3 &= \big(a^3a^2a^1\big)\big[r^{-1} \star 1\big]_{K,I}
\end{aligned}
\tag{8.8.30}
$$

The composite

$$g_1^{-1}f_3g_3^{-1}f_2g_2^{-1}f_1 \colon R_1 \longrightarrow R_1$$

in (8.8.28) is equal to going one round clockwise in the (K,I)-entry in (8.8.27), starting and ending at R_1.

- Using the notation in (8.7.6), the morphism

$$
\begin{array}{c}
R_1 = \Bigg[\bigoplus_{j=1}^{n} B_{kj}A_{ji}\Bigg]_{\mathsf{lt}}\Bigg[\Bigg\{\bigoplus_{j'=1}^{n'} B'_{k'j'}A'_{j'i'}\Bigg\}_{\mathsf{lt}}\Bigg\{\bigoplus_{j''=1}^{n''} B''_{k''j''}A''_{j''i''}\Bigg\}_{\mathsf{lt}}\Bigg] \\
\Big\downarrow h_1 \\
R_2 = \Bigg[\bigoplus_{\overline{j}=1}^{\overline{n}} \big(B_{kj}(B'_{k'j'}B''_{k''j''})\big)\big(A_{ji}(A'_{j'i'}A''_{j''i''})\big)\Bigg]_{\mathsf{lt}}
\end{array}
$$

has the form $(\alpha^{\pm\oplus}, \alpha^{\pm\otimes}, \xi^{\otimes}, \delta^r, \delta^l)$. In h_1, the following statements hold.

  - The morphisms involving δ^r and δ^l bring the three sums to the front.
  - The morphisms involving $\alpha^{\pm\otimes}$ and $\xi^{\otimes}$ permute

$$(B_{kj}A_{ji})(B'_{k'j'}A'_{j'i'})(B''_{k''j''}A''_{j''i''}) \quad \text{to} \quad \big(B_{kj}(B'_{k'j'}B''_{k''j''})\big)\big(A_{ji}(A'_{j'i'}A''_{j''i''})\big)$$

in each summand.

- The morphisms involving $\alpha^{-\oplus}$ move the additive brackets to match with those in R_2.

- The morphism

$$\begin{array}{c} R_2 = \Big[\bigoplus_{\overline{j}=1}^{\overline{n}} \big(B_{kj}(B'_{k'j'}B''_{k''j''})\big)\big(A_{ji}(A'_{j'i'}A''_{j''i''})\big)\Big]_{\mathsf{lt}} \\ \Big\downarrow h_2 \\ R_3 = \Big[\bigoplus_{\overline{j}=1}^{\overline{n}} \big(B_{kj}(B'_{k'j'}B''_{k''j''})\big)\big((A_{ji}A'_{j'i'})A''_{j''i''}\big)\Big]_{\mathsf{lt}} \end{array}$$

has the form $(\alpha^{-\otimes})$ with one copy of

$$A_{ji}(A'_{j'i'}A''_{j''i''}) \xrightarrow{\alpha^{-\otimes}} (A_{ji}A'_{j'i'})A''_{j''i''}$$

in each summand.

To show that the diagram (8.8.28) is commutative, we use the set

$$X = \left\{0^X, 1^X, a_{je}, a'_{j'e'}, a''_{j''e''}, b_{cj}, b'_{c'j'}, b''_{c''j''}\right\}$$

of formal variables, with the indices as in (8.8.22) and (8.8.29), and the function $\varphi\colon X \longrightarrow \mathsf{Ob}(\mathsf{C})$ defined as follows.

$$\varphi(x) = \begin{cases} \mathbb{0} & \text{if } x = 0^X. \\ \mathbb{1} & \text{if } x = 1^X. \\ A_{je} & \text{if } x = a_{je}. \\ A'_{j'e'} & \text{if } x = a'_{j'e'}. \\ A''_{j''e''} & \text{if } x = a''_{j''e''}. \\ B_{cj} & \text{if } x = b_{cj}. \\ B'_{c'j'} & \text{if } x = b'_{c'j'}. \\ B''_{c''j''} & \text{if } x = b''_{c''j''}. \end{cases}$$

By the description above of the objects and morphisms in (8.8.28), there exists a diagram in $\mathsf{Gr}(X)$ below in which the value in C of each path is the corresponding edge in (8.8.28).

(8.8.31)

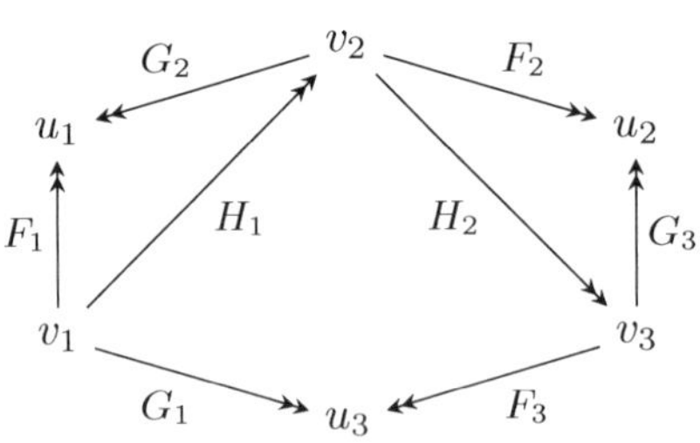

In the diagram (8.8.31), for $1 \leq t \leq 3$, the following statements hold.

- The images of v_t and u_t under φ are, respectively, R_t and D_t in (8.8.28). For example, the vertex

$$v_2 = \Big[\bigoplus_{\overline{j}=1}^{\overline{n}} \big(b_{kj}(b'_{k'j'}b''_{k''j''})\big)\big(a_{ji}(a'_{j'i'}a''_{j''i''})\big)\Big]_{\mathsf{lt}}.$$

- The images of F_t and G_t under φ are, respectively, f_t and g_t.
- The images of H_1 and H_2 under φ are, respectively, h_1 and h_2.

Moreover, by (8.2.4), (8.2.10), (8.3.5), Lemma 8.6.16, and (8.8.30), the following statements hold.

- Each prime edge in F_1 is an identity or contains $\alpha^{\pm\oplus}$, $\lambda^{-\oplus}$, $\rho^{-\oplus}$, $\xi^{\oplus}$, $\alpha^{\pm\otimes}$, $\rho^{-\otimes}$, $\xi^{-\otimes}$, $\rho^{-\bullet}$, δ^l, or δ^r.
- Each prime edge in F_2 is an identity or contains $\lambda^{-\oplus}$, $\rho^{-\oplus}$, $\alpha^{-\otimes}$, $\lambda^{-\otimes}$, $\lambda^{-\bullet}$, or δ^l.
- Each prime edge in F_3 is an identity or contains $\alpha^{\pm\oplus}$, $\lambda^{-\oplus}$, $\rho^{-\oplus}$, $\xi^{\oplus}$, $\alpha^{\pm\otimes}$, $\lambda^{-\otimes}$, $\lambda^{-\bullet}$, or δ^r.
- Each prime edge in G_1 is an identity or contains $\alpha^{-\oplus}$, $\lambda^{-\oplus}$, $\rho^{-\oplus}$, $\alpha^{\pm\otimes}$, $\lambda^{-\otimes}$, $\xi^{-\otimes}$, $\lambda^{-\bullet}$, δ^l, or δ^r.
- Each prime edge in G_2 is an identity or contains $\lambda^{-\oplus}$, $\rho^{-\oplus}$, $\rho^{-\otimes}$, $\rho^{-\bullet}$, or δ^l.
- Each prime edge in G_3 is an identity or contains $\alpha^{\pm\oplus}$, $\lambda^{-\oplus}$, $\rho^{-\oplus}$, $\xi^{\oplus}$, $\alpha^{\otimes}$, $\rho^{-\otimes}$, $\rho^{-\bullet}$, or δ^r.
- Each prime edge in H_1 is an identity or contains $\alpha^{-\oplus}$, $\alpha^{\pm\otimes}$, $\xi^{\otimes}$, δ^l, or δ^r.
- Each prime edge in H_2 is an identity or contains $\alpha^{-\otimes}$.

In (8.8.31), the vertex v_2 is regular as in Definition 3.1.25, so the vertex v_1 is also regular by Lemma 3.1.29. Since C is assumed to be tight, the Coherence Theorem 3.9.1 implies that in each of the following three cases, the two paths have the same value in C.

- F_1 and $(G_2, H_1)\colon v_1 \twoheadrightarrow u_1$.
- G_1 and $(F_3, H_2, H_1)\colon v_1 \twoheadrightarrow u_3$.
- F_2 and $(G_3, H_2)\colon v_2 \twoheadrightarrow u_2$.

This implies that the diagram (8.8.28) is commutative. This, in turn, implies that the diagram (8.8.27) is commutative. □

Lemma 8.8.32. *$a^{\boxtimes}$ in Definition 8.8.1 is a strong transformation.*

Proof. We have proved all the necessary assertions.

- Lemma 8.8.5 shows that $a^{\boxtimes}$ in (8.8.3) is a natural isomorphism.
- Lemma 8.8.11 shows that $a^{\boxtimes}$ satisfies the lax unity axiom (6.2.15).
- Lemmas 8.8.17 and 8.8.26 show that $a^{\boxtimes}$ satisfies the lax naturality axiom (6.2.16).

Therefore, $a^{\boxtimes}$ is a strong transformation. □

The Right Adjoint of the Monoidal Associator.

Definition 8.8.33. With respect to the pseudofunctor $(\boxtimes, \boxtimes^2, \boxtimes^0)$ in Definition 8.6.19 and Lemma 8.7.31 , define the data of a lax transformation

$$\boxtimes(1 \times \boxtimes) \xrightarrow{a^{\boxtimes\bullet}} \boxtimes(\boxtimes \times 1)$$

as follows.

Component 1-Cells: For each triple of objects $(m, n, p) \in (\mathsf{Mat}^{\mathsf{C}})^3$, define

$$(8.8.34)\qquad \big(m \boxtimes (n \boxtimes p)\big) = mnp \xrightarrow{a^{\boxtimes\bullet}_{m,n,p}} mnp = \big((m \boxtimes n) \boxtimes p\big)$$

as the identity matrix $\mathbb{1}^{mnp} \in \mathsf{Mat}^{\mathsf{C}}_{mnp,mnp}$ in (8.1.6).

Component 2-Cells: For each triple of 1-cells

$$\big(A = (A_{i'i}), B = (B_{j'j}), C = (C_{k'k})\big) \in \mathsf{Mat}^{\mathsf{C}}_{m,m'} \times \mathsf{Mat}^{\mathsf{C}}_{n,n'} \times \mathsf{Mat}^{\mathsf{C}}_{p,p'},$$

define the component 2-cell

$$a^{\boxtimes^\bullet}_{A,B,C} \in \mathsf{Mat}^{\mathsf{C}}_{mnp,m'n'p'}\Big([(A \boxtimes B) \boxtimes C]a^{\boxtimes^\bullet}_{m,n,p}; a^{\boxtimes^\bullet}_{m',n',p'}[A \boxtimes (B \boxtimes C)]\Big)$$

as the following vertical composite.

(8.8.35)

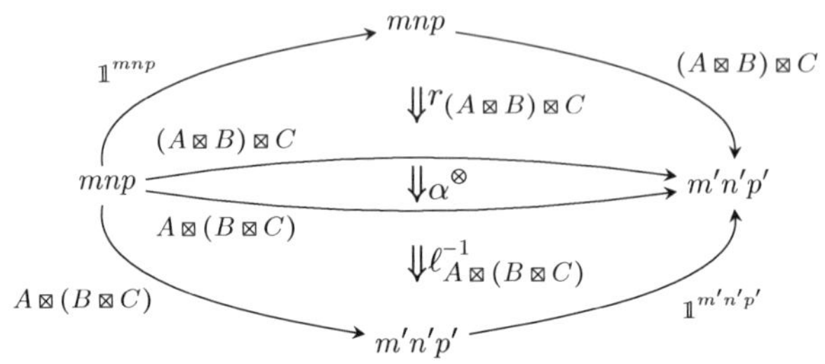

- ℓ is the base left unitor in $\mathsf{Mat}^{\mathsf{C}}$ in (8.2.2).
- r is the base right unitor in $\mathsf{Mat}^{\mathsf{C}}$ in (8.2.8).
- $\alpha^{\otimes}$ is the 2-cell with $((i'-1)n'p' + (j'-1)p' + k', (i-1)np + (j-1)p + k)$-entry the structure morphism in C,

$$\big(A_{i'i} \otimes B_{j'j}\big) \otimes C_{k'k} \xrightarrow{\alpha^{\otimes}_{A_{i'i},B_{j'j},C_{k'k}}} A_{i'i} \otimes \big(B_{j'j} \otimes C_{k'k}\big)$$

  for $1 \le i \le m$, $1 \le j \le n$, $1 \le k \le p$, $1 \le i' \le m'$, $1 \le j' \le n'$, and $1 \le k' \le p'$.

This finishes the definition of $a^{\boxtimes^\bullet}$. ◇

Lemma 8.8.36. *$a^{\boxtimes^\bullet}$ in Definition 8.8.33 is a strong transformation.*

Proof. The proofs of Lemmas 8.8.5, 8.8.11, 8.8.17, and 8.8.26 for $a^{\boxtimes}$ apply to $a^{\boxtimes^\bullet}$ with only cosmetic changes. □

The Unit and the Counit of the Monoidal Associator. Recall

- the identity strong transformation in Lemma 6.2.17,
- the horizontal composition of lax transformations in Definition 6.2.20,
- modification in Definition 6.3.1, and
- the base left unitor ℓ in (8.2.2).

Next we define the unit and the counit for $(a^{\boxtimes}, a^{\boxtimes^\bullet})$.

Definition 8.8.37. For the strong transformations

$$\boxtimes(\boxtimes \times 1) \underset{a^{\boxtimes^\bullet}}{\overset{a^{\boxtimes}}{\rightleftarrows}} \boxtimes(1 \times \boxtimes)$$

in Definitions 8.8.1 and 8.8.33, define the data

$$1_{\boxtimes(\boxtimes\times 1)} \xrightarrow{\eta^a} a^{\boxtimes^\bullet}a^{\boxtimes} \qquad a^{\boxtimes}a^{\boxtimes^\bullet} \xrightarrow{\varepsilon^a} 1_{\boxtimes(1\times\boxtimes)}$$

as consisting of the component 2-cells

$$(8.8.38)\qquad \begin{array}{c} \big(1_{\boxtimes(\boxtimes\times 1)}\big)_{(m,n,p)} = \mathbb{1}^{mnp} \\ \eta^a_{(m,n,p)} = \Big\downarrow \ell^{-1}_{\mathbb{1}^{mnp}} \\ (a^{\boxtimes\bullet}a^{\boxtimes})_{(m,n,p)} = \mathbb{1}^{mnp}\mathbb{1}^{mnp} \end{array} \qquad\qquad \begin{array}{c} (a^{\boxtimes}a^{\boxtimes\bullet})_{(m,n,p)} = \mathbb{1}^{mnp}\mathbb{1}^{mnp} \\ \varepsilon^a_{(m,n,p)} = \Big\downarrow \ell_{\mathbb{1}^{mnp}} \\ \big(1_{\boxtimes(1\times\boxtimes)}\big)_{(m,n,p)} = \mathbb{1}^{mnp} \end{array}$$

in $\mathsf{Mat}^{\mathsf{C}}_{mnp,mnp}$ for each triple of objects $(m,n,p) \in (\mathsf{Mat}^{\mathsf{C}})^3$. ◇

Lemma 8.8.39. *η^a and ε^a in Definition 8.8.37 are invertible modifications.*

Proof. Each component 2-cell of each of η^a and ε^a is invertible by Lemma 8.2.1. We will show that η^a is a modification; the proof for ε^a is obtained by a slight adjustment.

By (6.2.18) applied to $1_{\boxtimes(\boxtimes\times 1)}$ and (6.2.21) applied to $a^{\boxtimes\bullet}a^{\boxtimes}$, the modification axiom (6.3.2) for $\eta^a\colon 1_{\boxtimes(\boxtimes\times 1)} \longrightarrow a^{\boxtimes\bullet}a^{\boxtimes}$ states that, for each triple of 1-cells

$$(A,B,C) \in \mathsf{Mat}^{\mathsf{C}}_{m,m'} \times \mathsf{Mat}^{\mathsf{C}}_{n,n'} \times \mathsf{Mat}^{\mathsf{C}}_{p,p'},$$

the following equality of pasting diagrams in $\mathsf{Mat}^{\mathsf{C}}_{q,q'}$ holds, with $q = mnp$ and $q' = m'n'p'$.

(8.8.40)

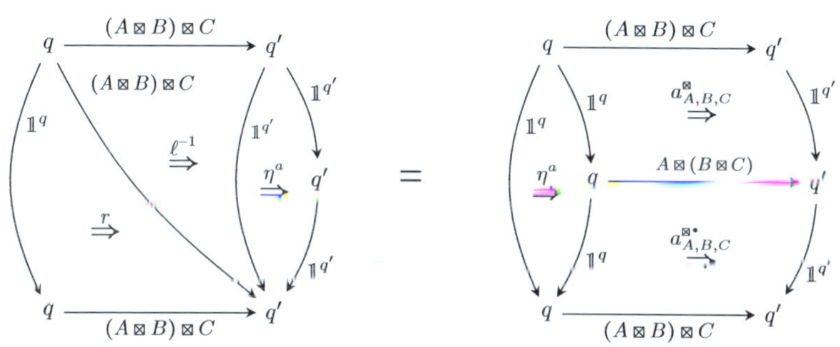

On the right-hand side of (8.8.40), by Convention 6.2.12, there are

- one instance of the base associator a in (8.3.2) and
- two instances of a^{-1}

that are not explicitly displayed. If $qq' = 0$, then $\mathsf{Mat}^{\mathsf{C}}_{q,q'}$ is the terminal category, and the equality in (8.8.40) holds.

Suppose $qq' > 0$. The proof for this case proceeds along the lines of the proof of Lemma 8.8.26. Due to the presence of both the base associator a and its inverse, a typical entry of (8.8.40) involves $\delta^{\pm r}$ and $\delta^{\pm l}$. Therefore, when the entries of the 2-cells in (8.8.40) are realized as paths in $\mathsf{Gr}(X)$, the diagram will need to be further subdivided before the Coherence Theorem 3.9.1 can be applied. The detailed argument is given below.

The two pasting diagrams in (8.8.40) are the two composites in $\mathsf{Mat}^{\mathsf{C}}_{q,q'}$ below.

(8.8.41)
$$\begin{array}{ccc}
 & ((A \boxtimes B) \boxtimes C)(\mathbb{1}^q \mathbb{1}^q) & \\
{}^{1 \star \eta^a \,=\, 1 \star \ell^{-1}} \nearrow & & \searrow {}^{a^{-1}} \\
((A \boxtimes B) \boxtimes C)\mathbb{1}^q & & \big[((A \boxtimes B) \boxtimes C)\mathbb{1}^q\big]\mathbb{1}^q \\
\downarrow {\scriptstyle r} & & \downarrow {\scriptstyle a^{\boxtimes \bullet} \star 1} \\
(A \boxtimes B) \boxtimes C & & \big[\mathbb{1}^{q'}(A \boxtimes (B \boxtimes C))\big]\mathbb{1}^q \\
\downarrow {\scriptstyle \ell^{-1}} & & \downarrow {\scriptstyle a} \\
\mathbb{1}^{q'}((A \boxtimes B) \boxtimes C) & & \mathbb{1}^{q'}\big[(A \boxtimes (B \boxtimes C))\mathbb{1}^q\big] \\
\downarrow {\scriptstyle \eta^a \star 1 \,=\, \ell^{-1} \star 1} & & \downarrow {\scriptstyle 1 \star a^{\boxtimes}} \\
(\mathbb{1}^{q'} \mathbb{1}^{q'})((A \boxtimes B) \boxtimes C) & \xleftarrow{\;a^{-1}\;} & \mathbb{1}^{q'}\big[\mathbb{1}^{q'}((A \boxtimes B) \boxtimes C)\big]
\end{array}$$

To show that each entry of (8.8.41) is a commutative diagram in C, consider the following indices and abbreviations.

(8.8.42)
$$\begin{array}{ll}
1 \le i, s_1, t_1 \le m & 1 \le i', u_1, v_1 \le m' \\
1 \le j, s_2, t_2 \le n & 1 \le j', u_2, v_2 \le n' \\
1 \le k, s_3, t_3 \le p & 1 \le k', u_3, v_3 \le p' \\
Q = (i-1)np + (j-1)p + k & Q' = (i'-1)n'p' + (j'-1)p' + k' \\
S = (s_1-1)np + (s_2-1)p + s_3 & U = (u_1-1)n'p' + (u_2-1)p' + u_3 \\
T = (t_1-1)np + (t_2-1)p + t_3 & V = (v_1-1)n'p' + (v_2-1)p' + v_3 \\
\bigoplus_{s=1}^{q} = \bigoplus_{s_1=1}^{m} \bigoplus_{s_2=1}^{n} \bigoplus_{s_3=1}^{p} & \bigoplus_{u=1}^{q'} = \bigoplus_{u_1=1}^{m'} \bigoplus_{u_2=1}^{n'} \bigoplus_{u_3=1}^{p'} \\
\bigoplus_{t=1}^{q} = \bigoplus_{t_1=1}^{m} \bigoplus_{t_2=1}^{n} \bigoplus_{t_3=1}^{p} & \bigoplus_{v=1}^{q'} = \bigoplus_{v_1=1}^{m'} \bigoplus_{v_2=1}^{n'} \bigoplus_{v_3=1}^{p'}
\end{array}$$

We also use

- the entrywise decomposition $a = a^4 a^3 a^2 a^1$ in (8.3.5),
- the definition of $a^{\boxtimes} = \ell^{-1} \alpha^{-\otimes} r$ in (8.8.3),
- the definition of $a^{\boxtimes \bullet} = \ell^{-1} \alpha^{\otimes} r$ in (8.8.35), and
- concatenation for $\otimes$ and the shorthand $a^{-4} = (a^4)^{-1}$.

The (Q', Q)-entry of (8.8.41) is equivalent to the following diagram in C.
(8.8.43)

$$Y_1 = \left[\bigoplus_{t=1}^{q}\big((A_{i't_1}B_{j't_2})C_{k't_3}\big)\left[\bigoplus_{s=1}^{q}\mathbb{1}^q_{T,S}\mathbb{1}^q_{S,Q}\right]_{\mathsf{lt}}\right]_{\mathsf{lt}} \xrightarrow{\;a^{-4}\;} \left[\bigoplus_{t=1}^{q}\left[\bigoplus_{s=1}^{q}\big((A_{i't_1}B_{j't_2})C_{k't_3}\big)\big(\mathbb{1}^q_{T,S}\mathbb{1}^q_{S,Q}\big)\right]_{\mathsf{lt}}\right]_{\mathsf{lt}} = Z_1$$

$$Y_2 = \left[\bigoplus_{t=1}^{q}\big((A_{i't_1}B_{j't_2})C_{k't_3}\big)\mathbb{1}^q_{T,Q}\right]_{\mathsf{lt}} \xrightarrow{\;1\star\ell^{-1}\;} Y_1$$

$$Z_2 = \left[\bigoplus_{s=1}^{q}\left[\bigoplus_{t=1}^{q}\big((A_{i't_1}B_{j't_2})C_{k't_3}\big)\mathbb{1}^q_{T,S}\right]_{\mathsf{lt}}\mathbb{1}^q_{S,Q}\right]_{\mathsf{lt}} \xrightarrow{\;a^3a^2a^1\;} Z_1$$

$$Y_2 \xrightarrow{\;r\;} Y_3 = (A_{i'i}B_{j'j})C_{k'k}$$

$$Z_2 \xrightarrow{\;(\ell^{-1}\alpha^{\otimes}r)\star 1\;} \left[\bigoplus_{s=1}^{q}\left[\bigoplus_{v=1}^{q'}\mathbb{1}^{q'}_{Q',V}\big(A_{v_1s_1}(B_{v_2s_2}C_{v_3s_3})\big)\right]_{\mathsf{lt}}\mathbb{1}^q_{S,Q}\right]_{\mathsf{lt}} = Z_3$$

$$Y_3 \xrightarrow{\;\ell^{-1}\;} Y_4 = \left[\bigoplus_{u=1}^{q'}\mathbb{1}^{q'}_{Q',U}\big((A_{u_1i}B_{u_2j})C_{u_3k}\big)\right]_{\mathsf{lt}}$$

$$Z_3 \xrightarrow{\;a^3a^2a^1\;} \left[\bigoplus_{v=1}^{q'}\left[\bigoplus_{s=1}^{q}\mathbb{1}^{q'}_{Q',V}\big\{\big(A_{v_1s_1}(B_{v_2s_2}C_{v_3s_3})\big)\mathbb{1}^q_{S,Q}\big\}\right]_{\mathsf{lt}}\right]_{\mathsf{lt}} = Z_4$$

$$Y_4 \xrightarrow{\;\ell^{-1}\star 1\;} Y_5 = \left[\bigoplus_{u=1}^{q'}\left[\bigoplus_{v=1}^{q'}\mathbb{1}^{q'}_{Q',V}\mathbb{1}^{q'}_{V,U}\right]_{\mathsf{lt}}\big((A_{u_1i}B_{u_2j})C_{u_3k}\big)\right]_{\mathsf{lt}}$$

$$Z_5 = \left[\bigoplus_{v=1}^{q'}\mathbb{1}^{q'}_{Q',V}\left[\bigoplus_{s=1}^{q}\big(A_{v_1s_1}(B_{v_2s_2}C_{v_3s_3})\big)\mathbb{1}^q_{S,Q}\right]_{\mathsf{lt}}\right]_{\mathsf{lt}} \xrightarrow{\;a^{-4}\;} Z_4$$

$$Y_5 \xrightarrow{\;a^3a^2a^1\;} Y_6 = \left[\bigoplus_{v=1}^{q'}\left[\bigoplus_{u=1}^{q'}\mathbb{1}^{q'}_{Q',V}\big\{\mathbb{1}^{q'}_{V,U}\big((A_{u_1i}B_{u_2j})C_{u_3k}\big)\big\}\right]_{\mathsf{lt}}\right]_{\mathsf{lt}}$$

$$Z_5 \xrightarrow{\;1\star(\ell^{-1}\alpha^{-\otimes}r)\;} \left[\bigoplus_{v=1}^{q'}\mathbb{1}^{q'}_{Q',V}\left[\bigoplus_{u=1}^{q'}\mathbb{1}^{q'}_{V,U}\big((A_{u_1i}B_{u_2j})C_{u_3k}\big)\right]_{\mathsf{lt}}\right]_{\mathsf{lt}} = Z_6$$

$$Z_6 \xrightarrow{\;a^{-4}\;} Y_6$$

To show that the diagram (8.8.43) is commutative when suitably subdivided, we use the set of formal variables

$$X = \left\{0^X, 1^X, a_{v_1s_1}, b_{v_2s_2}, c_{v_3s_3}\right\},$$

with the indices as in (8.8.42), and the function $\varphi\colon X \longrightarrow \mathsf{Ob}(\mathsf{C})$ defined as follows.

$$\varphi(x) = \begin{cases} \mathbb{0} & \text{if } x = 0^X. \\ \mathbb{1} & \text{if } x = 1^X. \\ A_{v_1s_1} & \text{if } x = a_{v_1s_1}. \\ B_{v_2s_2} & \text{if } x = b_{v_2s_2}. \\ C_{v_3s_3} & \text{if } x = c_{v_3s_3}. \end{cases}$$

For $1 \le l \le 6$,

- $y_l \in X^{\mathsf{fr}}$ is obtained from Y_l in (8.8.43) by replacing the symbols

$$(A, B, C, \mathbb{1}^q, \mathbb{1}^{q'}) \quad \text{with} \quad (a, b, c, \delta^X, \delta^X),$$

where δ^X is as in (8.4.6), and
- $z_l \in X^{\mathsf{fr}}$ is obtained in a similar way from Z_l in (8.8.43).

For example,

$$y_1 = \left[\bigoplus_{t=1}^{q}\big((a_{i't_1}b_{j't_2})c_{k't_3}\big)\left[\bigoplus_{s=1}^{q}\delta^X_{T,S}\delta^X_{S,Q}\right]_{\mathsf{lt}}\right]_{\mathsf{lt}}.$$

Using (8.2.4), (8.2.10), (8.3.5), and the notation in (8.7.6), there is a diagram in $\mathsf{Gr}(X)$ below in which the value in C of each boundary path is the corresponding edge in (8.8.43).

(8.8.44)

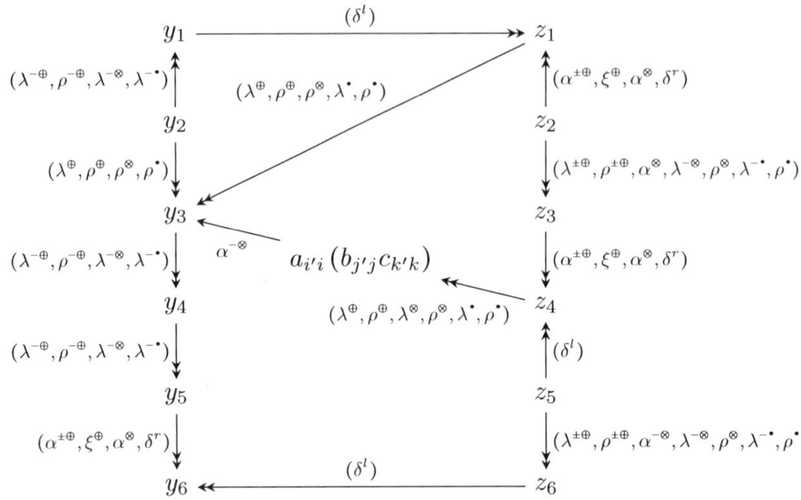

Consider the diagram (8.8.44).

- As an example of the notation, in the lower left path $y_5 \twoheadrightarrow y_6$, each prime edge is an identity or contains $\alpha^{\pm\oplus}$, $\xi^{\oplus}$, $\alpha^{\otimes}$, or δ^r.
- In the path $z_1 \twoheadrightarrow y_3$, the following statements hold.
  (i) $\rho^{\otimes}$, $\lambda^{\bullet}$, and $\rho^{\bullet}$ are used to reduce each summand

$$\big((a_{i't_1}b_{j't_2})c_{k't_3}\big)\big(\delta^X_{T,S}\delta^X_{S,Q}\big)$$

  in z_1 to

$$\begin{cases} 0^X & \text{if } T \neq S \text{ or } S \neq Q, \text{ and} \\ y_3 = (a_{i'i}b_{j'j})\,c_{k'k} & \text{if } T = S = Q. \end{cases}$$

  (ii) $\lambda^{\oplus}$ and $\rho^{\oplus}$ are then used to remove all the 0^X's.
- The path

$$z_4 \twoheadrightarrow a_{i'i}\,(b_{j'j}c_{k'k})$$

  is defined similarly.

Since the vertex y_3 is regular as in Definition 3.1.25, each vertex in (8.8.44) is regular by Lemma 3.1.29. Since C is assumed to be tight, the Coherence Theorem 3.9.1 implies that the image of (8.8.44) in C is a commutative diagram. This implies that (8.8.43) is commutative. This in turn implies that (8.8.41) is commutative. □

The Triangle Identities of the Monoidal Associator.

Lemma 8.8.45. *The quadruple* $(a^{\boxtimes}, a^{\boxtimes\bullet}, \eta^a, \varepsilon^a)$ *satisfies the triangle identities.*

Proof. We will show the left triangle identity in (6.3.10). The proof for the right triangle identity is obtained by a slight adjustment.

The left triangle identity states the commutativity of the left diagram below.

(8.8.46)
$$
\begin{array}{ccccc}
a^{\boxtimes}1_{\boxtimes(\boxtimes\times 1)} & & \xrightarrow{\;r_{a^{\boxtimes}}\;} & & a^{\boxtimes} \\
{\scriptstyle 1_{a^{\boxtimes}} * \eta^{a}}\downarrow & & & & \uparrow{\scriptstyle \ell_{a^{\boxtimes}}} \\
a^{\boxtimes}(a^{\boxtimes\bullet}a^{\boxtimes}) & & & & 1_{\boxtimes(1\times\boxtimes)}a^{\boxtimes} \\
& {\scriptstyle a^{-1}_{a^{\boxtimes},a^{\boxtimes\bullet},a^{\boxtimes}}}\searrow & & \nearrow{\scriptstyle \varepsilon^{a} * 1_{a^{\boxtimes}}} & \\
& & (a^{\boxtimes}a^{\boxtimes\bullet})a^{\boxtimes} & &
\end{array}
\qquad
\begin{array}{ccccc}
\mathbb{1}^{q}\mathbb{1}^{q} & & \xrightarrow{\;r_{\mathbb{1}^{q}}\;} & & \mathbb{1}^{q} \\
{\scriptstyle 1 \star \ell^{-1}_{\mathbb{1}^{q}}}\downarrow & & & & \uparrow{\scriptstyle \ell_{\mathbb{1}^{q}}} \\
\mathbb{1}^{q}(\mathbb{1}^{q}\mathbb{1}^{q}) & & & & \mathbb{1}^{q}\mathbb{1}^{q} \\
& {\scriptstyle a^{-1}_{\mathbb{1}^{q},\mathbb{1}^{q},\mathbb{1}^{q}}}\searrow & & \nearrow{\scriptstyle \ell_{\mathbb{1}^{q}} \star 1} & \\
& & (\mathbb{1}^{q}\mathbb{1}^{q})\mathbb{1}^{q} & &
\end{array}
$$

Consider the left diagram in (8.8.46).

- $1_{a^{\boxtimes}} * \eta^{a}$ and $\varepsilon^{a} * 1_{a^{\boxtimes}}$ are horizontal composites of modifications in (6.3.5).
- The component 2-cells of $\ell_{a^{\boxtimes}}$, $r_{a^{\boxtimes}}$, and $a^{-1}_{a^{\boxtimes},a^{\boxtimes\bullet},a^{\boxtimes}}$ are computed in the matrix bicategory $\mathsf{Mat}^{\mathsf{C}}$ using, respectively, the base left unitor in (8.2.2), the base right unitor in (8.2.8), and the inverse of the base associator in (8.3.2).
- The composite of any two consecutive edges is the vertical composite of modifications in (6.3.4).

To show that the left diagram in (8.8.46) is commutative, it suffices to show that the two composites have the same component 2-cells.

In other words, it suffices to show that, for each triple of objects $(m,n,p) \in (\mathsf{Mat}^{\mathsf{C}})^{3}$, the right diagram in (8.8.46) is commutative in $\mathsf{Mat}^{\mathsf{C}}_{q,q}$ with $q = mnp$. This is equivalent to showing that, for $1 \leq s,t \leq q$, the (s,t)-entry of the right diagram in (8.8.46) is commutative in C. Using the entrywise factorization $a = a^{4}a^{3}a^{2}a^{1}$ in (8.3.5) and with $\otimes$ abbreviated to concatenation, the (s,t)-entry of the right diagram in (8.8.46) is equivalent to the following diagram in C.

(8.8.47)
$$
\begin{array}{ccc}
\Big[\bigoplus_{i=1}^{q} \mathbb{1}^{q}_{si}\mathbb{1}^{q}_{it}\Big]_{\mathsf{lt}} & \xleftarrow{\;(r^{-1}_{\mathbb{1}^{q}})_{st}\;} & \mathbb{1}^{q}_{st} \\
{\scriptstyle (1\star\ell^{-1}_{\mathbb{1}^{q}})_{st}}\downarrow & & \downarrow{\scriptstyle (\ell^{-1}_{\mathbb{1}^{q}})_{st}} \\
\Big[\bigoplus_{i=1}^{q} \mathbb{1}^{q}_{si}\Big[\bigoplus_{j=1}^{q} \mathbb{1}^{q}_{ij}\mathbb{1}^{q}_{jt}\Big]_{\mathsf{lt}}\Big]_{\mathsf{lt}} & & \Big[\bigoplus_{j=1}^{q} \mathbb{1}^{q}_{sj}\mathbb{1}^{q}_{jt}\Big]_{\mathsf{lt}} \\
{\scriptstyle a^{-4}}\downarrow & & \downarrow{\scriptstyle (\ell^{-1}_{\mathbb{1}^{q}}\star 1)_{st}} \\
\Big[\bigoplus_{i=1}^{q}\Big[\bigoplus_{j=1}^{q} \mathbb{1}^{q}_{si}(\mathbb{1}^{q}_{ij}\mathbb{1}^{q}_{jt})\Big]_{\mathsf{lt}}\Big]_{\mathsf{lt}} & \xleftarrow{\;a^{3}a^{2}a^{1}\;} & \Big[\bigoplus_{j=1}^{q}\Big[\bigoplus_{i=1}^{q} \mathbb{1}^{q}_{si}\mathbb{1}^{q}_{ij}\Big]_{\mathsf{lt}}\mathbb{1}^{q}_{jt}\Big]_{\mathsf{lt}}
\end{array}
$$

To show that (8.8.47) is commutative, we use

- the set $X = \{0^{X}, 1^{X}\}$ and
- the function $\varphi \colon X \longrightarrow \mathsf{Ob}(\mathsf{C})$ defined by $\varphi(0^{X}) = \mathbb{0}$ and $\varphi(1^{X}) = \mathbb{1}$.

By (8.2.4), (8.2.10), (8.3.5), and the notation in (8.4.6) and (8.7.6), there exists a diagram in $\mathsf{Gr}(X)$ below in which the value in C of each path is the corresponding edge in (8.8.47).

$$\begin{array}{ccc}
\Big[\bigoplus_{i=1}^{q} \delta^X_{si}\delta^X_{it}\Big]_{\mathsf{lt}} & \xleftarrow{(\lambda^{-\oplus},\rho^{-\oplus},\rho^{-\otimes},\rho^{-\bullet})} & \delta^X_{st} \\
\Big\downarrow{\scriptstyle (\lambda^{-\oplus},\rho^{-\oplus},\lambda^{-\otimes},\lambda^{-\bullet})} & & \Big\downarrow{\scriptstyle (\lambda^{-\oplus},\rho^{-\oplus},\lambda^{-\otimes},\lambda^{-\bullet})} \\
\Big[\bigoplus_{i=1}^{q} \delta^X_{si}\Big[\bigoplus_{j=1}^{q} \delta^X_{ij}\delta^X_{jt}\Big]_{\mathsf{lt}}\Big]_{\mathsf{lt}} & & \Big[\bigoplus_{j=1}^{q} \delta^X_{sj}\delta^X_{jt}\Big]_{\mathsf{lt}} \\
\Big\downarrow{\scriptstyle (\delta^l)} & & \Big\downarrow{\scriptstyle (\lambda^{-\oplus},\rho^{-\oplus},\lambda^{-\otimes},\lambda^{-\bullet})} \\
\Big[\bigoplus_{i=1}^{q}\Big[\bigoplus_{j=1}^{q} \delta^X_{si}(\delta^X_{ij}\delta^X_{jt})\Big]_{\mathsf{lt}}\Big]_{\mathsf{lt}} & \xleftarrow{(\alpha^{\pm\oplus},\xi^{\oplus},\alpha^{\otimes},\delta^r)} & \Big[\bigoplus_{j=1}^{q}\Big[\bigoplus_{i=1}^{q} \delta^X_{si}\delta^X_{ij}\Big]_{\mathsf{lt}}\delta^X_{jt}\Big]_{\mathsf{lt}}
\end{array} \tag{8.8.48}$$

For example, in the bottom horizontal path, each prime edge is an identity or contains $\alpha^{\pm\oplus}$, $\xi^{\oplus}$, $\alpha^{\otimes}$, or δ^r.

- If $s \neq t$, then $\delta^X_{st} = 0^X$. Proposition 3.5.33 implies that the image of (8.8.48) is a commutative diagram in C.
- If $s = t$, then $\delta^X_{ss} = 1^X$, which is regular in the sense of Definition 3.1.25. Since C is assumed to be tight, the Coherence Theorem 3.9.1 implies that the image of (8.8.48) is a commutative diagram in C.

Therefore, the right, hence also the left, diagram in (8.8.46) is commutative. □

Lemma 8.8.49. *The quadruple $(a^{\boxtimes}, a^{\boxtimes\bullet}, \eta^a, \varepsilon^a)$ with*

- *$a^{\boxtimes}$ in Definition 8.8.1,*
- *$a^{\boxtimes\bullet}$ in Definition 8.8.33, and*
- *η^a and ε^a in Definition 8.8.37*

is an adjoint equivalence.

Proof. We already proved the following statements.

- $a^{\boxtimes}$ is a strong transformation by Lemma 8.8.32.
- $a^{\boxtimes\bullet}$ is a strong transformation by Lemma 8.8.36.
- η^a and ε^a are invertible modifications by Lemma 8.8.39.
- The triangle identities are satisfied by Lemma 8.8.45.

This is sufficient by Explanation 6.4.6. □

8.9. The Monoidal Unitors

We continue to assume that C is an arbitrary tight symmetric bimonoidal category as in Definition 2.1.2. We are in the process of constructing a monoidal bicategory structure, in the sense of Definition 6.4.1, on the matrix bicategory $\mathsf{Mat}^{\mathsf{C}}$. In this section, we define the monoidal unitors $(\ell^{\boxtimes}, \ell^{\boxtimes\bullet}, \eta^{\ell}, \varepsilon^{\ell})$ and $(r^{\boxtimes}, r^{\boxtimes\bullet}, \eta^{r}, \varepsilon^{r})$.

Recall

- the identity strict functor $1_{\mathsf{B}} \colon \mathsf{B} \longrightarrow \mathsf{B}$ for each bicategory B in Example 6.2.10;
- the strictly unitary pseudofunctor

$$(1_{\boxtimes}, 1^2_{\boxtimes}, 1^0_{\boxtimes}) \colon \mathbb{1} \longrightarrow \mathsf{Mat}^{\mathsf{C}}$$

in Definition 8.5.1 and Lemma 8.5.2; and

- the pseudofunctor
$$(\boxtimes, \boxtimes^2, \boxtimes^0) \colon \mathsf{Mat}^{\mathsf{C}} \times \mathsf{Mat}^{\mathsf{C}} \longrightarrow \mathsf{Mat}^{\mathsf{C}}$$
in Definition 8.6.19 and Lemma 8.7.31.

With the composite of pseudofunctors in Definition 6.2.6, we consider the following two composites.

$$\mathsf{Mat}^{\mathsf{C}} \underset{1_{\mathsf{Mat}^{\mathsf{C}}} \times 1_{\boxtimes}}{\overset{1_{\boxtimes} \times 1_{\mathsf{Mat}^{\mathsf{C}}}}{\rightrightarrows}} (\mathsf{Mat}^{\mathsf{C}})^2 \xrightarrow{\ \boxtimes\ } \mathsf{Mat}^{\mathsf{C}}$$

As discussed in Explanation 6.4.6, the left monoidal unitor $(\ell^{\boxtimes}, \ell^{\boxtimes\bullet}, \eta^{\ell}, \varepsilon^{\ell})$ consists of the following data.

(i) $\ell^{\boxtimes}$ and $\ell^{\boxtimes\bullet}$ are strong transformations as in Definition 6.2.14 as follows.
$$\boxtimes(1_{\boxtimes} \times 1_{\mathsf{Mat}^{\mathsf{C}}}) \underset{\ell^{\boxtimes\bullet}}{\overset{\ell^{\boxtimes}}{\rightleftarrows}} 1_{\mathsf{Mat}^{\mathsf{C}}}$$
These strong transformations are decorated with $\boxtimes$ to avoid confusion with the base left unitor (8.2.2) in $\mathsf{Mat}^{\mathsf{C}}$.

(ii) η^{ℓ} and ε^{ℓ} are invertible modifications as in Definition 6.3.1 as follows.
$$1_{\boxtimes(1_{\boxtimes} \times 1_{\mathsf{Mat}^{\mathsf{C}}})} \xrightarrow{\ \eta^{\ell}\ } \ell^{\boxtimes\bullet}\ell^{\boxtimes}$$
$$\ell^{\boxtimes}\ell^{\boxtimes\bullet} \xrightarrow{\ \varepsilon^{\ell}\ } 1_{1_{\mathsf{Mat}^{\mathsf{C}}}}$$

Moreover, these data are required to satisfy the triangle identities (6.3.10). The left monoidal unitor is defined in Definitions 8.9.1, 8.9.5, and 8.9.8.

The right monoidal unitor $(r^{\boxtimes}, r^{\boxtimes\bullet}, \eta^{r}, \varepsilon^{r})$ consists of the following data.

(i) $r^{\boxtimes}$ and $r^{\boxtimes\bullet}$ are strong transformations as follows.
$$\boxtimes(1_{\mathsf{Mat}^{\mathsf{C}}} \times 1_{\boxtimes}) \underset{r^{\boxtimes\bullet}}{\overset{r^{\boxtimes}}{\rightleftarrows}} 1_{\mathsf{Mat}^{\mathsf{C}}}$$
These strong transformations are decorated with $\boxtimes$ to avoid confusion with the base right unitor (8.2.8) in $\mathsf{Mat}^{\mathsf{C}}$.

(ii) η^{r} and ε^{r} are invertible modifications as follows.
$$1_{\boxtimes(1_{\mathsf{Mat}^{\mathsf{C}}} \times 1_{\boxtimes})} \xrightarrow{\ \eta^{r}\ } r^{\boxtimes\bullet}r^{\boxtimes}$$
$$r^{\boxtimes}r^{\boxtimes\bullet} \xrightarrow{\ \varepsilon^{r}\ } 1_{1_{\mathsf{Mat}^{\mathsf{C}}}}$$

Moreover, these data are required to satisfy the triangle identities (6.3.10). The right monoidal unitor is defined in Definitions 8.9.14, 8.9.17, and 8.9.20.

The Left Monoidal Unitor. First we define the left adjoint of the left monoidal unitor.

Definition 8.9.1. Define the data of a lax transformation
$$\boxtimes(1_{\boxtimes} \times 1_{\mathsf{Mat}^{\mathsf{C}}}) \xrightarrow{\ \ell^{\boxtimes}\ } 1_{\mathsf{Mat}^{\mathsf{C}}}$$
as follows.

Component 1-Cells: For each object $m \in \mathsf{Mat}^{\mathsf{C}}$, define

$$m = 1 \boxtimes m \xrightarrow{\ \ell^{\boxtimes}_m\ } (1_{\mathsf{Mat}^{\mathsf{C}}})\, m = m \tag{8.9.2}$$

as the identity matrix $\mathbb{1}^m \in \mathsf{Mat}^{\mathsf{C}}_{m,m}$ in (8.1.6).

Component 2-Cells: For each 1-cell $A = (A_{ji}) \in \mathsf{Mat}^{\mathsf{C}}_{m,n}$, define

$$\ell^{\boxtimes}_A \in \mathsf{Mat}^{\mathsf{C}}_{m,n}\Big((1_{\mathsf{Mat}^{\mathsf{C}}} A)\, \ell^{\boxtimes}_m ; \ell^{\boxtimes}_n \,(\boxtimes(1_{\boxtimes} \times 1_{\mathsf{Mat}^{\mathsf{C}}}))\,(A)\Big)$$

as the following vertical composite 2-cell.

(8.9.3)

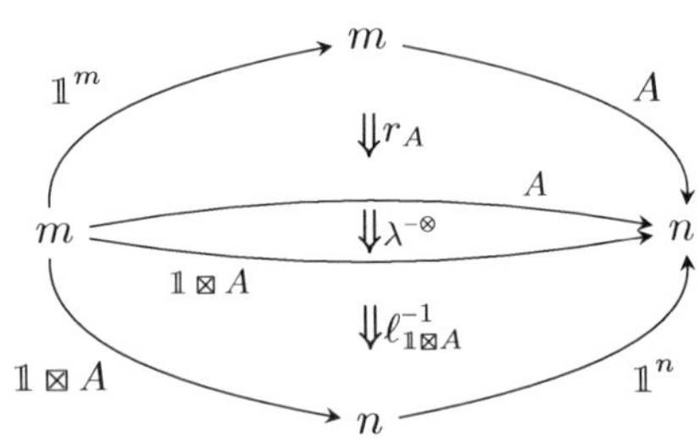

- $\mathbb{1} \boxtimes A$ is the scalar product in (8.6.2). This makes sense because the identity matrix $\mathbb{1}^1 \in \mathsf{Mat}^{\mathsf{C}}_{1,1}$ has the unique entry $\mathbb{1} \in \mathsf{C}$, and
$$(\boxtimes(1_{\boxtimes} \times 1_{\mathsf{Mat}^{\mathsf{C}}}))\,(A) = \mathbb{1}^1 \boxtimes A = \mathbb{1} \boxtimes A = \big(\mathbb{1} \otimes A_{ji}\big)$$
by the definition (8.6.3) of the matrix tensor product.
- ℓ is the base left unitor in $\mathsf{Mat}^{\mathsf{C}}$ in (8.2.2).
- r is the base right unitor in $\mathsf{Mat}^{\mathsf{C}}$ in (8.2.8).
- $\lambda^{-\otimes}$ is the 2-cell with (j,i)-entry the structure morphism in C,
$$A_{ji} \xrightarrow{\ \lambda^{-\otimes}_{A_{ji}}\ } \mathbb{1} \otimes A_{ji}$$
for $1 \le i \le m$ and $1 \le j \le n$.

This finishes the definition of $\ell^{\boxtimes}$. ◇

Explanation 8.9.4. The matrix $\mathbb{1} \boxtimes A \in \mathsf{Mat}^{\mathsf{C}}_{m,n}$ is obtained from A by replacing the (j,i)-entry A_{ji} with the product $\mathbb{1} \otimes A_{ji}$ for each $1 \le i \le m$ and $1 \le j \le n$. ◇

Next we define the right adjoint of the left monoidal unitor.

Definition 8.9.5. Define the data of a lax transformation

$$1_{\mathsf{Mat}^{\mathsf{C}}} \xrightarrow{\ \ell^{\boxtimes\bullet}\ } \boxtimes(1_{\boxtimes} \times 1_{\mathsf{Mat}^{\mathsf{C}}})$$

as follows.

Component 1-Cells: For each object $m \in \mathsf{Mat}^{\mathsf{C}}$, define

$$m = (1_{\mathsf{Mat}^{\mathsf{C}}})\, m \xrightarrow{\ \ell^{\boxtimes\bullet}_m\ } 1 \boxtimes m = m \tag{8.9.6}$$

as the identity matrix $\mathbb{1}^m \in \mathsf{Mat}^{\mathsf{C}}_{m,m}$ in (8.1.6).

Component 2-Cells: For each 1-cell $A = (A_{ji}) \in \mathsf{Mat}^{\mathsf{C}}_{m,n}$, define

$$\ell^{\boxtimes\bullet}_A \in \mathsf{Mat}^{\mathsf{C}}_{m,n}\Big((\boxtimes(1_{\boxtimes} \times 1_{\mathsf{Mat}^{\mathsf{C}}}))(A)\ell^{\boxtimes\bullet}_m ; \ell^{\boxtimes\bullet}_n (1_{\mathsf{Mat}^{\mathsf{C}}} A) \Big)$$

as the following vertical composite 2-cell.

(8.9.7)

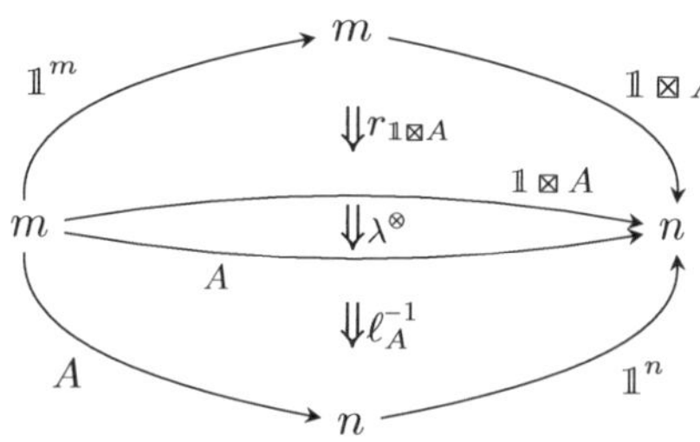

In this vertical composite, $\lambda^{\otimes}$ is the 2-cell with (j,i)-entry the structure morphism in C,

$$\mathbb{1} \otimes A_{ji} \xrightarrow{\lambda^{\otimes}_{A_{ji}}} A_{ji}$$

for $1 \le i \le m$ and $1 \le j \le n$.

This finishes the definition of $\ell^{\boxtimes\bullet}$. ◇

Next we define the unit and the counit for $(\ell^{\boxtimes}, \ell^{\boxtimes\bullet})$.

Definition 8.9.8. Define the data

$$1_{\boxtimes(1_{\boxtimes}\times 1_{\mathsf{Mat}^{\mathsf{C}}})} \xrightarrow{\eta^{\ell}} \ell^{\boxtimes\bullet}\ell^{\boxtimes} \qquad \ell^{\boxtimes}\ell^{\boxtimes\bullet} \xrightarrow{\varepsilon^{\ell}} 1_{1_{\mathsf{Mat}^{\mathsf{C}}}}$$

as consisting of the component 2-cells

$$\begin{array}{ccc} \left(1_{\boxtimes(1_{\boxtimes}\times 1_{\mathsf{Mat}^{\mathsf{C}}})}\right)_m = \mathbb{1}^m & & (\ell^{\boxtimes}\ell^{\boxtimes\bullet})_m = \mathbb{1}^m\mathbb{1}^m \\ \eta^{\ell}_m = \Big\downarrow \ell^{-1}_{\mathbb{1}^m} & & \varepsilon^{\ell}_m = \Big\downarrow \ell_{\mathbb{1}^m} \\ (\ell^{\boxtimes\bullet}\ell^{\boxtimes})_m = \mathbb{1}^m\mathbb{1}^m & & \left(1_{1_{\mathsf{Mat}^{\mathsf{C}}}}\right)_m = \mathbb{1}^m \end{array}$$

in $\mathsf{Mat}^{\mathsf{C}}_{m,m}$ for each object $m \in \mathsf{Mat}^{\mathsf{C}}$. ◇

Lemma 8.9.9. *The quadruple $(\ell^{\boxtimes}, \ell^{\boxtimes\bullet}, \eta^{\ell}, \varepsilon^{\ell})$ in Definitions 8.9.1, 8.9.5, and 8.9.8 is an adjoint equivalence.*

Proof. The statement of the lemma means the following three assertions.

(1) $\ell^{\boxtimes}$ and $\ell^{\boxtimes\bullet}$ are strong transformations.
(2) η^{ℓ} and ε^{ℓ} are invertible modifications.
(3) The triangle identities (6.3.10) are satisfied.

The naturality of $\ell^{\boxtimes}_A$ and $\ell^{\boxtimes\bullet}_A$ with respect to A follows from the naturality of $\lambda^{\oplus}$, $\rho^{\oplus}$, $\lambda^{\otimes}$, $\rho^{\otimes}$, $\lambda^{\bullet}$, and $\rho^{\bullet}$ in C.

Taking advantage of the similarity between $(\ell^{\boxtimes}, \ell^{\boxtimes\bullet}, \eta^{\ell}, \varepsilon^{\ell})$ and $(a^{\boxtimes}, a^{\boxtimes\bullet}, \eta^{a}, \varepsilon^{a})$ in Definitions 8.8.1, 8.8.33, and 8.8.37, the above assertions are proved by simpler versions of the proofs of Lemmas 8.8.5, 8.8.11, 8.8.17, 8.8.26, 8.8.39, and 8.8.45. Since we already showed all the detailed argument in Section 8.8, below we only state the equalities that one has to check. In marginal cases with $\mathsf{Mat}^{\mathsf{C}}_{m,0}$ or $\mathsf{Mat}^{\mathsf{C}}_{0,m}$, which are both the terminal category, the equality holds automatically. In every other case, one has to show that some diagram of 2-cells in $\mathsf{Mat}^{\mathsf{C}}$ is commutative. In each entry, the diagram is realized as paths in $\mathsf{Gr}(X)$ for some set X of formal variables and some function $\varphi\colon X \longrightarrow \mathsf{Ob}(\mathsf{C})$. The desired commutativity in C then follows from either

- Proposition 3.5.33, if one vertex has the same support as 0^X, or
- the Coherence Theorem 3.9.1 in every other case, where the vertices are regular.

Analogous to (8.8.12), the lax unity axiom (6.2.15) for $\ell^{\boxtimes}$ states the following equality of pasting diagrams in $\mathsf{Mat}^{\mathsf{C}}_{m,m}$ for $m \geq 0$.

(8.9.10)

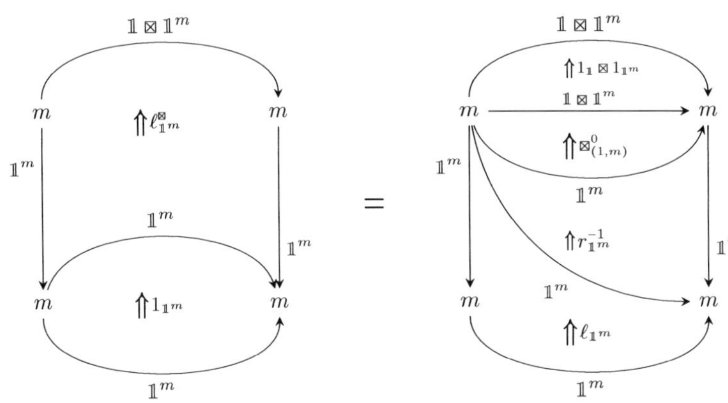

Consider (8.9.10).

- $\ell^{\boxtimes}_{\mathbb{1}^m} = \ell^{-1}_{\mathbb{1}\boxtimes\mathbb{1}^m} \lambda^{-\otimes} r_{\mathbb{1}^m}$.
- Each diagonal entry of $\mathbb{1} \boxtimes \mathbb{1}^m$ is $\mathbb{1} \otimes \mathbb{1}$, and each off-diagonal entry is $\mathbb{1} \otimes \mathbb{0}$.
- $\boxtimes^0_{(1,m)}$ is
  - $\lambda^{-\otimes}_{\mathbb{1}}\colon \mathbb{1} \longrightarrow \mathbb{1} \otimes \mathbb{1}$ in each diagonal entry and
  - $\rho^{-\bullet}_{\mathbb{1}}\colon \mathbb{0} \longrightarrow \mathbb{1} \otimes \mathbb{0}$ in each off-diagonal entry.
- $1_{\mathbb{1}} \boxtimes 1_{\mathbb{1}^m}$ is entrywise an identity morphism in C.

The lax unity axiom for $\ell^{\boxtimes\bullet}$ is analogous to (8.9.10).

Analogous to (8.8.16), the lax naturality axiom (6.2.16) for $\ell^{\boxtimes}$ states the following pasting diagram equality in $\mathsf{Mat}^{\mathsf{C}}_{m,p}$ for 1-cells $(A,B) \in \mathsf{Mat}^{\mathsf{C}}_{m,n} \times \mathsf{Mat}^{\mathsf{C}}_{n,p}$.

(8.9.11)

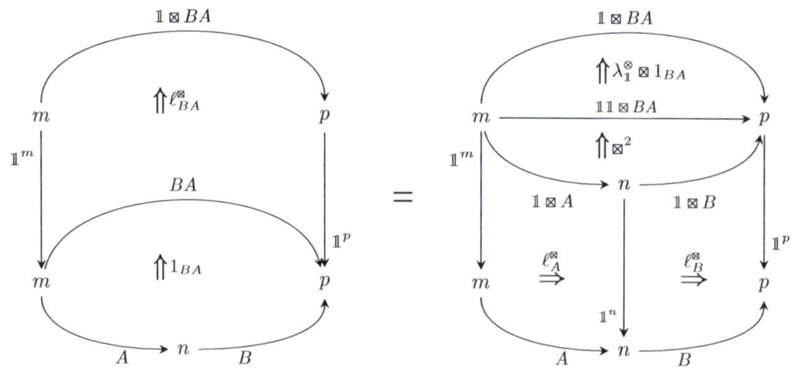

Consider (8.9.11).

- $\ell^{\boxtimes}_{BA} = \ell^{-1}_{\mathbb{1}\boxtimes BA}\lambda^{-\otimes} r_{BA}$, $\ell^{\boxtimes}_{A} = \ell^{-1}_{\mathbb{1}\boxtimes A}\lambda^{-\otimes} r_{A}$, and $\ell^{\boxtimes}_{B} = \ell^{-1}_{\mathbb{1}\boxtimes B}\lambda^{-\otimes} r_{B}$.
- $\boxtimes^2$ is either an identity morphism or as in (8.6.20).
- For $1 \le i \le m$ and $1 \le k \le p$, the (k,i)-entry of $\mathbb{1}\mathbb{1} \boxtimes BA$ is

$$\big(\mathbb{1}\mathbb{1} \boxtimes BA\big)_{ki} = (\mathbb{1} \otimes \mathbb{1}) \otimes (BA)_{ki} \in \mathsf{C}.$$

The lax naturality axiom for $\ell^{\boxtimes\bullet}$ is analogous to (8.9.11).

Analogous to (8.8.40), the modification axiom (6.3.2) for η^{ℓ} states that for each 1-cell $A \in \mathsf{Mat}^{\mathsf{C}}_{m,n}$, the following pasting diagram equality holds.

(8.9.12)

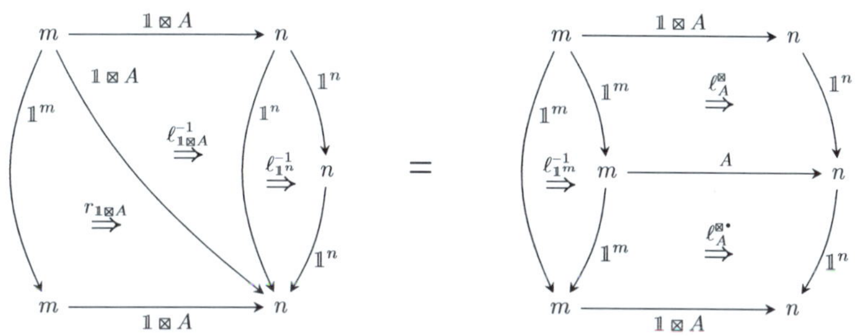

On the right-hand side, $\ell^{\boxtimes\bullet}_{A} = \ell^{-1}_{A}\lambda^{\otimes} r_{\mathbb{1}\boxtimes A}$. The modification axiom for ε^{ℓ} is analogous to (8.9.12).

Analogous to (8.8.46), the left triangle identity in (6.3.10) for $(\ell^{\boxtimes}, \ell^{\boxtimes\bullet}, \eta^{\ell}, \varepsilon^{\ell})$ states the commutativity of the left diagram below.

(8.9.13)

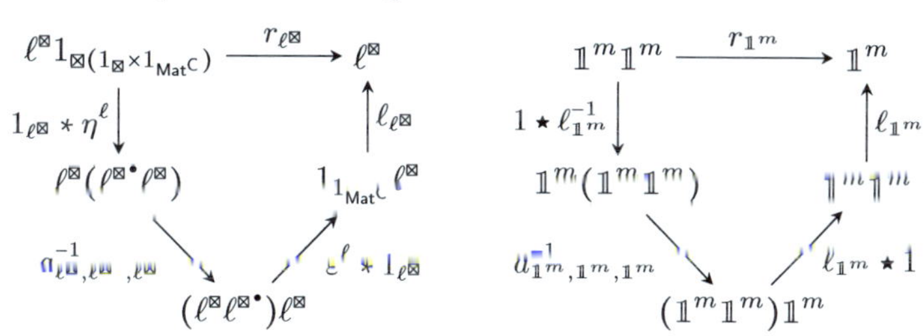

At each object $m \in \mathsf{Mat}^{\mathsf{C}}$, the left diagram in (8.9.13) yields the right diagram in $\mathsf{Mat}^{\mathsf{C}}_{m,m}$. The right triangle identity in (6.3.10) is analogous to (8.9.13). □

The Right Monoidal Unitor. Next we define the right monoidal unitor in $\mathsf{Mat}^{\mathsf{C}}$, which is similar to the left monoidal unitor.

Definition 8.9.14. Define the data of a lax transformation

$$\boxtimes(1_{\mathsf{Mat}^{\mathsf{C}}} \times 1_{\boxtimes}) \xrightarrow{r^{\boxtimes}} 1_{\mathsf{Mat}^{\mathsf{C}}}$$

as follows.

Component 1-Cells: For each object $m \in \mathsf{Mat}^{\mathsf{C}}$, define

$$m = m \boxtimes 1 \xrightarrow{r^{\boxtimes}_m} (1_{\mathsf{Mat}^{\mathsf{C}}})\, m = m \tag{8.9.15}$$

as the identity matrix $\mathbb{1}^m \in \mathsf{Mat}^{\mathsf{C}}_{m,m}$ in (8.1.6).

Component 2-Cells: For each 1-cell $A = (A_{ji}) \in \mathsf{Mat}^{\mathsf{C}}_{m,n}$, define

$$r^{\boxtimes}_A \in \mathsf{Mat}^{\mathsf{C}}_{m,n}\Big((1_{\mathsf{Mat}^{\mathsf{C}}} A)\, r^{\boxtimes}_m ; r^{\boxtimes}_n \, (\boxtimes(1_{\mathsf{Mat}^{\mathsf{C}}} \times 1_{\boxtimes}))\,(A)\Big)$$

as the following vertical composite 2-cell.

(8.9.16)

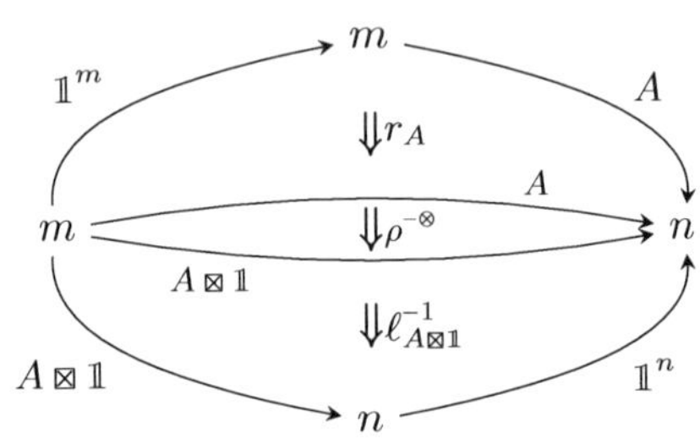

- $A \boxtimes \mathbb{1}$ is the matrix tensor product in (8.6.3) with

$$A \boxtimes \mathbb{1} = \big(A_{ji} \otimes \mathbb{1}\big) = (\boxtimes(1_{\mathsf{Mat}^{\mathsf{C}}} \times 1_{\boxtimes}))\,(A).$$

  Here and in what follows, the identity matrix $\mathbb{1}^1 \in \mathsf{Mat}^{\mathsf{C}}_{1,1}$ is abbreviated to $\mathbb{1}$.
- $\rho^{-\otimes}$ is the 2-cell with (j,i)-entry the structure morphism in C,

$$A_{ji} \xrightarrow{\rho^{-\otimes}_{A_{ji}}} A_{ji} \otimes \mathbb{1}$$

  for $1 \le i \le m$ and $1 \le j \le n$.

This finishes the definition of $r^{\boxtimes}$. ◇

Definition 8.9.17. Define the data of a lax transformation

$$1_{\mathsf{Mat}^{\mathsf{C}}} \xrightarrow{r^{\boxtimes\bullet}} \boxtimes(1_{\mathsf{Mat}^{\mathsf{C}}} \times 1_{\boxtimes})$$

as follows.

Component 1-Cells: For each object $m \in \mathsf{Mat}^{\mathsf{C}}$, define

$$m = (1_{\mathsf{Mat}^{\mathsf{C}}})\, m \xrightarrow{\;r^{\boxtimes\bullet}_m\;} m \boxtimes 1 = m \tag{8.9.18}$$

as the identity matrix $\mathbb{1}^m \in \mathsf{Mat}^{\mathsf{C}}_{m,m}$ in (8.1.6).

Component 2-Cells: For each 1-cell $A = (A_{ji}) \in \mathsf{Mat}^{\mathsf{C}}_{m,n}$, define

$$r^{\boxtimes\bullet}_A \in \mathsf{Mat}^{\mathsf{C}}_{m,n}\Big(\big(\boxtimes(1_{\mathsf{Mat}^{\mathsf{C}}} \times 1_{\boxtimes})\big)(A) r^{\boxtimes\bullet}_m ; r^{\boxtimes\bullet}_n (1_{\mathsf{Mat}^{\mathsf{C}}} A) \Big)$$

as the following vertical composite 2-cell.

(8.9.19)

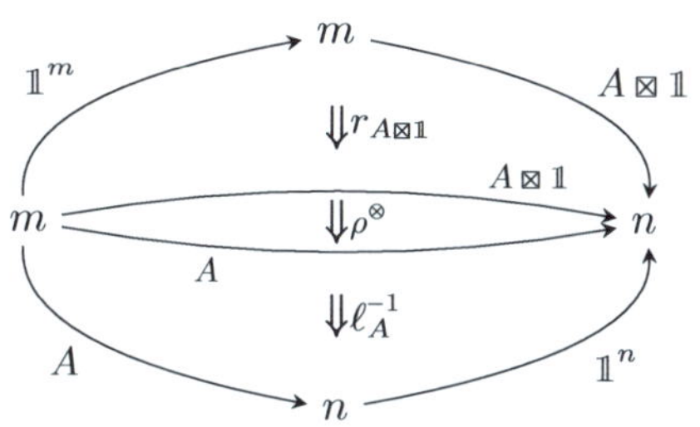

In this vertical composite, $\rho^{\otimes}$ is the 2-cell with (j,i)-entry the structure morphism in C,

$$A_{ji} \otimes \mathbb{1} \xrightarrow{\;\rho^{\otimes}_{A_{ji}}\;} A_{ji}$$

for $1 \le i \le m$ and $1 \le j \le n$.

This finishes the definition of $r^{\boxtimes\bullet}$. ◇

Definition 8.9.20. Define the data

$$1_{\boxtimes(1_{\mathsf{Mat}^{\mathsf{C}}} \times 1_{\boxtimes})} \xrightarrow{\;\eta^r\;} r^{\boxtimes\bullet} r^{\boxtimes} \qquad r^{\boxtimes} r^{\boxtimes\bullet} \xrightarrow{\;\varepsilon^r\;} 1_{1_{\mathsf{Mat}^{\mathsf{C}}}}$$

as consisting of the component 2-cells

$$\begin{array}{ccc} \big(1_{\boxtimes(1_{\mathsf{Mat}^{\mathsf{C}}} \times 1_{\boxtimes})}\big)_m = \mathbb{1}^m & & (r^{\boxtimes} r^{\boxtimes\bullet})_m = \mathbb{1}^m \mathbb{1}^m \\ \eta^r_m = \Big\downarrow \ell^{-1}_{\mathbb{1}^m} & & \varepsilon^r_m = \Big\downarrow \ell_{\mathbb{1}^m} \\ (r^{\boxtimes\bullet} r^{\boxtimes})_m = \mathbb{1}^m \mathbb{1}^m & & \big(1_{1_{\mathsf{Mat}^{\mathsf{C}}}}\big)_m = \mathbb{1}^m \end{array}$$

in $\mathsf{Mat}^{\mathsf{C}}_{m,m}$ for each object $m \in \mathsf{Mat}^{\mathsf{C}}$. ◇

Lemma 8.9.21. *The quadruple $(r^{\boxtimes}, r^{\boxtimes\bullet}, \eta^r, \varepsilon^r)$ in Definitions 8.9.14, 8.9.17, and 8.9.20 is an adjoint equivalence.*

Proof. The statement of the lemma means the following three assertions.

(1) $r^{\boxtimes}$ and $r^{\boxtimes\bullet}$ are strong transformations.
(2) η^r and ε^r are invertible modifications.
(3) The triangle identities (6.3.10) are satisfied.

The naturality of $r^{\boxtimes}_A$ and $r^{\boxtimes\bullet}_A$ with respect to A follows from the naturality of $\lambda^{\oplus}$, $\rho^{\oplus}$, $\lambda^{\otimes}$, $\rho^{\otimes}$, $\lambda^{\bullet}$, and $\rho^{\bullet}$ in C.

The rest of the proof is similar to those in Section 8.8, in particular Lemmas 8.8.5, 8.8.11, 8.8.17, 8.8.26, 8.8.39, and 8.8.45. As in Lemma 8.9.9, below we only state the equalities that one has to check.

Analogous to (8.8.12) and (8.9.10), the lax unity axiom (6.2.15) for $r^{\boxtimes}$ states the following pasting diagram equality in $\mathsf{Mat}^{\mathsf{C}}_{m,m}$ for $m \geq 0$.

(8.9.22)
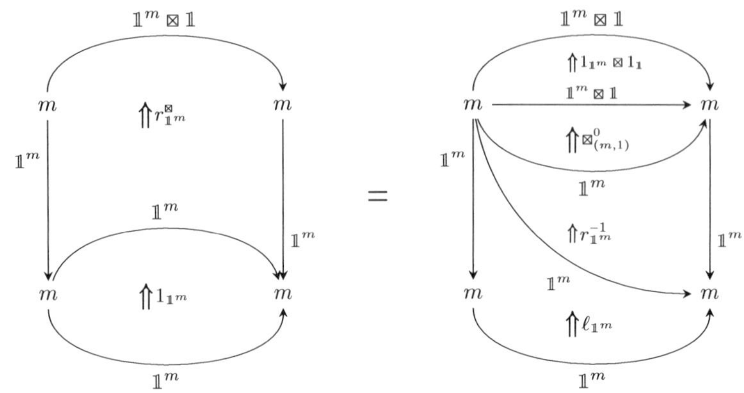

Consider (8.9.22).

- Each diagonal entry of $\mathbb{1}^m \boxtimes \mathbb{1}$ is $\mathbb{1} \otimes \mathbb{1}$, and each off-diagonal entry is $\mathbb{0} \otimes \mathbb{1}$.
- $\boxtimes^0_{(m,1)}$ is
  - $\lambda^{-\otimes}_{\mathbb{1}} : \mathbb{1} \longrightarrow \mathbb{1} \otimes \mathbb{1}$ in each diagonal entry and
  - $\lambda^{-\bullet}_{\mathbb{1}} : \mathbb{0} \longrightarrow \mathbb{0} \otimes \mathbb{1}$ in each off-diagonal entry.
- $1_{\mathbb{1}^m} \boxtimes 1_{\mathbb{1}}$ is entrywise an identity morphism in C.

The lax unity axiom for $r^{\boxtimes\bullet}$ is analogous to (8.9.22).

Analogous to (8.8.16) and (8.9.11), the lax naturality axiom (6.2.16) for $r^{\boxtimes}$ states the following pasting diagram equality in $\mathsf{Mat}^{\mathsf{C}}_{m,p}$ for 1-cells $(A,B) \in \mathsf{Mat}^{\mathsf{C}}_{m,n} \times \mathsf{Mat}^{\mathsf{C}}_{n,p}$.

(8.9.23)
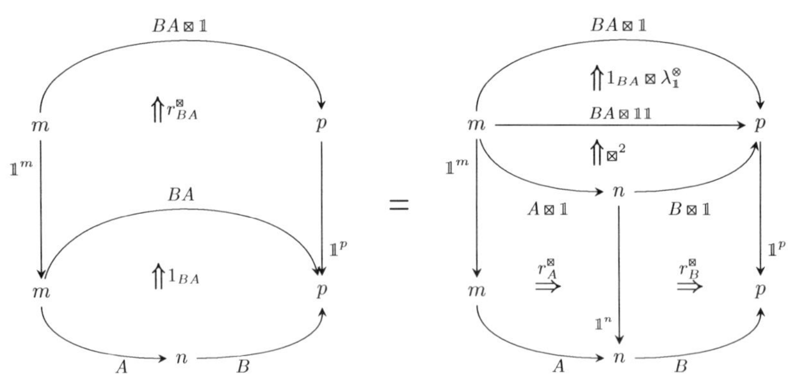

In (8.9.23), for $1 \leq i \leq m$ and $1 \leq k \leq p$, the (k,i)-entry of $BA \boxtimes \mathbb{1}\mathbb{1}$ is

$$\big(BA \boxtimes \mathbb{1}\mathbb{1}\big)_{ki} = (BA)_{ki} \otimes (\mathbb{1} \otimes \mathbb{1}) \in \mathsf{C}.$$

The lax naturality axiom for $r^{\boxtimes\bullet}$ is analogous to (8.9.23).

Analogous to (8.8.40) and (8.9.12), the modification axiom (6.3.2) for η^r states that for each 1-cell $A \in \mathsf{Mat}^{\mathsf{C}}_{m,n}$, the following pasting diagram equality holds.

(8.9.24)

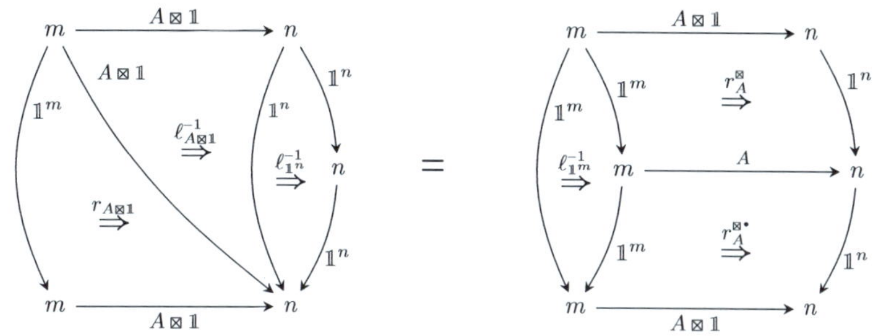

The modification axiom for ε^r is analogous to (8.9.24).

Analogous to (8.8.46) and (8.9.13), the left triangle identity in (6.3.10) for the data $(r^{\boxtimes}, r^{\boxtimes \bullet}, \eta^r, \varepsilon^r)$ states the commutativity of the left diagram below.

(8.9.25)

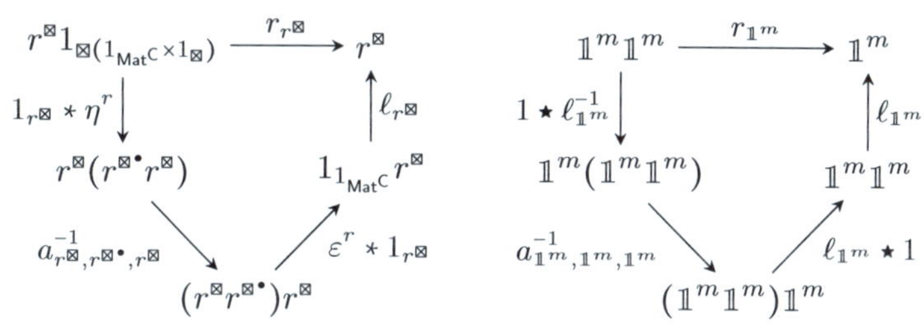

At each object $m \in \mathsf{Mat}^{\mathsf{C}}$, the left diagram in (8.9.25) yields the right diagram in $\mathsf{Mat}^{\mathsf{C}}_{m,m}$. The right triangle identity in (6.3.10) is analogous to (8.9.25). □

8.10. The Pentagonator

For a tight symmetric bimonoidal category C, we continue the construction of a monoidal bicategory structure, in the sense of Definition 6.4.1, on the matrix bicategory $\mathsf{Mat}^{\mathsf{C}}$ in Theorem 8.4.12. In this section, we construct the pentagonator in $\mathsf{Mat}^{\mathsf{C}}$.

Definition 8.10.1. Define π as consisting of the 2-cells

$$\pi_{m,n,p,q} \in \mathsf{Mat}^{\mathsf{C}}_{s,s} \quad \text{with} \quad m,n,p,q \geq 0 \quad \text{and} \quad s = mnpq$$

given by the composite of the following pasting diagram in $\mathsf{Mat}^{\mathsf{C}}_{s,s}$.

(8.10.2)

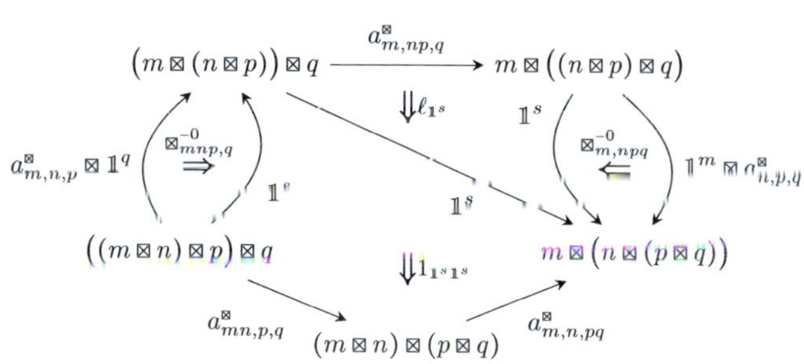

This finishes the definition of π. ◇

Explanation 8.10.3. Consider the pasting diagram in (8.10.2).

(1) By Definition 8.6.19, each vertex is the object $s \in \mathsf{Mat}^{\mathsf{C}}$.

(2) Each instance of $a^{\boxtimes}$ is a component 1-cell of the monoidal associator in (8.8.2). In particular,

$$a^{\boxtimes}_{m,n,p} = \mathbb{1}^{mnp}, \qquad a^{\boxtimes}_{n,p,q} = \mathbb{1}^{npq}, \quad \text{and}$$
$$a^{\boxtimes}_{mn,p,q} = a^{\boxtimes}_{m,np,q} = a^{\boxtimes}_{m,n,pq} = \mathbb{1}^{s}.$$

As in (8.6.10), each diagonal entry of the 1-cell $\mathbb{1}^{mnp} \boxtimes \mathbb{1}^{q}$ is $\mathbb{1} \otimes \mathbb{1}$, and each off-diagonal entry is $\mathbb{1} \otimes \mathbb{0}$, $\mathbb{0} \otimes \mathbb{1}$, or $\mathbb{0} \otimes \mathbb{0}$. A similar description applies to the 1-cell $\mathbb{1}^{m} \boxtimes \mathbb{1}^{npq}$.

(3) By Convention 6.2.12, the domain 1-cell is the horizontal composite

$$\Big(\big(\mathbb{1}^{m} \boxtimes \mathbb{1}^{npq}\big)\mathbb{1}^{s}\Big)\big(\mathbb{1}^{mnp} \boxtimes \mathbb{1}^{q}\big).$$

The codomain 1-cell is $\mathbb{1}^{s}\mathbb{1}^{s}$.

(4) Each instance of $\boxtimes^{0}$ is a component 2-cell of the lax unity constraint of the monoidal composition in (8.6.9), with $\boxtimes^{-0} = (\boxtimes^{0})^{-1}$. If $s > 0$, then each entry of each $\boxtimes^{-0}$ is
   - $\lambda^{\otimes}_{\mathbb{1}} \colon \mathbb{1} \otimes \mathbb{1} \longrightarrow \mathbb{1}$ for a diagonal entry and
   - $\rho^{\bullet}_{\mathbb{1}} \colon \mathbb{1} \otimes \mathbb{0} \longrightarrow \mathbb{0}$, $\lambda^{\bullet}_{\mathbb{1}} \colon \mathbb{0} \otimes \mathbb{1} \longrightarrow \mathbb{0}$, or $\lambda^{\bullet}_{\mathbb{0}} \colon \mathbb{0} \otimes \mathbb{0} \longrightarrow \mathbb{0}$ for an off-diagonal entry.

(5) $\ell_{\mathbb{1}^{s}} \colon \mathbb{1}^{s}\mathbb{1}^{s} \longrightarrow \mathbb{1}^{s}$ is the base left unitor in (8.2.2).

(6) $1_{\mathbb{1}^{s}\mathbb{1}^{s}}$ is the identity 2-cell of $\mathbb{1}^{s}\mathbb{1}^{s}$.

(7) Each entry of each component 2-cell of π is a composite of morphisms in C, each being an iterated sum and product of identity morphisms and one component of $\lambda^{\oplus}$, $\rho^{\oplus}$, $\lambda^{\otimes}$, $\lambda^{\bullet}$, or $\rho^{\bullet}$.

Moreover, each $\pi_{m,n,p,q} \in \mathsf{Mat}^{\mathsf{C}}_{s,s}$ is a well-defined 2-cell by the Bicategorical Pasting Theorem in [**JY21**, 3.6.6]. It states that each pasting diagram in a bicategory has a unique composite, once a bracketing is chosen for the (co)domain composite 1-cell. ◇

Lemma 8.10.4. *In Definition 8.10.1, π is an invertible modification.*

Proof. Each component 2-cell of π is a composite of invertible 2-cells by (8.2.2) and (8.6.9). To verify the modification axiom (6.3.2) for π, consider arbitrary 1-cells

$$m \xrightarrow{B^1} m' \qquad n \xrightarrow{B^2} n' \qquad p \xrightarrow{B^3} p' \qquad q \xrightarrow{B^4} q'$$

in $\mathsf{Mat}^{\mathsf{C}}$ with $s = mnpq$ and $s' = m'n'p'q'$. If either $s = 0$ or $s' = 0$, then $\mathsf{Mat}^{\mathsf{C}}_{s,s'}$ is the terminal category, and the modification axiom holds automatically.

Suppose $s, s' > 0$. We use the abbreviations

$$B^{12} = B^{1} \boxtimes B^{2} \qquad B^{2(34)} = B^{2} \boxtimes (B^{3} \boxtimes B^{4})$$
$$B^{[(12)3]4} = \big[(B^{1} \boxtimes B^{2}) \boxtimes B^{3}\big] \boxtimes B^{4} \qquad B^{(12)(34)} = (B^{1} \boxtimes B^{2}) \boxtimes (B^{3} \boxtimes B^{4})$$
$$a^{\boxtimes}_{2,3,4} = a^{\boxtimes}_{B^2,B^3,B^4} \qquad a^{\boxtimes}_{1,23,4} = a^{\boxtimes}_{B^1,B^2\boxtimes B^3,B^4}$$
$$\boxtimes^{-2} = (\boxtimes^{2})^{-1}$$

and so forth. Using (6.2.18), (8.6.20), and (8.8.3), the modification axiom (6.3.2) for the 1-cells (B^1, B^2, B^3, B^4) states the following pasting diagram equality in $\mathsf{Mat}^{\mathsf{C}}_{s,s'}$.

(8.10.5)

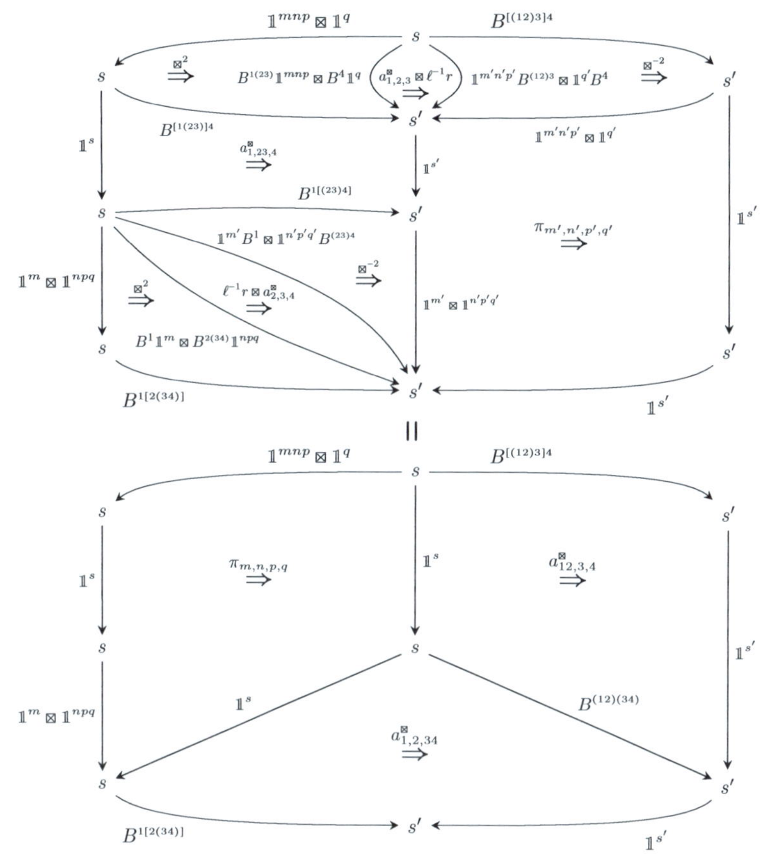

Consider (8.10.5).

- The common domain 1-cell is

$$B^{1[2(34)]}\Big[\big((\mathbb{1}^{m} \boxtimes \mathbb{1}^{npq})\mathbb{1}^{s}\big)\big(\mathbb{1}^{mnp} \boxtimes \mathbb{1}^{q}\big)\Big]. \tag{8.10.6}$$

  The common codomain 1-cell is

$$\big(\mathbb{1}^{s'}\mathbb{1}^{s'}\big)B^{[(12)3]4}. \tag{8.10.7}$$

- In the bottom pasting diagram, the following statements hold.
  - There is one instance of a^{-1} after $1_{B^{1[2(34)]}} * \pi_{m,n,p,q}$, with a the base associator in (8.3.2).
  - There is one instance of a after $a^{\boxtimes}_{1,2,34} * 1_{\mathbb{1}^s}$.
  - There is one instance of a^{-1} after $1_{\mathbb{1}^{s'}} * a^{\boxtimes}_{12,3,4}$.

  These a and a^{-1} are not explicitly displayed in the pasting diagram.
- In the top pasting diagram, the following statements hold.
  - There are multiple instances of the base associator and its inverse that are not explicitly displayed.
  - ℓ is the base left unitor in (8.2.2), and r is the base right unitor in (8.2.8).

It suffices to prove the equality (8.10.5) in each entry. As in the proofs in Section 8.8, in each entry, each of the two composite 2-cells in (8.10.5) is a composite of morphisms in C. The resulting diagram is the image in C of some diagram in $\mathsf{Gr}(X)$. The main subtlety comes from

- $\boxtimes^2$, which contains δ^{-r} and δ^{-l} by Lemma 8.6.16, and
- the base associator $a = a^4a^3a^2a^1$ in (8.3.5), which contains δ^r (8.3.6) in a^1 and δ^{-l} (8.3.8) in a^4.

Since the elementary edges δ^l and δ^r do not have formal inverses as in Definition 3.1.6, only the *inverses* of $\boxtimes^2$ and a^4 are represented as paths in $\mathsf{Gr}(X)$, creating a zigzag. Before the Coherence Theorem 3.9.1 can be applied, the diagram must first be subdivided such that each subdiagram consists of two parallel paths. This subdivision process is analogous to those in the proofs of Lemmas 8.4.9 and 8.8.26. The detailed argument is given below.

Consider the following indices and abbreviations.

$$\begin{array}{llll} 1 \le i \le m & 1 \le j \le n & 1 \le k \le p & 1 \le l \le q \\ 1 \le i' \le m' & 1 \le j' \le n' & 1 \le k' \le p' & 1 \le l' \le q' \end{array}$$
$$I = (i-1)npq + (j-1)pq + (k-1)q + l$$
$$I' = (i'-1)n'p'q' + (j'-1)p'q' + (k'-1)q' + l'$$

In the (I', I)-entry, the two pasting diagrams in (8.10.5) yield the outer diagram in C below, which we will explain in detail.

(8.10.8)

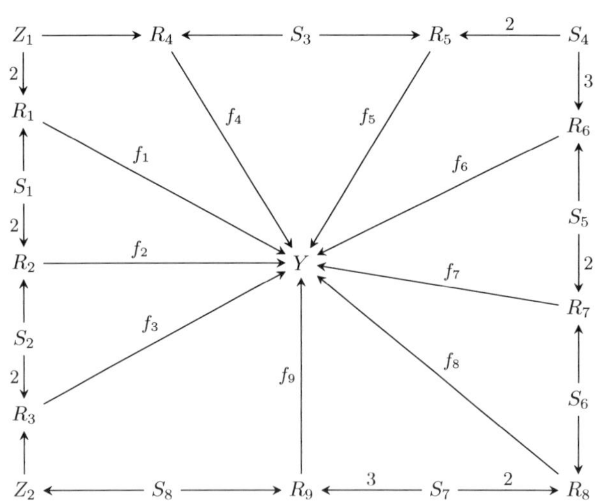

Consider the diagram (8.10.8).

- Z_1 is the (I', I)-entry of the common domain (8.10.6).
- Z_2 is the (I', I)-entry of the common codomain (8.10.7).
- Each arrow is an isomorphism built from the structure morphisms in C. The number 2 or 3 decorating an arrow means that it is a composite of that many morphisms.
- With the appropriate inverses taken into account, the left zigzag from Z_1 to Z_2 is the (I', I)-entry of the bottom pasting diagram in (8.10.5).

- The top-right-bottom zigzag, from Z_1 to S_4 to R_8 to Z_2, is the (I', I)-entry of the top pasting diagram in (8.10.5).
- Each R_i for $1 \le i \le 9$ is the codomain of an instance of $a^{-4} = (a^4)^{-1}$ (8.3.5), or of a morphism induced by a^{-4}.
- The object Y is the (I', I)-entry of $B^{[(12)3]4}$, that is,

$$Y = \big[(B^1_{i'i} \otimes B^2_{j'j}) \otimes B^3_{k'k}\big] \otimes B^4_{l'l}. \tag{8.10.9}$$

Next we describe explicitly the objects and the morphisms in (8.10.8). Then we will explain how to realize the diagram using paths in $\mathsf{Gr}(X)$ such that, by Theorem 3.9.1, each of the 9 subdiagrams is commutative in C. We will use the following abbreviations.

$$\begin{aligned} \mathbb{1}^{m,npq} &= \mathbb{1}^m \boxtimes \mathbb{1}^{npq} & \qquad \mathbb{1}^{m',n'p'q'} &= \mathbb{1}^{m'} \boxtimes \mathbb{1}^{n'p'q'} \\ \mathbb{1}^{mnp,q} &= \mathbb{1}^{mnp} \boxtimes \mathbb{1}^q & \qquad \mathbb{1}^{m'n'p',q'} &= \mathbb{1}^{m'n'p'} \boxtimes \mathbb{1}^{q'} \end{aligned} \tag{8.10.10}$$

Since we have specified the (I', I)-entry in (8.10.5), to save space, we will omit the subscript (I', I) in most objects and also the subscripts in π and $a^{\boxtimes}$. In displaying the objects R_i, we abbreviate $\otimes$ to concatenation. In the following diagrams, $\star$ is the matrix product of morphisms in (8.1.5), and $a = a^4a^3a^2a^1$ refers to an entry of the base associator in (8.3.5). As mentioned above, we will use $\boxtimes^{-2} = (\boxtimes^2)^{-1}$ and $a^{-4} = (a^4)^{-1}$.

The left zigzag in (8.10.8) from Z_1 to Z_2 is the following zigzag in C.

$$\begin{array}{ccc}
R_1 = \left[\bigoplus_{t=1}^{s}\left[\bigoplus_{u=1}^{s} B^{1[2(34)]}_{I't}\big(\mathbb{1}^s_{tu}\mathbb{1}^s_{uI}\big)\right]_{\mathrm{lt}}\right]_{\mathrm{lt}} & \xleftarrow{\;a^{-4}\;} & B^{1[2(34)]}(\mathbb{1}^s\mathbb{1}^s) \\
\uparrow a^3a^2a^1 & & \uparrow 1\star\pi \\
S_1 = \big(B^{1[2(34)]}\mathbb{1}^s\big)\mathbb{1}^s & & B^{1[2(34)]}\big[(\mathbb{1}^{m,npq}\mathbb{1}^s)\mathbb{1}^{mnp,q}\big] = Z_1 \\
\downarrow a^{\boxtimes}\star 1 & & \\
\big(\mathbb{1}^{s'}B^{(12)(34)}\big)\mathbb{1}^s & & (\mathbb{1}^{s'}\mathbb{1}^{s'})B^{[(12)3]4} = Z_2 \\
\downarrow a^3a^2a^1 & & \downarrow a^3a^2a^1 \\
R_2 = \left[\bigoplus_{t'=1}^{s'}\left[\bigoplus_{u=1}^{s} \mathbb{1}^{s'}_{I't'}\big(B^{(12)(34)}_{t'u}\mathbb{1}^s_{uI}\big)\right]_{\mathrm{lt}}\right]_{\mathrm{lt}} & & \left[\bigoplus_{t'=1}^{s'}\left[\bigoplus_{u'=1}^{s'} \mathbb{1}^{s'}_{I't'}\big(\mathbb{1}^{s'}_{t'u'}B^{[(12)3]4}_{u'I}\big)\right]_{\mathrm{lt}}\right]_{\mathrm{lt}} = R_3 \\
\uparrow a^{-4} & & \uparrow a^{-4} \\
S_2 = \mathbb{1}^{s'}\big(B^{(12)(34)}\mathbb{1}^s\big) & \xrightarrow{\;1\star a^{\boxtimes}\;} & \mathbb{1}^{s'}\big(\mathbb{1}^{s'}B^{[(12)3]4}\big)
\end{array} \tag{8.10.11}$$

As discussed in the bullet point following (8.10.7), the zigzag (8.10.11) is equivalent to the (I', I)-entry of the bottom pasting diagram in (8.10.5).

The top zigzag in (8.10.8) from Z_1 to S_4 is the following zigzag in C.

(8.10.12)

$$R_5 = \Big[\bigoplus_{v=1}^{s}\Big\{\bigoplus_{t=1}^{s}\Big[\bigoplus_{u=1}^{s} B^{1[2(34)]}_{l't}\big(\mathbb{1}^{m,npq}_{tu}\mathbb{1}^{s}_{uv}\big)\Big]_{\mathrm{lt}}\Big\}_{\mathrm{lt}}\mathbb{1}^{mnp,q}_{vl}\Big]_{\mathrm{lt}}$$

$$S_3 = \big[B^{1[2(34)]}(\mathbb{1}^{m,npq}\mathbb{1}^{s})\big]\mathbb{1}^{mnp,q} \xrightarrow{[\oplus_v a^{-4}1]_{\mathrm{lt}}} R_5 \xleftarrow{[\oplus_v (a^3a^2a^1)1]_{\mathrm{lt}}} \big[(B^{1[2(34)]}\mathbb{1}^{m,npq})\mathbb{1}^{s}\big]\mathbb{1}^{mnp,q}$$

$$S_3 \xrightarrow{a^3a^2a^1} R_4 = \Big[\bigoplus_{t=1}^{s}\Big[\bigoplus_{u=1}^{s} B^{1[2(34)]}_{l't}\big((\mathbb{1}^{m,npq}\mathbb{1}^{s})_{tu}\mathbb{1}^{mnp,q}_{ul}\big)\Big]_{\mathrm{lt}}\Big]_{\mathrm{lt}} \xleftarrow{a^{-4}} Z_1 = B^{1[2(34)]}\big[(\mathbb{1}^{m,npq}\mathbb{1}^{s})\mathbb{1}^{mnp,q}\big]$$

$$S_4 = \big[(B^1\mathbb{1}^m \boxtimes B^{2(34)}\mathbb{1}^{npq})\mathbb{1}^{s}\big]\mathbb{1}^{mnp,q} \xrightarrow{[\boxtimes^{-2}\star 1]\star 1} \big[(B^{1[2(34)]}\mathbb{1}^{m,npq})\mathbb{1}^{s}\big]\mathbb{1}^{mnp,q}$$

In (8.10.12), $\boxtimes^{-2}$ comes from the lower left $\boxtimes^2$ in the top pasting diagram in (8.10.5). The instances of a^{-1} before $\boxtimes^{-2}$ are needed because of the bracketing of the domain 1-cell (8.10.6).

The right zigzag in (8.10.8) from S_4 to R_8 is the following zigzag in C.

(8.10.13)

$$S_4 = \big[(B^1\mathbb{1}^m \boxtimes B^{2(34)}\mathbb{1}^{npq})\mathbb{1}^{s}\big]\mathbb{1}^{mnp,q} \xrightarrow{[(\ell^{-1}r\boxtimes a^{\boxtimes})\star 1]\star 1} \big[(\mathbb{1}^{m'}B^1 \boxtimes \mathbb{1}^{n'p'q'}B^{(23)4})\mathbb{1}^{s}\big]\mathbb{1}^{mnp,q}$$

$$\xrightarrow{[\boxtimes^{-2}\star 1]\star 1} \big[(\mathbb{1}^{m',n'p'q'}B^{1[(23)4]})\mathbb{1}^{s}\big]\mathbb{1}^{mnp,q} \xrightarrow{[\oplus_v (a^3a^2a^1)1]_{\mathrm{lt}}} R_6$$

$$R_6 = \Big[\bigoplus_{v=1}^{s}\Big\{\bigoplus_{t'=1}^{s'}\Big[\bigoplus_{t=1}^{s} \mathbb{1}^{m',n'p'q'}_{l't'}\big(B^{1[(23)4]}_{t't}\mathbb{1}^{s}_{tv}\big)\Big]_{\mathrm{lt}}\Big\}_{\mathrm{lt}}\mathbb{1}^{mnp,q}_{vl}\Big]_{\mathrm{lt}}$$

$$R_6 \xleftarrow{[\oplus_v a^{-4}1]_{\mathrm{lt}}} \big[\mathbb{1}^{m',n'p'q'}(B^{1[(23)4]}\mathbb{1}^{s})\big]\mathbb{1}^{mnp,q} = S_5$$

$$S_5 \xrightarrow{[1\star a^{\boxtimes}]\star 1} \big[\mathbb{1}^{m',n'p'q'}(\mathbb{1}^{s'}B^{[1(23)]4})\big]\mathbb{1}^{mnp,q} \xrightarrow{[\oplus_v a^{-4}1]_{\mathrm{lt}}} R_7$$

$$R_7 = \Big[\bigoplus_{v=1}^{s}\Big\{\bigoplus_{t'=1}^{s'}\Big[\bigoplus_{u'=1}^{s'} \mathbb{1}^{m',n'p'q'}_{l't'}\big(\mathbb{1}^{s'}_{t'u'}B^{[1(23)]4}_{u'v}\big)\Big]_{\mathrm{lt}}\Big\}_{\mathrm{lt}}\mathbb{1}^{mnp,q}_{vl}\Big]_{\mathrm{lt}}$$

$$S_6 = \big[(\mathbb{1}^{m',n'p'q'}\mathbb{1}^{s'})B^{[1(23)]4}\big]\mathbb{1}^{mnp,q} \xrightarrow{[\oplus_v (a^3a^2a^1)1]_{\mathrm{lt}}} R_7$$

$$S_6 \xrightarrow{a^3a^2a^1} R_8 = \Big[\bigoplus_{t'=1}^{s'}\Big[\bigoplus_{t=1}^{s} \big(\mathbb{1}^{m',n'p'q'}\mathbb{1}^{s'}\big)_{l't'}\big(B^{[1(23)]4}_{t't}\mathbb{1}^{mnp,q}_{tl}\big)\Big]_{\mathrm{lt}}\Big]_{\mathrm{lt}}$$

In (8.10.13), $\ell^{-1}r\boxtimes a^{\boxtimes}$, $\boxtimes^{-2}$, and $a^{\boxtimes}$ are from the left side of the top pasting diagram in (8.10.5).

The bottom zigzag in (8.10.8) from R_8 to Z_2 is the following zigzag in C.

(8.10.14)

$$\begin{array}{ccc}
Z_2 = (\mathbb{1}^{s'}\mathbb{1}^{s'})B^{[(12)3]4} & & \left[\bigoplus_{t'=1}^{s'}\left[\bigoplus_{t=1}^{s}(\mathbb{1}^{m',n'p'q'}\mathbb{1}^{s'})_{I't'}(B^{[1(23)]4}_{t't}\mathbb{1}^{mnp,q}_{tI})\right]_{\mathrm{lt}}\right]_{\mathrm{lt}} = R_8 \\
\uparrow{\scriptstyle \pi\star 1} & & \uparrow{\scriptstyle a^{-4}} \\
S_8 = \left[(\mathbb{1}^{m',n'p'q'}\mathbb{1}^{s'})\mathbb{1}^{m'n'p',q'}\right]B^{[(12)3]4} & & (\mathbb{1}^{m',n'p'q'}\mathbb{1}^{s'})(B^{[1(23)]4}\mathbb{1}^{mnp,q}) \\
\downarrow{\scriptstyle a^3a^2a^1} & & \uparrow{\scriptstyle 1\star\boxtimes^{-2}} \\
& & (\mathbb{1}^{m',n'p'q'}\mathbb{1}^{s'})(B^{1(23)}\mathbb{1}^{mnp}\boxtimes B^4\mathbb{1}^q) = S_7 \\
R_9 = \left[\bigoplus_{t'=1}^{s'}\left[\bigoplus_{u'=1}^{s'}(\mathbb{1}^{m',n'p'q'}\mathbb{1}^{s'})_{I't'}(\mathbb{1}^{m'n'p',q'}_{t'u'}B^{[(12)3]4}_{u'I})\right]_{\mathrm{lt}}\right]_{\mathrm{lt}} & & \downarrow{\scriptstyle 1\star(a^{\boxtimes}\boxtimes\ell^{-1}r)} \\
\uparrow{\scriptstyle a^{-4}} & & \\
(\mathbb{1}^{m',n'p'q'}\mathbb{1}^{s'})(\mathbb{1}^{m'n'p',q'}B^{[(12)3]4}) & \xleftarrow{1\star\boxtimes^{-2}} & (\mathbb{1}^{m',n'p'q'}\mathbb{1}^{s'})(\mathbb{1}^{m'n'p'}B^{(12)3}\boxtimes\mathbb{1}^{q'}B^4)
\end{array}$$

In (8.10.14), from R_8 to Z_2, the 2-cells $\boxtimes^{-2}$, $a^{\boxtimes}\boxtimes\ell^{-1}r$, $\boxtimes^{-2}$, and π are from the top and the right parts of the top pasting diagram in (8.10.5).

The morphisms $f_i\colon R_i \longrightarrow Y$ for $1 \le i \le 9$ in (8.10.8) are defined as follows. To define the morphism

$$R_2 = \left[\bigoplus_{t'=1}^{s'}\left[\bigoplus_{u=1}^{s}\mathbb{1}^{s'}_{I't'}(B^{(12)(34)}_{t'u}\mathbb{1}^{s}_{uI})\right]_{\mathrm{lt}}\right]_{\mathrm{lt}} \xrightarrow{f_2} \left[(B^1_{i'i}B^2_{j'j})B^3_{k'k}\right]B^4_{l'l} = Y,$$

first observe that the summands in R_2 are as follows.

$$(8.10.15)\qquad \mathbb{1}^{s'}_{I't'}(B^{(12)(34)}_{t'u}\mathbb{1}^{s}_{uI}) = \begin{cases} \mathbb{0}(B^{(12)(34)}_{t'u}\mathbb{0}) & \text{if } t' \neq I' \text{ and } u \neq I. \\ \mathbb{1}(B^{(12)(34)}_{t'u}\mathbb{0}) & \text{if } t' = I' \text{ and } u \neq I. \\ \mathbb{0}(B^{(12)(34)}_{t'u}\mathbb{1}) & \text{if } t' \neq I' \text{ and } u = I. \\ \mathbb{1}\big([(B^1_{i'i}B^2_{j'j})(B^3_{k'k}B^4_{l'l})]\mathbb{1}\big) & \text{if } t' = I' \text{ and } u = I. \end{cases}$$

With the notation in (8.7.6) for morphisms, f_2 is defined as the following composite.

$$\begin{array}{ccc}
R_2 & \xrightarrow{\quad f_2 \quad} & \left[(B^1_{i'i}B^2_{j'j})B^3_{k'k}\right]B^4_{l'l} = Y \\
\Big\downarrow & & \uparrow{\scriptstyle \alpha^{-\otimes}} \\
 & \xrightarrow{(\lambda^{\oplus},\rho^{\oplus},\lambda^{\otimes},\rho^{\otimes},\lambda^{\bullet},\rho^{\bullet})} & (B^1_{i'i}B^2_{j'j})(B^3_{k'k}B^4_{l'l})
\end{array}$$

In the morphism f_2, the following statements hold.

- $(\lambda^{\oplus},\rho^{\oplus},\lambda^{\otimes},\rho^{\otimes},\lambda^{\bullet},\rho^{\bullet})$ is a composite of morphisms, each being an iterated sum and product of identity morphisms and one component of one of the indicated structure morphisms in C.
- $\lambda^{\bullet}$ and $\rho^{\bullet}$ are applied in the first three cases of (8.10.15), when at least one of $\mathbb{1}^{s'}_{I't'}$ and $\mathbb{1}^{s}_{uI}$ is $\mathbb{0}$, to reduce that summand to $\mathbb{0}$.
- $\lambda^{\otimes}$ and $\rho^{\otimes}$ are applied in the last case of (8.10.15), when both $\mathbb{1}^{s'}_{I't'}$ and $\mathbb{1}^{s}_{uI}$ are $\mathbb{1}$, to reduce that summand to $(B^1_{i'i}B^2_{j'j})(B^3_{k'k}B^4_{l'l})$.
- $\lambda^{\oplus}$ and $\rho^{\oplus}$ are applied to additively remove all the $\mathbb{0}$'s.
- $\alpha^{-\otimes}$ moves the multiplicative brackets to match with those in Y.

The other 8 morphisms $f_i: R_i \longrightarrow Y$ are defined in almost the same way as f_2. Each such f_i is a composite consisting of

- a morphism of the form $(\lambda^{\oplus}, \rho^{\oplus}, \lambda^{\otimes}, \rho^{\otimes}, \lambda^{\bullet}, \rho^{\bullet})$ that reduces R_i to $B^{1234}_{I'I}$ and
- a morphism of the form

$$(\alpha^{-\otimes}): B^{1234}_{I'I} \longrightarrow Y$$

that moves the multiplicative brackets to match with those in Y.

Here the superscript in $B^{1234}_{I'I}$ has the bracketing as it appears in R_i. For example, $B^{1234}_{I'I}$ is $B^{[(12)3]4}_{I'I} = Y$ in R_3 and R_9. This finishes the description of the diagram (8.10.8).

To realize the diagram (8.10.8) using paths, we use the set of formal variables

$$X = \left\{0^X, 1^X, b^1_{c'c}, b^2_{d'd}, b^3_{e'e}, b^4_{g'g}\right\} \tag{8.10.16}$$

with the indices as follows.

$$\begin{array}{cccc} 1 \le c \le m & 1 \le d \le n & 1 \le e \le p & 1 \le g \le q \\ 1 \le c' \le m' & 1 \le d' \le n' & 1 \le e' \le p' & 1 \le g' \le q' \end{array}$$

Also define the function $\varphi: X \longrightarrow \mathsf{Ob(C)}$ as follows.

$$\varphi(x) = \begin{cases} \mathbb{0} & \text{if } x = 0^X. \\ \mathbb{1} & \text{if } x = 1^X. \\ B^1_{c'c} & \text{if } x = b^1_{c'c}. \\ B^2_{d'd} & \text{if } x = b^2_{d'd}. \\ B^3_{e'e} & \text{if } x = b^3_{e'e}. \\ B^4_{g'g} & \text{if } x = b^4_{g'g}. \end{cases}$$

To define the desired paths in $\mathsf{Gr}(X)$, in each of (8.10.11), (8.10.12), (8.10.13), and (8.10.14), we perform the following two steps.

(1) In each object, replace
  - $\left\{B^1_{c'c}, B^2_{d'd}, B^3_{e'e}, B^4_{g'g}\right\}$ with the variables $\left\{b^1_{c'c}, b^2_{d'd}, b^3_{e'e}, b^4_{g'g}\right\}$ and
  - $\{\mathbb{0}, \mathbb{1}\}$ in each identity matrix with the variables $\{0^X, 1^X\}$.

(2) Replace each morphism by a corresponding path in $\mathsf{Gr}(X)$ using
  - (8.3.12) for (a^1, a^2, a^3, a^{-4}),
  - Examples 8.2.5 and 8.2.11 for ℓ, r, and $a^{\boxtimes} = \ell^{-1}\alpha^{-\otimes}r$ (8.8.3),
  - Lemma 8.6.16 for $\boxtimes^{-2}$, and
  - Example 8.2.5 and (8.6.9) for π in (8.10.2).

Moreover, perform the same procedure on the object Y to obtain the element

$$y = \left[\left(b^1_{i'i} \otimes b^2_{j'j}\right) \otimes b^3_{k'k}\right] \otimes b^4_{l'l} \in X^{\mathsf{fr}}$$

and on the morphisms f_i for $1 \le i \le 9$ to obtain 9 corresponding paths in $\mathsf{Gr}(X)$.

After the procedure in the previous paragraph, the resulting diagram D in $\mathsf{Gr}(X)$ has precisely the same shape as (8.10.8), with each edge replaced by a path in $\mathsf{Gr}(X)$. Moreover, the image of D in C is the diagram (8.10.8). The vertex y is regular in the sense of Definition 3.1.25. By Lemma 3.1.29, each vertex in D is regular. Since C is assumed to be tight, the Coherence Theorem 3.9.1 implies that the image of D in C, that is, the diagram (8.10.8), is commutative. This finishes the proof of the equality (8.10.5). $\square$

8.11. The 2-Unitors

For a tight symmetric bimonoidal category C as in Definition 2.1.2, we are in the process of constructing a monoidal bicategory structure as in Definition 6.4.1 on the matrix bicategory $\mathsf{Mat}^{\mathsf{C}}$. In Section 8.10, we constructed the pentagonator π in $\mathsf{Mat}^{\mathsf{C}}$. In this section, we construct the 2-unitors in $\mathsf{Mat}^{\mathsf{C}}$.

- The middle 2-unitor μ is in Definition 8.11.1.
- The left 2-unitor $\lambda^{\boxtimes}$ is in Definition 8.11.6.
- The right 2-unitor $\rho^{\boxtimes}$ is in Definition 8.11.11.

The left and the right 2-unitors are decorated with $\boxtimes$ to avoid confusion with the structure morphisms in the tight symmetric bimonoidal category C.

The Middle 2-Unitor.

Definition 8.11.1. Define μ as consisting of the 2-cells

$$\mu_{m,n} \in \mathsf{Mat}^{\mathsf{C}}_{mn,mn} \quad \text{with} \quad m,n \geq 0$$

given by the composite of the following pasting diagram in $\mathsf{Mat}^{\mathsf{C}}_{mn,mn}$.

(8.11.2)

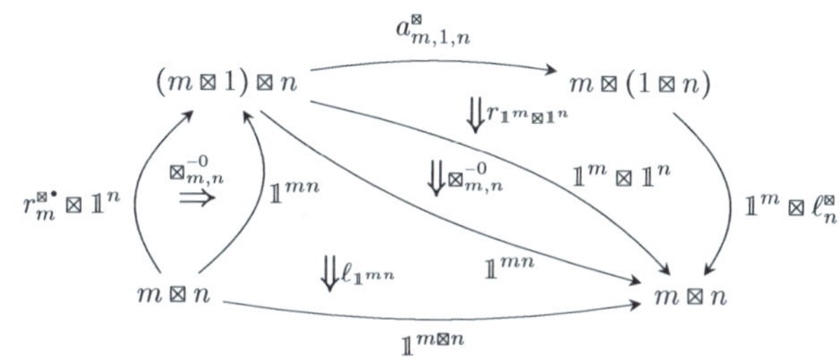

This finishes the definition of μ. $\diamond$

Explanation 8.11.3. Consider the pasting diagram in (8.11.2).

(1) By Definition 8.6.19, each vertex is the object $mn \in \mathsf{Mat}^{\mathsf{C}}$.

(2) By Convention 6.2.12, (8.8.2), (8.9.2), and (8.9.18), the domain 1-cell is the horizontal composite

$$\Big((\mathbb{1}^m \boxtimes \ell^{\boxtimes}_n)a^{\boxtimes}_{m,1,n}\Big)\big(r^{\boxtimes\bullet}_m \boxtimes \mathbb{1}^n\big) = \Big((\mathbb{1}^m \boxtimes \mathbb{1}^n)\mathbb{1}^{mn}\Big)\big(\mathbb{1}^m \boxtimes \mathbb{1}^n\big).$$

The codomain 1-cell is $\mathbb{1}^{m \boxtimes n} = \mathbb{1}^{mn}$.

(3) $\boxtimes^0$ is as in (8.6.9), and $\boxtimes^{-0} = (\boxtimes^0)^{-1}$.

(4) $r_{\mathbb{1}^m \boxtimes \mathbb{1}^n} \colon (\mathbb{1}^m \boxtimes \mathbb{1}^n)\mathbb{1}^{mn} \longrightarrow \mathbb{1}^m \boxtimes \mathbb{1}^n$ is the base right unitor in (8.2.8).

(5) $\ell_{\mathbb{1}^{mn}} \colon \mathbb{1}^{mn}\mathbb{1}^{mn} \longrightarrow \mathbb{1}^{mn}$ is the base left unitor in (8.2.2).

(6) Each entry of $\mu_{m,n}$ is a composite of morphisms in C, each being an iterated sum and product of identity morphisms and one component of $\lambda^{\oplus}$, $\rho^{\oplus}$, $\lambda^{\otimes}$, $\rho^{\otimes}$, $\lambda^{\bullet}$, or $\rho^{\bullet}$. $\diamond$

Lemma 8.11.4. *In Definition 8.11.1, μ is an invertible modification.*

Proof. Each component 2-cell $\mu_{m,n} \in \mathsf{Mat}^{\mathsf{C}}_{mn,mn}$ is well defined by the Bicategorical Pasting Theorem in [**JY21**, 3.6.6]. It is a composite of invertible 2-cells by (8.2.2), (8.2.8), and (8.6.9).

To verify the modification axiom (6.3.2) for μ, consider arbitrary 1-cells

$$(A,B) \in \mathsf{Mat}^{\mathsf{C}}_{m,m'} \times \mathsf{Mat}^{\mathsf{C}}_{n,n'}.$$

If one of m, m', n, and n' is 0, then $\mathsf{Mat}^{\mathsf{C}}_{mn,m'n'}$ is the terminal category, and the modification axiom holds automatically.

Suppose $m, m', n, n' > 0$. We use the abbreviation

$$\mathbb{1}^{m,n} = \mathbb{1}^m \boxtimes \mathbb{1}^n$$

in (8.10.10), $\boxtimes^2$ (8.6.20), $a^{\boxtimes}$ (8.8.3), $\ell^{\boxtimes}$ (8.9.3), and $r^{\boxtimes\bullet}$ (8.9.19). The modification axiom (6.3.2) for μ for the 1-cells (A,B) states the following pasting diagram equality in $\mathsf{Mat}^{\mathsf{C}}_{mn,m'n'}$.

(8.11.5)

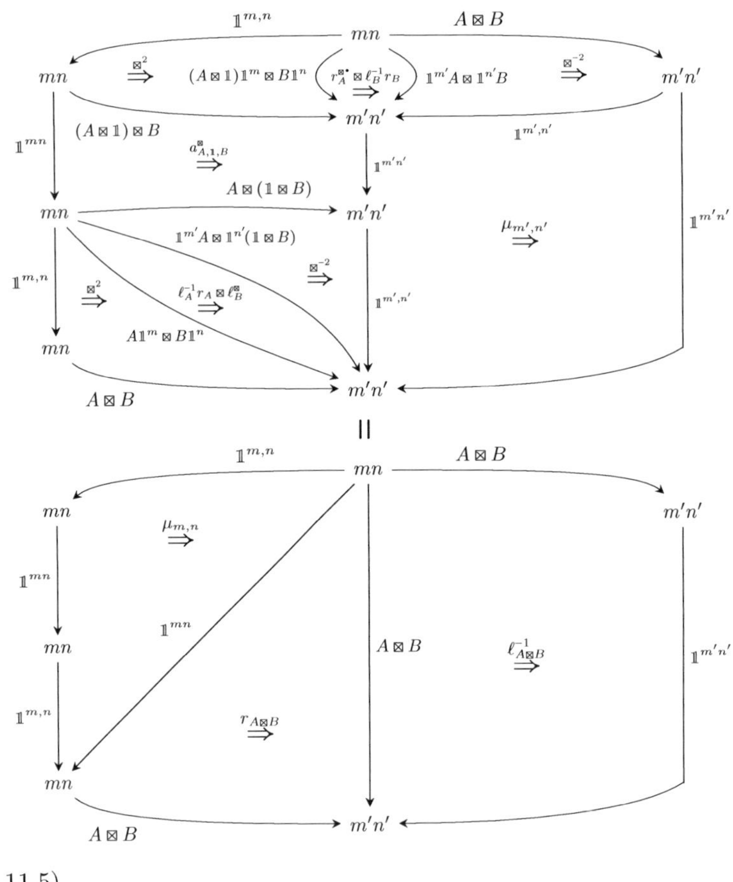

Consider (8.11.5).

- The common domain 1-cell is

$$(A \boxtimes B)\Big[\big((\mathbb{1}^m \boxtimes \mathbb{1}^n)\mathbb{1}^{mn}\big)\big(\mathbb{1}^m \boxtimes \mathbb{1}^n\big)\Big].$$

  The common codomain 1-cell is

$$\mathbb{1}^{m'n'}(A \boxtimes B).$$

- In the top pasting diagram, there are multiple instances of the base associator (8.3.2) and its inverse that are not explicitly displayed.

The proof of the equality (8.11.5) is adapted from that of (8.10.5), which is the modification axiom for the pentagonator π. Since we already showed all the detail in the proof of Lemma 8.10.4, we safely skip the detail here and only outline the remaining steps.

(1) First restrict to a typical (I', I)-entry with the following indices and abbreviations.

$$\begin{aligned} &1 \le i \le m \qquad 1 \le j \le n & &1 \le i' \le m' \qquad 1 \le j' \le n' \\ &I = (i-1)n + j & &I' = (i'-1)n' + j' \end{aligned}$$

In the (I', I)-entry, the two pasting diagrams in (8.11.5) yield a zigzag diagram in C analogous to the diagram (8.10.8), with the object Y in (8.10.9) replaced by the object

$$(A \boxtimes B)_{I'I} = A_{i'i} \otimes B_{j'j} \in \mathsf{C}.$$

The reason that the outer diagram consists of zigzags is that instances of $\boxtimes^2$ and a^4 in (8.3.5) are expressed using their respective inverses, that is., $\boxtimes^{-2}$ and a^{-4}.

(2) The set X of formal variables in (8.10.16) is replaced by

$$X = \left\{0^X, 1^X, a_{c'c}, b_{d'd}\right\}.$$

The function $\varphi: X \longrightarrow \mathsf{Ob}(\mathsf{C})$ is defined as follows.

$$\varphi(x) = \begin{cases} \mathbb{0} & \text{if } x = 0^X. \\ \mathbb{1} & \text{if } x = 1^X. \\ A_{c'c} & \text{if } x = a_{c'c}. \\ B_{d'd} & \text{if } x = b_{d'd}. \end{cases}$$

(3) The procedure in the proof of Lemma 8.10.4 after (8.10.16) yields a diagram D in $\mathsf{Gr}(X)$. Each subdiagram in D consists of two parallel paths with common codomain

$$y = a_{i'i} \otimes b_{j'j} \in X^{\mathsf{fr}}.$$

The image of D in C is the diagram in step (1).

(4) Since C is assumed to be tight, the Coherence Theorem 3.9.1 is applicable in each subdiagram in D because the central vertex y is regular, which implies that each vertex in D is regular.

The commutativity of D in C implies the equality (8.11.5) in the (I', I)-entry. $\square$

The Left 2-Unitor.

Definition 8.11.6. Define $\lambda^{\boxtimes}$ as consisting of the 2-cells

$$\lambda^{\boxtimes}_{m,n} \in \mathsf{Mat}^{\mathsf{C}}_{mn,mn} \quad \text{with} \quad m,n \geq 0$$

given by the composite of the following pasting diagram in $\mathsf{Mat}^{\mathsf{C}}_{mn,mn}$.

(8.11.7)

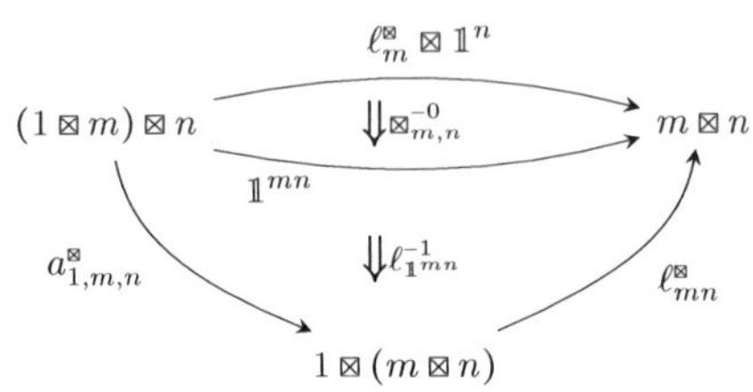

This finishes the definition of $\lambda^{\boxtimes}$. ◇

Explanation 8.11.8. Consider the pasting diagram in (8.11.7).

(1) By Definition 8.6.19, each vertex is the object $mn \in \mathsf{Mat}^{\mathsf{C}}$.
(2) By (8.8.2) and (8.9.2), the domain 1-cell is

$$\ell^{\boxtimes}_m \boxtimes 1\!\!1^n = 1\!\!1^m \boxtimes 1\!\!1^n.$$

The codomain 1-cell is the horizontal composite

$$\ell^{\boxtimes}_{mn} a^{\boxtimes}_{1,m,n} = 1\!\!1^{mn} 1\!\!1^{mn}.$$

(3) $\boxtimes^0$ is as in (8.6.9), and $\boxtimes^{-0} = (\boxtimes^0)^{-1}$.
(4) $\ell_{1\!\!1^{mn}} \colon 1\!\!1^{mn} 1\!\!1^{mn} \longrightarrow 1\!\!1^{mn}$ is the base left unitor in (8.2.2).
(5) Each entry of $\lambda^{\boxtimes}_{m,n}$ is a composite of morphisms in C, each being an iterated sum and product of identity morphisms and one component of $\lambda^{-\oplus}$, $\rho^{-\oplus}$, $\lambda^{\pm\otimes}$, $\lambda^{\pm\bullet}$, or $\rho^{\bullet}$. ◇

Lemma 8.11.9. *In Definition 8.11.6, $\lambda^{\boxtimes}$ is an invertible modification.*

Proof. Each component 2-cell $\lambda^{\boxtimes}_{m,n} \in \mathsf{Mat}^{\mathsf{C}}_{mn,mn}$ is well defined by the Bicategorical Pasting Theorem in [**JY21**, 3.6.6]. It is a composite of invertible 2-cells by (8.2.2) and (8.6.9).

To verify the modification axiom (6.3.2) for $\lambda^{\boxtimes}$, consider arbitrary 1-cells

$$(A,B) \in \mathsf{Mat}^{\mathsf{C}}_{m,m'} \times \mathsf{Mat}^{\mathsf{C}}_{n,n'}.$$

If one of m, m', n, and n' is 0, then $\mathsf{Mat}^{\mathsf{C}}_{mn,m'n'}$ is the terminal category, and the modification axiom holds automatically.

Suppose $m, m', n, n' > 0$. With the notation in the proof of Lemma 8.11.4, the modification axiom (6.3.2) for $\lambda^{\boxtimes}$ for the 1-cells (A, B) states the following pasting diagram equality in $\mathsf{Mat}^{\mathsf{C}}_{mn,m'n'}$.

(8.11.10)

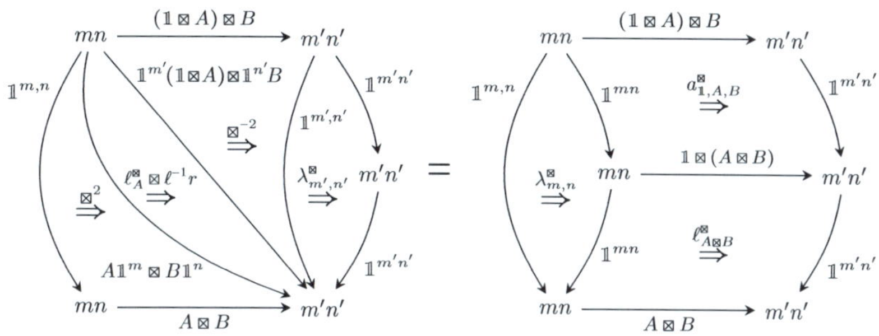

Consider (8.11.10).

- The common domain 1-cell is
$$(A \boxtimes B)(\mathbb{1}^m \boxtimes \mathbb{1}^n).$$
- The common codomain 1-cell is
$$\big(\mathbb{1}^{m'n'}\mathbb{1}^{m'n'}\big)\big[(\mathbb{1} \boxtimes A) \boxtimes B\big].$$
- In the right pasting diagram, there are instances of the base associator (8.3.2) and its inverse that are not explicitly displayed.

The proof of the equality (8.11.10) is adapted from that of (8.10.5), which is the modification axiom for the pentagonator π, in the proof of Lemma 8.10.4. The remaining steps are outlined in the proof of Lemma 8.11.4. □

The Right 2-Unitor.

Definition 8.11.11. Define $\rho^{\boxtimes}$ as consisting of the 2-cells
$$\rho^{\boxtimes}_{m,n} \in \mathsf{Mat}^{\mathsf{C}}_{mn,mn} \quad \text{with} \quad m, n \geq 0$$
given by the composite of the following pasting diagram in $\mathsf{Mat}^{\mathsf{C}}_{mn,mn}$.

(8.11.12)

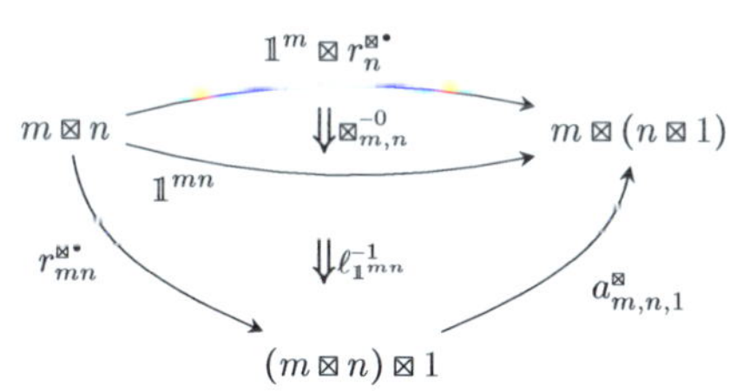

This finishes the definition of $\rho^{\boxtimes}$. ◇

Explanation 8.11.13. Consider the pasting diagram in (8.11.12).

(1) By Definition 8.6.19, each vertex is the object $mn \in \mathsf{Mat}^{\mathsf{C}}$

(2) By (8.8.2) and (8.9.18), the domain 1-cell is
$$\mathbb{1}^m \boxtimes r^{\boxtimes\bullet}_n = \mathbb{1}^m \boxtimes \mathbb{1}^n.$$
The codomain 1-cell is the horizontal composite
$$a^{\boxtimes}_{m,n,1} r^{\boxtimes\bullet}_{mn} = \mathbb{1}^{mn}\mathbb{1}^{mn}.$$

(3) $\rho^{\boxtimes}_{m,n} = \lambda^{\boxtimes}_{m,n}$ in (8.11.7). However, the modification axioms for $\rho^{\boxtimes}$ and $\lambda^{\boxtimes}$ are different because their (co)domain strong transformations are different. ◇

Lemma 8.11.14. *In Definition 8.11.11, $\rho^{\boxtimes}$ is an invertible modification.*

Proof. This proof is almost identical to that of Lemma 8.11.9, so we only write down the modification axiom (6.3.2) for $\rho^{\boxtimes}$ for the 1-cells (A, B) below.

(8.11.15)

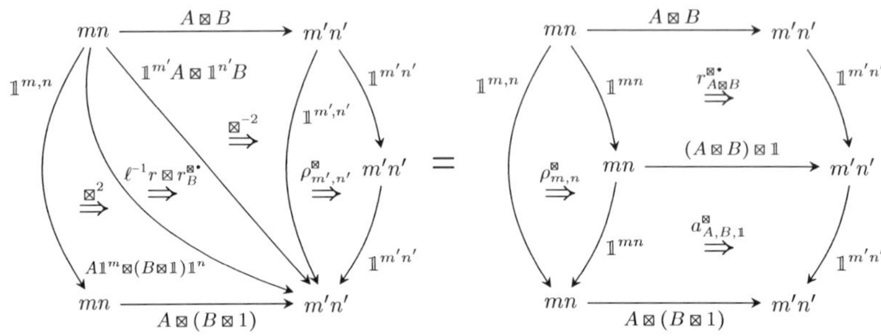

The proof of the equality (8.11.15) is adapted from that of (8.10.5) in the proof of Lemma 8.10.4. The remaining steps are outlined in the proof of Lemma 8.11.4. □

8.12. The Matrix Monoidal Bicategory

For a tight symmetric bimonoidal category C, in this section, we show that $\mathsf{Mat}^{\mathsf{C}}$ is a monoidal bicategory. So far in this chapter, we have defined the following data.

- The matrix bicategory $\mathsf{Mat}^{\mathsf{C}}$ is in Definition 8.4.11 and Theorem 8.4.12.
- The monoidal identity $(1_{\boxtimes}, 1^2_{\boxtimes}, 1^0_{\boxtimes})$ is in Definition 8.5.1 and Lemma 8.5.2.
- The monoidal composition $(\boxtimes, \boxtimes^2, \boxtimes^0)$ is in Definitions 8.6.1 and 8.6.19 and Lemma 8.7.31.
- The monoidal associator $(a^{\boxtimes}, a^{\boxtimes\bullet}, \eta^a, \varepsilon^a)$ is in Section 8.8.
- The left monoidal unitor $(\ell^{\boxtimes}, \ell^{\boxtimes\bullet}, \eta^{\ell}, \varepsilon^{\ell})$ and the right monoidal unitor $(r^{\boxtimes}, r^{\boxtimes\bullet}, \eta^r, \varepsilon^r)$ are in Section 8.9.
- The pentagonator π is in Section 8.10.
- The middle 2-unitor μ, the left 2-unitor $\lambda^{\boxtimes}$, and the right 2-unitor $\rho^{\boxtimes}$ are in Section 8.11.

To show that $\mathsf{Mat}^{\mathsf{C}}$ is a monoidal bicategory, it remains to check the three axioms in Definition 6.4.1.

Lemma 8.12.1. $\mathsf{Mat}^{\mathsf{C}}$ *satisfies the non-abelian 4-cocycle condition* (6.4.2).

Proof. Consider arbitrary objects $n_1, n_2, n_3, n_4, n_5 \in \mathsf{Mat}^{\mathsf{C}}$ with $N = n_1n_2n_3n_4n_5$. We use the abbreviations

$$\begin{aligned}
&\mathbb{1}^{(123,4),5} = \left(\mathbb{1}^{n_1n_2n_3} \boxtimes \mathbb{1}^{n_4}\right) \boxtimes \mathbb{1}^{n_5} && \mathbb{1}^{1,2345} = \mathbb{1}^{n_1} \boxtimes \mathbb{1}^{n_2n_3n_4n_5}\\
&\mathbb{1}^{1,(234,5)} = \mathbb{1}^{n_1} \boxtimes \left(\mathbb{1}^{n_2n_3n_4} \boxtimes \mathbb{1}^{n_5}\right) && \mathbb{1}^{123,45} = \mathbb{1}^{n_1n_2n_3} \boxtimes \mathbb{1}^{n_4n_5}\\
&a^{\boxtimes}_{1,2,345} = a^{\boxtimes}_{\mathbb{1}^{n_1},\mathbb{1}^{n_2},\mathbb{1}^{n_3n_4n_5}} && a^{-\boxtimes}_{1,234,5} = \left(a^{\boxtimes}_{\mathbb{1}^{n_1},\mathbb{1}^{n_2n_3n_4},\mathbb{1}^{n_5}}\right)^{-1}\\
&\pi_{1,23,4,5} = \pi_{n_1,n_2n_3,n_4,n_5} && \boxtimes^{-0}_{1,2} = \left(\boxtimes^0_{n_1,n_2}\right)^{-1}
\end{aligned} \tag{8.12.2}$$

and so forth. The non-abelian 4-cocycle condition (6.4.2) for $\mathsf{Mat}^{\mathsf{C}}$ for the objects $(n_1,\ldots,n_5)$ states the following pasting diagram equality in $\mathsf{Mat}^{\mathsf{C}}_{N,N}$.

(8.12.3)

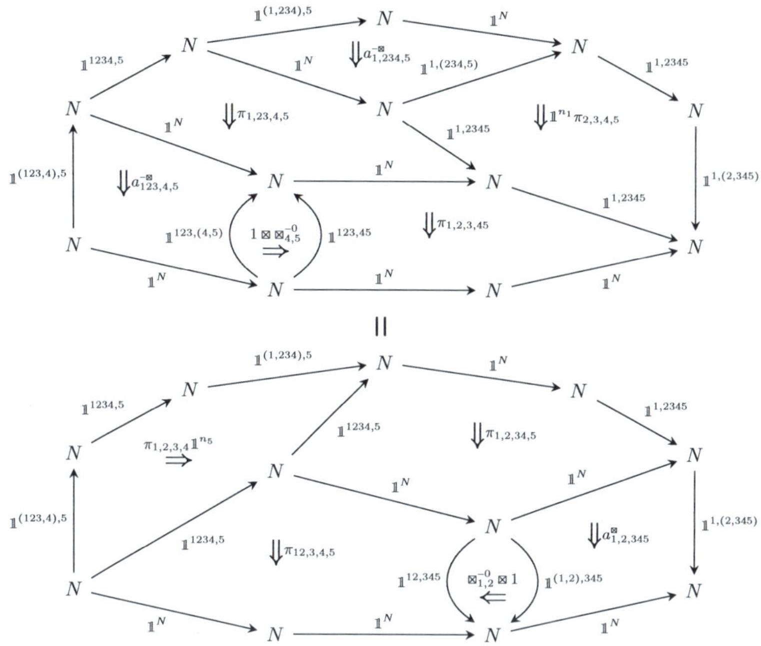

Consider (8.12.3).

- The common (co)domain 1-cell has the left normalized bracketing (5.2.13) by Convention 6.2.12.
- As discussed in Explanation 6.4.7, both $\mathbb{1}^{n_1}\pi_{2,3,4,5}$ and $\pi_{1,2,3,4}\mathbb{1}^{n_5}$ involve $\boxtimes^{\pm 2}$ in (8.6.20) and the base left unitor ℓ in (8.2.2).
- There are many instances of the base associator in (8.3.2) and its inverse that are not explicitly displayed in each pasting diagram.

If $N = 0$, then $\mathsf{Mat}^{\mathsf{C}}_{N,N}$ is the terminal category, and the equality (8.12.3) holds automatically.

The proof of the equality (8.12.3) for the case $N > 0$ is adapted from those of Lemmas 8.8.11 and 8.10.4. Since we already showed all the detail in those lemmas, here we only outline the remaining steps.

(1) First restrict to a typical (j,i)-entry with $1 \le i,j \le N$. In the (j,i)-entry, the two pasting diagrams in (8.12.3) yield a zigzag diagram in C analogous to the diagram (8.10.8). The object Y in (8.10.9) is now replaced by the object

$$\begin{cases} \mathbb{1} \in \mathsf{C} & \text{if } i = j \text{ and} \\ \mathbb{0} \in \mathsf{C} & \text{if } i \neq j. \end{cases}$$

The diagram involves zigzags because instances of $\boxtimes^2$ and a^1 in (8.3.5) are expressed using their respective inverses, that is, $\boxtimes^{-2}$ and a^{-1}.

(2) The set X in (8.10.16) and the function $\varphi\colon X \longrightarrow \mathsf{Ob}(\mathsf{C})$ are now defined as follows.

- $X = \{0^X, 1^X\}$.
- $\varphi(0^X) = \mathbb{0}$ and $\varphi(1^X) = \mathbb{1}$.

(3) The procedure in the proof of Lemma 8.10.4 after (8.10.16) yields a diagram D in $\mathsf{Gr}(X)$. Each subdiagram in D consists of two parallel paths with common codomain

$$\begin{cases} 1^X & \text{if } i = j \text{ and} \\ 0^X & \text{if } i \neq j. \end{cases}$$

The image of D in C is the diagram in step (1).

(4) If $i = j$, then the regularity of 1^X and Lemma 3.1.29 imply that each vertex in D is regular. Since C is assumed to be tight, the Coherence Theorem 3.9.1 implies that each subdiagram in D is commutative in C.

(5) If $i \neq j$, then each vertex in D has the same support as 0^X. Proposition 3.5.33 implies that each subdiagram in D is commutative in C.

The commutativity of D in C implies the equality (8.12.3) in the (j,i)-entry. □

Lemma 8.12.4. $\mathsf{Mat}^{\mathsf{C}}$ *satisfies the left normalization axiom* (6.4.3).

Proof. Consider arbitrary objects $n_1, n_2, n_3 \in \mathsf{Mat}^{\mathsf{C}}$ with $N = n_1 n_2 n_3$. We use the abbreviations in (8.12.2) and the following.

(8.12.5) $$\lambda^{\boxtimes}_{2,3} = \lambda^{\boxtimes}_{n_2,n_3} \qquad \mu_{1,2} = \mu_{n_1,n_2} \qquad \mu_{1,23} = \mu_{n_1,n_2n_3}$$

The left normalization axiom (6.4.3) for $\mathsf{Mat}^{\mathsf{C}}$ for the objects (n_1, n_2, n_3) states the following pasting diagram equality in $\mathsf{Mat}^{\mathsf{C}}_{N,N}$.

(8.12.6)

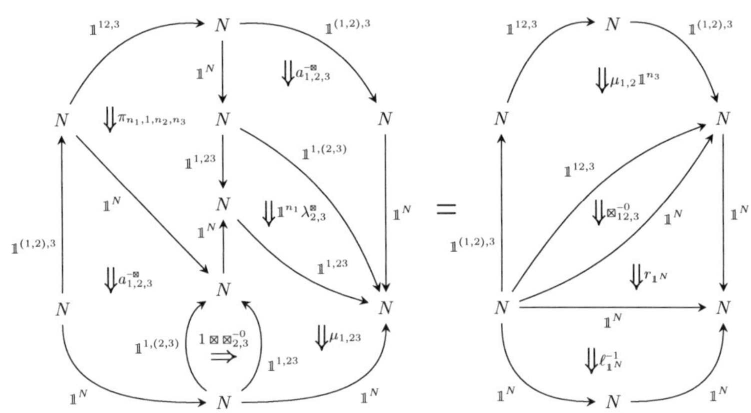

Consider (8.12.6).

- The common domain 1-cell has the left normalized bracketing (5.2.13) by Convention 6.2.12.
- As discussed in Explanation 6.4.7, both $\mathbb{1}^{n_1}\lambda^{\boxtimes}_{2,3}$ and $\mu_{1,2}\mathbb{1}^{n_3}$ involve $\boxtimes^{\pm 2}$ in (8.6.20) and the base left unitor ℓ in (8.2.2).
- There are instances of the base associator in (8.3.2) and its inverse that are not explicitly displayed.

If $N = 0$, then $\mathsf{Mat}^{\mathsf{C}}_{N,N}$ is the terminal category, and the equality (8.12.6) holds automatically. The proof of the equality (8.12.6) for the case $N > 0$ is adapted from those of Lemmas 8.8.11 and 8.10.4, using the steps (1)–(5) outlined in the proof of Lemma 8.12.1. □

Lemma 8.12.7. $\mathsf{Mat}^{\mathsf{C}}$ *satisfies the right normalization axiom* (6.4.4).

Proof. Consider arbitrary objects $n_1, n_2, n_3 \in \mathsf{Mat}^{\mathsf{C}}$ with $N = n_1n_2n_3$. We use the abbreviations in (8.12.2), (8.12.5), and $\rho^{\boxtimes}_{1,2} = \rho^{\boxtimes}_{n_1,n_2}$. The right normalization axiom (6.4.4) for $\mathsf{Mat}^{\mathsf{C}}$ for the objects (n_1, n_2, n_3) states the following pasting diagram equality in $\mathsf{Mat}^{\mathsf{C}}_{N,N}$.

(8.12.8)

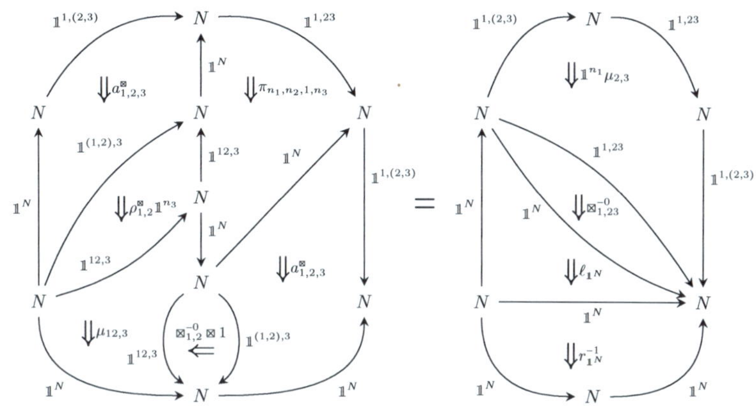

Consider (8.12.8).

- The common domain 1-cell has the left normalized bracketing (5.2.13) by Convention 6.2.12.
- As discussed in Explanation 6.4.7, both $\rho^{\boxtimes}_{1,2}\mathbb{1}^{n_3}$ and $\mathbb{1}^{n_1}\mu_{2,3}$ involve $\boxtimes^{\pm 2}$ in (8.6.20) and the base left unitor ℓ in (8.2.2).
- There are instances of the base associator in (8.3.2) and its inverse that are not explicitly displayed.

If $N = 0$, then $\mathsf{Mat}^{\mathsf{C}}_{N,N}$ is the terminal category, and the equality (8.12.8) holds automatically. The proof of the equality (8.12.8) for the case $N > 0$ is adapted from those of Lemmas 8.8.11 and 8.10.4, using the steps (1)–(5) outlined in the proof of Lemma 8.12.1. □

We now assemble the results so far in this chapter to obtain the following.

Theorem 8.12.9. *For each tight symmetric bimonoidal category* C*, the matrix bicategory* $\mathsf{Mat}^{\mathsf{C}}$ *is a monoidal bicategory.*

Proof. As noted in the beginning of this section, all the monoidal bicategorical data on $\mathsf{Mat}^{\mathsf{C}}$ have already been constructed in the previous sections. The three axioms in Definition 6.4.1 are proved in Lemmas 8.12.1, 8.12.4, and 8.12.7. □

8.13. The Braiding

For each tight symmetric bimonoidal category C, in Theorem 8.12.9, we observed that $\mathsf{Mat}^{\mathsf{C}}$ is a monoidal bicategory. The goal for the rest of this chapter is to show that this structure on $\mathsf{Mat}^{\mathsf{C}}$ extends to a *symmetric* monoidal bicategory as in Definition 6.5.9. Recall the following concepts.

- A symmetric monoidal bicategory is a sylleptic monoidal bicategory as in Definition 6.5.7 that satisfies the triple braid axiom (6.5.10).
- A sylleptic monoidal bicategory is a braided monoidal bicategory as in Definition 6.5.3 together with an invertible modification ν, which is called the syllepsis, that satisfies the two axioms in Definition 6.5.7.

Therefore, our next task is to construct a braided monoidal bicategory structure on $\mathsf{Mat}^{\mathsf{C}}$. In this section, we construct the braiding $(\beta, \beta^{\bullet}, \eta^{\beta}, \varepsilon^{\beta})$ in the monoidal bicategory $\mathsf{Mat}^{\mathsf{C}}$.

From Explanation 6.5.6, the braiding consists of the following data.

(i) β and $\beta^{\bullet}$ are strong transformations as in

$$\boxtimes \underset{\beta^{\bullet}}{\overset{\beta}{\rightleftarrows}} \boxtimes\tau$$

in which $\tau\colon (\mathsf{Mat}^{\mathsf{C}})^2 \longrightarrow (\mathsf{Mat}^{\mathsf{C}})^2$ switches the two arguments.

(ii) η^{β} and ε^{β} are invertible modifications as follows.

$$1_{\boxtimes} \xrightarrow{\eta^{\beta}} \beta^{\bullet}\beta \qquad \beta\beta^{\bullet} \xrightarrow{\varepsilon^{\beta}} 1_{\boxtimes\tau}$$

Moreover, these data satisfy the triangle identities (6.3.10).

Motivation 8.13.1. Let us motivate the construction of the braiding β. For complex matrices A and B, the two matrix tensor products $A \otimes B$ and $B \otimes A$ are equal up to a permutation of the entries. A permutation matrix is obtained from an identity matrix by permuting its columns. It has precisely one instance of 1 in each row and in each column, and every other entry is 0. There are permutation matrices P^1 and P^2 such that

$$A \otimes B = P^2(B \otimes A)P^1. \tag{8.13.2}$$

Here P^1 permutes the columns of $B \otimes A$, and P^2 permutes the rows.

The identity matrix $\mathbb{1}^n \in \mathsf{Mat}^{\mathsf{C}}_{n,n}$ makes sense in the matrix bicategory, and so do permutation matrices. For 1-cells A and B in $\mathsf{Mat}^{\mathsf{C}}$, the matrix tensor products $A \boxtimes B$ and $B \boxtimes A$, with entries in (8.6.4), differ by

- a permutation of the entries as in the previous paragraph and
- an entrywise multiplicative symmetry

$$A_{ji} \otimes B_{lk} \xrightarrow{\xi^{\otimes}} B_{lk} \otimes A_{ji}.$$

However, even with the $\xi^{\otimes}$ taken into account, the equality (8.13.2) does not hold strictly in $\mathsf{Mat}^{\mathsf{C}}$ in general. The reason is that $\mathbb{0}$ and $\mathbb{1}$ in C are not strict zero and strict unit, but only up to natural isomorphisms as in Definition 2.1.2. The analogue of the equality (8.13.2) is an invertible 2-cell $\beta_{A,B}$ that is a component 2-cell of the braiding β. Component 1-cells of β are analogues of the permutation matrices P^1 and P^2 in (8.13.2). ◇

The rest of this section is organized as follows.

- Before we define the braiding, first we define row permutation, column permutation, and permutation matrices in $\mathsf{Mat}^{\mathsf{C}}$ in Definition 8.13.3, and discuss some of their properties.
- The left adjoint β, the right adjoint $\beta^{\bullet}$, the unit η^{β}, and the counit ε^{β} are in Definitions 8.13.22, 8.13.30, 8.13.35, and 8.13.38.

Permutation Matrices. We now begin by defining row permutation, column permutation, and permutation matrices in the setting of $\mathsf{Mat}^{\mathsf{C}}$. Recall that Σ_m denotes the symmetric group on m letters.

Definition 8.13.3. For $m, n \geq 0$, suppose $\sigma \in \Sigma_m$, $\theta \in \Sigma_n$, and $A = (A_{ji}) \in \mathsf{Mat}^{\mathsf{C}}_{m,n}$.

- The *row permutation of A by θ*, denoted by ${}_\theta A$, is obtained from A by permuting its n rows by θ. Its entries are determined by

$$({}_\theta A)_{\theta(j),i} = A_{ji} \tag{8.13.4}$$

  for $1 \leq i \leq m$ and $1 \leq j \leq n$.
- The *column permutation of A by σ*, denoted by A^σ, is obtained from A by permuting its m columns by σ. Its entries are determined by

$$(A^\sigma)_{j,\sigma(i)} = A_{ji}. \tag{8.13.5}$$

- Define the *permutation matrix of σ* as

$$\mathbb{1}^\sigma = (\mathbb{1}^m)^\sigma. \tag{8.13.6}$$

  It is the column permutation of the $m \times m$ identity matrix $\mathbb{1}^m$ in (8.1.6) by σ. Its entries are given by

$$\mathbb{1}^\sigma_{ij} = \begin{cases} \mathbb{1} & \text{if } j = \sigma(i) \text{ and} \\ \mathbb{0} & \text{if } j \neq \sigma(i). \end{cases}$$

The notation in (8.13.4) and (8.13.5) also apply if A is a 2-cell in $\mathsf{Mat}^{\mathsf{C}}_{m,n}$. ◇

To define the (co)unit of the braiding, we will use the following preliminary observation. The transpose of a matrix A is denoted by A^T.

Lemma 8.13.7. *In the setting of Definition 8.13.3, the following statements hold.*

(1) *For $\sigma' \in \Sigma_m$ and $\theta' \in \Sigma_n$, the following equalities hold.*

$$\begin{aligned} {}_{\theta'}({}_\theta A) &= {}_{\theta'\theta} A \\ (A^\sigma)^{\sigma'} &= A^{\sigma'\sigma} \end{aligned} \tag{8.13.8}$$

(2) $(\mathbb{1}^\sigma)^T = \mathbb{1}^{\sigma^{-1}}$.

(3) *For $1 \leq i \leq m$, the ith diagonal entry in $\mathbb{1}^{\sigma^{-1}}\mathbb{1}^\sigma$ is*

$$\left(\mathbb{1}^{\sigma^{-1}}\mathbb{1}^\sigma\right)_{ii} = \Bigg\{ \underbrace{\Bigg[\bigoplus_{k=1}^{\sigma^{-1}(i)-1} \mathbb{0}\mathbb{0} \Bigg]}_{\text{empty if } \sigma^{-1}(i)=1} \oplus (\mathbb{1}\mathbb{1}) \oplus \underbrace{\Bigg[\bigoplus_{k=\sigma^{-1}(i)+1}^{m} \mathbb{0}\mathbb{0} \Bigg]}_{\text{empty if } \sigma^{-1}(i)=m} \Bigg\}_{\mathsf{lt}}. \tag{8.13.9}$$

(4) *Each off-diagonal entry in $\mathbb{1}^{\sigma^{-1}}\mathbb{1}^\sigma$ is a left normalized sum of m objects as in either*

$$\begin{gathered} \big[\mathbb{0}\mathbb{0} \oplus \cdots \oplus \mathbb{1}\mathbb{0} \oplus \cdots \oplus \mathbb{0}\mathbb{1} \oplus \cdots \oplus \mathbb{0}\mathbb{0}\big]_{\mathsf{lt}} \quad \text{or} \\ \big[\mathbb{0}\mathbb{0} \oplus \cdots \oplus \mathbb{0}\mathbb{1} \oplus \cdots \oplus \mathbb{1}\mathbb{0} \oplus \cdots \oplus \mathbb{0}\mathbb{0}\big]_{\mathsf{lt}}. \end{gathered} \tag{8.13.10}$$

In each case, there is one instance of each of $\mathbb{1}\mathbb{0}$ and $\mathbb{0}\mathbb{1}$, and every other summand is $\mathbb{0}\mathbb{0}$.

Proof. The equalities in (8.13.8) follow from the definitions (8.13.4) and (8.13.5).

For assertion (2), the transpose $(\mathbb{1}^\sigma)^T$ has $\mathbb{1}$ precisely in the $(\sigma(i), i)$-entries for $1 \le i \le m$, that is, the $(i, \sigma^{-1}(i))$-entries. These entries are where $\mathbb{1}^{\sigma^{-1}}$ has $\mathbb{1}$.

The last two assertions follow from assertion (2) and (8.1.4). $\square$

Component 2-cells of the braiding β involve the natural isomorphisms in Lemmas 8.13.12 and 8.13.15 below.

Motivation 8.13.11. Suppose for the moment that $\mathbb{0}$ and $\mathbb{1}$ in C are strict zero and strict unit. This is true, for example, if C is a right bipermutative category or a left bipermutative category as in Definitions 2.5.2 and 2.5.11. Then the matrix product $A\mathbb{1}^\sigma$ is equal to A^σ, the column permutation of A by σ. In the general case, when $\mathbb{0}$ and $\mathbb{1}$ are not strict zero and strict unit, $A\mathbb{1}^\sigma$ and A^σ are connected by a natural isomorphism as in the next lemma. Similar remarks apply to $\mathbb{1}^\theta A$ and ${}_{\theta^{-1}}A$. $\diamond$

Lemma 8.13.12. *For $\sigma \in \Sigma_m$ and $A = (A_{ji}) \in \mathsf{Mat}^{\mathsf{C}}_{m,n}$, there is a natural isomorphism*

$$A\mathbb{1}^\sigma \xrightarrow[\cong]{r^\sigma_A} A^\sigma \tag{8.13.13}$$

in $\mathsf{Mat}^{\mathsf{C}}_{m,n}$.

Proof. If either m or n is 0, then $\mathsf{Mat}^{\mathsf{C}}_{m,n}$ is the terminal category, and r^σ_A is the identity morphism of the empty matrix.

For $m, n > 0$, $1 \le i \le m$, and $1 \le j \le n$, the (j,i)-entry of $A\mathbb{1}^\sigma$ is as follows.

$$\begin{aligned}(A\mathbb{1}^\sigma)_{ji} &= \Big[\bigoplus_{k=1}^{m} A_{jk}\mathbb{1}^\sigma_{ki}\Big]_{\mathsf{lt}} \\ &= \Bigg\{\underbrace{\Big[\bigoplus_{k=1}^{\sigma^{-1}(i)-1} A_{jk}\mathbb{0}\Big]}_{\text{empty if } \sigma^{-1}(i)=1} \oplus \big(A_{j,\sigma^{-1}(i)}\mathbb{1}\big) \oplus \underbrace{\Big[\bigoplus_{k=\sigma^{-1}(i)+1}^{m} A_{jk}\mathbb{0}\Big]}_{\text{empty if } \sigma^{-1}(i)=m}\Bigg\}_{\mathsf{lt}}\end{aligned}$$

This object is isomorphic to A^σ_{ji} via the following composite in C.

$$\begin{array}{ccc} (A\mathbb{1}^\sigma)_{ji} & \xrightarrow{(r^\sigma_A)_{ji}} & A^\sigma_{ji} \\ \Big\downarrow{\scriptstyle (\rho^\bullet\oplus\cdots\oplus\rho^\bullet\oplus\rho^\otimes\oplus\rho^\bullet\oplus\cdots\oplus\rho^\bullet)_{\mathsf{lt}}} & & \Big\| \\ \Big(\underbrace{\mathbb{0}\oplus\cdots\oplus\mathbb{0}\oplus}_{\sigma^{-1}(i)-1 \text{ copies of } \mathbb{0}} A_{j,\sigma^{-1}(i)} \underbrace{\oplus\mathbb{0}\oplus\cdots\oplus\mathbb{0}}_{m-\sigma^{-1}(i) \text{ copies of } \mathbb{0}}\Big)_{\mathsf{lt}} & \xrightarrow{Z} & A_{j,\sigma^{-1}(i)} \end{array} \tag{8.13.14}$$

If $m = 1$, then Z is the identity morphism. If $m > 1$, then Z is a composite of $m-1$ morphisms, each being the sum of identity morphisms and one component of $\lambda^\oplus$ or $\rho^\oplus$.

The naturality of r^σ_A with respect to A follows from the naturality of $\rho^\bullet$, $\rho^\otimes$, $\lambda^\oplus$, and $\rho^\oplus$, and the functoriality of $\oplus$ in C. $\square$

In the next lemma, note that ${}_{\theta^{-1}}A$ is the row permutation of A by θ^{-1}.

Lemma 8.13.15. *For $\theta \in \Sigma_n$ and $A = (A_{ji}) \in \mathsf{Mat}^{\mathsf{C}}_{m,n}$, there is a natural isomorphism*

$$\mathbb{1}^{\theta} A \xrightarrow[\cong]{\ell^{\theta}_A} {}_{\theta^{-1}}A \tag{8.13.16}$$

in $\mathsf{Mat}^{\mathsf{C}}_{m,n}$.

Proof. If either m or n is 0, then $\mathsf{Mat}^{\mathsf{C}}_{m,n}$ is the terminal category, and ℓ^{θ}_A is the identity morphism of the empty matrix.

For $m,n > 0$, $1 \le i \le m$, and $1 \le j \le n$, the (j,i)-entry of $\mathbb{1}^{\theta}A$ is as follows.

$$\begin{aligned}(\mathbb{1}^{\theta}A)_{ji} &= \Big[\bigoplus_{l=1}^{n} \mathbb{1}^{\theta}_{jl}A_{li}\Big]_{\mathsf{lt}}\\ &= \Big\{\underbrace{\Big[\bigoplus_{l=1}^{\theta(j)-1} \mathbb{0}A_{li}\Big]}_{\text{empty if } \theta(j)=1} \oplus \big(\mathbb{1}A_{\theta(j),i}\big) \oplus \underbrace{\Big[\bigoplus_{l=\theta(j)+1}^{n} \mathbb{0}A_{li}\Big]}_{\text{empty if } \theta(j)=n}\Big\}_{\mathsf{lt}}\end{aligned}$$

This object is isomorphic to $({}_{\theta^{-1}}A)_{ji}$ via the following composite in C.

$$\begin{array}{ccc} (\mathbb{1}^{\theta}A)_{ji} & \xrightarrow{(\ell^{\theta}_A)_{ji}} & ({}_{\theta^{-1}}A)_{ji} \\ {\scriptstyle (\lambda^{\bullet}\oplus\cdots\oplus\lambda^{\bullet}\oplus\lambda^{\otimes}\oplus\lambda^{\bullet}\oplus\cdots\oplus\lambda^{\bullet})_{\mathsf{lt}}}\big\downarrow & & \big\| \\ \Big(\underbrace{\mathbb{0}\oplus\cdots\oplus\mathbb{0}\oplus}_{\theta(j)-1 \text{ copies of } \mathbb{0}} A_{\theta(j),i} \underbrace{\oplus\mathbb{0}\oplus\cdots\oplus\mathbb{0}}_{n-\theta(j) \text{ copies of } \mathbb{0}}\Big)_{\mathsf{lt}} & \xrightarrow{Z} & A_{\theta(j),i} \end{array} \tag{8.13.17}$$

If $n = 1$, then Z is the identity morphism. If $n > 1$, then Z is a composite of $n-1$ morphisms, each being the sum of identity morphisms and one component of $\lambda^{\oplus}$ or $\rho^{\oplus}$.

The naturality of ℓ^{θ}_A with respect to A follows from the naturality of $\lambda^{\bullet}$, $\lambda^{\otimes}$, $\lambda^{\oplus}$, and $\rho^{\oplus}$, and the functoriality of $\oplus$ in C. □

Remark 8.13.18. In Lemma 8.13.12, if σ is the identity permutation id_m, then $\mathbb{1}^{\mathrm{id}_m} = \mathbb{1}^m$, and $r^{\mathrm{id}_m} = r$, which is the base right unitor in (8.2.8). In Lemma 8.13.15, if θ is the identity permutation id_n, then $\mathbb{1}^{\mathrm{id}_n} = \mathbb{1}^n$, and $\ell^{\mathrm{id}_n} = \ell$, which is the base left unitor in (8.2.2). ◇

Denote by $\tau_{m,n} \in \Sigma_{mn}$ the permutation in (2.4.5). It rearranges n intervals of m objects each to m intervals of n objects each. Note that

$$(\tau_{m,n})^{-1} = \tau_{n,m}. \tag{8.13.19}$$

The next lemma makes precise the two bullet points in Motivation 8.13.1. Namely, the matrix tensor products $A \boxtimes B$ and $B \boxtimes A$ differ by a column permutation, a row permutation, and a multiplicative symmetry in each entry.

Lemma 8.13.20. *Suppose $A \in \mathsf{Mat}^{\mathsf{C}}_{m,m'}$ and $B \in \mathsf{Mat}^{\mathsf{C}}_{n,n'}$ are either both 1-cells or both 2-cells. Then there are equalities*

$$\begin{aligned}(B \boxtimes A)^{\tau_{m,n}}_{i'+(j'-1)m',\, j+(i-1)n} &= B_{j'j} \otimes A_{i'i}\\ \big({}_{\tau_{n',m'}}(A \boxtimes B)\big)_{i'+(j'-1)m',\, j+(i-1)n} &= A_{i'i} \otimes B_{j'j}\end{aligned}$$

for $1 \le i \le m$, $1 \le j \le n$, $1 \le i' \le m'$, and $1 \le j' \le n'$.

Proof. The first equality follows from (2.4.5), (8.13.5), and (8.6.4) as follows.

$$\begin{aligned}(B\boxtimes A)^{\tau_{m,n}}_{i'+(j'-1)m',j+(i-1)n} &= (B\boxtimes A)^{\tau_{m,n}}_{i'+(j'-1)m',\tau_{m,n}(i+(j-1)m)}\\ &= (B\boxtimes A)_{i'+(j'-1)m',i+(j-1)m}\\ &= B_{j'j}\otimes A_{i'i}\end{aligned}$$

The second equality follows similarly from (2.4.5), (8.13.4), and (8.6.4) as follows.

$$\begin{aligned}\Big({}_{\tau_{n',m'}}(A\boxtimes B)\Big)_{i'+(j'-1)m',j+(i-1)n} &= \Big({}_{\tau_{n',m'}}(A\boxtimes B)\Big)_{\tau_{n',m'}(j'+(i'-1)n'),j+(i-1)n}\\ &= (A\boxtimes B)_{j'+(i'-1)n',j+(i-1)n}\\ &= A_{i'i}\otimes B_{j'j}\end{aligned}$$

This finishes the proof of the lemma. □

Explanation 8.13.21. There is a more conceptual way to represent the two matrices in Lemma 8.13.20. Suppose $A_{\bullet i}\in\mathsf{Mat}^{\mathsf{C}}_{m',1}$ is the ith column of A for $1\le i\le m$. Then

$$(B\boxtimes A)^{\tau_{m,n}} = \big[\ B\boxtimes A_{\bullet 1}\ \big|\ \cdots\ \big|\ B\boxtimes A_{\bullet m}\ \big].$$

From left to right, the ith block is $B\boxtimes A_{\bullet i}$. In linear algebra, this column-wise matrix tensor product is an example of the Khatri-Rao product [**KR68**], which is related to the Tracy-Singh product [**Liu99, TS72**].

Similarly, suppose $B_{j'\bullet}\in\mathsf{Mat}^{\mathsf{C}}_{1,n}$ is the j'th row of B for $1\le j'\le n'$. Then

$${}_{\tau_{n',m'}}(A\boxtimes B) = \begin{bmatrix} A\boxtimes B_{1\bullet}\\ \hline \vdots \\ \hline A\boxtimes B_{n'\bullet}\end{bmatrix}.$$

From top to bottom, the j'th block is $A\boxtimes B_{j'\bullet}$. In linear algebra, this row-wise matrix tensor product is an example of the face product in [**Sly99**]. ◇

The Left Adjoint of the Braiding. Recall the monoidal composition $(\boxtimes,\boxtimes^2,\boxtimes^0)$ in Definition 8.6.19 and Lemma 8.7.31.

Definition 8.13.22. Define the data of a lax transformation

$$\boxtimes \xrightarrow{\ \beta\ } \boxtimes\tau$$

as follows, in which $\tau\colon(\mathsf{Mat}^{\mathsf{C}})^2\longrightarrow(\mathsf{Mat}^{\mathsf{C}})^2$ switches the two arguments.

Component 1-Cells: For each pair of objects $(m,n)\in(\mathsf{Mat}^{\mathsf{C}})^2$, define

$$mn = m\boxtimes n \xrightarrow{\ \beta_{m,n}\ } n\boxtimes m = nm$$

as the permutation matrix

$$\beta_{m,n} = \mathbb{1}^{\tau_{m,n}}\in\mathsf{Mat}^{\mathsf{C}}_{mn,nm} \tag{8.13.23}$$

in (8.13.6), with $\tau_{m,n}\in\Sigma_{mn}$ the permutation in (2.4.5).

Component 2-Cells: For 1-cells $A \in \mathsf{Mat}^{\mathsf{C}}_{m,m'}$ and $B \in \mathsf{Mat}^{\mathsf{C}}_{n,n'}$, define the component 2-cell

$$\beta_{A,B} \in \mathsf{Mat}^{\mathsf{C}}_{mn,n'm'}\big((B \boxtimes A)\beta_{m,n}; \beta_{m',n'}(A \boxtimes B)\big)$$

as the following vertical composite.

(8.13.24)

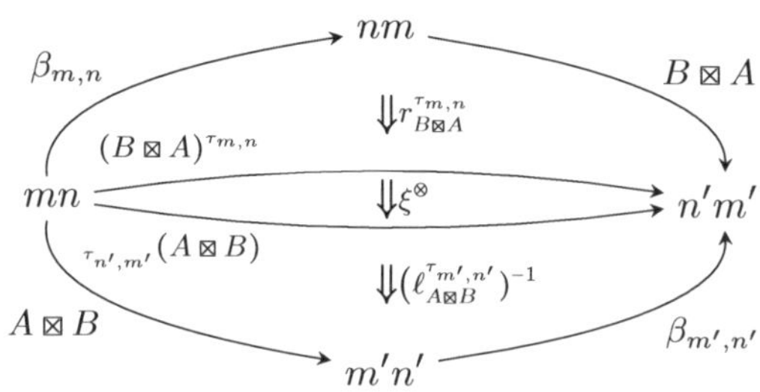

- $r^{\tau_{m,n}}_{B\boxtimes A}$ is the natural isomorphism in (8.13.13).
- $\ell^{\tau_{m',n'}}_{A\boxtimes B}$ is the natural isomorphism in (8.13.16), along with (8.13.19).
- $\xi^{\otimes}$ is the 2-cell with $\big(i' + (j'-1)m', j + (i-1)n\big)$-entry the structure morphism in C,

$$B_{j'j} \otimes A_{i'i} \xrightarrow{\xi^{\otimes}_{B_{j'j},A_{i'i}}} A_{i'i} \otimes B_{j'j}$$

  for $1 \le i \le m$, $1 \le j \le n$, $1 \le i' \le m'$, and $1 \le j' \le n'$.

This finishes the definition of β. ◇

Explanation 8.13.25. Consider Definition 8.13.22.

(1) The entries of $\beta_{m,n} = \mathbb{1}^{\tau_{m,n}}$ are determined as follows.

$$(\beta_{m,n})_{pq} = \begin{cases} \mathbb{1} & \text{if } (p,q) = \big(i + (j-1)m, j + (i-1)n\big) \\ & \quad \text{for some } 1 \le i \le m \text{ and } 1 \le j \le n, \text{ and} \\ \mathbb{0} & \text{otherwise.} \end{cases} \tag{8.13.26}$$

(2) By (8.13.14) and (8.13.17), each entry of $\beta_{A,B}$ in (8.13.24) is a composite of isomorphisms, each being an iterated sum of identity morphisms and at most one component of $\lambda^{\pm\oplus}$, $\rho^{\pm\oplus}$, $\lambda^{-\otimes}$, $\rho^{\otimes}$, $\xi^{\otimes}$, $\lambda^{-\bullet}$, or $\rho^{\bullet}$ in C.
(3) The 2-cell $\xi^{\otimes}$ in (8.13.24) is well defined by Lemma 8.13.20.
(4) There is a strong similarity between β and $a^{\boxtimes}$ in Definition 8.8.1. Since the proofs in Section 8.8 for $a^{\boxtimes}$ have all the detail, we will safely skip most of the detail below. ◇

Lemma 8.13.27. *In Definition 8.13.22, $\beta \colon \boxtimes \longrightarrow \boxtimes\tau$ is a strong transformation.*

Proof. Similar to Lemma 8.8.5, β in (8.13.24) is a natural isomorphism because the structure isomorphisms $\lambda^{\oplus}$, $\rho^{\oplus}$, $\lambda^{\otimes}$, $\rho^{\otimes}$, $\xi^{\otimes}$, $\lambda^{\bullet}$, and $\rho^{\bullet}$ in C are natural.

The lax unity axiom (6.2.15) for β states the following pasting diagram equality in $\mathsf{Mat}^{\mathsf{C}}_{mn,nm}$ for $m, n \geq 0$.

(8.13.28)

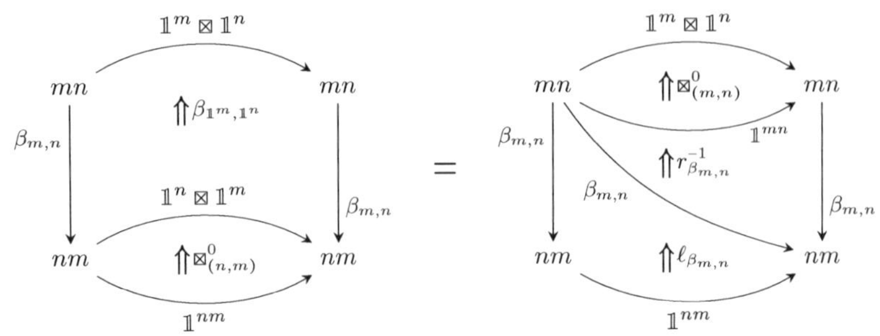

The proof of this equality is adapted from that of (8.8.12), which is the lax unity axiom for $a^{\boxtimes}$. Instead of diagonal entries, here we consider entries of the form

$$\big(i + (j-1)m, j + (i-1)n\big) \quad \text{for} \quad 1 \leq i \leq m \quad \text{and} \quad 1 \leq j \leq n.$$

These entries are where $\mathbb{1}$ appears in $\beta_{m,n}$ (8.13.26). Instead of off-diagonal entries, here we consider entries where $\mathbb{0}$ appears in $\beta_{m,n}$.

The lax naturality axiom (6.2.16) for β states that for 1-cells

$$m \xrightarrow{A} n \xrightarrow{B} p$$
$$m' \xrightarrow{A'} n' \xrightarrow{B'} p'$$

in $\mathsf{Mat}^{\mathsf{C}}$, the following pasting diagram equality in $\mathsf{Mat}^{\mathsf{C}}_{mm',p'p}$ holds.

(8.13.29)

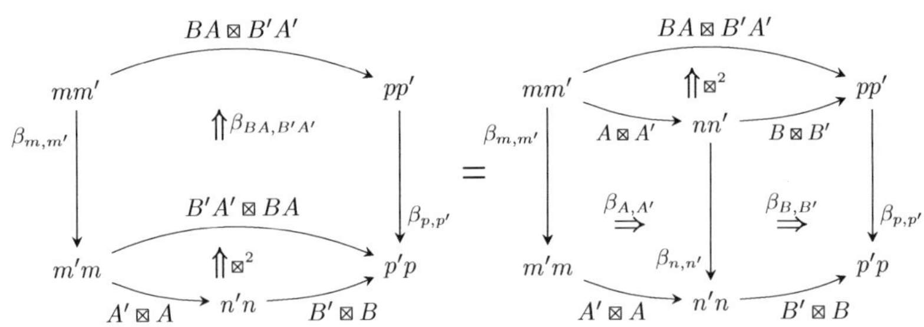

The proof of this equality is adapted from that of (8.8.16), which is the lax naturality axiom for $a^{\boxtimes}$, in Lemmas 8.8.17 and 8.8.26. Instead of the subdivided diagram (8.8.28), here we have a zigzag diagram analogous to (8.10.8). The object Y in (8.10.9) is replaced by the following object for $1 \leq i \leq m$, $1 \leq i' \leq m'$, $1 \leq k \leq p$, and $1 \leq k' \leq p'$.

$$\begin{aligned}
&\Big[\big((B' \boxtimes B)(A' \boxtimes A)\big)^{\tau_{m,m'}}\Big]_{(k+(k'-1)p,\, i'+(i-1)m')} \\
&= \Big[\big((B' \boxtimes B)(A' \boxtimes A)\big)^{\tau_{m,m'}}\Big]_{(k+(k'-1)p,\, \tau_{m,m'}(i+(i'-1)m))} \\
&= \big[(B' \boxtimes B)(A' \boxtimes A)\big]_{(k+(k'-1)p,\, i+(i'-1)m)} \\
&= \left[\bigoplus_{j'=1}^{n'} \bigoplus_{j=1}^{n} (B'_{k'j'} B_{kj})(A'_{j'i'} A_{ji})\right]_{\mathsf{lt}}
\end{aligned}$$

This object is a typical entry of the column permutation of $(B' \boxtimes B)(A' \boxtimes A)$ by $\tau_{m,m'}$. The three equalities above follow from, respectively, (2.4.5), (8.13.5), and (8.1.4) and (8.6.4). □

The Right Adjoint of the Braiding.

Definition 8.13.30. Define the data of a lax transformation

$$\boxtimes\tau \xrightarrow{\beta^{\bullet}} \boxtimes$$

as follows, in which $\tau\colon (\mathsf{Mat}^{\mathsf{C}})^2 \longrightarrow (\mathsf{Mat}^{\mathsf{C}})^2$ switches the two arguments.

Component 1-Cells: For each pair of objects $(m,n) \in (\mathsf{Mat}^{\mathsf{C}})^2$, define

$$nm = n \boxtimes m \xrightarrow{\beta^{\bullet}_{m,n}} m \boxtimes n = mn \tag{8.13.31}$$

as the permutation matrix

$$\beta^{\bullet}_{m,n} = \beta_{n,m} = \mathbb{1}^{\tau_{n,m}} \in \mathsf{Mat}^{\mathsf{C}}_{nm,mn}$$

in (8.13.6). Here $\tau_{n,m} = (\tau_{m,n})^{-1}$ with $\tau_{m,n} \in \Sigma_{mn}$ the permutation in (2.4.5).

Component 2-Cells: For 1-cells $A \in \mathsf{Mat}^{\mathsf{C}}_{m,m'}$ and $B \in \mathsf{Mat}^{\mathsf{C}}_{n,n'}$, define the component 2-cell

$$\beta^{\bullet}_{A,B} = \beta_{B,A} \in \mathsf{Mat}^{\mathsf{C}}_{nm,m'n'}\big((A \boxtimes B)\beta^{\bullet}_{m,n}; \beta^{\bullet}_{m',n'}(B \boxtimes A)\big).$$

This finishes the definition of $\beta^{\bullet}$. ◇

Explanation 8.13.32. Using (8.13.24) the component 2-cell $\beta^{\bullet}_{A,B} = \beta_{B,A}$ is the following vertical composite.

(8.13.33)

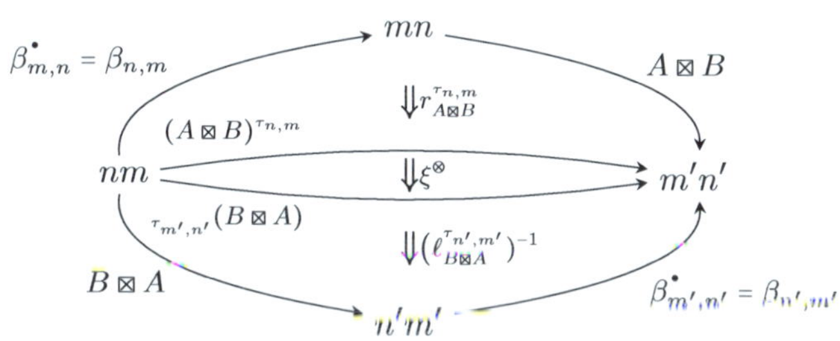

- $r^{\tau_{n,m}}_{A\boxtimes B}$ is the natural isomorphism in (8.13.13).
- $\ell^{\tau_{n',m'}}_{B\boxtimes A}$ is the natural isomorphism in (8.13.16), along with (8.13.19).
- $\xi^{\otimes}$ is the 2-cell with $(j' + (i'-1)n', i + (j-1)m)$-entry the structure morphism in C,

  $$A_{i'i} \otimes B_{j'j} \xrightarrow{\xi^{\otimes}_{A_{i'i},B_{j'j}}} B_{j'j} \otimes A_{i'i}$$

  for $1 \le i \le m$, $1 \le j \le n$, $1 \le i' \le m'$, and $1 \le j' \le n'$. This is well defined by Lemma 8.13.20.

By (8.13.14) and (8.13.17), each entry of $\beta^{\bullet}_{A,B}$ is a composite of isomorphisms, each being an iterated sum of identity morphisms and at most one component of $\lambda^{\pm\oplus}$, $\rho^{\pm\oplus}$, $\lambda^{-\otimes}$, $\rho^{\otimes}$, $\xi^{\otimes}$, $\lambda^{-\bullet}$, or $\rho^{\bullet}$ in C. ◇

Lemma 8.13.34. *In Definition 8.13.30, $\beta^{\bullet}\colon \boxtimes\tau \longrightarrow \boxtimes$ is a strong transformation.*

Proof. The proof of Lemma 8.13.27 applies here by switching symbols. In the lax unity axiom (8.13.28), m and n are switched. In the lax naturality axiom (8.13.29), the symbols (m,n,p,A,B) are switched with (m',n',p',A',B'). □

The Unit and the Counit of the Braiding. Recall that $\tau_{m,n} \in \Sigma_{mn}$ is the permutation in (2.4.5). The next two definitions refer to the strong transformations

$$\boxtimes \underset{\beta^{\bullet}}{\overset{\beta}{\rightleftarrows}} \boxtimes\tau$$

in Definitions 8.13.22 and 8.13.30.

Definition 8.13.35. Define the data

$$1_{\boxtimes} \xrightarrow{\eta^{\beta}} \beta^{\bullet}\beta$$

as consisting of the component 2-cells

$$\begin{array}{c} (1_{\boxtimes})_{(m,n)} = \mathbb{1}^{mn} \\ \big\downarrow \eta^{\beta}_{(m,n)} = \left(r^{\tau_{m,n}}_{\mathbb{1}^{\tau_{n,m}}}\right)^{-1} \\ (\beta^{\bullet}\beta)_{(m,n)} = \beta^{\bullet}_{m,n}\beta_{m,n} = \mathbb{1}^{\tau_{n,m}}\mathbb{1}^{\tau_{m,n}} \end{array} \tag{8.13.36}$$

in $\mathsf{Mat}^{\mathsf{C}}_{mn,mn}$ for $m,n \geq 0$, with r as in (8.13.13). ◇

Explanation 8.13.37. Consider (8.13.36).

- $\eta^{\beta}_{(m,n)}$ is well defined because $r^{\tau_{m,n}}_{\mathbb{1}^{\tau_{n,m}}}$ is a natural isomorphism

$$\mathbb{1}^{\tau_{n,m}}\mathbb{1}^{\tau_{m,n}} \xrightarrow{r^{\tau_{m,n}}_{\mathbb{1}^{\tau_{n,m}}}} \left(\mathbb{1}^{\tau_{n,m}}\right)^{\tau_{m,n}} = \left(\mathbb{1}^{mn}\right)^{\tau_{m,n}\tau_{n,m}} = \mathbb{1}^{mn}.$$

  - In the domain, the entries of $\mathbb{1}^{\tau_{n,m}}\mathbb{1}^{\tau_{m,n}}$ are described in (8.13.9) and (8.13.10).
  - In the codomain, the equalities follow from (8.13.8) and (8.13.19).
- The kth diagonal entry of $\eta^{\beta}_{(m,n)}$ has the following form.

$$\mathbb{1} \xrightarrow{(\lambda^{-\oplus},\rho^{-\oplus},\rho^{-\otimes},\rho^{-\bullet})} \left[(\beta^{\bullet}\beta)_{(m,n)}\right]_{kk}$$

  It is a composite of morphisms, each being an iterated sum of identity morphisms and at most one component of $\lambda^{-\oplus}$, $\rho^{-\oplus}$, $\rho^{-\otimes}$, or $\rho^{-\bullet}$.
- Similarly, each off-diagonal entry of $\eta^{\beta}_{(m,n)}$ has the form

$$\mathbb{0} \xrightarrow{(\lambda^{-\oplus},\rho^{-\otimes},\rho^{-\bullet})} \left[(\beta^{\bullet}\beta)_{(m,n)}\right]_{kl}$$

  for $1 \leq k \neq l \leq mn$. ◇

Definition 8.13.38. Define the data

$$\beta\beta^{\bullet} \xrightarrow{\varepsilon^{\beta}} 1_{\boxtimes\tau}$$

as consisting of the component 2-cells

(8.13.39)
$$\begin{array}{c} (\beta\beta^{\bullet})_{(m,n)} = \beta_{m,n}\beta^{\bullet}_{m,n} = \mathbb{1}^{\tau_{m,n}}\mathbb{1}^{\tau_{n,m}} \\ \Big\downarrow \varepsilon^{\beta}_{(m,n)} = r^{\tau_{n,m}}_{\mathbb{1}^{\tau_{m,n}}} \\ (1_{\boxtimes^{\tau}})_{(m,n)} = \mathbb{1}^{nm} = \mathbb{1}^{mn} \end{array}$$

in $\mathsf{Mat}^{\mathsf{C}}_{mn,mn}$ for $m,n \geq 0$, with r as in (8.13.13). ◇

Explanation 8.13.40. Consider (8.13.39).

- $\varepsilon^{\beta}_{(m,n)}$ is well defined by (8.13.8) and (8.13.19).
- The kth diagonal entry of $\varepsilon^{\beta}_{(m,n)}$ has the following form.

$$\big[(\beta\beta^{\bullet})_{(m,n)}\big]_{kk} \xrightarrow{(\lambda^{\oplus},\rho^{\oplus},\rho^{\otimes},\rho^{\bullet})} \mathbb{1}$$

- Each off-diagonal entry of $\varepsilon^{\beta}_{(m,n)}$ has the form

$$\big[(\beta\beta^{\bullet})_{(m,n)}\big]_{kl} \xrightarrow{(\lambda^{\oplus},\rho^{\otimes},\rho^{\bullet})} \mathbb{0}$$

for $1 \leq k \neq l \leq mn$. ◇

Lemma 8.13.41. *η^{β} and ε^{β} in, respectively, Definitions 8.13.35 and 8.13.38 are invertible modifications.*

Proof. Each component 2-cell of each of $\eta^{\beta}\colon 1_{\boxtimes} \longrightarrow \beta^{\bullet}\beta$ and $\varepsilon^{\beta}\colon \beta\beta^{\bullet} \longrightarrow 1_{\boxtimes^{\tau}}$ is invertible by (8.13.13).

The modification axiom (6.3.2) for η^{β} states that for each pair of 1-cells

$$(A,B) \in \mathsf{Mat}^{\mathsf{C}}_{m,m'} \times \mathsf{Mat}^{\mathsf{C}}_{n,n'}$$

the following pasting diagram equality holds in $\mathsf{Mat}^{\mathsf{C}}_{mn,m'n'}$.

(8.13.42)
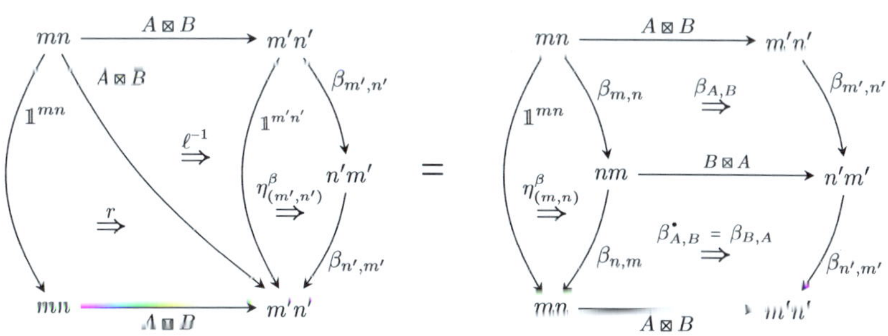

The proof of this equality is adapted from that of (8.8.40), the modification axiom for η^{a}. The proof of the modification axiom for ε^{β} is obtained similarly. □

Lemma 8.13.43. *The quadruple $(\beta, \beta^{\bullet}, \eta^{\beta}, \varepsilon^{\beta})$ satisfies the triangle identities.*

Proof. The left triangle identity (6.3.10) states the commutativity of the left diagram below.

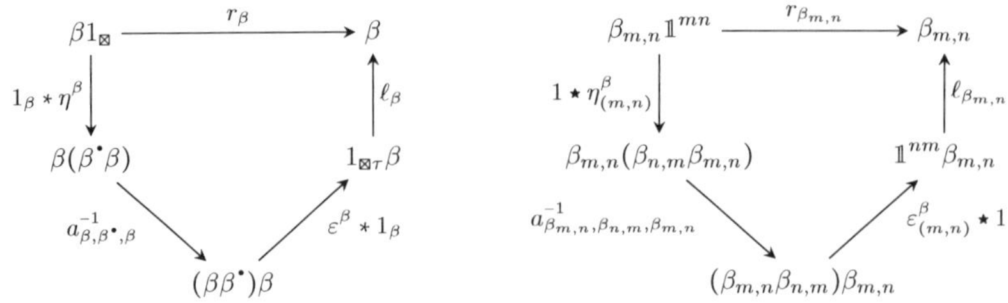

To show that the left diagram is commutative, it suffices to show that the two composites have the same component 2-cells. So it suffices to show that the right diagram is commutative for $m, n \geq 0$. The proof of the commutativity of the right diagram is adapted from that of (8.8.46), which is the left triangle identity for $a^{\boxtimes}$. The proof for the right triangle identity is obtained similarly. □

Lemma 8.13.44. *The quadruple* $(\beta, \beta^\bullet, \eta^\beta, \varepsilon^\beta)$ *with*

- β *in Definition 8.13.22,*
- $\beta^\bullet$ *in Definition 8.13.30,*
- η^β *in Definition 8.13.35, and*
- ε^β *in Definition 8.13.38*

is an adjoint equivalence.

Proof. This follows from Lemmas 8.13.27, 8.13.34, 8.13.41, and 8.13.43. □

8.14. The Matrix Braided Monoidal Bicategory

We continue to assume that C is a tight symmetric bimonoidal category. In Section 8.13, we constructed the braiding $(\beta, \beta^\bullet, \eta^\beta, \varepsilon^\beta)$. In this section, we construct the left hexagonator and the right hexagonator and observe that $\mathsf{Mat}^{\mathsf{C}}$ is a braided monoidal bicategory as in Definition 6.5.3.

- The left hexagonator is in Definition 8.14.9.
- The right hexagonator is in Definition 8.14.21.
- The verification of the braided monoidal bicategory axioms is in Theorem 8.14.26.

The Left Hexagonator. To define the left hexagonator, we will make use of an auxiliary 2-cell $h_{m|n,p}$ in Lemma 8.14.4 below.

Motivation 8.14.1. The left hexagonator $R_{-|--}$ in a braided monoidal bicategory is the 2-cell analogue of the left hexagon diagram in (II.1.3.17). With the braiding denoted by β, the left hexagonator compares $\beta_{A,B} \boxtimes 1_C$ and $1_B \boxtimes \beta_{A,C}$ with $\beta_{A,B\boxtimes C}$. In the context of the matrix bicategory $\mathsf{Mat}^{\mathsf{C}}$, the left hexagonator involves a comparison 2-cell

$$\left(\mathbb{1}^n \boxtimes \beta_{m,p}\right)\left(\beta_{m,n} \boxtimes \mathbb{1}^p\right) \xrightarrow{h_{m|n,p}} \beta_{m,np} \in \mathsf{Mat}^{\mathsf{C}}_{mnp,mnp}$$

for objects $m, n, p \in \mathsf{Mat}^{\mathsf{C}}$, that is, nonnegative integers. To define this comparison 2-cell, in Lemma 8.14.2, we first describe explicitly the three 1-cells involved. The definition of $h_{m|n,p}$ is given in the proof of Lemma 8.14.4 in terms of the structure morphisms in C. ◇

Lemma 8.14.2. *For $m,n,p \geq 0$, $1 \leq i \leq m$, $1 \leq j \leq n$, and $1 \leq k \leq p$, define the following integers.*

$$I = i + \big[k - 1 + (j-1)p\big]m \qquad J = k + \big[j - 1 + (i-1)n\big]p$$
$$J' = k + \big[i - 1 + (j-1)m\big]p$$

Then the following equalities hold for $1 \leq u, v \leq mnp$.

$$(8.14.3)\quad \begin{aligned}
(\beta_{m,np})_{uv} &= \begin{cases} \mathbb{1} & \text{if } (u,v) = (I,J) \text{ for some } (i,j,k), \text{ and} \\ \mathbb{0} & \text{otherwise.} \end{cases} \\
(\mathbb{1}^n \boxtimes \beta_{m,p})_{uv} &= \begin{cases} \mathbb{1} \otimes \mathbb{1} & \text{if } (u,v) = (I,J') \text{ for some } (i,j,k), \text{ and} \\ \mathbb{0} \otimes \mathbb{0},\ \mathbb{0} \otimes \mathbb{1}, \text{ or } \mathbb{1} \otimes \mathbb{0} & \text{otherwise.} \end{cases} \\
(\beta_{m,n} \boxtimes \mathbb{1}^p)_{uv} &= \begin{cases} \mathbb{1} \otimes \mathbb{1} & \text{if } (u,v) = (J',J) \text{ for some } (i,j,k), \text{ and} \\ \mathbb{0} \otimes \mathbb{0},\ \mathbb{0} \otimes \mathbb{1}, \text{ or } \mathbb{1} \otimes \mathbb{0} & \text{otherwise.} \end{cases}
\end{aligned}$$

Proof. If m, n, or p is 0, then $\mathsf{Mat}^{\mathsf{C}}_{mnp,mnp}$ is the terminal category, and the three matrices in (8.14.3) are all empty matrices. In this case, there is nothing to prove, so we assume that $m,n,p > 0$.

With $1 \leq j \leq n$ and $1 \leq k \leq p$, the sum $k + (j-1)p$ runs through $\{1, \ldots, np\}$. Therefore, the first equality in (8.14.3) is the result of applying (8.13.26) to $\beta_{m,np}$.

For the second equality, by (8.1.6) and (8.6.3), there is an equality as follows.

$$\mathbb{1}^n \boxtimes \beta_{m,p} = \begin{bmatrix} \mathbb{1} \boxtimes \beta_{m,p} & \cdots & \mathbb{0} \boxtimes \beta_{m,p} \\ \vdots & \ddots & \vdots \\ \mathbb{0} \boxtimes \beta_{m,p} & \cdots & \mathbb{1} \boxtimes \beta_{m,p} \end{bmatrix}$$

Among its n^2 blocks,

- each of the n diagonal blocks is $\mathbb{1} \boxtimes \beta_{m,p} \in \mathsf{Mat}^{\mathsf{C}}_{mp,mp}$, and
- each off-diagonal block is $\mathbb{0} \boxtimes \beta_{m,p} \in \mathsf{Mat}^{\mathsf{C}}_{mp,mp}$.

By (8.13.26), the (I,J')-entry in $\mathbb{1}^n \boxtimes \beta_{m,p}$ is the following object in the jth diagonal block $\mathbb{1} \boxtimes \beta_{m,p}$.

$$\begin{aligned}
(\mathbb{1}^n \boxtimes \beta_{m,p})_{I,J'} &= (\mathbb{1} \boxtimes \beta_{m,p})_{(i+(k-1)m,\, k+(i-1)p)} \\
&= \mathbb{1} \otimes \mathbb{1}
\end{aligned}$$

Every other entry in the jth diagonal block $\mathbb{1} \boxtimes \beta_{m,p}$ is $\mathbb{1} \otimes \mathbb{0}$. Each entry in each off-diagonal block $\mathbb{0} \boxtimes \beta_{m,p}$ is either $\mathbb{0} \otimes \mathbb{0}$ or $\mathbb{0} \otimes \mathbb{1}$. This proves the second equality in (8.14.3).

For the last equality in (8.14.3), by (8.6.3), $\beta_{m,n} \boxtimes \mathbb{1}^p$ has $(mn)^2$ blocks. By (8.13.26),

- its $\big(i + (j-1)m, j + (i-1)n\big)$-th block is $\mathbb{1} \boxtimes \mathbb{1}^p \in \mathsf{Mat}^{\mathsf{C}}_{p,p}$ for $1 \leq i \leq m$ and $1 \leq j \leq n$, and
- every other block is $\mathbb{0} \boxtimes \mathbb{1}^p \in \mathsf{Mat}^{\mathsf{C}}_{p,p}$.

In the $\big(i + (j-1)m, j + (i-1)n\big)$-th block of $\beta_{m,n} \boxtimes \mathbb{1}^p$, namely, $\mathbb{1} \boxtimes \mathbb{1}^p$,

- for $1 \leq k \leq p$, the (k,k)-entry is
$$(\beta_{m,n} \boxtimes \mathbb{1}^p)_{J',J} = \mathbb{1} \otimes \mathbb{1},$$
and
- every other entry is $\mathbb{1} \otimes \mathbb{0}$.

Each entry in every other block of $\beta_{m,n} \boxtimes \mathbb{1}^p$, namely, $\mathbb{0} \boxtimes \mathbb{1}^p$, is either $\mathbb{0} \otimes \mathbb{1}$ or $\mathbb{0} \otimes \mathbb{0}$. This proves the last equality in (8.14.3). □

Using Lemma 8.14.2, we now define the comparison 2-cell $h_{m|n,p}$.

Lemma 8.14.4. *For $m, n, p \geq 0$, there is an isomorphism*

$$(\mathbb{1}^n \boxtimes \beta_{m,p})(\beta_{m,n} \boxtimes \mathbb{1}^p) \xrightarrow{h_{m|n,p}} \beta_{m,np} \in \mathsf{Mat}^{\mathsf{C}}_{mnp,mnp} \tag{8.14.5}$$

that is defined entrywise by the structure morphisms in C.

Proof. If m, n, or p is 0, then $\mathsf{Mat}^{\mathsf{C}}_{mnp,mnp}$ is the terminal category, and $h_{m|n,p}$ is the identity morphism of the empty matrix. Next we assume that $m, n, p > 0$.

Using Lemma 8.14.2 and its notation, there are equalities as follows.

$$\begin{aligned}
&\Big[(\mathbb{1}^n \boxtimes \beta_{m,p})(\beta_{m,n} \boxtimes \mathbb{1}^p)\Big]_{I,J} \\
&= \Big[\bigoplus_{t=1}^{mnp} (\mathbb{1}^n \boxtimes \beta_{m,p})_{I,t} \otimes (\beta_{m,n} \boxtimes \mathbb{1}^p)_{t,J}\Big]_{\mathsf{lt}} \\
&= \Big[\underbrace{z_1 \oplus \cdots \oplus z_{J'-1}}_{\text{empty if } J'=1} \oplus (\mathbb{1}\mathbb{1})(\mathbb{1}\mathbb{1}) \oplus \underbrace{z_{J'+1} \oplus \cdots \oplus z_{mnp}}_{\text{empty if } J'=mnp}\Big]_{\mathsf{lt}}
\end{aligned}$$

Here each z_t for $1 \leq t \leq mnp$ with $t \neq J'$ has the form $(\delta_1\delta_2)(\delta_3\delta_4)$ such that

- each $\delta_i \in \{\mathbb{0}, \mathbb{1}\}$ and
- $(\delta_1, \delta_2) \neq (\mathbb{1}, \mathbb{1}) \neq (\delta_3, \delta_4)$.

The (I,J)-entry of $h_{m|n,p}$ is the following composite isomorphism in C.

$$\begin{array}{ccc}
\Big[(\mathbb{1}^n \boxtimes \beta_{m,p})(\beta_{m,n} \boxtimes \mathbb{1}^p)\Big]_{I,J} & \xrightarrow{(h_{m|n,p})_{I,J}} & (\beta_{m,np})_{I,J} \\
\Big\downarrow{\scriptstyle (\lambda^{\otimes},\lambda^{\bullet},\rho^{\bullet})} & & \Big\| \\
\Big(\underbrace{\mathbb{0} \oplus \cdots \oplus \mathbb{0}}_{J'-1 \text{ copies of } \mathbb{0}} \oplus \mathbb{1} \underbrace{\oplus \mathbb{0} \oplus \cdots \oplus \mathbb{0}}_{mnp-J' \text{ copies of } \mathbb{0}}\Big)_{\mathsf{lt}} & \xrightarrow{(\lambda^{\oplus},\rho^{\oplus})} & \mathbb{1}
\end{array} \tag{8.14.6}$$

Here we used the notation in (8.7.6) for morphisms. For example, $(\lambda^{\otimes}, \lambda^{\bullet}, \rho^{\bullet})$ is a composite of morphisms, each being an iterated sum and product of identity morphisms and at most one component of $\lambda^{\otimes}$, $\lambda^{\bullet}$, or $\rho^{\bullet}$.

If $(u,v) \neq (I,J)$ for any (i,j,k), then Lemma 8.14.2 implies that

$$\Big[(\mathbb{1}^n \boxtimes \beta_{m,p})(\beta_{m,n} \boxtimes \mathbb{1}^p)\Big]_{uv} = \big[y_1 \oplus \cdots \oplus y_{mnp}\big]_{\mathsf{lt}}.$$

Here each y_t for $1 \leq t \leq mnp$ has the form $(\delta_1\delta_2)(\delta_3\delta_4)$ with

- each $\delta_i \in \{\mathbb{0}, \mathbb{1}\}$ and
- $(\delta_1, \delta_2, \delta_3, \delta_4) \neq (\mathbb{1}, \mathbb{1}, \mathbb{1}, \mathbb{1})$.

The (u,v)-entry of $h_{m|n,p}$ is the following composite isomorphism in C.

(8.14.7)

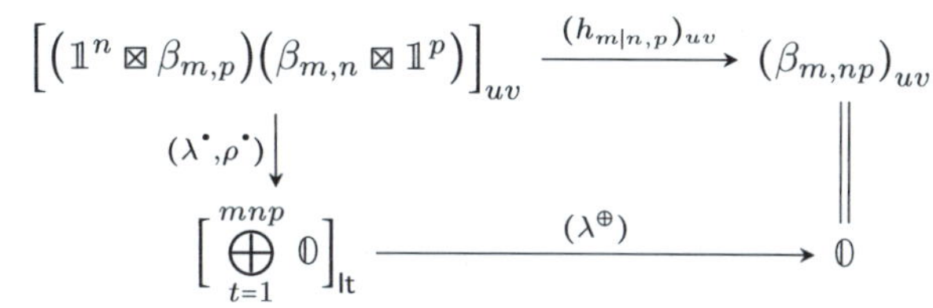

This finishes the construction of the isomorphism $h_{m|n,p}$. □

Explanation 8.14.8. While some choices are made in the definition of $h_{m|n,p}$—for example, in the order in which $\lambda^{\oplus}$ and $\rho^{\oplus}$ are applied in (8.14.6)—its value as a 2-cell in $\mathsf{Mat}^{\mathsf{C}}$ is unique. This follows from essentially the same argument as in Examples 8.1.14 and 8.2.5 using the set $X = \{0^X, 1^X\}$ of formal variables and the function $\varphi: X \longrightarrow \mathsf{Ob}(\mathsf{C})$ with $\varphi(0^X) = \mathbb{0}$ and $\varphi(1^X) = \mathbb{1}$.

- In the (I,J)-entry for some (i,j,k), this uniqueness follows from the regularity of 1^X, Lemma 3.1.29, and the Coherence Theorem 3.9.1.
- In every other entry, this uniqueness follows from Proposition 3.5.33. ◇

Using the comparison 2-cell $h_{m|n,p}$, we now define the left hexagonator in the matrix bicategory $\mathsf{Mat}^{\mathsf{C}}$.

Definition 8.14.9. Define $R_{-|--}$ as consisting of the 2-cells $R_{m|n,p} \in \mathsf{Mat}^{\mathsf{C}}_{mnp,npm}$ for $m,n,p \geq 0$ given by the following pasting diagram.

(8.14.10)

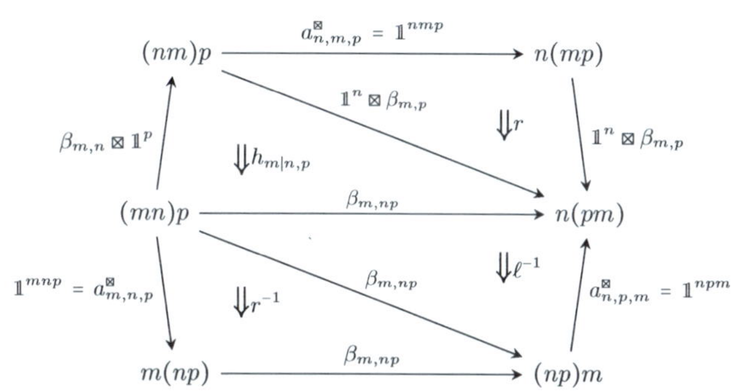

This finishes the definition of $R_{-|--}$. ◇

Explanation 8.14.11. Consider (8.14.10).

- By Convention 6.2.12, the (co)domain 1-cell has the left normalized bracketing.
- Each component 1-cell of β is a permutation matrix by (8.13.23).
- Each component 1-cell of $a^{\boxtimes}$ is an identity matrix by (8.8.2).
- The top 2-cell
$$r = r_{\mathbb{1}^n \boxtimes \beta_{m,p}} : \big(\mathbb{1}^n \boxtimes \beta_{m,p}\big)\mathbb{1}^{nmp} \longrightarrow \mathbb{1}^n \boxtimes \beta_{m,p}$$
is the base right unitor in (8.2.8).
- $h_{m|n,p}$ is the comparison 2-cell in (8.14.5).
- The 2-cell
$$\ell^{-1} = \ell^{-1}_{\beta_{m,np}} : \beta_{m,np} \longrightarrow \mathbb{1}^{npm}\beta_{m,np}$$
is the inverse of the base left unitor in (8.2.2).
- The bottom 2-cell $r^{-1} = r^{-1}_{\beta_{m,np}}$ is the inverse of the base right unitor.

- By (8.2.4), (8.2.10), (8.14.6), and (8.14.7), each entry of $R_{m|n,p}$ is a composite of morphisms, each being an iterated sum and product of identity morphisms and at most one component of $\lambda^{\pm\oplus}$, $\rho^{\pm\oplus}$, $\lambda^{\pm\otimes}$, $\rho^{\pm\otimes}$, $\lambda^{\pm\bullet}$, or $\rho^{\pm\bullet}$. ◇

Lemma 8.14.12. *In Definition 8.14.9, $R_{-|--}$ is an invertible modification.*

Proof. Each component 2-cell of $R_{-|--}$ is entrywise a composite of isomorphisms in C, so it is invertible.

Using the notation in the proof of Lemma 8.10.4, the modification axiom (6.3.2) for $R_{-|--}$ for the 1-cells

$$(B^1, B^2, B^3) \in \mathsf{Mat}^{\mathsf{C}}_{m,m'} \times \mathsf{Mat}^{\mathsf{C}}_{n,n'} \times \mathsf{Mat}^{\mathsf{C}}_{p,p'}$$

states the following pasting diagram equality in $\mathsf{Mat}^{\mathsf{C}}_{mnp,n'p'm'}$.

(8.14.13)

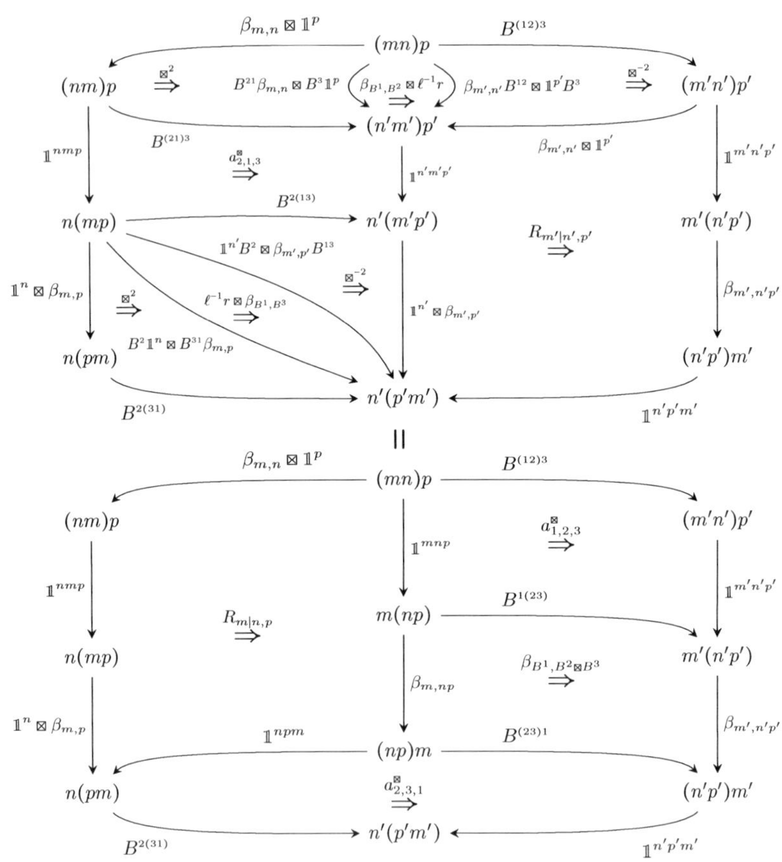

For example,

$$B^{2(31)} = B^2 \boxtimes (B^3 \boxtimes B^1)$$
$$a^{\boxtimes}_{2,1,3} = a^{\boxtimes}_{B^2,B^1,B^3}.$$

The proof of the equality (8.14.13) is adapted from that of (8.10.5), which is the modification axiom for the pentagonator π. When restricted to a typical entry, the two pasting diagrams in (8.14.13) yield a zigzag diagram analogous to (8.10.8). The object Y in (8.10.9) is now replaced by the corresponding entry of $B^{(12)3}$, which has the form

$$Y = \big(B^1_{i'i} \otimes B^2_{j'j}\big) \otimes B^3_{k'k}.$$

Since the proof of the equality (8.10.5) has all the detail, we safely skip the detail here. □

The Right Hexagonator. Similar to the left hexagonator $R_{-|--}$, to define the right hexagonator, we will make use of a comparison 2-cell that involves the 1-cells in the following lemma.

Lemma 8.14.14. *For $m,n,p \geq 0$, $1 \leq i \leq m$, $1 \leq j \leq n$, and $1 \leq k \leq p$, define the following integers.*

$$K = j + \big[i - 1 + (k-1)m\big]n \qquad J = k + \big[j - 1 + (i-1)n\big]p$$
$$K' = j + \big[k - 1 + (i-1)p\big]n$$

Then the following equalities hold for $1 \leq u, v \leq mnp$.

(8.14.15)

$$(\beta_{mn,p})_{uv} = \begin{cases} \mathbb{1} & \text{if } (u,v) = (K,J) \text{ for some } (i,j,k), \text{ and} \\ \mathbb{0} & \text{otherwise.} \end{cases}$$

$$(\mathbb{1}^m \boxtimes \beta_{n,p})_{uv} = \begin{cases} \mathbb{1}\otimes\mathbb{1} & \text{if } (u,v) = (K',J) \text{ for some } (i,j,k), \text{ and} \\ \mathbb{0}\otimes\mathbb{0},\ \mathbb{0}\otimes\mathbb{1},\ \text{or } \mathbb{1}\otimes\mathbb{0} & \text{otherwise.} \end{cases}$$

$$(\beta_{m,p} \boxtimes \mathbb{1}^n)_{uv} = \begin{cases} \mathbb{1}\otimes\mathbb{1} & \text{if } (u,v) = (K,K') \text{ for some } (i,j,k), \text{ and} \\ \mathbb{0}\otimes\mathbb{0},\ \mathbb{0}\otimes\mathbb{1},\ \text{or } \mathbb{1}\otimes\mathbb{0} & \text{otherwise.} \end{cases}$$

Proof. As in the proof of Lemma 8.14.2, we may assume that $m,n,p > 0$, since there is nothing to prove otherwise. With $1 \leq i \leq m$ and $1 \leq j \leq n$, the sum $j + (i-1)n$ runs through $\{1, \ldots, mn\}$. Therefore, the first equality in (8.14.15) is the result of applying (8.13.26) to $\beta_{mn,p}$.

The second equality in (8.14.15) is obtained from the second equality in (8.14.3) by switching (i,m) with (j,n). The third equality in (8.14.15) is obtained from the third equality in (8.14.3) by switching (j,n) with (k,p). □

Using Lemma 8.14.14, we now define the comparison 2-cell that will be used in the definition of the right hexagonator.

Lemma 8.14.16. *For $m,n,p \geq 0$, there is an isomorphism*

(8.14.17)
$$\big(\beta_{m,p} \boxtimes \mathbb{1}^n\big)\big(\mathbb{1}^m \boxtimes \beta_{n,p}\big) \xrightarrow{h_{m,n|p}} \beta_{mn,p} \in \mathsf{Mat}^{\mathsf{C}}_{mnp,mnp}$$

that is defined entrywise by the structure morphisms in C.

Proof. As in the proof of Lemma 8.14.4, we may assume that $m,n,p > 0$, since otherwise $h_{m,n|p}$ is the identity morphism of the empty matrix.

Using Lemma 8.14.14 and its notation, analogous to (8.14.6), the (K,J)-entry of $h_{m,n|p}$ is the following composite isomorphism in C.

$$\begin{array}{ccc}
\Big[(\beta_{m,p} \boxtimes \mathbb{1}^n)(\mathbb{1}^m \boxtimes \beta_{n,p})\Big]_{K,J} & \xrightarrow{(h_{m,n|p})_{K,J}} & (\beta_{mn,p})_{K,J} \\
\Big\downarrow{\scriptstyle(\lambda^{\otimes},\lambda^{\bullet},\rho^{\bullet})} & & \Big\| \\
\Big(\underbrace{\mathbb{0}\oplus\cdots\oplus\mathbb{0}\oplus}_{K'-1 \text{ copies of } \mathbb{0}} \ \mathbb{1} \ \underbrace{\oplus\mathbb{0}\oplus\cdots\oplus\mathbb{0}}_{mnp-K' \text{ copies of } \mathbb{0}}\Big)_{\mathsf{lt}} & \xrightarrow{(\lambda^{\oplus},\rho^{\oplus})} & \mathbb{1}
\end{array} \tag{8.14.18}$$

Suppose $(u,v) \neq (K,J)$ for any (i,j,k). Analogous to (8.14.7), the (u,v)-entry of $h_{m,n|p}$ is the following composite isomorphism in C.

$$\begin{array}{ccc}
\Big[(\beta_{m,p} \boxtimes \mathbb{1}^n)(\mathbb{1}^m \boxtimes \beta_{n,p})\Big]_{uv} & \xrightarrow{(h_{m,n|p})_{uv}} & (\beta_{mn,p})_{uv} \\
\Big\downarrow{\scriptstyle(\lambda^{\bullet},\rho^{\bullet})} & & \Big\| \\
\Big[\bigoplus_{t=1}^{mnp} \mathbb{0}\Big]_{\mathsf{lt}} & \xrightarrow{(\lambda^{\oplus})} & \mathbb{0}
\end{array} \tag{8.14.19}$$

This finishes the construction of the isomorphism $h_{m,n|p}$. □

Explanation 8.14.20. As in Explanation 8.14.8, $h_{m,n|p}$ is a well-defined 2-cell in $\mathsf{Mat}^{\mathsf{C}}_{mnp,mnp}$ by Proposition 3.5.33 and Theorem 3.9.1. ◇

We now define the right hexagonator in the matrix bicategory $\mathsf{Mat}^{\mathsf{C}}$.

Definition 8.14.21. Define $R_{-\,-|-}$ as consisting of the 2-cells $R_{m,n|p} \in \mathsf{Mat}^{\mathsf{C}}_{mnp,pmn}$ for $m,n,p \geq 0$ given by the following pasting diagram.

(8.14.22)

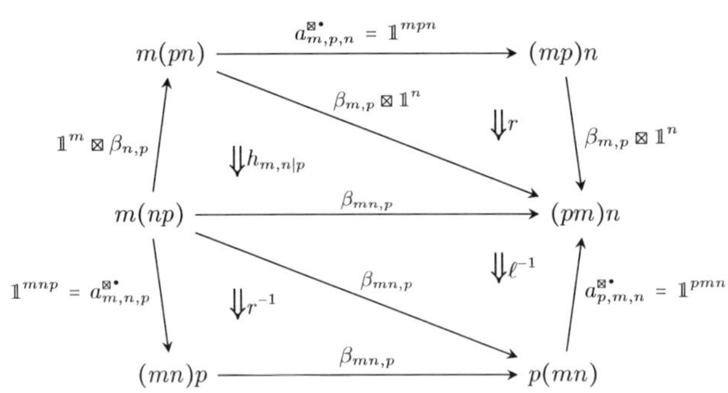

This finishes the definition of $R_{-\,-|-}$. ◇

Explanation 8.14.23. Consider (8.14.22).

- By Convention 6.2.12, the (co)domain 1-cell has the left normalized bracketing.
- $h_{m,n|p}$ is the comparison 2-cell in (8.14.17).
- $\ell = \ell_{\beta_{mn,p}}$ is the base left unitor (8.2.2).
- The top $r = r_{\beta_{m,p}\boxtimes\mathbb{1}^n}$ and the bottom $r = r_{\beta_{mn,p}}$ are the base right unitor (8.2.8).
- Each component 1-cell of β is a permutation matrix by (8.13.23).
- Each component 1-cell of $a^{\boxtimes\bullet}$ is an identity matrix by (8.8.34).

- By (8.2.4), (8.2.10), (8.14.18), and (8.14.19), each entry of $R_{m,n|p}$ is a composite of morphisms, each being an iterated sum and product of identity morphisms and at most one component of $\lambda^{\pm\oplus}$, $\rho^{\pm\oplus}$, $\lambda^{\pm\otimes}$, $\rho^{\pm\otimes}$, $\lambda^{\pm\bullet}$, or $\rho^{\pm\bullet}$. $\diamond$

Lemma 8.14.24. *In Definition 8.14.21, $R_{-,-|-}$ is an invertible modification.*

Proof. Each component 2-cell of $R_{-,-|-}$ is entrywise a composite of isomorphisms in C, so it is invertible.

Using the notation in the proof of Lemma 8.10.4, the modification axiom (6.3.2) for $R_{-,-|-}$ for the 1-cells

$$(B^1, B^2, B^3) \in \mathsf{Mat}^{\mathsf{C}}_{m,m'} \times \mathsf{Mat}^{\mathsf{C}}_{n,n'} \times \mathsf{Mat}^{\mathsf{C}}_{p,p'}$$

states the following pasting diagram equality in $\mathsf{Mat}^{\mathsf{C}}_{mnp,p'm'n'}$.

(8.14.25)

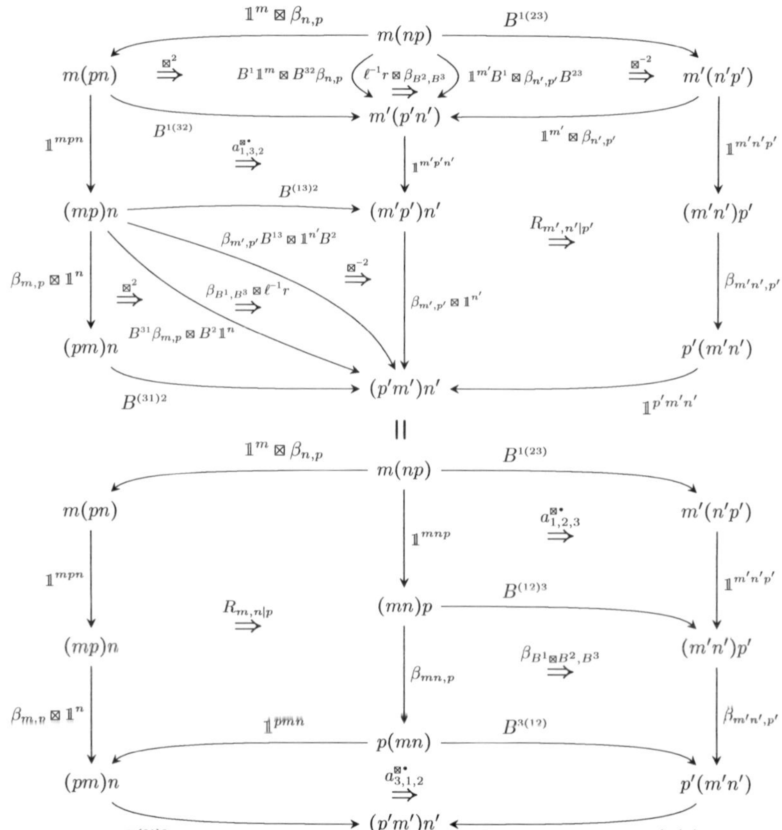

Analogous to (8.14.13), the proof of the equality (8.14.25) is adapted from the detailed proof of (8.10.5), which is the modification axiom for the pentagonator π. $\square$

The Axioms.

Theorem 8.14.26. *For each tight symmetric bimonoidal category C, the matrix bicategory $\mathsf{Mat}^{\mathsf{C}}$ equipped with*

- *the braiding $(\beta, \beta^{\bullet}, \eta^{\beta}, \varepsilon^{\beta})$ in Section 8.13,*

- *the left hexagonator* $R_{-|--}$ *in Definition 8.14.9, and*
- *the right hexagonator* $R_{--|-}$ *in Definition 8.14.21*

is a braided monoidal bicategory.

Proof. We already have the following.

- $\mathsf{Mat}^{\mathsf{C}}$ is a monoidal bicategory by Theorem 8.12.9.
- The braiding is an adjoint equivalence by Lemma 8.13.44.
- The left hexagonator and the right hexagonator are invertible modifications by, respectively, Lemmas 8.14.12 and 8.14.24.

It remains to check the four axioms in Definition 6.5.3.

For objects $m, n, p \in \mathsf{Mat}^{\mathsf{C}}$, the Yang-Baxter axiom states the following pasting diagram equality in $\mathsf{Mat}^{\mathsf{C}}_{mnp,pnm}$.

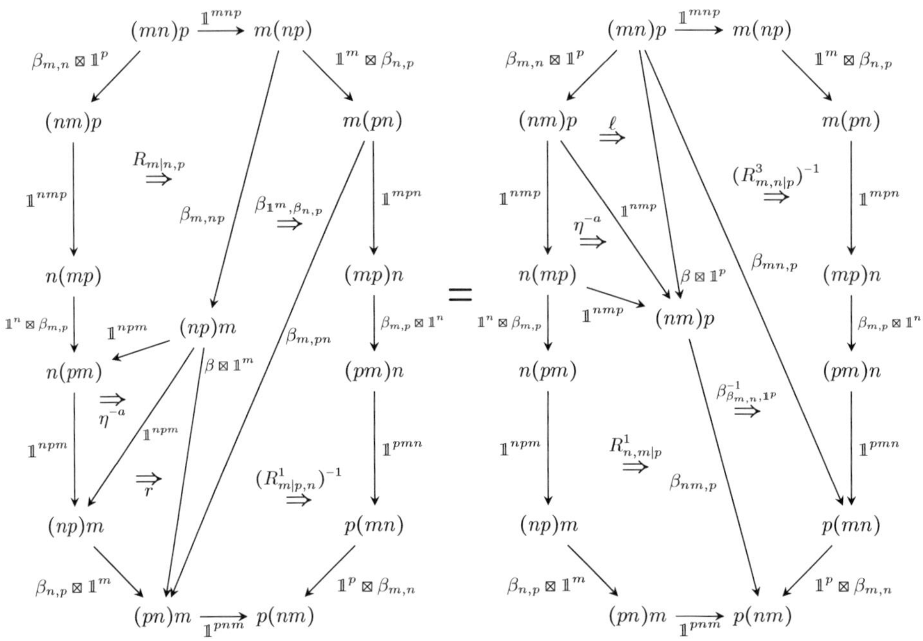

The proof of this equality is adapted from that of (8.12.3), which is the non-abelian 4-cocycle condition, in Lemma 8.12.1. The key points are as follows.

- Component 1-cells of $a^{\boxtimes}$ (8.8.2) and $a^{\boxtimes\bullet}$ (8.8.34) are identity matrices.
- Each component 1-cell of the braiding β (8.13.23) is a permutation matrix. It has precisely one $\mathbb{1}$ in each row and each column and $\mathbb{0}$ in every other entry.
- Each 2-cell in the Yang-Baxter axiom is entrywise defined by the structure morphisms in C.

These facts allow us to reuse steps (1)–(5) in the proof of Lemma 8.12.1. In the last two steps, instead of considering the two cases $i = j$ and $i \neq j$, we simply observe that by Lemma 3.1.29, all the vertices in the diagram D in $\mathsf{Gr}(X)$ have the same support, which is that of either 0^X or 1^X. Therefore, either Proposition 3.5.33 (for 0^X) or the Coherence Theorem 3.9.1 (for 1^X) implies the commutativity of D in C. This, in turn, proves the Yang-Baxter axiom.

The (3,1)-crossing, the (1,3)-crossing, and the (2,2)-crossing axioms are proved in the same way by following the steps in the proof of Lemma 8.12.1 as discussed in the previous paragraph. □

8.15. The Matrix Symmetric Monoidal Bicategory

For a tight symmetric bimonoidal category C as in Definition 2.1.2, we saw in Theorem 8.14.26 that the matrix bicategory $\mathsf{Mat}^{\mathsf{C}}$ is a braided monoidal bicategory. In this section, we finish the proof that $\mathsf{Mat}^{\mathsf{C}}$ is a symmetric monoidal bicategory as in Definition 6.5.9. Since a symmetric monoidal bicategory is a sylleptic monoidal bicategory as in Definition 6.5.7 with an extra axiom, our next task is to define the syllepsis that relates the braid-square to the identity.

Definition 8.15.1. Define the data of a modification ν with component 2-cells

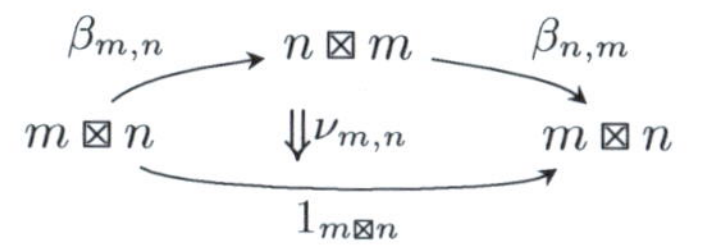

defined as

$$\beta_{n,m}\beta_{m,n} = \mathbb{1}^{\tau_{n,m}}\mathbb{1}^{\tau_{m,n}} \xrightarrow{\nu_{m,n} = r^{\tau_{m,n}}_{\mathbb{1}^{\tau_{n,m}}}} \mathbb{1}^{mn} \in \mathsf{Mat}^{\mathsf{C}}_{mn,mn}$$

for $m, n \geq 0$ and r as in (8.13.13). ◇

Explanation 8.15.2. In the setting of Definition 8.15.1, there are equalities

$$\nu_{m,n} = r^{\tau_{m,n}}_{\mathbb{1}^{\tau_{n,m}}} = \left(\eta^{\beta}_{(m,n)}\right)^{-1} = \varepsilon^{\beta}_{(n,m)}$$

with η^{β} the unit (8.13.36) and ε^{β} the counit (8.13.39) of the braiding. By (8.13.14) and the notation for morphisms in (8.7.6),

- each diagonal entry of $\nu_{m,n}$ has the form $(\lambda^{\oplus}, \rho^{\oplus}, \rho^{\otimes}, \rho^{\bullet})$, and
- each off-diagonal entry of $\nu_{m,n}$ has the form $(\lambda^{\oplus}, \rho^{\otimes}, \rho^{\bullet})$.

The entries of the domain 1-cell $\mathbb{1}^{\tau_{n,m}}\mathbb{1}^{\tau_{m,n}}$ are described in (8.13.9) and (8.13.10). ◇

Lemma 8.15.3. *ν in Definition 8.15.1 is an invertible modification.*

Proof. Each component 2-cell of ν is invertible by (8.13.13). The modification axiom (6.3.2) for ν is obtained from (8.13.42), which is the modification axiom for η^{β}, by pasting

- $\left(\eta^{\beta}_{(m,n)}\right)^{-1}$ on the left and
- $\left(\eta^{\beta}_{(m',n')}\right)^{-1}$ on the right

in each pasting diagram. □

We are now ready for the main result of this chapter.

Theorem 8.15.4 (Bicategorification)**.** *For each tight symmetric bimonoidal category C, the quintuple*

$$\left(\mathsf{Mat}^{\mathsf{C}}, \beta, R_{-|--}, R_{--|-}, \nu\right)$$

is a symmetric monoidal bicategory.

Proof. By Theorem 8.14.26, the monoidal bicategory $\mathsf{Mat}^{\mathsf{C}}$ in Theorem 8.12.9 becomes a braided monoidal bicategory when equipped with

- the braiding $(\beta, \beta^{\bullet}, \eta^{\beta}, \varepsilon^{\beta})$ in Section 8.13 and
- the hexagonators $R_{-|--}$ and $R_{--|-}$ in Definitions 8.14.9 and 8.14.21.

The syllepsis ν is an invertible modification by Lemma 8.15.3. It remains to check the two syllepsis axioms in Definition 6.5.7 and the triple braid axiom (6.5.10).

The (2,1)-syllepsis axiom is the next pasting diagram equality in $\mathsf{Mat}^{\mathsf{C}}_{mnp,mnp}$ for $m,n,p \geq 0$. Unlabeled edges are identity 1-cells.

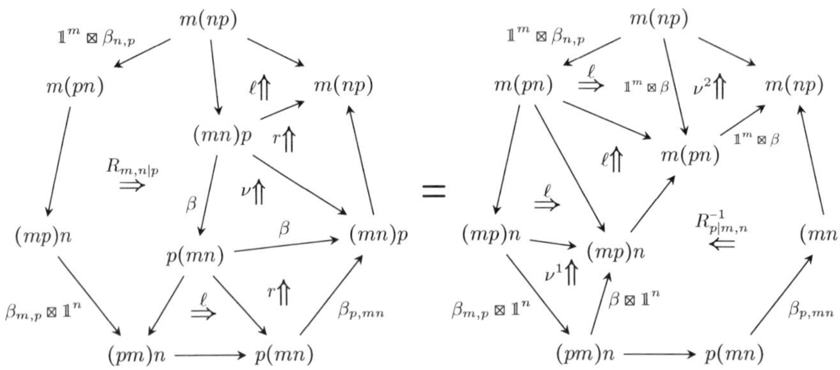

In these pasting diagrams, we used the fact that each component 2-cell of the counit ε^a (8.8.38) is an instance of the base left unitor ℓ (8.2.2), applied to an identity 1-cell. The proof of the above equality is adapted from that of (8.12.3), which is the non-abelian 4-cocycle condition, in Lemma 8.12.1 by following steps (1)–(5) there. As we pointed out in the proof of Theorem 8.14.26, the key points that allow us to reuse those steps are as follows.

- Component 1-cells of $a^{\boxtimes}$ (8.8.2) and $a^{\boxtimes\bullet}$ (8.8.34) are identity matrices.
- Each component 1-cell of the braiding β (8.13.23) is a permutation matrix, with precisely one $\mathbb{1}$ in each row and each column and $\mathbb{0}$ in every other entry.
- Each 2-cell in the (2,1)-syllepsis axiom is entrywise defined by the structure morphisms in C.

The (1,2)-syllepsis axiom is proved in the same way.

The triple braid axiom (6.5.10) states the following pasting diagram equality in $\mathsf{Mat}^{\mathsf{C}}_{mn,nm}$ for $m,n \geq 0$.

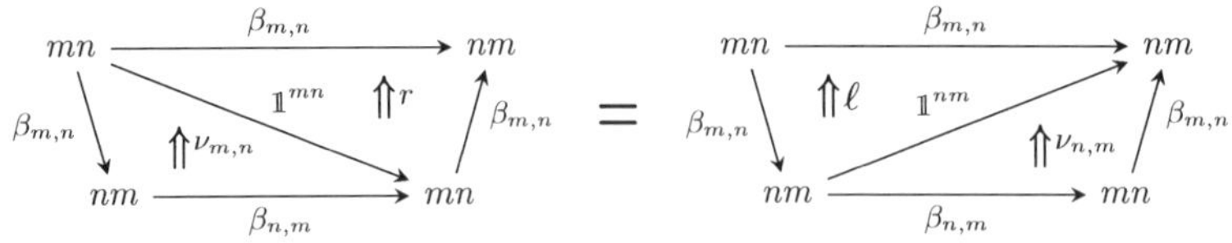

The proof of this equality is also adapted from that of (8.12.3) in Lemma 8.12.1 by following steps (1)–(5) there. In the last two steps, instead of considering the two cases $i = j$ and $i \neq j$, we simply observe that by Lemma 3.1.29, all the vertices in the diagram D in $\mathsf{Gr}(X)$ have the same support, which is that of either 0^X or 1^X. Therefore, either Proposition 3.5.33 (for 0^X) or the Coherence Theorem 3.9.1

(for 1^X) implies the commutativity of D in C. This in turn proves the triple braid axiom. □

Example 8.15.5. Theorem 8.15.4 applies to the tight symmetric bimonoidal categories below.

- Distributive symmetric monoidal categories by Proposition 2.3.2.
- Symmetric monoidal closed categories with finite coproducts by Example 2.3.3.
- The category of modules over a commutative ring by Example 2.3.4.
- Distributive categories by Example 2.3.5.
- Σ and Σ' by Propositions 2.4.8 and 2.4.23.
- Right bipermutative categories by Proposition 2.5.7.
- Left bipermutative categories by Proposition 2.5.16.
- The category $\mathsf{Vect}^{\mathbb{C}}$ of finite dimensional complex vector spaces in Example 8.4.13.

In particular, when applied to the last example $\mathsf{C} = \mathsf{Vect}^{\mathbb{C}}$, Theorem 8.15.4 says that 2-vector spaces, in the form $\mathsf{Mat}^{\mathsf{Vect}^{\mathbb{C}}}$, constitute a genuine symmetric monoidal bicategory in which no structure morphisms are identities. ◇

Bibliography and Indices

APPENDIX A

Open Questions

In this chapter, we discuss open questions related to the topics of this work. These open questions provide additional motivation for the main text.

A.1. Bimonoidal Categories

The following questions are about bimonoidal, symmetric bimonoidal, and braided bimonoidal categories in Definitions 2.1.2 and II.2.1.29.

Question A.1.1 (Functoriality of the Matrix Construction). In Theorem 8.15.4, we showed that, for each tight symmetric bimonoidal category C, the matrix construction $\mathsf{Mat}^{\mathsf{C}}$ is a symmetric monoidal bicategory. Denote by $\mathsf{Bi}^{\mathsf{tsy}}$ the full sub-2-category of the 2-category $\mathsf{Bi}^{\mathsf{sy}}$ in Proposition 7.1.7, with small *tight* symmetric bimonoidal categories as objects. Regard $\mathsf{Bi}^{\mathsf{tsy}}$ as a tricategory with only identity 3-cells. It is claimed in [**SP**∞] that small symmetric monoidal bicategories are the objects of a tricategory, denoted by SMB.

- Extend the assignment

$$\mathsf{C} \longmapsto \mathsf{Mat}^{\mathsf{C}}$$

to a trifunctor

$$\mathsf{Bi}^{\mathsf{tsy}} \longrightarrow \mathsf{SMB}.$$

Among other things, one should carefully verify the tricategory axioms for SMB. For a discussion of tricategories and a detailed verification of the tricategory of small bicategories, the reader is referred to [**JY21**, Ch. 11]. ◇

Question A.1.2 (Bimonoidal Bicategories). Taking the categorification from (commutative) rigs to (symmetric) bimonoidal categories one step further, we could ask about two different monoidal structures, $\boxplus$ and $\boxtimes$, on a bicategory, with $\boxplus$ symmetric.

- Define such a *(braided/sylleptic/symmetric) bimonoidal bicategory*, generalizing the (braided/sylleptic/symmetric) monoidal bicategories in Sections 6.4 and 6.5.
- For a tight symmetric bimonoidal category C, prove that the symmetric monoidal bicategory $\mathsf{Mat}^{\mathsf{C}}$ extends to a symmetric bimonoidal bicategory.
- For a tight braided bimonoidal category C, prove that the monoidal bicategory $\mathsf{Mat}^{\mathsf{C}}$ extends to a bimonoidal bicategory.

More precisely, in Theorem 8.15.4, the monoidal composition $\boxtimes$ in the symmetric monoidal bicategory $\mathsf{Mat}^{\mathsf{C}}$ involves the matrix tensor product in (8.6.3). There should be another symmetric monoidal bicategory structure on the matrix bicategory $\mathsf{Mat}^{\mathsf{C}}$ in Theorem 8.4.12, whose monoidal composition uses the matrix direct sum as in Example 2.5.9. These two symmetric monoidal structures on $\mathsf{Mat}^{\mathsf{C}}$ should make it into a symmetric bimonoidal bicategory. A similar discussion applies in the braided case, with Theorem II.8.4.7 showing that $\mathsf{Mat}^{\mathsf{C}}$ is a monoidal bicategory.

- Extend the tricategory SMB (Question A.1.1) to a tricategory SBB with small symmetric bimonoidal bicategories as objects.
- Construct a tricategory BB with small bimonoidal bicategories as objects.
- Extend the assignment $\mathsf{C} \longmapsto \mathsf{Mat}^{\mathsf{C}}$ in
  - Theorem 8.15.4 in the symmetric case to a trifunctor

$$\mathsf{Bi}^{\mathsf{tsy}} \longrightarrow \mathsf{SBB}$$

    and
  - Theorem II.8.4.7 in the braided case to an analogous trifunctor with codomain BB. ◇

Question A.1.3 (Bimonoidal Bicategorical Centers)**.** Extend the bimonoidal centers in Theorems II.4.4.3 and II.4.5.3 to the bimonoidal bicategorical setting in Question A.1.2. In other words:

- Extend the bimonoidal Drinfeld center in Theorem II.4.4.3 to a bimonoidal bicategory and show that it is a braided bimonoidal bicategory.
- Show that the center of a braided bimonoidal bicategory is a sylleptic bimonoidal bicategory.
- Extend the bimonoidal symmetric center in Theorem II.4.5.3 to a sylleptic bimonoidal bicategory and show that it is a symmetric bimonoidal bicategory.

There are simpler centers of Gray monoids [**BN96**], braided monoidal 2-categories, and sylleptic monoidal 2-categories [**Cra98**]. As in Theorem II.4.4.3, a tightness assumption is likely necessary for some of these center constructions. ◇

Question A.1.4 (Gray Rings and Bipermutative Gray Monoids)**.** Recall from Sections 6.6 and 6.7 that a Gray monoid is a 2-category equipped with a monoid structure under the Gray tensor product. A permutative Gray monoid is a Gray monoid equipped with a compatible Gray symmetry. Symmetric monoidal bicategories can be strictified to permutative Gray monoids; see [**GJO17b**] and the discussion near the end of Section 6.7.

- Define a *Gray ring* and a *bipermutative Gray monoid* that are analogous to, respectively, a right rigid bimonoidal category (Definition 5.5.8) and a right bipermutative category (Definition 2.5.2).
- Along the lines of Theorem 5.5.11, prove a strictification result from bimonoidal bicategories (Question A.1.2) to Gray rings.
- Along the lines of Theorem 5.4.6, prove a strictification result from symmetric bimonoidal bicategories to bipermutative Gray monoids.

A bipermutative Gray monoid should be a 2-category equipped with two compatible permutative Gray monoid structures, $\boxplus$ and $\boxtimes$, that interact via distributivity. In a Gray ring, $\boxtimes$ is a Gray monoid structure that is not assumed to be permutative. The following table summaries these (conjectural) concepts.

lax structure	strict structure	strictification
bicategories	2-categories	[**JY21**, 8.4.1]
monoidal bicategories	Gray monoids	[**GPS95, Gur13**]
symmetric monoidal bicategories	permutative Gray monoids	[**GJO17b**]
bimonoidal bicategories	Gray rings	conjecture
symmetric bimonoidal bicategories	bipermutative Gray monoids	conjecture

In each row, the left column can be strictified to the middle column. ◇

Question A.1.5 (Horizontal Bicategories of Double Categories)**.** A number of bicategories, including those of spans and bimodules, are the horizontal bicategories of some double categories.

- For a tight bimonoidal category C, is the matrix bicategory $\mathsf{Mat}^{\mathsf{C}}$ in Theorem 8.4.12 the horizontal bicategory of a double category?
- If so, does the symmetric monoidal bicategory in Theorem 8.15.4 arise from a symmetric monoidal structure on the double category?

See [**HS**∞] and [**JY21**, 12.3 and 12.4] for a discussion of (monoidal) double categories and their horizontal bicategories. ◇

Question A.1.6 (Braided Sheet Diagrams)**.** String diagrams are graphical reasoning tools in monoidal categories [**JS91a, Sel11**]. Sheet diagrams [**CDH**∞], which we mentioned in Notes 2.7.5 and 7.9.2 and Example 3.10.9, are their analogues for tight bimonoidal categories.

- Develop sheet diagrams for
  - symmetric bimonoidal categories (Definition 2.1.2) and
  - braided bimonoidal categories (Definition II.2.1.29).

This is, in fact, a coherence question with several parts. More precisely, a *bimonoidal signature* S consists of (i) a set of generating objects and (ii) a set of generating morphisms, each with (co)domain in the free $\{\oplus, \otimes\}$-algebra S^{fr} (Definition 3.1.2). Given a bimonoidal signature S, one first defines the appropriate braided bimonoidal sheet diagrams and topological deformations corresponding to the axioms of a braided bimonoidal category. Then one constructs a braided bimonoidal category S' with object set S^{fr} and, as morphisms, braided bimonoidal sheet diagrams modulo topological deformations. Finally, one proves that S' is braided bimonoidally equivalent to the free braided bimonoidal category on S. A similar discussion applies in the symmetric case.

In view of the results in Chapter II.3, braided bimonoidal sheet diagrams can be used as graphical reasoning tools in (i) quantum group theory and (ii) the Fibonacci and Ising anyons in topological quantum computation. Sheet diagrams for tight bimonoidal categories in [**CDH**∞] involve the symmetric monoidal string diagrams in [**JS91a**]. Braided bimonoidal sheet diagrams will likely involve both the symmetric monoidal (for the additive structure $\oplus$) and the braided monoidal (for the multiplicative structure $\otimes$) string diagrams in [**JS91a**, Ch. 2–3]. The Braided Bimonoidal Coherence Theorem II.5.4.4 will be needed to check the axioms in Definition II.2.1.29 for braided bimonoidal sheet diagrams. Similarly, Laplaza's Coherence Theorems 3.9.1 and 4.4.3 will be needed to check the axioms in Definition 2.1.2 for symmetric bimonoidal sheet diagrams.

The distributivity morphisms δ^l and δ^r in Definitions 2.1.2 and II.2.1.29 are not invertible in general.

- Is it possible to replace the tightness assumption—that is, the invertibility of δ^l and δ^r—in the sheet diagrams in [**CDH**∞] with flatness in Definitions 3.9.9 and II.5.4.5?

Related to Section 7.9, to replace tightness with the much weaker assumption of flatness, one would need to work directly with a flat (symmetric/braided) bimonoidal category and *avoid* using the Strictification Theorems 5.4.6, 5.4.7, 5.5.11, 5.5.12, II.6.3.6, and II.6.3.7. The reason is that each of these theorems requires the tightness assumption. See Question A.2.8 for further problems about sheet diagrams. $\diamond$

A.2. E_n-Monoidal Categories

The following questions are about the E_n-monoidal categories in Part II.2.

Question A.2.1 (Coherence of E_n-Monoidal Categories).

- Prove coherence theorems for ring categories (Definition II.9.1.2) along the lines of Theorems 3.10.7 and 4.5.8. Each such coherence theorem should say that any reasonable formal diagram in a ring category involving
$$(\oplus, \mathbb{0}, \xi^{\oplus}, \otimes, \mathbb{1}, \partial^l, \partial^r)$$
is commutative, with an assumption on either the common domain or the two paths.
- Prove coherence theorems for bipermutative categories (Definition II.9.3.2) along the lines of Theorems 3.9.1 and 4.4.3.
- Prove a coherence theorem for braided ring categories (Definition II.9.5.1) along the lines of Theorem II.5.4.4.
- More generally, prove a coherence theorem for E_n-monoidal categories (Definition II.10.7.2) along the lines of Theorem II.5.4.4. The coherence theorems for n-fold monoidal categories (Theorem II.10.6.8) should be relevant.

As in Theorems 3.9.1 and 4.4.3, one may need to assume a monomorphism or an epimorphism condition on the factorization morphisms ∂^l and ∂^r $\diamond$

Question A.2.2 (n-Monoidal Categories). In an n-fold monoidal category (Definition II.10.1.1), each monoidal structure $\otimes_i$ is strictly associative with a common strict unit $\mathbb{1}$. There is a more general concept called an *n-monoidal category* in [**AM10**, Def. 6.1, 7.1, and 7.24]. It allows each monoidal structure $\otimes_i$ to be nonstrict and distinct monoidal units.

- Describe the free n-monoidal category of a small category, along the lines of Proposition II.10.5.9 and Theorem II.10.5.18.
- Generalize the Coherence Theorem II.10.6.8 to n-monoidal categories.

In [**AM10**, Section 6.2], it is stated without detail that, in a 2-monoidal category, each formal diagram is commutative. It is stated there that this coherence result can be deduced from the work in [**Lew72**]. So one possible first step in answering these questions would be to prove in detail this coherence result for 2-monoidal categories. $\diamond$

Questions A.2.3 through A.2.7 below are all related to each other.

Question A.2.3 (Lax n-Fold Monoidal Categories). Between an n-fold monoidal category and an n-monoidal category (Question A.2.2) is a *lax n-fold monoidal category*. The latter allows each monoidal structure $\otimes_i$ to be nonstrict, and it assumes a common nonstrict monoidal unit $\mathbb{1}$. Analogous to Propositions II.10.1.14 and II.10.1.21, general nonstrict braided monoidal categories should be examples of lax 2-fold monoidal categories, and general nonstrict symmetric monoidal categories should be examples of lax n-fold monoidal categories for $n \geq 2$.

- Describe the free lax n-fold monoidal category of a small category, along the lines of Proposition II.10.5.9 and Theorem II.10.5.18.
- Extend the Coherence Theorem II.10.6.8 to lax n-fold monoidal categories.
- Can n-monoidal categories be strictified to lax n-fold monoidal categories?
- Can lax n-fold monoidal categories be strictified to n-fold monoidal categories?

There are two other possible variants of n-fold monoidal categories. The variant in [**For04**] corresponds to a lax n-fold monoidal category with a common *strict* monoidal unit. The variant in [**FSS07**] corresponds to an n-monoidal category with generally distinct but strict monoidal units. The following table summaries the strictness assumptions of n-fold monoidal categories and its four variants.

monoidal units	strict $\{\otimes_i\}_{1\leq i\leq n}$	nonstrict $\{\otimes_i\}_{1\leq i\leq n}$
common strict	n-fold monoidal	[**For04**]
common nonstrict		lax n-fold monoidal
distinct strict		[**FSS07**]
distinct nonstrict		n-monoidal [**AM10**]

The questions below refer to lax n-fold monoidal categories as defined above, with $\otimes_i$ generally nonstrict and a common nonstrict monoidal unit. ◇

Question A.2.4 (Unstable Periodic Table of Weak n-Categories)**.** The *periodic table* in [**BD98**] of k-tuply monoidal n-categories is a guiding principle for defining some versions of weak n-categories. In the $n = 1$ column in the periodic table, the values $k = 0, 1, 2$, and ≥ 3 correspond to, respectively, categories, monoidal categories, braided monoidal categories, and symmetric monoidal categories. On the other hand, by Propositions II.10.1.14 and II.10.1.21, braided strict monoidal categories and permutative categories are special examples of, respectively, 2-fold monoidal categories and k-fold monoidal categories for $k \geq 2$. Proposition II.10.2.8 has examples of 2-fold monoidal categories that are not braided strict monoidal categories. The k-fold monoidal category operad Mon^k (Proposition III.13.1.20) is an E_k-operad that parametrizes k-fold monoidal categories (Theorem III.13.2.1 and Proposition III.13.3.18).

- Construct an *unstable* periodic table in which the $n = 1$ column consists of lax k-fold monoidal categories (Question A.2.3) for $k \geq 1$.

The $n = 1$ and $n = 2$ columns of the unstable periodic table should look like this:

	$n = 1$	$n = 2$
$k = 0$	categories	bicategories
$k = 1$	monoidal categories	monoidal bicategories
$k \geq 2$	lax k-fold monoidal categories	k-fold monoidal bicategories

The unstable periodic table does *not* stabilize like the periodic table in [**BD98**], where the ($n = 1, k \geq 3$) entries are all symmetric monoidal categories. Moreover, the ($n = 2, k \geq 2$) column in the unstable periodic table contains yet-to-be defined *k-fold monoidal bicategories*.

- Prove that braided, sylleptic, and symmetric monoidal bicategories in Section 6.5 are examples of k-fold monoidal bicategories for, respectively, $k = 2, 3$, and ≥ 4. ◇

This question may be regarded as both (i) a litmus test for the correct definition of a k-fold monoidal bicategory and (ii) a conceptual unification of braided, sylleptic, and symmetric monoidal bicategories. Further examples of k-fold monoidal bicategories should arise from the matrix construction in Question A.2.7.

Question A.2.5 (Iterated Gray Monoids)**.** This is a variation of Question A.1.4.

- Analogous to k-fold monoidal categories (Definition II.10.1.1) for $k \geq 1$, define the concept of a *k-fold Gray monoid* that satisfies the following statements:
  - A Gray monoid is precisely a 1-fold Gray monoid.
  - A braided monoidal 2-category [**Cra98**] is an example of a 2-fold Gray monoid, analogous to Proposition II.10.1.14.
  - A sylleptic monoidal 2-category [**Cra98**] is an example of a 3-fold Gray monoid.
  - A permutative Gray monoid is an example of a k-fold Gray monoid for $k \geq 4$, analogous to Proposition II.10.1.21.
- Prove a strictification theorem from k-fold monoidal bicategories (Question A.2.4) to k-fold Gray monoids. This should fit into the following table of strictification theorems.

bicategories	2-categories	[**JY21**, 8.4.1]
monoidal bicategories	Gray monoids	[**GPS95, Gur13**]
braided monoidal bicategories	braided monoidal 2-categories	[**Gur11**]
k-fold monoidal bicategories ($k \geq 1$)	k-fold Gray monoids	conjecture
symmetric monoidal bicategories	permutative Gray monoids	[**GJO17b**]

In each row, the left column can be strictified to the middle column. For the symmetric case, see the discussion near the end of Section 6.7. ◇

Question A.2.6 (Laplaza E_n-Monoidal Categories)**.** The factorization morphisms $\{\partial^{l,i}, \partial^{r,i}\}_{1 \leq i \leq n}$ in an E_n-monoidal category (Definition II.10.7.2) go in the opposite direction as the distributivity morphisms in a bimonoidal category (Definition 2.1.2). Moreover, in an E_n-monoidal category, the monoidal structures $\oplus$ and $\{\otimes_i\}_{i=1}^n$ are all strict.

- Define a *Laplaza E_n-monoidal category* with
  - a generally nonstrict additive structure $(\oplus, \mathbb{0})$,
  - a lax n-fold monoidal structure $\big(\{\otimes_i\}_{i=1}^n, \{\eta^{i,j}\}\big)$ (Question A.2.3),
  - for each $1 \leq i \leq n$, a bimonoidal structure $(\oplus, \otimes_i, \lambda_i^\bullet, \rho_i^\bullet, \delta_i^l, \delta_i^r)$ (Definition 2.1.2), and
  - appropriate axioms relating the lax n-fold monoidal structure and the n bimonoidal structures, analogous to those in Definition II.10.7.2.

Laplaza E_n-monoidal categories should contain the following examples:

- An E_n-monoidal category with invertible factorization morphisms should be an example of a Laplaza E_n-monoidal category, analogous to Theorems II.9.1.15, II.9.3.7, and II.9.5.6.
- A Laplaza E_1-monoidal category should be precisely a bimonoidal category (Definition 2.1.2), analogous to Example II.10.7.13.

- A braided bimonoidal category (Definition II.2.1.29) should be an example of a Laplaza E_2-monoidal category with $\otimes_1 = \otimes_2$, analogous to Theorem II.10.8.1.
- A symmetric bimonoidal category should be an example of a Laplaza E_n-monoidal category for $n \geq 2$ with $\otimes_1 = \cdots = \otimes_n$, analogous to Theorem II.10.9.1.
- Similar to Theorem II.2.4.22, an abelian category with a compatible lax n-fold monoidal structure should be a Laplaza E_n-monoidal category.

Moreover:

- Prove a coherence theorem for Laplaza E_n-monoidal categories, along the lines of Theorem II.5.4.4. This will certainly involve the coherence theorem for lax n-fold monoidal categories in Question A.2.3.
- Prove a strictification theorem for *tight* Laplaza E_n-monoidal categories, along the lines of Theorems II.6.3.6 and II.6.3.7. Here *tight* means that all the distributivity morphisms, δ_i^l and δ_i^r, are natural isomorphisms.
- Is there an analogue of Baez's Conjecture (Theorems 7.8.1, 7.8.3, II.7.3.4, and II.7.3.6) for Laplaza E_n-monoidal categories? ◇

Question A.2.7 (Matrix Construction)**.** In Theorem 8.4.12 we showed that the matrix construction $\mathsf{Mat}^{\mathsf{C}}$ is a bicategory for each tight bimonoidal category C. Moreover, $\mathsf{Mat}^{\mathsf{C}}$ is (i) a monoidal bicategory if C is a tight braided bimonoidal category (Theorem II.8.4.7) and (ii) a symmetric monoidal bicategory if C is a tight symmetric bimonoidal category (Theorem 8.15.4).

- Show that the matrix construction $\mathsf{Mat}^{\mathsf{C}}$ of a tight Laplaza E_{k+1}-monoidal category (Question A.2.6) is a k-fold monoidal bicategory.

This question asks for a refinement of the table in the introduction of Chapter II.8 as follows.

tight – category C	– bicategory $\mathsf{Mat}^{\mathsf{C}}$	
bimonoidal	plain	8.4.12
braided bimonoidal	monoidal	II.8.4.7
Laplaza E_{k+1}-monoidal ($k \geq 1$)	k-fold monoidal	conjecture
symmetric bimonoidal	symmetric monoidal	8.15.4

The k-fold monoidal bicategories in the conjectural row refer to the $n = 2$ column in the unstable periodic table in Question A.2.4. Proving that $\mathsf{Mat}^{\mathsf{C}}$ is a k-fold monoidal bicategory will certainly involve the coherence theorem for Laplaza E_{k+1}-monoidal categories in Question A.2.6. The general picture of the table above is that it takes a sum $\oplus$ and a product $\otimes$ to construct the matrix bicategory $\mathsf{Mat}^{\mathsf{C}}$. So any further monoidal structures on the bicategory $\mathsf{Mat}^{\mathsf{C}}$ would have to come from further monoidal structures on C. ◇

Question A.2.8 (Higher Sheet Diagrams)**.** As in Question A.1.6, develop higher dimensional sheet diagrams for

- n-fold monoidal categories (Definition II.10.1.1),
- n-monoidal categories (Question A.2.2),
- lax n-fold monoidal categories (Question A.2.3),
- E_n-monoidal categories (Definition II.10.7.2), and
- Laplaza E_n-monoidal categories (Question A.2.6).

As discussed in Question A.1.6, each of these items is a coherence question with several parts. ◇

A.3. Enriched Monoidal Categories

The following questions are about the concepts in Chapters III.1, III.2, and III.3.

Question A.3.1 (Enriched Lax n-Fold Monoidal Categories)**.** In Section III.1.4, with V a braided monoidal category, we defined monoidal, braided monoidal, and symmetric monoidal V-categories, with the latter two assuming that V is symmetric. See Lemma III.1.3.23 and Explanation III.1.3.25.

- For $n \geq 2$ and V a symmetric monoidal category, extend the lax n-fold monoidal categories in Question A.2.3 to the V-enriched setting.
- Extend the results in Chapter III.2 to lax n-fold monoidal V-categories.

Analogous to Propositions II.10.1.14 and II.10.1.21, braided monoidal V-categories should be examples of lax 2-fold monoidal V-categories, and symmetric monoidal V-categories should be examples of lax n-fold monoidal V-categories for $n \geq 2$. Theorem II.10.4.5 says that small $(n+1)$-fold monoidal categories are precisely the monoids in the monoidal category MCat^n of small n-fold monoidal categories and n-fold monoidal functors.

- Extend Theorem II.10.4.5 to the V-enriched setting. ◇

Question A.3.2 (Centers and Enriched Centers)**.** The Drinfeld center of a monoidal category is a braided monoidal category (Theorem II.1.4.27), and the symmetric center of a braided monoidal category is a symmetric monoidal category (Proposition II.1.5.3). Moreover, the Drinfeld center and the symmetric center are generalized to (i) the bimonoidal setting in Theorems II.4.4.3 and II.4.5.3 and (ii) the ring categorical setting in Corollary II.9.6.1 and Theorem II.9.6.4.

- Define a center construction that sends an n-fold monoidal category (Definition II.10.1.1) to an $(n+1)$-fold monoidal category.

Theorem II.10.4.5 should be relevant.

- Repeat the previous question for
  - n-monoidal categories (Question A.2.2),
  - lax n-fold monoidal categories (Question A.2.3),
  - E_n-monoidal categories (Definition II.10.7.2),
  - Laplaza E_n-monoidal categories (Question A.2.6), and
  - lax n-fold monoidal V-categories (Question A.3.1).

As a special case of the last item, the Drinfeld center of a monoidal V-category with V *strict* is studied in [**KYZZ**∞, **KZ18**]. ◇

Question A.3.3 (Autonomous Enriched Monoidal Categories)**.** The definition of a monoidal V-category K in [**MP19**, 2.1] assumes that V is braided strict monoidal and K is strict monoidal, so it is the special case of Definition III.1.4.2 with both V and K strict. The main theorem in [**MP19**, 1.1] shows that, using their definition, there is a bijective correspondence between (i) some autonomous monoidal V-categories and (ii) some braided oplax monoidal functors from V to the Drinfeld center of an autonomous monoidal category. This bijective correspondence is extended to a 2-equivalence between 2-categories in [**Del**∞, 1.2]. In bicategorical language (Definition 6.3.9), *autonomy* means that each object $X \in \mathsf{K}$ is equipped with both a left adjoint X_* and a right adjoint X^* that satisfy the triangle identities. A (braided) monoidal functor (F, F^2, F^0) is *oplax* if its monoidal constraint F^2 and unit constraint F^0 go in the opposite directions as those in Definitions III.1.1.6 and III.1.1.17, with appropriately adjusted axioms. Discussion of autonomous monoidal categories can be found in [**FY92, JS91b, JS93**].

- Extend the 2-equivalence in [**Del**∞, 1.2] to general monoidal V-categories K, with V and K not necessarily strict, as in Definition III.1.4.2.

To extend this 2-equivalence to the general nonstrict case, the coherence and strictification results of enriched monoidal categories in Sections III.2.5 and III.2.6 will likely be necessary. $\diamond$

A.4. Homotopy Theory

Question A.4.1 (Homotopy Theory of Matrix Bicategories)**.** In Example 8.15.5, we listed some examples of tight symmetric bimonoidal categories C, to which the Bicategorification Theorem 8.15.4 may be applied to yield a symmetric monoidal bicategory $\mathsf{Mat}^{\mathsf{C}}$.

- What can be said about the homotopy theoretic properties of any of these symmetric monoidal bicategories?
- Consider the previous question for the symmetric bimonoidal bicategories in Question A.1.2.

For instance, for the finite ordinal category Σ in Section 2.4, Mat^{Σ} may be related to a remark in [**JO12**] about the multiplicative structure on the categorical model for the sphere spectrum. $\diamond$

Question A.4.2 (Categorical Model of BP)**.** The Brown-Peterson spectrum BP has an E_4 structure [**BM13**], but not an E_∞ structure at any prime [**Law18, Sen**∞]. By Corollary III.13.5.2, the Elmendorf-Mandell K-theory of each small E_n-monoidal category is an E_n-symmetric spectrum.

- Is there a small E_4-monoidal category (Definition II.10.7.2) whose K-theory is BP?

A positive answer to this question would provide a categorical model of BP. If Question A.5.7 has a positive answer for the 4-fold monoidal category operad Mon^4, then this question also has a positive answer, at least up to weak equivalences. $\diamond$

Question A.4.3 (Boardman-Vogt E_n-Operads)**.** In a symmetric monoidal category C, the commutative operad Com is the operad with each entry the monoidal unit $\mathbb{1}$ and structure morphisms given by the coherence isomorphism $\mathbb{1} \otimes \mathbb{1} \cong \mathbb{1}$. The algebras of Com are precisely commutative monoids. One model of an E_∞-operad (Definition III.11.6.1) is WCom, where W is the Boardman-Vogt W-construction. In the topological setting, it was introduced in [**BV73**]. For a conceptual presentation of the W-construction in terms of coends in a general symmetric monoidal category, see [**Yau20**, Ch. 6–7].

- Describe E_n-operads (Definition III.12.2.3) for $n \geq 1$ as a filtration of sub-operads of WCom.
- Compare these models of E_n-operads to
  - the n-fold monoidal category operad Mon^n (Definition III.13.1.12),
  - Berger's filtration [**Ber96**] of the Barratt-Eccles operad $N(E\mathsf{As})$ (Definition III.11.4.10),

- Smith's filtration [**Smi89**] of the Barratt-Eccles operad,
- Batanin's E_n-operads [**Bat07, Bat08**],
- Fiedorowicz's E_n-operads [**Fie∞b**], and
- the Fulton-MacPherson E_n-operads [**Fre17, FM94, GJ∞, Sal01**].

In [**Yau20**, 3.2.11 and 6.3.1], the W-construction of an operad in a symmetric monoidal category is defined entrywise as a coend indexed by a substitution category whose objects are trees and whose morphisms correspond to tree substitution. This applies, in particular, to the commutative operad Com. So describing E_n-operads as sub-operads of WCom would provide a combinatorial description of E_n-operads in terms of trees. ◇

A.5. Algebraic K-Theory

Question A.5.1 (Multifunctorial K-Theory of Pointed Multicategories)**.** Contrary to the claim in [**EM09**, Theorem 1.3], Elmendorf-Mandell J-theory J^{EM} (Definition III.10.3.25) does *not* extend to a multifunctor on all of $\mathsf{Multicat}_*$, the category of small pointed multicategories, but only to the full subcategory $\mathsf{Mod}^{\mathcal{M}\underline{1}}$ of left $\mathcal{M}\underline{1}$-modules, via the symmetric monoidal Cat_*-functor $\mathsf{J}^{\mathcal{T}}$ (Theorem III.10.3.17). Examples III.10.2.15 and III.10.2.16 present some small pointed multicategories that are not left $\mathcal{M}\underline{1}$-modules.

- Is there a K-theory multifunctor that is objectwise equivalent to Segal K-theory K^{Se} and extends to $\mathsf{Multicat}_*$ via the endomorphism multicategory End in Corollary III.5.3.9 and Definition III.6.5.1?

The key issue is about the monoidal units. In $\mathsf{Multicat}_*$ the smash unit is $\mathsf{S} = \mathsf{I} \coprod \mathsf{T}$ (Definition III.5.6.20), which is different from the monoidal unit $\mathcal{M}\underline{1}$ in $\mathsf{Mod}^{\mathcal{M}\underline{1}}$. Unlike Definition III.10.3.16 with $\mathsf{J}^{\mathcal{T}}(\mathcal{M}\underline{1})$, the object $\mathsf{J}^{\mathcal{T}}(\mathsf{S})$ is the terminal $\mathcal{G}_*$-category $*$. If the monoidal unit constraint $(\mathsf{J}^{\mathcal{T}})^0$ for $\mathsf{J}^{\mathcal{T}}$ were to be defined as the unique morphism $J \longrightarrow \mathsf{J}^{\mathcal{T}}(\mathsf{S}) = *$ to the terminal object, as stated in the last paragraph in [**EM09**, Section 5], then $\mathsf{J}^{\mathcal{T}}$ cannot satisfy the unity axioms (III.1.1.10). The reason is that a general left or right unit isomorphism for $\mathsf{J}^{\mathcal{T}}(-)$ does not factor through the zero morphism in $\mathcal{G}_*$-Cat. So with $J \longrightarrow *$ as the unit constraint, $\mathsf{J}^{\mathcal{T}}$ would not be a monoidal functor. ◇

Question A.5.2 (Comparison of K^{Se} and K^{EM} for $\mathcal{M}\underline{1}$-Modules)**.** The Segal and Elmendorf-Mandell K-theory constructions are defined as the following composites, respectively:

$$\mathsf{K}^{\mathsf{Se}} = \mathsf{K}^{\mathcal{F}} N_* \mathsf{J}^{\mathcal{M}} \mathsf{End} \quad \text{and} \quad \mathsf{K}^{\mathsf{EM}} = \mathsf{K}^{\mathcal{G}} N_* \mathsf{J}^{\mathcal{T}} \mathsf{End}.$$

The domain of $\mathsf{J}^{\mathcal{M}}$ is the category of small pointed multicategories, $\mathsf{Multicat}_*$, and the domain of $\mathsf{J}^{\mathcal{T}}$ is the category of left $\mathcal{M}\underline{1}$-modules within $\mathsf{Multicat}_*$. Thus both K^{Se} and K^{EM} can be expanded to $\mathsf{Mod}^{\mathcal{M}\underline{1}}$. We will write

$$\mathsf{Mod}^{\mathcal{M}\underline{1}} \xrightarrow{\tilde{\mathsf{K}}^{\mathsf{Se}} = \mathsf{K}^{\mathcal{F}} N_* \mathsf{J}^{\mathcal{M}}} \mathsf{SymSp} \quad \text{and} \quad \mathsf{Mod}^{\mathcal{M}\underline{1}} \xrightarrow{\tilde{\mathsf{K}}^{\mathsf{EM}} = \mathsf{K}^{\mathcal{G}} N_* \mathsf{J}^{\mathcal{T}}} \mathsf{SymSp}.$$

- Is there a (natural) level equivalence $\tilde{\mathsf{K}}^{\mathsf{Se}}\mathsf{P} \longrightarrow \tilde{\mathsf{K}}^{\mathsf{EM}}\mathsf{P}$ for each left $\mathcal{M}\underline{1}$-module P?

The level equivalence $\mathsf{K}^{\mathsf{Se}}\mathsf{C} \longrightarrow \mathsf{K}^{\mathsf{EM}}\mathsf{C}$ given in Theorem III.10.6.10 for each small permutative category C depends crucially on the adjunctions of Proposition III.10.6.7 and these, in turn, depend on Proposition III.8.5.4, which gives a strong symmetric monoidal adjunction

$$\mathsf{Cat}_*(a,\mathsf{C}) \underset{R}{\overset{L}{\rightleftarrows}} \mathsf{Multicat}_*(\mathcal{M}a,\mathsf{End}(\mathsf{C}))$$

for each small permutative category C and each pointed finite set a. Therefore the proof of Theorem III.10.6.10 does not immediately generalize to $\tilde{\mathsf{K}}^{\mathsf{Se}}$ and $\tilde{\mathsf{K}}^{\mathsf{EM}}$. ◇

Question A.5.3 (Homotopy Theory of $\mathcal{G}_*$-Objects)**.** Notes III.8.6.4 and III.8.6.6 summarize equivalences of homotopy categories from the work of [**BF78, Tho95, Man10**]:

$$Ho(\mathsf{PermCat}^{\mathsf{su}}) \simeq Ho(\Gamma\text{-}\mathsf{Cat}) \simeq Ho(\Gamma\text{-}\mathsf{sSet}) \simeq Ho(\mathsf{Sp}^{\mathbb{N}}_{\geq 0}),$$

where $\mathsf{Sp}^{\mathbb{N}}_{\geq 0}$ denotes the category of connective sequential spectra. The composite from the homotopy category of small permutative categories to that of connective spectra is induced by Segal K-theory, K^{Se}. Therefore, by Theorem III.10.6.10, the composite on homotopy categories is equal to that induced by Elmendorf-Mandell K-theory, K^{EM}. However, K^{EM} factors through $\mathcal{G}_*$-Cat and $\mathcal{G}_*$-sSet instead of Γ-Cat and Γ-sSet.

As diagram categories, $\mathcal{G}_*$-Cat and $\mathcal{G}_*$-sSet have Quillen model structures defined similarly to those of Γ-Cat and Γ-sSet. The functor

$$\wedge \colon \mathcal{G} \longrightarrow \mathcal{F}$$

is described in Definition III.9.1.15 and is used in the comparison of $\mathsf{K}^{\mathcal{F}}$ and $\mathsf{K}^{\mathcal{G}}$ in Proposition III.9.3.16. The induced functor from Γ-objects to $\mathcal{G}_*$-objects is symmetric monoidal by Theorem III.9.4.18.

- Are the induced functors

$$\Gamma\text{-}\mathsf{Cat} \xrightarrow{\wedge^*} \mathcal{G}_*\text{-}\mathsf{Cat} \quad\text{and}\quad \Gamma\text{-}\mathsf{sSet} \xrightarrow{\wedge^*} \mathcal{G}_*\text{-}\mathsf{sSet}$$

  (monoidal) Quillen equivalences?

Question A.5.7 poses a related question regarding categories of algebras over operads. ◇

Question A.5.4 (K-Theory of Matrix Bicategories)**.** There is a K-theory construction via the following composite on objects only.

$$\mathsf{Bi}^{\mathsf{tsy}} \xrightarrow{\mathsf{Mat}} \mathsf{SMB} \xrightarrow{\text{strictify}} \mathsf{PGray} \xrightarrow{K} \Gamma\mathsf{2Cat}$$

Mat is the matrix construction in Theorem 8.15.4 that sends a tight symmetric bimonoidal category to a symmetric monoidal bicategory. The middle arrow is the strictification of symmetric monoidal bicategories to permutative Gray monoids. The right arrow is the K-theory of permutative Gray monoids in [**GJO17b**].

- How does this compare with the K-theory of tight symmetric bimonoidal categories in [**BDR04**], which is defined using the direct sum instead of the tensor product of matrices?
- What extra structures on Γ-2-categories are there when it is the K-theory of the strictification of $\mathsf{Mat}^{\mathsf{C}}$ for some tight symmetric bimonoidal category C, such as those in Example 8.15.5 and $\mathsf{Vect}^{\mathbb{C}}_{\mathsf{c}}$ in Example 2.5.9? ◇

Question A.5.5 (K-Theory of Matrix Symmetric Bimonoidal Bicategories)**.** Repeat Question A.5.4 for

- the matrix symmetric bimonoidal bicategories in Question A.1.2 and
- the bipermutative Gray monoids in Question A.1.4. ◇

Question A.5.6 (K-Theory of Distortion Categories)**.** Recall the finite ordinal category Σ in Definition 2.4.1. Here we consider Σ as a permutative category with respect to its additive structure $\oplus$. Quillen's +-construction $(B\Sigma)^+$ of the classifying space $B\Sigma$ is the sphere spectrum by the Barratt-Priddy-Quillen Theorem [**BP72**]. A different way to say this is that the algebraic K-groups of Σ are the stable homotopy groups of the spheres.

- Can the algebraic K-groups of
  - the distortion category $\mathcal{D}$ in Section 4.2,
  - the additive distortion category $\mathcal{D}^{\mathsf{ad}}$ in Section 4.5, and
  - the braided distortion category $\mathcal{D}^{\mathsf{br}}$ in Section II.5.2

  be computed in similar terms?

By Examples III.11.3.18, III.11.6.13, and III.12.5.4, respectively, $\mathsf{K}^{\mathsf{EM}}\mathcal{D}^{\mathsf{ad}}$, $\mathsf{K}^{\mathsf{EM}}\mathcal{D}$, and $\mathsf{K}^{\mathsf{EM}}\mathcal{D}^{\mathsf{br}}$ are strict ring, E_∞-, and E_2-symmetric spectra. Moreover, each of the distortion categories $\mathcal{D}$, $\mathcal{D}^{\mathsf{ad}}$, and $\mathcal{D}^{\mathsf{br}}$ is a Grothendieck construction over Σ by, respectively, Propositions 4.6.5, 4.6.7, and II.5.5.3.

- Does that yield a computation of their $(B?)^+$ and algebraic K-groups? $\diamond$

Question A.5.7 (Lifting K-Theory Equivalences to Algebras)**.** The Segal K-theory functor in Definition III.8.5.1 induces an equivalence of homotopy categories, from permutative categories to connective symmetric spectra. See Question A.5.3 for further description and a related question about K^{EM}.

- Do the (Quillen) equivalences in Segal's K-theory lift to the categories of algebras over categorical operads, such as the E_2-operad Br and the E_n-operads Mon^n in Theorems III.12.2.4 and III.13.2.1?

For a categorical operad P and a Cat-enriched multicategory M, such as $\mathsf{PermCat}^{\mathsf{su}}$ in Section III.6.6, a P-*algebra in* M is defined as a Cat-enriched multifunctor

$$F\colon \mathsf{P} \longrightarrow \mathsf{M}.$$

This is equivalent to a Cat-enriched operad morphism

$$\mathsf{P} \longrightarrow \mathsf{End}(A)$$

to the Cat-enriched endomorphism operad of the object $A = F(*)$, with $*$ the unique object in the multicategory P. If the answer to this question is yes for a categorical operad P, then Segal's K-theory induces an equivalence between the homotopy categories of (i) P-algebras in permutative categories and (ii) P-algebras in connective symmetric spectra.

To answer this question, it is tempting to use [**WY19**, Th. 4.4 and 4.6], which give sufficient conditions under which a Quillen equivalence between monoidal model categories lifts to a Quillen equivalence between the categories of algebras over some colored operads. This will *not* work because the domain of Segal's K-theory is the multicategory $\mathsf{PermCat}^{\mathsf{su}}$ (Section III.6.6), which is not a symmetric monoidal category, hence also not a monoidal model category. $\diamond$

Bibliography

[Abe80] E. Abe, *Hopf algebras*, Cambridge Tracts in Mathematics, vol. 74, Cambridge University Press, Cambridge-New York, 1980, Translated from Japanese by Hisae Kinoshita and Hiroko Tanaka. (cit. on p. II.106).

[Ada95] J. F. Adams, *Stable homotopy and generalised homology*, Chicago Lectures in Mathematics, University of Chicago Press, Chicago, IL, 1995, Reprint of the 1974 original. (cit. on p. xvii).

[AM10] M. Aguiar and S. Mahajan, *Monoidal functors, species and Hopf algebras*, CRM Monograph Series, vol. 29, American Mathematical Society, Providence, RI, 2010. doi:10.1090/crmm/029 (cit. on pp. I.430, I.431, II.307, II.314, II.315, III.508, III.509).

[Art47] E. Artin, *Theory of braids*, Ann. of Math. **48** (1947), 101–126. doi:10.2307/1969218 (cit. on p. II.34).

[Awo10] S. Awodey, *Category theory*, second ed., Oxford Logic Guides, vol. 52, Oxford University Press, Oxford, 2010. (cit. on pp. xxx, I.18).

[BDR04] N. A. Baas, B. I. Dundas, and J. Rognes, *Two-vector bundles and forms of elliptic cohomology*, Topology, geometry and quantum field theory, London Math. Soc. Lecture Note Ser., vol. 308, Cambridge Univ. Press, Cambridge, 2004, pp. 18–45. doi:10.1017/CBO9780511526398.005 (cit. on pp. I.298, I.315, I.325, I.437, II.321, III.515).

[Bae18] J. C. Baez, *rig category*, 2018, https://ncatlab.org/nlab/show/rig+category. (cit. on pp. xvi, I.255).

[BD98] J. C. Baez and J. Dolan, *Categorification*, Higher category theory (Evanston, IL, 1997), Contemp. Math., vol. 230, Amer. Math. Soc., Providence, RI, 1998, pp. 1–36. doi:10.1090/conm/230/03336 (cit. on pp. I.431, II.315, III.509).

[BN96] J. C. Baez and M. Neuchl, *Higher-dimensional algebra I: Braided monoidal 2-categories*, Adv. Math. **121** (1996), no. 2, 196–244. doi:10.1006/aima.1996.0052 (cit. on pp. I.428, II.312, III.506).

[BS11] J. C. Baez and M. Stay, *Physics, topology, logic and computation: a Rosetta Stone*, New structures for physics, Lecture Notes in Phys., vol. 813, Springer, Heidelberg, 2011, pp. 95–172. doi:10.1007/978-3-642-12821-9_2 (cit. on p. I.45).

[BFSV03] C. Balteanu, Z. Fiedorowicz, R. Schwänzl, and R. Vogt, *Iterated monoidal categories*, Adv. Math. **176** (2003), no. 2, 277–349. doi:10.1016/S0001-8708(03)00065-3 (cit. on pp. xvi, xx, xxvi, II.259, II.260, II.262, II.264, II.293, II.294, II.295, II.306, II.307, III.476, III.485, III.488, III.502).

[BR20] D. Barnes and C. Roitzheim, *Foundations of stable homotopy theory*, Cambridge Studies in Advanced Mathematics, vol. 185, Cambridge University Press, 2020. (cit. on p. xxxii).

[BW12] M. Barr and C. Wells, *Category theory for computing science*, Repr. Theory Appl. Categ. (2012), no. 22, xviii+538, Corrected reprint. (cit. on p. III.345).

[BE74a] M. G. Barratt and P. J. Eccles, *Γ^+-structures. I. A free group functor for stable homotopy theory*, Topology **13** (1974), 25–45. doi:10.1016/0040-9383(74)90036-6 (cit. on p. III.447).

[BE74b] ———, *Γ^+-structures. II. A recognition principle for infinite loop spaces*, Topology **13** (1974), 113–126. doi:10.1016/0040-9383(74)90002-0 (cit. on p. III.447).

[BE74c] ———, *Γ^+-structures. III. The stable structure of $\Omega^\infty\Sigma^\infty A$*, Topology **13** (1974), 199–207. doi:10.1016/0040-9383(74)90011-1 (cit. on p. III.447).

[BP72] M. Barratt and S. Priddy, *On the homology of non-connected monoids and their associated groups*, Comment. Math. Helv. **47** (1972), 1–14. doi:10.1007/BF02566785 (cit. on pp. I.438, II.322, III.516).

[BM13] M. Basterra and M. A. Mandell, *The multiplication on BP*, J. Topol. **6** (2013), no. 2, 285–310. doi:10.1112/jtopol/jts032 (cit. on pp. I.435, II.319, III.513).

[Bat98] M. A. Batanin, *Monoidal globular categories as a natural environment for the theory of weak n-categories*, Adv. Math. **136** (1998), no. 1, 39–103. doi:10.1006/aima.1998.1724 (cit. on p. II.307).

[Bat07] ———, *Symmetrisation of n-operads and compactification of real configuration spaces*, Adv. Math. **211** (2007), no. 2, 684–725. doi:10.1016/j.aim.2006.07.022 (cit. on pp. I.436, II.307, II.320, III.514).

[Bat08] ———, *The Eckmann-Hilton argument and higher operads*, Adv. Math. **217** (2008), no. 1, 334–385. doi:10.1016/j.aim.2007.06.014 (cit. on pp. I.436, II.307, II.320, III.514).

[BM12] M. Batanin and M. Markl, *Centers and homotopy centers in enriched monoidal categories*, Adv. Math. **230** (2012), no. 4-6, 1811–1858. doi:10.1016/j.aim.2012.04.011 (cit. on p. II.307).

[BJP08a] H.-J. Baues, M. Jibladze, and T. Pirashvili, *Quadratic algebra of square groups*, Adv. Math. **217** (2008), no. 3, 1236–1300. doi:10.1016/j.aim.2007.08.007 (cit. on p. III.448).

[BJP08b] H.-J. Baues, M. Jibladze, and T. Pirashvili, *Third Mac Lane cohomology*, Math. Proc. Cambridge Philos. Soc. **144** (2008), no. 2, 337–367. doi:10.1017/S030500410700076X (cit. on p. III.448).

[BM11] H.-J. Baues and F. Muro, *The algebra of secondary homotopy operations in ring spectra*, Proc. Lond. Math. Soc. (3) **102** (2011), no. 4, 637–696. doi:10.1112/plms/pdq034 (cit. on p. III.448).

[Bec67] J. M. Beck, *Triples, algebras and cohomology*, Ph.D. thesis, Columbia University, 1967. (cit. on p. III.219).

[Bén65] J. Bénabou, *Catégories relatives*, C. R. Acad. Sci. Paris **260** (1965), 3824–3827. (cit. on p. I.207).

[Bén67] ———, *Introduction to bicategories*, Reports of the Midwest Category Seminar, Springer, Berlin, 1967, pp. 1–77. (cit. on p. I.207).

[Ber96] C. Berger, *Opérades cellulaires et espaces de lacets itérés*, Ann. Inst. Fourier (Grenoble) **46** (1996), no. 4, 1125–1157. (cit. on pp. I.435, II.307, II.319, III.447, III.473, III.501, III.513).

[BF02] C. Berger and B. Fresse, *Une décomposition prismatique de l'opérade de Barratt-Eccles*, C. R. Math. Acad. Sci. Paris **335** (2002), no. 4, 365–370. doi:10.1016/S1631-073X(02)02489-5 (cit. on p. III.447).

[BF04] ———, *Combinatorial operad actions on cochains*, Math. Proc. Cambridge Philos. Soc. **137** (2004), no. 1, 135–174. doi:10.1017/S0305004103007138 (cit. on p. III.447).

[BK00] A. J. Berrick and M. E. Keating, *Categories and modules with K-theory in view*, Cambridge Studies in Advanced Mathematics, vol. 67, Cambridge University Press, Cambridge, 2000. (cit. on pp. xxx, I.18, II.62).

[BKP89] R. Blackwell, G. M. Kelly, and A. J. Power, *Two-dimensional monad theory*, J. Pure Appl. Algebra **59** (1989), no. 1, 1–41. doi:10.1016/0022-4049(89)90160-6 (cit. on p. III.90).

[BG16] A. Blass and Y. Gurevich, *On quantum computation, anyons, and categories*, Martin Davis on computability, computational logic, and mathematical foundations, Outst. Contrib. Log., vol. 10, Springer, Cham, 2016, pp. 209–241. (cit. on p. II.106).

[BG20a] ———, *Braided distributivity*, Theoret. Comput. Sci. **807** (2020), 73–94. doi:10.1016/j.tcs.2019.11.027 (cit. on pp. xvi, xx, xxiii, xxiv, xxv, I.48, II.35, II.36, II.43, II.44, II.46, II.54, II.65, II.127, II.157).

[BG20b] ———, *Witness algebra and anyon braiding*, Math. Structures Comput. Sci. **30** (2020), no. 3, 234–270. doi:10.1017/s0960129520000055 (cit. on p. II.106).

[BM11] A. J. Blumberg and M. A. Mandell, *Derived Koszul duality and involutions in the algebraic K-theory of spaces*, J. Topol. **4** (2011), no. 2, 327–342. doi:10.1112/jtopol/jtr003 (cit. on p. III.411).

[Bly05] T. S. Blyth, *Lattices and ordered algebraic structures*, Universitext, Springer-Verlag London, Ltd., London, 2005. (cit. on p. II.307).

[BV73] J. M. Boardman and R. M. Vogt, *Homotopy invariant algebraic structures on topological spaces*, Lecture Notes in Mathematics, Vol. 347, Springer-Verlag, Berlin-New York, 1973. (cit. on pp. I.435, II.230, II.260, II.307, II.319, III.219, III.220, III.457, III.513).

[BO20] A. M. Bohmann and A. M. Osorno, *A multiplicative comparison of Segal and Waldhausen K-theory*, Math. Z. **295** (2020), no. 3-4, 1205–1243. doi:10.1007/s00209-019-02394-7 (cit. on pp. III.345, III.346, III.411).

[Bor94a] F. Borceux, *Handbook of categorical algebra 1: Basic category theory*, Encyclopedia of Mathematics and its Applications, vol. 50, Cambridge University Press, Cambridge, 1994. (cit. on pp. xxx, I.18).

[Bor94b] ———, *Handbook of categorical algebra 2: Categories and structures*, Encyclopedia of Mathematics and its Applications, vol. 51, Cambridge University Press, Cambridge, 1994. (cit. on pp. xxx, I.18, III.156, III.194, III.286, III.409).

[Bor94c] ———, *Handbook of categorical algebra 3: Categories of sheaves*, Encyclopedia of Mathematics and its Applications, vol. 52, Cambridge University Press, Cambridge, 1994. (cit. on pp. xxx, I.18).

[BF78] A. K. Bousfield and E. M. Friedlander, *Homotopy theory of Γ-spaces, spectra, and bisimplicial sets*, Geometric applications of homotopy theory (Proc. Conf., Evanston, Ill., 1977), II, Lecture Notes in Math., vol. 658, Springer, Berlin, 1978, pp. 80–130. (cit. on pp. I.437, II.321, III.287, III.315, III.515).

[BK72] A. K. Bousfield and D. M. Kan, *Homotopy limits, completions and localizations*, Lecture Notes in Mathematics, Vol. 304, Springer-Verlag, Berlin-New York, 1972. (cit. on p. III.501).

[BJS11] W. J. Bowman, R. P. James, and A. Sabry, *Dagger traced symmetric monoidal categories and reversible programming*, Proceedings of the 4th Workshop on Reversible Computation (RC 2011), 2011, available at https://www.williamjbowman.com/papers/index.html. (cit. on p. I.46).

[BP07] S. L. Braunstein and A. K. Pati, *Quantum information cannot be completely hidden in correlations: implications for the black-hole information paradox*, Phys. Rev. Lett. **98** (2007), no. 8, 080502, 4. doi:10.1103/PhysRevLett.98.080502 (cit. on p. I.46).

[Bro64] R. Brown, *Function spaces and product topologies*, Quart. J. Math. Oxford Ser. (2) **15** (1964), 238–250. doi:10.1093/qmath/15.1.238 (cit. on pp. III.286, III.473).

[Buc55] D. A. Buchsbaum, *Exact categories and duality*, Trans. Amer. Math. Soc. **80** (1955), 1–34. doi:10.2307/1990000 (cit. on p. II.69)

[CS16] J. Carette and A. Sabry, *Computing with semirings and weak rig groupoids*, Programming languages and systems, Lecture Notes in Comput. Sci., vol. 9632, Springer, Berlin, 2016, pp. 123–148. doi:10.1007/978-3-662-49498-1_6 (cit. on pp. I.20, I.45, I.46, I.47, I.48, I.49).

[CS02] W. Chachólski and J. Scherer, *Homotopy theory of diagrams*, Mem. Amer. Math. Soc. **155** (2002), no. 736, x+90. doi:10.1090/memo/0736 (cit. on p. III.501).

[CDH∞] C. Comfort, A. Delpeuch, and J. Hedges, *Sheet diagrams for bimonoidal categories*, 2020. arXiv:2010.13361v1 (cit. on pp. xvii, xxiii, I.49, I.131, I.293, I.429, I.430, II.313, II.314, III.507, III.508).

[Cor16] A. S. Corner, *Day convolution for monoidal bicategories*, Ph.D. thesis, University of Sheffield, 2016. (cit. on p. III.156).

[Cra98] S. E. Crans, *Generalized centers of braided and sylleptic monoidal 2-categories*, Adv. Math. **136** (1998), no. 2, 183–223. doi:10.1006/aima.1998.1720 (cit. on pp. I.253, I.428, I.432, II.312, II.316, III.506, III.510).

[Cru09] G. Cruttwell, *Normed Spaces and the Change of Base for Enriched Categories*, Ph.D. thesis, Dalhousie University, 2009. (cit. on pp. xxx, III.55, III.90).

[Cur71] E. B. Curtis, *Simplicial homotopy theory*, Advances in Math. **6** (1971), 107–209 (1971). doi:10.1016/0001-8708(71)90015-6 (cit. on pp. xxxii, III.286).

[Day70] B. Day, *On closed categories of functors*, Reports of the Midwest Category Seminar, IV, Lecture Notes in Mathematics, Vol. 137, Springer, Berlin, 1970, pp. 1–38. doi:10.1007/BFb0060437 (cit. on pp. xxx, I.18, III.139, III.156).

[DK69] B. Day and G. M. Kelly, *Enriched functor categories*, Reports of the Midwest Category Seminar, III, Lecture Notes in Mathematics, Vol. 106, Springer, Berlin, 1969, pp. 178–191. doi:10.1007/BFb0059139 (cit. on pp. xxx, I.18, III.156).

[DS97] B. Day and R. Street, *Monoidal bicategories and Hopf algebroids*, Adv. Math. **129** (1997), no. 1, 99–157. doi:10.1006/aima.1997.1649 (cit. on pp. III.55, III.156).

[Del∞] Z. Dell, *A characterization of braided enriched monoidal categories*, 2021. arXiv:2104.07747 (cit. on pp. I.434, I.435, II.318, II.319, III.512, III.513).

[Del20] A. Delpeuch, *A complete language for faceted dataflow programs*, Proceedings Applied Category Theory 2019 (ACT 2019), University of Oxford, UK, 15-19 July 2019 (J. Baez and B. Coecke, eds.), Electronic Proceedings in Theoretical Computer Science 323, 2020, pp. 1–14. doi:10.4204/EPTCS.323.1 (cit. on p. I.49).

[DP12] K. Došen and Z. Petrić, *Intermutation*, Appl. Categ. Structures **20** (2012), no. 1, 43–95. doi:10.1007/s10485-010-9228-x (cit. on p. II.306).

[Dri87] V. G. Drinfel'd, *Quantum groups*, Proceedings of the International Congress of Mathematicians, Vol. 1, 2 (Berkeley, Calif., 1986), Amer. Math. Soc., Providence, RI, 1987, pp. 798–820. (cit. on p. II.106).

[Dri89] ———, *Almost cocommutative Hopf algebras*, Algebra i Analiz **1** (1989), no. 2, 30–46. (cit. on p. II.106).

[Dug∞] D. Dugger, *A primer on homotopy colimits*, preprint available at https://pages.uoregon.edu/ddugger/hocolim.pdf. (cit. on pp. III.488, III.501).

[Dun97] G. Dunn, *E_n-ring categories*, J. Pure Appl. Algebra **119** (1997), no. 1, 27–45. doi:10.1016/S0022-4049(96)00015-1 (cit. on p. II.307).

[DS95] W. G. Dwyer and J. Spaliński, *Homotopy theories and model categories*, Handbook of algebraic topology, North-Holland, Amsterdam, 1995, pp. 73–126. doi:10.1016/B978-044481779-2/50003-1 (cit. on p. III.287).

[Ehr65] C. Ehresmann, *Catégories et Structures*, Dunod, Paris, 1965. (cit. on p. I.18).

[EK66] S. Eilenberg and G. M. Kelly, *Closed categories*, Proc. Conf. Categorical Algebra (La Jolla, Calif., 1965), Springer, New York, 1966, pp. 421–562. (cit. on pp. III.54, III.90).

[EML45] S. Eilenberg and S. Mac Lane, *General theory of natural equivalences*, Transactions of the American Mathematical Society **58** (1945), no. 2, 231–294. (cit. on pp. I.18, III.65).

[Elg21] J. Elgueta, *The groupoid of finite sets is biinitial in the 2-category of rig categories*, Journal of Pure and Applied Algebra **225** (2021), no. 11, 106738. doi:10.1016/j.jpaa.2021.106738 (cit. on pp. xxiii, I.48, I.203, I.259, I.292, I.293).

[EKMM97] A. D. Elmendorf, I. Kriz, M. A. Mandell, and J. P. May, *Rings, Modules, and Algebras in Stable Homotopy Theory*, Mathematical Surveys and Monographs, vol. 47, American Mathematical Society, Providence, RI, 1997, With an appendix by M. Cole. (cit. on pp. xvii, III.286).

[EM06] A. D. Elmendorf and M. A. Mandell, *Rings, modules, and algebras in infinite loop space theory*, Adv. Math. **205** (2006), no. 1, 163–228. doi:10.1016/j.aim.2005.07.007 (cit. on pp. xvi, xvii, xviii, xix, xx, xxvi, xxviii, xxix, I.49, I.173, I.204, II.229, II.230, II.231, II.232, II.244, II.247, II.257, II.258, II.261, II.307, III.250, III.253, III.314, III.317, III.345, III.346, III.347, III.389, III.408, III.409, III.410, III.413, III.423, III.431, III.440, III.445, III.447).

[EM09] ———, *Permutative categories, multicategories and algebraic K-theory*, Algebr. Geom. Topol. **9** (2009), no. 4, 2391–2441. doi:10.2140/agt.2009.9.2391 (cit. on pp. xvii, xviii, xix, xx, xxvi, xxviii, xxix, I.436, II.229, II.230, II.261, II.320, III.175, III.220, III.253, III.314, III.317, III.345, III.346, III.347, III.350, III.408, III.409, III.410, III.514).

[Eps66] D. B. A. Epstein, *Functors between tensored categories*, Invent. Math. **1** (1966), 221–228. doi:10.1007/BF01452242 (cit. on p. I.18).

[EGNO15] P. Etingof, S. Gelaki, D. Nikshych, and V. Ostrik, *Tensor categories*, Mathematical Surveys and Monographs, vol. 205, American Mathematical Society, Providence, RI, 2015. doi:10.1090/surv/205 (cit. on p. xxx).

[Fie∞b] Z. Fiedorowicz, *Constructions of E_n operads*, 1999. arXiv:math/9808089 (cit. on pp. I.436, II.307, II.320, III.514).

[Fie∞] ———, *The symmetric bar construction*, preprint available at `https://people.math.osu.edu/fiedorowicz.1/`. (cit. on pp. II.230, II.260, III.450, III.460).

[FSV13] Z. Fiedorowicz, M. Stelzer, and R. M. Vogt, *Homotopy colimits of algebras over Cat-operads and iterated loop spaces*, Adv. Math. **248** (2013), 1089–1155. doi:`10.1016/j.aim.2013.07.016` (cit. on pp. II.260, II.307).

[FV03] Z. Fiedorowicz and R. Vogt, *Simplicial n-fold monoidal categories model all loop spaces*, Cah. Topol. Géom. Différ. Catég. **44** (2003), no. 2, 105–148. (cit. on pp. II.260, II.307).

[For04] S. Forcey, *Enrichment over iterated monoidal categories*, Algebr. Geom. Topol. **4** (2004), 95–119. doi:`10.2140/agt.2004.4.95` (cit. on pp. xxx, I.431, II.306, II.315, III.54, III.509).

[FSS07] S. Forcey, J. Siehler, and E. S. Sowers, *Operads in iterated monoidal categories*, J. Homotopy Relat. Struct. **2** (2007), no. 1, 1–43, errata available at `http://www.math.uakron.edu/~sf34/`. (cit. on pp. I.431, II.271, II.306, II.307, II.315, III.509).

[FKLW03] M. H. Freedman, A. Kitaev, M. J. Larsen, and Z. Wang, *Topological quantum computation*, vol. 40, 2003, pp. 31–38. doi:`10.1090/S0273-0979-02-00964-3` (cit. on p. II.106).

[Fre17] B. Fresse, *Homotopy of operads and Grothendieck-Teichmüller groups. Part 1. The algebraic theory and its topological background*, Mathematical Surveys and Monographs, vol. 217, American Mathematical Society, Providence, RI, 2017. (cit. on pp. xxxi, I.436, II.307, II.320, III.434, III.450, III.460, III.461, III.464, III.473, III.514).

[FY92] P. Freyd and D. N. Yetter, *Coherence theorems via knot theory*, J. Pure Appl. Algebra **78** (1992), no. 1, 49–76. doi:`10.1016/0022-4049(92)90018-B` (cit. on pp. I.435, II.319, III.513).

[Fre03] P. J. Freyd, *Abelian categories*, Repr. Theory Appl. Categ. (2003), no. 3, 1–190. (cit. on pp. xxx, II.62, II.63).

[Fri12] G. Friedman, *Survey article: An elementary illustrated introduction to simplicial sets*, Rocky Mountain J. Math. **42** (2012), no. 2, 353–423. doi:`10.1216/RMJ-2012-42-2-353` (cit. on p. III.286).

[FP18] T. Fritz and P. Perrone, *Bimonoidal structure of probability monads*, Electronic Notes in Theoretical Computer Science **341** (2018), 121–149. (cit. on p. I.49).

[Ful97] W. Fulton, *Young Tableaux: With Applications to Representation Theory and Geometry*, London Mathematical Society Student Texts, vol. 35, Cambridge University Press, Cambridge, 1997. (cit. on p. II.307).

[FM94] W. Fulton and R. MacPherson, *A compactification of configuration spaces*, Ann. of Math. (2) **139** (1994), no. 1, 183–225. doi:`10.2307/2946631` (cit. on pp. I.436, II.307, II.320, III.514).

[GZ67] P. Gabriel and M. Zisman, *Calculus of fractions and homotopy theory*, Ergebnisse der Mathematik und ihrer Grenzgebiete, Band 35, Springer-Verlag New York, Inc., New York, 1967. (cit. on pp. xxxii, III.280, III.459, III.473).

[GH06] T. Geisser and L. Hesselholt, *On the K-theory of complete regular local $\mathbb{F}_p$-algebras*, Topology **45** (2006), no. 3, 475–493. doi:`10.1016/j.top.2005.09.002` (cit. on p. III.411).

[GJ∞] E. Getzler and J. D. Jones, *Operads, homotopy algebra and iterated integrals for double loop spaces*, 1994. arXiv:hep-th/9403055 (cit. on pp. I.436, II.307, II.320, III.514).

[GJ09] P. G. Goerss and J. F. Jardine, *Simplicial homotopy theory*, Modern Birkhäuser Classics, Birkhäuser Verlag, Basel, 2009, Reprint of the 1999 edition. doi:`10.1007/978-3-0346-0189-4` (cit. on pp. xxxii, III.260, III.286).

[GPS95] R. Gordon, A. J. Power, and R. Street, *Coherence for tricategories*, Mem. Amer. Math. Soc. **117** (1995), no. 558, vi+81. doi:`10.1090/memo/0558` (cit. on pp. I.429, I.432, II.313, II.316, III.55, III.507, III.510).

[Gra18] M. Grandis, *Category theory and applications: A textbook for beginners*, World Scientific Publishing Co. Pte. Ltd., Hackensack, NJ, 2018. doi:`10.1142/10737` (cit. on pp. xxx, I.18).

[Gra76] D. Grayson, *Higher algebraic K-theory: II (after Daniel Quillen)*, Algebraic K-theory (Proc. Conf., Northwestern Univ., Evanston, Ill., 1976), Springer, Berlin, 1976, pp. 217–240. Lecture Notes in Math., Vol. 551. doi:10.1007/BFb0080003 (cit. on pp. III.253, III.411).

[Gro57] A. Grothendieck, *Sur quelques points d'algèbre homologique*, Tohoku Math. J. (2) **9** (1957), 119–221. doi:10.2748/tmj/1178244839 (cit. on p. II.62).

[Gui10] B. J. Guillou, *Strictification of categories weakly enriched in symmetric monoidal categories*, Theory Appl. Categ. **24** (2010), No. 20, 564–579. (cit. on pp. I.204, I.293).

[GMMO∞] B. J. Guillou, J. P. May, M. Merling, and A. M. Osorno, *Multiplicative equivariant K-theory and the Barratt-Priddy-Quillen theorem*. arXiv:2102.13246 (cit. on p. III.411).

[Gur11] N. Gurski, *Loop spaces, and coherence for monoidal and braided monoidal bicategories*, Adv. Math. **226** (2011), no. 5, 4225–4265. doi:10.1016/j.aim.2010.12.007 (cit. on pp. I.432, II.316, III.510).

[Gur13] ———, *Coherence in three-dimensional category theory*, Cambridge Tracts in Mathematics, vol. 201, Cambridge University Press, Cambridge, 2013. doi:10.1017/CBO9781139542333 (cit. on pp. I.429, I.432, II.313, II.316, III.55, III.507, III.510).

[GJO17a] N. Gurski, N. Johnson, and A. M. Osorno, *Extending homotopy theories across adjunctions*, Homology Homotopy Appl. **19** (2017), no. 2, 89–110. doi:10.4310/HHA.2017.v19.n2.a6 (cit. on pp. III.315, III.350).

[GJO17b] ———, *K-theory for 2-categories*, Adv. Math. **322** (2017), 378–472. doi:10.1016/j.aim.2017.10.011 (cit. on pp. I.209, I.246, I.253, I.254, I.428, I.429, I.432, I.437, II.312, II.313, II.316, II.321, III.315, III.316, III.506, III.507, III.510, III.515).

[HRY15] P. Hackney, M. Robertson, and D. Yau, *Infinity properads and infinity wheeled properads*, Lecture Notes in Mathematics, vol. 2147, Springer, Cham, 2015. doi:10.1007/978-3-319-20547-2 (cit. on p. III.220).

[HS∞] L. Hansen and M. A. Shulman, *Constructing symmetric monoidal bicategories functorially*, 2019. arXiv:1910.09240 (cit. on pp. I.429, II.313, III.507).

[Hat02] A. Hatcher, *Algebraic topology*, Cambridge University Press, Cambridge, 2002. (cit. on p. III.473).

[Hin13] P. Hines, *Quantum speedup and categorical distributivity*, Computation, logic, games, and quantum foundations, Lecture Notes in Comput. Sci., vol. 7860, Springer, Heidelberg, 2013, pp. 122–138. doi:10.1007/978-3-642-38164-5_9 (cit. on pp. xvii, I.48, I.50).

[Hir03] P. S. Hirschhorn, *Model categories and their localizations*, Mathematical Surveys and Monographs, vol. 99, American Mathematical Society, Providence, RI, 2003. (cit. on pp. III.287, III.501).

[Hou07] R. Houston, *Linear logic without units*, Ph.D. thesis, University of Manchester, 2007. arXiv:1305.2231 (cit. on p. III.90).

[Hov99] M. Hovey, *Model categories*, Mathematical Surveys and Monographs, vol. 63, American Mathematical Society, Providence, RI, 1999. (cit. on pp. III.284, III.287).

[Hov01] ———, *Spectra and symmetric spectra in general model categories*, J. Pure Appl. Algebra **165** (2001), no. 1, 63–127. doi:10.1016/S0022-4049(00)00172-9 (cit. on pp. III.265, III.286).

[HSS00] M. Hovey, B. Shipley, and J. Smith, *Symmetric spectra*, J. Amer. Math. Soc. **13** (2000), no. 1, 149–208. doi:10.1090/S0894-0347-99-00320-3 (cit. on pp. xvii, III.265, III.286, III.287).

[JS12] R. P. James and A. Sabry, *Information effects*, ACM SIGPLAN Notices **47** (2012), no. 1, 73–84. (cit. on p. I.46).

[JP07] M. Jibladze and T. Pirashvili, *Third Mac Lane cohomology via categorical rings*, J. Homotopy Relat. Struct. **2** (2007), no. 2, 187–216. (cit. on p. III.448).

[JO12] N. Johnson and A. M. Osorno, *Modeling stable one-types*, Theory Appl. Categ. **26** (2012), No. 20, 520–537. (cit. on pp. I.435, II.319, III.513).

[JY21] N. Johnson and D. Yau, *2-Dimensional Categories*, Oxford University Press, New York, 2021. doi:10.1093/oso/9780198871378.001.0001 (cit. on pp. xxii, xxx, I.11, I.133, I.164, I.204, I.209, I.213, I.216, I.217, I.230, I.239, I.256, I.291, I.325, I.386, I.393, I.396, I.427, I.429, I.432, II.194, II.198, II.311, II.313, II.316, III.3, III.90, III.345, III.505, III.507, III.510).

[JS91a] A. Joyal and R. Street, *The geometry of tensor calculus. I*, Adv. Math. **88** (1991), no. 1, 55–112. doi:10.1016/0001-8708(91)90003-P (cit. on pp. I.49, I.131, I.429, II.313, III.507).

[JS91b] ———, *An introduction to Tannaka duality and quantum groups*, Category Theory (Como, 1990) (A. Carboni, M. C. Pedicchio, and G. Rosolini, eds.), Lecture Notes in Math., vol. 1488, Springer, Berlin, 1991, pp. 413–492. doi:10.1007/BFb0084235 (cit. on pp. I.435, II.319, III.513).

[JS91c] ———, *Tortile Yang-Baxter operators in tensor categories*, J. Pure Appl. Algebra **71** (1991), no. 1, 43–51. doi:10.1016/0022-4049(91)90039-5 (cit. on p. II.34).

[JS93] ———, *Braided tensor categories*, Adv. Math. **102** (1993), no. 1, 20–78. doi:10.1006/aima.1993.1055 (cit. on pp. I.18, I.435, II.17, II.19, II.34, II.156, II.319, III.513).

[JT91] A. Joyal and M. Tierney, *Strong stacks and classifying spaces*, Category Theory (Como, 1990) (A. Carboni, M. C. Pedicchio, and G. Rosolini, eds.), Lecture Notes in Math., vol. 1488, Springer, Berlin, 1991, pp. 213–236. doi:10.1007/BFb0084235 (cit. on p. III.284).

[KGB+19] A. R. Kalra, N. Gupta, B. K. Behera, S. Prakash, and P. K. Panigrahi, *Demonstration of the no-hiding theorem on the 5-qubit IBM quantum computer in a category-theoretic framework*, Quantum Inf. Process. **18** (2019), no. 6, Paper No. 170, 13. doi:10.1007/s11128-019-2288-4 (cit. on p. I.46).

[KV94] M. Kapranov and V. Voevodsky, *2-categories and Zamolodchikov tetrahedra equations*, Algebraic groups and their generalizations: quantum and infinite-dimensional methods (University Park, PA, 1991), Proc. Sympos. Pure Math., vol. 56, Amer. Math. Soc., Providence, RI, 1994, pp. 177–259. (cit. on pp. xvi, I.44, I.325).

[Kas95] C. Kassel, *Quantum groups*, Graduate Texts in Mathematics, vol. 155, Springer-Verlag, New York, 1995. doi:10.1007/978-1-4612-0783-2 (cit. on pp. xvii, II.34, II.106).

[KT08] C. Kassel and V. Turaev, *Braid groups*, Graduate Texts in Mathematics, vol. 247, Springer, New York, 2008, With the graphical assistance of Olivier Dodane. doi:10.1007/978-0-387-68548-9 (cit. on p. II.34).

[Kel55] J. L. Kelley, *General topology*, D. Van Nostrand Company, Inc., Toronto-New York-London, 1955. (cit. on pp. III.286, III.473).

[Kel64] G. M. Kelly, *On MacLane's conditions for coherence of natural associativities, commutativities, etc.*, J. Algebra **1** (1964), 397–402. (cit. on p. I.11).

[Kel69] ———, *Adjunction for enriched categories*, Reports of the Midwest Category Seminar, III, Lecture Notes in Mathematics, Vol. 106, Springer, Berlin, 1969, pp. 166–177. (cit. on p. III.156).

[Kel74] ———, *Coherence theorems for lax algebras and for distributive laws*, Category Seminar (Proc. Sem., Sydney, 1972/1973), Lecture Notes in Math., Vol. 420, 1974, pp. 281–375. (cit. on pp. I.133, I.169, II.34).

[Kel05] ———, *Basic concepts of enriched category theory*, Repr. Theory Appl. Categ. (2005), no. 10, vi+137, Reprint of the 1982 original (Cambridge Univ. Press, Cambridge). (cit. on pp. xxx, I.213, III.54, III.155, III.156, III.286).

[KML71a] G. M. Kelly and S. Mac Lane, *Coherence in closed categories*, J. Pure Appl. Algebra **1** (1971), no. 1, 97–140. doi:10.1016/0022-4049(71)90013-2 (cit. on p. III.155).

[KML71b] ———, *Erratum: "Coherence in closed categories"*, J. Pure Appl. Algebra **1** (1971), no. 2, 219. doi:10.1016/0022-4049(71)90019-3 (cit. on p. III.155).

[KML72] ———, *Closed coherence for a natural transformation*, Coherence in categories, Lecture Notes in Math., Vol. 281, 1972, pp. 1–28. doi:10.1007/BFb0059553 (cit. on p. III.155)

[KR68] C. G. Khatri and C. R. Rao, *Solutions to some functional equations and their applications to characterization of probability distributions*, Sankhya Ser. A **30** (1968), 167–180. (cit. on p. I.406).

[KYZZ∞] L. Kong, W. Yuan, Z.-H. Zhang, and H. Zheng, *Enriched monoidal categories I: centers*, 2021. arXiv:2104.03121 (cit. on pp. I.434, II.318, III.512).

[KZ18] L. Kong and H. Zheng, *Drinfeld center of enriched monoidal categories*, Adv. Math. **323** (2018), 411–426. doi:10.1016/j.aim.2017.10.038 (cit. on pp. I.434, II.318, III.55, III.90, III.512).

[Lac02] S. Lack, *Codescent objects and coherence*, J. Pure Appl. Algebra **175** (2002), no. 1-3, 223–241, Special volume celebrating the 70th birthday of Professor Max Kelly. doi:10.1016/S0022-4049(02)00136-6 (cit. on p. III.90).

[Lac10] ———, *A 2-categories companion*, Towards higher categories (J. Baez and J. May, eds.), IMA Vol. Math. Appl., vol. 152, Springer, New York, 2010, pp. 105–191. doi:10.1007/978-1-4419-1524-5_4 (cit. on p. I.209).

[LP17] V. T. Lahtinen and J. K. Pachos, *A short introduction to topological quantum computation*, SciPost Physics **3** (2017), 021. doi:10.21468/SciPostPhys.3.3.021 (cit. on p. II.106).

[Lam69] J. Lambek, *Deductive systems and categories II. Standard constructions and closed categories*, Category Theory, Homology Theory and their Applications, I (Battelle Institute Conference, Seattle, Wash., 1968, Vol. One), Springer, Berlin, 1969, pp. 76–122. (cit. on pp. III.219, III.220).

[Lap72a] M. L. Laplaza, *Coherence for distributivity*, Coherence in categories, Lectures Notes in Mathematics, Vol. 281, Springer-Verlag, Berlin-Heidelberg, 1972, pp. 29–65. (cit. on pp. xvi, xix, xxi, I.48, I.51, I.131, I.132, I.133, I.167).

[Lap72b] ———, *A new result of coherence for distributivity*, Coherence in categories, Lectures Notes in Mathematics, Vol. 281, Springer-Verlag, Berlin-Heidelberg, 1972, pp. 214–235. (cit. on pp. xvi, xix, xxii, I.48, I.135, I.167, I.168).

[Law18] T. Lawson, *Secondary power operations and the Brown-Peterson spectrum at the prime 2*, Ann. of Math. (2) **188** (2018), no. 2, 513–576. doi:10.4007/annals.2018.188.2.3 (cit. on pp. I.435, II.319, III.513).

[Lei∞] T. Leinster, *Basic bicategories*, 1998. arXiv:math/9810017 (cit. on p. I.209).

[Lei14] ———, *Basic category theory*, Cambridge Studies in Advanced Mathematics, vol. 143, Cambridge University Press, Cambridge, 2014. doi:10.1017/CBO9781107360068 (cit. on pp. xxx, I.18).

[Lew72] G. Lewis, *Coherence for a closed functor*, Coherence in categories, Lecture Notes in Math., Vol. 281, 1972, pp. 148–195. (cit. on pp. I.430, II.314, III.508).

[Lim59] E. L. Lima, *The Spanier-Whitehead duality in new homotopy categories*, Summa Brasil. Math. **4** (1959), 91–148 (1959). (cit. on p. III.287).

[Liu99] S. Liu, *Matrix results on the Khatri-Rao and Tracy-Singh products*, vol. 289, 1999, Linear algebra and statistics (Istanbul, 1997), pp. 267–277. doi:10.1016/S0024-3795(98)10209-4 (cit. on p. I.406).

[Lor21] F. Loregian, *(Co)end calculus*, London Mathematical Society Lecture Note Series, Cambridge University Press, 2021. (cit. on pp. xxx, I.18, III.156).

[ML63] S. Mac Lane, *Natural associativity and commutativity*, Rice Univ. Stud. **49** (1963), no. 4, 28–46. (cit. on pp. I.18, I.51).

[ML98] ———, *Categories for the working mathematician*, second ed., Graduate Texts in Mathematics, vol. 5, Springer-Verlag, New York, 1998. (cit. on pp. xxx, I.10, I.18, I.51, II.62, II.158, III.86, III.192, III.219).

[Maj91] S. Majid, *Representations, duals and quantum doubles of monoidal categories*, Proceedings of the Winter School on Geometry and Physics (Srní, 1990), no. 26, 1991, pp. 197–206. (cit. on pp. xvii, II.34).

[Maj93] ———, *Anyonic quantum groups*, Spinors, twistors, Clifford algebras and quantum deformations (Sobótka Castle, 1992), Fund. Theories Phys., vol. 52, Kluwer Acad. Publ., Dordrecht, 1993, pp. 327–336. (cit. on p. II.106).

[Maj95] ———, *Foundations of quantum group theory*, Cambridge University Press, Cambridge, 1995. doi:10.1017/CBO9780511613104 (cit. on p. II.106).

[MMSS01] M. A. Mandell, J. P. May, S. Schwede, and B. Shipley, *Model categories of diagram spectra*, Proc. London Math. Soc. (3) **82** (2001), no. 2, 441–512. doi:10.1112/S0024611501012692 (cit. on pp. xvii, III.286).

[Man10] M. A. Mandell, *An inverse K-theory functor*, Doc. Math. **15** (2010), 765–791. (cit. on pp. I.437, II.321, III.314, III.315, III.316, III.350, III.515).

[Man12] O. Manzyuk, *Closed categories vs. closed multicategories*, Theory Appl. Categ. **26** (2012), 132–175. (cit. on p. III.220).

[MSS02] M. Markl, S. Shnider, and J. Stasheff, *Operads in algebra, topology and physics*, Mathematical Surveys and Monographs, vol. 96, American Mathematical Society, Providence, RI, 2002. (cit. on p. xxxi).

[May72] J. P. May, *The geometry of iterated loop spaces*, Lectures Notes in Mathematics, Vol. 271, Springer-Verlag, Berlin-New York, 1972. (cit. on pp. xxxi, II.230, II.260, II.307, III.219, III.286, III.457, III.473).

[May74] ———, *E_∞ spaces, group completions, and permutative categories*, New developments in topology (Proc. Sympos. Algebraic Topology, Oxford, 1972), London Math. Soc. Lecture Note Ser., No. 11, 1974, pp. 61–93. (cit. on p. II.260).

[May77] ———, *E_∞ ring spaces and E_∞ ring spectra*, Lecture Notes in Mathematics, Vol. 577, Springer-Verlag, Berlin-New York, 1977, With contributions by Frank Quinn, Nigel Ray, and Jørgen Tornehave. (cit. on pp. xvi, I.48, I.49, I.171, I.173, I.203, I.204, III.411).

[May78] ———, *The spectra associated to permutative categories*, Topology **17** (1978), no. 3, 225–228. doi:10.1016/0040-9383(78)90027-7 (cit. on p. III.314).

[May82] ———, *Multiplicative infinite loop space theory*, J. Pure Appl. Algebra **26** (1982), no. 1, 1–69. doi:10.1016/0022-4049(82)90029-9 (cit. on p. III.411).

[May92] ———, *Simplicial objects in algebraic topology*, Chicago Lectures in Mathematics, University of Chicago Press, Chicago, IL, 1992, Reprint of the 1967 original. (cit. on pp. xxxii, III.286).

[May99] ———, *A concise course in algebraic topology*, Chicago Lectures in Mathematics, University of Chicago press, Chicago, IL, 1999. (cit. on pp. xxxii, III.286, III.287, III.473).

[May09a] ———, *The construction of E_∞ ring spaces from bipermutative categories*, New topological contexts for Galois theory and algebraic geometry (BIRS 2008), Geom. Topol. Monogr., vol. 16, Geom. Topol. Publ., Coventry, 2009, pp. 283–330. doi:10.2140/gtm.2009.16.283 (cit. on p. III.411).

[May09b] ———, *What are E_∞ ring spaces good for?*, New topological contexts for Galois theory and algebraic geometry (BIRS 2008), Geom. Topol. Monogr., vol. 16, Geom. Topol. Publ., Coventry, 2009, pp. 331–365. doi:10.2140/gtm.2009.16.331 (cit. on p. III.411).

[May09c] ———, *What precisely are E_∞ ring spaces and E_∞ ring spectra?*, New topological contexts for Galois theory and algebraic geometry (BIRS 2008), Geom. Topol. Monogr., vol. 16, Geom. Topol. Publ., Coventry, 2009, pp. 215–282. doi:10.2140/gtm.2009.16.215 (cit. on p. III.411).

[MP12] J. P. May and K. Ponto, *More concise algebraic topology: Localization, completion, and model categories*, Chicago Lectures in Mathematics, University of Chicago Press, Chicago, IL, 2012. (cit. on pp. xxxii, III.277, III.279, III.280, III.281, III.282, III.285, III.286, III.287, III.488, III.501).

[MS06] J. P. May and J. Sigurdsson, *Parametrized homotopy theory*, Mathematical Surveys and Monographs, vol. 132, American Mathematical Society, Providence, RI, 2006. doi:10.1090/surv/132 (cit. on p. III.286).

[MS76] D. McDuff and G. Segal, *Homology fibrations and the "group-completion" theorem*, Invent. Math. **31** (1975/76), no. 3, 279–284. doi:10.1007/BF01403148 (cit. on p. II.260).

[Mil66] R. J. Milgram, *Iterated loop spaces*, Ann. of Math. (2) **84** (1966), 386–403. doi:10.2307/1970453 (cit. on pp. II.307, III.502).

[Mil20] H. Miller, *Handbook of homotopy theory*, CRC Press/Chapman and Hall Handbooks in Mathematics Series, CRC Press, Boca Raton, FL, 2020. (cit. on p. xxxii).

[Mil71] J. Milnor, *Introduction to algebraic K-theory*, Princeton University Press, Princeton, N.J.; University of Tokyo Press, Tokyo, 1971, Annals of Mathematics Studies, No. 72. (cit. on p. xxxii).

[MS74] J. W. Milnor and J. D. Stasheff, *Characteristic classes*, Princeton University Press, Princeton, N. J.; University of Tokyo Press, Tokyo, 1974, Annals of Mathematics Studies, No. 76. (cit. on p. III.432).

[Mit64] B. Mitchell, *The full imbedding theorem*, Amer. J. Math. **86** (1964), 619–637. doi:10.2307/2373027 (cit. on p. II.63).

[Mit65] ———, *Theory of Categories*, Pure and Applied Mathematics, Vol. XVII, Academic Press, New York-London, 1965. (cit. on pp. xxx, I.18, II.62, II.63).

[MT10] I. Moerdijk and B. Toën, *Simplicial methods for operads and algebraic geometry*, Advanced Courses in Mathematics. CRM Barcelona, Birkhäuser/Springer Basel AG, Basel, 2010, Edited by Carles Casacuberta and Joachim Kock. doi:10.1007/978-3-0348-0052-5 (cit. on pp. III.219, III.220).

[Mon93] S. Montgomery, *Hopf algebras and their actions on rings*, CBMS Regional Conference Series in Mathematics, vol. 82, Published for the Conference Board of the Mathematical Sciences, Washington, DC; by the American Mathematical Society, Providence, RI, 1993. doi:10.1090/cbms/082 (cit. on p. II.106).

[MP19] S. Morrison and D. Penneys, *Monoidal categories enriched in braided monoidal categories*, Int. Math. Res. Not. IMRN (2019), no. 11, 3527–3579. doi:10.1093/imrn/rnx217 (cit. on pp. I.434, II.318, III.55, III.512).

[Mun00] J. R. Munkres, *Topology*, Prentice Hall, Inc., Upper Saddle River, NJ, 2000. (cit. on p. III.473).

[MT07] F. Muro and A. Tonks, *The 1-type of a Waldhausen K-theory spectrum*, Adv. Math. **216** (2007), no. 1, 178–211. doi:10.1016/j.aim.2007.05.008 (cit. on p. III.448).

[NSS+08] C. Nayak, S. H. Simon, A. Stern, M. Freedman, and S. Das Sarma, *Non-abelian anyons and topological quantum computation*, Rev. Modern Phys. **80** (2008), no. 3, 1083–1159. doi:10.1103/RevModPhys.80.1083 (cit. on p. II.106).

[Pac12] J. K. Pachos, *Introduction to topological quantum computation*, Cambridge University Press, Cambridge, 2012. doi:10.1017/CBO9780511792908 (cit. on pp. II.106, II.107).

[PP11] P. Panangaden and E. O. Paquette, *A categorical presentation of quantum computation with anyons*, New structures for physics, Lecture Notes in Phys., vol. 813, Springer, Heidelberg, 2011, pp. 983–1025. doi:10.1007/978-3-642-12821-9_15 (cit. on p. II.106).

[Pin14] C. C. Pinter, *A book of set theory*, Dover, New York NY, 2014. (cit. on p. III.282).

[Pro13] T. U. F. Program, *Homotopy type theory—univalent foundations of mathematics*, The Univalent Foundations Program, Princeton, NJ; Institute for Advanced Study (IAS), Princeton, NJ, 2013. (cit. on p. I.46).

[Qui67] D. G. Quillen, *Homotopical algebra*, Lecture Notes in Mathematics, No. 43, Springer-Verlag, Berlin-New York, 1967. (cit. on p. III.287).

[Qui73] ———, *Higher algebraic K-theory I: Higher K-theories*, Proc. Conf., Battelle Memorial Inst., Seattle, Wash., (1972), Lecture Notes in Math., Vol. 341, 1973, pp. 85–147. (cit. on pp. xxxii, III.253, III.411).

[Rad93] D. E. Radford, *Minimal quasitriangular Hopf algebras*, J. Algebra **157** (1993), no. 2, 285–315. doi:10.1006/jabr.1993.1102 (cit. on p. II.106).

[Rez∞] C. Rezk, *A model category for categories*, preprint available at http://www.math.uiuc.edu/~rezk. (cit. on p. III.284).

[Ric10] B. Richter, *An involution on the K-theory of bimonoidal categories with anti-involution*, Algebraic & Geometric Topology **10** (2010), no. 1, 315–342. (cit. on pp. xvi, xx, II.43, II.134, II.231, II.258).

[Ric20] ———, *From categories to homotopy theory*, Cambridge Studies in Advanced Mathematics, vol. 188, Cambridge University Press, Cambridge, 2020. (cit. on pp. xxxii, III.156, III.265, III.286).

[Rie11] E. Riehl, *Algebraic model structures*, New York J. Math. **17** (2011), 173–231. (cit. on p. III.287).

[Rie14] ———, *Categorical homotopy theory*, New Mathematical Monographs, vol. 24, Cambridge University Press, Cambridge, 2014. doi:10.1017/CBO9781107261457 (cit. on pp. xxxii, III.156, III.286, III.501).

[Rie16] ———, *Category theory in context*, Courier Dover Publications, 2016. (cit. on pp. xxx, I.18, III.194, III.219, III.220, III.286).

[Rom17] S. Roman, *An introduction to the language of category theory*, Compact Textbooks in Mathematics, Birkhäuser/Springer, Cham, 2017. doi:10.1007/978-3-319-41917-6 (cit. on pp. xxx, I.18).

[Ros95] J. Rosenberg, *Algebraic K-theory and its applications*, Graduate Texts in Mathematics, vol. 147, Springer Science & Business Media, 1995. (cit. on p. xxxii).

[Sal01] P. Salvatore, *Configuration spaces with summable labels*, Cohomological methods in homotopy theory (Bellaterra, 1998), Progr. Math., vol. 196, Birkhäuser, Basel, 2001, pp. 375–395. (cit. on pp. I.436, II.307, II.320, III.514).

[SPK11] J. R. Samal, A. K. Pati, and A. Kumar, *Experimental test of the quantum no-hiding theorem*, Phys. Rev. Lett. **106** (2011), 080401. doi:10.1103/PhysRevLett.106.080401 (cit. on p. I.46).

[SP∞] C. Schommer-Pries, *The classification of two-dimensional extended topological field theories*, 2011. arXiv:1112.1000 (cit. on pp. I.253, I.254, I.427, II.311, III.505).

[Sch03] B. S. W. Schröder, *Ordered Sets: An Introduction*, Birkhäuser Boston, Inc., Boston, MA, 2003. doi:10.1007/978-1-4612-0053-6 (cit. on p. II.307).

[Sch72] H. Schubert, *Categories*, Springer-Verlag, New York-Heidelberg, 1972, Translated from German by Eva Gray. (cit. on pp. xxx, I.18, II.62, II.63).

[Seg74] G. Segal, *Categories and cohomology theories*, Topology **13** (1974), 293–312. doi:10.1016/0040-9383(74)90022-6 (cit. on pp. xvii, xxviii, II.260, III.289, III.292, III.314, III.315).

[Sel11] P. Selinger, *A survey of graphical languages for monoidal categories*, New structures for physics, Lecture Notes in Phys., vol. 813, Springer, Heidelberg, 2011, pp. 289–355. doi:10.1007/978-3-642-12821-9_4 (cit. on pp. I.49, I.429, II.313, III.507).

[Sen∞] A. Senger, *The Brown-Peterson spectrum is not $E_{2(p^2+2)}$ at odd primes*, 2017. arXiv:1710.09822 (cit. on pp. I.435, II.319, III.513).

[SV02] A. Shen and N. K. Vereshchagin, *Basic set theory*, Student Mathematical Library, vol. 17, American Mathematical Society, Providence, RI, 2002, Translated from the 1999 Russian edition by Shen. doi:10.1090/stml/017 (cit. on p. III.282).

[Shu∞a] M. A. Shulman, *Homotopy limits and colimits and enriched homotopy theory*, 2009. arXiv:math/0610194 (cit. on pp. III.156, III.501).

[Shu∞b] ———, *Set theory for category theory*, 2008. arXiv:0810.1279 (cit. on pp. III.65, III.90).

[Sim11] H. Simmons, *An introduction to category theory*, Cambridge University Press, Cambridge, 2011. doi:10.1017/CBO9780511863226 (cit. on pp. xxx, I.18).

[Sly99] V. I. Slyusar, *A family of face products of matrices and its properties*, Kibernet. Sistem. Anal. (1999), no. 3, 43–49, 189. doi:10.1007/BF02733426 (cit. on p. I.406).

[Smi89] J. H. Smith, *Simplicial group models for $\Omega^n S^n X$*, Israel J. Math. **66** (1989), no. 1-3, 330–350. doi:10.1007/BF02765902 (cit. on pp. I.436, II.307, II.320, III.447, III.501, III.513).

[Sta15] S. Staton, *Algebraic effects, linearity, and quantum programming languages*, ACM SIGPLAN Notices **50** (2015), no. 1, 395–406. (cit. on p. I.49).

[Ste67] N. E. Steenrod, *A convenient category of topological spaces*, Michigan Math. J. **14** (1967), 133–152. (cit. on pp. III.286, III.473).

[Sto79] R. Steiner, *A canonical operad pair*, Math. Proc. Cambridge Philos. Soc. **86** (1979), no. 3, 443–449. doi:10.1017/S0305004100056292 (cit. on p. III.473).

[Swe69] M. E. Sweedler, *Hopf algebras*, Mathematics Lecture Note Series, W. A. Benjamin, Inc., New York, 1969. (cit. on p. II.106).

[Tho95] R. W. Thomason, *Symmetric monoidal categories model all connective spectra*, Theory Appl. Categ. **1** (1995), No. 5, 78–118. (cit. on pp. I.437, II.321, III.316, III.515).

[TS72] D. S. Tracy and R. P. Singh, *A new matrix product and its applications in partitioned matrix differentiation*, Statistica Neerlandica **26** (1972), 143–157. doi:10.1111/j.1467-9574.1972.tb00199.x (cit. on p. I.406).

[TTWL08] S. Trebst, M. Troyer, Z. Wang, and A. W. W. Ludwig, *A short introduction to fibonacci anyon models*, Progress of Theoretical Physics Supplement **176** (2008), 384–407. (cit. on p. II.106).

[Vog73] R. M. Vogt, *Homotopy limits and colimits*, Math. Z. **134** (1973), 11–52. doi:10.1007/BF01219090 (cit. on p. III.501).

[Vog77] ———, *Commuting homotopy limits*, Math. Z. **153** (1977), no. 1, 59–82. doi:10.1007/BF01214734 (cit. on p. III.501).

[Wal85] F. Waldhausen, *Algebraic K-theory of spaces*, Algebraic and geometric topology (New Brunswick, N.J., 1983), Lecture Notes in Math., vol. 1126, Springer, Berlin, 1985, pp. 318–419. doi:10.1007/BFb0074449 (cit. on pp. xxxii, III.411).

[Wan10] Z. Wang, *Topological quantum computation*, CBMS Regional Conference Series in Mathematics, vol. 112, Published for the Conference Board of the Mathematical Sciences, Washington, DC; by the American Mathematical Society, Providence, RI, 2010. doi:10.1090/cbms/112 (cit. on p. II.106).

[Wei13] C. A. Weibel, *The K-book: An introduction to algebraic K-theory*, Graduate Studies in Mathematics, vol. 145, American Mathematical Society Providence, RI, 2013. (cit. on pp. xxxii, III.411).

[WY18] D. White and D. Yau, *Bousfield localization and algebras over colored operads*, Appl. Categ. Structures **26** (2018), no. 1, 153–203. doi:10.1007/s10485-017-9489-8 (cit. on pp. III.219, III.287).

[WY19] ———, *Homotopical adjoint lifting theorem*, Appl. Categ. Structures **27** (2019), no. 4, 385–426. doi:10.1007/s10485-019-09560-2 (cit. on pp. I.438, II.322, III.351, III.516).

[WY20] ———, *Right Bousfield localization and operadic algebras*, Tbilisi Mathematical Journal, Special Issue (HomotopyTheorySpectra - 2020) (2020), 71–118. (cit. on p. III.287).

[Yau∞] D. Yau, *Infinity operads and monoidal categories with group equivariance*, 2019. arXiv:1903.03839 (cit. on pp. II.34, II.156, III.447, III.473, III.474).

[Yau16] ———, *Colored operads*, Graduate Studies in Mathematics, vol. 170, American Mathematical Society, Providence, RI, 2016. (cit. on pp. xxxi, III.197, III.219, III.447).

[Yau18] ———, *Operads of wiring diagrams*, Lecture Notes in Mathematics, vol. 2192, Springer, Cham, 2018. doi:10.1007/978-3-319-95001-3 (cit. on p. III.219).

[Yau20] ———, *Homotopical quantum field theory*, World Scientific, Singapore, 2020. (cit. on pp. I.435, I.436, II.319, II.320, III.219, III.250, III.447, III.513, III.514).

[YJ15] D. Yau and M. W. Johnson, *A foundation for PROPs, algebras, and modules*, Mathematical Surveys and Monographs, vol. 203, American Mathematical Society, Providence, RI, 2015. doi:10.1090/surv/203 (cit. on p. III.250).

[Yon60] N. Yoneda, *On Ext and exact sequences*, J. Fac. Sci. Univ. Tokyo Sect. I **8** (1960), 507–576 (1960). (cit. on p. I.9).

[Zak18] I. Zakharevich, *The category of Waldhausen categories is a closed multicategory*, New directions in homotopy theory, Contemp. Math., vol. 707, Amer. Math. Soc., Providence, RI, 2018, pp. 175–194. doi:10.1090/conm/707/14259 (cit. on pp. III.220, III.411).

List of Main Facts

Part 1. Symmetric Bimonoidal Categories

Chapter 1. Basic Category Theory

(1.1.11) An adjunction satisfies the triangle identities.
(p. I.7) A functor is an equivalence if and only if it is fully faithful and essentially surjective.
(1.1.14) Left adjoints preserve colimits. Right adjoints preserve limits.
(1.2.1) A monoidal category satisfies the unity axiom and the pentagon axiom.
(1.2.7) A monoidal category satisfies the left and the right unity properties.
(1.3.3) **Mac Lane's Coherence Theorem.** Any two words of the same length in a monoidal category are connected by a unique canonical map.
(1.3.5) **Mac Lane's Strictification Theorem.** Each monoidal category is adjoint equivalent to a strict monoidal category via strong monoidal functors.
(1.3.8) **Symmetric Coherence Theorem.** Any two permuted words of the same length in a symmetric monoidal category are connected by a unique permuted canonical map.
(1.3.10) **Symmetric Strictification Theorem.** Each symmetric monoidal category is adjoint equivalent to a permutative category via strong symmetric monoidal functors.
(1.3.12) **Epstein's Coherence Theorem.** For each (symmetric) monoidal functor $F\colon \mathsf{C} \longrightarrow \mathsf{D}$ and F-iterates $G, H\colon \mathsf{C}^n \longrightarrow \mathsf{D}$, there exists at most one F-coherent map $G \longrightarrow H$.

Chapter 2. Symmetric Bimonoidal Categories

(2.1.2) A symmetric bimonoidal category has two symmetric monoidal structures, left/right multiplicative zero natural isomorphisms, and left/right distributivity natural monomorphisms, and satisfies 24 axioms.
(2.1.2) A bimonoidal category is defined in the same way as a symmetric bimonoidal category, but without the multiplicative symmetry $\xi^{\otimes}$ and the two axioms that involve $\xi^{\otimes}$. So a bimonoidal category is defined by the other 22 axioms.
(2.1.32) There is a tight symmetric bimonoidal category $\mathsf{Vect}^{\mathbb{C}}$ of finite dimensional complex vector spaces.
(2.2.13) Half of the 24 axioms in a symmetric bimonoidal category are formal consequences of the others.
(2.2.14) One axiom is redundant in a bimonoidal category, which is, therefore, determined by 21 axioms.
(2.3.2) Each distributive symmetric monoidal category yields a tight symmetric bimonoidal category, whose sum is the coproduct.

(2.3.3–2.3.5) Symmetric monoidal closed categories with finite coproducts, the category of modules over a commutative ring, and distributive categories are examples of tight symmetric bimonoidal categories.
(2.4.8) The category Σ of finite ordinals and permutations is a tight symmetric bimonoidal category.
(2.4.23) The variant Σ' of Σ is a tight symmetric bimonoidal category.
(2.5.7) Each right bipermutative category is a tight symmetric bimonoidal category.
(2.5.8–2.5.9) Σ' and $\mathsf{Vect}_{\mathsf{c}}^{\mathbb{C}}$ are right bipermutative categories.
(2.5.16) Each left bipermutative category is a tight symmetric bimonoidal category.
(2.5.17) Σ is a left bipermutative category.
(2.6.2) There is a symmetric bimonoidal groupoid Π with syntax of finite types as objects and Π-terms and Π-combinators as morphisms.

Chapter 3. Coherence of Symmetric Bimonoidal Categories

(3.1.6) In the elementary graph $\mathsf{Gr}^{\mathsf{el}}(X)$, δ^l and δ^r do not have formal inverses.
(3.1.8) Each prime edge involves at most one nonidentity elementary edge.
(3.1.9) The graph $\mathsf{Gr}(X)$ consists of the vertex set X^{fr} and prime edges.
(3.1.14) Each functor $\varphi: X \longrightarrow \mathsf{Ob}(\mathsf{C})$ extends additively and multiplicatively to a graph morphism $\varphi: \mathsf{Gr}(X) \longrightarrow \mathsf{C}$.
(3.1.18) The value in C of a path in $\mathsf{Gr}(X)$ is the composite of the images of its constituent prime edges under φ.
(3.1.25) An element in X^{fr} is regular if it has the same support as a formal polynomial whose monomials are distinct in the strict $\{\oplus, \otimes\}$-algebra X^{st}, and whose factors in each monomial are distinct elements in X.
(3.1.29) Any two elements in X^{fr} connected by a path in $\mathsf{Gr}(X)$ have the same support, and one of them is regular if and only if the other one is regular.
(3.2.15) For an element in X^{fr}, the size is equal to the rank if and only if it is a sum, with each summand either in X or a product of two elements in X.
(3.3.6) Each element in X^{fr} has a 0^X-reduction.
(3.3.11–3.3.12) Any two 0^X-reductions of an element in X^{fr} have the same codomain and the same value in a symmetric bimonoidal category.
(3.5.32) Each path in $\mathsf{Gr}(X)$ has a 0^X-reduction.
(3.5.33) Any two parallel paths in $\mathsf{Gr}(X)$ whose domain has the same support as 0^X have the same value in a symmetric bimonoidal category.
(3.6.5) An element in X^{fr} is a polynomial if and only if it is δ-reduced.
(3.6.9) Each element in X^{fr} has a δ-reduction.
(3.7.19) Each path in $\mathsf{Gr}(X)$ that does not contain 0^X has a $(0^X, \delta)$-reduction.
(3.8.5–3.8.7) Each element in X^{fr} has a 1^X-reduction. If the original element is δ-reduced, then all of its 1^X-reductions have the same codomain and the same value in a symmetric bimonoidal category.
(3.8.14) Each $(0^X, \delta)$-free path in $\mathsf{Gr}(X)$ whose (co)domain is δ-reduced has a 1^X-reduction.
(3.9.1) **Laplaza's First Coherence Theorem.** In each symmetric bimonoidal category C satisfying a monomorphism assumption, any two parallel paths in $\mathsf{Gr}(X)$ with a regular domain have the same value in C.
(3.9.9–3.9.10) Theorem 3.9.1 applies to symmetric bimonoidal categories that are flat, in particular, tight.

(3.10.7) **Bimonoidal Coherence Theorem**. In each bimonoidal category C satisfying a monomorphism assumption, any two parallel paths in $\mathsf{Gr}^{\mathsf{ns}}(X)$ with a nonsymmetric regular domain have the same value in C.
(3.10.8) Theorem 3.10.7 applies to flat, in particular, tight, bimonoidal categories.

Chapter 4. Coherence of Symmetric Bimonoidal Categories II

(4.2.1) In the distortion category, each object is a finite sequence of nonnegative integers, and each morphism is a finite sequence of permutations.
(4.2.5) The distortion category is a groupoid.
(4.2.12) The additive structure of the distortion category is a permutative category.
(4.2.19) The multiplicative structure of the distortion category is a permutative category.
(4.2.29) The distortion category is a left bipermutative category.
(4.3.1) For a path in $\mathsf{Gr}(X)$, its distortion is defined as its value in the distortion category.
(4.4.3) **Laplaza's Second Coherence Theorem**. In each symmetric bimonoidal category C satisfying a monomorphism assumption, any two parallel paths in $\mathsf{Gr}(X)$ with the same distortion have the same value in C.
(4.4.5) Theorem 4.4.3 applies to symmetric bimonoidal categories that are flat, in particular, tight.
(4.5.2) In the additive distortion category, each object is a finite sequence of nonnegative integers, and each morphism is a permutation.
(4.5.6) The additive distortion category is a groupoid and a tight bimonoidal category. It faithfully embeds into the distortion category.
(4.5.7) For a path in $\mathsf{Gr}^{\mathsf{ns}}(X)$, its additive distortion is defined as its value in the additive distortion category.
(4.5.8) **Bimonoidal Coherence Theorem II**. In each bimonoidal category C satisfying a monomorphism assumption, any two parallel paths in $\mathsf{Gr}^{\mathsf{ns}}(X)$ with the same additive distortion have the same value in C.
(4.5.9) Theorem 4.5.8 applies to flat, in particular, tight, bimonoidal categories.
(4.6.5) The distortion category $\mathcal{D}$ is isomorphic to the Grothendieck construction $\int_{\Sigma} F$.
(4.6.7) The additive distortion category $\mathcal{D}^{\mathsf{ad}}$ is isomorphic to the Grothendieck construction $\int_{\Sigma} F^{\mathsf{ad}}$.

Chapter 5. Strictification of Tight Symmetric Bimonoidal Categories

(5.1.1) A symmetric bimonoidal functor is a functor equipped with two symmetric monoidal functor structures, and satisfies two axioms.
(5.1.10) There is a 1-category $\mathsf{Bi}^{\mathsf{sy}}$ of small symmetric bimonoidal categories and symmetric bimonoidal functors.
(5.1.11) Each symmetric monoidal functor between distributive symmetric monoidal categories induces a symmetric bimonoidal functor.
(5.1.15–5.1.16) Σ and Σ' are isomorphic via symmetric bimonoidal functors.
(5.3.9) Each tight symmetric bimonoidal category C has an associated right bipermutative category A, whose objects are formal polynomials in the objects in C.
(5.4.6–5.4.7) **Bipermutative Strictification Theorems**. Each tight symmetric bimonoidal category is adjoint equivalent to a right/left bipermutative category via symmetric bimonoidal functors.

(5.5.1) A bimonoidal functor is a functor equipped with an additive symmetric monoidal functor structure and a multiplicative monoidal functor structure that satisfies four axioms.
(5.5.4) There is a 1-category Bi of small bimonoidal categories and bimonoidal functors.
(5.5.10) Each tight bimonoidal category C has an associated right rigid bimonoidal category A, whose objects are formal polynomials in the objects in C.
(5.5.11–5.5.12) **Rigid Strictification Theorems**. Each tight bimonoidal category is adjoint equivalent to a right/left rigid bimonoidal category via bimonoidal functors.

Part 2. Bicategorical Aspects of Symmetric Bimonoidal Categories

Chapter 6. Definitions from Bicategory Theory

(6.1.2) A bicategory has objects, (identity) 1-cells, (identity) 2-cells, vertical and horizontal compositions, an associator, and two unitors, and satisfies the unity axiom and the pentagon axiom.
(6.1.8) A 2-category is a bicategory whose associator and unitors are identities.
(6.1.10) A 2-category can be described by data and axioms.
(6.1.11) A locally small 2-category is precisely a Cat-category.
(6.1.16) A monoidal category is a one-object bicategory.
(6.2.1) A lax functor has an object assignment, local functors, and two laxity constraints, and satisfies the lax associativity axiom and the lax unity axioms.
(6.2.11) There is a 1-category Bicat with small bicategories as objects and lax functors as morphisms.
(6.2.14) A lax transformation has component 1-cells and natural component 2-cells, and satisfies the lax unity axiom and the lax naturality axiom.
(6.2.26) There is a 2-category $\mathsf{2Cat}$ of small 2-categories, 2-functors, and 2-natural transformations.
(6.3.1) A modification has component 2-cells and satisfies the modification axiom.
(6.3.7) For bicategories B and B' with $\mathsf{Ob}(\mathsf{B})$ a set, there is a bicategory $\mathsf{Bicat}(\mathsf{B}, \mathsf{B}')$ with lax functors $\mathsf{B} \longrightarrow \mathsf{B}'$ as objects, lax transformations as 1-cells, and modifications as 2-cells. It is a 2-category if B' is a 2-category. It contains a full subbicategory $\mathsf{Bicat}^{\mathsf{ps}}(\mathsf{B}, \mathsf{B}')$ with pseudofunctors as objects and strong transformations as 1-cells.
(6.3.9) An adjunction in a bicategory consists of two 1-cells and two 2-cells, and satisfies two triangle identities.
(6.4.1) A monoidal bicategory has a base bicategory, a monoidal composition, a monoidal identity, a monoidal associator, two monoidal unitors, a pentagonator, and three 2-unitors, and satisfies the non-abelian 4-cocycle condition and two normalization axioms.
(6.5.3) A braided monoidal bicategory is a monoidal bicategory equipped with a braiding and two hexagonators, and satisfies four axioms.
(6.5.7) A sylleptic monoidal bicategory is a braided monoidal bicategory equipped with a syllepsis that satisfies two axioms.
(6.5.9) A symmetric monoidal bicategory is a sylleptic monoidal bicategory that satisfies the triple braid axiom.
(6.6.12) The 1-category $\mathsf{2Cat}$ equipped with the Gray tensor product is a symmetric monoidal closed category Gray.

(6.6.13) A Gray monoid is a monoid in Gray.
(6.7.1) A permutative Gray monoid is a Gray monoid equipped with a Gray symmetry that satisfies three axioms.
(6.7.16) A permutative 2-category is a monoid in $(\mathsf{2Cat}, \times)$ that is equipped with a symmetry 2-natural isomorphism and that satisfies the same three axioms as for permutative Gray monoids.

Chapter 7. Baez's Conjecture

(7.1.2) A bimonoidal natural transformation is a natural transformation that is a monoidal natural transformation for each of the additive structure and the multiplicative structure.
(7.1.7) There is a 2-category $\mathsf{Bi}^{\mathsf{sy}}$ of small symmetric bimonoidal categories, symmetric bimonoidal functors, and bimonoidal natural transformations.
(7.1.9) Each monoidal natural transformation between symmetric monoidal functors between distributive symmetric monoidal categories is a bimonoidal natural transformation.
(7.2.9) For each symmetric bimonoidal category C, there is a strong symmetric monoidal functor $F_{\oplus}: \Sigma \longrightarrow \mathsf{C}$ between the additive structures.
(7.3.28) For each flat symmetric bimonoidal category C, there is a symmetric monoidal functor $F_{\otimes}: \Sigma \longrightarrow \mathsf{C}$ between the multiplicative structures.
(7.4.4) For each flat symmetric bimonoidal category C, $F: \Sigma \longrightarrow \mathsf{C}$ is a robust symmetric bimonoidal functor.
(7.5.8) Epstein's Coherence Theorem 1.3.12 has a bimonoidal analogue.
(7.6.2–7.6.3) For a symmetric bimonoidal category C and robust symmetric bimonoidal functors $G, H: \Sigma \longrightarrow \mathsf{C}$, there is at most one bimonoidal natural transformation $G \longrightarrow H$, which must be invertible if it exists.
(7.7.9) For each flat symmetric bimonoidal category C and robust symmetric bimonoidal functor $G: \Sigma \longrightarrow \mathsf{C}$, there exists a unique bimonoidal natural transformation $\theta: F \longrightarrow G$, which is, moreover, invertible.
(7.8.1) **Baez's Conjecture.** Σ is a lax bicolimit of the 2-functor $\varnothing \longrightarrow \mathsf{Bi}^{\mathsf{fsy}}_{\mathsf{r}}$.
(7.8.3) **Baez's Conjecture, Ver. 2.** Σ' is a lax bicolimit of the 2-functor $\varnothing \longrightarrow \mathsf{Bi}^{\mathsf{fsy}}_{\mathsf{r}}$.

Chapter 8. Symmetric Monoidal Bicategorification

(8.1.1) For a category C, $\mathsf{Mat}^{\mathsf{C}}_{m,n}$ has $n \times m$ matrices of objects in C as objects and $n \times m$ matrices of morphisms in C as morphisms.
(8.1.8) The matrix product is a functor.
(8.2.2) There is a natural isomorphism $\ell_A: \mathbb{1}^n A \xrightarrow{\cong} A$ for each flat bimonoidal category C and $A \in \mathsf{Mat}^{\mathsf{C}}_{m,n}$.
(8.2.8) There is a natural isomorphism $r_A: A\mathbb{1}^m \xrightarrow{\cong} A$ for each flat bimonoidal category C and $A \in \mathsf{Mat}^{\mathsf{C}}_{m,n}$.
(8.3.1) There is a natural isomorphism $a: (CB)A \xrightarrow{\cong} C(BA)$ for each tight bimonoidal category C.
(8.4.12) For each tight bimonoidal category C, $\mathsf{Mat}^{\mathsf{C}}$ is a bicategory.
(8.4.14) $\mathsf{Mat}^{\mathsf{C}}_{n,n}$ is a monoidal category.
(8.6.7) The matrix tensor product is a functor.
(8.7.31) The triple $(\boxtimes, \boxtimes^2, \boxtimes^0)$ on $\mathsf{Mat}^{\mathsf{C}}$ is a pseudofunctor.
(8.8.49) The quadruple $(u^{\boxtimes}, u^{\boxtimes\bullet}, \eta^u, \varepsilon^u)$ is an adjoint equivalence.

(8.9.9) The quadruple $(\ell^{\boxtimes}, \ell^{\boxtimes\bullet}, \eta^{\ell}, \varepsilon^{\ell})$ is an adjoint equivalence.
(8.9.21) The quadruple $(r^{\boxtimes}, r^{\boxtimes\bullet}, \eta^{r}, \varepsilon^{r})$ is an adjoint equivalence.
(8.10.4) π is an invertible modification.
(8.11.4) μ is an invertible modification.
(8.11.9) $\lambda^{\boxtimes}$ is an invertible modification.
(8.11.14) $\rho^{\boxtimes}$ is an invertible modification.
(8.12.9) For each tight symmetric bimonoidal category C, $\mathsf{Mat}^{\mathsf{C}}$ is a monoidal bicategory.
(8.13.13) For $\sigma \in \Sigma_m$, there is a natural isomorphism $r^{\sigma}_{A}\colon A\mathbb{1}^{\sigma} \xrightarrow{\cong} A^{\sigma}$ for $A \in \mathsf{Mat}^{\mathsf{C}}_{m,n}$.
(8.13.16) For $\theta \in \Sigma_n$, there is a natural isomorphism $\ell^{\theta}_{A}\colon \mathbb{1}^{\theta}A \xrightarrow{\cong} {}_{\theta^{-1}}A$ for $A \in \mathsf{Mat}^{\mathsf{C}}_{m,n}$.
(8.13.20) The matrix tensor products $A \boxtimes B$ and $B \boxtimes A$ differ by a column permutation, a row permutation, and a multiplicative symmetry in each entry.
(8.13.44) The quadruple $(\beta, \beta^{\bullet}, \eta^{\beta}, \varepsilon^{\beta})$ is an adjoint equivalence.
(8.14.12) $R_{-|--}$ is an invertible modification.
(8.14.24) $R_{--|-}$ is an invertible modification.
(8.14.26) For each tight symmetric bimonoidal category C, $\mathsf{Mat}^{\mathsf{C}}$ is a braided monoidal bicategory.
(8.15.4) **Bicategorification Theorem.** For each tight symmetric bimonoidal category C, $\mathsf{Mat}^{\mathsf{C}}$ is a symmetric monoidal bicategory.
(8.15.5) Coordinatized 2-vector spaces $2\mathsf{Vect}_{\mathsf{c}} = \mathsf{Mat}^{\mathsf{Vect}^{\mathbb{C}}}$ form a symmetric monoidal bicategory.

Part II.1. Braided Bimonoidal Categories

Chapter II.1. Preliminaries on Braided Structures

(1.1.1) The braid group B_n on n strings is generated by $s_1, \ldots, s_{n-1}$ and subject to two braid relations.
(1.1.9) Sum braids generalize block sums of permutations.
(1.1.12) Each braid has an underlying permutation.
(1.1.20) Block braids generalize block permutations.
(1.2.4) Elementary block braids generalize interval-swapping permutations.
(1.2.14) Elementary block braids are compatible with sum braids.
(1.2.16) Elementary block braids satisfy the hexagon axioms.
(1.3.15) A braided monoidal category is a monoidal category equipped with a braiding that satisfies two hexagon axioms.
(1.3.18) A braided monoidal functor is defined in the same way as a symmetric monoidal functor.
(1.3.21) In each braided monoidal category, the left unit isomorphism uniquely determines the right unit isomorphism, and vice versa, via the braiding.
(1.3.28, 1.3.31) Each braided monoidal category satisfies the third Reidemeister move.
(1.3.36) A symmetric monoidal category is precisely a braided monoidal category whose braiding satisfies the symmetry axiom.
(1.4.27) The Drinfeld center of a monoidal category is a braided monoidal category.
(1.5.3) The symmetric center of a braided monoidal category is a symmetric monoidal category.

(1.6.3) **Braided Coherence Theorem**. Two braided canonical maps with the same (co)domain in a braided monoidal category are equal if their underlying braids are equal.
(1.6.5) **Braided Strictification Theorem**. Each braided monoidal category is adjoint equivalent to a braided strict monoidal category via strong braided monoidal functors.

Chapter II.2. Braided Bimonoidal Categories

(2.1.29) A braided bimonoidal category is a category equipped with an additive symmetric monoidal structure, a multiplicative braided monoidal structure, left/right multiplicative zero natural isomorphisms, and left/right distributivity natural monomorphisms, and satisfies twelve Laplaza's axioms and two additional axioms involving the braiding.
(2.1.37) Tight braided bimonoidal categories are equivalent to BD categories in the sense of Blass and Gurevich.
(2.2.1) Each braided bimonoidal category satisfies all 24 Laplaza axioms.
(2.2.3) A symmetric bimonoidal category is precisely a braided bimonoidal category whose braiding satisfies the symmetry axiom.
(2.3.2) In an Ab-category, composition with a zero morphism yields a zero morphism, and composition commutes with taking the additive inverse of a morphism.
(2.3.7 (1)) For any two objects in an Ab-category, a product, a coproduct, and a direct sum are equivalent.
(2.3.7 (2)) In an Ab-category, the direct sum morphism $f \oplus f'$ can be characterized in terms of the inclusions.
(2.3.7 (3)) In an Ab-category, the sum morphism $f + g$ factors as $\nabla_B(f \oplus g)\Delta_A$.
(2.3.7 (4)) A functor between Ab-categories whose domain has all direct sums is an additive functor if and only if it preserves direct sums.
(2.3.12) For any two objects in a preadditive category, the zero morphism is the unique morphism that factors through the zero object.
(2.3.15) An abelian category is an Ab-category with a zero object, a direct sum for any two objects, and a (co)kernel for each morphism, such that (i) each monomorphism is a kernel and (ii) each epimorphism is a cokernel.
(2.3.17) Each abelian category has all finite (co)limits, with (co)products given by direct sums.
(2.4.22) An abelian category with a compatible braided monoidal structure is a tight braided bimonoidal category.
(2.5.1–2.5.2) An abelian category with a compatible (symmetric) monoidal structure is a tight (symmetric) bimonoidal category.

Chapter II.3. Applications to Quantum Groups and Topological Quantum Computation

(3.1.19) A braided bialgebra is a bialgebra equipped with an R-matrix that satisfies two axioms. A symmetric bialgebra is a braided bialgebra in which the inverse of the R-matrix is its opposite.
(3.1.27–3.1.30) Each cocommutative bialgebra is a symmetric bialgebra with the R-matrix $1 \otimes 1$. Examples include group bialgebras, the universal enveloping bialgebra of a Lie algebra, and Sweedler's 4-dimensional non-(co)commutative bialgebra.
(3.1.33) Each anyonic quantum group is a braided bialgebra.

(3.2.6) The category of left modules over each bialgebra is a monoidal category under the tensor product.
(3.2.12) The category of left modules over each braided bialgebra is a braided monoidal category.
(3.2.13) The category of left modules over each symmetric bialgebra is a symmetric monoidal category.
(3.2.19) The category of left modules over each bialgebra is a tight bimonoidal category. The braided and the symmetric analogues are also true.
(3.3.27) The Fibonacci anyons form a monoidal category.
(3.4.5) The Fibonacci anyons form a braided monoidal category.
(3.4.13) The Fibonacci anyons form a tight braided bimonoidal category.
(3.5.27) The Ising anyons form a monoidal category.
(3.6.7) The Ising anyons form a braided monoidal category.
(3.6.14) The Ising anyons form a tight braided bimonoidal category.

Chapter II.4. Bimonoidal Centers

(4.2.6) The additive structure of the bimonoidal Drinfeld center is a symmetric monoidal category.
(4.3.3) The multiplicative structure of the bimonoidal Drinfeld center is a braided monoidal category.
(4.4.3) For each tight bimonoidal category, the bimonoidal Drinfeld center is a tight braided bimonoidal category.
(4.5.3) For each braided bimonoidal category, the bimonoidal symmetric center is a symmetric bimonoidal category.

Chapter II.5. Coherence of Braided Bimonoidal Categories

(5.1.2) A left permbraided category has an additive permutative structure, a multiplicative braided strict monoidal structure, and identities for $\lambda^{\bullet}$, $\rho^{\bullet}$, δ^{l}, $\xi^{\otimes}_{-,\mathbb{0}}$, and $\xi^{\otimes}_{\mathbb{0},-}$, and satisfies four braided bimonoidal category axioms.
(5.1.8) Each left bipermutative category is a left permbraided category.
(5.1.10) Each left permbraided category is a tight braided bimonoidal category.
(5.1.11) A right permbraided category has an additive permutative structure, a multiplicative braided strict monoidal structure, and identities for $\lambda^{\bullet}$, $\rho^{\bullet}$, δ^{r}, $\xi^{\otimes}_{-,\mathbb{0}}$, and $\xi^{\otimes}_{\mathbb{0},-}$, and satisfies four braided bimonoidal category axioms.
(5.1.17) Each right bipermutative category is a right permbraided category.
(5.1.19) Each right permbraided category is a tight braided bimonoidal category.
(5.2.7) The braided distortion category is a groupoid.
(5.2.13) The additive structure of the braided distortion category is a permutative category.
(5.2.21) The multiplicative structure of the braided distortion category is a braided strict monoidal category.
(5.2.28) In the braided distortion category, the right distributivity morphism δ^{r} has identity braid components.
(5.2.30) The braided distortion category is a left permbraided category.
(5.2.33–5.2.34) The braided distortion category is a tight braided bimonoidal category and satisfies all 24 Laplaza axioms.
(5.3.14) For a path in $\mathsf{Gr}(X)$, its value in a braided bimonoidal category is the composite of the images of its constituent prime edges.

(5.3.15) The braided distortion of a path in $\mathsf{Gr}(X)$ is its value in the braided distortion category.
(5.4.4) **Braided Bimonoidal Coherence Theorem**. In each braided bimonoidal category C satisfying a monomorphism assumption, any two parallel paths in $\mathsf{Gr}(X)$ with the same braided distortion have the same value in C.
(5.4.6) Theorem II.5.4.4 applies to flat, in particular, tight, braided bimonoidal categories.
(5.5.3) The braided distortion category $\mathcal{D}^{\mathsf{br}}$ is isomorphic to the Grothendieck construction $\int_\Sigma F^{\mathsf{br}}$.

Chapter II.6. Strictification of Tight Braided Bimonoidal Categories

(6.1.1) A braided bimonoidal functor is a functor equipped with an additive symmetric monoidal structure and a multiplicative braided monoidal structure, and satisfies two axioms.
(6.1.10) There is a category $\mathsf{Bi}^{\mathsf{br}}$ with small braided bimonoidal categories as objects and braided bimonoidal functors as morphisms.
(6.1.12) Each braided monoidal functor that is also an additive functor between abelian categories with a compatible braided monoidal structure canonically extends to a braided bimonoidal functor.
(6.1.15) Each symmetric monoidal functor that is also an additive functor between abelian categories with a compatible symmetric monoidal structure canonically extends to a symmetric bimonoidal functor.
(6.2.39) Each tight braided bimonoidal category has a canonically associated right permbraided category.
(6.3.6–6.3.7) **Permbraided Strictification**. Each tight braided bimonoidal category is adjoint equivalent to a right/left permbraided category via braided bimonoidal functors.

Chapter II.7. The Braided Baez Conjecture

(7.1.4) There is a 2-category $\mathsf{Bi}^{\mathsf{br}}$ of small braided bimonoidal categories, braided bimonoidal functors, and bimonoidal natural transformations.
(7.1.7) Each monoidal natural transformation between braided monoidal functors that are also additive functors between abelian categories with a compatible braided monoidal structure is a bimonoidal natural transformation.
(7.2.4) For each braided bimonoidal category C, there is a strong symmetric monoidal functor $F_\oplus\colon \Sigma \longrightarrow \mathsf{C}$ between the additive structures.
(7.2.9) For each flat braided bimonoidal category C, there is a braided monoidal functor $F_\otimes\colon \Sigma \longrightarrow \mathsf{C}$ between the multiplicative structures.
(7.2.11) For each flat braided bimonoidal category C, $F\colon \Sigma \longrightarrow \mathsf{C}$ is a robust braided bimonoidal functor.
(7.3.4) **Braided Baez Conjecture**. Σ is a lax bicolimit of the 2-functor $\varnothing \longrightarrow \mathsf{Bi}^{\mathsf{fbr}}_{\mathsf{r}}$.
(7.3.6) **Braided Baez Conjecture, Ver. 2**. Σ' is a lax bicolimit of the 2-functor $\varnothing \longrightarrow \mathsf{Bi}^{\mathsf{fbr}}_{\mathsf{r}}$.

Chapter II.8. Monoidal Bicategorification

(8.1.13) For each tight braided bimonoidal category C, $\mathsf{Mat}^{\mathsf{C}}$ is a bicategory.
(8.4.7) For each tight braided bimonoidal category C, $\mathsf{Mat}^{\mathsf{C}}$ is a monoidal bicategory.

Part II.2. E_n-Monoidal Categories

Chapter II.9. Ring, Bipermutative, and Braided Ring Categories

(9.1.15) Tight ring categories form a subclass of tight bimonoidal categories.
(9.1.19) Right and left rigid bimonoidal categories are tight ring categories.
(9.1.20) Each tight ring category is adjoint equivalent to a right, respectively, left, rigid bimonoidal category.
(9.2.14) Each small permutative category has an endomorphism ring category.
(9.2.20) Each small permutative category has an endomorphism tight ring category.
(9.3.7) Each tight bipermutative category yields a tight symmetric bimonoidal category.
(9.3.12) Right and left bipermutative categories are tight bipermutative categories.
(9.3.13) Each tight bipermutative category is adjoint equivalent to a right, respectively, left, bipermutative category.
(9.4.7) In a bipermutative category, about half of the ring category axioms are redundant.
(9.5.4) A bipermutative category is a braided ring category whose braiding satisfies the symmetry axiom.
(9.5.5) In a braided ring category, about half of the ring category axioms are redundant.
(9.5.6) Tight braided ring categories form a subclass of tight braided bimonoidal categories.
(9.5.10) Right and left permbraided categories are tight braided ring categories.
(9.5.11) Each tight braided ring category is adjoint equivalent to a right, respectively, left, permbraided category.
(9.6.1) The bimonoidal Drinfeld center of a tight ring category is a tight braided ring category.
(9.6.4) The symmetric center of a braided ring category with left factorization a natural epimorphism is a bipermutative category.

Chapter II.10. Iterated and E_n-Monoidal Categories

(10.1.9) A 1-fold monoidal category is a strict monoidal category.
(10.1.14) Braided strict monoidal categories form a subclass of 2-fold monoidal categories.
(10.1.21) Permutative categories form a subclass of n-fold monoidal categories for each $n \geq 2$.
(10.2.3) A totally ordered set with a least element forms a permutative category with identity symmetry.
(10.2.8) A totally ordered monoid whose unit is also the least element forms a 2-fold monoidal category.
(10.3.7) A 1-fold monoidal functor is a strictly unital monoidal functor.
(10.3.11) A braided strictly unital monoidal functor is a 2-fold monoidal functor.
(10.3.15) A symmetric strictly unital monoidal functor is an n-fold monoidal functor for each $n \geq 2$.
(10.3.20) The composite of two n-fold monoidal functors is an n-fold monoidal functor.
(10.4.2) MCat^n is a monoidal category.
(10.4.5) Monoids in MCat^n are small $(n+1)$-fold monoidal categories.

(10.4.13) A morphism of monoids in MCat^n is an $(n+1)$-fold monoidal functor with the last monoidal constraint the identity.
(10.5.9) $\mathsf{FMon}^n\colon \mathsf{Cat} \longrightarrow \mathsf{MCat}^n_{\mathsf{st}}$ is the left adjoint of the forgetful functor.
(10.5.18) $\mathsf{FMon}^n(\mathsf{C})$ decomposes into a coproduct $\coprod_{k\geq 0} \mathsf{Mon}^n(k) \times_{\Sigma_k} \mathsf{C}^{\times k}$.
(10.5.26) $\coprod_{k\geq 0} \mathsf{Mon}^n(k)/\Sigma_k$ is the free n-fold monoidal category on one object.
(10.5.28) There are evaluation functors $\theta_k\colon \mathsf{Mon}^n(k) \times_{\Sigma_k} \mathsf{C}^{\times k} \longrightarrow \mathsf{C}$ for each small n-fold monoidal category C.
(10.6.8 (1)) Each morphism set in $\mathsf{Mon}^n(k)$ has at most one morphism.
(10.6.8 (2)) There exists a morphism $A \longrightarrow B \in \mathsf{Mon}^n$ if and only if $a \otimes_i b \in A$ implies either $a \otimes_j b \in B$ for some $j \geq i$ or $b \otimes_j a \in B$ for some $j > i$.
(10.6.9) In each n-fold monoidal category, each formal diagram built from identity morphisms, the exchanges $\{\eta^{i,j}\}_{i<j}$, the monoidal products $\{\otimes_i\}_{i=1}^n$, and composites is commutative.
(10.7.13) An E_1-monoidal category is a ring category.
(10.8.1) Braided ring categories form a subclass of E_2-monoidal categories.
(10.9.1) Bipermutative categories form a subclass of E_n-monoidal categories for each $n \geq 2$.
(10.10.2) Each small category has a free E_n-monoidal category.

Part III.1. Enriched Monoidal Categories and Multicategories

Chapter III.1. Enriched Monoidal Categories

(1.1.31) **Mac Lane's Coherence Theorem.** Any two words of the same length in a monoidal category are connected by a unique canonical map.
(1.1.32) **Mac Lane's Strictification Theorem.** Each monoidal category is adjoint equivalent to a strict monoidal category via strong monoidal functors.
(1.1.38) **Braided Coherence Theorem.** Two braided canonical maps with the same (co)domain in a braided monoidal category are equal if their underlying braids are equal.
(1.1.39) **Braided Strictification Theorem.** Each braided monoidal category is adjoint equivalent to a braided strict monoidal category via strong braided monoidal functors.
(1.1.41) **Symmetric Coherence Theorem.** Any two permuted words of the same length in a symmetric monoidal category are connected by a unique permuted canonical map.
(1.1.42) **Symmetric Strictification Theorem.** Each symmetric monoidal category is adjoint equivalent to a permutative category via strong symmetric monoidal functors.
(1.1.44) **Epstein's Coherence Theorem.** For each (symmetric) monoidal functor $F\colon \mathsf{C} \longrightarrow \mathsf{D}$ and F-iterates $G, H\colon \mathsf{C}^n \longrightarrow \mathsf{D}$, there exists at most one F-coherent map $G \longrightarrow H$.
(1.2.1) For a monoidal category V, a V-category has hom objects in V satisfying associativity and unity axioms.
(1.2.4) A V-functor satisfies composition and identity axioms.
(1.2.7) A V-natural transformation satisfies a naturality axiom.
(1.2.13) There is a 2-category formed by small V-categories, V-functors, and V-natural transformations.
(1.2.16) The opposite of a V-category is defined if V is braided monoidal.
(1.3.3) The tensor product of V-categories is defined if V is braided monoidal.

(1.3.6) The tensor product of V-categories is 2-functorial.
(1.3.35) The underlying 1-category of V-Cat is monoidal if V is braided, and is symmetric monoidal if V is symmetric.
(1.4.2) A monoidal V-category has associator and unitor V-natural transformations that satisfy unity axioms and a pentagon axiom.
(1.4.7) Composition in a monoidal V-category has an enriched interchange.
(1.4.10) The definition of braided monoidal V-category requires that V be symmetric monoidal.
(1.4.10) A braided monoidal V-category has a V-natural braiding satisfying two hexagon axioms.
(1.4.13) A symmetric monoidal V-category is a braided monoidal V-category satisfying an additional symmetry axiom.
(1.4.17) A monoidal V-functor satisfies associativity and unity axioms.
(1.4.18) A braided monoidal V-functor satisfies a braid axiom. A symmetric monoidal V-functor is a braided monoidal V-functor whose domain and codomain are symmetric monoidal V-categories.
(1.4.22) A monoidal V-natural transformation satisfies monoidal naturality and monoidal unity axioms.
(1.4.25) There are 2-categories formed by each of: monoidal V-categories, braided monoidal V-categories, and symmetric monoidal V-categories with, in each case, the corresponding V-functors and V-natural transformations.
(1.5.1) We use the term Cat-monoidal 2-category to indicate monoidal V-categories when V = Cat.
(1.5.2) The underlying 1-category of a plain/braided/symmetric Cat-monoidal 2-category has the corresponding structure as a 1-category.
(1.5.3) A Cat-monoidal 2-category has a strict form of the data and axioms for a monoidal bicategory. Similar statements hold for the braided and symmetric cases.
(1.5.4) With the Cartesian product, Cat is a symmetric Cat-monoidal 2-category.
(1.5.5) For a braided monoidal category V, V-Cat is a Cat-monoidal 2-category. If V is symmetric, then so is V-Cat.

Chapter III.2. Change of Enrichment

(2.1.2) Change of enrichment along a monoidal functor U is 2-functorial.
(2.1.7) The functor from V-Cat to Cat that takes underlying categories is injective on 2-cells.
(2.2.7) The assignment $E\colon \mathsf{V} \longmapsto \mathsf{V}\text{-}\mathsf{Cat}$ is 2-functorial with respect to monoidal functors and monoidal natural transformations.
(2.2.11) For small monoidal V, there is a 2-equivalence between V-Cat and $(\mathsf{V}_{\mathsf{st}})$-Cat.
(2.3.7) If U is braided moniodal, then the change of enrichment induced by U is a Cat-monoidal 2-functor. If U is symmetric, then so is the change of enrichment.
(2.3.9) The assignment $E\colon \mathsf{V} \longmapsto \mathsf{V}\text{-}\mathsf{Cat}$ of braided monoidal categories to Cat-monoidal 2-categories is 2-functorial. A similar result holds for symmetric monoidal categories with E producing symmetric Cat-monoidal categories.
(2.3.16) For small symmetric monoidal V, there is a symmetric Cat-monoidal 2-equivalence between V-Cat and $(\mathsf{V}_{\mathsf{st}})$-Cat.
(2.4.10, 2.4.15) For braided monoidal $U\colon \mathsf{V} \longrightarrow \mathsf{W}$, change of enrichment along U induces 2-functors between the 2-categories of monoidal V- and W-categories. If U is symmetric, then a similar result holds for braided and symmetric monoidal V- and W-categories.

(2.4.17) The underlying category of a monoidal V-category is monoidal. Similar statements hold for braided and symmetric cases, and for functors and natural transformations.
(2.5.1) Given the data of a monoidal V-category, the enriched monoidal category axioms are satisfied if and only if the underlying data satisfy the ordinary monoidal category axioms. Similar results hold for the braided and symmetric monoidal cases, and also for functors and natural transformations.
(2.5.6) **Enriched (Braided/Symmetric) Monoidal Coherence Theorem.** Any two V-words of the same length in a monoidal V-category are connected by a unique canonical V-map. Similar coherence results hold in the braided and symmetric cases.
(2.5.8) **Enriched Epstein's Coherence Theorem.** For each (symmetric) monoidal V-functor $F\colon \mathsf{K} \longrightarrow \mathsf{L}$, and F-iterates $G, H\colon \mathsf{K}^{\otimes n} \longrightarrow \mathsf{L}$, there exists at most one F-coherent map $G \longrightarrow H$.
(2.6.1) **Enriched Monoidal Strictification Theorem.** Each monoidal V-category is adjoint V-equivalent to a strict monoidal V-category via strong monoidal V-functors.
(2.6.3) **Enriched Braided Strictification Theorem.** Each braided monoidal V-category is adjoint V-equivalent to a braided strict monoidal V-category via strong braided monoidal V-functors.
(2.6.4) **Enriched Symmetric Strictification Theorem.** Each symmetric monoidal V-category is adjoint V-equivalent to a strict monoidal V-category via strong symmetric monoidal V-functors.

Chapter III.3. Self-Enrichment and Enriched Yoneda

(3.1.11) Each symmetric monoidal closed V has a canonical enrichment over itself, $\underline{\mathsf{V}}$.
(3.2.1, 3.2.2) A category enriched over symmetric monoidal closed V has co/represented V-functors $\mathcal{Y}^X$ and $\mathcal{Y}_Y$ to $\underline{\mathsf{V}}$.
(3.3.2) The self-enriched category $\underline{\mathsf{V}}$ is symmetric monoidal as a V-category.
(3.3.4) The standard enrichment of a symmetric monoidal functor is symmetric monoidal in the enriched sense.
(3.4.12) **V-Yoneda Bijection Theorem.** For each V-functor $F\colon \mathsf{C} \longrightarrow \underline{\mathsf{V}}$ and each $X \in \mathsf{C}$, there is a bijection of sets $\mathsf{V}\text{-nat}(\mathcal{Y}^X, F) \cong \mathsf{V}(\mathbb{1}, FX)$. For each V-functor $G : \mathsf{C}^{\mathsf{op}} \longrightarrow \underline{\mathsf{V}}$ and each $Y \in \mathsf{C}$, there is a bijection of sets $\mathsf{V}\text{-nat}(\mathcal{Y}_Y, G) \cong \mathsf{V}(\mathbb{1}, GY)$.
(3.5.1) A V-coend is initial among V-cowedges. A V-end is terminal among V-wedges.
(3.5.5) If V is cocomplete, then V-coends are computed by a coequalizer in V. If V is complete, then V-ends are computed by an equalizer in V.
(3.5.12) For V complete symmetric monoidal closed, the mapping object for V-functors to $\underline{\mathsf{V}}$ is given by a V-end.
(3.6.9) **V-Yoneda Lemma.** For a V-functor $F\colon \mathsf{C} \longrightarrow \underline{\mathsf{V}}$ with C small, there is a V-natural isomorphism

$$F \xrightarrow{\cong} \mathsf{Map}(\mathcal{Y}^{(-)}, F).$$

(3.7.3) For V complete and cocomplete symmetric monoidal closed, the Day convolution and hom diagram are given by a V-coend and a V-end, respectively.

(3.7.8) **V-Yoneda Density Theorem**. For a V-functor $X: \mathcal{D} \longrightarrow \underline{\mathsf{V}}$ with $\mathcal{D}$ small, there is a V-natural isomorphism

$$\int^x \mathcal{D}(x,-) \otimes X_x \xrightarrow{\cong} X.$$

(3.7.13) There is an isomorphism $\mathsf{Map}(X,Y) \cong \mathsf{Hom}(X,Y)_e$, for V-diagrams X and Y.

(3.7.22) **Day Convolution Theorem**. For a small symmetric monoidal V-category $\mathcal{D}$, the category of $\mathcal{D}$-shaped diagrams in V is symmetric monoidal closed with the Day convolution product, internal hom, and monoidal unit $J = \mathcal{Y}^e$.

(3.7.28) Precomposition with a symmetric monoidal V-functor induces a symmetric monoidal functor between diagram V-categories.

(3.8.4) Change of enrichment along a symmetric monoidal functor $U: \mathsf{V} \longrightarrow \mathsf{W}$ induces a symmetric monoidal functor from $\mathcal{D}$-V to $\mathcal{D}_U$-W.

(3.9.3) If C is tensored and cotensored over V, then $X \otimes -$ and $Y^{(-)}$ extend uniquely to V-functors that are V-adjoint to the respective co/represented V-functors.

(3.9.8) If (F, U) is an adjunction of monoidal functors between symmetric monoidal closed categories and if F^2 is invertible, then F transfers tensor and cotensor structure over its codomain to corresponding structure over its domain.

(3.9.15) The symmetric monoidal closed diagram category $\mathcal{D}$-V is enriched, tensored, and cotensored over V.

Chapter III.4. Pointed Objects, Smash Products, and Pointed Homs

(4.1.6) Smash product with respect to a terminal object T is given by a pushout from a monoidal product.

(4.2.1) Pointed hom with respect to a terminal object T is given by a pullback from an internal hom.

(4.1.5, 4.1.8, 4.2.3) Suppose C is complete and cocomplete symmetric monoidal closed. Then C_* is complete and cocomplete symmetric monoidal closed with respect to the smash product and pointed hom.

(4.3.11) For a small symmetric monoidal category $\mathcal{D}$ with a null object, its pointed unitary enrichment over $(\mathsf{V}_*, \wedge, E)$ is given by taking wedge sums of E over nonzero morphisms in $\mathcal{D}$.

(4.3.19) Assuming the basepoint of V is terminal and the basepoint of $\mathcal{D}$ is null, there is an equivalence of categories between pointed functors from $\mathcal{D}$ to V_* and V_*-enriched functors from the pointed unitary enrichment of $\mathcal{D}$ to the self enrichment of V_*.

(4.3.37) The category of pointed diagrams $\mathcal{D}_*$-V is complete and cocomplete symmetric monoidal closed. Moreover, it is enriched, tensored, and cotensored over V_*.

Chapter III.5. Multicategories

(5.1.2) A multicategory has objects, n-ary operations, symmetric group actions, colored units, and composition that are subject to axioms for symmetry, associativity, unity, and equivariance.

(5.1.2) An operad is a multicategory with one object.

(5.1.11) Each small permutative category C has an endomorphism multicategory with the same objects and with n-ary operations given by morphisms out of n-fold sums in C.

(5.1.12) A multifunctor satisfies axioms for symmetric group action, units, and composition.
(5.1.17) A multinatural transformation satisfies a naturality condition.
(5.1.20) There is a 2-category consisting of small multicategories, multifunctors, and multinatural transformations.
(5.1.21) The initial operad I has a single object and only one operation, which is the unit on its one object.
(5.2.1, 5.2.2) The terminal multicategory T has a single object and a single n-ary operation for each $n \geq 0$. The terminal multicategory is also known as the commutative operad, Com.
(5.3.9) Taking endomorphism operads gives a 2-functor from $\mathsf{PermCat}^{\mathsf{su}}$ to $\mathsf{Multicat}_*$.
(5.4.1) A monad consists of an endofunctor together with multiplication and unit natural transformations such that the associativity and unit diagrams commute.
(5.4.2) A monad algebra consists of an object and a structure morphism such that associativity and unity diagrams commute.
(5.4.13) **Beck's Precise Tripleability Theorem.** An adjunction $L \dashv U$ is strictly monadic if and only if U strictly creates coequalizers for parallel pairs f, g for which (Uf, Ug) has a split coequalizer.
(5.4.18) If T is a monad on a complete and cocomplete category, and if T preserves filtered colimits, then the category of T-algebras is complete and cocomplete.
(5.5.1) A multigraph consists of vertices and multiedges.
(5.5.4, 5.5.9) The forgetful functor from small multicategories to small multigraphs has a left adjoint.
(5.5.11) The category of small multicategories is strictly monadic over the category of small multigraphs.
(5.5.14) The category of small multicategories is complete and cocomplete.
(5.6.9) The sharp product of multicategories is generated by operations $\phi \times d$ and $c \times \psi$ subject to symmetry and compatibility axioms.
(5.6.14) The tensor product of multicategories is generated by those of the sharp product, and subject to an additional interchange relation.
(5.6.14) A multifunctor out of a tensor product of multicategories consists of an assignment on objects that is multifunctorial in each variable separately and that preserves the interchange relation.
(5.7.2, 5.7.4) The internal hom for multicategories has operations given by transformations that satisfy a naturality condition.
(5.7.14) The category of small multicategories is complete and cocomplete symmetric monoidal closed with monoidal product given by the tensor product and closed structure given by the internal hom.
(5.7.22) The category of small pointed multicategories is complete and cocomplete symmetric monoidal closed with monoidal product given by the smash product and closed structure given by the pointed hom.
(5.7.23) The symmetric monoidal structure on $\mathsf{Multicat}_*$ does not restrict along End to a symmetric monoidal structure on $\mathsf{PermCat}^{\mathsf{su}}$.

Chapter III.6. Enriched Multicategories

(6.1.1) A V-enriched multicategory has n-ary operation objects, symmetric group action, colored units, and composition morphisms in V. These satisfy axioms given by commutative diagrams in V for symmetric group action, associativity, unity, and equivariance.

(6.1.8) A V-enriched operad is a V-multicategory with one object.
(6.1.9) An object of an enriched multicategory has a V-enriched endomorphism operad.
(6.1.10) A V-enriched multifunctor satisfies axioms given by commutative diagrams in V for symmetric group action, units, and composition.
(6.1.14) An algebra c over a V-enriched operad P is given by a V-enriched operad morphism $\mathsf{P} \longrightarrow \mathsf{End}(c)$.
(6.1.15) A V-enriched multinatural transformation satisfies a V-naturality diagram in V.
(6.1.18) There is a 2-category consisting of small V-enriched multicategories together with V-enriched multifunctors and multinatural transformations.
(6.2.9) For a symmetric monoidal functor $U\colon \mathsf{V} \longrightarrow \mathsf{W}$, change of enrichment along U provides a 2-functor from small V-enriched multicategories to small W-enriched multicategories.
(6.3.3, 6.3.6) A symmetric monoidal V-category has a V-enriched endomorphism multicategory with V-objects of n-ary operations given by morphism objects out of n-fold left normalized products in K.
(6.3.10) A symmetric monoidal V-functor induces a V-enriched multifunctor between V-enriched endomorphism multicategories of its domain and codomain.
(6.4.3) The tensor product is a Cat-enriched symmetric monoidal product for the 2-category of small multicategories.
(6.4.4) The smash product is a Cat-enriched symmetric monoidal product for the 2-category of small pointed multicategories.
(6.4.5) Each of Multicat and $\mathsf{Multicat}_*$ has the structure of a Cat-enriched multicategory induced by the tensor and smash product, respectively.
(6.5.1) The Cat-enriched multicategory structure on $\mathsf{Multicat}_*$ induces a corresponding structure on $\mathsf{PermCat}^{\mathsf{su}}$.
(6.5.4) Multilinear functors of permutative categories consist of functors out of a Cartesian product together with linearity constraints. They are subject to axioms for unity, constraint unity, constraint associativity, constraint symmetry, and constraint 2-by-2.
(6.5.11) Multilinear transformations between multilinear functors satisfy multilinearity conditions with respect to linearity constraints and identities.
(6.5.10, 6.5.13) The categories of n-ary operations in $\mathsf{PermCat}^{\mathsf{su}}$ are canonically isomorphic to the corresponding categories of n-linear functors and transformations.
(6.6.13) The Cat-enriched multicategory structure of $\mathsf{PermCat}^{\mathsf{su}}$ is described explicitly in terms of multilinear functors and transformations.

Part III.2. Algebraic K-Theory

Chapter III.7. Homotopy Theory Background

(7.1.16) The geometric realization of the standard n-simplex is the topological n-simplex.
(7.1.19) The category of simplicial sets is symmetric monoidal closed with the monoidal product given by the levelwise Cartesian product.
(7.1.23) The category of pointed simplicial sets is symmetric monoidal closed with monoidal product given by the levelwise smash product.
(7.2.4) The nerve of a small category is a simplicial set with p-simplices given by strings of p composable arrows.

(7.2.5 (1)) A natural transformation between functors induces a simplicial homotopy on nerves.
(7.2.5 (2)) An adjunction of functors induces a simplicial homotopy equivalence on nerves.
(7.3.7) The category of symmetric sequences is symmetric monoidal closed with monoidal product given by Day convolution.
(7.4.5) The category of symmetric spectra is the category of left modules over the symmetric sphere.
(7.4.6) A symmetric spectrum consists of a symmetric sequence with structure morphisms satisfying unity, associativity, and equivariance axioms.
(7.5.5) The category of symmetric spectra is complete and cocomplete.
(7.6.1) The smash product of symmetric spectra is given by a coequalizer of actions by S.
(7.6.8) The internal hom for symmetric spectra is given by an equalizer of actions by S.
(7.6.15) The category of symmetric spectra is complete and cocomplete symmetric monoidal closed.
(7.8.8) Every level equivalence of symmetric spectra is a stable equivalence.

Chapter III.8. Segal K-Theory of Permutative Categories

(8.1.8) A Γ-object in a pointed category C is a pointed functor from $\mathcal{F}$ to C.
(8.2.6) The construction $\mathsf{K}^{\mathcal{F}}$ is a functor from Γ-simplicial sets to symmetric spectra.
(8.3.13) For a small permutative category C, there are three variant constructions of Γ-categories, $\mathsf{C}^{\mathcal{F}} = \mathsf{C}^{\mathcal{F}}_{\cong}$, $\mathsf{C}^{\mathcal{F}}_{\mathrm{lax}}$, and $\mathsf{C}^{\mathcal{F}}_{\mathrm{co}}$.
(8.3.21) For a small permutative category C, each of the Γ-simplicial sets $N_*\mathsf{C}^{\mathcal{F}}_{\cong}$, $N_*\mathsf{C}^{\mathcal{F}}_{\mathrm{lax}}$, and $N_*\mathsf{C}^{\mathcal{F}}_{\mathrm{co}}$ is special, and all three are levelwise weakly-equivalent.
(8.4.5) The partition multicategory $\mathcal{M}\underline{1}$ has two objects, $\varnothing$ and $\{1\}$, with operations given by partitions.
(8.4.7) The partition multicategory $\mathcal{M}$ defines a pointed functor from $\mathcal{F}^{\mathrm{op}}$ to $\mathsf{Multicat}_*$.
(8.4.8, 8.4.10) For a small permutative category C, there is an isomorphism of Γ-categories $\mathsf{J}^{\mathsf{Se}}\mathsf{C} \cong \mathsf{C}^{\mathcal{F}}_{\mathrm{lax}}$.

Chapter III.9. Categories of $\mathcal{G}_*$-Objects

(9.1.7) The objects of $\mathcal{G}$ are tuples of objects of $\mathcal{F}$ subject to certain basepoint identifications.
(9.1.15) Smash product of pointed finite sets defines a strict symmetric monoidal functor from $\mathcal{G}$ to $\mathcal{F}$.
(9.2.1) A $\mathcal{G}_*$-object in a pointed category C is a pointed functor from $\mathcal{G}$ to C.
(9.2.15) If C is complete and cocomplete symmetric monoidal closed with terminal basepoint, then the category for $\mathcal{G}_*$-objects in C is complete and cocomplete symmetric monoidal closed with monoidal product given by Day convolution.
(9.2.19) The nerve induces a symmetric monoidal sSet_*-functor from small $\mathcal{G}_*$-categories to $\mathcal{G}_*$-simplicial sets.
(9.3.16) The construction $\mathsf{K}^{\mathcal{G}}$ is a functor from $\mathcal{G}_*$-simplicial sets to symmetric spectra whose restriction along $\wedge^*$ is equal to $\mathsf{K}^{\mathcal{F}}$.
(9.4.9) The functor $\mathsf{K}^{\mathcal{G}}$ is a unital symmetric monoidal sSet-functor.
(9.4.18) The functors $\wedge^*$ and $\mathsf{K}^{\mathcal{F}}$ are symmetric monoidal sSet_*-functors.

Chapter III.10. Elmendorf-Mandell K-Theory of Permutative Categories

(10.1.6) The partition multicategory $\mathcal{M}$ is a symmetric monoidal functor, with monoidal constraint given by the partition product $\prod$.
(10.1.14) The category of left $\mathcal{M}\underline{1}$-modules is complete and cocomplete.
(10.1.15) The partition products $\prod_{\underline{1},b}$ and $\prod_{b,\underline{1}}$ are isomorphisms.
(10.1.28) The 2-category of left $\mathcal{M}\underline{1}$-modules in $\mathsf{Multicat}_*$ is a full sub-2-category and the smash product over $\mathcal{M}\underline{1}$ is isomorphic to that of $\mathsf{Multicat}_*$.
(10.1.36) The category of left $\mathcal{M}\underline{1}$-modules is symmetric monoidal in the Cat_*-enriched sense.
(10.2.14) If C is a small permutative category, $\mathsf{End}(\mathsf{C})$ has a canonical left $\mathcal{M}\underline{1}$-module structure. Taking this structure, End factors through $\mathsf{Mod}^{\mathcal{M}\underline{1}}$.
(10.2.15, 10.2.16) The category of left $\mathcal{M}\underline{1}$-modules is a proper subcategory of $\mathsf{Multicat}_*$. In particular, the monoidal unit S is not an $\mathcal{M}\underline{1}$-module.
(10.2.17) The symmetric monoidal structure on $\mathsf{Mod}^{\mathcal{M}\underline{1}}$ does not restrict to $\mathsf{PermCat}^{\mathsf{su}}$.
(10.2.22) The symmetric monoidal structure on $\mathsf{Mod}^{\mathcal{M}\underline{1}}$ is closed.
(10.3.3) The smash product of partition multicategories, $\mathcal{T}$, defines a pointed functor from $\mathcal{G}^{\mathsf{op}}$ to $\mathsf{Mod}^{\mathcal{M}\underline{1}}$.
(10.3.7) The functor $\mathcal{T}$ is strictly unital strong symmetric monoidal.
(10.3.13) The monoidal constraint for the partition J-theory $\mathsf{J}^{\mathcal{T}}$ uses the inverse monoidal constraint for $\mathcal{T}$.
(10.3.17) The partition J-theory $\mathsf{J}^{\mathcal{T}}$ is a symmetric monoidal Cat_*-functor.
(10.3.25) Elmendorf-Mandell J-theory $\mathsf{J}^{\mathsf{EM}} = \mathsf{J}^{\mathcal{T}} \circ \mathsf{End}$ is a Cat-enriched multifunctor.
(10.3.32) Elmendorf-Mandell K-theory $\mathsf{K}^{\mathsf{EM}} = \mathsf{K}^{\mathcal{G}} N_* \mathsf{J}^{\mathsf{EM}}$ is a sSet-enriched multifunctor.
(10.3.33) The multifunctor K^{EM} preserves enriched operad actions.
(10.4.18) For a small permutative category C there are three variant constructions of $\mathcal{G}_*$-categories, $\mathsf{C}^{\mathcal{G}} = \mathsf{C}^{\mathcal{G}}_{\mathrm{lax}}$, $\mathsf{C}^{\mathcal{G}}_{\cong}$, and $\mathsf{C}^{\mathcal{G}}_{\mathrm{co}}$.
(10.5.1) For a small permutative category C, there is an isomorphism of $\mathcal{G}_*$-categories $\mathsf{J}^{\mathsf{EM}}\mathsf{C} \cong \mathsf{C}^{\mathcal{G}}_{\mathrm{lax}}$.
(10.6.10) There is a level equivalence of symmetric spectra $\mathsf{K}^{\mathsf{Se}}\mathsf{C} \longrightarrow \mathsf{K}^{\mathsf{EM}}\mathsf{C}$ for each small permutative category C. It is natural with respect to strictly unital symmetric monoidal functors.
(10.7.16) There is a level equivalence $\mathsf{K}^{\mathsf{EM}}_{\cong}\mathsf{C} \longrightarrow \mathsf{K}^{\mathsf{EM}}_{\mathsf{lax}}$ for each small permutative category C.
(10.7.19 (4)) There is a level equivalence $\mathsf{K}^{\mathsf{EM}}_{\cong} \longrightarrow \mathsf{K}^{\mathsf{EM}}_{\mathsf{co}}\mathsf{C}$ for each small permutative category C.
(10.7.22, 10.7.27) The $\mathcal{G}_*$-category morphisms $\mathsf{C}^{\mathcal{G}}_{\cong} \longrightarrow \mathsf{C}^{\mathcal{G}}$ and $\mathsf{C}^{\mathcal{G}}_{\cong} \longrightarrow \mathsf{C}^{\mathcal{G}}_{\mathsf{co}}$ are components of Cat-enriched multinatural transformations.

Chapter III.11. K-Theory of Ring and Bipermutative Categories

(11.1.4) As is an operad.
(11.1.7) As is generated by id_0 and id_2, which are subject to unity and associativity relations.
(11.1.15) As is the operad for monoids.
(11.2.16) As detects ring category structures on small permutative categories.

(11.3.2) A strict ring symmetric spectrum is a symmetric spectrum equipped with multiplication and unit morphisms in symmetric sequences that satisfy compatibility, associativity, and unity axioms.
(11.3.13–11.3.15) The sphere spectrum, the suspension spectrum of a monoid in pointed simplicial sets, and the Eilenberg-Mac Lane spectrum of a ring are strict ring symmetric spectra.
(11.3.16) $\mathsf{K}^{\mathsf{EM}}\mathsf{C}$ is a strict ring symmetric spectrum for each small ring category C.
(11.3.17) For each small permutative category, the K-theory of its (tight) endomorphism ring category is a strict ring symmetric spectrum.
(11.3.18) The K-theory of the additive distortion category is a strict ring symmetric spectrum.
(11.3.19) For each small tight bimonoidal category, the K-theory of its right/left rigid bimonoidal strictification is a strict ring symmetric spectrum.
(11.4.7) The translation category functor E is a right adjoint.
(11.4.11) Each morphism in the Barratt-Eccles operad decomposes into a categorical composite of ϕv with v a permutation and ϕ an operadic composite of one $\tau\colon \mathrm{id}_2 \longrightarrow (1,2)$ and identity morphisms.
(11.4.14) The Barratt-Eccles operad is generated by two objects and one isomorphism, which are subject to relations that are formally identical to those of a permutative category.
(11.4.26) The Barratt-Eccles operad is the Cat-enriched operad for permutative categories.
(11.5.5) The Barratt-Eccles operad detects bipermutative category structures on small permutative categories.
(11.6.3) The Barratt-Eccles operad is an E_∞-operad.
(11.6.6 (1)) Each commutative monoid in SymSp is an E_∞-symmetric spectrum.
(11.6.6 (2)) Each E_∞-symmetric spectrum via the Barratt-Eccles operad has a strict ring structure.
(11.6.7) The symmetric sphere is a commutative monoid in SymSp.
(11.6.9) The suspension spectrum of a commutative monoid in sSet_* is a commutative monoid in SymSp.
(11.6.10) The Eilenberg-Mac Lane spectrum of a commutative ring is a commutative monoid in SymSp.
(11.6.12) $\mathsf{K}^{\mathsf{EM}}\mathsf{C}$ is an E_∞-symmetric spectrum for each small bipermutative category C.
(11.6.13) The K-theory of each small right/left bipermutative category is an E_∞-symmetric spectrum. For example, this applies to the finite ordinal category Σ, its variant Σ', $\mathsf{Vect}_{\mathsf{c}}^{\mathbb{C}}$, and the distortion category.
(11.6.14) For each small tight symmetric bimonoidal category, the K-theory of its right/left bipermutative strictification is an E_∞-symmetric spectrum. For example, this applies to small distributive symmetric monoidal categories, the symmetric bimonoidal groupoid Π, and the bimonoidal symmetric center of a small tight braided bimonoidal category.
(11.6.15) For each small braided ring category whose left factorization morphism is a natural epimorphism, the K-theory of its symmetric center is an E_∞-symmetric spectrum.

Chapter III.12. K-Theory of Braided Ring Categories

(12.1.10) The braid operad is a Cat-enriched operad.
(12.2.4) The braid operad is an E_2-operad.
(12.3.6) Each morphism in the braid operad admits a categorical decomposition into isomorphisms of the form ϕv with v a permutation and ϕ an operadic composite of one $s_1^{\pm 1}\colon \mathrm{id}_2 \longrightarrow (1,2)$ and identity morphisms.
(12.3.10) The braid operad is generated by two objects and one isomorphism, which are subject to relations that are formally identical to those of a braided strict monoidal category.
(12.3.22) The braid operad is the Cat-enriched operad for braided strict monoidal categories.
(12.4.5) The braid operad detects braided ring category structures on small permutative categories.
(12.5.2 (1)) An E_∞-symmetric spectrum via the Barratt-Eccles operad has an E_2-structure.
(12.5.2 (2)) An E_2-symmetric spectrum via the braid operad has a strict ring structure.
(12.5.3) $\mathsf{K}^{\mathsf{EM}}\mathsf{C}$ is an E_2-symmetric spectrum for each small braided ring category C.
(12.5.4) The K-theory of the braided distortion category is an E_2-symmetric spectrum.
(12.5.5) For each small tight ring category, the K-theory of its bimonoidal Drinfeld center is an E_2-symmetric spectrum.
(12.5.6) For each small tight braided bimonoidal category, the K-theory of its right/left permbraided strictification is an E_2-symmetric spectrum. For example, this applies to a small abelian category with a compatible braided monoidal structure, Fibonacci anyons, Ising anyons, and the bimonoidal Drinfeld center of a small tight bimonoidal category.

Chapter III.13. K-Theory of E_n-Monoidal Categories

(13.1.20) Mon^n is a Cat-enriched operad.
(13.1.23) Mon^1 is the associative operad.
(13.2.1) Mon^n is an E_n-operad.
(13.3.3) Mon^n is generated by the objects $\mathbb{1}$ and $\{1 \otimes_i 2\}_{i=1}^n$ and the exchange morphisms $\{\eta_{1,2,3,4}^{i,j}\}_{1 \le i < j \le n}$, which are subject to relations that are formally identical to those of an n-fold monoidal category.
(13.3.18) Mon^n is the Cat-enriched operad for n-fold monoidal categories.
(13.4.12) Mon^n detects E_n-monoidal category structures on small permutative categories.
(13.5.1) The canonical Cat-enriched operad morphism $\mathsf{As} \longrightarrow E\mathsf{As}$ factors through Mon^n. As a result, an E_∞-symmetric spectrum via the Barratt-Eccles operad induces an E_n-structure. An E_{n+1}-structure via Mon^{n+1} induces an E_n-structure.
(13.5.2) $\mathsf{K}^{\mathsf{EM}}\mathsf{C}$ is an E_n-symmetric spectrum for each small E_n-monoidal category C.
(13.5.3) For each small category, the K-theory of its free E_n-monoidal category is an E_n-symmetric spectrum.

List of Notations

Part 1

Chapter 3

Chapter 4

F^{ad}	I.166	functor $\Sigma^{\mathsf{op}} \longrightarrow \mathsf{Cat}$
$\mathbb{N}^{\times n}$	I.166	n-fold Cartesian product of $\mathbb{N}$

Chapter 5

$(F, F_{\oplus}^2, F_{\oplus}^0, F_{\otimes}^2, F_{\otimes}^0)$	I.173	a (symmetric) bimonoidal functor
$F_{\oplus}$	I.174	additive structure $(F, F_{\oplus}^2, F_{\oplus}^0)$
$F_{\otimes}$	I.174	multiplicative structure $(F, F_{\otimes}^2, F_{\otimes}^0)$
$\mathsf{Bi}^{\mathsf{sy}}$	I.177	category of small symmetric bimonoidal categories
A	I.180	associated right bipermutative category
$(-)_{\mathsf{rt}}$, $(-)_{\mathsf{lt}}$	I.182	right/left normalized bracketing
π	I.182	$\mathsf{Ob}(\mathsf{A}) \longrightarrow \mathsf{Ob}(\mathsf{C})$ and $\mathsf{A} \longrightarrow \mathsf{C}$
$\cong_{\mathsf{ML}}^{\oplus}$	I.183	a Mac Lane coherence isomorphism
$\cong_{\mathsf{Lap}}$	I.184	a Laplaza coherence isomorphism
$\cong_{\mathsf{Lap}}^{-1}$	I.184	inverse of a Laplaza coherence isomorphism
ι	I.193	functor $\mathsf{C} \longrightarrow \mathsf{A}$
Bi	I.198	category of small bimonoidal categories
A	I.198	associated right rigid bimonoidal category

Part 2
Chapter 6

$\mathsf{Ob}(\mathsf{B})$	I.209	objects in a bicategory
$\Rightarrow$	I.210	a 2-cell
1_f	I.210	identity 2-cell of f
1_X	I.210	identity 1-cell of X
gf, $\beta * \alpha$	I.210	horizontal composition
a	I.210	associator
ℓ, r	I.210	left and right unitors
Cat	I.213	2-category of small categories, functors, and natural transformations
MCat	I.213	2-category of small monoidal categories
SMCat	I.214	2-category of small symmetric monoidal categories
$\Sigma\mathsf{C}$	I.214	one-object bicategory of a monoidal category C
Bimod	I.214	bicategory with bimodules as 1-cells
(F, F^2, F^0)	I.214	a lax functor
1_{B}	I.217	identity strict functor of B
Bicat	I.217	category of small bicategories and lax functors
$\mathsf{Bicat}^{\mathsf{ps}}$	I.217	wide subcategory of Bicat with pseudofunctors
α_X, α_f	I.218	component 1-/2-cells of a lax transformation α
1_F	I.219	identity strong transformation of F
$\beta\alpha$	I.220	horizontal composite of lax transformations
Γ_X	I.221	a component 2-cell of a modification Γ
$\Omega\Gamma$	I.222	vertical composite of modifications
$\Gamma' * \Gamma$	I.222	horizontal composite of modifications
$\mathsf{Bicat}(\mathsf{B}, \mathsf{B}')$	I.222	bicategory of lax functors/transformations and modifications
$\mathsf{Bicat}^{\mathsf{ps}}(\cdot, \cdot)$	I.223	$\mathsf{Bicat}(\cdot, \cdot)$ with pseudofunctors and strong transformations
$f \dashv g$, $(f, g, \eta, \varepsilon)$	I.224	an adjunction in a bicategory
$f^{\bullet}$	I.224	an adjoint of f
B^n	I.225	$\mathsf{B} \times \cdots \times \mathsf{B}$ with n copies of B
$(\boxtimes, \boxtimes^2, \boxtimes^0)$	I.225	monoidal composition
$(1_{\boxtimes}, 1_{\boxtimes}^2, 1_{\boxtimes}^0)$	I.225	monoidal identity
$(a, a^{\bullet}, \eta^a, \varepsilon^a)$	I.225	monoidal associator
$(\ell, \ell^{\bullet}, \eta^{\ell}, \varepsilon^{\ell})$	I.225	left monoidal unitor
$(r, r^{\bullet}, \eta^r, \varepsilon^r)$	I.225	right monoidal unitor

Chapter 7

Chapter 8

Part II.1
Chapter II.1

Chapter II.2

Chapter II.3

$\mathsf{Vect}^{\Bbbk}$	II.67	category of $\Bbbk$-vector spaces
$A^{\otimes n}$	II.67	$A \otimes \cdots \otimes A$ with n copies of A and $A^{\otimes 0} = \Bbbk$
$A^{\oplus n}$	II.67	$A \oplus \cdots \oplus A$ with n copies of A and $A^{\oplus 0} = \mathbb{0}$
$\mathsf{Vect}^{\Bbbk}_{\otimes}$	II.67	$\mathsf{Vect}^{\Bbbk}$ with the tensor product
$\mathsf{Vect}^{\Bbbk}_{\oplus}$	II.67	$\mathsf{Vect}^{\Bbbk}$ with the direct sum
$(A, \mu, \eta, \Delta, \varepsilon)$	II.68	a bialgebra
$\sum_i s'_i \otimes s''_i$	II.69	an element in $A^{\otimes 2}$
S^{op}	II.69	$\xi^{\otimes} S$
S_{12}, S_{13}, S_{23}	II.69	elements obtained from $S \in A^{\otimes 2}$ by inserting 1
$\sum_{(x)} x^{(1)} \otimes x^{(2)}$	II.69	Sweedler's notation for comultiplication
Δ^{op}	II.70	opposite comultiplication $\xi^{\otimes}\Delta$
$\Bbbk G$	II.71	group bialgebra of G
$U\mathfrak{g}$	II.72	universal enveloping bialgebra of $\mathfrak{g}$
H_4	II.72	Sweedler's 4-dimensional bialgebra
$\mathbb{C}\mathbb{Z}_n$	II.74	anyonic quantum groups
$\mathsf{Mod}(A)_{\otimes}$	II.76	$\mathsf{Mod}(A)$ with the tensor product
$\mathsf{Mod}(A)_{\oplus}$	II.79	$\mathsf{Mod}(A)$ with the direct sum
$\mathsf{Vect}^{\mathbb{C}}_{\mathsf{sk}}$	II.81	abelian category with linear maps $\mathbb{C}^m \longrightarrow \mathbb{C}^n$ as morphisms
$\mathcal{F}^{\mathsf{any}}$	II.82	Fibonacci anyons
$\mathbb{0} = (0; 0)$	II.82	additive zero in $\mathcal{F}^{\mathsf{any}}$
$\mathbb{1} = (1; 0)$	II.82	vacuum in $\mathcal{F}^{\mathsf{any}}$
$\tau = (0; 1)$	II.82	non-abelian anyon in $\mathcal{F}^{\mathsf{any}}$
q	II.85	reciprocal of the golden ratio
z	II.87	$e^{3\pi i/5}$
$\mathcal{I}^{\mathsf{any}}$	II.91	Ising anyons
$\mathbb{0} = (0; 0; 0)$	II.91	additive zero in $\mathcal{I}^{\mathsf{any}}$
$\mathbb{1} = (1; 0; 0)$	II.91	vacuum in $\mathcal{I}^{\mathsf{any}}$
$\sigma = (0; 1; 0)$	II.91	non-abelian anyon in $\mathcal{I}^{\mathsf{any}}$
$\psi = (0; 0; 1)$	II.91	fermion in $\mathcal{I}^{\mathsf{any}}$
w	II.102	$e^{\pi i/8}$

Chapter II.4

$\overline{\mathsf{C}}^{\mathsf{bi}}$	II.112	bimonoidal Drinfeld center
$(\overline{\oplus}, \overline{\mathbb{0}}, \alpha^{\overline{\oplus}}, \lambda^{\overline{\oplus}}, \rho^{\overline{\oplus}}, \xi^{\overline{\oplus}})$	II.112	additive structure of $\overline{\mathsf{C}}^{\mathsf{bi}}$
$(\overline{\otimes}, \overline{\mathbb{1}}, \alpha^{\overline{\otimes}}, \lambda^{\overline{\otimes}}, \rho^{\overline{\otimes}}, \xi^{\overline{\otimes}})$	II.112	multiplicative structure of $\overline{\mathsf{C}}^{\mathsf{bi}}$
$\overline{\lambda}^{\bullet}, \overline{\rho}^{\bullet}$	II.112	multiplicative zeros of $\overline{\mathsf{C}}^{\mathsf{bi}}$
$\overline{\delta}^{l}, \overline{\delta}^{r}$	II.112	distributivity morphisms of $\overline{\mathsf{C}}^{\mathsf{bi}}$
$(A; \beta^{A})$	II.112	an object in $\overline{\mathsf{C}}^{\mathsf{bi}}$
$\mathsf{C}^{\mathsf{sym}}$	II.124	bimonoidal symmetric center

Chapter II.5

$\mathcal{D}^{\mathsf{br}}$	II.135	braided distortion category
$\underline{r}$	II.135	an object in $\mathcal{D}^{\mathsf{br}}$
$\varnothing$	II.135	empty sequence in $\mathcal{D}^{\mathsf{br}}$
$\underline{\sigma}$	II.135	a morphism in $\mathcal{D}^{\mathsf{br}}$
$\int_{\mathsf{C}} F$	II.154	Grothendieck construction of $F \colon \mathsf{C}^{\mathsf{op}} \longrightarrow \mathsf{Cat}$
$\mathcal{B}$	II.154	braid category
F^{br}	II.155	functor $\Sigma^{\mathsf{op}} \longrightarrow \mathsf{Cat}$

Chapter II.6

$(F, F^2_\oplus, F^0_\oplus, F^2_\otimes, F^0_\otimes)$	II.160	a braided bimonoidal functor
$F_\oplus$	II.160	additive structure $(F, F^2_\oplus, F^0_\oplus)$
$F_\otimes$	II.160	multiplicative structure $(F, F^2_\otimes, F^0_\otimes)$
$\mathsf{Bi}^{\mathsf{br}}$	II.162	category of small braided bimonoidal categories
$(-)_{\mathsf{rt}}$	II.165	right normalized bracketing
$\underline{a}$	II.165	an object in A
π	II.165	realization function $\mathsf{Ob}(\mathsf{A}) \longrightarrow \mathsf{Ob}(\mathsf{C})$
$\cong^{\oplus}_{\mathsf{ML}}$	II.166	a Mac Lane coherence isomorphism
$\cong_{\mathsf{Lap}}$	II.168	a Laplaza coherence isomorphism
A	II.176	associated right permbraided category
π	II.177	functor $\mathsf{A} \longrightarrow \mathsf{C}$
ι	II.177	functor $\mathsf{C} \longrightarrow \mathsf{A}$

Chapter II.7

$\mathsf{Bi}^{\mathsf{br}}$	II.184	2-category of small braided bimonoidal categories
$\mathsf{Bi}^{\mathsf{fbr}}_{\mathsf{r}}$	II.185	full sub-2-category of $\mathsf{Bi}^{\mathsf{br}}$ with flat objects and robust 1-cells
Σ	II.187	category of $n \geq 0$ and permutations
$(-)_{\mathsf{lt}}$	II.188	left normalized bracketing
$\overline{n}$	II.188	left normalized sum of n copies of $\mathbb{1}$
θ^G	II.193	unique bimonoidal natural transformation $F \longrightarrow G$

Chapter II.8

$(\mathsf{B}, 1, c, a, \ell, r)$	II.199	a bicategory
$\mathsf{Mat}^{\mathsf{C}}_{m,n}$	II.200	category of $n \times m$ matrices in C
$A = (A_{ji})$	II.200	a matrix with (j,i)-entry A_{ji}
$\mathbb{0}_{m,n}$	II.201	0 matrix in $\mathsf{Mat}^{\mathsf{C}}_{m,n}$
BA	II.201	matrix product
$g \star f$	II.201	matrix product of morphisms
$\mathbb{1}^n$	II.201	$n \times n$ identity matrix
ζ^ℓ, ζ^r, ℓ, r, a	II.202	natural isomorphisms in $\mathsf{Mat}^{\mathsf{C}}$
$(1_\boxtimes, 1^2_\boxtimes, 1^0_\boxtimes)$	II.208	monoidal identity
$A \boxtimes B$	II.209	matrix tensor product
$(\boxtimes, \boxtimes^2, \boxtimes^0)$	II.211	monoidal composition
$(a^\boxtimes, a^{\boxtimes\bullet}, \eta^a, \varepsilon^a)$	II.217	monoidal associator
$(\ell^\boxtimes, \ell^{\boxtimes\bullet}, \eta^\ell, \varepsilon^\ell)$	II.219	left monoidal unitor
$(r^\boxtimes, r^{\boxtimes\bullet}, \eta^r, \varepsilon^r)$	II.221	right monoidal unitor
π	II.221	pentagonator
μ	II.222	middle 2-unitor
$\lambda^\boxtimes$	II.222	left 2-unitor
$\rho^\boxtimes$	II.223	right 2-unitor

Part II.2

Chapter II.9

$(\mathsf{C}, \oplus, \mathbb{0}, \xi^\oplus, \otimes, \mathbb{1}, \partial^l, \partial^r)$	II.232	a ring category
∂^l, ∂^r	II.233	left and right factorization morphisms
$\mathsf{End}(A)$	II.237	endomorphism rig of a monoid
$\mathsf{Perm}^{\mathsf{su}}(\mathsf{C};\mathsf{D})$	II.238	category of strictly unital symmetric monoidal functors
$\mathsf{Perm}^{\mathsf{su}}(\mathsf{C};\mathsf{C})$	II.239	endomorphism ring category
$(\boxplus, \mathbb{0}, \xi^\boxplus)$	II.239	additive structure of $\mathsf{Perm}^{\mathsf{su}}(\mathsf{C};\mathsf{C})$
$\mathsf{Perm}^{\mathsf{sug}}(\mathsf{C};\mathsf{C})$	II.243	a full subcategory of $\mathsf{Perm}^{\mathsf{su}}(\mathsf{C};\mathsf{C})$

$(\mathsf{C}, \oplus, \mathbb{0}, \xi^{\oplus}, \otimes, \mathbb{1}, \partial^l, \partial^r)$	II.244	a bipermutative category
$(\mathsf{C}, \oplus, \mathbb{0}, \xi^{\oplus}, \otimes, \mathbb{1}, \partial^l, \partial^r)$	II.252	a braided ring category
$\overline{\mathsf{C}}^{\mathsf{bi}}$	II.256	bimonoidal Drinfeld center
$\mathsf{C}^{\mathsf{sym}}$	II.256	symmetric center

Chapter II.10

$(\mathsf{C}, \{\otimes_i\}, \mathbb{1}, \{\eta^{i,j}\})$	II.264	an n-fold monoidal category
$\otimes_i$	II.264	ith product
$\eta^{i,j}$	II.264	(i,j)-exchange
$(\mathsf{C}, \otimes_1, \otimes_2, \mathbb{1}, \eta)$	II.266	a 2-fold monoidal category
$\leq$	II.269	a partial ordering
$x < y$	II.270	$x \leq y$ and $x \neq y$
$\max(x, y)$	II.270	maximum of x and y
$(M, \mu, \mathbb{1}, \leq)$	II.270	a totally ordered monoid
$(M, \max, \mu, \mathbb{1}, \eta)$	II.271	2-fold monoidal category of $(M, \mu, \mathbb{1}, \leq)$ with $\mathbb{1}$ the least element
$(F, F_1^2, \ldots, F_n^2)$	II.273	an n-fold monoidal functor
F_i^2	II.273	ith monoidal constraint
MCat^n	II.278	category of small n-fold monoidal categories and functors
$\mathsf{MCat}^n_{\mathsf{sg}}$	II.278	wide subcategory of MCat^n with strong n-fold monoidal functors
$\mathsf{MCat}^n_{\mathsf{st}}$	II.278	wide subcategory of MCat^n with strict n-fold monoidal functors
$(\mathsf{MCat}^n, \times, \mathbb{1})$	II.279	monoidal category of small n-fold monoidal categories
(A, μ, η)	II.280	a monoid in a monoidal category
$(\varepsilon, \{\varepsilon_i^2\}_{1 \leq i \leq n})$	II.280	unit n-fold monoidal functor
$(\otimes_{n+1}, \{\eta^{i,n+1}\}_{1 \leq i \leq n})$	II.280	multiplication n-fold monoidal functor
U	II.284	forgetful functor $\mathsf{MCat}^n_{\mathsf{st}} \longrightarrow \mathsf{Cat}$
$\mathsf{FMon}^n(\mathsf{C})$	II.284	free n-fold monoidal category of C
$\overline{F}$	II.286	$\mathsf{FMon}^n(F)$ for a functor F
$\mathsf{Mon}^n(k)$	II.288	a full subcategory of $\mathsf{FMon}^n\{1, \ldots, k\}$
ϕ_{C}	II.290	isomorphism $\coprod_{k \geq 0} \mathsf{Mon}^n(k) \times_{\Sigma_k} \mathsf{C}^{\times k} \longrightarrow \mathsf{FMon}^n(\mathsf{C})$
$\mathsf{FMon}^n(\mathbb{1})$	II.293	free n-fold monoidal category on one object
θ_k	II.293	evaluation functor $\mathsf{Mon}^n(k) \times_{\Sigma_k} \mathsf{C}^{\times k} \longrightarrow \mathsf{C}$
$R_{a,b}$	II.293	restriction functor $\mathsf{Mon}^n(k) \longrightarrow \mathsf{Mon}^n(\{a,b\})$
$a \otimes_i b \in A$	II.293	$R_{a,b}(A) = a \otimes_i b \in \mathsf{Mon}^n(\{a,b\})$
$\eta^{i,j}$	II.297	(i,j)-exchange in an E_n-monoidal category
$\otimes_i, \partial^{l,i}, \partial^{r,i}$	II.297	ith product and left/right factorization morphisms
$\mathsf{FE}^n(\mathsf{C})$	II.303	free E_n-monoidal category of C

Part III.1
Chapter III.1

$(\mathsf{C}, \otimes, \mathbb{1}, \alpha, \lambda, \rho)$	III.4	monoidal category
(F, F^2, F^0)	III.5	monoidal functor
ξ	III.6	braiding
$[-,-]$	III.9	internal hom
C_{st}	III.10	strictification of C
$(\mathsf{V}, \otimes, \mathbb{1}, \alpha, \lambda, \rho, \xi)$	III.13	monoidal or braided monoidal category
C	III.13	V-category
m	III.13	composition in C
i_X	III.13	identity of X
F	III.14	V-functor
θ	III.14	V-natural transformation
$\phi\theta$	III.15	vertical composition
$\theta' * \theta$	III.15	horizontal composition

$\theta * 1_H$, $1_K * \theta$	III.16	whiskering
V-Cat	III.17	2-category of small V-categories
$L \dashv R$, η, ε	III.17	V-adjunction
C^{op}	III.17	opposite V-category
V-$\mathsf{Cat}^{\mathsf{co}}$	III.19	V-Cat with 2-cells reversed
ξ_{mid}	III.22	middle four interchange isomorphisms
$\mathsf{C} \otimes \mathsf{D}$	III.22	tensor product of V-categories
$\mathbb{I}$	III.25	unit V-category
$\ell^{\otimes}$, $\rho^{\otimes}$	III.26	left and right unitor for $\otimes$
$a^{\otimes}$	III.27	associator for $\otimes$
(V, ξ)	III.30	symmetric monoidal category
$\beta^{\otimes}$	III.30	braiding for $\otimes$
$\gamma_{\mathsf{C},\mathsf{D}}$	III.33	V-functor $\mathsf{C}^{\mathsf{op}} \otimes \mathsf{D}^{\mathsf{op}} \longrightarrow (\mathsf{D} \otimes \mathsf{C})^{\mathsf{op}}$
$(\mathsf{K}, \boxtimes, \mathrm{I}^{\boxtimes}, a^{\boxtimes}, \ell^{\boxtimes}, r^{\boxtimes})$	III.38	monoidal V-category
$a_1^{\boxtimes}$, $a_1^{-\boxtimes}$	III.41	a mate of $a^{\boxtimes}$ and its inverse
$(\mathsf{K}, \beta^{\boxtimes})$	III.42	braided or symmetric monoidal V-category
$\ell_1^{\boxtimes}$, $r_1^{\boxtimes}$	III.43	a mate of $\ell^{\boxtimes}$ and a mate of $r^{\boxtimes}$.
(F, F^2, F^0)	III.44	monoidal, braided monoidal, or symmetric monoidal V-functor
θ	III.48	monoidal V-natural transformation
V-MCat	III.50	2-category of small monoidal V-categories
V-BMCat	III.50	2-category of small braided monoidal V-categories
V-SMCat	III.50	2-category of small symmetric monoidal V-categories
$F \dashv G$	III.50	monoidal V-adjunction
$(\overline{\mathsf{K}}, \overline{\boxtimes})$	III.51	rotation of K
$(\mathsf{V}$-$\mathsf{Cat}, \otimes, \mathbb{I})$	III.54	Cat-monoidal 2-category of small V-categories

Chapter III.2

$U\colon \mathsf{V} \longrightarrow \mathsf{W}$	III.57	monoidal functor between monoidal categories
$(-)_U$	III.57	change of enrichment 2-functor
C_0	III.60	underlying category of C
$\mathsf{D}(X, -)$, $\mathsf{D}(-, Y)$	III.61	co/represented V-functors
$\mathsf{C}_\mu\colon \mathsf{C}_U \longrightarrow \mathsf{C}_T$	III.62	enriched functor induced by monoidal natural μ
E	III.65	enrichment 2-functor $\mathsf{V} \longmapsto \mathsf{V}$-$\mathsf{Cat}$
$T \dashv U$	III.67	monoidal adjunction
$U\colon \mathsf{V} \longrightarrow \mathsf{W}$	III.68	braided or symmetric monoidal functor
$(-)^2_U$	III.68	monoidal constraint for $(-)_U$
BMCat	III.71	2-category of small braided monoidal categories
SMCat	III.71	2-category of small symmetric monoidal categories
$\mathsf{CM2Cat}$	III.71	2-category of small Cat-monoidal 2-categories
$\mathsf{SCM2Cat}$	III.71	2-category of small symmetric Cat-monoidal 2-categories
$T \dashv U$	III.73	braided monoidal adjunction
$(\mathbb{I}, a, \ell, r, \beta)$	III.75	braided monoidal data of either V-Cat or W-Cat
K_{st}	III.86	enriched strictifiction of K

Chapter III.3

V	III.91	symmetric monoidal closed category
eval, coeval	III.91	evaluation and coevaluation
$\underline{\mathsf{V}}$	III.92	canonical self-enrichment of V
$\mathcal{Y}^X\colon \mathsf{C} \longrightarrow \underline{\mathsf{V}}$	III.95	corepresented V-functor
$\mathcal{Y}_Y\colon \mathsf{C}^{\mathsf{op}} \longrightarrow \underline{\mathsf{V}}$	III.95	represented V-functor
$\theta^{\perp}\colon \mathbb{1} \longrightarrow [P, Q]$	III.96	adjoint to $\theta \circ \lambda$
$(\boxtimes, \mathrm{I})$	III.103	monoidal product and unit for $\underline{\mathsf{V}}$

$\kappa\colon [\mathbb{1}, W] \longrightarrow W$	III.108	composite of $\rho_{[\mathbb{1},W]}$ and eval
$\mathsf{V}\text{-nat}(F,G)$	III.109	set of V-natural transformations $F \longrightarrow G$
$\overline{\eta}_Y\colon \mathsf{C}(X,Y) \longrightarrow FY$	III.109	composite induced by $\eta \in \mathsf{V}(\mathbb{1}, FX)$
$\widetilde{\eta}\colon \mathcal{Y}^X \longrightarrow F$	III.109	V-natural transformation $\widetilde{\eta}_Y = (\overline{\eta}_Y)^{\perp}$
$\theta_0\colon \mathbb{1} \longrightarrow FX$	III.109	composite induced by V-natural $\theta\colon \mathcal{Y}^X \longrightarrow F$
$\theta \longmapsto \theta_0$, $\eta \longmapsto \widetilde{\eta}$	III.110	V-Yoneda bijection
(X,ζ)	III.113	V-cowedge of F
$(\int^{a\in\mathsf{A}} F(a,a),\omega)$	III.113	V-coend of F
(Y,δ)	III.113	V-wedge of F
$(\int_{a\in\mathsf{A}} F(a,a),\sigma)$	III.114	V-end of F
$\mathsf{Map}(F,G)$	III.116	mapping object between V-functors
$\theta^{\#}\colon Q \longrightarrow [P,R]$	III.116	transform of $\theta\colon P \longrightarrow [Q,R]$
C-V	III.118	mapping V-category of V-functors $\mathsf{C} \longrightarrow \underline{\mathsf{V}}$
m^{Map}	III.119	composition in mapping V-category
$\mathcal{Y}^{(-)}\colon \mathsf{C} \longrightarrow (\mathsf{C}\text{-}\mathsf{V})^{\mathsf{op}}$	III.121	Yoneda V-functor
$\mathsf{Map}(-,F)$	III.123	V-functor $(\mathsf{C}\text{-}\mathsf{V})^{\mathsf{op}} \longrightarrow \underline{\mathsf{V}}$ represented by F
$\phi_X\colon FX \longrightarrow \mathsf{Map}(\mathcal{Y}^X, F)$	III.125	morphisms induced by $F^{\#}_{X,Z}$
$\phi^{\perp}\colon F \longrightarrow \mathsf{Map}(\mathcal{Y}^{(-)}, F)$	III.127	V-Yoneda isomorphism for F
$\mathcal{D}\text{-}\mathsf{V} = \mathsf{V}\text{-}\mathsf{Cat}(\mathcal{D},\underline{\mathsf{V}})$	III.132	category of $\mathcal{D}$-shaped diagrams in V
$X \otimes Y$	III.132	Day convolution of $\mathcal{D}$-shaped diagrams X and Y
$\mathsf{Hom}_{\mathcal{D}}(X,Y)$	III.132	hom diagram from X to Y
ψ_s	III.133	morphisms induced by adjoints of $X_{x,s}$
$\psi^{\perp}$	III.133	V-Yoneda density isomorphism
ϕ_s, $\phi^{\perp}$	III.134	V-Yoneda isomorphism for X
$\gamma_{a,c;b}$, $\gamma_{b;a,c}$	III.134	density isomorphisms
$J = \mathcal{Y}^e$	III.135	unit diagram for Day convolution
α, λ, ρ, ξ	III.135	associativity, units, and symmetry for Day convolution
$F^*\colon \mathcal{E}\text{-}\mathsf{V} \longrightarrow \mathcal{D}\text{-}\mathsf{V}$	III.141	change-of-shape induced by $F\colon \mathcal{D} \longrightarrow \mathcal{E}$
$\mathsf{ev}_e\colon \mathcal{D}\text{-}\mathsf{V} \longrightarrow \mathsf{V}$	III.142	evaluation at e
$U_*\colon \mathcal{D}\text{-}\mathsf{V} \longrightarrow \mathcal{D}_U\text{-}\mathsf{W}$	III.143	change of target induced by $U\colon \mathsf{V} \longrightarrow \mathsf{W}$
$\mathcal{D}_{\mathbb{1}}$	III.145	unitary enrichment of $\mathcal{D}$
$X \otimes A$, X^A	III.147	tensoring and cotensoring of C over V
$L_e\colon \mathsf{V} \longrightarrow \mathcal{D}\text{-}\mathsf{V}$	III.153	symmetric monoidal functor $L_e A = A \otimes J$

Chapter III.4

T, C_*	III.158	terminal object T and category of objects under T in C
$\iota_X\colon T \longrightarrow X$	III.158	basepoint of pointed object X
X_+	III.158	X with disjoint basepoint
$X \vee Y$	III.158	wedge product of X and Y
$\wedge$, $E = \mathbb{1}_+$	III.159	smash product and smash unit
$\mathsf{Hom}_*(X,Y)$	III.165	pointed hom
T, T^{V}	III.167	terminal object of V
VCat, $\mathsf{V}_*\mathsf{Cat}$	III.167	2-categories of small V-categories, respectively small pointed V-categories
$(\mathcal{D},\boxdot,e,\mathrm{T})$	III.168	symmetric monoidal category with null object T
$X^{\flat}$	III.170	punctured set
$\widehat{\mathcal{D}}$	III.170	pointed unitary enrichment of $\mathcal{D}$
$(\widehat{\mathcal{D}})\text{-}(\mathsf{V}_*) = \mathsf{V}_*\mathsf{Cat}(\widehat{\mathcal{D}},\underline{\mathsf{V}_*})$	III.174	V_*-enriched $\widehat{\mathcal{D}}$-shaped diagrams
$\mathcal{D}\text{-}\mathsf{V}$	III.174	pointed diagrams from $\mathcal{D}$ to V
(L_e, ev_e)	III.174	strong symmetric monoidal adjunction

Chapter III.5

$\mathsf{Prof}(C)$	III.177	C-profiles
$\langle c\rangle = (c_1, \ldots, c_n)$	III.177	C-profile of length n
$\oplus$	III.177	concatenation of profiles
$(\langle c\rangle; c')$	III.177	C-profile $\langle c\rangle$ and element $c' \in C$
$(\mathsf{M}, \gamma, 1)$	III.178	multicategory, composition, and unit
$\mathsf{M}(\langle c\rangle; c')$	III.178	n-ary operations with input profile $\langle c\rangle$ and output c'
$\langle c\rangle\sigma$	III.178	right permutation of $\langle c\rangle$ by σ
1_c	III.178	c-colored unit
$\mathsf{End}(X)$	III.181	endomorphism operad of X
$\mathsf{End}(\mathsf{C})$	III.181	endomorphism multicategory of C
$F\colon \mathsf{M} \longrightarrow \mathsf{N}$	III.181	multifunctor from M to N
$\alpha\colon F \longrightarrow G$	III.182	multinatural transformation from F to G
$\beta\alpha$	III.183	vertical composition
$\alpha' * \alpha$	III.183	horizontal composition
$\mathsf{Multicat}$	III.183	2-category of small multicategories
I	III.183	initial operad
$\mathsf{T} = \mathsf{Com}$	III.183	terminal multicategory; commutative operad
$(\mathsf{M}, *^{\mathsf{M}}, \iota^{\mathsf{M}})$	III.186	pointed multicategory
$\mathsf{Multicat}_*$	III.186	2-category of small pointed multicategories
$\mathsf{PermCat}^{\mathsf{su}}$	III.187	2-category with strictly unital symmetric monoidal functors
(T, μ, η)	III.190	monad T with multiplication μ and unit η
$\mathsf{Alg}(T)$	III.190	category of T-algebras
Δ_W	III.191	constant functor at W
$\mathsf{Vt}X$	III.196	vertices of a multigraph X
$X(\langle c\rangle; c')$	III.196	multiedges with source $\langle c\rangle$ and target c'
MGraph, MGraph^C	III.196	category of small multigraphs; subcategory with vertex set C
$\mathsf{Multicat}^C$	III.197	subcategory of multicategories with object set C
(L^C, U^C)	III.197	left adjoint to forgetful $U^C\colon \mathsf{Multicat}^C \longrightarrow \mathsf{MGraph}^C$
f_0^*	III.197	change of object/vertex set by function $f_0\colon C \longrightarrow D$
(L, U)	III.198	left adjoint to forgetful $U\colon \mathsf{Multicat} \longrightarrow \mathsf{MGraph}$
$\langle c\rangle \otimes \langle d\rangle$, $\langle c\rangle \otimes^{\mathsf{t}} \langle d\rangle$	III.202	product and transposed product of tuples
$\xi^{\otimes}$	III.202	transpose permutation
$X \& Y$	III.202	product of multigraphs
$\mathsf{M} \# \mathsf{N}$	III.203	sharp product of multicategories
$\phi \otimes \psi$, $\phi \otimes^{\mathsf{t}} \psi$	III.204	product and transposed product of operations
$\xi^{\otimes}$	III.204	transposition bijection for operations
$\mathsf{M} \otimes \mathsf{N}$	III.204	Boardman-Vogt tensor product of multicategories
$\wedge$, $\mathsf{S} - \mathsf{I}_+$	III.207	smash product and smash unit for pointed multicategories
$\mathsf{M} \vee \mathsf{N}$	III.207	wedge product of pointed multicategories
$\langle Fc\rangle$, $\langle Fc\rangle^{\mathsf{t}}$, $\langle F\rangle\phi$	III.207	applying a tuple of operations $\langle F\rangle$ to a profile $\langle c\rangle$ and operation ϕ
$\mathsf{Hom}(\mathsf{M}, \mathsf{N})$	III.207	internal hom for multicategories
$\alpha\colon \langle F\rangle \longrightarrow G$	III.207	transformation in $\mathsf{Hom}(\mathsf{M}, \mathsf{N})$
$\mathsf{Hom}_*(\mathsf{M}, \mathsf{N})$	III.217	pointed hom multicategory

Chapter III.6

$(\mathsf{V}, \otimes, \xi)$	III.222	permutative category, monoidal product, and symmetry
$(\mathsf{M}, \gamma, 1)$	III.222	V-enriched multicategory, composition, and unit
$\mathsf{M}(\langle c\rangle; c')$	III.222	n-ary operation object with input profile $\langle c\rangle$ and output c'
$\langle c\rangle\sigma$	III.222	right permutation of $\langle c\rangle$ by σ
1_c	III.222	c-colored unit
$\mathsf{End}(c)$	III.224	endomorphism V-enriched operad of c

$F: \mathsf{M} \longrightarrow \mathsf{N}$	III.225	V-enriched multifunctor from M to N
P	III.226	V-enriched operad
(c, θ)	III.226	P-algebra
$\alpha: F \longrightarrow G$	III.226	V-enriched multinatural transformation
$\beta\alpha$	III.226	vertical composition
$\alpha' * \alpha$	III.227	horizontal composition
$\mathsf{V}\text{-}\mathsf{Multicat}$	III.227	2-category of small V-enriched multicategories
$U: \mathsf{V} \longrightarrow \mathsf{W}$	III.228	symmetric monoidal functor between permutative categories
M_U	III.228	W-enriched multicategory with enrichment via U
F_U	III.229	W-enriched multifunctor with enrichment via U
α_U	III.229	W-enriched multinatural transformation with enrichment via U
$(-)_U$	III.230	change of enrichment 2-functor
$\boxtimes\langle X \rangle$	III.230	left normalized product
$\mathsf{End}(\mathsf{K})$	III.231	V-enriched endomorphism multicategory
$\mathsf{End}(X)$	III.233	endomorphism operad of $X \in \mathsf{K}$
$\mathsf{End}(F)$	III.234	V-enriched multifunctor induced by symmetric monoidal V-functor F
$\theta \otimes \omega$	III.235	tensor product of multinatural transformations
$\mathsf{Multicat}$, $\mathsf{Multicat}_*$	III.236	endomorphism multicategories of $(\mathsf{Multicat}, \otimes)$ and $(\mathsf{Multicat}_*, \wedge)$
$\mathsf{PermCat}^{\mathsf{su}}$	III.237	Cat-enriched sub-multicategory of $\mathsf{Multicat}_*$
$\langle X \circ_i X_i' \rangle$	III.237	tuple $\langle X \rangle$ with ith entry X_i'
$\langle X \circ_i X_i' \circ_k X_k' \rangle$	III.237	tuple $\langle X \rangle$ with ith entry X_i' and kth entry X_k'
$F: \mathsf{C}_1 \times \cdots \times \mathsf{C}_n \longrightarrow \mathsf{D}$	III.237	n-linear functor
F_i^2	III.237	ith linearity constraint
$\alpha: F \longrightarrow F'$	III.240	n-linear transformation
F^σ	III.241	image of F under σ-action
α^σ	III.242	image of α under σ-action

Part III.2
Chapter III.7

$\underline{n}$	III.254	totally ordered set $\{0 < 1 < \ldots < n\}$
Δ	III.254	category of ordinals $\underline{n}$ and order-preserving functions
d^i, s^i	III.254	coface and codegeneracy morphisms
X_n, d_i, s_i	III.254	images of $\underline{n}$, d^i, and s^i under simplicial C-object X
sC	III.254	category of simplicial objects in C
sSet	III.254	category of simplicial sets
sCat	III.254	category of simplicial small categories
$*$	III.255	one-point simplicial set
$\Delta^n = \Delta(-, \underline{n})$	III.255	standard n-simplex
$\iota^n = 1_{\underline{n}}$	III.255	fundamental simplex
$\partial\Delta^n$, Λ_k^n	III.255	boundary and k-horn of Δ^n
$\Delta^?: \Delta \longrightarrow \mathsf{sSet}$	III.256	functor $\underline{n} \longmapsto \Delta^n$
S^1	III.256	simplicial circle
Top	III.256	category of compactly generated weak Hausdorff spaces
$\triangle^n$	III.256	topological n-simplex
$\lvert X \rvert$	III.256	geometric realization of simplicial set X
$\mathsf{Sing}: \mathsf{Top} \longrightarrow \mathsf{sSet}$	III.256	total singular complex
$\mathsf{Map}(X, Y)$, $X \otimes A$, X^A	III.257	mapping, tensor, and cotensor objects
$\mathsf{Hom}(X, Y)$	III.258	internal hom for simplicial sets
sC_*	III.259	pointed simplicial objects in C
sSet_*, $\wedge$, $\mathsf{Hom}_{\mathsf{sSet}_*}$	III.259	pointed simplicial sets, smash product, and internal hom
$N\mathsf{C} \in \mathsf{sSet}$	III.260	nerve of a small category C
$N(\Sigma G)$	III.262	simplicial bar construction on group G

Chapter III.8

$\mathsf{C}^{\mathcal{F}} = \mathsf{C}^{\mathcal{F}}_{\cong}, \mathsf{C}^{\mathcal{F}}_{\mathrm{lax}}, \mathsf{C}^{\mathcal{F}}_{\mathrm{co}}$	III.300	strong, lax, and colax Segal Γ-categories of C
$P_\nu, Q_\nu, \nu \in \{\cong, \mathrm{lax}, \mathrm{co}\}$	III.301	adjoint functors comparing Segal Γ-categories
$\mathcal{M}a$	III.304	partition multicategory of a
$2^{a^\flat}$	III.304	set of basepoint-free subsets of a
$\iota_{\langle s\rangle}$	III.304	unique operation in $(\mathcal{M}a)(\langle s\rangle ; t)$ if $\langle s\rangle$ is a partition of t
$\mathcal{M}\underline{1}$	III.305	partition multicategory of $\underline{1}$
$\mathsf{J}^{\mathcal{M}} = \mathsf{Multicat}_*(\mathcal{M}, -)$	III.309	$\mathcal{M}$-partition J-theory
$\mathsf{J}^{\mathsf{Se}}\mathsf{C} = \mathsf{J}^{\mathcal{M}}(\mathsf{End}(\mathsf{C}))$	III.309	Segal J-theory of C
$\mathsf{K}^{\mathsf{Se}} = \mathsf{K}^{\mathcal{F}} N_* \mathsf{J}^{\mathsf{Se}}$	III.309	Segal K-theory functor

Chapter III.9

$\overline{p} = \underline{p}^\flat$	III.319	unpointed finite set $\{1, \ldots, n\}$
$\mathcal{F}^{(q)}$	III.319	smash powers with additional basepoint for $q = 0$
$f_*: \mathcal{F}^{(q)} \longrightarrow \mathcal{F}^{(p)}$	III.320	reindexing functor associated to injection f
$\mathcal{G}, *$	III.320	indexing category and basepoint object for K^{EM}
$(f, \langle\psi\rangle): \langle\underline{n}\rangle \longrightarrow \langle\underline{m}\rangle$	III.320	objects and morphism in $\mathcal{G}$
$\oplus: \mathcal{G} \times \mathcal{G} \longrightarrow \mathcal{G}$	III.323	concatenation product in $\mathcal{G}$
$\gamma_{q,q'}, \xi$	III.324	block permutations and symmetry for concatenation product
$\wedge: \mathcal{G} \longrightarrow \mathcal{F}$	III.325	functor induced by smash product of pointed finite sets
$X: (\mathcal{G}, *) \longrightarrow (\mathsf{C}, *)$	III.326	$\mathcal{G}_*$-object in C
$\mathcal{G}_*$-C	III.326	category of $\mathcal{G}_*$-objects in C
$\mathcal{G}^\flat(\langle\underline{a}\rangle ; \langle\underline{b}\rangle)$	III.327	subset of nonzero morphisms from $\langle\underline{a}\rangle$ to $\langle\underline{b}\rangle$
$\widehat{\mathcal{G}}$	III.327	pointed unitary enrichment of $\mathcal{G}$
$X \wedge Y$, $\mathsf{Hom}_*(X, Y)$, J	III.327	Day convolution, hom diagram, and unit diagram for $\mathcal{G}_*$-C
$L_{\langle\rangle} \dashv \mathsf{ev}_{\langle\rangle}$	III.328	evaluation at empty tuple, and left adjoint $L_{\langle\rangle}$
$i_p: \mathcal{F}^{(p)} \longrightarrow \mathcal{G}$	III.331	inclusion of length-p tuples
$c_p: \mathcal{F}^p \longrightarrow \mathcal{F}^{(p)}$	III.332	p-fold canonical projection
$\wedge^*: \Gamma\text{-}\mathsf{sSet} \longrightarrow \mathcal{G}_*\text{-}\mathsf{sSet}$	III.332	functor induced by $\wedge: \mathcal{G} \longrightarrow \mathcal{F}$
$\eta_{\underline{m},\langle\underline{n}\rangle,x}, \eta_{\underline{m},\langle\underline{n}\rangle}, \eta_{\langle\underline{m}\rangle,\langle\underline{n}\rangle}$	III.332	structure morphisms for definition of $\mathsf{K}^{\mathcal{G}}$
$\mathsf{K}^{\mathcal{G}} X$	III.333	K-theory symmetric spectrum of $\mathcal{G}_*$-simplicial set X
$(\mathsf{K}^{\mathcal{G}} X)_{p,m} = X(\underline{m}^{\oplus p})_m$	III.336	m-simplices of $(\mathsf{K}^{\mathcal{G}} X)_p$

Chapter III.10

$\prod_{a,b}$	III.351	partition product for pointed finite sets a and b
$(\mathsf{Mod}^{\mathcal{M}\underline{1}}, \wedge, \mathcal{M}\underline{1})$	III.361	Cat_*-enriched category of left $\mathcal{M}\underline{1}$-modules
$\mathcal{T}\langle a\rangle = \wedge_k \mathcal{M}a_k$	III.371	smash product of partition multicategories
$\mathsf{J}^{\mathcal{T}} = \mathsf{Multicat}_*(\mathcal{T}, -)$	III.373	$\mathcal{T}$-partition J-theory of $\mathcal{M}\underline{1}$-modules
$\mathsf{J}^{\mathsf{EM}}\mathsf{C} = \mathsf{J}^{\mathcal{T}}(\mathsf{End}(\mathsf{C}))$	III.378	Elmendorf-Mandell J-theory of C
$\mathsf{K}^{\mathsf{EM}} = \mathsf{K}^{\mathcal{G}} N_* \mathsf{J}^{\mathsf{EM}}$	III.379	Elmendorf-Mandell K-theory multifunctor
$(C, \rho) = \{C_{\langle s\rangle}, \rho_{\langle s\rangle;k,t,u}\}$	III.380	$\langle\underline{n}\rangle$-system in C with gluing morphisms $\rho_{\langle s\rangle;k,t,u}$
$\{\alpha_{\langle s\rangle}\}$	III.381	morphism of $\langle\underline{n}\rangle$-systems
$\mathsf{C}^{\mathcal{G}}_{\mathrm{lax}}\langle\underline{n}\rangle, \mathsf{C}^{\mathcal{G}}_{\cong}\langle\underline{n}\rangle, \mathsf{C}^{\mathcal{G}}_{\mathrm{co}}\langle\underline{n}\rangle$	III.382	lax, strong, and colax Elmendorf-Mandell $\mathcal{G}_*$-categories of C
$\tilde{f}, \langle\tilde{\psi}\rangle$	III.383	functors associated to morphism $(f, \langle\psi\rangle)$ in $\mathcal{G}_*$.
$\mathsf{C}^{\mathcal{G}} = \mathsf{C}^{\mathcal{G}}_{\mathrm{lax}}, \mathsf{C}^{\mathcal{G}}_{\cong}, \mathsf{C}^{\mathcal{G}}_{\mathrm{co}}$	III.386	lax, strong, and colax Elmendorf-Mandell $\mathcal{G}_*$-categories of C
$\mathsf{K}^{\mathsf{EM}}_{\cong}\mathsf{C}, \mathsf{K}^{\mathsf{EM}}_{\mathrm{co}}\mathsf{C}$	III.388	strong and colax Elmendorf-Mandell K-theory of C
$(-)^{\mathcal{G}}$	III.388	Elmendorf-Mandell Cat-enriched multifunctor
$\prod: \mathcal{T}\langle a\rangle \longrightarrow \mathcal{M}(\wedge_k a_k)$	III.392	partition product for tuples
$\mathcal{P}\langle a\rangle$	III.394	product and disjoint basepoint
i, j	III.394	pointed multifunctors comparing $\mathcal{T}$, $\mathcal{P}$, and $\mathcal{M}$
$\widetilde{(-)}$	III.399	comparison functor for $\mathsf{C}^{\mathcal{G}}\langle\underline{n}\rangle$ and $\mathsf{C}^{\mathcal{G}}_{\cong}\langle\underline{n}\rangle$
$L^{\mathsf{C}}_{\langle\underline{n}\rangle}, R^{\mathsf{C}}_{\langle\underline{n}\rangle} = \widetilde{(-)}$	III.399	adjoint comparison functors
$I^{\mathsf{C}}_{\langle\underline{n}\rangle}$	III.402	full subcategory inclusion from $\mathsf{C}^{\mathcal{G}}_{\cong}\langle\underline{n}\rangle$ to $\mathsf{C}^{\mathcal{G}}\langle\underline{n}\rangle$

Chapter III.11

Chapter III.12

Chapter III.13

$\mathsf{hocolim}_{\mathsf{C}} F$	III.488	homotopy colimit of a functor $F\colon \mathsf{C} \longrightarrow \mathsf{Top}$
α_F	III.488	canonical map from the homotopy colimit to the colimit
$*$	III.488	constant functor $\mathsf{Mon}^n(k) \longrightarrow \mathsf{Top}$

Appendix A

$\mathsf{Bi}^{\mathsf{tsy}}$	I.427	full sub-2-category of $\mathsf{Bi}^{\mathsf{sy}}$ with tight objects
SMB	I.427	tricategory of symmetric monoidal bicategories
SBB	I.428	tricategory of symmetric bimonoidal bicategories
BB	I.428	tricategory of bimonoidal bicategories
$(B\Sigma)^+$	I.438	Quillen's +-construction of $B\Sigma$

Index

N

Q

Selected Published Titles in This Series

285 **Niles Johnson and Donald Yau,** Bimonoidal Categories, E_n-Monoidal Categories, and Algebraic K-Theory: Volume III: From Categories to Structured Ring Spectra, 2024

284 **Donald Yau,** Bimonoidal Categories, E_n-Monoidal Categories, and Algebraic K-Theory: Volume II: Braided Bimonoidal Categories with Applications, 2024

283 **Donald Yau,** Bimonoidal Categories, E_n-Monoidal Categories, and Algebraic K-Theory: Volume I: Symmetric Bimonoidal Categories and Monoidal Bicategories, 2024

281 **Horst R. Thieme,** Discrete-Time Dynamics of Structured Populations and Homogeneous Order-Preserving Operators, 2024

280 **Tadashi Ochiai,** Iwasawa Theory and Its Perspective, Volume 2, 2024

279 **Ellen Elizabeth Eischen, Wee Teck Gan, Aaron Pollack, and Zhiwei Yun,** Automorphic Forms Beyond GL_2, 2024

278 **Delaram Kahrobaei, Ramón Flores, Marialaura Noce, Maggie E. Habeeb, and Christopher Battarbee,** Applications of Group Theory in Cryptography, 2024

277 **George Metcalfe, Francesco Paoli, and Constantine Tsinakis,** Residuated Structures in Algebra and Logic, 2023

276 **Balázs Bárány, Károly Simon, and Boris Solomyak,** Self-similar and Self-affine Sets and Measures, 2023

275 **Alekos Vidras and Alain Yger,** Multidimensional Residue Theory and Applications, 2023

274 **Willi Freeden and M. Zuhair Nashed,** Recovery Methodologies: Regularization and Sampling, 2023

273 **Gennadiy Feldman,** Characterization of Probability Distributions on Locally Compact Abelian Groups, 2023

272 **Tadashi Ochiai,** Iwasawa Theory and Its Perspective, Volume 1, 2023

271 **Hideto Asashiba,** Categories and Representation Theory, 2022

270 **Leslie Hogben, Jephian C.-H. Lin, and Bryan L. Shader,** Inverse Problems and Zero Forcing for Graphs, 2022

269 **Ralph S. Freese, Ralph N. McKenzie, George F. McNulty, and Walter F. Taylor,** Algebras, Lattices, Varieties, 2022

268 **Ralph S. Freese, Ralph N. McKenzie, George F. McNulty, and Walter F. Taylor,** Algebras, Lattices, Varieties, 2022

267 **Dragana S. Cvetković Ilić,** Completion Problems on Operator Matrices, 2022

266 **Kate Juschenko,** Amenability of Discrete Groups by Examples, 2022

265 **Nabil H. Mustafa,** Sampling in Combinatorial and Geometric Set Systems, 2022

264 **J. Scott Carter and Seiichi Kamada,** Diagrammatic Algebra, 2021

263 **Vugar E. Ismailov,** Ridge Functions and Applications in Neural Networks, 2021

262 **Ragnar-Olaf Buchweitz,** Maximal Cohen–Macaulay Modules and Tate Cohomology, 2021

261 **Shiri Artstein-Avidan, Apostolos Giannopoulos, and Vitali D. Milman,** Asymptotic Geometric Analysis, Part II, 2021

260 **Lindsay N. Childs, Cornelius Greither, Kevin P. Keating, Alan Koch, Timothy Kohl, Paul J. Truman, and Robert G. Underwood,** Hopf Algebras and Galois Module Theory, 2021

259 **William Heinzer, Christel Rotthaus, and Sylvia Wiegand,** Integral Domains Inside Noetherian Power Series Rings, 2021

258 **Pramod N. Achar,** Perverse Sheaves and Applications to Representation Theory, 2021

257 **Juha Kinnunen, Juha Lehrbäck, and Antti Vähäkangas,** Maximal Function Methods for Sobolev Spaces, 2021

For a complete list of titles in this series, visit the AMS Bookstore at **www.ams.org/bookstore/survseries/**.